SUSTAINABLE MANAGEMENT OF LAND RESOURCES

An Indian Perspective

SUSTAINABLE MANAGEMENT OF LAND RESOURCES

An Indian Perspective

Edited by

G. P. Obi Reddy, PhD

N. G. Patil, PhD

Arun Chaturvedi, PhD

AAP | APPLE ACADEMIC PRESS

Apple Academic Press Inc.
3333 Mistwell Crescent
Oakville, ON L6L 0A2 Canada

Apple Academic Press Inc.
9 Spinnaker Way
Waretown, NJ 08758 USA

Exclusive worldwide distribution by CRC Press, a member of Taylor & Francis Group
No claim to original U.S. Government works
Printed in the United States of America on acid-free paper
International Standard Book Number-13: 978-1-77188-517-1 (Hardcover)
International Standard Book Number-13: 978-1-315-36556-5 (eBook)

Library and Archives Canada Cataloguing in Publication

Sustainable management of land resources : an Indian perspective / edited by G.P. Obi Reddy, PhD, N.G. Patil, PhD, Arun Chaturvedi, PhD.
Includes bibliographical references and index.
Issued in print and electronic formats.
ISBN 978-1-77188-517-1 (hardcover).--ISBN 978-1-315-36556-5 (PDF)
1. Land use--India--Management. 2. Land use mapping--India. I. Chaturvedi, Arun, author, editor II. Patil, N. G., author, editor III. Obi Reddy, G. P., author, editor
HD876.5.S87 2017 333.73'130954 C2017-900854-4 C2017-900855-2

Library of Congress Cataloging-in-Publication Data

Names: Obi Reddy, G. P., editor. | Patil, N. G., editor. | Chaturvedi, Arun, editor.
Title: Sustainable management of land resources : an Indian perspective / editors, G.P. Obi Reddy, PhD, N.G. Patil, PhD, Arun Chaturvedi, PhD.
Description: Oakville, ON ; Waretown, NJ : Apple Academic Press, 2017. |
Includes bibliographical references and index.
Identifiers: LCCN 2017003365 (print) | LCCN 2017009612 (ebook) | ISBN 9781771885171 (hardcover : alk. paper) | ISBN 9781315365565 (ebook)
Subjects: LCSH: Land use--India. | Land use mapping--India. | Land degradation--Control--India.
Classification: LCC HD876.5 .S87 2017 (print) | LCC HD876.5 (ebook) | DDC 333.730954--dc23
LC record available at https://lccn.loc.gov/2017003365

Apple Academic Press also publishes its books in a variety of electronic formats. Some content that appears in print may not be available in electronic format. For information about Apple Academic Press products, visit our website at **www.appleacademicpress.com** and the CRC Press website at **www.crcpress.com**

CONTENTS

LIST OF CONTRIBUTORS

B. K. Agarwal
Chief Scientist cum Professor, Department of Soil Science and Agricultural Chemistry, BAU, Ranchi – 834 006, Jharkhand, India

A. B. Age
Assistant Professor, Department of Soil Science and Agricultural Chemistry, Dr. Panjabrao Deshmukh Krishi Vidyapeeth, Akola – 444 104, India

Arvind Kumar
Assistant Professor, Birsa Agricultural University, Rachi-834 006, India

V. Suresh Babu
Associate Professor, National Institute of Rural Development and Panchayati Raj (NIRD&PR), Hyderabad – 500 030, India

T. Banerjee
Principal Scientist, ICAR-National Bureau of Soil Survey and Land Use Planning, Regional Centre, Kolkata – 700 091, India

B. P. Bhaskar
Principal Scientist, ICAR-National Bureau of Soil Survey and Land Use Planning, Amravati Road, Nagpur – 440 033, India

T. Bhattacharyya
Principal Scientist and former Head, ICAR-National Bureau of Soil Survey and Land Use Planning, Amravati Road, Nagpur – 440 033, India

P. Chandran
Principal Scientist, ICAR-National Bureau of Soil Survey and Land Use Planning, Amravati Road, Nagpur – 440 033, India

S. Chattaraj
Scientist, ICAR – National Bureau of Soil Survey and Land Use Planning, Amravati Road, Nagpur – 440 033, India

S. Chatterji
Pr. Scientist, ICAR – National Bureau of Soil Survey and Land Use Planning, Amravati Road, Nagpur – 440 033, India

A. Chaturvedi
Principal Scientist and Head, ICAR-National Bureau of Soil Survey and Land Use Planning, Amravati Road, Nagpur – 440 033, India

S. N. Das
Director, Maharashtra Remote Sensing Applications Centre, Nagpur, Maharashtra – 440 010, India

A. S. Dhawan
Director of Extension Education, Vasantrao Naik Marathwada Agricultural University, Parbhani – 431 402, India

P. N. Dubey
Senior Scientist, ICAR-National Research Centre on Seed Spices, Ajmer – 305 206, India

V. V. Gabhane
Associate Professor, AICRP for Dryland Agriculture, Dr. Panjabrao Deshmukh Krishi Vidyapeeth, Akola – 444 104, India

S. S. Gaikwad
Assistant Technical Officer, ICAR-National Bureau of Soil Survey and Land Use Planning, Amravati Road, Nagpur – 440 033, India

A. S. Gajare
Research Scholar, Department of Soil Science and Agricultural Chemistry, Dr. Panjabrao Deshmukh Krishi Vidyapeeth, Akola – 444 104, India

S. K. Gangopadhyay
Principal Scientist, ICAR-National Bureau of Soil Survey and Land Use Planning, Regional Centre, Kolkata – 700 091, India

M. M. Ganvir
Assistant Professor, College of Agriculture, Dr. Panjabrao Deshmukh Krishi Vidyapeeth, Akola – 444 104, India

Rupali Ghogare
PG Student, Dept. of Soil Science and Agricultural Chemistry, Dr. Panjabrao Deshmukh Krishi Vidyapeeth, Akola – 444 104, India

S. S. Girhe
Research Scholar, Department of Geology, RTM, Nagpur University, Nagpur – 440 001, India

J. D. Giri
Senior Scientist, ICAR-National Bureau of Soil Survey and Land Use Planning, Amravati Road, Nagpur – 440 033, India

U. D. Ikhe
Senior Research Fellow, Department of Soil Science and Agricultural Chemistry, Dr. Panjabrao Deshmukh Krishi Vidyapeeth, Akola – 444 104, India

S. M. Jadhao
Assistant Professor, Dr. Panjabrao Deshmukh Krishi Vidyapeeth, Akola – 444 104, India

A. K. Joshi
General Manager, Regional Remote Sensing Centre-Central, NRSC, ISRO, Amravati Road, Nagpur, Maharashtra – 440 033, India

P. K. Joshi
Principal Scientist, ICAR-Central Soil Salinity Research Institute, Zarifa Farm, Karnal, Haryana – 132 001, India

Prakash R. Kadu
Professor, Department of Soil Science and Agricultural Chemistry, Dr. Panjabrao Deshmukh Krishi Vidyapeeth, Akola – 444 104, India

B. Kalaiselvi
Scientist, ICAR-National Bureau of Soil Surve y and Land Use Planning, Regional Centre, Bangalore – 560 024, India

R. N. Katkar
Associate Professor, Department of Soil Science and Agricultural Chemistry, Dr. Panjabrao Deshmukh Krishi Vidyapeeth, Akola – 444 104, India

V. K. Kharche
Professor, Department of Soil Science and Agricultural Chemistry, Dr. Panjabrao Deshmukh Krishi Vidyapeeth, Akola – 444 104, India

N. M. Konde
Assistant Professor, Department of SSAC, PG Institute, Dr. Panjabrao Deshmukh Krishi Vidyapeeth, Akola – 444 104, India

K. R. Kranthi
Director, ICAR-Central Institute for Cotton Research, Nagpur – 400 010, India

K. S. Anil Kumar
Principal Scientist, ICAR-National Bureau of Soil Survey and Land Use Planning, Regional Centre, Bangalore – 560 024, India

Nirmal Kumar
Scientist, ICAR-National Bureau of Soil Survey and Land Use Planning, Amravati Road, Nagpur – 440 033, India

S. C. Ramesh Kumar
Principal Scientist, National Bureau of Soil Survey and Land Use Planning Regional Centre, Bangalore – 560 024, India

Suresh Kumar
Scientist–SG and Head, Agriculture and Soils Department, Indian Institute of Remote Sensing, 4-Kalidas Road, Dehradun – 248 001, India

S. R. Lakhe
Research Associate, Department of Soil Science and Agricultural Chemistry, Dr. Panjabrao Deshmukh Krishi Vidyapeeth, Akola – 444 104, India

M. Lalitha
Scientist, ICAR-National Bureau of Soil Survey and Land Use Planning, Regional Centre, Bangalore – 560 024, India

A. K. Maji
Ex-Director (Act.), National Bureau of Soil Survey & Land Use Planning, (ICAR), Regional Centre, Delhi – 110 012, India

D. V. Mali
Assistant Professor, Department of Soil Science and Agricultural Chemistry, Dr. Panjabrao Deshmukh Krishi Vidyapeeth, Akola – 444 104, India

A. K. Mandal
Principal Scientist, ICAR-Central Soil Salinity Research Institute, Zarifa Farm, Karnal, Haryana – 132 001, India

C. Mandal
Principal Scientist, National Bureau of Soil Survey & Land Use Planning (ICAR), Amravati Road, Nagpur – 440 033, India

Jugal Kishore Mani
Scientist, Regional Remote Sensing Centre-Central, NRSC, ISRO, Amravati Road, Nagpur, Maharashtra – 440 033, India

B. S. Manjare
Assistant Professor, Department of Geology, RTM, Nagpur University, Nagpur – 440 001, India

B. P. Meena
Scientist, ICAR-Indian Institute of Soil Science, Babibagh, Bhopal – 462 038, India

A. R. Mhaske
Associate Professor, College of Agriculture, Dr. Panjabrao Deshmukh Krishi Vidyapeeth, Nagpur – 444 104, India

B. K. Mishra
Senior Scientist, ICAR-National Research Centre on Seed Spices, Ajmer – 305206, India

S. Mukhopadhyay
Senior Scientist, ICAR-National Bureau of Soil Survey and Land Use Planning, Regional Centre, Kolkata – 700 091, India

M. S. S. Nagaraju
Principal Scientist, ICAR-National Bureau of Soil Survey and Land Use Planning, Amravati Road, Nagpur – 440 033, India

M. B. Nagdeve
Chief Scientist, AICRP for Dryland Agriculture, Dr. Panjabrao Deshmukh Krishi Vidyapeeth, Akola – 444 104, India

M. V. S. Naidu
Professor, S. V. Agricultural College, ANGRAU, Tirupathi – 517 502, India

K. M. Nair
Principal Scientist, ICAR-National Bureau of Soil Survey and Land Use Planning, Regional Centre, Bangalore – 560 024, India

R. K. Naitam
Scientist, ICAR-National Bureau of Soil Survey and Land Use Planning, Regional Centre, Udaipur – 313 001, India

D. C. Nayak
Principal Scientist & Head, ICAR-National Bureau of Soil Survey and Land Use Planning, Regional Centre, Kolkata – 700 091, India

Dhanashree Pable
PhD Scholar, ICAR-National Bureau of Soil Survey and Land Use Planning, Amravati Road, Nagpur – 440 033, India

M. A. Padhye
Research Scholar, Department of Geology, RTM, Nagpur University, Nagpur – 440 001, India

N. G. Patil
Principal Scientist, ICAR-National Bureau of Soil Survey and Land Use Planning, Amravati Road, Nagpur 440 033, India

S. M. Patil
M.Sc. Scholar, Dr. Panjabrao Deshmukh Krishi Vidyapeeth, Akola – 444 104, India

Sidharam Patil
Research Scholar, Department of Soil Science and Agricultural Chemistry, University of Agricultural Sciences, Bangalore – 560 065, India

M. C. Patnaik
Principal Scientist, Agricultural Research Institute, Jayshankar Telangana State Agricultural University, Rajendranagar, Hyderabad – 500 030, India

A. R. Pimpale
Assistant Professor, Section of Agricultural Engineering, College of Agriculture, Nagpur – 440 001, India

Jambhulkar Priti
Research Scholar, Department of Geology, RTM, Nagpur University, Nagpur – 440 001, India

G. Rajanikanth
Associate Professor, National Institute of Rural Development and Panchayati Raj (NIRD&PR), Hyderabad – 500 030, India

P. B. Rajankar
Associate Scientist, Maharashtra Remote Sensing Applications Centre, Nagpur – 440 010, India

M. Rajkumar
Principal Scientist, Fruit Research Station, Sangareddy, Medak, Telangana – 502 001

V. Ramamurthy
Principal Scientist, National Bureau of Soil Survey and Land Use Planning, Regional Centre, Bangalore – 560 024, India

I. K. Ramteke
Scientific Associate, Maharashtra Remote Sensing Applications Centre, Nagpur – 440 010, India

A. R. Reddy
Principal Scientist, ICAR-Agricultural Technology Application Research Institute, Zone V, Hyderabad – 500 059, India

G. P. Obi Reddy
Principal Scientist, ICAR-National Bureau of Soil Survey and Land Use Planning, Amravati Road, Nagpur – 440 033, India

S. K. Reza
Senior Scientist, ICAR-National Bureau of Soil Survey and Land Use Planning, Regional Centre, Kolkata – 700 091, India

A. K. Sahoo
Principal Scientist, ICAR-National Bureau of Soil Survey and Land Use Planning, Regional Centre, Kolkata – 700 091, India

Nisha Sahu
Scientist, ICAR-National Bureau of Soil Survey and Land Use Planning, Amravati Road, Nagpur – 440 033, India

A. K. Sarkar
Ex-Dean (Agriculture), Birsa Agricultural University, Ranchi – 834 006, Jharkhand, India

Dipak Sarkar
Ex-Director, National Bureau of Soil Survey & Land Use Planning (ICAR), Amravati Road, Nagpur – 440 033, India

Vandana Satish
Senior Technical Assistant, ICAR-Central Institute for Cotton Research, Nagpur – 400 010, India

T. K. Sen
Principal Scientist, ICAR-National Bureau of Soil Survey and Land Use Planning, Amravati Road, Nagpur – 440 033, India

D. K. Shahi
Chairman, Department of Soil Science and Agricultural Chemistry, Birsa Agricultural University, Ranchi Jharkhand, India

D. K. Sharma
Director, ICAR-Central Soil Salinity Research Institute, Zarifa Farm, Karnal, Haryana – 132 001, India

A. O. Shirale
Scientist, ICAR-Indian Institute of Soil Science, Babibagh, Bhopal – 462 038, India

B. Singh
Director, ICAR-National Research Centre on Seed Spices, Ajmer – 305 206, India

Ranbir Singh
Principal Scientist, ICAR-Central Soil Salinity Research Institute, Zarifa Farm, Karnal, Haryana – 132 001, India

S. K. Singh
Director, ICAR-National Bureau of Soil Survey and Land Use Planning, Amravati Road, Nagpur – 440 033, India

A. K. Srivastava
Principal Scientist, ICAR – Central Citrus Research Institute, Nagpur, Maharashtra – 440 010, India

Rajeev Srivastava
Principal Scientist and Head, ICAR-National Bureau of Soil Survey and Land Use Planning, Amravati Road, Nagpur – 440 033, India

M. Subbarao
Principal Scientist & Head, Agricultural Research Station, Perumallapalle – 517 505, ANGRAU, Tirupathi, Andhra Pradesh, India

S. M. Taley
Professor and Head, Department of SWC Engineering, Dr. Panjabrao Deshmukh Krishi Vidyapeeth, Akola – 444 104, India

A. B. Turkhede
Agronomist, AICRP for Dryland Agriculture, Dr. Panjabrao Deshmukh Krishi Vidyapeeth, Akola – 444 104, India

M. Umadevi
Director, Water Technology Centre, PJTSTAU, Rajendranagar, Hyderabad – 500 030, India

B. Vajantha
Scientist, Agricultural Research Station, Perumallapalle – 517 505, Chittoor, India

A. O. Varghese
Scientist, Regional Remote Sensing Centre-Central, NRSC, ISRO, Amravati Road, Nagpur, Maharashtra – 440 033, India

M. V. Venugopalan
Principal Scientist, ICAR-Central Institute for Cotton Research, Nagpur – 400 010, India

S. B. Wadatkar
Head, Department of Irrigation and Drainage Engineering, Dr. PDKV, Akola – 444 104, India

M. S. Yadav
Chief Technical Officer, ICAR-Central Institute for Cotton Research, Nagpur – 400 010, India

R. S. Zadode
PhD Scholar, Department of Agronomy, Dr. Panjabrao Deshmukh Krishi Vidyapeeth, Akola – 444 104, India

LIST OF ABBREVIATIONS

AESR	Agro-Ecological Sub Region
AFSIS	Africa Soil Information Service
ALTM	Airborne Laser Terrain Mapper
AMSR-E	Advanced Microwave Scanning Radiometer on Earth Observing System
ANN	Artificial Neural Network
ANOVA	Analysis of Variance
API	Application Programming Interface
ASRIS	Australian Soil Resource Information System
ASSOD	Soil Degradation in South and Southeast Asia
ASTER	Advanced Spaceborne Thermal Emission and Reflection Radiometer
AVHRR	Advanced Very High Resolution Radiometer
AWC	Available Water Capacity
AWiFS	Advanced Wide Field Sensor
BPL	Below Poverty Line
BOD	Biological Oxygen Demand
BS	Base Saturation
CANSIS	Canadian Soil Information System
CACP	Commission for Agricultural Costs and Prices
CAPE	Crop Acreage and Production Estimation
CAZRI	Central Arid Zone Research Institute
CER	Crop Equivalent Rating
CEC	Cation Exchange Capacity
CIA	Chemical Index of Alteration
CIW	Chemical Weathering Index
CLI	Composite Land Index
COD	Chemical Oxygen Demand
COLE	Coefficient of Linear Extensibility
CPCB	Central Pollution Control Board
CROPMON	Crop Monitoring and Production Forecast Program
CSM	Soil-Landscape Model
CSSRI	Central Soil Salinity Research Institute

CSWCR&TI	Central Soil Water Conservation and Training Institute
CTI	Compound Topographic Index
DBMS	Data Base Management System
DDP	Desert Development Program
DEMs	Digital Elevation Models
DES	Directorate of Economic and Statistics
DED	Digital Elevation Data
DOLR	Department of Land Resources
DRIS	Diagnosis and Recommendation Integrated System
DTD	Digital Terrain Data
DTM	Digital Terrain Model
DSM	Digital Soil Mapping
EC	Electrical Conductivity
EMP	Exchangeable Magnesium Percentage
EOS	Earth Observing System
ERS	European Remote Sensing
DSS	Decision Support System
ESP	Exchangeable Sodium Percentage
EUSIS	European Soil Information System
EVI	Enhanced Vegetation Index
FAO	Food and Agriculture Organization
FCC	False Color Composite
FCN	Four Closest Neighbors
FSI	Forest Survey of India
FYM	Farm Yard Manure
GAMs	General Additive Models
GCPs	Ground Control Points
GCS	Global Coordinate System
GDEM	Global Digital Elevation Model
GDP	Gross Domestic Product
GEAC	Genetic Engineering Approval Committee
GEOSS	Global Earth Observation System of Systems
GIS	Geographic Information System
GLASOD	Global Assessment of Land Degradation
GLMs	Generalized Linear Models
GML	Geography Markup Language
GPP	Gross Primary Productivity
GPS	Global Positioning System

GSP	Global Soil Partnership
GUI	Graphical User Interface
HANTS	Harmonic Analysis of Time Series
IBDLH	Integrated Basin Development and Livelihood Program
ICAR	Indian Council of Agricultural Research
ICT	Information and Communication Technologies
IDW	Inverse Distance Weight
IGP	Indo-Gangetic Alluvial Plain
IIHR	Indian Institute of Horticulture Research
IRNSS	Indian Regional Navigational Satellite System
INARIS	Integrated National Agricultural Resource Information System
INM	Integrated nutrient management
IRS	Indian Remote Sensing
ISODATA	Iterative Self Organizing Data Analysis Technique
IUSS	International Union of Soil Science
JMF	Joint Membership Function
KML	Keyhole Markup Language
LACA	Large Area Crop Acreage
LADA	Land Degradation Assessment in Dry lands
LAI	Leaf Area Index
LCC	Lambert Conformal Conic
LACIE	Large Area Crop Inventory Experiment
LGP	Length of growing period
LRI	Land Resource Inventory
LSA	Land Suitability Analysis
LSD	Least Significant Differences
LTFE	Long Term Fertilizer Experiments
LUC	Land Use Capability classification
LUE	Light Use Efficiency
LUP	Land Use Planning
MARS	Monitoring Agriculture through Remote Sensing
MAST	Mean Annual Soil Temperature
MCDM	Multi-criterion Decision Making
MCE	Multi-Criterion Evaluation
MDS	Minimum Datasets
MGNREGS	Mahatma Gandhi National Rural Employment Guarantee Scheme

MODIS	Moderate Resolution Imaging Spectroradiometer
MoRD	Ministry of Rural Development
MRT	MODIS Reprojection Tool
MSL	Mean Sea Level
MSP	Minimum Support Price
MSST	Mean Summer Soil Temperature
MWST	Mean Winter Soil Temperature
MXL	Maximum Likelihood
NAIP	National Agriculture Innovation Project
NARP	National Agricultural Research Project
NAAS	National Academy of Agricultural Sciences
NBSS&LUP	National Bureau of Soil Survey and Land Use Planning
NCSS	National Cooperative Soil Survey
NDSI	Normalized Differential Salinity Index
NDVI	Normalized Difference Vegetation Index
NEE	Net Ecosystem Exchange
NIV	Nutrient Index Value
NPP	Net Primary Production
NRCS	Natural Resources Conservation Service
NRSC	National Remote Sensing Centre
NSDB	National Soil Database
NTFP	Non Timber Forest Produce
OC	Organic Carbon
PAR	Photosynthetically Active Radiation
PAT	Polygon Attribute Table
PAWC	Plant Available Water Capacity
PC	Pedogenic Carbonates
PCA	Principal Component Analysis
PET	Potential Evapotranspiration
PLUP	Participatory Land Use Planning
RDF	Recommended Doses of Fertilizers
RISAT-1	Radar Imaging Satellite-1
RSC	Residual Sodium Carbonate
RWW	Recycled Waste Water
SAR	Synthetic Aperture Radar
SAT	Semi-Arid Tropics
SCS	Soil Control Section
SMCS	Soil Moisture Control Section

SMR	Soil Moisture Regime
SDSS	Spatial Decision Support Systems
sHC	Saturated Hydraulic Conductivity
SIS	Soil Information System
SLUB	State Land Use Board
SOC	Soil Organic Carbon
SOI	Survey of India
SOTER	Soil and Terrain Digital Database
SPI	Stream Power Index
SRM	Soil Resource Mapping
SRTM	Shuttle Radar Topographic Mission
SSD	Sub-Surface Drainage
SSNM	Site Specific Nutrient Management
SSURGO	Soil Survey Geographic Data Base
STATSGO	State Soil Geographic
SYI	Sustainable Yield Index
TDSW	Treated Domestic Sewage Water
TM	Thematic Mapper
TMWW	Treatment Municipal Waste Water
TWI	Terrain Wetness Index
TPI	Topographic Position Index
TSPs	Temporal Spectral Profile
UNEP	United Nations Environmental Program
USLE	Universal Soil Loss Equation
UTM	Universal Transverse Mercator
VI	Vegetation Index
WCS	Web Catalogue Services
WFS	Web Feature Services
WIP	Weathering index of Parker
WISE	World Inventory of Soil Emission Potential
WMS	Web Map Services
WMTS	Web Map Tile Service
WPS	Web Processing Service
XML	Extensible Markup Language

ABOUT THE EDITORS

G. P. Obi Reddy, PhD

Principal Scientist, ICAR-National Bureau of Soil Survey and Land Use Planning (NBSS&LUP), Nagpur; Honorary Secretary, Indian Society of Soil Survey and Land Use Planning, Nagpur, India

G. P. Obi Reddy, PhD, is a Principal Scientist in the Division of Remote Sensing Applications, ICAR-National Bureau of Soil Survey and Land Use Planning (NBSS&LUP), Nagpur, India. His areas of specialization include remote sensing and GIS applications in geomorphology, digital terrain analysis, landform mapping, soil-landscape modeling, land use/land cover studies, watershed characterization, land resource management, and design and development of soil information systems. He teaches courses on geomorphology, remote sensing and GIS applications in land resource management. He has published more than 70 research articles in peer-reviewed national and international journals. Some of his other publications include *Soil Erosion of Goa, Degraded and Wastelands of India-Status and Spatial Distribution, Acid Soils of India: Their Extent and Spatial Variability, Assessment of Soil Loss for Prioritization of Sub-watersheds: A Remote Sensing and GIS Approach,* and *Remote Sensing and GIS in Digital Terrain Analysis and Soil-Landscape Modeling.* In recognition of his outstanding contributions in applications of geospatial technologies and tools in management of soil resources database of national importance, the Indian Society of Remote Sensing conferred on him the 2007 Indian National Geospatial Award and 2013 National Geospatial Award for Excellence. He obtained his doctoral degree in Geography from Sri Krishnadevaraya University, Anantapur, India, and later he joined the Agriculture Research Service (ARS).

N. G. Patil, PhD

Principal Scientist, Indian Council of Agricultural Research;
Editor-in-Chief, Indian Society of Soil Survey and Land Use Planning

N. G. Patil, PhD, is a Principal Scientist whose specialization is soil and water conservation engineering. He joined Agriculture Research Service in January 1992 and since then he has been an active researcher in different institutions of the Indian Council of Agricultural Research. He has more than 24 years of research experience covering different facets of natural resources management. He has worked extensively in the field of soil hydraulics and agriculture water management and land use planning. His contributions in building the Geo-referenced Soil Information System of the country through the development of pedotransfer functions are well recognized. His research contributions have been published in highly reputed international and national journals. He also serves as an editor and reviewer for many international journals published from USA, Europe, Africa, and Asia. Currently he is an Editor-in-Chief of the Indian Society of Soil Survey and Land Use Planning. He obtained a BTech (agricultural engineering) from Mahatma Phule Agricultural University, Rahuri (Maharashtra) in 1988. He also studied at the Indian Institute of Technology, Kharagpur, for masters degree in technology during 1988–1990.

Arun Chaturvedi, PhD
Principal Scientist and Head, Division of Land Use Planning, ICAR-National Bureau of Soil Survey and Land Use Planning, Nagpur, India; President, Indian Society of Soil Survey and Land Use Planning

Arun Chaturvedi, PhD, is a Principal Scientist and Head, Division of Land Use Planning at ICAR-National Bureau of Soil Survey and Land Use Planning, Nagpur, India. His specialization is in land use planning, remote sensing, and GIS. He is an Honorary Fellow of the Bhoo-Vigyan Vikas Foundation, New Delhi, and was a member of UNEP-CSE committee for "Alternative Approaches to Urban Development" during 1984–1985, as well as a member of the DSTE Committee for project formulation and evaluation. He worked as member of the Working Group for the UNESCO-NCERT Panel on Educational Films on Environment during June 1986. He has received several awards, including the Bhoo-Vigyan Sammaan from the Bhoo-Vigyan Vikas Foundation during 2005; the CLUMA Award during 2005 from the Center for Land Use and Management, Hyderabad; and a NAGI Citation during 2000 from the National Association of Geographers, India. He has held many important roles, including founding Joint Secretary of the Indian Society of Soil Survey and Land Use Planning (1986–1988); Honorary Secretary of the Indian Society of Soil Survey and Land Use Planning (2003–2005); Chairman of the Commission on Digital Mapping Techniques, NAGI (2002–2006); Vice President of the National Association of Geographers, India (1999–2000); President of the Indian Society of Soil Survey and Land Use Planning; and Vice President of the Soil Conservation Society of India (present). He has three books and more than 50 research papers to his credit.

FOREWORD

Land and soil resources, in recent years, are under tremendous pressure with highly competing demands from rising population not only to meet the demands of food, fodder and fuel but also due to increasing claims on land for settlements, urban growth, industrial expansion and infrastructure development. As a consequence, per capita arable land availability has registered a serious declining trend. Fertile and productive lands are being progressively converted from crop lands to industrial and other uses, whereas, marginal lands with impoverished soils are being tilled for agricultural crops. Hence, utilization and management of land resources, according to the capability, is necessary to ensure that land resources are utilized to the best advantage in an enduring manner, particularly with respect to enhancing agricultural productivity, keeping in view the prevailing ecological conditions and population pressure in the region.

In this context, the availability of information on extent and spatial distribution of land resources forms is vital for local, regional, national and global planning processes towards optimizing land use and to maintain a sound ecological balance. Thus, land resource inventory, mapping and management are necessary to deal with different issue like land use planning, food security, environmental protection, sustainability of land resources and climate change. It is a matter of great concern to all of us that various types of degradation *viz.,* water erosion, wind erosion, physical and chemical deterioration, flooding, salinization and alkalinization are affecting our land resources. Advanced geospatial tools like remote sensing and Geographic Information System (GIS) applications have immense potential in land resources inventory, mapping and generation of databases for better planning, management, monitoring and implementing the land use plans more efficiently at different levels.

In this context, it gives me an immense pleasure to note that the Editors have brought out this publication on *Sustainable Management of Land Resources: An Indian Perspective.* It is an outcome of the intensive deliberations held in the National Conference on *Sustainable Management of Land Resources for Livelihood Security* during January 28–30, 2015 at ICAR-NBSS&LUP, Nagpur. The contributions are grouped into four sections namely land resource inventory and characterization; geospatial technologies in land resource mapping and management; soil nutrient status and management; land use planning and livelihood security. I appreciate and congratulate the editors of the volume for this excellent achievement.

The publication is extremely valuable for students, researchers, planners and policy makers. I hope it will be highly beneficial in better understanding of the land resources and to formulate future research, development and extension programs in optimizing land resources for sustainable agricultural land use planning.

—Alok K. Sikka
Ex-DDG (NRM)
ICAR, New Delhi

PREFACE

The depletion of land resources is the greatest challenge for the mankind in this millennium. The shrinking land resources, weather aberrations, deterioration of land quality, globalization and liberalization of market economies have become intertwined to influence sustainable management of land resources and hence land use plans. With an increased human impact, natural resources are under great pressure. Issues concerning to land use have gained importance in recent times as the increasing human and livestock population and diversified human demands have further intensified the competition for the finite land resources. The neglect and deterioration of the land resources is impacting the livelihood and food security. Sustaining the productivity of soil and water resources in the next century is an important national as well as global goal. This will require scientists to translate their knowledge and experience into implementable policy decisions. Since India is agrarian economy, the land resources assume a more critical role affecting the livelihood of vast majority of populace in the country.

In order to deliberate various emerging issues of land resource management, National Seminar was organized on Sustainable Management of Land Resources for Livelihood Security at National Bureau for Soil Survey and Land Use Planning (NBSSLUP), Nagpur in collaboration with Indian Council of Agricultural Research (ICAR), New Delhi and National Bank of Agriculture and Rural Development (NABARD), Mumbai during January 28–30, 2015. Besides, two lead chapters, this volume contains twenty eight contributions developed as chapters from the selected research papers presented in the National Seminar. The contributions were grouped under four sections namely land resource inventory and characterization, geospatial technologies in land resource mapping and management, soil nutrient status and management and land use planning and livelihood security.

We are highly thankful to Dr. Alok K. Sikka, Hon'ble Ex-Deputy Director General (NRM), ICAR, New Delhi for his valuable guidance and support. We are extremely thankful to Dr. S.K. Singh, Director, ICAR-NBSS&LUP, Nagpur, for his valuable suggestions, support and encouragement. As editors, we would like to profusely thank Dr. G.S. Sidhu, Former Principal Scientist & Head, ICAR-NBSS&LUP, Delhi, Dr. L.G.K. Naidu, Former

Principal Scientist & Head, ICAR-NBSS&LUP, Bengaluru, Dr. A.K. Maji, Former Director (Act.), ICAR-NBSS&LUP, Nagpur, Dr. S. Vadivelu, Former Principal Scientist and Head, ICAR-NBSS&LUP, Bengaluru for their valuable support and the help extended in finalizing the contributions. We express our thanks to Mrs. Manisha, Office Assistant, ISSLUP for her secretarial assistance.

We firmly believe that this publication will be highly useful for the researchers, academicians, extension workers, policymakers, planners, officials of land resources survey, planning and management institutions/agencies/departments.

—G. P. Obi Reddy, PhD
N. G. Patil, PhD
Arun Chaturvedi, PhD

CHAPTER 1

LAND RESOURCE INVENTORY, MAPPING AND MANAGEMENT: AN INDIAN PERSPECTIVE

G. P. OBI REDDY,[1] S. K. SINGH,[2] N. G. PATIL,[1] and A. CHATURVEDI[3]

[1]*Principal Scientist, ICAR-National Bureau of Soil Survey and Land Use Planning, Amravati Road, Nagpur – 440 033, India*

[2]*Director, ICAR-National Bureau of Soil Survey and Land Use Planning, Amravati Road, Nagpur – 440 033, India*

[3]*Principal Scientist and Head, ICAR-National Bureau of Soil Survey and Land Use Planning, Amravati Road, Nagpur – 440 033, India*

CONTENTS

ABSTRACT

Land resource inventory and mapping play a vital role in resource planning and management to assess its potential and limitations for wide range of land use options and formulate sustainable land use plans to meet the ever increasing demand for food, fodder and fuel production. Further, land resource inventory is necessary to deal with the issues of sustainable land resource management and land use planning, food security and assess the impact of climate change on soil resources and their sustainability. Detailed terrain analysis and landform mapping derived from ancillary data and satellite data analysis form a base map in land resource inventory. The detailed landform maps with spatial variations of the terrain features helps in soil survey and mapping the soils and finalize the soil-mapping units in the region. The conventional field investigation in land resource inventory is becoming increasingly unaffordable in terms of financial cost, time and data deliverability. The integrated remote sensing and Geographic Information System (GIS) applications have immense potential in land resources inventory, mapping and generation of spatial databases for better planning, management, monitoring and implementing the land use plans more efficiently at different levels. In India, ICAR-National Bureau of Soil Survey and Land Use Planning (NBSS & LUP), the premier institute in soil resource inventory and mapping is being used various satellite remote sensing products in soil resource inventory and mapping at different scales ranges from 1:250,000 to 1:4,000 scale depending upon the objectives and scale of mapping. The advent of new age Information and Communication Technologies (ICTs), especially, personal computers, the Internet and mobile technologies have immense potential in inventory, mapping, collection, storage, processing, transmission and presentation of land resource information in multiple formats. These advancements provide accurate, timely, relevant information in cost effective and time efficient manner on real time basis. The information generated through land resource inventory on climate, soils and water resources, cropping systems, land use pattern, production and productivity, vegetation, socio-economic profile of the region, etc., could be effectively used to assess land capability, land irrigability, crop suitability, delineation of land management units and evaluate the alternative land use options.

1.1 INTRODUCTION

Land is a delineable area of the earth's terrestrial surface, encompassing all attributes of the biosphere immediately above or below this surface, including those of the near-surface climate, the soil and terrain forms, the surface hydrology, the near-surface sedimentary layers and associated groundwater reserve, the plant and animal populations, the human settlement pattern and physical results of past and present human activity (FAO, 1995). The ever increasing demand for food, fodder and fuel production could be met through systematic survey of the land resources, evaluating their potentials and limitations for wide range of land use options and formulating sustainable land use plans. Information on the nature, extent, and spatial distribution of land resources is pre-requisite to develop rational land use plans for agriculture, forestry, irrigation, drainage, etc. Land resource inventory provides an insight into the potentialities and limitation of land for its effective planning, utilization and management. The soil, as a major subsystem of land, is changing with time consequent upon changes in its environment or in management. Rapid increase in human population has increased the stress on land resources, including the soil. Soil degradation shows its impact on agricultural production and adversely affects other interrelated natural resources (Doran and Parkin, 1994, 1996; Karlen et al., 2001). Soil survey provides an accurate and scientific inventory of different soils, their kind and nature, and extent of distribution so that one can make prediction about their characters and potentialities (Manchanda et al., 2002). Land resource inventory also helps to assess the land degradation status and delineate vulnerability zones for sustainable land resource management. Hence, land resource inventory and mapping is pre-requisite to generate reliable information on potentials and limitations of land resources for various land resource developments activities. Utilization and management of land resources, according to its capability, is necessary to ensure that land resources are utilized to the best advantage in an enduring manner, particularly with respect to enhance agricultural productivity, keeping in view the prevailing ecological conditions and population pressure in a region.

Remote sensing and GIS technologies play an important role in the areas of land resource inventory, mapping and management at different scales. Remotely sensed data, in particular, provide an unparalleled view of the Earth with synoptic or periodic observations for land resource inventory, monitoring and management. A wide variety of satellite remote sensing data from LANDSAT-ETM+, IRS-IC, IRS-ID, IRS-P6, Cartosat-I & II, Quickbird and Google are now available for generation of spatial database on land resources for various applications. GIS provides wide range of analytical tools to analyze topology or spatial aspects and attributes of the geographical data (Burrough, 1986). With the increasing demand for mapping, evaluation, monitoring and management of land resources, GIS has become an efficient and inevitable platform. The system basically involves four main functions *viz*., data acquisition, data storage and retrieval, manipulation and analysis and output generation. The integrated remote sensing and GIS applications are rapidly advancing in land resources inventory, mapping and generation of databases for better planning, management, monitoring and implementing the land use plans more efficiently at different levels. However, remotely sensed data coupled with field observations and contemporary technologies provide more realistic datasets of land resources as compared to remotely sensed data alone. Besides database derived from satellite data, socio-economic profile of the region is important for integrated land resources management.

1.2 LAND RESOURCE INVENTORY AND MAPPING

Land resource inventory is necessary to deal with the issues of land use planning, food security, environmental problem, sustainability of land resources and climate change impact on soil resources and their sustainability. Land use planning and management decisions at any scale need information about the soil system and its relationships with the environment (Maji et al., 2001). Furthermore, the characterization and investigation of the spatial distribution of soils and their properties gain importance due to the increasing need of knowledge on soil resources, triggered by their importance in the environmental well-being and agricultural activities. The conventional field investigation and laboratory analysis of soils at every site is becoming increasingly unaffordable in terms of financial cost, time and data deliverability. Advances in space technology opened promising application possibilities of remote sensing in land resource inventory and mapping. Traditionally, remotely sensed imageries have been used to support segmentation of the landscape into rather homogeneous soil–landscape units for which soil composition can be established by sampling. Since its inception, ICAR-National Bureau of Soil Survey and Land Use Planning

(NBSS & LUP), Nagpur, the premier institute in soil resource inventory and mapping in India is using used various satellite remote sensing products in soil resource inventory and mapping at different scales to generate viable land use plans in the country. Conjunctive use of remote sensing data and collateral information like climate, lithology, physiography etc., enable the surveyor to map the soils at different scales, ranging from 1:250,000 to 1:50,000 with the abstraction level of subgroups/association thereof and association of families. High-resolution stereo data from the satellites like Cartosat-1 are found to be more useful for generation of information on soil resources at 1:10,000 scales for micro-level land-use planning. The use of remote sensing and GIS has been found to be highly helpful for time efficient and cost effective land resource inventory and mapping.

1.2.1 DIGITAL TERRAIN ANALYSIS AND MODELING

Terrain characterization and landform analysis are prerequisite for precise soil resources mapping (Fairbridge, 1968; Reddy et al., 1999). The utility of soil–land resource information for proper agricultural land use was proposed by Dumanski et al. (1987). Digital terrain modeling is a technique for deriving spatially explicit, quantitative measures of the shape character of topography (Wilson and Gallant, 2000a). Space technology plays a significant role in digital terrain analysis and modeling in replacing the qualitative and characterization of topography. Stereo-pair satellite data has its comparative advantages in quantitative measurement of elevation, enables to derive any other terrain attribute quantitatively, enables to visualize topography in more realistic way than ever before, and enables to store, update, proliferate and manipulate topographic data digitally (Li et al., 2005; Moore et al., 1993; Wilson and Gallant, 2000b). Due to the fact that topography influences endogenic and exogenic soil forming factors and processes, it plays crucial role in the spatial distribution of soils and their properties (Schaetzl and Anderson, 2005). Reddy et al. (2014) developed seamless mosaic of Shuttle Radar Topographic Mission (SRTM) digital elevation data (90 m) for India to analyze and characterize the selected geomorphological parameters namely elevation, slope, aspect, hill shade, plane curvature, profile curvature, total curvature, flow direction, flow accumulation and topographic wetness index. Integrated remote sensing and GIS techniques especially use of digital elevation models (DEMs) at different resolutions enable us to analyze the elevation, slope and image characteristics to delineate the distinct terrain features, such as drainage pattern, geology, structures, geomorphological units (Manjare et al., 2017) and these parameters could be used in the model to develop various terrain based applications.

1.2.2 MAPPING AND ANALYSIS OF LANDFORMS

Detailed landform map, derived from topographical sheets and satellite data analysis through visual interpretation, can be prepared and digitized to form a base map in soil survey. The detailed landform maps show spatial variations of the terrain features, which in turn, help in soil survey and mapping the soils and finalize the soil-mapping units in the region. Soil mapping units, forming the skeleton of the final soil data base, are then labeled and related by appropriate GIS functions to tabular data base containing essential soil characteristics, to form an integrated soil coverage, comprising both spatial and descriptive information of soil mapping unit. The resulting coverage is then analyzed using the analytical capabilities of GIS and tabular database, numerous thematic and derivative maps could be developed to use in various planning process of land resource inventory and mapping. In recent times, satellite remote sensing and GIS techniques are being effectively used in determining the quantitative description of the terrain features, geomorphological mapping (Maji et al., 2004; Reddy et al., 2002, 2004; Reddy and Maji, 2003), soil resource mapping (Reddy et al., 2013; Sahu et al., 2014, 2015; Sarkar et al., 2006; Vishakha et al., 2013;). The systematic analysis of various terrain parameters in GIS helps to understand the soil resources distribution, watershed prioritization, planning and management at watershed level. GIS techniques proved to be a competent tool in morphometric analysis as it not only upgrade and monitor parameters, but also permits the analysis of drainage information in association with other resources and environmental parameters (Sahu et al., 2017).

1.2.3 MAPPING AND ANALYSIS OF LAND USE SYSTEMS

Land use system is composed of one land utilization type practiced on one land unit, complex forms of land use systems are aggregations of different single land use systems (Driessen, 1997). The spatio-temporal dimensions of land use systems must be understood to survey, describe and classify in solving specific land resources management problems (de Bie, 2000). Land use systems are dynamic, the specifications of both land unit and land utilization type change over time. Land use system analysis must account for the dynamics of the system (Driessen and Konijn, 1992). Land use systems analysis aims to cover the successful management of resources to satisfy human needs without degrading the natural resource base and environment (Duivenbooden, 1995). Scientific knowledge of the physical extent, character and consequences of land transformation serves as the foundation for any assessment (Liu et al., 2002). The use of remotely sensed data facilitates the synoptic analysis of land use systems, and change at local to regional scale over time to assess temporal and spatial changes in terrestrial biological productivity (Mani et al., 2017). Multi temporal remote sensing data are widely acknowledged as having significant advantages over single date imagery (Townshend et al., 1985) for studying dynamic phenomena of land use systems. Recent advances in remote sensing and GIS technologies have emerged to meet ever increasing demand for generation of precise and timely information on various cropping systems, which in turn could be effectively used to map, monitor and manage the land use systems. High spatial resolution remote sensing data provides valuable information on location, spatial distribution and extent of land use systems. With increase in spatial, spectral and temporal resolutions of the instruments, the satellite data becomes effective source for accurate land use systems mapping. The satellite remote sensing data with their repetitive nature have proved to be quite useful in mapping land use systems and changes with time (Pimple et al., 2017). GIS techniques having the capability of data storage, retrieval and analysis can play an important role in land use systems studies and development. An integrated GIS-based approach in analysis of land suitability units in association with the prevailing land use systems offers significant advantages to identify potential areas and optimum allocation of land for particular crop.

1.2.4 SOIL–LANDSCAPE MODELING

Soil–landscape modeling is a quantitative approach to analyze and predict the spatial distribution of soil properties based on the variability of environmental correlates, particularly topographic and hydrologic parameters (Moore et al., 1991) derived from DEM. In any soil resources inventory, the primary objective is the delineation of meaningful map units with a minimum of within unit variance from the virtually continuous variability of soils on the landscape (Young et al., 1991). However, naturally soils appear to be more of continuous variables than discrete objects (Qi et al., 2006). Therefore, their conceptualization as discrete objects involves uncertainties. Such variables are best predicted through fuzzy logic approach so that the uncertainties related to conceptualization can be reduced. Application of fuzzy logic requires establishment of knowledge bases or models for the fuzzy membership criteria. For digital soil mapping, the fuzzy logic approach uses the principle that a spatial unit, e.g., a pixel can contain soil, which cannot be exclusively classified into one soil class. Membership values in fuzzy mapping can be determined using deterministic or empirical approaches. Several freely available DEMs are very useful to derive various topographic parameters, which facilitates soil–landscape modeling. They provide large opportunities for predictive soil mapping to generate accurate spatially explicit soil maps. Soil surveyors follows soil–landscape model to capture the relationship between soil and their formative environment. The spatial extents of these soil formative environments are interpreted using satellite data to delineate soil–landscape units known as physiographic units for soil mapping. In digital soil mapping, laboratory analyzed soil data and remote sensing data are integrated with statistical methods to infer spatial pattern and distribution of soil properties (Suresh Kumar et al., 2017). Various approaches such as multivariate statistical analysis, geostatistics, artificial neural network (ANN), expert systems, etc., have emerged as advanced tools in digital soil mapping.

1.3 LAND RESOURCE INVENTORY AND MAPPING IN INDIA

In the last three decades increasing emphasis was laid on inventory, mapping and characterization of soils for developing rational and scientific land use planning in India. This calls for comprehensive knowledge on soil resources in terms of types of soils, their spatial extent, physical and chemical properties and capabilities/limitations. In recent years, the advancements in the field of remote sensing and GIS augmented the soil resource inventory and mapping activities. The remote sensing technology is found to be more efficient and economical than conventional survey (Reddy et al., 2013; Srivastava and Saxena, 2003). Depending upon the scale of mapping, Landsat-MSS/TM, SPOT, IRS P6-LISS-III, and Cartosat-1, etc., could be employed to map the soils. The integrated use of GIS and remote sensing is of immense help in time saving and efficient land resource inventory and mapping. GIS applications in soil survey and mapping could be used in pre-survey landform mapping and post survey database generation and analysis. The soil information generated through surveys and laboratory analysis could be interpreted for various purposes like assessment of land capability, land irrigability, crop suitability, management and prioritization of watersheds (Ali and Kotb, 2010).

1.3.1 SOIL RESOURCE INVENTORY AND MAPPING AT NATIONAL LEVEL

In India, as a part of soil resource inventory and mapping at national scale, the first attempt was made by Raychaudhuri and Mathur (1954), who have divided India into 16 major soil regions and 108 minor basic soil regions by integrating the effects of climate, vegetation, topography on soil formation. Further, Raychaudhuri et al. (1963) have divided India into 27 major soil groups. However, Govindarajan (1971) categorized the soils of India into 23 major groups. Later the concept of identifying and describing soil series was adopted. Murthy et al. (1982) published the Benchmark soils of India-morphology, characterization and classification by describing 64 soil series for resource management. NBSS & LUP compiled the soil survey work conducted by different organizations at different scales and additional information was obtained through rapid reconnaissance survey (whatever needed) for extrapolation and subsequently 1:1 m draft soil map sheet of the country (371 great group association) has been prepared (Staff, NBSS & LUP, 2000). These sheets were further progressively reduced to 1:6.3 m scale. In the process cartographic and categoric generalization was made consistent with the scale and 103 suborder associations have been delineated. As a nodal organization, ICAR-NBSS & LUP has developed soil resource maps of different states on 1:250,000 scale and generalized soil map of India on 1:1 m scale by following unified approach and methodology. The soil map of India on 1:1 m scale contains 1649 mapping units composed of sub-group association. These units are arranged as per

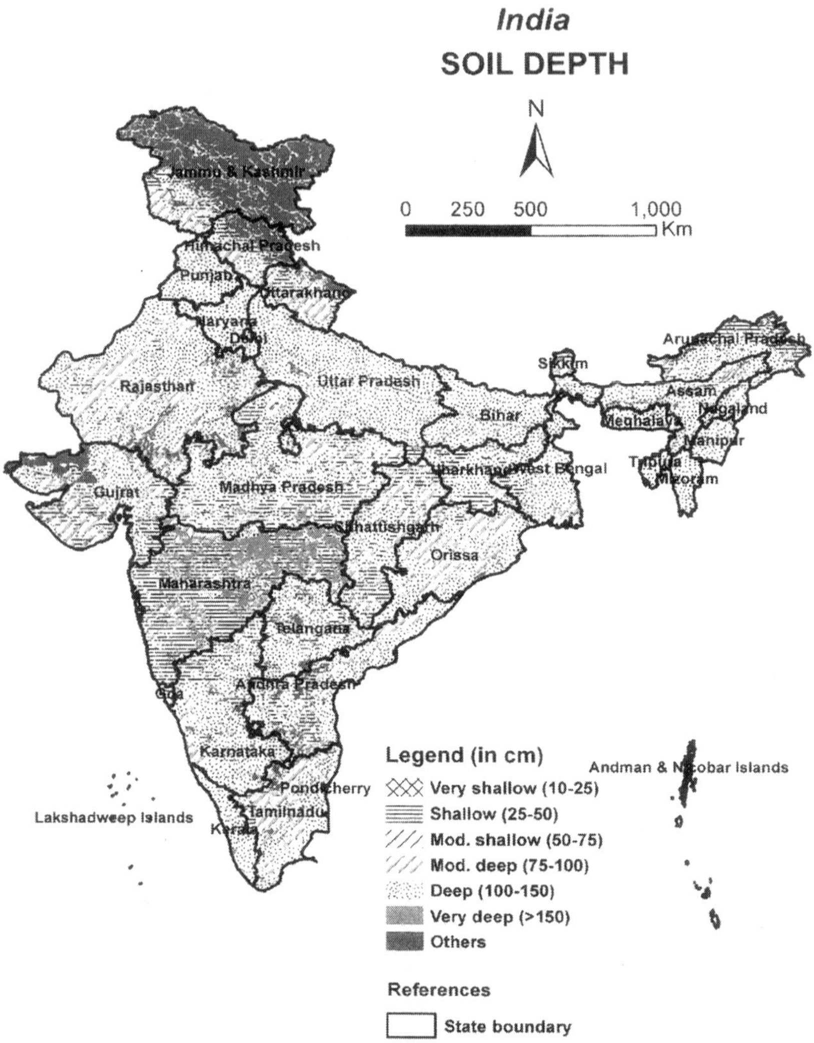

FIGURE 1.1A Soil depth map of India (Source: Staff, NBSS & LUP, 2000).

the physiographic region. Besides taxonomic classification, these mapping units are also described in terms of their depth, drainage, texture, calcareousness, topography of the landforms and phases like slope, erosion, stoniness, rockiness, salinity, sodicity and flooding, etc. (Staff NBSS & LUP, 2000). The soil depth and surface texture maps of India are shown in Figures 1.1A and 1.1B.

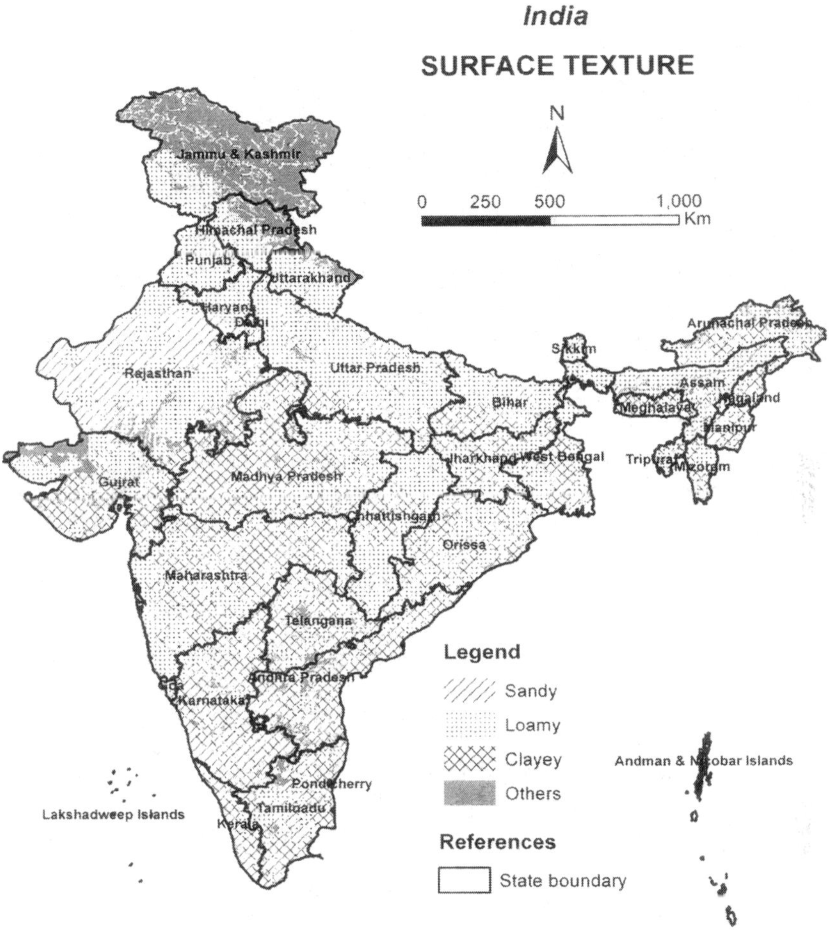

FIGURE 1.1B Surface texture map of India (Source: Staff, NBSS & LUP, 2000).

1.3.2 SOIL RESOURCE INVENTORY AND MAPPING AT STATE LEVEL

ICAR-NBSS & LUP under the Soil Resource Mapping project mapped the soils of different states of India at 1:250,000 scale through systematic soil survey following a three-tier approach. The approach includes (a) image interpretation, (b) field survey and laboratory investigation, and (c) cartography and printing (Sehgal et al., 1987). Eleven broad physiographic regions

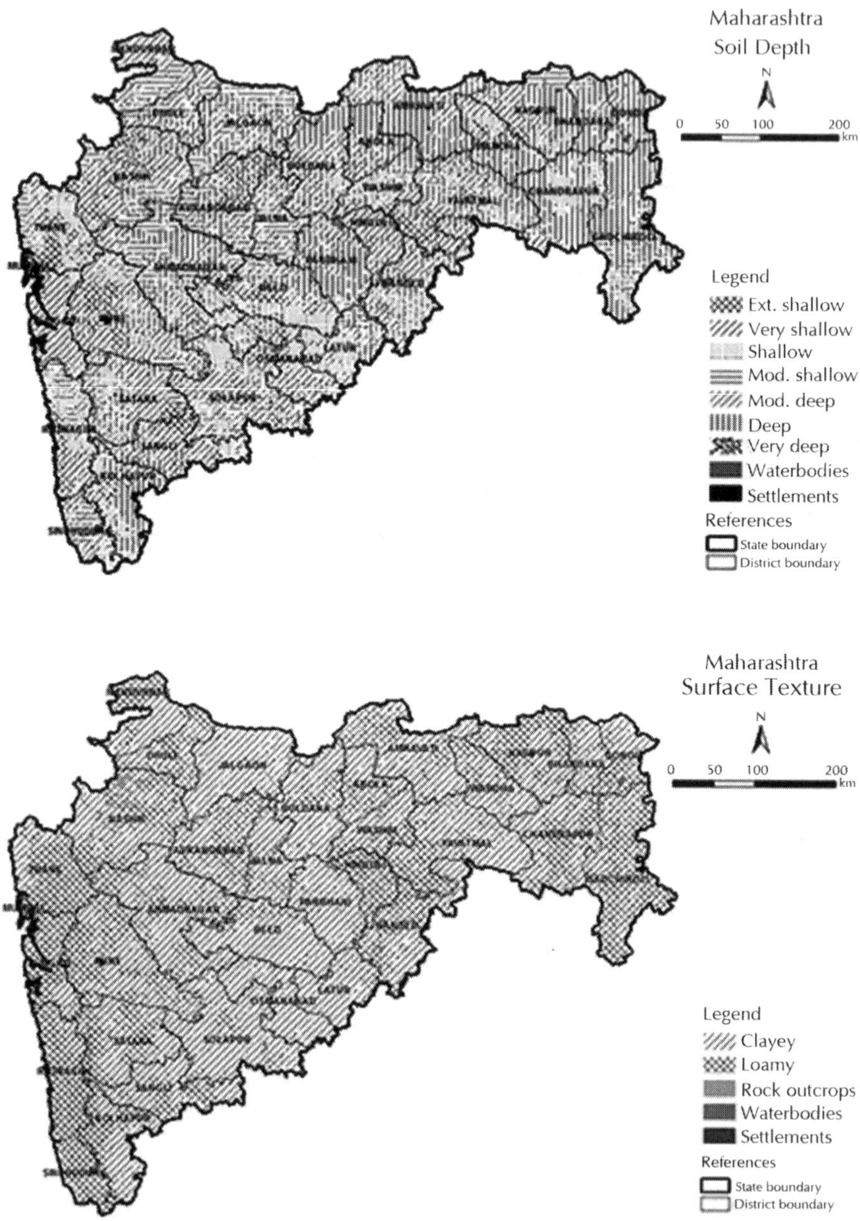

FIGURE 1.2 Soil depth and surface texture maps of Maharashtra state India (Source: Challa et al., 1995).

of India were delineated based on the geology, geomorphology and agro-climatic conditions. The second and third level sub divisions were delineated by considering the regional physiography that expresses the integrated effect of geology, terrain and environmental conditions. Further, subdivisions were made based on the differences in local conditions (landscape elements) interpreted from 1:250,000 topographic maps in conjunction with Landsat images printed as overlays at 1:250,000 scale to match the toposheets. The soil-physiography relationship was established in each photo-interpretation unit in each sheet and thereafter in each state. Soil mapping units were classified at Soil Taxonomy (Soil Survey Staff, 1975) family level, for the dominant (>50% areal extent), sub- dominant (>35% and 50% aerial extent) and included (<15% areal extent) soils. The resulting map legend key is a soil association number, representing the association of two soil families (dominant and subdominant) with phases. The soil map formed thc basic spatial database, however, the attribute database contain various parameters of soil site characteristics, soil physical, morphological and chemical properties of both dominant and sub-dominant soils for each map unit (Soil family association). The state wise attribute database on soil site characteristics, physical, morphological and soil chemical properties have been used to generate various thematic maps and generation of area statistics. As an example, soil depth and surface texture maps of Maharashtra (Challa et al., 1995) are shown in Figure 1.2.

1.3.3 SOIL RESOURCE INVENTORY AND MAPPING AT DISTRICT LEVEL

Reconnaissance soil survey was carried for the priority districts in India by ICAR-NBSS & LUP, as per the procedure outlined by AISLUS (1971) to provide information on natural resources with particular emphasis on soils. Soil surveys at district level include use of remote sensing data with adequate ground truth on 1:50,000 scale to prepare a physiography map. The delineation of physiographic units involves study of soil profiles, soil classification and correlation, soil–physiography relationship for soil map generation, laboratory characterization, soil survey interpretations and generation of thematic maps. Soil resource inventory and mapping at district level for the selected districts has been completed for land use planning on 1:50,000 scale using IRS P6-LISS III data and ancillary datasets. Imageries were visually interpreted in conjunction with Survey of India (SOI) toposheets to delineate photomorphic units. Based on spectral signature (tone, texture, pattern), geology, slope and land use, four major physiograhic units were identified in the district. The physiographic/landform delineations were made to use as a base for the study of soils in several transects covering all the landforms. A minimum of three or more soil profiles were studied in each physiographic unit. The initial legend prepared based on the relationship established between physiography and soils in various transects. This was progressively developed and refined as more data was accumulated from the transects and random locations. Soil maps at district level shows associations of soil series as mapping units (1, 2, 3) were generated on 1:50,000 scale for the district. The soil survey information enables preparation of soil, land use, hydrology and other interpretative maps for land use planning and land resources management activities by the line departments at district level. Soil depth and surface texture maps generated through reconnaissance soil survey of Raisen district, Madhya Pradesh carried by NBSS & LUP, (2002) on 1:50,000 scale are shown in Figure 1.3.

1.3.4 SOIL RESOURCE INVENTORY AND MAPPING AT BLOCK LEVEL

Large scale soil resource data and information helps to formulate site-specific land use planning and resource conservation policies to sustain agricultural production system and ensure food security in the country. High-resolution satellite imageries provide a synoptic view of landscapes critical for integrating soil landscape patterns with observable or measurable soil properties that can vary in both space and time. Use of digital imaging and associated geospatial information for characterizing and mapping soils is expanding rapidly with the advent of new sensors, aircraft and satellite platforms, ortho-rectification techniques, mathematical models for integrating disparate spatial data sources, and visualization of soil properties using conventional and web-enabled technologies. As a pilot study, NBSS & LUP and its five regional centers located in different parts of India selected one block in each agro-ecological regions of India for soil resource inventory and characterization. In this effort, IRS P6 LISS-IV satellite data has been extensively used in soil resource inventory and characterization. The study demonstrates that the space technology will be of immense help to develop the farm level

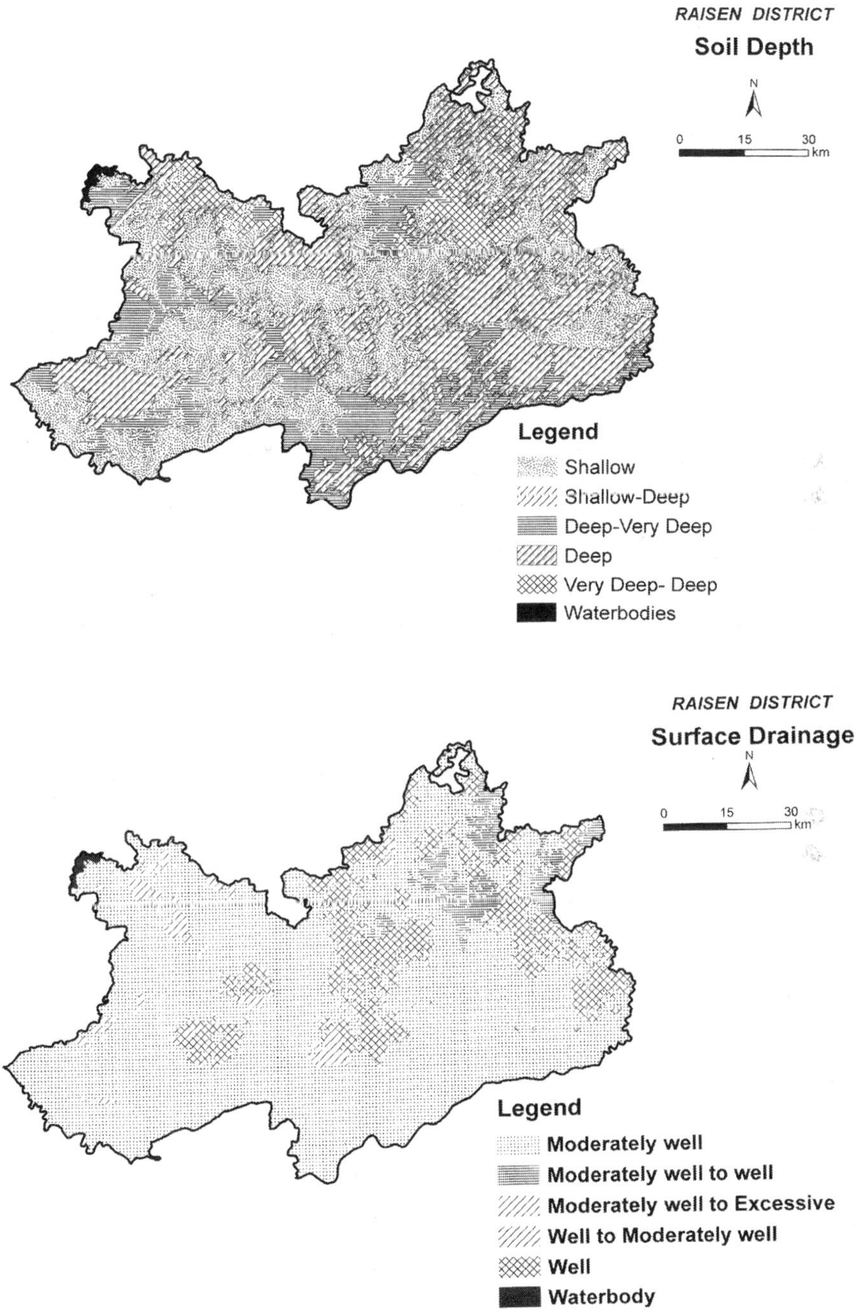

FIGURE 1.3 Soil depth and surface texture maps of Raisen district, Madhya Pradesh (Source: NBSS & LUP, 2002).

resource data in crop planning to maximize the crop productivity. Satellite data from IRS-P6-LISS III, CARTOSAT-1, 2 and IKONOS have been found useful for detailed characterization of soils (Das et al., 2009) and to generate thematic maps such as drainage, soils, land use and land cover, physiography, slope and soil erosion using GIS technique and suggest alternate land use options.

1.3.5 SOIL RESOURCE INVENTORY AND MAPPING AT VILLAGE LEVEL

The aim of detailed soil survey is to delineate homogenous areas, which respond or expected to respond similarly to a given level of management. Soil resource inventory at village level involves the study of all the site characteristics such as slope, erosion, drainage, salinity, occurrence of rock fragments, etc., and soil characteristics such as depth, texture, color, texture, coarse fragments, porosity and soil reaction, etc., in detail to delineate homogenous areas based on their similarity or difference into various management units. Normally for village level land resource inventory cadastral maps on 1:4,000 to 1:10,000 scale with field boundaries and survey numbers form the most suitable base maps. The high resolution satellite data like IRS-P6 LISS-IV fused with Cartosat-I data could be used as base maps in conjunction with cadastral maps to carry out the land resource inventory at village level. Bhaskar et al. (2017) carried out a detailed soil resource inventory of Lagadwal village, Sakri tehsil, Dhule district of Maharashtra, India representing semi-arid ecosystem of Western India. Soil resource inventory was conducted at 1:10,000 scale using latest technologies of remote sensing and GIS, to identify the problems and potentials of the soils and to suggest management practices to develop appropriate and sustainable production systems that deal with the problems of food security and rural livelihoods. Naitam et al. (2017) studied the swell- shrink soils of the Ramagarh village of Purna valley in Amravati district, Maharashtra in semi-arid region of Central India for their morphological, physical and chemical characteristics. They reported that soils in the Ramagarh village are very deep, clay in texture, alkaline in reaction, calcareous in nature and low to medium organic carbon content.

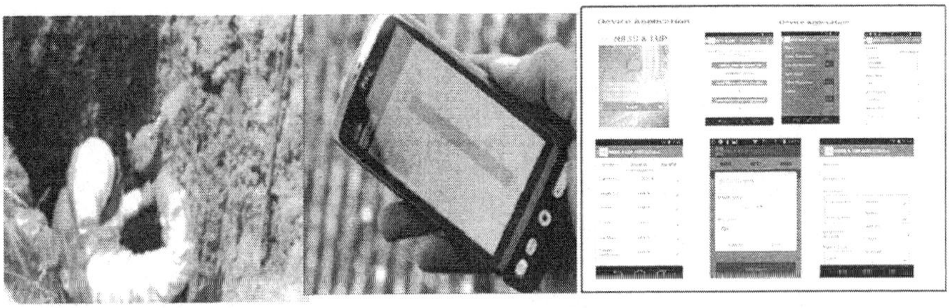

FIGURE 1.4 Android based GPS device for collection and transformation of soil survey data from the field.

1.3.6 ADVANCES IN LAND RESOURCE INVENTORY AND MAPPING

The conventional land resource inventory obviously time consuming and requires hard copy of survey maps for navigation in the field. Accurate and inexpensive methods for measuring soil properties are required to enhance interpretation of land resource maps and improve planning for precision farming strategies. Advances in geospatial and mobile technologies paved the way to find out the innovative ways in land resources inventory and mapping. The advent of new age Information and Communication Technologies (ICTs), especially, personal computers, the internet and mobile technologies has provided a much wider choice in inventory, mapping, collection, storage, processing, transmission and presentation of land resource information in multiple formats to meet the requirements. These advancements provide accurate, timely, relevant information and services to the surveyors, thereby facilitating an ecosystem for cost effective and time efficient services in recording the data in the field and transferring the same to central server on real time basis. ICAR-NBSS & LUP developed Android based GPS device for collection and transformation of soil survey data from field to laboratory on real-time basis and the same is shown in Figure 1.4.

1.4 LAND RESOURCE PLANNING AND MANAGEMENT

Planning and management of land resources including its utilization and conservation is of crucial importance in the context of growing population pressure, increased demand for food, fodder and fuelwood, and competing demands from different sectors like industrial, urban and recreation development. Generation of land resource information system is of great use to assess the current status and utilization potential of land resources and identify local specific alternate land use management practices. Remote sensing and GIS technologies greatly assist in land evaluation and monitoring at different scales (national, regional, provincial/district, and village/block level) for proposing suitable land use planning. Land resources appraisal linked with the agro-ecological zones in GIS environment greatly enhanced the capability to develop alternate land use plan scenarios, management and decision support systems and improved interfaces to promote use of such systems together by scientists, development administrators and land users (Velayutham et al., 1999). Furthermore, derivative information, such as land capability, land irrigability, erodibility, reclamability, and suitability for different crops, which in turn enable preparing the optimal land use plans and in taking up site-specific land reclamation measures.

1.4.1 DEVELOPMENT OF LAND RESOURCE INFORMATION SYSTEM

The availability of accurate and timely information on land resources enables and rationalizes the decision-making processes of planners and managers in a cost effective and time efficient manner. Remote sensing and GIS have immense potential in development of Land Resource Information System (LRIS) to store, query, update, manage and visualize the spatial and non-spatial information. Development of systematic databases in GIS framework on land resources eliminate redundancies, duplication of efforts enforce consistency and standards to develop cross-domain applications for effective utilization of land resources. The available hard copies of land resource maps could be digitized to generate seamless mosaic, standardized spatial and non-spatial data using various capabilities of GIS. The soil maps and attribute database thus produced through surveys were digitized in GIS for generation of digital soils resource information system (Bhattacharyya et al., 2010; Maji et al., 2001, Reddy et al., 2014, 2014a). This activity encompasses creation and updating of the soil resource database (spatial and non-spatial). Soil mapping units, forming the skeleton of the final soil data base, are then labeled and related by appropriate GIS functions to tabular database containing essential soil characteristics, to form an integrated soil coverage, comprising both spatial and descriptive information's of soil mapping unit.

The soil resource maps at state level have been developed at 1:250,000 scale to provide information up to subgroup level classification of the soils. Maps, thus produced, can be utilized for district level information dissemination. For selected districts district soil information system has been developed and the methodology includes digitization of base maps, georeferencing, edge matching/mosaicing, topology building, attribution of polygons, linking non-spatial data, rasterization, thematic layer generation, area analysis, map composition and output (Reddy et al., 2014b). Various soil characteristics arranged in attribute database are linked to the district wise master soil layer and reclassified to produce different thematic maps. Maps, thus produced, help us in understanding the spatial distribution and nature of a particular theme character which, in turn, helps as diagnostics of land use planning parameters. The non-spatial (attribute) data are linked with soil layers to generate various derivative thematic maps, like soil depth, soil drainage, soil erosion, slope, soil reaction (pH), particle size class, salinity, sodicity, etc. Further, LRIS could be enriched with DEMs and satellite remote sensing data. (Reddy at al., 2017) made an attempt to demonstrate the capabilities of geospatial technologies in development of Soil Information System (SIS) and prototype Geoportal to store, query, update, manage and visualize the spatial and non-spatial soil information of India and provide robust web map services.

1.4.2 ASSESSMENT OF POTENTIAL AND LIMITATIONS OF LAND RESOURCES

Land resource inventory and mapping plays a vital role in assessment of its potential and limitations for wide range of land use options and sustainable land use planning. Land capability refers to the ability of land to support a type of land use without causing damage. Land capability classification is an interpretative grouping made primarily for conservation of land resources, which emphasizes the intensity of a particular piece of land it can be used with minimum hazards (Klingebiel and Montogomery, 1961). It plays an important role in land use planning to show the relative suitability of soils for cultivation of crops, pastures and forestry in addition to focusing the problems. Class I, II, III, and IV lands are used to represent arable and class V, VI, VII, and VIII lands are used to represent non-arable. These are based on the effects of combination of climate and permanent soil characteristics, risks of soil damage, existing land use, productive capacity and management requirements. Land capability classes are divided into subclass that represents groups of soils having the same kind of limitations for agricultural use. Four kinds of limitations are recognized at subclass level *viz*., 'e' for water or wind erosion; 'w' for drainage problem, wetness or overflow; 's' for soil limitations affecting the plant growth and 'c' for climatic limitations (AISLUS, 1971). Land irrigability classification is an interpretative grouping based on soil and land characteristics, which indicate relative suitability of land for irrigation as well as predicted behavior of soils under irrigation. The factors like soil depth, texture, moisture holding capacity, soil surface cover, soil salinity and alkalinity hazards, drainage, soil erosion, groundwater table and landform decides the irrgability class of soil unit (AISLUS, 1971). Lands suitable for irrigation are grouped under classes 1 to 4 according to their limitations. Lands not suitable for irrigation are grouped under class 5 and 6. Land irrigability classes have subclasses to indicate their dominant limitations for irrigation purposes. Three subclasses are denoted by 's' for soil limitations such as heavy clay or sandy texture, soil depth and gravel/stones, 'd' for drainage problems and 't' for limitations of topography.

1.4.3 ASSESSMENT OF LAND DEGRADATION FOR MITIGATION

Land degradation affects about 2.6 billion people in more than a hundred countries and over 33% of the Earth's land surface (Adams and Eswaran, 2000). Combating land degradation is critical for ensuring sustainability of agriculture to support current and future demands in crop and livestock production. The immediate causes of land degradation are inappropriate land use that leads to degradation of soil, water and vegetative cover and loss of both soil and vegetative biological diversity, affecting ecosystem structure and functions (Snel and Bot, 2003). Degraded lands are more susceptible to the adverse effects of climatic change such as increased temperature and changes in land use (Dubey et al., 2017). Remote sensing has been successfully utilized in land degradation assessment and monitoring over a range of spatial and temporal scales (Jafari et al., 2008). Ostir et al. (2003) pointed out that remote sensing has developed as an important tool for assessment and monitoring of vegetation, and erosion. It can provide calibrated, quantitative, repeatable and cost effective information for large areas and can be related to the field data (Jafari et al., 2008; Pickup, 1989; Tueller, 1987). Different developmental agencies faced problems in implementing plans to rehabilitate degraded lands owing to non-consistent area statistics of the waste lands and also because of incomplete information available on their extent and location. However, harmonized estimates of degraded and waste-lands in India shows that about 120.72 Mha, area is under different kinds of land degradation (ICAR and NAAS, 2010). These datasets available at district level provide a new dimension towards attempting rehabilitation of problematic lands of India. In trans- Gangetic Plains of India, temporal remote sensing data has been used to map and assess the land degradation and understand the causes and consequences to develop the location specific database and mitigation strategies using the latest geospatial technologies. Soil salinity/alkalinity and poor quality groundwater are major constraints causing reduced productivity in the arid and semiarid region of India (Mandal, et al., 2017). Based on the soil physico-chemical characteristics and the quality of groundwater, suitable reclamation and management options could be suggested. Thus, land resource inventory helps to assess the land degradation status and vulnerability of different physiographic units. Based on climate, terrain and soil characteristics, the land degradation status and vulnerability could be evaluated and categorized into different vulnerability zones (Anil Kumar et al., 2017).

1.4.4 PRIORITIZATION OF AREAS FOR LAND RESOURCE MANAGEMENT

Watershed Development Program is one of the major initiatives in the country towards conservation of soil and water resources in the rainfed area for enhancing agricultural production, and to ensure livelihood security to rural communities. Such endeavor needs scientific databases for evolving suitable strategies for planning. Watershed, as normally defined, is a natural hydrological entity that covers a specific area within which the entire rainfall runoff, ultimately passes through a specific channel at a particular point. The information about the land resources of the watershed is very important for watershed management planning. The various parameters of the watershed, (i.e., stream network (drainage), physiography, land use, vegetation/forest cover and snow cover can be mapped and monitored using remote sensing data. Remote sensing and GIS technologies could be effectively used in strategic planning adopted for watershed development comprises delineation, codification and prioritization of watersheds, detailed soil resource inventory of very high and high priority watersheds in the catchments, treatment of very high and high priority watersheds, evaluation and monitoring of the impact of the treatment. In recent years, high resolution satellite remote sensing techniques proved to be of immense value for evaluation of land resources and land use systems at watershed level and to prepare land management strategies and monitoring changes at regular intervals (Reddy et al., 2008). Analysis of satellite data for drainage, lithology, geomorphology, soils and land use/land cover aspects in conjunction with collateral data, facilitate effective evaluation of status of soil loss at watershed level (Reddy and Sarkar, 2012). This is essential for the conservation and management of land and water resources of watersheds for optimum productivity. In general, remote sensing and GIS technologies have immense potential in watershed characterization soil, land use/land cover mapping, erosion intensity mapping and identification of erosion prone areas (Srinivas et al., 2002; Reddy et al., 2004), potentials and management requirements of various watersheds, evolving water conservation strategies, selection of sites for the construction of check dams/reservoirs on streams evaluation and monitoring of the impact of the treatment and suggesting sites for rain water harvesting structures.

1.4.5 LAND EVALUATION FOR SUSTAINABLE LAND USE PLANNING

The detailed spatial information on land resources is essential for many land use applications (Burrough, 1989; Maji et al., 2001, 2002; Reddy et al., 2014a). Land suitability evaluation forms a pre- requisite for land use planning (Sys et al., 1991). Rossiter (1996) opined that land evaluation using a scientific procedure is essential in assessing the potential and constraints of a given land parcel for agricultural purposes. The reliable information on the landforms, soil site, physical and chemical characteristics has been recognized as an important requirement in land suitability evaluation for different crops (NBSS & LUP, 1994). Land evaluation is concerned with the assessment of land performance for specific land utilization purposes and provides a rational basis for taking land-use decisions based on analysis of relations between the land use and land, giving estimates of required inputs and predicted outputs (FAO, 1985; Sys et al., 1991). The principle objective of soil-suitability evaluation is to select the optimum land use for each defined land unit taking into account both physical, socio-economic considerations and the conservation of environmental resources for sustainable use (FAO, 1983). The physical land suitability evaluation is a prerequisite for land use planning and development (Sys, 1985; Van Ranst et al., 1996). It provides information on the constraints and opportunities for the use of the land and therefore guides decisions on optimal utilization of land resources (FAO, 1984). The FAO (1976) framework for land suitability involves the construction of matching tables or transfer functions, and subsequent calculations of suitability. Overlay and modeling techniques in GIS help in assessing suitability of crops. The procedure follows multi-criteria based decision making with or without weightage of different parameters and evaluates the conditions to generate suitability maps (Maji et al., 2005; Walke et al., 2014). The evaluation of soil resources and assessment of soil related constraints enable to manage the soils for sustainable crop production. The database on soil properties particularly through soil resource inventories is important tool for practical purpose of assessing suitability of each soil unit for agricultural crops. The information on climate, soils and water resources, crops, and cropping system, land use pattern, production and productivity, vegetation, socio-economic profile, etc., form the base to evaluate the land suitability (Gangopadhyay et al., 2017; Kadu and Karche, 2017; Kumar et al., 2017) and alternative land use options. Decision Support Systems could be developed to assist land managers in land resources planning, allocation and management at different levels.

1.5 CONCLUSIONS

Land resource inventory and mapping play a pivotal role in assessment of its potential and limitations for wide range of land use options and sustainable land use planning. The availability of wide ranges of satellite remote sensing datasets has immense potential to generate spatial database on land resources for planning and management. The integrated remote sensing and GIS applications are rapidly advancing in land resources inventory, mapping and generation of databases for better planning, management, monitoring and implementing the land use plans more efficiently at different levels. Besides database derived from satellite data, socio-economic profile of the people serves as vital input for integrated land resources management. Land resource inventory helps to assess the land degradation status and identify the different vulnerability zones for resource management. The availability of accurate and timely information through land resource information system enables and rationalizes the decision-making processes of planners and managers in a cost effective and time efficient manner. Such systems have immense potential in identification of erosion prone areas, evolving water conservation strategies and monitoring of the impact of the land management interventions on the health of land resources. The information generated through land resource inventory and mapping on climate, soils and water resources, crops, and cropping system, land use pattern, production and productivity, vegetation, socio-economic profile of the region could be effectively used to evaluate the land suitability. Integrated remote sensing and GIS technologies could be effectively used in precisely identification of erosion prone areas, develop soil and water conservation strategies and monitoring the impact of the treatments on regular basis.

KEYWORDS

- **Digital Terrain Analysis**
- **Land Resource Information System**
- **Land Resource Inventory**
- **Land Resource Management**
- **Land Use Systems**
- **Soil–Landscape Modeling**
- **Sustainable Land Use Planning**

REFERENCES

Adams, C. R., & Eswaran, H. (2000). Global land resources in the context of food and environmental security, *In*: *Advances in Land Resources Management for the 20th Century*. (S. P. Gawande et al., Eds.), New Delhi: pp. 35–50.

AISLUS. (1971). All India Soil and Land Use Survey, Soil Survey Manual, Indian Agricultural Research Institute (IARI) Publ. New Delhi.

Ali, R. R., & Kotb, M. M. (2010). Use of satellite data and GIS for Soil Mapping and Capability Assessment. *Nature and Science*, *8*(8), 104–115.

Anil Kumar, K. S. Lalitha, M. Patil, S., Kalaiselvi, B., Nair, K. M., & Singh, S. K. (2017). Assessment of Land Degradation Vulnerability: A Case Study from Part of Western

Ghats and West Coast of India, *In*: *Sustainable Management of Land Resources – An Indian Perspective,* Reddy, G. P. O., Patil, N. G., & Chaturvedi, A. (Eds.); Apple Academic Press, Canada, pp. 213-238.

Bhaskar, B. P., Chaturvedi, A., Bhattacharyya, T., & Gaikwad, S. S. (2017). Land Resource Inventory and Evaluation for Agricultural Land Use Planning in Semi-Arid Ecosystem of Western India, *In*: *Sustainable Management of Land Resources – An Indian Perspective*, Reddy, G. P. O., Patil, N. G., & Chaturvedi, A. (Eds.); Apple Academic Press, Canada, pp. 47-76.

Bhattacharyya, T., Sarkar, D., Pal, D. K., Mandal, C., Baruah, U., Telpande, B., & Vaidya, P. H. (2010). Soil information system for resource management – Tripura as a case study, *Current Science*, *99*(9), 1208–1217.

Burrough, P. A. (1986). '*Principles of Geographical Information System for Land Resource Assessment,* ' Monograph.

Burrough, P. A. (1989). Fuzzy mathematical methods for soil survey and land evaluation. *Journal of Soil Science, 40*, 477–492.

Challa, O., Vadivelu, S., & Sehgal, J. (1995). *Soils of Maharashtra for Optimizing Land Use.* NBSS Publ. 54b, NBSS & LUP, Nagpur. p.112.

Das, D. K., Bandyopadhyay S., Chakraborty D., & Srivastava, R. (2009). Application of modern techniques in characterization and management of soil and water resources. *Journal of Indian Society of Soil Science, 57*(4), 445–460.

de Bie, C. A. J. M. (2000). *Comparative Performance Analysis of Agro-Ecosystems*, ITC Dissertation No. 75, 232.

Doran, J. W., & Parkin, T. B. (1994). *Defining and Assessing Soil Quality*, SSSA, Madison, WI, *35*, pp. 1–22.

Doran, J. W., & Parkin, T. B. (1996). *Qualitative Indicators of Soil Quality: A Minimum Dataset*, SSSA, Madison, WI, *49*, pp. 25–37.

Driessen, P. M. (1997). Biophysical sustainability of land use systems, *ITC Journal, 3/4*, 243–247.

Driessen, P. M., & Konijn, N. T. (1992). *Land Use Systems Analysis,* Department of Soil Science and Geology, Wageningen Agricultural University, The Netherlands.

Dubey, P. N., Bhaskar, B. P., Chandran, P., Singh, B., & Mishra, B. K. (2017). Major and Trace Element Geochemistry in Ferruginous Soils Developed Under Hot Humid Malabar Region, India, *In*: *Sustainable Management of Land Resources – An Indian Perspective*, Reddy, G. P. O., Patil, N. G., & Chaturvedi, A. (Eds.); Apple Academic Press, Canada, pp. 239-270.

Duivenbooden, N. (1995). *Land use systems analysis as a tool in land use planning with special reference to North and West African agro-ecosystems,* Landbouwuniversiteit te Wageningen.

Dumanski, J., Phipps, M., & Huffman, E. (1987). A study of relationships between soil survey data and agricultural land use using information theory. *Canadian Journal of Soil Science,* 67, 95–102.

Fairbridge, R. W. (1968). *The Encyclopedia of Geomorphology*. Reinhold Bk., NY, 1295 pp.

FAO. (1976). A framework for land evaluation. FAO soils bulletin 32, Rome, Italy: FAO.

FAO. (1984). Guidelines: Land evaluation for forestry. FAO, Rome, Italy, *Soils Bull., 48*, 8–42.

FAO. (1985). Guidelines: Land evaluation for irrigated agriculture. Soils Bulletin 55, Rome, Italy: FAO. 231 pp. S590. F68 no. 55.

FAO. (1995). Land and Water Bulletin 2. Planning for Sustainable Use of Land Resources: Towards a New Approach. Rome.

FAO. (1983). Guidelines: Land evaluation for rainfed agriculture. FAO, Rome, Italy, *Soils Bull*, *52*, 11–54.

Gangopadhyay, S. K., Reza, S. K., Mukhopadhyay, S., Nayak, D. C., & Singh, S. K. (2017). Characterization of Coastal Soils for Enhancement of Productivity and Livelihood Security: A Case Study from Coastal Plains of West Bengal, India, *In*: *Sustainable Management of Land Resources – An Indian Perspective*, Reddy, G. P. O., Patil, N. G., & Chaturvedi, A. (Eds.); Apple Academic Press, Canada, pp. 175-211.

Govindarajan, (1971). Soil Map of India (1:7 m). *In: Review of Soil Research in India* (J. S. Kanwar and Raychaudhuri, Eds.).

ICAR and NAAS. (2010). Degraded and Wastelands of India Status and Spatial Distribution, Indian Council of Agricultural Research and National Academy of Agricultural Science, New Delhi, p. 158.

Jafari, R., Lewis, M. M., & Ostendorf, B. (2008). An image-based diversity for assessing land degradation in an arid environment in South Australia. *Journal of Arid Environment. 72*, 1282–1293.

Kadu, P. R., & Kharche, V. K. (2017). Evaluation of Shrink-Swell Soils in Semi-Arid Region of Central India for Soil Resource Management and Sustainable Agriculture, *In*: *Sustainable Management of Land Resources – An Indian Perspective*, Reddy, G. P. O., Patil, N. G., & Chaturvedi, A. (Eds.); Apple Academic Press, Canada, pp. 105-143.

Karlen, D. L., Andrews, S. S., & Doran, J. W. (2001). Soil quality, current concepts and applications. *Advances in Agronomy*, *74*, 1–40.

Klingebiel, A. A., & Montogomery, P. H. (1961). Land Capability Classification, U.S. Department of Agricultural Handbook No. 210.

Kumar, N., Reddy, G. P. O., Chatterji, S., Srivastava, R., & Singh, S. K. (2017). Land Suitability Evaluation for Soybean Using Temporal Satellite and GIS: A Case Study from Central India, *In*: *Sustainable Management of Land Resources – An Indian Perspective*, Reddy, G. P. O., Patil, N. G., & Chaturvedi, A. (Eds.); Apple Academic Press, Canada, pp. 387-410.

Li, Z., Zhu, Q., & Gold, C. (2005). *Digital Terrain Modeling: Principles and Methodology*. CRC Press, Boca Raton.

Liu, S. H., Sylvia, P., & Li, X. B. (2002). 'Spatial patterns of urban land use growth in Beijing,' *Journal of Geographical Sciences 12*(3), 266–274.

Maji, A. K. Srinivas, C. V., Dubey, P. N., Reddy, G. P. O., Kamble, K. H., & Velayutham, M. (2002). Soil Resources Information of Nagaland in GIS for Land use planning in Mountainous Region, *GIS India*, *11*(4), 13–16.

Maji, A. K., Nayak, D. C., Krishna, N. D. R., Srinivas, C. V., Kamble, K., Reddy, G. P. O., & Velayutham, M. (2001). Soil Information System of Arunachal Pradesh in a GIS Environment for Land Use Planning, *International Journal of Applied Earth Observation and Geoinformatics, 3*(1), 69–77.

Maji, A. K., Reddy, G. P. O., Tamgadge, D. B., & Gajbhiye, K. S. (2005). Spatial Modeling for Crop Suitability Analysis Using AGROMA GIS software, *Asian Journal of Geoinformatics*, *5*(3), 47–56.

Maji, A. K., Reddy, G. P. O., Thayalan, S., & Walke, N. J. (2004). Characterization and Classification of Landforms and Soils over Basaltic Terrain in Sub-humid Tropics of Central India, *Journal of Indian Society of Soil Science, 53*(2), 154–162.

Manchanda, M. L., Kudrat, M., & Tiwari, A. K. (2002). Soil survey and mapping using remote sensing, *Tropical Ecology, 43*(1), 61–74.

Mandal, A. K., Singh, R., Joshi, P. K., & Sharma, D. K. (2017). Mapping and Characterization of Salt Affected Soils for Reclamation and Management: A Case Study from Trans-Gangetic Plains of India, *In*: *Sustainable Management of Land Resources – An Indian Perspective*, Reddy, G. P. O., Patil, N. G., & Chaturvedi, A. (Eds.); Apple Academic Press, Canada, pp. 145-173.

Mani, J. K., Varghese, A. O., & Joshi, A. K. (2017). Assessment of Gross Primary Productivity in Semi Arid Agricultural Region of Central India using Temporal MODIS Data, *In*: *Sustainable Management of Land Resources – An Indian Perspective*, Reddy, G. P. O., Patil, N. G., & Chaturvedi, A. (Eds.); Apple Academic Press, Canada, pp. 365-386.

Manjare, B. S. Jambhulkar, P., Padhye, M. A., & Girhe, S. S. (2017). Digital Terrain Analysis and Geomorphological Mapping Using Remote Sensing and GIS: A Case Study from Central India, *In*: *Sustainable Management of Land Resources – An Indian Perspective*, Reddy, G. P. O., Patil, N. G., & Chaturvedi, A. (Eds.); Apple Academic Press, Canada, pp. 327-345.

Moore, I. D., Grayson R. B., & Ladson, A. R. (1991). Digital terrain modeling: a review of hydrological, geomorphologic and biological applications, *Hydrological Processes, 5*, 3–30.

Moore, I. D., Grayson, R. B., & Ladson, A. R. (1993). Digital terrain modeling: a review of hydrological, geomorphological, and biological applications. Terrain Analysis and Distributed Modeling in Hydrology. *In*: *Advances in Hydrological Processes* (K. J. Beven and I. D. Moore, Eds.), John Wiley & Sons, Inc, Chichester.

Murthy, R. S. Hirekerur, L. R., Deshpande, S. B., & Venkat Rao, B. V. (1982). Benchmark Soils of India-morphology, characteristics and classification for resource management, NBSS & LUP, Nagpur, 374 p.

Naitam, R. K., Kharche, V. K., Gabhane, V. V., Bhattacharyya, T., Taley, S. M., & Konde, N. M. (2017). Characterization and Classification of Soils of Purna Valley in Semi-Arid Region of Central India, *In*: *Sustainable Management of Land Resources – An Indian Perspective*, Reddy, G. P. O., Patil, N. G., & Chaturvedi, A. (Eds.); Apple Academic Press, Canada, pp. 77-103.

NBSS & LUP. (2002). Soils of Raisen district, Madhya Pradesh, NBSS Publ. No. 511, p.58.

NBSS & LUP. (1994). Soil site suitability criteria for different crops. *In*: *Proceedings of National Meet on Soil Site Suitability Criteria for Different Crops*, Feb, 7–8, 1994, 31.

Ostir, K., Veljanovski, T., Podobnikar, T., & Stancic, Z. (2003). Application of satellite remote sensing in natural hazard management: The Mount Mangart landslide case study. *International Journal of Remote Sensing 24*(20), 3983–4002.

Pickup, G. (1989). New land degradation survey techniques for arid Australia: problems and prospects. *Australian Rangeland Journal, 11*, 74–82.

Pimpale, A. R., Rajankar, P. B., Wadatkar, S. B., & Ramteke, I. K. (2017). Time Series Satellite Data and GIS for Crop Acreage Estimation: A Case Study from Central India, *In*: *Sustainable Management of Land Resources – An Indian Perspective*, Reddy, G. P. O., Patil, N. G., & Chaturvedi, A. (Eds.); Apple Academic Press, Canada, pp. 347-364.

Qi, F., Zhu, A. X., Harrower, M., & Burt, J. E. (2006). Fuzzy soil mapping based on prototype category theory. *Geoderma, 136*(3–4), 774–787.

Raychaudhuri, S. P., & Mathur, L. M. (1954). *Bull. N. Inst. Sci.*

Raychaudhuri, S. P. Agarwal, R. P. Dutta Biswas, N. R., Gupta, S. P., & Thomas P. K. (1963). Soils of India, Indian Council of Agricultural Research, New Delhi.

Reddy, G. P. O., & Maji, A. K. (2003). Delineation and Characterization of Geomorphological features in a part of Lower Maharashtra Metamorphic Plateau, using IRS-ID LISS-III data, *Journal of the Indian Society of Remote Sensing, 31*(4), pp. 241–250.

Reddy, G. P. O., & Sarkar Dipak (2012). Assessment of Soil Loss for Prioritization of Sub-watersheds – A Remote sensing and GIS Approach, NBSS Publ. No. 137, NBSS & LUP, Nagpur, pp. 55.

Reddy, G. P. O., Maji, A. K., Chary, G. R., Srinivas, C. V., Tiwary, P., & Gajbhiye, K. S. (2004). GIS and Remote sensing Applications in Prioritization of River sub basins using Morphometric and USLE Parameters: A Case Study. *Asian Journal of Geoinformatics, 4*(4), 35–49.

Reddy, G. P. O., Maji, A. K., Srinivas, C. V., & Velayutham, M. (2002). Geomorphological analysis for inventory of degraded lands in a river basin of basaltic terrain, using remote sensing data and Geographical Information Systems. *Journal of the Indian Society of Remote Sensing 30*, 15–31.

Reddy, G. P. O., Mondal C., Srivastava R. Bhattacharyya, T. Naidu, L. G. K., Sidhu, G. S. Baruah U, Singh R. S., Kumar Nirmal, Singh S. K., & Sarkar, D. (2014). *Development of Indian Soil Information System – A Geoportal Project Report,* pp. 34.

Reddy, G. P. O., Nagaraju, M. S. S., Ramteke, I. K., & Sarkar, D. (2013). Terrain Characterization for soil resource mapping in part of semi-tract of Central India using high resolution satellite data and GIS, *Journal of the Indian Society of Remote Sensing, 41*, 331–343.

Reddy, G. P. O., Sarkar, D., Mandal, C., Srivastava, R., Bhattacharyya, T., Naidu, L. G. K., Sidhu, G. S., Baruah, U., Singh, S. K., Singh, R. S., Nair, K. M., Sen, T. K., Chandran, P., Sahoo, A. K., Srinivas, S., Kumar, N., & Chavan, S. (2014a). Soil Resource Database and Information System, *In: Geospatial Technology for Integrated Natural Resources Management* (P. S. Roy and R. S. Dwivedi, Eds.), Yes Dee Publishing Pvt. Ltd, Chennai. pp. 370–406.

Reddy, G. P. O., Shekinah, D. E., Maurya, U. K., Thayalan, S., Prasad, J., Ray, S. K., & Bhaskar, B. P. (1999). Landscape-soil relationship in part of Bazargaon plateau, Maharashtra, *The Geographical Review, 63*(3), pp. 280–291.

Reddy, G. P. O., Singh S. K., Mondal C., Srivastava R., Bhattacharyya, T., Naidu, L. G. K., Sidhu, G. S., Baruah U., Singh R. S., Nirmal, K., & Sarkar, D. (2014b). Development of District Soil Information System (DSIS) on 1:50,000 Scale (50 Districts), NBSS & LUP, Nagpur, Project Report, pp. 160.

Reddy, G. P. O., Singh, S. K., Kumar, N., Mondal, C., Srivastava, R., Maji, A. K., & Sarkar, D. (2017). Geospatial Technologies in Development of Soil Information System and Prototype GeoPortal, *In*: *Sustainable Management of Land Resources – An Indian Perspective*, Reddy, G. P. O., Patil, N. G., & Chaturvedi, A. (Eds.); Apple Academic Press, Canada, pp. 411-442.

Reddy, M. G. R., Reddy, G. P. O., Maji, A. K., & Rao, K. N. (2004). Landscape Analysis for Pedo-Geomorphological Characterization in part of Basaltic Terrain, Central India using Remote Sensing and GIS, *Journal of the Indian Society of Remote Sensing, 31*(4), 271–282.

Reddy. G. P. O., Maji, A. K., Nagaraju, M. S. S., Thayalan, S., & Ramamurthy, V. (2008). Ecological evaluation of land resources and land use systems for sustainable development at watershed level in different agro-ecological zones of Vidarbha region, Maharashtra using Remote sensing and GIS techniques, Project Report, NBSS & LUP, Nagpur. 270p.

Rossiter, D. G. (1996). A theoretical framework for land evaluation, *Geoderma, 72*, 165–202.

Sahu, N., Reddy, G. P. O., Kumar, N., Nagaraju, M. S. S., Srivastava, R., & Singh, S. K. (2014). Characterization of landforms and land use/land Cover in basaltic terrain using IRS-P6 LISS-IV and Cartosat-1 DEM data: A case study, *Agropedology, 24*, 166–178.

Sahu, N., Reddy, G. P. O., Kumar, N., Nagaraju, M. S. S., Srivastava, R., & Singh, S. K. (2017). Morphometric Analysis using GIS Techniques: A Case Study from Basaltic Terrain of Central India, *In*: *Sustainable Management of Land Resources – An Indian Perspective*, Reddy, G. P. O., Patil, N. G., & Chaturvedi, A. (Eds.); Apple Academic Press, Canada, pp. 301-326.

Sahu, N., Reddy. G. P. O. Kumar, N., & Nagaraju, M. S. S. (2015). High resolution remote sensing. GPS and GIS in soil resource mapping and characterization: A Review, *Agricultural Reviews, 36*(1), 14–25.

Sarkar, D., Gangopadhyay, S. K., & Sahoo, A. K. (2006). Soil resource appraisal towards land use planning using satellite remote sensing and GIS- a case study in Patloinala micro-watershed, District Puruliya, West Bengal. *Journal of the Indian Society of Remote Sensing, 34*, 245–260.

Schaetzl, R. J., & Anderson, S. (2005). *Soils: Genesis and Geomorphology*. Cambridge University Press, Cambridge, XIII, 817, pp.

Sehgal, J., Saxena, R. K., & Vadivelu, S. (1987). Field Manual, Soil Resource Mapping of different states in India. NBSS Publ. 13, NBSS & LUP, Nagpur, India.

Snel, M., & Bot, A. (2003). *Some Suggested Indicators for Land Degradation Assessment of Drylands* (draft paper). Food and Agriculture Organization, Rome.

Soil Survey Staff. (1975). Soil Taxonomy. A Basic System of Soil Classification for Making and Interpreting Soil Surveys. *Agricultural Handbook*, 436 (Washington DC).

Srinivas, C. V., Maji, A. K., Reddy, G. P. O., & Chary, G. R. (2002). Assessment of Soil Erosion Using Remote sensing and GIS in Nagpur District, Maharashtra, for Prioritization and Delineation of Conservation Units, *Journal of the Indian Society of Remote Sensing, 30*(4), 197–211.

Srivastava, R., & Saxena R. K. (2003). Technique of large scale soil mapping in basaltic terrain using satellite remote sensing data. *International Journal of Remote Sensing 24*(21), 4187–4194.

Staff, NBSS & LUP. (2000). Soils of India, NBSS Publ. 94. National Bureau of Soil Survey and Land Use Planning, Nagpur, India pp. 130+11 sheets maps.

Suresh Kumar (2017). Geospatial Tools and Techniques in Land Resources Inventory, *In*: *Sustainable Management of Land Resources – An Indian Perspective*, Reddy, G. P. O., Patil, N. G., & Chaturvedi, A. (Eds.); Apple Academic Press, Canada, pp. 273-300.

Sys, C. (1985). Land evaluation. Algemeen Bestuur vande Ontwikkelingss, Ghent, Belgium: International Training Centre for Post-Graduate Soil Scientists. State University of Ghent.

Sys, C., Van Ranst, E., & Debaveye, J. (1991). Land evaluation Part I and II – crop requirement. Belgium General Administration for Development Cooperation. Agricultural Publications No. 7.

Townshend, J. R. G., Goff, T. E., & Tucker, C. J. (1985). Multi-temporal dimensionality of images of normalized difference vegetation index at continental scales, *IEEE Transactions* on Geosciences and *Remote Sensing,* GE, *23*, 888–895.

Tueller, P. T. (1987). Remote sensing science applications in arid environment. *Remote Sensing of Environment*, *23*, 143–154.

Van Ranst, E., Tang, H., Groenemans, R., & Sinthurahat, S. (1996). Application of fuzzy logic to land suitability for rubber production in peninsular Thailand. *Geoderma, 70*, 1–19.

Velayutham, M. (1999). Available soil information and the need for a systematic classification of soils of India. *Journal of Indian Society of Soil Science, 48*(4), 683–689.

Vishakha, T. D., Reddy. G. P. O., Maji, A. K., & Ramteke, I. K. (2013). Characterization of Landforms and Soils in Complex Geological Formations – A Remote Sensing and GIS Approach, *Journal of The Indian Society of Remote Sensing*, *41*(1), 91–104.

Walke, N., Reddy, G. P. O., Maji, A. K., & Thayalan, S. (2012). GIS-based multi-criteria overlay analysis in soil- suitability evaluation for cotton: A case study in the black soil region of central India. *Computers and Geosciences, 41*, 108–118.

Wilson, D. J., & Gallant, J. C. (2000a). Digital terrain analysis. *In*: *Principles and Applications* (Wilson, D. J., & Gallant, J. C., Eds.), Terrain Analysis: John Willey & Sons, Inc., New York, pp. 1–27.

Wilson, P. J., & Gallant, J. C. (2000b). Secondary topographic attributes. *In*: Terrain Analysis: Principles and Applications (Wilson P. J., & Gallant J. C., Eds.), John Willey & Sons, Inc., New York, pp. 87–131.

Young, F. J., Maatta, J. M., & Hammer, R. D. (1991). Confidence intervals for soil properties within map units. *In: Spatial Variability of Soils and Landforms*. SSSA Special Publication, *28*, 213–230.

CHAPTER 2

SOIL MANAGEMENT AND LAND USE PLANNING: AN INDIAN PERSPECTIVE

N. G. PATIL,[1] S. K. SINGH,[2] G. P. OBI REDDY,[1] and A. CHATURVEDI[3]

[1]*Principal Scientist, ICAR-National Bureau of Soil Survey and Land Use Planning, Amravati Road, Nagpur – 440 033, India*

[2]*Director, ICAR-National Bureau of Soil Survey and Land Use Planning, Amravati Road, Nagpur – 440 033, India*

[3]*Principal Scientist and Head, ICAR-National Bureau of Soil Survey and Land Use Planning, Amravati Road, Nagpur – 440 033, India*

CONTENTS

ABSTRACT

Soil surveys generate detailed information about the soils that could be used for number of applications like land use planning (LUP), soil quality assessment and management, nutrient mapping, analysis of spatial distribution, etc. Intertwined with bio-physical, socio-economic and environmental information, it is used for visualizing potential scenarios of different land uses asides evaluating the prevailing land uses. The administrators, researchers, planners and farmers are greatly aided by modern tools like Geographic information system (GIS), Global Positioning System (GPS) and fast processing computer aids to arrive at decisions that are beneficial for sustainable natural resources management. However, in India paucity of data and variation in scales, acquisition methods, consistency, lack of skills and many such constraints confine LUP research to an academic exercise. Before commencement of an inclusive LUP, the focus must shift from soil based LUP to resource based LUP with due considerations to other sources of livelihood like water, forest, common property resources, etc. Current status of research is also reviewed including the articles included in section III and IV of this compilation/book. There are articles ranging from soil nutrient management in GIS to on field evaluation results of applying water tank silt in croplands for improving nutrient and physical status of the soils. Crop specific soil management practices are also reported in two different chapters. Similarly there are two reports on site specific nutrient application derived from spatial information/distribution in semi-arid part of the country. It is argued that research efforts are limited and skilled manpower is required for promoting LUP and soil management research in India.

2.1 INTRODUCTION

Soil survey information is used for many different applications such as land use planning, nutrient mapping, recreational planning, infrastructure planning, land use management or framing policies, etc. Administrators, land users and planners often combine the soil data with climate, environment, socio-economic factors, prevailing policies and development goals to analyze and simulate/test decisions for sustainable utilization of natural resources. Assessment of soil and land resources is the first step in any land related planning process. World Bank estimates (2016) show that per capita arable land in India is 0.12 ha in 2015. In a densely populated (204 million population) state like Uttar Pradesh, it declines to 0.08 ha. Although agriculture accounts for only about 14% of the Gross Domestic Product, it is still the main source of livelihood for the majority of the rural population, and provides the basis of food security for the nation (DOLR 2013). In a country with enormous variety of natural resources and equally diverse socio-economic factors affecting land use decisions, land use planning is indeed a complex task. Earlier two sections of this book dealt with land resources inventory, the methods used and modern tools/techniques. However, it is essential to recapitulate land evaluation here because method of land evaluation is often chosen according to the purpose of Land Use Planning (LUP).

LUP research work in India is hampered by lack of institutional support and other constraints like non-availability or availability of relevant data at varied scales (e.g., land records, soil information, water resources). There is inadequate co-ordination amongst the agencies engaged in acquiring related data because of which integration of bio-physical and socio-economic data, livestock data, demographic information, and environment information becomes difficult. Objective land evaluation based on scientific information is vital to decision making and deciding priorities for a specified land use but it is difficult to achieve under such constraints.

Land evaluation in India so far has been confined to crop suitability evaluation and academic exercise such as generating irrigation suitability, erosion maps, etc. In the developed world where land evaluation is now a complex process, no generic models of land use planning are developed. As knowledge, societal needs, data availability and environmental conditions evolve, it is an illusion to claim that land units, ecosystem attributes and objectives can be defined only once and for all. A specific, closed land evaluation system, developed for a certain objective and for well-defined criteria, will become irrelevant for land use planning with time (Meyer et al., 2013). It is apparent that future land evaluation systems in India need to be open ended.

In natural resources based development planning, watershed is a preferred unit. Soil maps, however may transcend the watershed boundaries, therefore, land evaluation may not always intersect with the two main components of LUP (i.e., soil and water). LUP at national scale helps in setting national priorities for development, whereas local level LUP could play a major role in livelihood of individual or a community. Land evaluation is applicable both in areas where there is strong competition between existing land uses in highly populated zones as well as in zones that are largely undeveloped (FAO, 2007). Land ownership in India is complex. Government is an owner in all the forest lands while individual farmer owns agricultural land. Then there are other issues like tenancy, *Jaminadari*, absentee landlords discussion about which is out of the scope of this article but suffice to state that stakeholders like state and individual farmers perceive LUP with sometimes completely opposing objectives. The people living in and around forest depend heavily on non-timber forest produce for their living but have limited access to the forest lands as regulated by the state. The farmer on the other hand is completely at liberty to grow the crop of his/her choice in accordance with the market forces or family needs. Thus LUP in India is mostly academic exercise with limited advisory role.

2.1.1 LAND USE PLANNING AT NATIONAL LEVEL

Based on the concept of length of growing period (LGP), which is an index of crop production that considers soil moisture availability for the crop, the country was divided into 20 Agro-ecological (AER) regions. Later the AERs were further subdivided into 60 sub-regions (Figure 2.1). Utility of subzonation has been demonstrated in estimating carbon soil carbon status and available potassium in Indo Gangetic Plains and Black Soils Region of the country. The AESR is currently being modified based on greater climate and soil data that became available recently. These regions have assisted decision makers and planners in setting development priorities as well managing natural resources.

2.1.2 LAND USE PLANNING AT REGIONAL LEVEL

Based on the soil survey information, soil maps of major states of the country have been generated by the bureau. The soil data were further analyzed to form research bulletins, articles, etc., depicting LUP, for e.g., land use plan for development of Bundelkhand region, benchmark shrink-swell soils of India, Red and lateritic soils of India. Soil moisture map of India is another example of regional LUP. The maps are accompanied by suggested land use plans and also highlight the problems related to soils and land use.

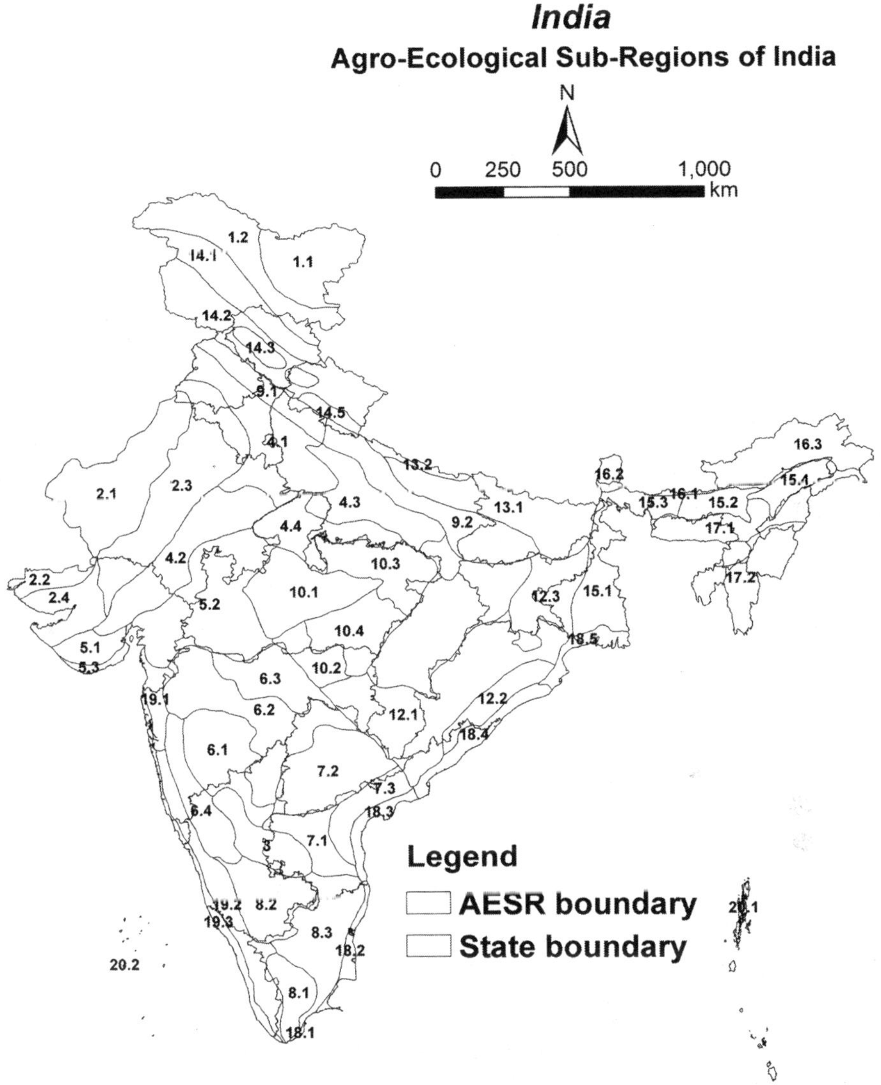

FIGURE 2.1 Agro-Ecological Subregions of India (Source: Velayutham et al., 1999).

2.1.3 LAND USE PLANNING AT DISTRICT LEVEL

Similar to regional level plans, district soil survey reports and land use plans have been published by the bureau. Districts like Yavatmal were surveyed at

1:50,000 scale and hence these reports have detailed information on suggestive land use options that could be implemented in the study district. Jalna, Medak, Meerut, Mysore, Udaipur, Wardha are among the districts across India that have been studied for LUP. Recent developments in LUP for district level studies emphasize decision support systems in GIS environment.

2.1.4 LAND USE PLANNING AT WATERSHED/VILLAGE LEVEL

Village level or watershed level studies require soil information at larger scale like 1:10,000 or 1:5000 scale. The bureau has conducted many studies that investigate efficient cropping systems, utilization of common property resources, livestock management in the realm of LUP. These studies highlight the importance of local issues like socio-economic condition of the village and family preferences for land use. It is beyond the purview of this article to discuss at length about LUP at different levels in India primarily because the articles in the section III and IV do not contain any report on such studies. It is however pertinent to note that the subsequent chapters are manifestation of the LUP studies conducted at different levels.

2.2 CURRENT NEEDS AND FUTURE DEMANDS

It is projected that (3.5% GDP growth scenario), demand for food grains (including feed, seed, wastage and export) in India will rise to 256 mt by the year 2020. Land use planning will pertinently target a per hectare yield of 2.7 tons for rice and 3.1 tons for wheat. Keeping this in view, the land use plans have to be made for achieving higher productivity and sustainable food security. The National Bureau of Soil Survey and Land Use Planning (NBSS & LUP) is the sole research organization actively engaged in LUP research. Work done at NBSS and LUP and by extension in India could be broadly divided into four categories in a chronological order (Patil et., 2015).

1. During the early years, zonation of the country based on length of growing period criteria was focused.
2. Development of soil suitability criteria for major crops of the country followed by crop experiments to evaluate the developed criteria including soil attributes
3. LUP at different planning levels/units like village, watershed, district based on soil distribution, topography and climate.

4. Customized LUP like identifying suitable area for commercial crops—rubber, tea, etc.

Based on the articles appearing in this book and review of literature, following inferences could be drawn (Patil et al., 2015).

A paradigm shift in land use planning from edaphic factors to resources based planning is required to include water resources, livestock, common property resources, socio-economics, and eco-services in LUP. A protocol for data collection on these factors to be developed and followed for national, regional and micro level land use planning. Georeferencing of all the data to be made mandatory.

Provide legislative support to protect most fertile lands/most suitable lands for agriculture in national interest. Such lands (highly productive) to be delineated to facilitate decisions after appropriate evaluation of potential loss. Eco-sensitive zones to be delineated and evaluated for the ecological services they provide.

Encourage policy that aims at protection and optimal utilization of common property resources such as pasture lands, water bodies, community forests, etc. Develop LUPs encompassing growth of non-timber forest produce species on common lands.

Development of National Decision Support System (DSS) that simulates different scenarios at national scale. All major crops to be included in DSS.

National priorities for land use and crop choice need to be defined and mechanism to be set up to implement decisions taken in national interest.

Economic evaluation of eco-services to be made an integral part of Land Use Planning.

Raising awareness levels, human resources development for LUP research to be accorded priority. Geo-portal will overcome the inadequacies partially by providing soil data. Other data also need to be made accessible.

All future studies and policy analyses to include water resources data as an integral part. A paradigm shift in approach would also be required as input costs differ across regions. Globally the work in LUP could be divided into distinct phases: (i) LUP based mostly on edaphic factors, then (ii) participatory land use planning implemented at smaller levels like watershed, village or cluster of villages; (iii) the third phase in developed world is now incorporating environment concerns and ways to address it though LUP. In India, the LUP research is still limited to soil centric evaluation and possibilities of alternate crops. All the three phases could be observed in the inclusions. For instance an article on soil evaluation to grow cotton represents typical first phase approach (Pable et al., 2017). Another article reports second phase (Ramamurthy et al., 2017) of LUP (i.e., participatory LUP work in India. and there is article focusing on necessity of paradigm shift (Patil et al., 2015) to incorporate non-edaphic factors (third phase). There is also an interesting article that shows how technological advances like genetically modified cotton seed (Bt cotton) is changing crop decisions (Venugopalan et al., 2017). Bt hybrids along with associated agro-techniques propelled India to become the second largest producer (390.9 lakh bales), consumer (306 lakh bales) and the largest exporter (118 lakh bales) of raw cotton in 2013. The land use changes accompanying Bt-cotton revolution is the unprecedented increase in cotton acreage from 7.8 million ha in 2002 to 12.6 million ha in 2014. Other structural changes include the spread of intra-hirsutum hybrids to 92% in 2012, a concomitant decline in the area of Diploid and Egyptian cotton from 31% to 3%. This caused a glut of long staple cotton and a shortage of both short staple cotton and extra long staple cotton. The gain in area under cotton in the post-Bt era is at the expense of pulses, oilseeds and coarse cereals that are vital for our food security. Cereal crops and traditionally drought resistant crops have been replaced by cotton in many parts of the country. The pattern is disturbing and a warning to the planners of possible crop dynamics triggered by market forces and technologies.

Traditional LUP work is found in an article on the nutrient optima-based productivity zoning for fertilizer use in citrus (Srivastava and Das, 2017). Application of geospatial tools like GPS and GIS coupled with diagnosis and recommendation integrated system (DRIS) based nutrient diagnostics has substantially aided in developing rationale of fertilizer use using Site Specific Nutrient Management (SSNM) concept in crops like citrus, avocado, coconut, olive, etc., which could well be expanded to other perennial crops. The last article (Suresh Babu and Rajanikanth, 2017) on effect of employment guarantee scheme of the Central Government attempts to evaluate its impact on land use pattern and natural resources management in drought prone area.

2.3 NUTRIENT STATUS AND MANAGEMENT

For monitoring the impact of present land use and management on soil health and to develop sustainable alternate land use plan, mapping of nutrient status and stock on real time is imperative. This kind of mapping also helps to understand the contribution of soils towards the stability of ecosystem and climate change. NBSS & LUP has completed GPS based nutrient mapping status for the state of Assam, Jharkhand, Tripura and West Bengal in GIS environment. Soil nutrient maps have thus been generated indicating spatial distribution of nutrients.

The fertility status of the soils mainly depends on the nature of vegetation, climate, and topography, texture of soil and decomposition rate of organic matter. Optimum productivity of any cropping systems depends on adequate supply of plant nutrients. NBSS & LUP has completed nutrient mapping in selected states like West Bengal, Jharkhand, Assam and Tripura and prepared GIS based district soil nutrient maps (organic carbon, available N, P, K, S and available Fe, Mn, Zn, Cu, B and Mo). Soil nutrient mapping based on geo-referenced soil sampling, laboratory analysis, database structuring in GIS and subsequent interpolation maps provide information regarding soil nutrients status at district level, which enable to identify the site specific nutrient deficiencies for site specific nutrient management in agricultural planning and attaining the highest fertilizer use efficiency. In climate change mitigation strategies, assessment and site-specific integrated nutrient management improvement of nutrient use efficiency play its important role. About 47% of the nitrogen applied (36 out 78 million tons) is lost annually to the environment through leaching, erosion, runoff, and gaseous emissions (Roy et al., 2002). Nitrogen recovery for rainfed crops is about 20–30%, while irrigated crops recover only 30–40% of applied nitrogen (Roberts, 2008). By increasing nutrient use efficiency through better timing of fertilizer and organic input application, precision, and effectiveness through improved placement of appropriate quantities of applied inputs.

2.4 ADVANCED TOOLS IN SOIL NUTRIENTS MANAGEMENT

Part III and IV of this book has 14 chapters related to application of soil information in enhancing agricultural productivity, assessing, monitoring soil nutrient status and strategies to sustain the productivity and nutrient status. Research results on use of geospatial tools like GIS, GPS and other modern tools in soil management have been discussed by the researchers for fertility monitoring; there are two articles reporting development of nutrient indices (Gajare and Dhawan, 2017; Katkar et al., 2017) derived from the soil information/properties. These two articles emphasize importance of site-specific nutrient application derived from spatial information/distribution in semi-arid part of the country. This region depends heavily on rainfall for growing seasonal crops, therefore agricultural productivity is constrained by soil moisture availability in time and space. In soybean growing soils, depletion of available phosphorus soils is notable. The other report (Sahoo et al., 2017) on soil fertility status of Jamtara district of Jharkhand elaborates evaluation and mapping using GIS environment and based on the georeferenced database generated through the analysis of seven thousand two hundred nine (7209) composite surface (0–20 cm) soil samples collected at 500 m grid using GPS. The soils in general are low in organic matter content and available phosphorus along with moderate contents of available potassium and sulfur as well as deficiency of zinc and boron certainly reflects in general poor fertility status of the district. The next two articles (Shirale et al., 2017; Vajantha et al., 2017) discuss crop specific management interventions to overcome soil constraints. In Ashwagandha (*Withania somnifera*) growing soils of Ananthapur district (semi-arid region) low to medium soil OC (0.42 to 0.64%), low available N (185 to 242 kg ha^{-1}), medium available P (12.15 to 18.24 kg ha^{-1}), medium to high available K (235 to 310 kg ha^{-1}), and sufficient available S (10.34 to 18.22 mg kg^{-1}) is reported.

2.5 SOIL QUALITY MANAGEMENT AND ISSUES

In soil management research reports, a case study (Shirale et al., 2017) from Central India (Purna Valley) has interesting findings on shrink-swell soils. The major problems of these soils are native salinity/sodicity, poor hydraulic conductivity; compact, dense subsoil and incomplete leaching of salts from soil due to severe drainage impairments. Field experiments on cotton followed by green gram in *kharif* and chickpea in *rabi* season are reported. The pH of experimental sites varied from 8.27 to 8.34, the cation exchange capacity between 52.17–53.47 cmol (p^+) kg^{-1}) and Exchangeable Sodium Percentage between 10.39 and 11.29. Significant improvement in chemical properties of soils was observed under gypsum indicating reduction in pH from 8.31 to 8.19 and ESP from average initial of 11.15–8.11, but simultaneously significant reduction was also observed under organic amendments like sunhemp in terms of reduction in pH upto 8.25 and ESP upto 8.82. Among the crop residues and green manures sunhemp in situ green manuring showed the highest significant decrease in exchangeable Na followed by sunhemp and cowpea. The crop residues and green manures although slow in reclamation were found effective in improving chemical properties of sodic soils gradually in addition to more carbon sequestration. In another field study (Gabhane et al., 2017) on cotton crop in *Vertisols* land configurations comprising of flat bed, ridges and furrows, opening of furrow after two rows and opening of furrow after each row were evaluated. Significantly higher seed cotton yield was recorded in land treatment-opening of furrow after each row followed by ridges and furrows which were at par with each other. The chapter recommends integrated application of 50% recommended dose of fertilizer + farm yard manure+ phosphate solubilizing bacteria + *Azotobacter.*

2.5.1 IMPROVING PHYSICAL PROPERTIES

During the last two decades, many small and minor irrigation tanks have been built as a part of watershed development in India. These tanks often lose their storage capacity due to siltation. The trapped silt reportedly contains significant amount of nutrients (Kharche et al., 2017) and improve texture of surface soils when used in *Vertisols*. Its application also improved saturated hydraulic conductivity, soil water retention and there was slight decrease in bulk density. The section has solitary report on possible use of sewage water for irrigation after Phytorid treatment (Mhaske et al., 2017). The use of the treated domestic wastewater showed improvement in the physicochemical properties of the soil, along with the nutrient status as compared to the application of well water. The content of micronutrients and heavy metals in soil due to treated domestic wastewater irrigation were well below the safe limits in soil. These findings provide practically feasible recommendations for peri-urban agriculture where treatment cost could be offset by location advantage of direct selling to the urban consumer.

2.6 CAPACITY BUILDING IN LAND USE PLANNING

Apart from the constraints in LUP research discussed above, lack of skills/ trained manpower is acutely felt in the country. Only one organization (NBSS & LUP) pursuing LUP research with a small base of 10-12 scientists located at six different places is an inadequate pool of skilled persons for a country of sub-continental size. Equally important is to identify vulnerable communities like rural landless people or communities dependent on livestock or common property resources. Adaptation involves combinations of strategy, policy, institutional and technical options that require a wide range of skills and multidisciplinary actions, including ecosystem-based and livelihood approaches. Particular attention has to be given to the most vulnerable groups and communities, e.g., those in fragile environments such as drylands, mountain areas, lakes and coastal zones (FAO, 2009a), as well as those disadvantaged by socio-economic factors such as land ownership, gender, caste and age constraints. Effective adaptation involves creating the capacity to cope with more frequent, increasingly difficult conditions and gradual climate changes, even without being able to anticipate their precise nature. Under such circumstances, the focus will be on decision-making and capacity development that strengthen institutions, social learning, iterative planning, innovation and development processes. This means taking a "no regrets" approach, promoting adaptive actions that will be beneficial even if future impacts are uncertain and climate change threats do not occur exactly as anticipated (FAO, 2009b). Institutional support should be provided to assess the climate change impacts and vulnerabilities, disaster risk management, sustainable land, water and biodiversity management, strengthening institutions and policies for adaptation, developing and disseminating technologies, practices and processes for adaptation and accessing potential sources of adaptation financing (FAO, 2009a).

2.7 CONCLUSIONS

Numbers of issues in LUP were identified. The most important issue is the paradigm shift required from LUP based on edaphic factors to more inclusive LUP based on water, livestock, forest and common property resources. It is anticipated that many problems like scale limitations in acquisition of varied data, bringing out uniformity and consistency, co-ordination amongst different related agencies, we argue that the beginning has to be made to make LUP in India better understood by administrators, planners and stakeholders alike.

KEYWORDS

- **Land Use Planning**
- **Soil Nutrient Status**
- **GIS**
- **Natural Resources Management**

REFERENCES

DOLR. (2013). National Land Utilization Policy-framework for land use planning and management. Last accessed in April 2016 at http://dolr.nic.in/dolr/downloads/PDFs.

FAO. (2007). Land Evaluation-Towards a Revised Framework. Food and Agriculture Organization, Rome.

FAO. (2009a). Coping with Changing Climate: Considerations for Adaptation and Mitigation in Agriculture. Food and Agriculture Organization, Rome.

FAO. (2009b). Profile for Climate Change, Food and Agriculture Organization, Rome.

Gabhane, V. V., Rupali, G., Nagdeve, M. B., Ganvir, M. M., & Turkhede, A. B. (2017). Impact of Land Configuration and Integrated Nutrient Management on Productivity of Rainfed Cotton in Vertisols of Central India, *In: Sustainable Management of Land Resources – An Indian Perspective,* Reddy, G. P. O., Patil, N. G., & Chaturvedi, A. (Eds.); Apple Academic Press, Canada, pp. 555-579.

Gajare, A. S., & Dhawan, A. S. (2017). Assessment of Soil Fertility Status in Soybean Growing Soils: A Case Study from Semi-arid Region of Central India, *In: Sustainable Management of Land Resources – An Indian Perspective,* Reddy, G. P. O., Patil, N. G., & Chaturvedi, A. (Eds.); Apple Academic Press, Canada, pp. 461-483.

Katkar, R. N., Kharche, V. K., Lakhe, S. R., Ikhe, U. D., Age, A. B., & Mali, D. V. (2017). Geo-Referenced Soil Fertility Monitoring for Optimized Fertilizer Use: A Case Study from Semi-arid Region of Western India, *In: Sustainable Management of Land Resources – An Indian Perspective,* Reddy, G. P. O., Patil, N. G., & Chaturvedi, A. (Eds.); Apple Academic Press, Canada, pp. 445-460.

Kharche, V. K., Patil, S. M., Mali, D. V., Jadhao, S. M., Shirale, A. O., & Katkar, R. N. (2017). Impact of Tank Silt on Soil Quality and Crop Productivity in Rainfed Areas: A Case Study from Central India, *In: Sustainable Management of Land Resources – An Indian Perspective,* Reddy, G. P. O., Patil, N. G., & Chaturvedi, A. (Eds.); Apple Academic Press, Canada, pp. 581-600.

Meyer, A. D., Estrella, R., Jacxsens, P., Deckers, J., Rompaey, A. V., & Orshoven, J. V. (2013). A conceptual framework and its software implementation to generate spatial decision support systems for land use planning. *Land Use Policy*, 35, 271–282.

Mhaske, A. R., Taley, S. M., & Katkar, R. N. (2017). Impact of Treated Domestic Sewage Water Irrigation on Soil Properties, Maize Yield and Plant Uptake: A Case Study from Nagpur City, Central India, *In: Sustainable Management of Land Resources – An Indian Perspective,* Reddy, G. P. O., Patil, N. G., & Chaturvedi, A. (Eds.); Apple Academic Press, Canada, pp. 601-621.

Pable, D., Chatterji S., Sen T. K., Venugopalan M. V., & Giri J. D. (2017). Land Evaluation for Rainfed Cotton: A Case Study from Central India, *In: Sustainable Management of Land Resources – An Indian Perspective,* Reddy, G. P. O., Patil, N. G., & Chaturvedi, A. (Eds.); Apple Academic Press, Canada, pp. 699-716.

Patil, N. G., Chaturvedi, A., & Singh, S. K. (2015) Land Use Planning in India: Past and Future. *Agropedology 25*(1), 1–19.

Ramamurthy, V., Singh, S. K., Chattaraj, S., Reddy, G. P. O., & Ramesh Kumar, S. C. (2017). Land Resource Inventory towards Village Level Agricultural Land Use Planning, *In: Sustainable Management of Land Resources – An Indian Perspective,* Reddy, G. P. O., Patil, N. G., & Chaturvedi, A. (Eds.); Apple Academic Press, Canada, pp. 643-668.

Roberts, T. (2008). Improving nutrient use efficiency. *Turkish Journal of Agriculture and Forestry 32*, 177–182.

Roy, R. N., Misra, R. V., & Montanez, A. (2002). Decreasing reliance on mineral nitrogen – yet more food. *Ambio, 31*(2), 177–183.

Sahoo, A. K., Singh, S. K., Nayak, D. C., Mukhopadhyay, S., Banerjee, T., Sarkar, D., Sarkar, A. K., Agarwal, B. K., & Shahi, D. K. (2017). Soil Acidity and Poor Nutrient Status – Emerging Issues for Agricultural Land Use Planning in Jamtara District of Jharkhand, *In: Sustainable Management of Land Resources – An Indian Perspective,* Reddy, G. P. O., Patil, N. G., & Chaturvedi, A. (Eds.); Apple Academic Press, Canada, pp. 485-510.

Shirale, A. O., Kharche, V. K., Zadode, R. S., Katkar, R. N., Meena, B. P., & Age, A. B. (2017). Management of Sodic Black Calcareous Soils: A Case Study from Central India, *In: Sustainable Management of Land Resources – An Indian Perspective,* Reddy, G. P. O., Patil, N. G., & Chaturvedi, A. (Eds.); Apple Academic Press, Canada, pp. 531-554.

Srivastava, A. K., & Das, S. N. (2017). Nutrient Optima-Based Productivity Zonality and Rationale of Fertilizer Use in Citrus, *In: Sustainable Management of Land Resources – An Indian Perspective,* Reddy, G. P. O., Patil, N. G., & Chaturvedi, A. (Eds.); Apple Academic Press, Canada, pp. 717-738.

Suresh Babu, V., & Rajanikanth, G. (2017). Impact of Mahatma Gandhi Nregs on Land Use Pattern and Natural Resource Management in Drought Prone Rayalaseema Region of Andhra Pradesh, India, *In: Sustainable Management of Land Resources – An Indian Perspective,* Reddy, G. P. O., Patil, N. G., & Chaturvedi, A. (Eds.); Apple Academic Press, Canada, pp. 739-767.

Vajantha, B., Umadevi, M., Patnaik, M. C., Rajkumar, M., Subbarao, M., & Naidu, M. V. S. (2017). Spatial Distribution of Available Nutrients in Ashwagandha (Withania Somnifera) Grown Soils of Farmer Fields in Andhra Pradesh, *In: Sustainable Management of Land Resources – An Indian Perspective,* Reddy, G. P. O., Patil, N. G., & Chaturvedi, A. (Eds.); Apple Academic Press, Canada, pp. 511-529.

Velayutham, M., Mandal, D. K., Mandal, C., & Sehgal, J. (1999). Agro-Ecological Subregions of India for Planning and Development, NBSS & LUP Publ. No. 35, National Bureau of Soil Survey and Land Use Planning, Nagpur, p. 327.

Venugopalan, M. V., Reddy, A. R., Kranthi, K. R., Yadav, M. S., Vandana, S., & Pable, D. (2017). A Decade of Bt-Cotton in India: Land Use Changes and Other Socio-Economic Consequences, *In: Sustainable Management of Land Resources – An Indian Perspective,* Reddy, G. P. O., Patil, N. G., & Chaturvedi, A. (Eds.); Apple Academic Press, Canada, pp. 669-698.

World Bank, URL: http://data.worldbank.org/indicator/AG.LND.ARBL.HA.PC. Accessed on April 2016.

PART I

LAND RESOURCE INVENTORY AND CHARACTERIZATION

CHAPTER 3

LAND RESOURCE INVENTORY AND EVALUATION FOR AGRICULTURAL LAND USE PLANNING IN SEMI-ARID ECOSYSTEM OF WESTERN INDIA

B. P. BHASKAR,[1] A. CHATURVEDI,[2] T. BHATTACHARYYA,[3] and S. S. GAIKWAD[4]

[1]*Principal Scientist, ICAR National Bureau of Soil Survey and Land Use Planning, Amravati Road, Nagpur – 440 033, India*

[2]*Principal Scientist and Head, ICAR-National Bureau of Soil Survey and Land Use Planning, Amravati Road, Nagpur – 440 033, India*

[3]*Principal Scientist and Former Head, ICAR-National Bureau of Soil Survey and Land Use Planning, Amravati Road, Nagpur – 440 033, India*

[4]*Assistant Technical Officer, ICAR-National Bureau of Soil Survey and Land Use Planning, Amravati Road, Nagpur – 440 033, India*

CONTENTS

ABSTRACT

The agricultural suitability model considering soil, landscape and climate criteria was used to develop strategic development plans for selected commodities in Lagadwal village, Sakri tehsil of Dhule district representing semi-arid ecosystem of Western India. In the study, an effort was made to evaluate basaltic land resources of Lagadwal village through detailed soil survey and photo-pedological interview walks and discussions with farmers. The landscape was divided into 25 soil mapping units based on geology and topography that are common to the extent of common broad land uses at similar level of performance. These soil-mapping units were used as base for discussing and characterizing the capabilities for a range of land uses and intended uses under consideration. The dryland agriculture with soybean, sorghum and pearl millet crops on low quality agricultural lands of Lagadwal village has not changed little over decades. Accordingly, the visual mosaics of landscape photographs were arranged on six themes related to soil – landscape units (soil depth, soil surface unevenness, subsoil compaction, forest cover and top soil erosion) and eight themes on management options (grass field borders, multispecies buffer, ground cover crops, minimum tillage, diverse rotations, strip grasses, agroforestry and crop intensity) and opinions were sought so as to develop linkages for thematic based agro-programs such as erosion control, increasing crop diversity and raising crops at landscape level. The study advocates socio-agro-ecological approach linking of esthetic value of natural resources as per the socio-economic needs of rural communities.

3.1 INTRODUCTION

The extent, distribution and exploitation of land are factors that have long been identified as fundamental influences on agricultural development paths and poverty reduction (Binswanger and Pingali, 1988; Jayne et al., 2014). There is an evidence of rising rural population density combined with land use policy choices that profoundly affect the structure of farming systems and the economy of rural communities in terms of shrinking farm size (83% of farmers have land holdings less than 2 ha and they cultivate nearly 41% of the arable land) (Sharma, 2011), monocultivation of farms contributing to land degradation and unsustainable forms of agricultural intensification (over-exploitation of groundwater: 75% blocks in Punjab, 59% in Rajasthan, 37% in Karnataka and Tamil Nadu (GOI, 2011) and changes in land allocation (plateauing of the productivity in major crops is a matter of concern and efforts are needed to step up crop productivity). India's landmass is only 2.4% of the global total, but it is currently home to 16.7% of the world's population (James, 2011). Conventionally, land use surveys are undertaken by conjunctive use of cadastral and topographical maps accompanied by questionnaires to collect the data on various aspects of land utilization. But usually this process is found to be time consuming (Sharma et al., 1984) and thus the data and maps become outdated by the time they are published and circulated (Gautam and Narayan, 1983) due to very dynamic nature of changes in utilization pattern (Saxena et al., 1983). The decision making about land use allocation are more urgent in rural areas because of loss of agricultural land, population pressure and livelihood involving greater economy between uses (Verheye, 1997). Since, early times, land evaluation has evolved with the increasing use of non soil factors (Van Diepen, et al., 1991) but later on the edaphic land evaluation methods such as the land capability classification system (Klingebiel and Montgomery, 1961) or the United States Bureau of Reclamation (USBR) land classification for irrigated land use were used widely. The mathematical models, known as parametric indices, began to be applied to determine soil production capacity. Amongst these indices, the Storie index stands out (Storie, 1933). Other indices are described in Boixadera and Porta (1991) and Pierce, et al. (1983). Owing to the diversity of the existing land-evaluation systems, in 1976 the Food and Agriculture Organization (FAO) published A Framework for Land Evaluation with further modifications in later publications considering land qualities in each unit with the requirements of each land use type (FAO, 1983, 1985a, 1985b, 1993). The earlier work done on land resource inventory and land evaluation for rural agricultural planning at global level and Indian scenario is discussed below.

3.1.1 LAND RESOURCE CHARACTERIZATION AND EVALUATION: GLOBAL PERSPECTIVE

Site-specific soils mapping is often considered to be synonymous with Order 1 soil surveys completed by the National Cooperative Soil Survey (NCSS). Minimum delineations are very small (perhaps less than 1 ha). This detailed soil-site information can be used for designing experimental agriculture plots, preparing management plans for public lands, and other uses requiring detailed and very precise knowledge of the soils and their variability. Field procedures permit observations of soil boundaries throughout their length and soils in each delineation are identified by transecting or traversing. Mapping units are mostly consociations with some complexes and are phases of soil series or are miscellaneous areas. Base map for detail survey is generally at 1:12,000 scale (1″ = 1000) or a larger (Soil Survey Division Staff, 1993). The large-scale soil survey system in Hungary consists of four main parts: (1) genetic soil map, indicating soil taxonomy units, and the parent material; (2) thematic soil maps on the most important physical and chemical soil properties; (3) thematic maps, indicating recommendations for rational land use, cropping pattern, amelioration, tillage practice and fertilization; (4) explanatory booklets, including a short review on the physiographical conditions; description of soils, recommendations for their rational utilization; field description of soil profiles; results of field observations or measurements and data of laboratory analyzes (Sarkadi and Varallay, 1989). These maps were used widely and successfully in Hungary and became an easily applicable scientific basis of intensive, large-scale agricultural production in spite of the fact that generally these maps were not published in printed form and are available only as manuscripts at the given farming units or at the Plant and Soil Conservation Stations (Pasztor et al., 2002). The large scale soil map on 1:10,000 in Estonia serves as a major important document with details of soil texture and reaction by layers, but also about the thickness and characterization of the epipedon, quality indices for soil assessment, classes of stoniness, prevalent fractions of stones and erosion risk. Special maps of agronomical status, with a list of proper measures for improvement of soils and their associations, should form a regular component of large-scale mapping (Reintam et al., 2003). The land resource inventory and mapping in New Zealand suggested the smallest area of land that can be treated or managed at farm level between 1:5000 and 1:15,000 scale and considered five physical factors such as rock type, soil type, slope angle, erosion type and vegetation (Lynn et al., 2006). The standard method for attributing and evaluating conventional land resource survey maps can be used for strategic decisions about the management, development and conservation of land resources. The land evaluation rules can be auto generated if populated with the consistently structured soil and landscape information (Schoknecht et al., 2004). The standards described are similar to the land suitability assessment

(FAO, 1983) and land capability assessment (Klingbiel and Montgomery, 1961). The large scale land resource information offers opportunities for improving techniques that utilize digital elevation models (DEMs), remotely sensed data, climate information or crop yield information and impacts of seasonal and long-term climatic change (van Gool et al., 2004).

The best known parameteric systems for rating the quality of land is storie index (Riquier, 1970; Sys and Vereheye, 1975). Productivity ratings (based on physical and chemical properties of soils, Huddleston, 1982) are reflection of real value for agriculture or forestry (Miller, 1984) but later recognizing the importance of management in obtaining economic yields, CER (crop equivalent rating) values were used as index of evaluating soils in Minnesota (Robert, 1989; Rust et al., 1984). These mapping units are evaluated as per the land evaluation classification for different kinds of land use systems (FAO, 1983; van Diepen et al., 1991). When the objective is to design a model for all the land uses present in the rural environment (agriculture, woodlands, recreation, etc.), it is necessary to apply a land-evaluation method in which the land uses are very specific and are defined with great accuracy. This condition excludes the capability systems. In addition, in the framework of land-use planning, land evaluation should not be confined to assessing the physical characteristics, it should also consist the analysis of physical suitability, economic viability, social consequences, and environmental impact produced (Riveira and Maseda, 2006). Using the simple landscape-based presentations of interpreted data (Drohan et al., 2010) and linked with conceptual toposequence models, key soil indicators important for crop production at local level were proposed (Grealish et al., 2015). In land evaluation studies, the agricultural landscape photographs in conjunction with soil surveys were used to measure the perception of farmer's about the changes in agri landscapes (Eija et al., 2014; Tvet et al., 2009). The relevance and feasibility of landscape planning in addressing and resolving the problems of rural communities in Ukaraine was highlighted and advocated to develop landscape plans (Rudenko et al., 2014). Later on, the strong links of rural development with landscape planning was discussed in length by Rega (2014).

3.1.2 LAND RESOURCE CHARACTERIZATION AND EVALUATION: INDIAN PERSPECTIVE

India supports approximately 16% of the world population and 20% of its livestock on 2.5% of its geographical area, making its environment a highly stressed and vulnerable (Maria, 2003). The pressure on land has led to soil erosion, waterlogging, salinity, nutrient depletion, lowering of the groundwater table, and soil pollution—largely a consequence of thoughtless human intervention. The extent of land degradation of 45% of the total geographical area (Sehgal and Abrol, 1994) and in the past few years it is alarming. Agricultural activities that cause land degradation include shifting cultivation without adequate fallow periods, absence of soil conservation measures, cultivation of fragile lands, unbalanced fertilizer use, faulty planning or management of irrigation. Improper agricultural practices are usually observed under constraints of saturation of good lands and population pressure leading to cultivation of 'too shallow' or 'too steep' soils and plowing of fallow land before it has recovered its fertility. Overgrazing and over-extraction of green fodder lead to forest degradation through decreased vegetative regeneration, compaction of soil, and reduced infiltration and vulnerability to erosion. Annual environmental costs for India in 1995 were estimated at US$ 9.7 billion, of which surface water pollution, land degradation and deforestation contributed 84% (Maria, 2003). About 58% of households in the villages are marginal farmers, having less than 1 ha of land and 18% having less than 2 ha. In the Indian context, rural development assumes special significance for two important reasons. Firstly, about two thirds of the population still lives in villages and there cannot be any progress as long as rural areas remain backward. Secondly, the backwardness of the rural sector would be a major impediment to the overall progress of the economy. Ministry of Rural Development (MoRD), Govt. of India on National Land Reform Policy (2013) laid stress on a detailed and comprehensive survey and documentation of the existing land area, its use, titles, and all the information relating to the land, so that its possible potential use can scientifically be determined. A creative integration of the state plans with national perspective of development plan that will result in a comprehensive national land use plan that not only takes into account the development needs of the country but also the best utilization of a particular type of land.

The potentials and constraints of basaltic landscape evaluations for agriculture and or rural livelihood in parts of Maharashtra is well recognized (Bhaskar et al., 2011). It was reported that high resolution (PAN and LISS III) satellite sensor data can be registered with cadastral maps of remarkable accuracy and the cadastral information in the form of maps and records can be updated. For village-level planning, large-scale soil maps are very useful because individual soil series and their phases can be delineated Rao et al. (1996). The large scale mapping using IRS-1C PAN merged data of November and December was used to delineate physiography-land use map for basatilc terrain of Junewani village in Maharashtra by Srivastava and Saxena (2004). Similar approach was followed but with IRS-P6-LISS IV data in mini-watersheds of Upper Subarnrekha sub-catchment Jharkhand India (Londhe and Nathawat, 2010) and using IRS-Resourcesat-1 LISS-IV and Cartosat-1 fused data in Mohammad village of Nalgonda district, Andhra Pradesh (Wadodkar and Ravishankar, 2011). The standard method for generation of land resource survey maps with land evaluation rules can be auto generated if populated with the consistently structured soil and landscape information by using IRS-Resourcesat-1 LISS-IV and Cartosat-1 fused datasets. In recent times, the necessity of detailed soil survey and soil mapping at large scales at village/watershed level for holistic farm planning taking into account climate landscape-soil characteristics (Soil Health Card) and socio-economic conditions of the farm holdings linked to Kisan card credit support is therefore obvious, urgent and cannot brook any further delay (Velayutham, 2012). The integrated rural developmental programs need set of mutually supportive components such as land system surveys by landscape analysis on satellite imagery followed by characterization of land mapping units from field observations in combination of agronomic packages with credit and development of conservation schemes that comes under the domain of sustainable development. The recent development of Geographic Information System (GIS) enables these data to be displayed instantly in a variety of ways, as single parameter maps or in any desired combination (catena concept in prediction of soil distribution on landscape). The profound implications of these developments for natural resources surveys and their applications have yet to be realized.

3.1.3 IMPORTANCE OF THE STUDY

Dhule district in Maharashtra accounts for the highest percentage of scheduled tribes in the state. Out of the total district population of 20.50 lakhs, scheduled tribe population is 8.31 lakhs (40.53%) (Walters, 2013). The cultivated area is 4.64 lakh ha of total area (806,300 ha) with 65% of area under food crops and 56% of land holding less than 2 ha. The agricultural landscape pictures in rural sectors of this part of Maharashtra offers an opportunity to look into the reflective realties of drought and difficulties faced by farmers to rise farm productivity with rudimentary farming systems. Farming systems vary with agro-ecological conditions and no single technical fix will work as a magical cure for improving farm productivity. The policies impose the basic obligations on farmers to keep the open and well managed but agri-developmental programs are poorly implemented with multiple aims in the farming. The photo-elicitation techniques were used to give the experiences and perspectives of local farmers on various scenarios depicted in photographs (Cambell, 2006) and effectively used in land use planning exercises through visual preference surveys (Cambell, 2006; Van Auken, et al., 2010). Hence, the present study aims at integrating soil – land information system with visual perspectives and experiences of farmer's on various land use themes and possible linkages to be needed for implementation of land developmental programs at local level. The objectives were: (i) to characterize, classify and map the soils; (ii) to evaluate soils as per FAO framework for local grown crops; and (iii) to integrate soil–landscape system for evaluating sustainability of current land use and of suggesting alternate land use options for crop developmental programs at landscape level.

3.2 STUDY AREA

The agricultural landscape in Lagadwal village, Sakri tehsil of Dhule district, Maharashtra is located between 21°08′01″ N latitude and 74°06′51″ E longitude and dominated by dykes and residual hills of the Sahyadri Spurs with shallow to stony soils (Figure 3.1). The major part of the area is covered by basaltic flows commonly known as 'Deccan Traps' intruded by dykes of Upper Cretaceous-Lower Eocene age. The Deccan Trap includes several flows of Basalt, which are supposed to have extruded from fissure volcanoes that includes the "pahoehoe" and the "aa" types of flows, the former being very common (Deolankar, 1980). The water bearing strata occur below 30 m depth, beneath the redbole and dense massive basalt exhibits semiconfined to confined conditions. This geological structure in rural landscapes are subjected to severe to very severe on steep slopes running towards south and culminate to pinnacles. The tunnel, gully and landslip erosional features are common with duplex soil association with hard and compact subsoils, which will favor the lateral movement of water. The climate is dry with mean annual rainfall of 576 mm, of which 88% of annual rainfall is received during the south-west monsoon. This area has hot semiarid ecosystem with a length of growing period of 120–150 days (Agro-ecoregion of

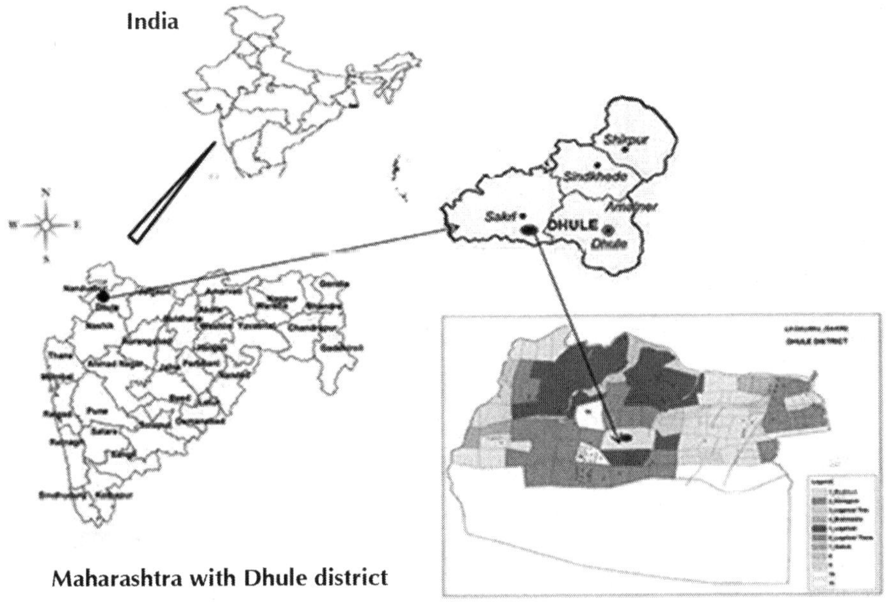

FIGURE 3.1 Location map of the study area.

K4Dd4-Velayutham et al., 1999). The cultivated crops in the village are sorghum (Sorghum *bicolor*), pearl millet (Pennisetum *graminae*), maize (Zea *mays*), wheat (Triticum *aestivum*), sugar cane (Saccharum *officinarum*), chilies (Capsicum *annuum*) and brinjal (Solanum *melogena*). The natural vegetation consists of trees *viz*., Gunj (Abrus *precatorius*), Hivvar (Aegle *marmelos*), Piwala (dhotra – Argemona *mexicana),* Neem (Azadirachta *indica*), Katesawar (Bombax *ceiba*), Rui (Calotropis *gigantean*), Dhak (Butea *monosperma*), Bahawa (Cassia *fistula*), Sisam (Dalbergia *sissoo),* Lokhadi (Ehretia *laevis*), Ghute umbar (Ficus *heterophylla),* Umbar (Ficus *racemosa*), Sabar (Euphorbia *neriifolia)* (Patil and Patil, 2007).

3.3 METHODOLOGY

The two-step methodological approach was adopted in this study involving preliminary landscape analysis at farm scale in terms of topography, geology and land use. The first step was to carry out detailed soil survey on 1:10,000 scale using cadastral map covering 664 ha of land in Lagadwal village of Sakri tehsil, Dhule district, Maharashtra. The intensive field traverses were made to check field boundaries and to acquaint with landscape patterns. Nine representative soil transects with fifty seven soil profiles were studied in different locations and recorded latitude, longitude and elevation of each site with the help of hand held Global Positioning System (GPS). Morphological descriptions of each pedon (Schoeneberger et al., 2002) were recorded. These soils were classified upto subgroup level in the soil orders of Entisols, Inceptisols, Alfisols and Vertisols as per Soil Survey Staff (2010). The soil map was generated with mapping units defined as phases of each series in GIS environment with ARCINFO *Ver.* 8. The land capability classification for arable lands (Klingebiel and Montgomery, 1961) treatment oriented capability classification for soil-water conservation measures (Sheng, 1972) and soil-site suitability for crops was done as per Sys et al. (1991).

The field survey was supported by an interview with the local farmers to pinpoint the landscape conservation to landscape change and the agricultural typology at farm level. Each georeference point is linked with photographical file so as to obtain description of landscape and identification of land use change at farm level. In the second step, the landscape photographs taken during survey were analyzed to form personal construct theory for deriving land use plans (Dalton and Dunnett, 1990). The personal construct theory was made through photo-elicitation process with selected landscape scenes that offered an opportunity to connect practice and theory. These highly personal 'constructs' were based on individual experiences and deeply held values in informal, conversational ways with the farmers to sort out photographic prints into themes or categories. The farmers then prioritized, which of the categories was most important and ranked the photographs for their significance within each category. This process of sorting the photographs seemed to organize their thoughts around the task. This involved grouping of photographs under thematic headings and ordering each group into levels of significance. The effectiveness of photo voice interviews were expressed in agricultural landscapes of steep hills of South Eastern Australia (Beilin, 2005), farming ecosystem services of multifunctional agriculture in Iowa (Drake Larsen, 2011) and in range land management practices of South Africa (Ming Kong, 2012). An open-ended question asking farmers to make a list of key landscape indicators that he/she envisions was obtained and then list any changes that are needed to improve land productivity were solicited. The visual preferences of landscape components such as soil depth, soil surface unevenness, slope, level of subsoil compaction, condition of forest cover and top soil erosion were identified and analyzed to do whole farming in accordance with soil capabilities and connecting soil units with management options like grass field borders, multispecies buffer, ground cover crops, minimum tillage, diverse crop rotations, strip grasses on the contour, agroforestry and increase of crop variety and diversity of species as mentioned in literature. To alleviate negative externalities of conventional approaches on rural agricultural landscapes and to enhance crop productivity, the approaches like personal interviews with photographs were employed (Atwell et al., 2010; Nassauer et al., 2011). Nineteen farmers above 45 years were selected (six in high hills, 5 in mid hills and 8 in low hills) to collect their perspectives and experiences on agricultural practices during transect walks. The results of photovoice and expert land evaluation results were integrated with local knowledge to draw thematic based agri-planning at landscape level. The schematic methodology is given in Figure 3.2.

3.4 RESULTS AND DISCUSSION

3.4.1 SOIL–LANDSCAPE SYSTEMS

A soil landscape is a mapping unit that has recognizable and specifiable topographic and soil properties which can be meaningfully represented on a

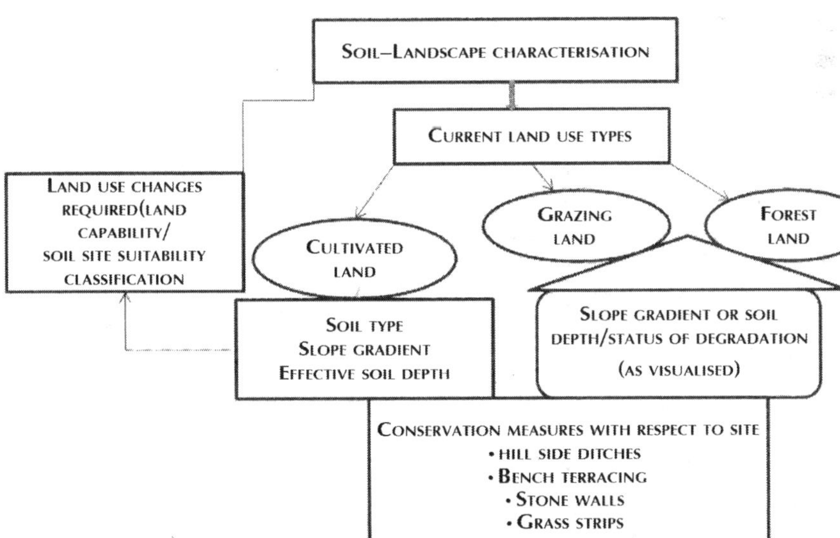

FIGURE 3.2 Flow chart of conservation measures at landscape level based on slope and land capability.

map and described by concise statements (Tulau, 1997). Soil landscapes are not uniform in terrain and soil characteristics, but have a definable pattern of variation. For instance, the Lagadwal landscape ranges from gently sloping ridge crests to very steep side slopes and narrow drainage lines (Figure 3.3). There is considerable variation in slope gradient, soil depth, drainage characteristics and hence the properties which determine capability. The village covers 664 ha of land having high hills in southern part (>620 m above mean sea level) with distinct steep slopes running towards north direction and covers more than 306.3 ha. The steep side slopes cover 10% of total area with narrow incised drainage lines (46% of area) supporting extremely shallow Budkhed (P1) and Lagadwal series (P2). The mid hills (600 to 620 m) covers 13% of area with four land forms *viz.*, broad crests, rounded crests and side slopes, gently undulating uplands and lower slopes. These land forms have moderately shallow Lagadwal thana series (P3) on crests and side slopes, Gaikot (P4) and Lagadwal tola (P5) in foot slopes. The low hills (580 to 600 m) have ridge lines and drainage depressions covering 40% of area with soil associations of moderately deep Navgaon (P6) and Brahmasila series (P7). These landscapes show varying degrees of sheet erosion on undulating plains, gullies near drainage lines and rocky exposures, stone cover on hill tops and ridges, crests and side slopes. These landscapes suggest themselves

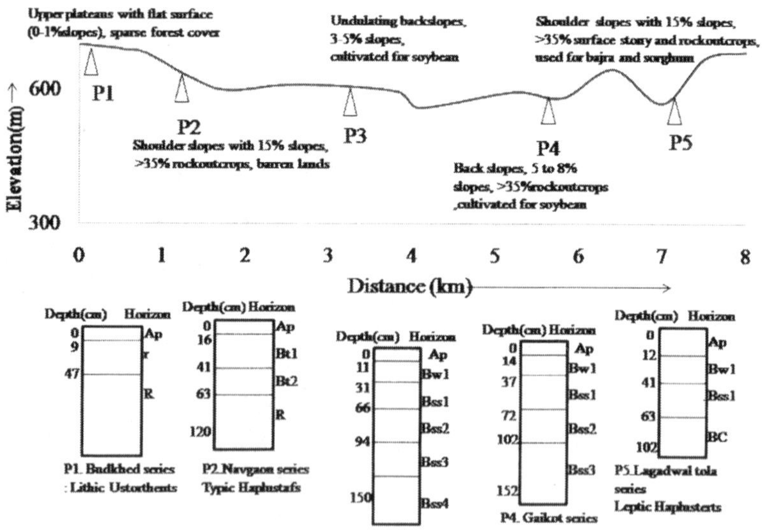

FIGURE 3.3 Schematic soil–landscape diagram of Lagadwal village.

as measures to estimate the impacts of current land use activities on steep lands posing questions such as (i) Is production sufficient to subsistence or profit? (ii) Is the ground cover capable of maintaining good soil conditions? and (iii) Is the soil erosion a cause of declining crop productivity under semi arid climate?

Soil typology is usually defined as a portion of soil cover with diagnostic characteristics resulting from similar process of soil genesis and location of soil typology is known through soil mapping unit delineation. Basaltic terrain is common in this part of Maharashtra and has been in the focus of tribal dominated and developmental programs linked with dry land agricultural crops in the past 20 years. The high basalt rises with thin forest cover (> 620 m above sea level) have moderately deep *Typic Haplustepts* to moderately shallow *Typic Ustorthents* (Budkhed series) and are used for the cultivation of pearl millet and black gram in patches. The mid hills (600 to 620 m) have moderately shallow *Typic Ustorthents* (Lagadwal series) under pearl millet, groundnut, redgram cultivation on crests and *Lithic Ustorthents* with forest cover on side slopes to very deep *Typic or Chromic Haplusterts* (Brahmasila and Gaikot series) in footslopes used for sorghum, maize and sugarcane. The moderately deep *Typic Haplustalfs* on 3–6% slopes are used for cultivation of soybean and redgram. The low hills (580 to 600 m) have moderately deep *Typic Haplustepts* (Lagadwal thana series) and *Typic Haplustalfs* (Navgaon series). These soils are used for cultivation of soybean, onion and sugar cane, very deep *Typic or Leptic Haplusterts* are put under wheat, maize and soybean. The red soil series namely Budkhed, Lagadwal, Lagadwal thana and Navgaon are in association with black soil series such as Gaikot, Brahmasila and Lagadwal tola in mid and low hill s of Lagadwal village. Such kind of "red-black" soil catena with "red soils" on the higher positions of the landscape, and black vertisols in the lower positions were reported (Bhattacharyya et al., 1993; Kantor and Schwertmann, 1974). These kinds of soil–landscapes are put under dryland agriculture during *kharif* season starting from June to September and for *rabi* crops (wheat, maize and groundnut) from November to February, wherever irrigation facilities are available.

3.4.2 SOIL RESOURCES

The distribution of soil series in the study area as shown in the soil map consists of generic characteristics of seven soil series at 23 phases depicted as mapping units (Figure 3.4 and Table 3.1). The mapping units is defined as

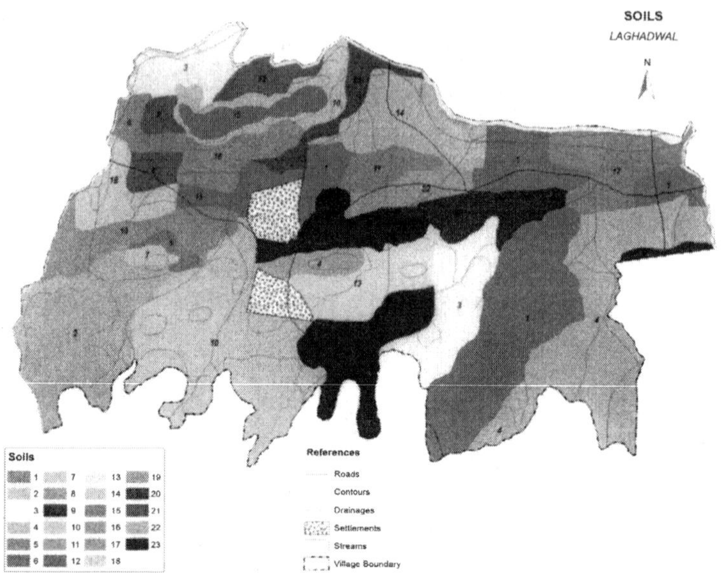

FIGURE 3.4 Soil map of Lagadwal Village, Sakri tehsil of Dhule district, Maharashtra.

per the guidelines of AISLUS (1970) considering texture, series name, erosion and slope class as symbolized: iB2F (where i = sandy clay, B = budkhed series, 2 = moderate erosion and F = >15% slope). It is observed that only three mapping units occupied more than 10% of total area in the village *viz.,* iB3F (96.7 ha, 14.6%), mN2B (78.1 ha, 11.8%) and mN2C (94 ha, 14.2%). Regarding depth classes (Sehgal et al., 1987), the soil series is arranged as follows: extremely shallow – Budkhed and Lagadwal, shallow – Lagadwal thana, moderately shallow – Lagadwal tola and Navgaon, deep – Gaikot and very deep – Brahmasila. Considering the slope >15% not suitable for mechanization and stony features in high, mid and low hills, the 10 mapping units are classified under non arable covering 272.08 ha (40.98%).

3.4.3 PHYSICAL AND CHEMICAL CHARACTERISTICS OF SOILS

The inherent soil properties embodied in land capability classification have a determining role in land use suitability for crops on comprehensive range of soil types and attributes examined. This determines their likely success on a particular soil type and the practical management. The basic physical

TABLE 3.1 Area and Description of Soil Mapping Units

S.No.	Mapping Unit	Mapping Unit Description	Area	
			(ha)	%
1	iB3F	Budkhed: sandy clay, severe erosion with 15–25% slope, moderately gravelly	96.7	14.6
2	mB2E	Budkhed: gravelly clay, moderately erosion with 3–5% slope, very gravelly	54.8	8.3
3	iB3E	Budkhed: sandy clay, severe erosion with 10–15% slope, slightly gravelly	4.7	0.7
4	mN2B	Navagaon: clayey, moderate erosion with 1–3% slope	78.1	11.8
5	mN3E	Navagaon: clayey, severe erosion with 10–15% slope, moderately gravelly	3.8	0.6
6	mN2C	Navagaon: clayey, moderately erosion with 3–5% slope, slightly gravelly	10.9	1.7
7	iN3F	Navagaon: sandy clay, severe erosion with 15–25% slope, very gravelly with wide gullies	3.0	0.5
8	hN2C	Navagaon: sandy clay loam, moderately erosion with 3–5% slope, slightly gravelly	7.8	1.2
9	mN2E	Navagaon: clayey, moderate erosion with 10–15% slope, narrow gullies	46.6	7.0
10	mN2C	Navagaon: clayey, moderate erosion with 3–5% slope, narrow gullies	94.0	14.2
11	mL2B	Lagadwal: clayey, moderately erosion with 1–3% slope, moderately gravelly, poorly terraced	10.4	1.6
12	hL2C	Lagadwal: sandy clay loam, moderate erosion with 3–5% slope, very gravelly, poorly bunded and single cropped	8.3	1.3
13	mL3E	Lagadwal: clayey, severe erosion with 10–15% slope, very gravelly, thin forest	24.2	3.6
14	mL2C	Lagadwal: clayey, moderate erosion with 3–5% slope, moderate gravelly, double cropped	30.4	4.6
15	mL2B	Lagadwal: clayey, moderate erosion with 1–3% slope, terraced and double cropped	21.0	3.2
16	hL2D	Lagadwal: sandy clay loam, moderate erosion with 5–10% slope, moderately gravelly, terraced and single cropped	15.8	2.4
17	mBr2B	Brahmasila: clayey, moderate erosion with 1–3% slope, double cropped	25.0	3.8
18	mGa2B	Gaikot: clayey, moderate erosion with 1–3% slope, poorly bunded and single cropped	15.8	2.4

TABLE 3.1 (Continued)

S.No.	Mapping Unit	Mapping Unit Description	Area (ha)	%
19	mGa2C	Gaikot: clayey, moderate erosion with 3–5% slope, moderate gravelly, poor terraced and single cropped	24.0	3.6
20	mLt1B	Lagadwal tola: clayey, slightly erosion with 1–3% slope, double cropped	18.7	2.8
21	hLt2B	Lagadwal tola: sandy clay loam, moderate erosion with 1–3% slope, double cropped	11.9	1.8
22	mLth2C	Lagadwal thana: clayey, moderate erosion with 3–5% slope, slightly gravelly, poorly bunded land includes newly open lands occurring on different slopes	11.5	1.7
23	mLth2E	Lagadwal thana: clayey, moderately erosion with 10–15% slope, narrow gullies, poorly bunded land, slightly gravelly	6.6	1.0
Total			**663.8**	**100.0**

and chemical characteristics of seven series used in crop evaluation are presented in Table 3.2. The Budkhed series (P1) is extremely shallow with clay loam texture (clay content of 29–35%) and fine loamy particle size class. The surface horizon is neutral with 1.05% of organic carbon and DTPA extractable zinc of 0.38 mg/kg. The exchangeable complex is dominated by Ca (19.2 cmol/kg), Mg (10.1 cmol/kg) and K (1.2 cmol/kg) in surface horizon. The Navgaon series (P2) is well drained, moderately shallow, slightly acid to neutral and fine loamy (clay of 9.92 to 24.9%). This soil has 18.8 to 27.6 cmol/kg exchangeable Ca, 10.9 to 14.3 cmol/kg Mg and irregular distribution of DTPA extractable Zn of 0.48 to 0.98 mg/kg. The Lagadwal series (P3) is slightly acid with fine loamy particle size class (clay of 19 to 26%), exchangeable Ca of 21 to 32.3 cmol/kg and DTPA extractable Zn of 0.72 mg/kg. Brahmasila series (P4) is very deep and fine loamy textured (clay content in control section (25–100 cm) is 29/97%, Soil Survey Staff, 2010). This soil has neutral surface layer with moderately alkaline slicken-sided zones having ex. Ca to ex. Mg ratio more than 1 with its contents varying from 19 to 32 cmol/kg. These soils have DTPA extractable Zn below critical limit (0.8 mg/kg, Lindsay and Norvell, 1978). The Gakot series (P5) is very deep with neutral to slightly alkaline and moderately alkaline slicken-sided zones. This soil has more than 1% of organic

TABLE 3.2 Physical and Chemical Characteristics of Soils Under Study

Horizon	Depth (cm)	Particle size distribution (%)			pH	EC (dSm^{-1})	OC (%)	Exchangeable bases (cmol/kg)				Available K (mg/kg)	DTPA Extractable (mg/kg)			
		Sand	Silt	Clay				Ca	Mg	K	Na		Fe	Mn	Cu	Zn
P1. Budkhed series																
Ap	0–9	25.1	39.6	35.2	6.5	0.054	1.05	19.2	10.1	1.20	0.39	44.8	28.08	52.8	6.92	0.38
r	9–47	23.9	46.3	29.7	6.2	0.092	1.03	23.2	18.0	0.51	0.43	114.2	44.66	61.5	18.0	0.92
P2. Navgaon series																
Ap	0–16	29.88	60.2	9.9	6.2	0.75	1.13	18.8	10.9	0.53	0.39	118.7	64.4	100.5	18.5	0.98
Bt1	16–41	17.23	57.87	24.9	6.6	0.041	1.07	22.2	10.9	0.56	0.43	125.4	51.92	69.8	13.0	0.48
Bt2	41–63	22.85	56.75	20.4	6.8	0.039	0.79	27.6	14.3	0.51	0.43	114.2	25.94	110.0	12.5	0.88
P3. Lagadwal series																
Ap	0–8	51.47	38.73	9.80	6.7	0.12	0.99	21.7	11.9	0.48	0.52	107.5	54.8	69.2	14.0	0.72
P4. Brahmasila series																
Ap	0–11	16.87	49.58	33.55	7.20	0.045	0.568	26.5	18.9	0.75	0.71	33.6	52.4	29.4	5.9	0.48
Bw1	11–31	15.35	45.01	39.64	7.22	0.042	0.549	32.1	21.2	0.78	0.85	34.94	48.3	31.2	8.8	0.38
Bss1	31–66	13.50	61.3	25.20	8.17	0.128	0.176	30.1	20.4	0.81	0.73	36.28	46.4	30.4	7.3	0.36
Bss2	66–94	24.11	40.43	35.46	8.24	0.129	0.098	29.6	19.6	0.78	0.69	34.94	45.5	29.8	10.2	0.28
Bss3	94–120	34.07	32.05	33.88	8.30	0.117	0.078	24.5	18.7	0.62	0.53	27.77	50.6	32.6	11.2	0.43
Bss4	120–150	22.22	48.02	29.76	6.34	0.089	2.262	27.3	18.0	0.52	0.72	23.29	48.8	30.7	8.8	0.40

TABLE 3.2 (Continued)

Horizon	Depth (cm)	Particle size distribution (%)			pH	EC (dSm^{-1})	OC (%)	Exchangeable bases (cmol/kg)				Available K (mg/kg)	DTPA Extractable (mg/kg)			
		Sand	Silt	Clay				Ca	Mg	K	Na		Fe	Mn	Cu	Zn
P5. Gaikot series																
Ap	0–14	12.95	42.6	44.5	3.9	0.036	1.07	44.3	36.9	0.89	0.69	39.87	24.36	66.2	16.2	1.2
Bw1	14–37	22.9	31.1	46.0	7.2	0.049	1.08	35.0	42.9	1.09	0.75	48.83	77.42	49.6	14.2	1.04
Bss1	37–72	10.3	43.8	45.9	7.3	0.057	1.18	35.3	39.6	1.02	0.59	26.43	66.7	47.4	15.2	1.0
Bss2	72–102	33.6	24.83	41.6	7.6	0.087	1.1	37.2	48.4	0.86	0.72	38.52	60.28	27.4	13.6	0.78
Bss3	102–152	14.0	44.7	41.5	7.8	0.052	0.29	37.8	41.9	0.92	0.62	41.21	37.6	43.8	4.8	0.44
P6. Lagadwal tola																
Ap	0–12	34.53	35.44	30.03	7.05	0.035	0.98	25.6	15.4	0.59	0.55	26.43	48.6	30.6	11.0	0.38
Bw1	12–41	21.75	58.08	25.17	7.38	0.046	0.99	28.9	19.4	0.99	0.68	41.66	46.7	33.7	12.3	0.42
Bss1	41–63	39.20	25.11	35.69	7.58	0.039	0.53	30.1	20.1	0.84	0.69	37.63	50.8	30.1	14.5	0.52
Bc	63–102	28.19	39.03	32.78	7.16	0.084	0.71	32.1	12.8	0.82	0.72	36.73	50.2	28.2	10.6	0.32
P7. Lagadwal thana series																
Ap	0–16	45.89	39.28	14.97	6.45	0.084	0.97	21.1	12.91	0.33	0.39	73.92	47.86	62.6	8.00	0.84
Bw1	16–32	34.70	50.26	19.49	6.79	0.058	0.71	27.2	14.08	0.38	0.48	85.12	41.14	50.8	8.32	0.36
Bw2	32–65	44.21	40.82	26.12	7.09	0.063	0.614	30.3	14.33	0.25	0.52	56.0	22.86	16.8	3.46	0.34

carbon throughout depth except in Bss2 layer (72–102 cm) and an increase of exchangeable magnesium from 36.9 to 48.4 cmol/kg with DTPA extractable zinc (0.44 mg/kg) in that particular layer. Lagadwal tola series (P6) is very deep, fine loamy textured (clay in control section is 32.01%) and neutral to slightly acid with an exchangeable Ca of 25 to 32 cmol/kg, Mg of 12.8 to 20.1cmol/kg. The DTPA extractable Zn in this soil is 0.38 to 0.52 mg/kg which is below critical limit. Lagadwal thana series (P7) has loamy texture with sand content of 35 to 46%, silt of 39 to 50% and clay of 15 to 16%. The particle size class is defined as fine loamy at family level as clay content is 21.74%. This soil has slightly acid to neutral with 0.96 to 0.61% of organic carbon. The calcium is dominant on exchange complex with its values ranging from 21.1 to 30.3 cmol/kg. Next to calcium, magnesium is dominant with its contents from 12.9 to 17.49 cmol/kg, Na of 0.39 to 0.55 cmol/kg and K of 0.25 to 0.56 cmol/kg. Among DTPA extractable elements, zinc in subsoils is in deficient range (0.34 to 0.36 mg/kg).

3.4.4 LAND CAPABILITY ANALYSIS

Using land capability classification system, the area has been divided into thirteen mapping units covering 58.99% of total area (391.72 ha) under arable class. Out of 58.99% of arable land, 14.67% of land is classified as Class II land whereas 19.78% of land is classified as Class III and 24.54% as Class IV land (Table 3.3). The Class IV lands (24.54%) have severe limitations of moderate to steep slopes with shallow infertile soils. These lands are capable of supporting grazing and forest land use. It was apparent then 'Treatment-oriented' land capability classification was proposed for steep lands to put under intensive agriculture with prescribed conservation practices (Sheng, 1972) and summary of results presented in Table 3.4. In this system of classification, slope and soil depth as major factors plus considering other soil limiting factors (stony, wet, severe erosion) are considered to classify each piece of land with conservation treatment (s) to follow in the study area. The summits under high hills cover 8.3% of total area with extremely shallow Budkhed series which needs stone wall construction while side slopes (14.6% of total area) and narrow shoulder slopes (6.7%) with Lagadwal series needs hill side ditches. Likewise for mid hills, the crest slopes having moderately shallow Navgaon (36.9% of total area) and Lagadwal thana (1.7% of total area), the suggested conservation methods

TABLE 3.3 Extent and Percentage of Arable Soil Mapping Units Under the Study

Land capability class for arable units	Soil mapping unit	Area ha.	Area %	Suitable crops
IIet: Agriculturally versatile, but requiring a higher level of inputs to achieve high productivity. Slope is greater, soils more variable, and the growing season is 90 to 120 days under rainfed but extended to 6 to 8 months if irrigation water is available. Suitable for high production extensive cropping with irrigation.	mBh2D	25.03	6.39	Sorghum, Soybean, Wheat Sunflower
	mG2D	15.81	4.04	
	mG2C	23.95	6.11	
	iN1A	32.68	8.34	
	Total	**97.47**	**14.67**	
IIIes: Sound grazing and moderate cropping land but limited in versatility. Growing season can be limited to approximately <90 days months due to dryness. With high inputs, moderate to high animal production may be achieved, and moderate cropping yields can be achieved using high inputs and minimum tillage techniques.	mL1B	18.73	4.78	Sorghum, Soybean, Wheat
	mN2C	112.6	28.75	
	Total	**131.33**	**19.78**	
IVst: Capable of supporting grazing under moderate to low stocking rates where clearing has occurred. Slopes are moderate to steep, with shallow infertile soils that need care in their management. Fertility levels are generally low. Unsuited to cropping either because of limitations due to slope, drainage, lack of topsoil depth, weaker structure, low water holding capacity or presence of rock. High inputs may not be economic. Erosion hazard is high. Forest is often the best and most stable form of land use	eN2C	78.06	19.93	Sorghum, Pearl millet
	eN3E	3.77	0.96	
	iN2 F	2.96	0.76	
	iN2E	13.94	3.56	
	mL3E	24.19	6.18	
	hL2B	11.91	3.04	
	hB2D	28.09	7.17	
	Total	**162.92**	**24.54**	

are hill side ditches and stone wall construction. In foot slopes on low hills (3.0% of total area) and gently undulating uplands on mid hills (6.0% of total area), bench terraces and broad base terraces are suggested.

TABLE 3.4 Treatment Land Capability Classification for Conservation Plans

Landscape components	**High hills (>620 m)**			**Mid hills (620–600 m)**			**Low hills (<600 m)**	
Landforms	Summits	Steep side slopes	Narrow shoulder slopes	Broad crest slopes	Round crest side slopes	Gently undulating uplands	Ridge lines	Drainage depressions/ Foot slopes
Slope (%)	1–3	15–25	3–8	3–8	8–15	3–8	3–5	1–3 0–1
Soil series	Budkhed		Lagadwal	Navgaon	Lagadwal thana	Gaikot	Lagadwal tola	Brahmasila
Soil taxonomy	*Lithic Ustorthents*		*Typic Ustorthents*	*Typic Haplustalfs*	*Typic Haplustepts*	*Typic Haplusterts*	*Leptic Haplusterts*	*Chromic Haplusterts*
Area (ha)	57.78	96.7	44.7	244.2	18.1	39.8	30.6	25.0
Area (%)	8.3	14.6	6.7	36.9	1.7	6.0	4.6	3.8
Stoniness /erosion	Strongly gravelly, moderately to severely erosion		Slightly gravelly, moderate erosion	Slightly gravelly, slight to moderate erosion	Moderately gullied, slightly gravelly	Slightly eroded, moderately gravelly	Moderately eroded, moderately gravelly	Slightly eroded
Slope land capability	Class-5/6 (land for conservation and reserves)			Class-3 (A3-intensive soil-water conservation)		Class-2 (A2-moderate soil-water conservation)		Class-1 (A1-unrestricted agricultural use)
Soil-site suitability	Marginally suitable for soybean-wheat			Marginally suitable for wheat, moderately suitable for soybean-sorghum		Moderately suitable for soybean-wheat		Suitable for sorghum-soybean-wheat
Conservation methods	Stone wall, Hill side ditches			Hill side ditches, stone walls		Bench terrace, grass strip, stone wall		bench terraces and broad based terraces

3.4.5 SOIL-SITE SUITABILITY FOR CROPS

The soil-site suitability evaluation shows that Budkhed and Lagadwal series on high hills are marginally suitable for soybean and wheat and these units are suitable for grazing and forestry. The land unit covers 23% of total land. Navgaon and Lagadwal thana series on mid hills cover 38.6% of area that needs intensive soil water conservation measures despite their suitability is moderate for soybean and sorghum. Lagadwal village comes under rainfall scarcity zone of Sakri tehsil with mean annual rainfall of 541 mm. The main livelihood for rural communities is from livestock mainly of cattle, goat and sheep. This village has 41% non-arable land mostly suitable for forestry and grazing purposes. These non-arable lands are now put under cultivation by clearing forest due to moderate and low grazing stocks of natural vegetation.

3.4.6 VISUAL STUDIES OF FIELD PHOTOS

The visual studies of photographs provide a straight view on variety of scenes for comparison and on site evaluation (Kellomaki and Savolainen, 1984; Shuttleworth, 1980). The visual studies of photographs show that tribal communities in Lagadwal village distinguish six landscape indicators such as soil depth exposed on the landscapes and the farming activities of current production mandate (sowing row crops on high hilly terrains with distinct visual signs of erosion and strong stoniness) and supported conservation values for changing landscapes (subsoil hardening). The land limitation may be purely physical such as a shallow depth of soil which is adequate for grazing but not for cropping under low erratic rainfall zone.

The Lagadwal village consists of rolling hills under mixed arable/livestock farming. The agricultural landscapes are subjected to seasonal changes through seasonal production cycles of soybean – sorghum – pearl millet, wheat – maize – sugar cane – vegetable based systems. The poor crop stand and yield of soybean in mid hills and sorghum in high hills clearly indicate the difficulties to work on stony soils (>50% surface stone cover) with lack of machinery skills to work on these landscapes. It is clear from the photographs that the inhabitant tribal farmers are not able to perceive the subtle changes in agricultural landscapes due to their faulty farming practices such as felling of trees in high hills, poor bunding signs of high degree of dissections all along the drainage lines, subsoil hardening and opening of landscapes for agriculture. The conservation of agricultural mosaics (mainly soybean on broad crest slopes of mid hills, drainage floors in lowhill regions for paddy and wheat through the combination of drainage and leveling with contour bunds) is now regarded as critical needs of landscape as expressed within local discourse as negative nature (eroded crest slopes of high hills, money driven production systems, illegal logging of trees, lack of green cover that livestock eats and lack of scenic value) with barren landscapes symbolizing the connection of farmer's role in nurturing the soil. The soil erosion is accentuated by poor practices of farmers on steep hillsides expose to the direct impact of raindrops on clearing and burning, planting with little attempts at soil preservation. Planting is done up and down the hillsides, so that the inter-row is left bare providing ideal conditions for rill erosion down the hill. There is also no attempt to cover the soil by using mulches or cover crops particularly during the inter-crop period. There was some intercropping seen, but unfortunately this was also planted down the hill. The use of any form of barriers to slow down or stop the movement of soil is also minimal.

3.4.7 FARMERS' PERSPECTIVES ON LAND MANAGEMENT OPTIONS

The matrix diagram is made to integrate landscape elements with indicators and management options for sustainable agriculture in the hill lands of Lagadwal village with two distinct color boxes indicating the change of themes to work out on these landscapes in consultation with local farmers (Table 3.5). The farmers perceptions on 6 soil–landscape views and eight management options related to agronomic values are selected based on farmers opinion/review of literature. The rural landscapes in the region are less attractive with greater homogeneity of agricultural landscapes and aridity syndrome. The farmers at high hills are of opinion that the major constraints for farming are unevenness of land surface, effective soil depth, strong slopes and poor condition of forest cover that needs multispecies buffer zones and agroforestry. At mid hills, all farmers agreed upon the poor condition of forest cover but lower level of subsoil compaction problems with limited options for diverse crop rotations and increase of crop variety and diversity. At low hills, the farmers expressed that they are not aware of

TABLE 3.5 Identification of Themes and Change in Level of Details

	High hills					Mid hills					Low hills								
	Number of farmers interviewed																		
Indicators	1	2	3	4	5	6	7	8	9	10	11	12	13	14	15	16	17	18	19
1. Soil depth(cm)	+	0	0	+	–	+	–	+	0	0	0	–	0	–	+	–	0	0	–
2. Soil surface unevenness	+	+	+	–	+	+	+	–	+	–	+	–	0	0	–	+	–	+	–
3. Slope/aspect	+	0	0	+	0	0	0	0	+	0	+	–	0	–	–	0	–	0	–
4. Level of subsoil compaction	–	0	–	0	0	–	–	0	–	0	–	–	0	–	0	–	–	–	–
5. Condition of forest (poor)	+	+	+	+	+	+	+	+	+	+	+	0	+	–	0	+	0	–	+
6. Top soil erosion (rills >50%)	0	–	+	–	–	+	0	–	+	0	–	–	–	–	–	–	–	–	–
Management options (themes added)																			
1. Grass field borders	–	0	–	+	0	–	–	+	+	–	0	–	+	–	0	–	–	–	–
2. Multispecies buffer (shrubs/ trees)	–	+	+	+	0	0	–	–	+	0	0	+	0	–	–	+	–	–	–
3. Ground cover crops	–	–	0	–	–	0	–	0	0	–	–	–	–	–	–	–	–	–	–
4. Minimum tillage	–	–	–	–	–	0	–	0	–	–	0	–	–	0	–	–	0	–	–
5. Diverse crop rotations	–	–	–	–	–	0	–	0	–	–	0	–	0	–	–	0	–	0	–
6. Strip of grasses on the contour	–	+	–	–	–	0	0	0	+	–	–	+	–	0	0	–	+	–	–
7. Agroforestry	0	0	+	0	0	+	0	0	+	+	–	–	–	–	–	–	–	–	–
8. Increase of crop variety and diversity of species	–	0	0	–	0	–	0	–	0	–	–	0	+	+	0	0	+	+	0

The direction of change in level of details: + = high level of details, 0 = same level of details, – = lower level of details, light gray = one theme added, dark gray = three or more themes added.

subsoil compaction and top soil erosion but opted for cover crops and grass field borders. They are not aware of strip or grasses on contours as conservation measure to check top soil erosion at low hills.

3.5 RECOMMENDATIONS

The soil and land resource information at farm level is generated based on scientific understanding of soil–landscape mapping techniques and evaluating the land units according to its capabilities/suitabilities to locally adopted crops. To reverse aridity syndrome and low attractive agricultural landscapes of Lagadwal village, the study advocates two means of landscape research activities such as linking of esthetic valuation of natural resources as per the socio-economic needs of rural communities and emphasizes the role of local population on conserving rural landscapes. With reference to the linking of aesthic value of natural resources, the agricultural landscapes should be defined with a holistic approach by integrating agri-environmental issues with social preferences and perceptions on landscape changes. Therefore, landscape planning for rural agricultural development should be based on socio – ecological approach. Regarding the role of local communities in preserving agri – landscapes, a participatory mode of understanding and sharing of traditional wisdom to improve crop productivity should be resilient and adapt to a changing climate.

3.6 CONCLUSIONS

The conclusions drawn from the land resource survey in hilly basaltic terrain of Lagadwal village in Western Khandesh, Maharashtra shows seven soil series identified and used for deriving soil map containing 23 soil mapping units. Based on land capability analysis, thirteen soil mapping units are classified as arable covering 391.72 ha (58.99% of total area). Further it is estimated that 14.67% of land is grouped under Class II, 19.78% of land as Class-III and 24.54% as Class IV lands with natural constraints of stoniness, hard and compact subsoils, effective rooting depth, slope and low crop, productivity. The study area is a rainfall scarcity zone with limited crop options for suitability evaluation. The suitability analysis shows that sorghum, soybean, wheat and sunflower are good options for Class II and III lands of low hills and mid hills whereas sorghum and pearl millet for Class

IV lands. Based on slope and depth parameters as per treatment oriented land capability classification, the conservation measures like stone wall and hill side ditches are suggested for high and mid hills where slope is above 15%, whereas broad base terraces and grass strips are suggested in low hills. The visual studies of field photo views of experienced farmers on landscape (>45 years) pave a way forward for socio-agroecological approach at land level and is the need of hour to revert the aridity syndrome of agricultural landscapes in the region.

KEYWORDS

- **Conservation Plans**
- **Lagadwal**
- **Land Capability**
- **Photo Elicitation**
- **Soil Survey**

REFERENCES

AISLUS. (1970). *Soil Survey Manual*, IARI, New Delhi, pp. 13–61.

Atwell R. C., Schulte L. A., & Westphal L. M. (2010). How to build multifunctional agricultural landscapes in the U.S. Corn Belt: Add perennials and partnerships. *Land Use Policy, 27*, 1082–1090.

Beilin, R. (2005). Photoelicitation and the agricultural landscape: seeing and telling about farming, community and place. *Visual Studies*, *20*(1), 55–68.

Bhaskar, B. P., Dipak Sarkar, Bobade, S. V., Gaikwad, M. S., Gaikwad, S. S., Nimkar, A. M., Anantwar, S. G., Patil, S. V., & Tapas Bhattacharyya. (2011). Land resource evaluation for optimal land use plans in cotton growing Yavatmal district, Maharasahtra. *The Ecoscan, 1*, 251–259.

Bhattacharyya, T., Pal, D. K., & Deshpande, S. B. (1993). Genesis and transformation of minerals in the formation of red (Alfisols) and black (Inceptisols and Vertisols) soils on Deccan basalt in the Western Ghats, India. *Journal of Soil Science, 44*(1), 159–171.

Binswanger, H., & Ruttan, V. (1978). Induced innovation: technology, institutions and development. Johns Hopkins University Press, Baltimore, MD.

Boixadera. J., & Porta, J. (1991). Soil Information and Cadastral Evaluation. *Method of Index Value* (Spanish Ministry of Economy and Finance, Madrid) (in Spanish).

Campbell, D. C. (2006). A proposed approach for employing resident perceptions to help define and measure rural character. Paper presented at the 12th International Symposium on *Society and Natural Resources*. Vancouver, BC, 6 June. 2006.

Dalton, P., & Dunnett, G. (1990). *A Psychology for Living: Personal Construct Theory for Professionals and Clients*. Wiley and Sons, Chichester.

Deolankar, S. B. (1980). *The Deccan Basalts of Maharashtra, India—Their Potential as Aquifiers*. Groundwater. Lecture, Department of Geology, University of Poona, March, 434–437.

Drake Larsen, G. L. (2011). *Farming for Ecosystem Services: A Case Study of Multifunctional Agriculture in Iowa*, USA. M.Sc. thesis. Iowa State University, USA. pp. 1–125.

Drohan, P. J., Havlin, J. L., Megonigal, J. P., & Cheng, H. H. (2010). The Dig it Smithsonian soils exhibition: lessons learned and goals for the future. *Soil Science Society of America Journal, 74*, 697–705.

Eija, P., Grammatikopoulou, I., Timo Hurme, T., Soini, K., & Uusitalo, M. (2014). Assessing the Quality of Agricultural Landscape Change with Multiple Dimensions. *Land. 3*, 598–616.

FAO. (1983). *Guidelines: Land Evaluation for Rainfed Agriculture*. FAO Soils Bulletin. 52.

FAO. (1985a). *Guidelines: Land Evaluation for Irrigation.* FAO Soils Bulletin no. 55 (FAO, Rome).

FAO. (1985b). Evaluación de Tierras con Fines Forestales. Estudio FAO, Montes 48 (FAO, Rome).

FAO. (1993). *Guidelines for Land-Use Planning.* FAO Development Series (FAO, Rome).

Gautam, N. C., & Narayan L. R. A. (1983). Landsat MSS data for Land Use and Land Cover Inventory and Mapping: A case study of Andhra Pradesh. *Journal of Indian Society of Remote Sensing*, 11(3), 15–28.

Government of India. (2011). *Mid-Term Appraisal Eleventh Five Year Plan 2007–2012,* Planning Commission, Government of India, Oxford University Press, New Delhi.

Grealish, G. J., Fitzpatrick, and R. W., Hutson, J. L. (2015). Soil survey data rescued by means of user friendly soil identification keys and toposequence models to deliver soil information for improved soil management. *Geo Res Journal* 6, 81–91.

Huddleston, J. H. (1982). *Agricultural Productivity Rating for Soils of the Willamette Valley.* Oregon State University. Extension Service. EC1105, 32p.

James, K. S. (2011). India's demographic change: opportunities and challenges. *Science Journal, 333*, 576–580.

Jayne, T. S., Chamberlin, J., & Headey, D. D. (2014). Land pressures: the evolution of farming systems and development status in Africa—A synthesis. *Food Policy, 48*, 1–17.

Kantor, W., & Schwertmann, U. (1974). Mineralogy and genesis of clays in red-black soil toposequences on basic igneous rocks in Kenya. *European Journal of Soil Science. 25*(1), 67–78.

Kellomaki, S., & Savolainan, R. (1984). The scenic value of the forest landscape as assessed in the field and the laboratory. *Landscape Planning, 11,* 97–107.

Klingebiel, A. A., & Mongomery, P. H. (1961). Land capability classification. *Agri. Handbook. 210*. USDA Soil Conservation Service.

Lindsay, W. L., & Norvell, W. A. (1978). Development of DTPA soil test for zinc, iron, manganese and copper. *Soil Science Society of American Journal, 42*, 421–428.

Londhe, S., & Nathawat, M. S. (2010). Large scale soil mapping techniques for granitic terrain using high resolution satellite data. *Trends in Soil and Plant Sciences, 1*(1), 19–31.

Lynn, I. H., Manderson, A. K., Page, M. J., Hamsworth, G. R., Eyles, G. O., Douglas, G. B., Mackay, A. D., & Newsome, P. J. F. (2006). Land Use Capability Survey Handbook – A New Zealand Handbook for the Classification of Land. 3rd edition, Hamilton, Agri Research, Lincoln, Land Care Research, Lower Hutt, GNS Science, 163p.

Maria, A. (2003). *The Costs of Water Pollution in India,* Paper Presented at the conference on Market Development of Water & Waste Technologies through Environmental Economics, 30th–31st October 2003, New Delhi, available at http://www.cerna.ensmp.fr/cerna_globalization/Documents/maria, Delhi.

Miller, G. A. (1984). *Corn Suitability Rating: An Index to Soil Productivity.* Iowa State University. PM1168.

Ming Kong, T. M. (2012). *Understanding Land Management and Desertification in the South African Kalahari with Local Knowledge and Perspectives.* PhD thesis. University of Arizona. pp. 1–225.

Ministry of Rural Development. (2013). *Draft Report on National Land Reform Policy—For Discussion Purpose Only.* Department of Land resources. Government of India. pp. 1–35.

Nassauer, J. I., Dowdell, J. A., Wang, Z., McKahn, D., Chilcott, B., Kling, C. L., & Secchi, S. (2011). Iowa farmers' responses to transformative scenarios for Corn Belt agriculture. *Journal of Soil and Water Conservation, 66,* 18A–24A.

Pasztor, J. Szabo, J., & Bakacsi, Z. C. (2002). GIS Processing of Large-scale Soil Maps in Hungary. Agrokemia Es Talajtan. *51,* 273–282.

Patil, S. L., & Patil, D. A. (2007). Ethnomedicinal plants of Dhule district, Maharashtra. Natural Product *Radiance, 6*(2), 148–151.

Pierce, F. J., Larson, W. E., Dowdy, R. H., & Graham, W. A. P. (1983). Productivity of soils: Assessing long-term changes due to erosion. *Journal of Soil and Water Conservation,* 38, 39–44.

Rao, D. P., Guatham, N. C., Nagaraja, R., & Ramamohan P. (1996). IRS-IC applications in land use mapping and planning, *Current Science, 70,* 575–581.

Rega, C. (2014). *Landscape Planning and Rural Development: Key Issues and Options Towards Integration.* Springer, XIII, 147p.

Reintam, L. L., Kull, A., Palang, H., & Rooma, I. (2003). Large-scale soil maps and a supplementary database for land use planning in Estonia. *Journal* of *Plant Nutrition and Soil Science, 166,* 223–225.

Riquier, J., Bramao, D., & Cornet, J. P. (1970). A new system of soil appraisal in terms of actual and potential productivity. AGL/TESR/70/6. FAO, Rome, 35p.

Riveria, I. S., & Maseda, R. C. (2006). A review of rural land use planning models. *Environment* and *Planning B*: *Planning and Design, 33,* 165–183.

Robert, P. C. (1989). Land evaluation at farm scale using soil survey information systems. In: *Land Qualities in Space and Time.* J. Bouma and A. K. Brecht (Eds.). Pudoc, Wageningen, pp. 299–311.

Rudenko, L., Maruniak, E., & Lisovskiy, S. (2014). Landscape planning for Ukrainian rural communities: challenges, outputs, *Geoadria, 19*(2), 191–204.

Rust, R. H., Hanson, L. D., & Anderson, J. L. (1984). *Productivity Factors and Crop Ratings for Soils of Minnesota.* AG-BU-2199, Minnesota Extension Service, University of Minnesota. St. Paul. 57p.

Sarkadi, J., & Varallay, G. Y. (1989). Advisory system for mineral fertilization based on large-scale landsite maps. *Agrokémia* és *Talajtan. 38,* 775–789.

Saxena, R. K., Pofali R. M., & Hirekerur L. R. (1983). Preparation of optimal Land use Maps for agricultural development using remote sensing techniques, *Journal of Indian Society of Remote Sensing, 11*(3), 29–38.

Schoeneberger, P. J., Wysocki, D. A., Benham, E. C., & Broderson, W. D. (2002). *Field Book for Describing and Sampling Soils,* Version 2.0. Natural Resources Conservation Service, National Soil Survey Center, Lincoln, NE.

Schoknecht, N., Tille, P., & Purdie, B. (2004). Soil–landscape mapping in south-western Australia. Overview of methodology and outputs. Department of Agriculture, Resource Management *Technical Report,* 280.

Sehgal, J., & Abrol, I. P. (1994). *Soil Degradation in India: Status and Impact.* Oxford and IBH Publishing Co. Pvt Ltd., New Delhi.

Sehgal, J. L., Saxena, R. K., & Vadivelu, S. (1987). *Field Manual. Soil Resource Mapping of Different States in India.* NBSS & LUP, Nagpur. pp. 73.

Sharma, K. P., Jain S. C., & Garg P. K. (1984). Monitoring land use and landcover changes using Landsat images, *Journal of Indian Society of Remote Sensing, 12*(2), 65–70.

Sharma, V. P. (2011). India's agricultural development under the new economic regime: perspective and strategy for the 12th five year plan. 7th *Annual conference on Indian society of Agricultural Economics*, University of Agricultural Sciences, Dharwad, 3–5th, November, 1–49 pp.

Shcng, T. C. (1972). A treatment oriented land capability classification for hilly land of humid tropics. In: Technical Report No. TA 3112. *Food and Agriculture Organization*, United Nations. Rome, Italy.

Shuttleworth, S. (1980). The use of photographs as an environmental presentation medium in landscape studies. *Journal of Environmental Management, 11*, 61–76.

Soil Survey Staff. (1993). Soil Survey Manual. Soil Conservation Service. U.S. Department of Agriculture Handbook, 18.

Soil Survey Staff. (2010). *Keys to Soil Taxonomy.* Eleventh edition. United States Department of Agrculture. Natural Resopurce Conservation Service.

Srivastava, R., & Saxena, R. K. (2004). Technique of large-scale soil mapping in basaltic terrain using satellite remote sensing data. *International Journal of Remote Sensing, 25*(4), 679–688.

Storie, R. (1933). An index for rating the agricultural values of soils. California Agricultural Experimental Station, USA, Bulletin No. 526, 25 pp.

Sys, C., & Vereheye, W. (1975). Principles of land classification in arid and semiarid areas (revised). Publ. intern. Training Centre Post Graduate Soil Science, Ghent, 42p.

Sys, C., Van Ranst, E., & Debaveye, J. (1991). Land evaluation, Part 1: Principles in land evaluation and crop production calculations. International Training Centre for Post Graduate Soil Scientists, University of Ghent, Belgium, pp. 265.

Tulau, M. J. (1997). *Soil Landscapes of the Cooma 1, 100,000 Sheet Map.* Department of Conservation & Land Management, Sydney.

Tveit, M. S. (2009). Indicators of visual scale as predictors of landscape preference. A comparison between groups. *Journal of Environmental Management, 90*, 2882–2888.

Van Aukena, P. M., Frisvoll, S. J., Susan I., & Stewart, S. I. (2010). Visualizing community: using participant-driven photo-elicitation for research and application. *Local Environment, 15*(4), 373 –388.

Van Diepen, C. A., Van Keulen, H., Wolf, J., & Berkhout, J. A. A. (1991). Land evaluation: from intuition to quantification. In: *Advances in Soil Science*. B. A. Stewart (ed.), Springer, New York, 139–204 pp.

van Gool, D., White, P., Schoknecht, N., Bell, R., & Vance, W. (2004). Land evaluation for pulse production in WA in Agribusiness Crop Updates, pp. 60–62.

Velayutham, M. (2012). National Soil Information System (NASIS) and Land Resource Mapping for Perspective Land Use Planning and Pragmatic Farm level Planning. *Madras Agricultural Journal, 99*(4–6), 147–154.

Velayutham, M., Mandal, D. K., Mandal, C., & Sehgal, J. (1999). *Agro-Ecological Subregions of India for Planning and Development.* NBSS and LUP, Publ. No. 35, 372p.

Verheye, W. H. (1997). Land use planning and national soils policies. *Agricultural Systems*, *53*, 161–174.

Wadodkar, M., & Ravishankar, T. (2011). Soil resource database at village level foe developmental planning, *Journal of the Indian Society of Remote Sensing, 31*(1), 43–57.

Walters, S. (2013). *A Sociolinguistic Profile of the Bhils of Northern Dhule District.* SIL Electronic Survey Report, pp. 26.

CHAPTER 4

CHARACTERIZATION AND CLASSIFICATION OF SOILS OF THE PURNA VALLEY IN THE SEMI-ARID REGION OF CENTRAL INDIA

R. K. NAITAM,[1] V. K. KHARCHE,[2] V. V. GABHANE,[3] T. BHATTACHARYYA,[4] S. M. TALEY,[5] and N. M. KONDE[6]

[1]*Scientist, ICAR-National Bureau of Soil Survey and Land Use Planning, Regional Centre, Udaipur – 313 001, India*

[2]*Professor, Department of SSAC, PG Institute, Dr. Panjabrao Deshmukh Krishi Vidyapeeth, Akola – 444 104, India*

[3]*Associate Professor, AICRP for Dryland Agriculture, Dr. Panjabrao Deshmukh Krishi Vidyapeeth, Akola – 444 104, India*

[4]*Principal Scientist and Former Head, Division of Soil Resource Studies, ICAR-National Bureau of Soil Survey and Land Use Planning, Nagpur – 440 033, India*

[5]*Professor and Head, Department of SWC Engineering, Dr. Panjabrao Deshmukh Krishi Vidyapeeth, Akola – 444 104, India*

[6]*Assistant Professor, Department of SSAC, PG Institute, Dr. Panjabrao Deshmukh Krishi Vidyapeeth, Akola – 444 104, India*

CONTENTS

ABSTRACT

Detailed information on soils is necessary for agricultural planning and management. Keeping this in view the swell-shrink soils of the Ramagarh village of Purna valley in Amravati district, Maharashtra in semi-arid region of Central India were studied for their morphological, physical and chemical characteristics. Soils in the Ramagarh village are very deep, dark grayish brown to very dark grayish brown in color, clay in texture and exhibits medium, moderate, sub angular blocky structure in the surface layers and the subsoil horizons had medium, weak to strong angular blocky structure. Soils of the study area are alkaline in reaction, calcareous in nature and had low to medium organic carbon content. In general the pH, $CaCO_3$ and exchangeable sodium percentage (ESP) increase with depth in all the soils. Because of high smectitic clay content and ESP down the profile, these soils have impeded drainage and show ponding of water in the rainy season. The soils of the uplands are classified as *Sodic Haplusterts* and low land soils belong to *Typic Haplusterts* category at sub group level. The study suggests that the micro topographic differences can modify the important properties of soils in a village with Ustic soil moisture regime. The study indicates that these soils need to be classified based on measurable properties like hydraulic conductivity and the level of ESP used for categorizing sodic black soils in semi-arid Indian continent should be brought down.

4.1 INTRODUCTION

Soil is the most precious natural resource of any nation and its judicious management is of paramount importance. Growing human population and consequently increase in food requirements compel us to look for more land resources. Till now, the increasing demands of production were being met by putting more arable lands under cultivation. The per capita availability of land in India decreased from 0.48 ha in 1951 to 0.15 ha in the year 2000 and is likely to reduce further to 0.08 ha by the year 2020 because of demographic pressure. Besides reduction in land area, there is decline in land quality, what we call land degradation, either quantitatively/qualitatively or both as a result of processes, such as soil erosion by water and wind, salinization, sodification, water logging, depletion of plant nutrients, depletion of soil structure, desertification and pollution. Approximately 952 m ha of land worldwide are affected by salinity and alkalinity, of which 7.1 m ha of land are in India. Saline and alkaline soil occurs extensively in the arid and semiarid regions of India (Abrol et al., 1988). Plant growth is inhibited under such conditions due to unfavorable pH, excessive salt, imbalance of nutrients and poor soil structure. Irrigation-induced soil sodicity creates problems for cultivation and also in growth performance. Salt toxicity is one of the major edaphic factors limiting the salinized and/or sodic soils throughout the world. Apart from natural salinization, human-induced secondary salinization occurs frequently as a consequence of overproduction and irrigation caused by improper management of irrigation facilities, poor soil internal drainage conditions and unsuitable quality of irrigation water (Liang et al., 2005).

Black soils (Vertisols and vertic intergrades) occur widely in many parts of the world including India. They occupy an area of 72.9 m ha in India, 35.5% of which are in the state of Maharashatra. The earlier work carried out in the area (Kadam et al., 2013; Padekar, 2014) indicated that many soils in this valley are non-saline, the EC being less than 2 dS m^{-1} and the ESP less than 15. However, these soils have not been categorized in appropriate class because they show deterioration at ESP values much lower than 15. They are prone to waterlogging and show severe problems of drainage. It is, therefore, necessary to study the properties of these soils with their spatial distribution in the valley and their deterioration at different ESP in order to decide appropriate site-specific management options. The cracking clay soils in the present study area are peculiar and prone to degradation problems and detrimental for crop production. These soils being different from the salt affected alluvial soils found elsewhere in the country need to be studied for their detailed morphological, physical and chemical characterization. For proper agricultural planning and management of natural resources the detailed information on soils and their quality attributes is essential.

4.1.1 GLOBAL SCENARIO

Salt affected soils occur in the arid and semi-arid regions of the world where evapo-transpiration greatly exceeds precipitation. Agricultural production in the arid and semiarid region of the world is limited by poor water resources, limited rainfall and the detrimental effects associated with an excess of soluble salts, constrained to a localized area or sometimes extending over the whole of the basin. In order to minimize vagaries of arid weather, to bring more land under irrigation, and produce and stabilize greater yields per unit area, numerous water development projects have been commissioned over the world. Extension of irrigation to the arid and semi arid regions, however, usually have led to an increase in the area under shallow water tables and to intensify and expanding the hazards of salinity and sodicity. This is because irrigation water brings in additional salts and releases immobilized salts in the soil through evapotranspiration and concentrating dissolved salts in the soil solution. Fertilizers and decaying organic matter also serve as additional salt source. The relative significance of each source in contributing soluble salts depends on the natural drainage condition, soil properties, ground water quality, irrigation water quality, and management practices followed for crop production.

Salt-affected soils occur in all continents and under almost all climatic conditions. Their distribution, however, is relatively more extensive in the arid and semi-arid regions compared to the humid regions. There are extensive areas of salt-affected soils on all the continents but their extent and distribution has not been studied in detail. In some countries even the existence of these soils was reported only through a survey or the pressing demand for agricultural utilization of a region. The first attempt to compile information on the extent of salt-affected soils on a worldwide basis was made by F. Massoud based on the FAO/UNESCO (1974) Soil Map of the World. Information in respect of countries in Europe accounting 50.8 m ha is based on publications by Szabolcs (1974). Globally, salt affected soils cover an area of 952 m ha (Table 4.1) of land in the world under various degrees of deterioration due to excessive accumulation of salts in the soils (Szabolcs, 1974). Any attempt to increase food production in coming years must pay adequate attention to the improvement of existing salt affected soils with little or no production and to prevent further deterioration of productive soils through these degradation processes.

TABLE 4.1 Global Distribution of Salt Affected Soils

Continents	Salinity (m ha)	Alkalinity (m ha)	Area (m ha)
North America	0.2	9.6	15.8
Mexico and Central America	1.9	–	1.9
South America	69.4	59.6	129.0
Europe	–	50.8	50.8
Africa	53.5	26.9	80.4
South Asia	83.3	1.8	85.1
North and Central Asia	91.6	120.1	211.7
South-East Asia	19.9	–	19.9
Australia	17.4	340.0	357.4
Total	343.2	608.3	952.0

Source: FAO/UNESO (1974), World soil map; Szabolcs (1974).

4.1.2 INDIAN SCENARIO

There is wide difference in the estimates about the area of degraded land in the country made by different agencies. The earliest assessment of the land degradation in India made by the National Commission on Agriculture at 148 m ha followed by 175 m ha by the Ministry of Agriculture (Soil and Water Conservation Division). The National Wasteland Development Board estimated an area of 123 m ha under wastelands. The National Bureau of Soil Survey and Land Use Planning (NBSS & LUP) estimated an area of 187 m ha as degraded lands in 1994 following GLASOD methodology (Sehgal and Abrol, 1994) and later based on harmonization of datasets, it was estimated that about 120 m ha is affected by various kinds of degradation (ICAR and NAAS, 2010).

The recent database about the distribution of salt affected soils in various states in India is given by Central Soil Salinity Research Institute, Karnal (Mandal et al., 2010) and they estimated an area at 6.7 m ha presented state wise in Table 4.2. In Maharashtra, salt affected soils occupy an area of 6.07 lakh ha, found in Western Maharashtra and Vidarbaha region covering Ahmednagar, Jalgon, Akola, Sholapur, Buldana, Amravati, Nasik, Kolhapur and Aurangabad districts (Mandal et al., 2010). The Purna Valley, part of Payanghat Plain, is an oval- shaped basin covering part of Amravati, Akola and Buldana district on both sides of the river Purna around 10–45 km in

TABLE 4.2 Distribution of Salt Affected (Saline and Sodic Soils) Soils in India

State	Area (ha)
Andhra Pradesh	274,207
Andaman & Nicobar Islands	77,000
Bihar	153,153
Gujarat	2,222,000
Haryana	232556
Karnataka	150029
Kerala	20000
Madhya Pradesh	139720
Maharashtra	606759
Orissa	147138
Punjab	151717
Rajasthan	374942
Tamil Nadu	368015
Uttar Pradesh	1368960
West Bengal	441272
Total	6727468

Source: Mandal et al. (2010).

width and 150 km in length in Vidarbha region of Maharashtra under the semi-arid ecosystem of Central India. It covers a large area extending from 20°7′ to 21°31′ N, and 75°58′ to 77°53′ E. The Purna Valley is a faulted basin filled with sediments derived entirely from the Deccan basalts. The total thickness of these deposits is up to 420 m (Adyalkar, 1963). The river and its tributaries form a parallel to sub parallel drainage system. The whole area looks flat with only very gentle undulations here and there. The elevation of the alluvial plain varies from 274 to 395 m above mean sea level. This tract spreads on both sides of Purna River influencing about 892 villages, covering about 4692 sq.km area with mainly plain topography in Vidarbha region of Maharashtra (Shirale, 2013). The soils are black cotton, fine textured, swell-shrink soils with imperfect to poor drainage and having high water holding capacity.

Vertisols are dark montmorillonite-rich clays with characteristic shrinking/swelling properties, have high clay content (>30% up to at least 50 cm from the surface) and produce typical cracks (at least 1 cm wide reaching a depth of 50 cm or more). In India, Vertisols cover about 73 m ha area, 80% of which lies in the states of Maharashtra, Madhya Pradesh, Gujarat and Andhra Pradesh, 13% in Karnataka and Tamil Nadu and the rest in the other adjoining states. They are important soils, yielding high to moderate agricultural production in Madhya Pradesh, Maharashtra, Andhra Pradesh and Northern Karnataka. These soils occur extensively in the state of Maharashtra, occupying 36% of the 73 m ha area. These soils occur in the lower piedmont plains or valleys or in micro depressions, and are developed in the alluvium of weathered Deccan basalt (Pal and Deshpande, 1987).

4.1.3 IMPORTANCE OF THE STUDY

The poor structural stability of the Vertisols particularly during the monsoon season renders the agricultural activities very difficult; the low Saturated Hydraulic Conductivity (sHC) causes waterlogging. As a result vast lands remain vacant particularly during monsoon season. The earlier studies on the soils of Purna Valley showed that the considerable portion of these soils is reported to have developed salinity and waterlogging (Adyalkar, 1963; Puranik et al., 1972; Sagare et al., 1991; Tanpure, 1971). However, the presence of salinity is not always perceived because soil surface lacks salt-efflorescence (Kadu et al., 1993; Magar, 1990; Nimkar et al., 1992). But the soils remain waterlogged after rains, and this disrupts considerably the scheduling of sowing of *kharif* crops. As a consequence farmers are unable to maintain a sustainable production of agricultural crops. While studying the factors and processes of soil degradation in Purna valley Balpande et al. (1996) observed that there is development of sub soils sodicity with pedon depth and semi-arid climate is the prime factor responsible for depletion of Ca^{2+} ions from the soil solution resulting in precipitation of $CaCO_3$ and development of subsoil sodicity. On the other hand, Vaidya and Pal (2002) had reported that micro topography is a prime factor responsible for the natural chemical degradation of these soils. As a matter of fact, Purna valley covers large area and different researchers had their findings in the different parts of the valley. However, there are no detailed studies on soil resources of a village or micro- watershed level in the region. Hence, the present investigation is undertaken with the objective to characterize and classify the soils of the Ramagarh village of Purna valley region for successful land use planning and sustainable management of natural resources.

4.2 STUDY AREA

The study area comprises central part of the Purna valley in Vidarbha region of Central India. Ramagarh village was selected for present study, which is located between 77°12′36″ to 77°13′50″' E longitude and 20°52′46″ to 20°53′59″ N latitudes in Daryapur tehsil of Amravati district of Maharashtra covering an area of 324 ha (Figure 4.1). The mean elevation of the village ranges from 250 to 286 m above the mean sea level (MSL). This is 18 km away from Daryapur and 50 km away from Akola. The area is characterized by hot summer and a dry weather conditions except, during the south west monsoon season and thus represents a tropical sub humid dry to semi-arid dry climate. The study area has monsoonal climate, beginning from June or July through September which receives 85–95% of the total annual rainfall of 700–975 mm. However, the district experiences an erratic rainfall pattern with low as 600 to as high as 1100 mm. This is followed by a dry season from October to May or June. April and May are the hottest months with mean monthly temperature of 32.5 and 35.2°C respectively. December and January are the coolest months with monthly temperature of 22°C. The length of growing period in the area is 152 days. The soils have a Typic Tropo-ustic moisture regime (Van Wambeke, 1985). The soil temperature regime is hyperthermic. Majority percentage of cultivated land in *kharif* is under Soybean (*Glycin max*), greengram (*Phaseolus aurens*) and cotton (*Gossypium spp.)* as principal crops. Legumes like pigeon pea (*Cajanous cajan*), black gram (*Phaseolus mungo*) and cowpea (*Vigna catiang*) are also grown. Chickpea (*Cicer arietinum*) is dominant crop in *rabi* season under residual soil moisture and/or with some protective irrigations. The natural vegetation of the area comprises of dry deciduous tree species and grasses. The dominant tree species are babul (*Accacia arabica*), ber (*Ziziphus jujube*), palas (*Butea frondosa*), neem (*Azadiracta indica*), rui (*Calotropic giganda*), kans (*Saccharum spontaneum*) and dub (*Cynadon dactylon*).

4.3 METHODOLOGY

The information about land resources of the Ramagarh village was interpreted and studied from the collateral maps. The cadastral map of the village on 1:8,000 scale showed the field boundaries with survey number of each field, the other permanent details like habitation, roads, farm ponds, community ponds and stream, etc., were also obtained from the same. In

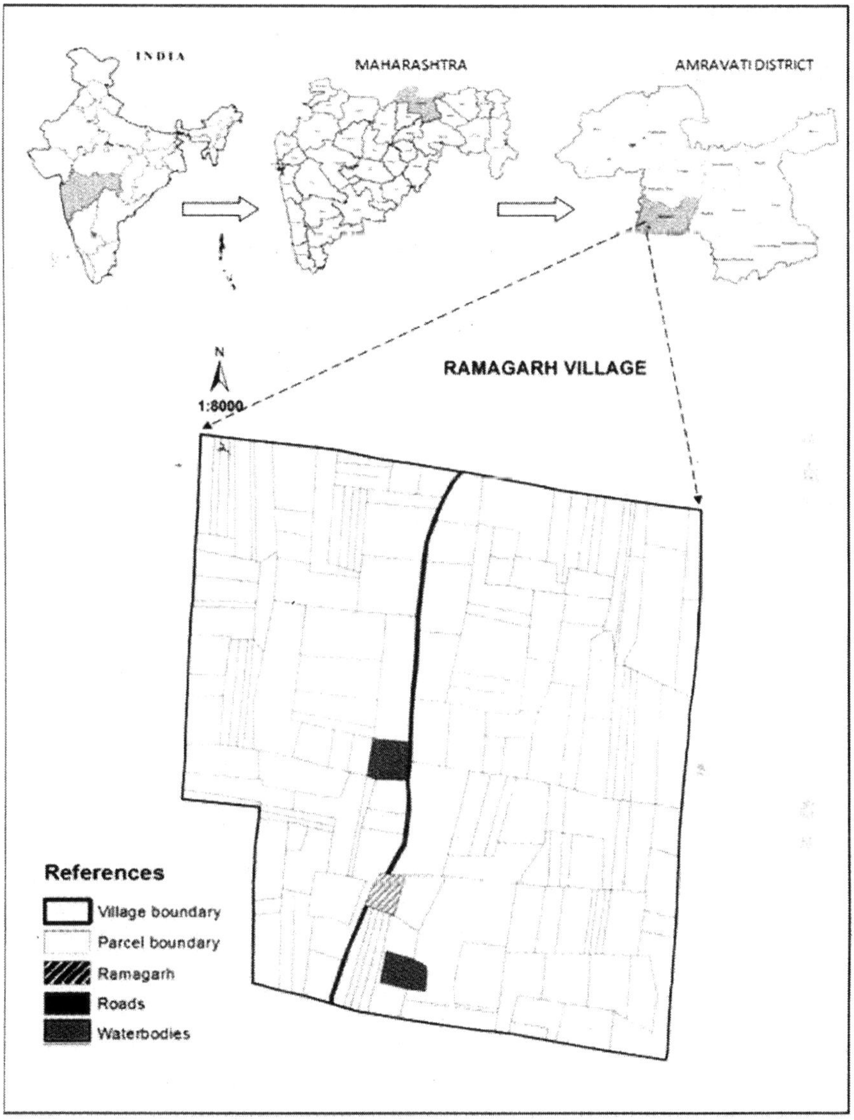

FIGURE 4.1 Location map of Study area.

addition to this the Survey of India (SOI) toposheet No. 55 H/1 (1:50,000 scale) was used to collect topographic information for landform analysis. Simultaneously the latitude and longitude of the study area were recorded using portable hand held Garmin GPS instrument for georeferencing of the study area. Then the cadastral map was scanned and georeferenced with the toposheet using maximum GCP and digitized in Arc GIS Ver. 10 GIS software. The detailed soil survey of the village was carried out using cadastral map as a base map on 1:8000 scale.

On the basis of soil variability in terms of changes in surface characteristics, break in slope, degree of erosion and other micro features of the eight representative sites for profile were recorded and soil-site characteristics were noted in the field. The soil profiles of 1.2 meter length, 1 m wide and 1.5 m deep were dug at all the selected sites. Soil site characteristics of the pedons were studied in the field as per the methodology laid out in the soil survey manual (Soil Survey Staff, 1998). Observations like cracking depth, slickensides and other micro-features were also recorded. Horizon wise soil samples were collected, air dried and processed as per the standard methodology and stored in plastic bottles for laboratory analysis.

Physical properties of the soils, such as particle size distribution were determined by the international pipette method (Klute, 1986). The bulk density was determined by clod coating method (Black and Hartge, 1986). The sHC was measured by constant head method described by Klute and Dirksen (1986). The coefficient of linear extensibility (COLE) was estimated by following the method of Schafer and Signer (1976), and is defined as the ratio of difference between the moist length (Lm, length of soil clod at 33 kPa) and dry length of soil clod to its dry length (Ld, length of soil clod when dry (room temperature). The moisture retention and release behavior within the available range of 33 kPa to 1500 kPa were measured on less than 2 mm size soil sample using pressure plate membrane apparatus as per method outlined by Richards (1954).

Chemical properties like pH and Ec of the soil suspension (1:2 soil: water ratio) was determined by the methodology of Jackson (1973). For the determination of soil organic carbon (SOC), the modified Walkley and Black wet oxidation method was used (Walkley and Black, 1934; Jackson, 1973). The free calcium carbonate was determined by rapid titration method (Piper, 1966). The exchangeable cations, cation exchange capacity (CEC) of soils were determined by methods outlined by Richards (1954).

4.4 RESULTS AND DISCUSSION

4.4.1 MORPHOLOGICAL PROPERTIES

All the soils were very deep (>150 cm), clayey in texture and dark brown (10YR 3/3) to dark grayish brown (10 YR 3/2) in color (Table 4.3). It is

TABLE 4.3 Morphological Properties of Soils

Horizon	Depth (cm)	Boundary D	Boundary T	Matrix color	Tex.	Structure	Consistence D	Consistence M	Consistence W	Nodules S	Nodules Q	Roots S	Roots Q	Eff.	SS/ Pf	Other features Cracks
Pedon:1 Fine, smectitic, hyperthermic (calc.), Typic Haplusterts																
Ap	0–16	c	s	10YR3/2	C	m2sbk	h	fr	vsvp	F	m	vf,f	m	es	-	2 to 3 cm wide cracks extending up to 45 cm depth.
Bw1	16–40	c	s	10YR3/2	C	m2sbk	h	fr	vsvp	f,m	m,c	vf,f	m	ev	-	
Bw2	40–64	g	w	10YR3/2	C	m2abk	h	fi	vsvp	m	c	f	f	ev	Pf	
Bss1	64–99	g	w	10YR3/2	C	m3abk	–	fi	vsvp	m	c	f	f	ev	SS	1.5 to1 cm wide cracks extending up to 100 cm depth
Bss2	99–130	c	s	10YR3/1	C	m3abk	–	fi	vsvp	m	c	c	f	ev	SS	
Bss3	130–160	–	–	10YR3/1	C	m3abk	–	fi	vsvp	m,c	c	-	-	ev	SS	
Pedon:2 Fine, smectitic, hyperthermic (calc.), Typic Haplusterts																
Ap	0–18	c	s	10YR3/2	C	m2sbk	h	fr	vsvp	vf,f	m	vf,f	m,c	es	-	2 to 3 cm wide cracks extending up to 20 cm depth.
Bw1	18–46	g	w	10YR3/2	C	m2sbk	h	fr	vsvp	vf,f	m	vf,f	m	es	-	
Bw2	46–70	g	w	10YR3/2	C	m2abk	h	fi	vsvp	F	m	f	m	ev	Pf	
Bss1	70–99	g	s	10YR3/1	C	m3abk	–	fi	vsvp	F	m	f	m	ev	SS	1.5 to1 cm wide cracks extending up to 90 cm depth.
Bss2	99–128	g	s	10YR3/1	C	m3abk	–	fi	vsvp	F	m	f	f	ev	SS	
Bss3	128–157	–	–	10YR3/1	C	m2abk	–	fi	vsvp	m	m	-	-	ev	SS	

TABLE 4.3 (Continued)

Horizon	Depth (cm)	Boundary D	Boundary T	Matrix color	Tex.	Structure	Consistence D	Consistence M	Consistence W	Nodules S	Nodules Q	Roots S	Roots Q	Eff.	SS/ Pf	Other features Cracks
Pedon:3 Fine, smectitic, hyperthermic (calc.), Typic Haplusterts																
Ap	0–19	c	s	10YR3/3	C	m2sbk	h	fr	vsvp	vf,f	m	vf,f	m	es	-	2 to 3 cm wide cracks extending up to 40 cm depth
Bw1	19–49	g	w	10YR3/2	C	m2sbk	h	fr	vsvp	F	m	vf	m	es	-	
Bw2	49–82	g	w	10YR3/2	C	m2abk	vh	fr	vsvp	m	c	vf	m	es	Pf	
Bss1	82–109	g	w	10YR3/1	C	m3abk	–	fi	vsvp	m	c	f	m	ev	SS	1.5 to1 cm wide cracks extending up to 90 cm depth.
Bss2	109–135	g	s	10YR3/1	C	m3abk	–	fi	vsvp	m	c	m	f	ev	SS	
Bss3	135–160	–	–	10YR3/2	C	m2abk	–	fi	vsvp	C	c	-	-	ev	SS	
Pedon:4 Fine, smectitic, hyperthermic (calc.), Sodic Haplusterts																
Ap	0–18	c	s	10YR3/2	C	m2sbk	h	fr	vsvp	vf,f	m	vf,f	m	es	-	1.5 to 2 cm wide cracks extending up to 80 cm depth
Bw1	18–45	c	s	10YR3/2	C	m2sbk	h	fr	vsvp	vf,f	m	vf,f	m	ev	-	
Bw2	45–90	g	w	10YR3/1	C	m2abk	h	fr	vsvp	m	c	f	m	ev	Pf	
Bss1	90–121	g	w	10YR3/1	C	m2abk	–	fi	vsvp	m	c	c	f	ev	SS	
Bss2	121–140	g	w	10YR3/1	C	m3abk	–	fi	vsvp	m	c	c	f	ev	SS	
Bss3	140–160	–	–	10YR3/1	C	m3abk	–	fi	vsvp	m	c			ev	SS	
Pedon:5 Fine, smectitic, hyperthermic (calc.), Typic Haplusterts																
Ap	0–18	C	s	10YR3/2	c	m2sbk	h	fr	vsvp	vf, f	m	vf,f	m	es	-	2 to 3 cm wide cracks extending up to 40 cm depth.
Bw1	18–39	C	s	10YR3/2	c	m2sbk	h	fr	vsvp	vf, f	m	vf,f	m	es	-	
Bw2	39–74	G	w	10YR3/1	c	m2abk	h	fr	vsvp	vf, f	m	f	m	es	Pf	
Bss1	74–100	C	s	10YR3/1	c	m3abk	vh	fi	vsvp	vf, f	m	m	f	ev	SS	1.0 to 0.5 cm wide cracks extending up to 100 cm depth.
Bss2	100–131	G	w	10YR3/1	c	m3abk	vh	fi	vsvp	m	m	vf,f	f	ev	SS	
Bss3	131–162	–	–	10YR3/3	c	m1abk	–	fi	vsvp	c	c	-	-	ev	Pf	

Pedon:6 Fine, smectitic, hyperthermic (calc.), Typic Haplusterts																
Ap	0–16	C	s	10YR3/2	c	m2sbk	h	fr	vsvp	vf, f	m	vf,f	m	es	-	2 to 3 cm wide cracks extending up to 18 cm depth.
Bw1	16–47	G	w	10YR3/2	c	m2sbk	h	fr	vsvp	vf, f	m	vf,f	m	es	-	
Bw2	47–80	G	w	10YR3/2	c	m2sbk	h	fr	vsvp	m	c	f	c	ev	Pf	
Bss1	80–110	G	w	10YR3/1	c	m2abk	–	fi	vsvp	m	c	f	c	ev	SS	1.5 to1 cm wide cracks extending up to 110 cm depth.
Bss2	110–135	G	s	10YR3/1	c	m3abk	–	fi	vsvp	m	c	f	c	ev	SS	
Bss3	135–170	–	–	10YR3/2	c	m3abk	–	fi	vsvp	m	c	-	-	ev	SS	
Pedon:7 Fine, smectitic, hyperthermic (calc.), Typic Haplusterts																
Ap	0–16	C	s	10YR3/2	c	m2sbk	h	fr	vsvp	vf, f	m	vf,m	f,c	es	-	2 to 3 cm wide cracks extending up to 40 cm depth
Bw	16–43	C	w	10YR3/2	c	m2sbk	h	fr	vsvp	f, m	m	vf,f	m	ev	Pf	
Bss1	43–80	C	w	10YR3/2	c	m2abk	h	fi	vsvp	f, m	m	f	c	ev	SS	
Bss2	80–103	C	w	10YR3/1	c	m3abk	–	fi	vsvp	f, m	m	f	c	ev	SS	1.5 to1 cm wide cracks extending up to 100 cm depth.
Bss3	103–132	G	s	10YR3/1	c	m3abk	–	fi	vsvp	f, m	m	f	c	ev	SS	
Bss4	132–157	–	–	10YR3/3	c	m2abk	–	fi	vsvp	m, c	m	-	-	ev	SS	
Pedon:8 Fine, smectitic, hyperthermic (calc.), Typic Haplusterts																
Ap	0–18	C	s	10YR3/2	c	m2sbk	h	fr	vsvp	f, m	m	vf,f	m	es	-	2 to 3 cm wide cracks extending up to 40 cm depth
Bw1	18–41	C	s	10YR3/2	c	m2sbk	h	fr	vsvp	f, m	m	f	m	es	-	
Bw2	41–57	G	w	10YR3/3	c	m2abk	vh	fi	vsvp	m, c	c	f	m	es	Pf	
Bss1	57–86	G	w	10YR3/3	c	m2abk	–	fi	vsvp	m	c	m	m	ev	SS	1.0 to 0.5 cm wide cracks extending up to 80 cm depth
Bss2	86–117	C	w	10YR3/2	c	m3abk	–	fi	vsvp	m	c	f	m	ev	SS	
Bss3	117–154	–	–	10YR3/6	c	m1abk	–	fi	vsvp	C	m			ev	Pf	

Tex. – Texture; Eff. – Effervescence with dil. HCl.

observed that all the soils exhibit a hue of 10YR, a value of 3 and chroma ranging from 1 to 6 corresponding to very dark gray to dark yellowish brown colored soils. The subsurface horizons in all pedons are very dark grayish brown (10 YR 3/2) to very dark gray (10 YR 3/1) in color except Bss3 horizon of pedon 8, which is dark yellowish brown (10 YR 3/6) in color. This may be due to presence of more $CaCO_3$ in diffused form with the depth. The dark color of these soils may be attributed to humus and minerals like titaniferous magnetite (Zonn, 1986). All the soils in the Ramagarh village exhibit medium, moderate, sub angular blocky structure in the surface layers and hard (dry) and friable (moist) consistence. However, the subsoil horizons had medium, weak to strong angular blocky structure. In the subsoil, the structure consists of intersecting slickensides, forming parallelepipeds with their long axes at 30–45° from the horizontal. This separates into strong, coarse angular blocks with shiny pressure faces and firm (moist) and very sticky and very plastic (moist) consistence. Common, many, few fine and very fine sized roots were observed in the surface layers. The number of roots decreased with depth. Below 100 cm depth there are only very few to few, fine and medium roots. The soils of Ramagarh village are calcareous in nature and showed strong effervescence (with 10% HCl) in the surface horizons and it was violent in rest of the profile, which is attributed to the presence of diffuse powdery form of $CaCO_3$ (Balpande, 1993). Calcium carbonate concretions are observed throughout the depth in all the soils.

The slickensides were commonly found in the soils of Ramagarh village. The slickensides appear first at the depth of 43 to 74 cm from the surface and extend beyond 155 cm if there are no restricting layers like Ck horizons. These slickensides form an angle of about 35° to 70° with the horizon. All the soils exhibit cracks measuring more than 2 cm wide at the surface during dry season. These cracks separate soil mass into number of polyhedrons. These cracks were extended up to slickensides zone in pedon 2, 3 4 and 8. On the other hand the cracks were extended through whole of the slickensides zone in pedon 1, 5, 6 and 7 (Table 4.3).

4.4.2 PHYSICAL PROPERTIES

The soils are clayey in texture and the clay content varies from 57.0 to 67.3% and it increases with depth in all the pedons, it might be due to downward translocation of finer particles from the surface layers. A similar trend was also reported by earlier workers (Balpande, 1993; Kadam, 2011; Kadu, 1991; Magar, 1990). The very high clay content of these soils can be attributed to their formation from basaltic parent material (Pal and Deshpande, 1987). The bulk density was quite variable in different horizons and varied from 1.44 to 2.07 Mg m^{-3}. Similar observations were recorded by Nimkar (1990) and Padekar (2014) while studying the soils of Purna valley. It was relatively lower in the surface horizons and increases with depth in all the soils that may be due to comparatively more organic matter in the surface horizons and higher swelling pressure and compaction caused due to smectitic clay content in the subsoil (Ahuja et al., 1988). The sHC of the soils of village varies from 0.20 to 7.56 mm hr^{-1} and rapidly decreased with depth in all the pedons except pedons 5 and 8 where it shows slight increase within the 50 cm over the surface layers. The data (Table 4.4) indicate imperfect to poor internal drainage condition of these soils. In general, it is noticed that the sHC is less in the subsurface horizons than that in surface layer. This might be due to the compactness of the subsurface layers and presence of excess of sodium salts indicated by high ESP in the subsurface horizons. Considerable decrease in sHC with increasing depth was also observed in deep black soils by Bharambe et al. (1986), Kadu (1991), Nimkar et al. (1992) and Balpande (1993). It is generally observed that the soils which have ESP $\geq$ 5 have low sHC value (Pal et al., 2000) indicating poor internal drainage condition. The value of the COLE varies from 0.19 to 0.26. On the basis of categorization of COLE value, all the soils under the study fall into very high shrink swell soils category (Nayak et al., 2006). The mean weight diameter was found to be varied from 0.42 to 0.98 mm in different horizons of the soil profile except pedon 1 and 2 and it was decreased down the depth. The downward decrease in mean weight diameter can be attributed to subsoil sodicity in these soils (Table 4.4). The gravimetric water retention at 33 kPa and 1500 kPa tension indicated that the values of AWC range from 5.0 to 35.1% and increase with depth in all the soils. The moisture retention and release functions in the soils of Purna valley indicated that, in general, the subsoil retained more moisture than the surface soils at the given tensions. This effect may be due to higher Na^+ saturation in the subsurface layers (Balpande, 1993).

4.4.3 CHEMICAL PROPERTIES

The soils are slight to moderately alkaline in reaction and pH value varies from 8.4 to 9.4, which increases down the depth in the soil profile (Table 4.5). The

TABLE 4.4 Physical Properties of Soils

Horizon	Depth (cm)	Sand	Silt	Clay	BD ($Mg\ m^{-3}$)	sHC ($mm\ hr^{-1}$)	MWD (mm)	COLE ($cm\ cm^{-1}$)	Water retention (%)		
		%							33 kPa	1500 kPa	AWC
Pedon: 1 *Fine, smectitic, hyperthermic (calc.), Typic Haplusterts*											
Ap	0–16	5.6	37.4	57.0	1.66	3.18	0.74	0.19	32.7	22.3	10.4
Bw1	16–40	7.1	34.5	58.5	1.71	0.97	0.42	0.21	37.0	18.5	18.6
Bw2	40–64	6.2	35.4	58.5	1.72	0.79	0.84	0.20	37.2	21.7	15.5
Bss1	64–99	6.4	34.4	59.3	1.88	0.73	0.78	0.19	39.1	22.2	16.9
Bss2	99–130	5.0	35.5	59.6	1.85	0.50	0.96	0.22	38.5	22.7	15.8
Bss3	130–160	5.6	33.8	60.6	1.87	0.42	0.98	0.25	40.7	23.1	17.6
Pedon: 2 *Fine, smectitic, hyperthermic (calc.), Typic Haplusterts*											
Ap	0–18	4.6	30.8	64.6	1.52	3.18	0.83	0.22	52.5	22.9	29.6
Bw1	18–46	5.4	29.5	65.1	1.56	1.41	0.85	0.22	40.3	23.6	16.7
Bw2	46–70	2.5	34.5	63.1	1.61	0.91	0.88	0.23	41.4	25.0	16.4
Bss1	70–99	3.4	29.4	67.3	1.59	0.91	0.87	0.25	60.9	25.8	35.1
Bss2	99–128	3.1	32.2	64.7	1.68	0.79	0.90	0.24	45.0	26.3	18.7
Bss3	128–157	2.3	36.6	61.1	1.64	0.60	0.88	0.23	51.2	26.2	25.0
Pedon: 3 *Fine, smectitic, hyperthermic (calc.), Typic Haplusterts*											
Ap	0–19	6.9	33.8	59.4	1.67	5.77	0.67	0.24	35.6	23.8	11.8
Bw1	19–49	9.8	28.7	61.5	1.63	2.51	0.89	0.24	34.8	22.9	11.9
Bw2	49–82	7.5	32.0	60.6	1.68	1.42	0.75	0.25	35.4	23.9	11.5
Bss1	82–109	5.5	31.5	63.1	1.68	0.91	0.63	0.26	37.6	24.4	13.2

Bss2	109–135	3.0	32.5	64.6	1.95	0.68	0.58	0.25	39.3	26.0	13.3
Bss3	135–160	4.8	30.5	64.7	1.87	0.64	0.54	0.26	39.4	25.8	13.5
Pedon: 4 *Fine, smectitic, hyperthermic (calc.), Sodic Haplusterts*											
Ap	0–18	7.1	32.3	60.6	1.56	3.23	0.89	0.21	35.1	23.0	12.1
Bw1	18–45	7.3	35.7	57.0	1.91	0.86	0.88	0.20	33.6	21.8	11.9
Bw2	45–90	5.6	32.9	61.5	1.91	0.51	0.70	0.22	34.5	24.3	10.3
Bss1	90–121	4.5	32.3	63.2	1.95	0.31	0.65	0.24	38.3	26.6	11.8
Bss2	121–140	4.5	32.5	63.1	2.07	0.48	0.68	0.25	41.3	27.0	14.4
Bss3	141–160	2.9	38.2	58.9	2.06	0.45	0.72	0.24	45.6	29.0	16.6
Pedon: 5 *Fine, smectitic, hyperthermic (calc.), Typic Haplusterts*											
Ap	0–18	6.0	36.0	58.0	1.48	6.12	0.90	0.19	37.3	30.0	7.3
Bw1	18–39	5.4	34.1	60.6	1.60	3.52	0.91	0.21	38.5	25.6	12.9
Bw2	39–74	6.5	31.6	61.9	1.66	7.56	0.88	0.22	44.0	25.5	18.5
Bss1	74–100	5.4	33.2	61.4	1.75	0.99	0.76	0.21	39.2	29.6	9.6
Bss2	100–131	4.7	33.1	62.2	1.72	0.98	0.72	0.24	43.0	28.7	14.3
Bss3	131–162	5.7	34.7	59.5	1.76	0.79	0.72	0.25	43.8	25.5	18.3
Pedon: 6 *Fine, smectitic, hyperthermic (calc.), Typic Haplusterts*											
Ap	0–16	6.4	31.7	62.0	1.47	3.57	0.91	0.22	40.9	35.8	5.0
Bw1	16–47	5.8	31.3	62.9	1.49	2.50	0.88	0.22	42.2	24.8	17.4
Bw2	47–80	6.3	30.0	63.7	1.75	1.41	0.83	0.24	41.8	25.8	16.0
Bss1	80–110	3.9	33.8	62.3	1.79	0.52	0.86	0.25	44.5	28.6	15.9

TABLE 4.4 (Continued)

Horizon	Depth (cm)	Sand	Silt	Clay	BD ($Mg\ m^{-3}$)	sHC ($mm\ hr^{-1}$)	MWD (mm)	COLE ($cm\ cm^{-1}$)	Water retention (%)		
			%						33 kPa	1500 kPa	AWC
Bss2	110–135	5.2	32.4	63.4	1.87	0.53	0.78	0.24	46.7	28.7	18.0
Bss3	135–170	3.4	31.0	65.6	1.82	0.20	0.66	0.30	48.3	30.2	18.1
Pedon: 7 *Fine, smectitic, hyperthermic (calc.), Typic Haplusterts*											
Ap	0–16	5.4	33.8	60.8	1.44	3.11	0.88	0.23	40.4	21.9	18.5
Bw1	16–43	6.5	29.6	63.9	1.48	1.21	0.90	0.23	38.2	23.5	14.7
Bw2	43–80	5.5	33.9	60.6	1.62	0.70	0.91	0.24	39.7	22.8	16.9
Bss1	80–103	6.0	31.4	62.6	1.65	0.54	0.83	0.25	42.8	22.5	20.3
Bss2	103–132	6.3	31.2	62.5	1.79	0.53	0.78	0.26	44.1	26.1	18.0
Bss3	132–157	3.5	32.9	63.6	1.85	0.40	0.64	0.26	47.2	28.6	18.5
Pedon: 8 *Fine, smectitic, hyperthermic (calc.), Typic Haplusterts*											
Ap	0–18	2.9	35.4	61.7	1.65	5.76	0.53	0.19	43.5	23.3	20.2
Bw1	18–41	3.8	34.7	61.7	1.73	4.18	0.50	0.21	38.9	22.4	16.5
Bw2	41–57	3.2	33.4	63.4	1.72	6.85	0.56	0.22	40.1	23.0	17.1
Bss1	57–86	2.5	32.3	65.3	1.84	0.72	0.60	0.23	43.1	27.6	15.5
Bss2	86–117	3.0	34.7	62.3	1.79	0.50	0.86	0.24	44.1	25.3	18.8
Bss3	117–154	2.4	39.6	58.1	1.73	0.44	0.58	0.24	42.6	24.0	18.7

TABLE 4.5 Chemical Properties of Soils

Horizon	Depth (cm)	pH (1:2 H_2O)	EC (1:2 H_2O)	OC (g kg^{-1})	$CaCO_3$ (%)	Exchangeable bases Ca^{2+}	Mg^{2+}	Na^+	K^+	Sum	CEC	Clay CEC	BS	ESP	EMP	Ca/Mg
						cmol (p+) kg^{-1}							%			
Pedon:1 Fine, smectitic, hyperthermic (calc.), Typic Haplusterts																
Ap	0–16	8.6	0.23	6.7	6.5	37.2	17.6	2.3	1.1	58.3	54.5	95.6	106.9	4.3	32.3	2.1
Bw1	16–40	8.7	0.21	5.4	7.1	32.4	21.2	3.2	0.9	57.7	54.3	92.8	106.4	5.9	39.1	1.5
Bw2	40–64	9.0	0.23	4.8	7.5	28.0	22.4	4.4	0.8	55.6	52.5	89.7	105.8	8.3	42.6	1.2
Bss1	64–99	9.1	0.29	4.7	7.1	26.4	22.4	6.7	0.8	59.1	55.1	93.1	107.2	12.1	40.6	1.2
Bss2	99–130	9.2	0.29	4.9	7.1	27.2	22.4	7.2	0.8	58.8	53.7	90.2	109.5	13.4	41.7	1.2
Bss3	130–160	9.2	0.42	4.8	7.3	22.8	22.4	8.5	0.7	55.7	56.3	92.9	98.5	15.1	39.8	1.0
Pedon:2 *Fine, smectitic, hyperthermic (calc.), Typic Haplusterts*																
Ap	0–18	8.6	0.17	5.4	7.9	40.8	19.2	2.7	0.9	63.6	61.7	95.5	103.2	4.4	31.1	2.1
Bw1	18–46	8.7	0.15	5.1	8.6	36.0	24.8	2.6	0.7	64.1	61.7	94.7	104.0	4.2	40.2	1.4
Bw2	46–70	8.7	0.19	5.1	8.4	34.0	26.4	3.7	0.7	64.8	60.5	95.9	107.0	6.0	43.6	1.3
Bss1	70–99	8.8	0.20	5.4	7.9	31.6	28.8	3.3	0.8	64.5	65.4	97.2	98.6	5.0	44.0	1.1
Bss2	99–128	8.8	0.21	5.2	6.9	13.6	25.2	3.6	0.8	43.1	62.8	97.0	68.7	5.7	40.1	0.5
Bss3	128–157	8.8	0.23	3.8	8.9	19.2	32.0	2.9	0.8	54.9	53.1	86.9	103.3	5.4	60.2	0.6
Pedon:3 *Fine, smectitic, hyperthermic (calc.), Typic Haplusterts*																
Ap	0–19	8.8	0.21	5.8	8.2	37.6	16.8	2.1	1.2	57.7	56.5	95.2	102.0	3.7	29.7	2.2
Bw1	19–49	8.8	0.17	4.8	8.9	33.2	20.4	2.3	0.7	56.6	56.7	92.1	99.8	4.0	36.0	1.6
Bw2	49–82	8.9	0.19	4.8	8.4	32.0	26.0	2.5	0.7	61.2	59.5	98.1	102.9	4.2	43.7	1.2
Bss1	82–109	9.0	0.20	4.7	7.8	31.2	22.8	3.0	0.7	57.7	57.5	91.1	100.5	5.3	39.7	1.4

TABLE 4.5 (Continued)

Horizon	Depth (cm)	pH (1:2 H_2O)	EC (1:2 H_2O)	OC (g kg^{-1})	$CaCO_3$ (%)	Exchangeable bases Ca^{2+}	Mg^{2+}	Na^+	K^+	Sum	CEC	Clay CEC	BS	ESP	EMP	Ca/Mg
						cmol (p+) kg^{-1}							%			
Bss2	109–135	9.1	0.29	4.7	8.5	26.8	28.4	4.8	0.7	60.7	59.6	92.2	101.9	8.0	47.7	0.9
Bss3	135–160	9.1	0.35	4.1	9.1	21.6	27.2	4.7	0.7	54.2	56.0	86.5	96.8	8.4	48.6	0.8
Pedon:4 *Fine, smectitic, hyperthermic (calc.), Sodic Haplusterts*																
Ap	0–18	8.6	0.21	8.1	5.9	39.2	17.2	2.6	1.4	60.5	57.4	94.7	105.3	4.5	30.0	2.3
Bw1	18–45	9.0	0.27	5.3	6.8	37.6	13.6	4.4	0.8	56.5	54.2	95.0	104.2	8.2	25.1	2.8
Bw2	45–90	9.0	0.23	5.0	6.6	34.0	14.4	6.9	0.8	56.1	57.7	93.8	97.1	11.9	24.9	2.4
Bss1	90–121	9.3	0.32	4.8	6.4	30.8	17.6	11.6	0.7	60.7	59.7	94.5	101.6	19.4	29.5	1.8
Bss2	121–140	9.4	0.35	5.3	6.4	26.0	18.0	10.3	0.7	55.1	57.0	90.3	96.6	18.2	31.6	1.4
Bss3	141–160	9.3	0.47	4.9	7.3	23.6	21.6	9.6	0.7	55.6	55.6	94.3	100.0	17.4	38.9	1.1
Pedon:5 *Fine, smectitic, hyperthermic (calc.), Typic Haplusterts*																
Ap	0–18	8.5	0.26	6.5	6.8	39.2	13.6	0.8	1.4	54.9	55.9	96.6	98.3	1.4	24.3	2.9
Bw1	18–39	8.7	0.24	5.7	6.6	36.4	16.8	0.4	0.9	55.5	57.6	95.1	94.6	0.8	29.1	2.2
Bw2	39–74	8.9	0.29	5.4	6.8	31.2	18.4	1.6	0.9	52.1	53.1	85.8	98.0	2.9	34.6	1.7
Bss1	74–100	9.0	0.31	4.9	6.9	28.4	17.6	3.1	0.9	50.0	53.7	87.6	93.0	5.8	32.8	1.6
Bss2	100–131	9.2	0.35	4.5	7.8	23.6	21.6	4.0	0.8	50.0	50.3	80.8	99.5	8.0	43.0	1.1
Bss3	131–162	9.2	0.43	3.4	9.5	17.2	22.0	4.0	0.7	43.9	48.9	82.1	89.8	8.2	45.0	0.8
Pedon:6 *Fine, smectitic, hyperthermic (calc.), Typic Haplusterts*																
Ap	0–16	8.4	0.16	6.2	6.8	31.6	22.0	1.1	1.0	55.8	55.2	89.1	101.0	2.0	39.8	1.4
Bw1	16–47	8.5	0.21	5.7	7.5	28.8	24.8	0.9	0.9	55.4	57.9	92.1	95.6	1.5	42.8	1.2

Bw2	47–80	8.7	0.31	5.5	7.0	23.6	25.6	2.3	0.8	52.3	55.5	87.1	94.2	4.1	46.1	0.9
Bss1	80–110	8.7	0.38	5.1	7.4	21.6	27.6	4.6	0.8	54.6	54.9	88.0	99.6	8.4	50.3	0.8
Bss2	110–135	8.8	0.43	5.4	8.1	17.6	29.2	4.4	0.8	52.0	50.3	79.3	103.4	8.8	58.1	0.6
Bss3	135–170	8.8	0.56	5.2	7.7	16.0	29.6	6.0	0.8	52.4	55.6	84.7	94.3	10.8	53.3	0.5
Pedon:7 *Fine, smectitic, hyperthermic (calc.), Typic Haplusterts*																
Ap	0–16	8.7	0.19	7.1	7.3	36.0	16.8	1.8	1.3	55.9	55.6	91.4	100.6	3.3	30.2	2.1
Bw1	16–43	9.1	0.29	6.2	7.[illegible]	31.6	20.4	2.7	0.9	55.6	57.2	89.5	97.2	4.7	35.6	1.5
Bw2	43–80	9.3	0.30	6.0	7.1	30.4	20.8	3.0	0.9	55.1	53.7	88.7	102.5	5.5	38.7	1.5
Bss1	80–103	9.4	0.37	5.8	7.3	27.2	21.2	4.1	0.9	53.4	55.9	89.3	95.6	7.3	37.9	1.3
Bss2	103–132	9.4	0.45	5.5	7.7	24.8	21.6	4.8	0.8	52.0	57.0	91.1	91.2	8.4	37.9	1.1
Bss3	132–157	9.3	0.73	4.1	9.3	18.0	24.4	6.3	0.7	49.5	51.7	81.4	95.7	12.3	47.2	0.7
Pedon:8 *Fine, smectitic, hyperthermic (calc.), Typic Haplusterts*																
Ap	0–18	8.5	0.29	5.7	6.9	37.6	12.8	1.5	1.4	53.2	56.4	91.5	94.3	2.6	22.7	2.9
Bw1	18–41	8.7	0.23	5.4	7.6	35.2	15.6	1.5	1.1	53.3	53.8	87.3	99.1	2.7	29.0	2.3
Bw2	41–57	8.8	0.25	5.2	7.9	30.8	20.4	0.5	1.0	52.7	53.8	84.9	97.9	1.0	37.9	1.5
Bss1	57–86	9.0	0.28	5.1	7.5	29.2	23.2	1.2	1.0	54.6	57.7	88.5	94.5	2.1	40.2	1.3
Bss2	86–117	9.2	0.26	4.5	7.7	22.8	24.4	4.3	0.9	52.4	52.5	84.4	99.7	8.1	46.5	0.9
Bss3	117–154	9.2	0.43	2.7	9.9	12.8	24.0	3.8	0.8	41.4	43.4	74.7	95.5	8.8	53.3	0.5

EC of all the pedons is much less than 4 dS m^{-1}, indicating no hazards of salinity. The values of EC vary from 0.12- 0.47 dS m^{-1} in different horizons which increases down the depth in all profiles. The subsoil increase in EC indicates that salinization process is also operative in these soils which results in accumulation of salts in lower horizons. All the soils are calcareous in nature and the value of $CaCO_3$ varies from 5.9 to 9.9% in different horizons with a tendency to increase with depth. This may be due to semi-arid climatic condition, where the leaching of bicarbonates during rainy season from upper layers and subsequent precipitation triggers development of sodicity in subsoil of black soils (Balpande et al., 1996). These soils are impoverished of organic carbon and the SOC content ranges from 2.7 to 8.1 g kg^{-1} and it decreased with depth in all the pedons. The CEC of the soils was high and it varies from 43.4 to 65.4 cmol (p^+) kg^{-1}. It was very high due to predominance of smectitic mineralogy of the soils. The clay CEC values estimated on the basis of soil CEC and clay percentage ranged from 74.7 to 98.1 cmol (p^+) kg^{-1}.

Among the exchangeable cations, calcium is the dominant cation followed by magnesium, sodium or potassium in surface layers of all the soils. On the other hand magnesium is the dominant cation followed by calcium, sodium or potassium in sub surface layers of all the soils. The exchangeable Ca^{2+} and K^+ content decreased with the depth in all the soils however, exchangeable Mg^{2+} and Na^+ shows reverse trend in all pedons. The exchangeable Ca^{2+} content ranged from 12.8 to 40.8 cmol (p^+) kg^{-1}. On the contrary exchangeable Mg^{2+} varied from 12.8 to 32.0 cmol (p^+) kg^{-1}. The ESP ranges from 0.8 to 19.4 in different pedons and in general it increases with depth in the profile. This can be attributed to lower topographical situation of these soils formed in this valley, which favors accumulation of soils and subsequent sodification under the semi –arid climatic condition coupled with slow permeability of these soils. An increase in ESP with depth is general observation for black soils in the semi- arid region of the peninsular India (Nimkar et al., 1992). The exchangeable K^+ content ranged from 0.7 to 1.1 cmol (p^+) kg^{-1}. The exchangeable magnesium percentage (EMP) ranged from 22.7 to 60.2 and it increases with depth in all the soils. Similar trend of EMP with depth were also reported by Magar (1990) and Kadu (1991) for soils of the central and southwest part of the valley. This increase in EMP with depth causes structural deterioration under the specific conditions and results into reduction in sHC and increase in COLE values (Table 4.3). The Ca^{2+}/Mg^{2+} ratio varies from 0.5 to 2.9 and it decreases with depth in all the soils. The opposite depth function of exchangeable magnesium resulted in the reduction in Ca^{2+}/Mg^{2+} ratio in the sub soils and leads to impairment of sHC.

4.5 SOIL CLASSIFICATION

Based on morphometric, physical and chemical characteristics, the pedons were grouped into different taxa. All the pedons are very deep (>150 cm) with cracks, clay content more than 30% and slickensides (>25 cm thick zone) underlain by cambic horizon qualify for order Vertisols and meet the requirement for the subgroup *Typic Haplusterts* with very fine textural class. However, the ESP of pedon 4 soils within the 100 cm depth from the surface is more than 15%. Hence, these soils have been grouped into order Vertisols and sub group *sodic Haplusterts*. In view of ustic soil moisture regime for the region, all the soils qualify for Ustert suborder. In the entire horizons of all soils the clay CEC was found to be more than 74.7 cmol (p^+) kg^{-1}. Soil Taxonomy (Smith, 1986) advocates clay CEC limit of 16–24 cmol (p^+) kg^{-1} or less for a kaolinitic mineralogy class, and 24–45 cmol (p^+) kg^{-1} for soils of mixed mineralogy class and >45 cmol (p^+) kg^{-1} for soils of montmorillonitic mineralogy class at family level of soil classification. Thus, the mineralogy class of these soils is montmorillonitic as reported earlier by Pal and Deshpande (1987) through x-ray diffraction technique.

The landform-soil relationship was established after systematic study of soils in different landform units and it indicate the changes in important soil properties *viz*., profile development (morphological), physical and chemical properties with the variation in landform unit. Based on soil correlation, tentatively four soil series namely Ramagarh-1 (Rmg-1), Ramagarh-2 (Rmg-2), Ramagarh-3 (Rmg-3) and Ramagarh-4 (Rmg-4) were identified. Taking into account surface texture, slope, erosion and kind of degradation in terms of sodicity the four soil series are further subdivided into soil phases and mapped into seven soil units at the level of phases of soil series (Figure 4.2). Vertisols of arid and semi-arid climates contain more pedogenic carbonates (PC) in their soil control sections (SCSs) than those of sub humid climates lowlands (Vaidya and Pal, 2002). Based on information of related studies made earlier in the region it was observed that formation of PC is the prime chemical reaction responsible for the increase in pH, the decrease in the Ca/Mg ratio of exchange site with depth and in the development of subsoil sodicity and higher ESP values in the uplands than the lowlands in their soil control sections.

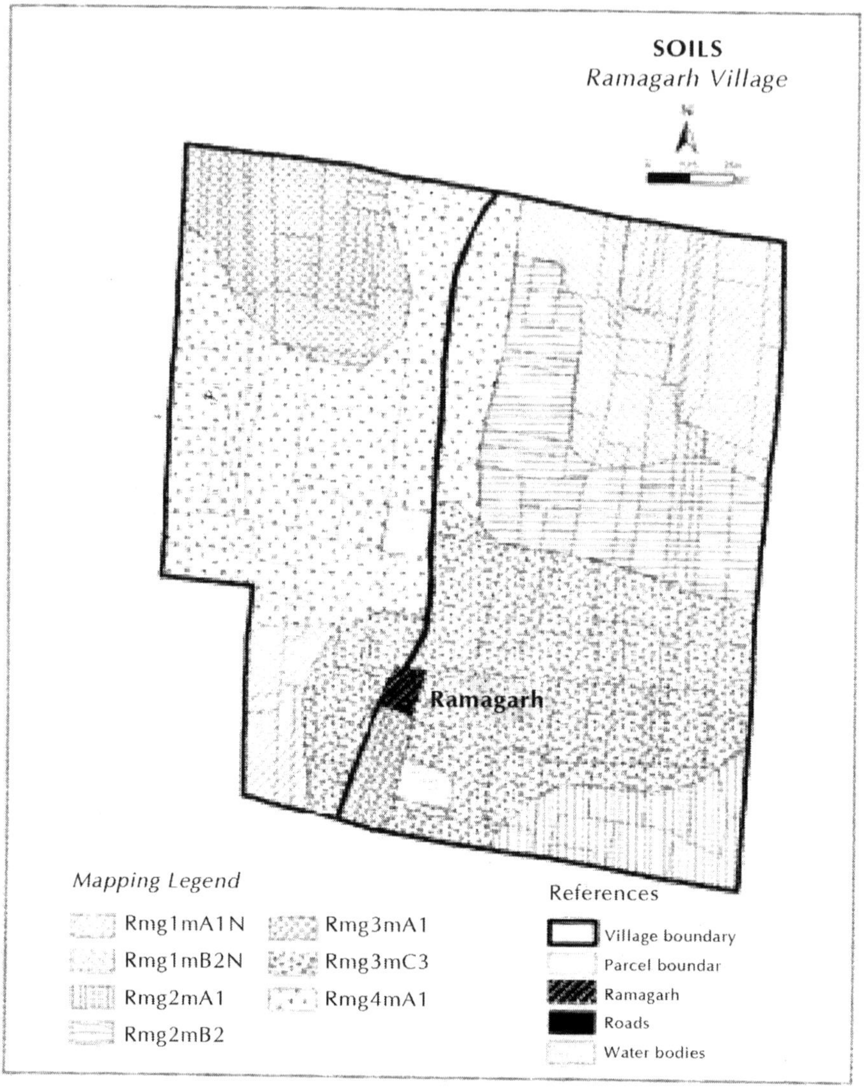

FIGURE 4.2 Soil map of the Ramagarh village.

4.6 CONCLUSIONS

From the present study it can be concluded that these soils formed in the basin or lower topographical position in the valley under semi-arid climate with high amount of smectitic clay. The soils are slight to moderately alkaline in reaction and pH value increases down the depth in the soil profile.

All the soils are calcareous and the $CaCO_3$ increases with depth. This may be due to leaching of bicarbonates during rainy season from the upper layers and their subsequent precipitation as carbonates in the lower layer during dry and hot periods indicates the presence of more lime concretions in the sub soil and resulted in the increase in pH in the subsoils. Exchangeable Ca^{2+} ions decreases with depth on one side and on the other hand exchangeable Mg^{2+} and Na^{+} increases with depth in all the soils. This opposite depth function of exchangeable magnesium causes reduction in the Ca/Mg ratio and concomitant increase in ESP in the sub soil deteriorate the soil structure and impaired the sHC of these soils which shows ponding of water during the rainy season. The reduction in mean weight diameter and sHC observed in the sub soil with concomitant increase in ESP is the cause of degradation of these soils and it also further becomes apparent that these adverse degradative processes occurs in these soils at much lower ESP values than 15. The study indicates that the sodic swell shrink soils need to be classified based on measurable properties like sHC and the level of ESP for categorizing sodic swell shrink soils in semi-aridic Indian continent should be brought down.

KEYWORDS

- **Purna Valley**
- **Salt Affected Soils**
- **Sodic Soils**
- **Swell-Shrink Soils**
- **Vertisols**

REFERENCES

Abrol, I. P., Yadav, J. S. P., & Massoud, F. I. (1988). Salt-affected soils and their management. *FAO Soils Bulletin in 39*, FAO, Rome.

Adyalkar, P. G. (1963). Paleogeography, nature and pattern of sedimentation and groundwater potentiality of the Purna Basin of Maharashtra. *Proceedings of National Institute of Science, India, 29*, 25–45.

Ahuja, L. R., Naney, J., Williams, W. R. D., & Ross, J. R. (1998). Vertical variability of soil properties in a small watershed. *Journal of Hydrology, 99*, 307–318.

Balpande, S. S., Deshpande, S. B., & Pal, D. K. (1996). Factors and processes of soil degradation in Vertisols of the Purna valley, Maharashtra. *Land Degradation and Development, 7*, 313–324.

Balpande, S. S. (1993). *Characteristics, genesis and degradation of Vertisols of the Purna valley, Maharashtra*. PhD Thesis, Panjabrao Krishi Vidyapeeth, Akola, Maharashtra.

Bharambe, P. R., Awasarmal, B. C., & Ghonsikar, C. P. (1986). Physical properties of soils in Jayakwadi and Purna commands. *Agricultural development of Jayakwadi and Purna command areas*. Publication of Marathwada Agricultural University, Parbhani, Maharashtra. pp. 137–144.

Black, G. R., & Hartge, K. H. (1986). Bulk density, *In*: *Method of Soil Analysis, Part I: Physical and Mineralogical Methods*; 2d ed. A. Klute (Ed.). American Society of Agronomy, Madison, WI, pp. 363–375.

FAO-UNESCO. (1974). Soil Map of the World. 1:5,000,000. Volume 1. Legend. UNESCO, Paris.

ICAR and NAAS. (2010). Degraded and wastelands of India: Status and spatial distribution. Project report, Indian Council of Agricultural Research (ICAR), New Delhi. p. 158.

Jackson, M. L. (1973). Soil Chemical Analysis, Prentice Hall of India Pvt. Ltd. New Delhi. pp. 10–114.

Kadam, Y. B. (2011). Decadal change in land degradation in Purna valley of Maharashtra. M.Sc. (Agri.) Thesis, Panjabrao Krishi Vidyapeeth, Akola, Maharashtra.

Kadam, Y. B., Kharche, V. K., Naitam, R. K., Katkar, R. N., & Konde. N. M. (2013). Characterization and classification of salt-affected soils of Purna valley in Vidarbha region of Maharashtra. *Journal of Soil Salinity and Water Quality, 5*(2), 125–135.

Kadu, P. R., Pal, D. K., & Deshpande, S. B. (1993). Effect of low exchangeable sodium on hydraulic conductivity and drainage in shrink-swell soils of the Purna Valley, Maharashtra. *Clay Research, 12*, 65–70.

Kadu, P. R. (1991). Physical and chemical properties affecting hydraulic conductivity of black soils of the Purna valley. M.Sc. (Agri.) Thesis, Panjabrao Krishi Vidyapeeth, Akola, Maharashtra.

Klute, A. (1986). Methods of Soil Analysis, Part 1. *In*: Physical and Mineralogical Methods; A. Klute Ed. Agronomy Monograph, 9, Part 1. 2nd Edison.

Klute, A., & Dirksen, C. (1986). Hydraulic conductivity and diffusivity laboratory methods in methods of soil analysis Part I. *In*: Klute, A. (Ed). Agron Monograph 9, Madison, Wisconsin. pp. 716–719.

Liang, Y. C., Jin, S., Microslav, N., Yu, P., Chen, W., & Jiang, Y. (2005). Organic manure stimulate biological activity and barley growth in soil subject to secondary salinization. *Soil Biology & Biochemistry, 37*(1), 185–195.

Magar, A. S. (1990). An Appraisal of the Nature of Salinity and Sodicity in Black Soils oflhe Puma Valley, M.Sc. (Ag.) Thesis, Panjabrao Krishi Vidyapeeth, Akola, Maharashtra, India.

Mandal, A. K., Sharma, R. C., Singh, Gurbachan. and Dagar, J. C. (2010). Computerized Database on Salt Affected Soil in India. Technical Bulletin No. CSSRI, Karnal, pp. 28.

Nayak, A. K., Chinchmalatpure, A. R., Rao, G. G., & Verma, A. K. (2006). Swell shrink potential of Vertisols in relation to clay content and exchangeable sodium under different ionic Environment. *Journal of Indian Society Soil Science, 54*, 1–5.

Nimkar, A. M. (1990). Evaluation of physical and chemical characteristics of soils of the Purna valley of Maharashtra. M.Sc. (Agri.) Thesis. Panjabrao Deshmukh Krishi Vidyapeeth, Akola, Maharashtra.

Nimkar, A. M., Deshpande, S. B., & Babrekar, P. G. (1992). Evaluation of a salinity problem in swell-shrink soils of a part of the Purna valley, Maharashtra. *Agropedology, 2*, 59–65.

Padekar, D. G. (2014). Soil quality as influenced by land use management with special reference to irrigation in selected tehsils of Amravati district, Maharashtra. PhD (Agri.) Thesis. Panjabrao Deshmukh Krishi Vidyapeeth, Akola, Maharashtra.

Pal, D. K., & Deshpande, S. B. (1987). Characteristics and genesis of minerals in some benchmark Vertisols of India. Pedologie (Ghent), *37*, 259–275.

Pal, D. K., Deshpande, S. B., Valayutham, M., Srivastava, P., & Durge, S. L. (2000). Climate change and polygenesis in Vertisols of the Purna valley (Maharashtra) and their management. NBSS Res. Bult. 83. NBSS & LUP, Nagpur, p. 35.

Piper, C. S. (1966). Soil and Plant Analysis. Hans Publishers, Bombay, pp. 368.

Puranik, R. B., Barde, N. K., & Ballal, D. K. (1972). Studies on some saline-alkali soils of Akola district, Maharashtra state, *PKV Research Journal, 1*, 51–58.

Richards, L. A. (1954). Diagnosis and Improvement of Saline and Alkali Soils. USDA Handbook No. 60, Oxford and IBH Publ. Co. Calcutta 160.

Sagare, B. N., Kalane, R. L., & Guhe, Y. S. (1991). Characterization of salt-affected Vertisols of Puma Valley in Vidarbha region. *Journal of Maharashtra Agricultural Universities, 16*, 310–312.

Schafer, W. M., & Singer, M. J. (1976). A new method of measuring shrink swell potential using pastes. *Soil Science Society America Journal, 40*, 805–806.

Sehgal, J., & Abrol, I. P. (1994). Soil degradation in India: status and impact. New Delhi: Oxford and IBH. 80 pp.

Shirale, A. O. (2013). Effect of crop residue and green manuring on soil properties and crop productivity in Purna valley of Vidarbha. PhD Thesis (Unpub). Dr. PDKV, Akola, M. S.

Smith, G. D. (1986). The Guy Smith interview: rationale for concepts in Soil Taxonomy. SMSS. Technical Mongraphy, No. 11, SMSS, SCS, USDA.

Soil Survey Staff (1998). Keys to Soil Taxonomy (8th Edn.). Soil Conservation Service, US Department of Agriculture, Washington, D.C., 1998.

Szabolcs, I. (1974). Salt affected soils in Europe. Martinus Nijhoff, Research Institute of Soil Science and Agricultural Chemistry, Academy of Sciences, Budapest.

Tanpure, V. D. (1971). A study of salinity in well waters of saline tract of Vidarbha. M.Sc. (Agri.) Thesis, Punjabrao Krishi Vidyapeeth, Akola, Maharashtra.

Vaidya, P. H., & Pal, D. K. (2002). Microtopography as a factor in the degradation of Vertisols in Central India. *Land Degradation and Development, 13*, 429–445.

Van Wambeke, A. (1985). Calculated Soil Moisture and Temperature Regimes of Asia. SMSS Technical Monograph No. 9. Cornell University, Ithaca, NY.

Walkley, A., & Black, C. A. (1934). An examination of degtjareff method for determining soil organic matter and a propose modification of chromic acid titration method. *Soil Science, 37*, 29–38.

Zonn, S. V. (1986). Tropical and Subtropical Soil Science (Translated from the Russian), Mir. Publishers, Moscow, p. 260.

CHAPTER 5

EVALUATION OF SHRINK-SWELL SOILS IN SEMI-ARID REGIONS OF CENTRAL INDIA FOR SOIL RESOURCE MANAGEMENT AND SUSTAINABLE AGRICULTURE

PRAKASH R. KADU and V. K. KHARCHE

Professor, Department of Soil Science and Agricultural Chemistry, Dr. Panjabrao Deshmukh Krishi Vidyapeeth, Akola – 444104, India

CONTENTS

ABSTRACT

The evaluation of soil resources and assessment of soil related constraints is the only proper way to manage the soils for sustainable crop production. The database on soil properties particularly measurable, obtained through soil resource studies is important tool for practical purpose of assessing suitability of each soil unit for agricultural crops. Therefore, the present study was undertaken to evaluate the shrink- swell soils of Chandrabhaga valley of central India for various crops. Various geomorphological processes operating on different landforms and the typifying pedons representing various landforms were studied. Most of the soils formed in basaltic alluvium were deep dark brown to very dark gray in color, had very high clay content, calcareous in nature, imperfect to moderately drained, low saturated hydraulic conductivity (sHC) (0.01 to 27.4 mm hr^{-1}). Lower values of sHC in some soils was due to higher exch. sodium percentage (>5) and low exch. Ca/Mg ratio (<1.5). The soils are neutral to strongly alkaline, non-saline with exchangeable sodium percent (ESP) well below 5% except in some irrigated soils. They have variable exchangeable Ca/Mg ratio (0.5 to 5.3) and exchangeable magnesium percentage (12.5 to 51.5). Although the thermal regime of the region is conducive for cropping throughout the year, the period of moisture availability (LGP) is quite variable from soil to soil. The exercise on land suitability evaluation indicated that soils have good potential for crop production because of their better rooting depth, high clay of smectitic nature associated with high cation exchange capacity (CEC) and plant available water capacity with a high base status. However, poor drainage appears to be the main parameter controlling the crop production in the area. The deep-rooted crops like cotton and pigeonpea can be successfully grown in well-developed Vertisols with sHC higher than 10 mm hr^{-1} while soybean, sorghum and groundnut can be grown on vertic intergrades with relatively higher sHC and better drainage. Thus the productivity of these soils is primarily governed by important parameters like sHC indicating that the determination of measured parameters in a particular crop-climatic-soil environment would be a prudent approach in the field of land evaluation instead of depending upon inferred parameters. sHC, ESP and Ca/Mg ratio can be regarded as the soil quality indices of Vertisols and associated vertic intergrades as they govern the productivity of these soils.

5.1 INTRODUCTION

Shrink-swell soils are a group of fine textured soils, most of which belong to the order Vertisols and vertic intergrades. The striking feature of these soils is that they are dominantly of smectitic mineralogy with a high coefficient of expansion and contraction, thereby setting up a steady process of churning within the pedons. Because of this behavior, these soils do not express a uniform horizonation as can be seen in other soils. However, this does not imply that they are of simple homogeneous soil systems. On the contrary, they exhibit extensive spatial and temporal variability in their properties and remain the most difficult land resource system in the world to be managed successfully (Coulombe et al., 1996).

Land use planning (LUP) is a systematic and iterative procedure carried out to create an enabling environment for sustainable development of land resources, which meet the need and demands of the people. Technically, it is the systematic assessment of land and water potential, alternatives for land use and economic and social conditions in order to select and adopt the best land-use options. Land evaluation is the ranking of the soil units on the basis of their capabilities to provide optimum returns per unit area besides conserving the natural resources for future use. The evaluation of land potential is a major tool in soil survey interpretation and natural resource management. The broad objective of land evaluation is to provide the integrated management of land and to provide information for the future allocation of land towards those uses that provide the greater sustainable benefits.

Land suitability classification is the assessment of the potential of a given type of land for a defined use. The categories recognized in land suitability classification (FAO, 1976) are order, classes, subclasses and units. There are only two orders suitable (S) and Non-suitable (N). The classes distinguished are S1 – highly suitable, S2 – moderately suitable, and S3 – marginal suitable. The subclasses reflect kinds of limitation as in land capability subclasses. The suitability units in a subclass differ in management requirements based on limitations. Depending upon the purpose, scale and intensity of study either all or limited number of categories may be adopted. The approach of suitability has been usefully employed for different crops by several workers in different areas (NBSS & LUP, 1994). There is a need to translate the innovative research in pedology, land capability assessment and soil survey interpretations for land resource managers and other end-users.

5.1.1 EVALUATION OF SHRINK-SWELL SOILS: GLOBAL PERSPECTIVE

Vertisols have attracted global attention in research, yielding a large volume of data on their properties and management (Coulombe et al., 1996; Mermut et al., 1996). The global area under vertisols is estimated to be approximately 308 m ha, covering nearly 2.23% of the global ice-free land area (USDASCS, 1994); however, the reliability of this estimate remains uncertain because several countries have not yet been included in the inventory (Coulombe et al., 1996). Vertisols and vertic intergrades occur in 80 countries, but more than 75% of the global vertisol area is contained in only 6 countries: India (25%), Australia (22%), Sudan (16%), the USA (6%), Chad (5%), and China (4%); (Dudal and Eswaran, 1988; Wilding and Coulombe, 1996). Because of their shrink–swell properties and stickiness, vertisols are known by a number of local regional and vernacular names (Dudal and Eswaran, 1988). Current agricultural land uses (edaphological) demonstrate that although vertisols are a relatively homogeneous soil group, they occur in a wide range of climatic environments globally and also show considerable variability in their uses and crop productivity (Pal et al., 2011).

5.1.2 EVALUATION OF SHRINK-SWELL SOILS: INDIAN PERSPECTIVE

Vertisols and associated intergrades occur extensively in the Peninsular India, covering about 72.9 m ha, accounting for 22.2% of the total geographical area of the country (Murthy et al., 1982), 28% of the world's dark clay soils and 40% of the total black soil in the semiarid tropics (Swindale and Miranda, 1984). In India about 80% of these soils form typical soilscape in the central Peninsula extending from 16° to 26° N latitudes and 73° to 81° E longitudes, covering parts of Maharashtra, Madhya Pradesh, Telangana, Andhra Pradesh and the Northern Karnataka. Majority of these soils are confined to the lower topographical situation of most of the river valleys. These soils are very important for agricultural production, as they are considered highly productive and sustainable resources, in terms of their deeper solum, high water holding capacity and nutrient status. Despite this fact, many researchers reported that the problems of internal drainage alongwith the accumulation of salts in the sub-surface horizons resulted in poor crop production (Balpande et al., 1996; Bharambe and Ghonsikar, 1985; Bhargava and Abrol, 1990; Kadu et al., 1993, 2003).

Chandrabhaga valley is one of such a river valleys covering the northwestern part of Nagpur district, Maharashtra State, India wherein the soils exhibit a relatively better drainability supporting good crops (except in areas of foot slope where the problems of waterlogging prevails), in contrast to the Vertisols of other river valleys such as Jayakwadi, Purna and elsewhere. In addition, the soils of this river valley are irrigated by wells without showing any sign of problem relating to salt encrustation, apparently indicating favorable cationic and anionic composition of the well waters. It is known that while a favorable cation composition of both the soil and irrigation water helps in raising good crops., unfavorable cations may lead to chemical degradation of this soil.

Kadu et al. (2003) attempted to identify bio-physical factors that limit the yield of deep-rooted crops (cotton) in 29 basaltic-alluvium vertisols of the Nagpur, Amravati and Akola districts in the Vidarbha region in central India. Natural soil degradation in terms of pedogenic carbonate formation and development of subsoil sodicity in the vertisols of the Purna Valley of Maharashtra, central India; (total area ~0.6 Mha) is triggered by the semi-arid climate, with an mean annual rainfall (MAR) of 875 mm, a tropustic moisture regime and a hyperthermic temperature regime (Balpande et al., 1996; Pal et al., 2000). Such soils have severe drainage problems, but in the Pedhi Watershed, in the adjacent east upland of the Purna valley (area ~45,000 ha), vertisols also have drainage problems. The area, however, has a higher MAR (975 mm) than the Purna valley and has similar moisture and temperature regimes to the Purna Valley. Vertisols are the dominant soil type in the watershed, but as a result of micro-topographic variation (0.5–5 m;), *Sodic Haplusterts* occur on micro-high (MH) positions and at a distance of approximately 6 km, while *Aridic Haplusterts* occur on micro-low (ML) positions (Vaidya and Pal, 2002). Cotton performs better in *Aridic Haplusterts* (ESP<5; 0.6–1.6 t/ha of seed+lint, yield obtained by the farmers following typical managements as detailed elsewhere (Kadu et al., 2003) than in *Aridic Haplusterts* (ESP > 5, but <15; 0.6–1.0 t/ha) and *Sodic Haplusterts* (ESP ≥ 15; 0.2–0.8 t/ha yield; Kadu et al., 2003). In view of the comparatively poor crop productivity of the *Aridic Haplusterts* with the *Sodic Haplusterts* with an ESP>5 but <15 (having no soil modifiers; Pal et al., 2006), *Aridic Haplusterts* were classified as *Sodic Haplusterts* (Balpande et al., 1996; Pal et al., 2000) because their resilience could be improved by appropriate management interventions to enhance crop productivity. Many productive vertisols under rainfed conditions have been rendered unproductive for agriculture under irrigated conditions in the longer-term. However, some zeolitic vertisols of the semi arid regions of western India have been irrigated through canals for the last twenty years to produce sugarcane.

These soils lack salt-efflorescence on the surface and are not waterlogged at present, suggesting that these soils are not degraded due to their better drainage. However, these soils are now *Sodic Haplusterts* in view of their pH, ECe and ESP values, but they have sHC >10 mmhr^{-1} (weighted mean in the 0–100 cm) (Pal et al. (2011).

5.1.3 IMPORTANCE OF THE STUDY

In view of the above observations, it was thought that an investigation should be carried out in parts of this river valley, constituting a sizable area namely, the Adasa watershed, which is in fact in between the Chandrabhaga and Kolar river basins to elucidate the precise pedogenetic processes that may be responsible for the formation of these soils with due importance to their geomorphic setting and the evolutionary history. Precise understanding on these aspects would help to explain the various soil parameters that are beneficial for raising good crops in general, and also their influence on impairing crop productivity at places, in the Adasa watershed. Furthermore, the investigation was carried out to examine such soil related constraints in the shrink-swell soils of Chandrabhaga valley in central India by adopting a pedo-edaphological approach. The management options formulated on the basis of identifying stress causing and yield controlling soil related attributes serve as a standard norms or guide for decision support system for management of soils and agricultural crop production in semi-arid parts of Central India. The study was undertaken with the objectives to establish the relationship between the environmental factors and soil properties, to study the detailed geomorphology, morphological, physical, chemical and mineralogical properties of soil, to identify the problems and potentials of the soils, to evaluate the soil site suitability for different crops, and to suggest the proper agricultural land use planning based on the available data and their interpretation.

5.2 STUDY AREA

The study area lies between 21°18′ 30″ to 21°22′ 20″ N latitude and 78° 50′ 40″ to 78°59′ 00″ E longitude with a total area of 60.8 sq km in Saoner tahsil of Nagpur district. The maximum elevation of 400 m above MSL is encountered on the extreme west of the watershed and the minimum elevation of 298 m above the MSL at the extreme east where the tributaries meet the Chandrabhaga River. Thus, the relief amplitude is about 102 m (Figure 5.1). The geology of the area consists of the Deccan trap, sandstone outcrops and colluvic-alluvial fills. The traps are basaltic composition and they occur in the western portion of the study area and sandstone outcrops are exposed as low mounds on lower central portion of the study area. The central and lower portion of the study area is occupied by colluvic-alluvial fills of recent origin. Geomorphologically, the study area consists of various landforms, which can be distinguished from their slope, form and composition, relief amplitude and position in toposequence. Denuded basaltic plateau occurs on the western most portion of the watershed and occupies the highest position. It is immediately followed by scarp slopes and colluvial foot slopes. Below this unit the main valley side slopes occur, which were subdivided as upper, middle and lower sectors based on their geomorphic position in a scqucncc. Denuded/isolated mounds are scattered in the lower part of the watershed. Valley floor occupies the lower position and it is subdivided into entrenched valley floors and narrow flood plain. The drainage network is mainly of dendritic pattern on plateau and sub parallel on valley slopes.

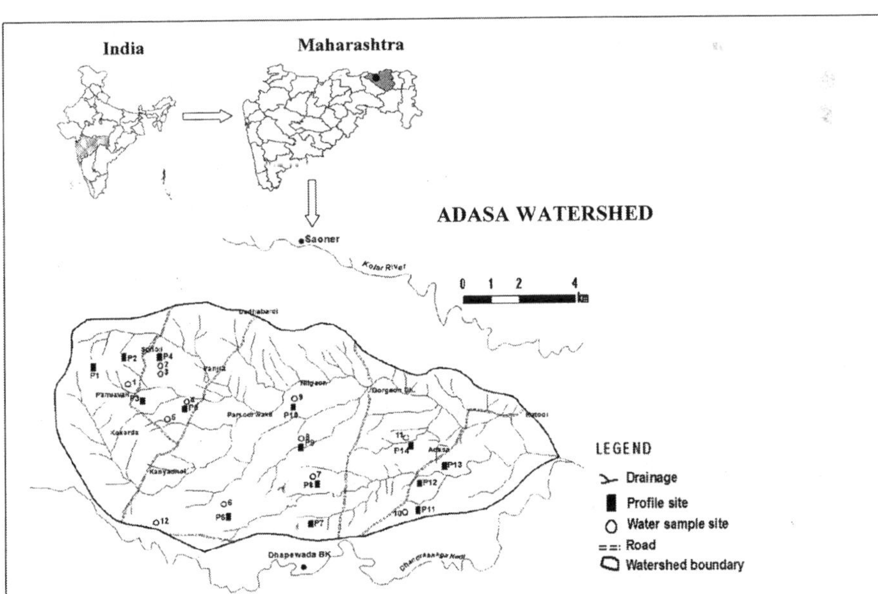

FIGURE 5.1 Location map of the study area showing soil and water sampling sites.

The climate is semiarid–subtropical with annual rainfall of 1060 mm. The maximum rainfall is received during the month of July (305.1 mm). The mean annual temperature is 26.9°C with mean maximum of 33.5 and mean minimum of 20.3°C. Highest relative humidity (79.5%) is observed in the month of August, while the lowest value (28.5%) has been recorded during May. The soils have ustic moisture and hyperthermic temperature regime. The Length of growing period (LGP) for the area is 160 days calculated by standard method given by FAO (1976) in which 100 mm of AWC is utilized by plants is considered, whereas, the different soils of watershed have variable LGP depending upon the actual AWC of the soil. The area is under rainfed *kharif* cultivation of cotton (*Gossypium sp*), sorghum (*Sorghum bicolor*), tur (*Cajanus cajan*), Soybean (*Glycine max*) black gram (*Phaseolus aureus*). In the *rabi* season wheat (*Triticum sp*) and gram (*Cicer arietinum*) are grown on black soils under stored soil moisture. But in some pockets orchards of mandarin (*Citrus reticulate*) and vegetables such as cabbage (*Brassica oleracea*), cauliflower (Brrassica botrytis) and tomato (*Lycopersicon esculentum*) are grown under irrigation.

5.3 METHODOLOGY

5.3.1 SOIL RESOURCE INVENTORY

The detailed characterization of soils of Adasa watershed, a part of Chandrabhaga valley in Central India with respect to their geomorphology, genesis and classification, physical, chemical and mineralogical properties of horizon wise soil samples and assessment of quality of irrigation water were accomplished. Based on the geomorphological analysis of the Adasa watershed, representative profile sites were selected and they were positioned in such way that they represent most of the landforms units identified. Various landforms were identified based on the slope, form and composition, relief amplitude and position in toposequence. In all fourteen pedons were selected on various landform units and they were examined for different morphological properties. The pedological diagnostic features were studied and horizonwise samples were collected. A detailed soil resource inventory for Adasa watershed was then prepared after classifying soils according to soil taxonomy (Soil Survey Staff, 1999).

5.3.2 SOIL ANALYSIS

The physical, chemical and mineralogical properties were assessed using standard procedures laid down by Richards (1954), Jackson (1979), Gardner, et al. (1984), Coughlan, et al. (1986), Gjems (1967) and land evaluation was carried out by the following approach of FAO with some modifications.

5.3.3 SOIL SITE-SUITABILITY EVALUATION

Land evaluation was carried out by following the approach of FAO (1976) and Sys et al. (1991) with some modifications. The crop requirements were modified for local conditions taking into account the available literature published on these soils in relation to crops grown and the actual experience of site specific conditions (Naidu et al., 2006; NBSS & LUP, 1994). This enabled identification of the soil attributes governing yield of important crops grown in the area that are important in the context of not only sustainable land management for enhancing soil quality but also useful for improving the crop productivity per unit area.

5.4 RESULTS AND DISCUSSION

5.4.1 MORPHOLOGICAL CHARACTERISTICS OF SOILS

For the present case study, fourteen pedons were examined covering different landform units mentioned in the section 5.2. Pedon 1 was situated on beveled crest and Pedon 2 on the dissected spur of the denuded plateau (Figure 5.2). Pedon 3 was situated on head slope and 4 on nose slope of foot slope units. Pedons 5 and 6 were situated on upper sectors and 7 to 10 on middle sectors of main valley side slopes whereas pedon 11 and 12 were located in lower sectors of main valley side slopes. Pedon 13 and 14 were situated near the sandstone mound, in which pedon 13 was on footslope and pedon 14 was on narrow valley floor. All the soils except pedon 1 and 2 were very deep and dark brown (10 YR 3/3 M) to very dark gray (10 YR 3/1 M) in color. The surface horizons of all the soils generally had subangular blocky structure and hard (dry) and friable (moist) consistency. In case of pedon 1 and 2 (located on plateau top), there was no effervescence with dilute HCl, whereas in case of pedons 3 to 14, the

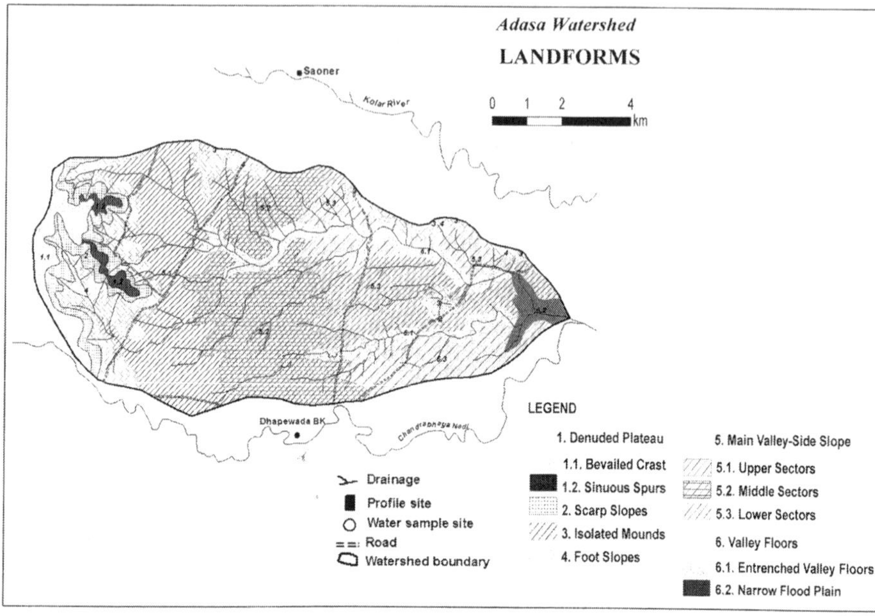

FIGURE 5.2 Landforms map of Adasa watershed.

effervescence was light in the surface horizons and strong to violent in subsurface horizons. All the soils except 1, 3, 13 and 14 showed most of the usual morphological characteristics of Vertisols, such as intersecting slickensides, forming parallelepipeds with their long axes tilted at 30–45° from the horizontal. There were shiny pressure faces and cracks penetrated to >50 cm. The detailed classification of these soils upto family level is presented in Table 5.1.

5.4.2 PHYSICAL PROPERTIES OF SOILS

The soils are mostly clayey and clay varied from 30.1 to 75.5%, whereas fine clay varied from 13.6 to 57.4% and it increases gradually throughout the depth. The silt content varied from 20.6 to 49.6 and sand content was below 8% in all the soils except pedons 13 and 14 located near the sandstone mound (5.5 to 47.1%). Bulk density at field capacity was 1.13 to 1.70 Mg m^{-3}. The saturated hydraulic conductivity of these soils ranged from 0.01 to 27.4 mm hr^{-1} for pedons 1 to 12 whereas pedon situated near sandstone

TABLE 5.1 Classification of Studied Soils of the Adasa Watershed

Pedon	Location	Classification at family level
Pedon 1	Panubali	Fine, montmorillonitic, hyperthermic family of *Lithic Ustorthents*
13	Sonpur	Fine, montmorillonitic, hyperthermic family of *Vertic Ustifluvents*
3	Panubali	Fine, montmorillonitic, hyperthermic family of *Vertic Endoaquepts*
14	Adasa	Fine, montmorillonitic, hyperthermic family of *Typic Haplumbrepts*
2	Sonali	Very fine, montmorillonitic, hyperthermic family of *Leptic Haplusterts*
Pedon 4	Pan-ubali	Very fine, montmorillonitic, hyperthermic family of *Typic Haplusterts*
5	Parsodi	
7	Dhapewada	
8	Dhapewada	
9	Dhapewada	
10	Nilgaon	
12	Adasa	
Pedon 6	Madhasawangi	Very fine, montmorillonitic, hyperthermic family of *Ustic Endoaquerts*
Pedon 11	Brahmpuri	Fine, montmorillonitic, hyperthermic family of *Typic Calciusterts*

mound had values ranged from 20.2 to 84.8 mm hr^{-1}. The coefficient of linear extensibility (COLE) varied from 0.06 to 0.28 cm^{-1} indicating a high shrink-swell potential of these soils. The moisture content at field capacity (33 kpa) and at permanent wilt capacity (1500 kpa) varied from 10.9 to 65.2 and 6.8 to 38.6%, respectively. The Available water capacity (AWC) of soils ranges between 103 to 298 mm m^{-1} and plant available water capacity (PAWC) 40 to 428 mm.

5.4.3 *CHEMICAL PROPERTIES OF SOILS*

The soils of Adasa watershed are slightly to strongly alkaline. The electrical conductivity values are less than 4 dS m^{-1}, indicating these soils are non saline. The organic carbon content of the soils varied from 0.04 to 1.11% and calcium carbonate content ranged between 0.8 to 28.2%. Calcium is the dominant exchangeable cation, followed by magnesium, sodium and potassium on the exchange complex in all the soils. The cation exchange capacity (CEC) and base saturation (BS) varied from 30.6 to 67.8 cmol (p+) kg^{-1} and 69 to 99%, respectively. The Ca/Mg ratio generally decreased with the depth indicating decrease of calcium with simultaneous increase of magnesium. The exchangeable sodium percentage (ESP) of these soils ranged between 0.2 and 17.9 and higher values were observed in the subsurface horizons of pedon 3, 5, 7 and 8. ESP less than 15 indicates the absence of alkalinity and sodium as per Richards (1954). Accordingly, these soils do not qualify as alkali soils except pedon 3 with 17.9 ESP. The exchangeable magnesium percentage (EMP) varied from 12.5 to 51.5 in these soils positive correlation was found between EMP and water dispersible clay and negative correlation between EMP and HC (Palaveyev and Penkov, 1990; Kadu et al., 2003).

5.4.4 MINERALOGICAL PROPERTIES OF SOILS

The total clay composed of smectite, vermiculite kaolin, mica, 1.0–1.4 nm minerals, small amount of chlorite, quartz and feldspars whereas the fine clay fraction composed of smectite as the dominant mineral with very minor amount of kaolin and mica. The semi-quantitative estimates of clay minerals of the total clay fraction indicated that the smectite was the dominant clay mineral amounting to 64–80%. The smectite was also dominant in pedons 13 (64%) and 14 (69%) situated near the sandstone mounds. The vermiculite, mica, kaolin and chlorite content ranged between 1 to 14, 4 to 8, 5 to 22, and 1 to 9%, respectively. The semi-quantitative estimate of clay mineral of fine clay fraction indicated that smectite was the dominant mineral (92 to 99%) followed by kaolin (1 to 8%).

5.4.5 SOIL-SITE SUITABILITY EVALUATION FOR DIFFERENT CROPS

Despite their enormous yield potential, Vertisols and vertic intergrades offer great challenge to both the scientists and the farmers for management and utilization. In depth study of their various properties, *viz.*, morphological, physical, chemical, mineralogical and their inter-relationship with landform setting may serve as a base to identify properties that are favoring or impairing the prospect of agricultural production.

Existing land use in the Adasa watershed studied by using satellite imagery with subsequent field review indicated that a major part of the area is under cultivation, which is mostly confined to soils of upper, middle and lower sectors of the main valley-side slopes. The soils developed on beveled crest and scarp slope are kept fallow and have scrub vegetation at places because of their skeletal nature. The isolated mounds are the lands without scrub, whereas the soils developed on the sinuous spurs are relatively well developed with better rooting depth and are cultivated in *kharif*. Among the watershed soils the double cropping is practiced in both upper and middle sectors of the main valley-side slopes, whereas, in lower sectors only *kharif* crop is grown. It is observed that the non-calcareous soils of upper and middle sectors with high clay (fine clay) and higher PAWC can sustain the *rabi* crop on residual soil moisture. In addition, the availability of irrigation water through wells provides an avenue for double cropping and perennial fruit trees (orange orchards).

Since every soil unit possesses a definite set of conditions to grow each crop and every plant species needs specific soil-site conditions for its optimum growth and yield, it is imperative to use the finite soil resource according to its capability or suitability for particular land use. This can be achieved only by critically evaluating the soil-site conditions vis-à-vis specific requirements of different crops (Naidu et al., 2006).

Soil-site suitability evaluation is the process to estimate the potential of a particular soil for designated uses. An intensive study of soil-site conditions in the Adasa watershed was, therefore, carried out in order to determine the suitability of some dominant crops of the area. The FAO Framework on Land Evaluation (FAO, 1976) suggests a crop specific suitability system. The basic feature is the comparison of the requirements of land use with the resources offered by the land. The framework sets out basic concepts and principles of land evaluation that are universally valid, applicable in any part of the world and at any level from global to single farm provided the crop requirement are defined taking into account the local conditions. An attempt has been made here to interpret the basic data on climate, soil, crops and qualities of land for determining soil-site suitability evaluation of the crops commonly cultivated in the area, such as sorghum, cotton, soybean, pigeonpea, and groundnut.

The literature on specific crop requirements forms an important part of suitability evaluation. This kind of information on the crops under study has been adopted from the proceedings of the National Meet on Soil-site Suitability Criteria for Different Crops organized at the National Bureau of Soil Survey and Land Use Planning, Nagpur (Naidu et al., 2006; NBSS & LUP, 1994). The suggested soil-site suitability criteria considered as a base was modified and refined, taking into account the local conditions.

The soils of Adasa watershed are endowed with relatively better PAWC, LGP, clay, organic matter and mineralogy. However, these soils have poor internal drainage (low HC), which is influenced by ESP, EMP, low exchangeable Ca/Mg and pedogenic $CaCO_3$ (Balpande et.al., 1996; Kadu et al., 2003). Hence, these properties need to be considered in the suitability criteria of different crops. Instead of considering the inferred parameter like drainage, the measured parameters like sHC have been used here. The sHC of <1 mm h^{-1} has been considered as the lower limit for satisfactory physical conditions (unsuitable) for crop growth (Balpande et al., 1996; Kadu et al., 1993), whereas HC 10 mm h^{-1} or more seems to be good for internal drainage and physical conditions, as it is evident from the good crop performance on these soils.

The CEC of these soils is high and hence does not pose any limitation for crop production in terms of nutrient and water retentions, but higher ratio of exchangeable Ca and Mg needs to be considered as criteria as it reflects on internal drainage and water retention of these soils. Therefore, exchangeable Ca/Mg ratio >2.5 is considered here as desirable, whereas <0.5 as a lower limit (Cockroft and Tisdall, 1978; Emerson and Bakker, 1973).

The ESP of >5 is considered as a critical limit for crop production for these soils (Balpande et al., 1996; Kadu et al., 1993, 2003; Northcote and Skene, 1972). Therefore, ESP >5 was considered as moderate limitation for all crops and >10 and >7.5 as severe limitation for cotton and pulses, respectively.

The existing climate and soil-site conditions (Tables 5.2 and 5.3) were compared with the criteria of each crop and based on the number of intensity of limitations, the overall suitability was determined (Sys, 1985; Sys et al., 1991). The method defines land classes with regard to number and intensity of limitations. The criteria of limitation method as given by Sys et al. (1991) is slightly modified because of the increased number of parameters in the present study for the suitability evaluation, and given below.

Land classes with regard to number and intensity of limitations:

Land	Classes	Definitions
S1	(Very suitable)	Land units with nil or upto 5 slight limitations.
S2	(Moderately suitable)	Land units with more than 5 slight limitations and/or no more than 4 moderate limitations.
S3	(Marginally suitable)	Land units with more than 4 moderate limitations and/or no more than 2 severe limitations.
N1	(Presently unsuitable; but potentially suitable)	Land units with more than 2 severe limitations and/or very severe limitations that can be corrected.
N2	(Unsuitable)	Land units having very severe limitations that cannot be corrected.

In addition, the suitability class was also derived based on the actual yields as suggested by FAO (1983). As suggested in the framework the yield is used as the criteria to confirm whether the methodology of land evaluation adopted is valid or not. This was based on the yield levels for the suitability classes as S1 – > 80%, S2 – 40 to 80%, S3 – 20 to 40% and N – < 20%. The yield reduction levels have been decided based on the optimum yield of the crop.

The prevailing climatic condition in watershed is given in Section 4.3, whereas the climatic characteristics in Table 5.2 are the actual computed values for the crop growing duration. However, no severe limitations of climate have been observed for growing these crops in the area. The LGP, which is based on both soil and climate has been found as major limitations for most of the crops.

TABLE 5.2 Climatic Characteristics of Different Crops in the Adasa Watershed

	Crops				
Characteristics	Cotton	Sorghum	Pigeonpea	Soybean	Groundnut
Rainfall during growing season	987	960	976	914	914
Mean temperature during growing season (°C)	26.1	28.1	27.1	28.7	28.7
Mean maximum temperature during growing season (°C)	31.7	32.7	32.2	32.9	32.9
Mean minimum temperature during growing season (°C)	20.6	23.5	22.0	24.4	24.4
Mean relative humidity in growing season (%)	65	70	66	71	71

TABLE 5.3 Soil-Site Characteristics of Studied Soils of the Adasa Watershed

Soil-site characteristics	Pedon 1	Pedon 2	Pedon 3	Pedon 4	Pedon 5	Pedon 6	Pedon 7	Pedon 8	Pedon 9	Pedon 10	Pedon 11	Pedon 12	Pedon 13	Pedon 14
• **LGP**	128	177	223	198	198	223	198	198	198	198	177	177	154	154
• **Site Characteristics**														
Slope (%)	1–3	0–1	3–5	3–5	1–3	1–3	1–3	1–3	1–3	3–5	1–3	1–3	3–5	3–5
Erosion	e2	e0	e2	e2	e0	e0	e1	e0	e0	e2	e1	e1	e2	e2
Saturated Hydraulic Conductivity (mmh^{-1})	27.4	14.5 to 20.2	0.01 to 3.8	14.2 to 20.2	3.3 to 8.8	4.4 to 7.4	14.5 to 20.3	2.2 to 6.5	5.8 to 14.3	9.4 to 16.4	4.7 to 12.2	6.4 to 17.8	20.2 to 84.8	20.3 to 54.8
Drainage	Well	Mod. Well	Very Poor	Mod. Well	Im-perfect	Im-perfect	Mod. Well	Im-perfect	Im-perfect	Mod. Well	Im-perfect	Mod. Well	Exces-sive	Well
Water Stagnation (days)	–	–	3–4	–	1–2	–	1–2	1–2	–	–	–	–	–	–
PAWC (mm)	40	146	281	220	217	298	198	234	239	213	175	175	112	103
Stoniness (Surface)	3–15	Nil	Nil	Nil	Nil	Nil	Nil	Nil	Nil	Nil	Nil	Nil	Nil	Nil
• **Soil Characteristics** *														
Texture	cl	c	c	c	c	c	c	c	c	c	c	c	c	cl
Coarse fragments (Vol%)														

within 50 cm	21	3	5	4	8	4	4	4	5	5	5	4	9	7
below 50 cm	–	3	8	6	10	6	7	8	8	11	8	3	12	8
Thickness (cm)	25	70	123	150	153	150	160	140	164	158	155	166	130	135
$CaCO_3$ (%)	–	–	9	4	3	2	3	4	3	2	7	2	11	4
Exch. Ca/Mg ratio	2.0	3.7	0.9	2.1	1.6	2.2	1.7	2.0	2.1	2.9	2.3	1.3	3.6	0.9
CEC (soil) Cmol(p^+)kg^{-1}	50	51	45	51	55	62	58	65	66	59	60	55	39	35
BS (%)	96	83	93	95	95	96	85	89	94	94	92	92	81	72
OM (%) (0–15 cm)	0.7	2.1	1.2	0.8	0.9	0.9	1.5	1.9	1.4	1.1	0.5	0.6	1.6	1.3
EC (dS m^{-1})	0.1	0.2	0.6	0.2	0.2	0.2	0.2	0.4	0.2	0.2	0.2	0.1	0.2	0.1
ESP	1.0	0.9	17.9	0.9	4.2	3.8	1.2	6.2	2.2	0.4	2.1	1.2	2.3	1.9
pH (1:2)	7.2	7.2	8.8	8.1	8.3	8.2	8.1	8.5	8.5	8.0	8.0	7.8	8.0	7.9

e0 – Nil erosion; e1 – Slight erosion; e2 – moderate erosion; e3 – severe erosion; cl – clay loam; c – clay; BS – Base Saturation; OM – organic matter; PAWC – Plant available water capacity; Weighted mean upto 1 m depth.

5.4.5.1 Soil-Site Suitability Evaluation for Sorghum

Sorghum is the main *kharif* crop of the region, which is found to be moderately suitable (S2) in most of the soils of the watershed (Tables 5.4 and 5.5). These soils possess moderate limitations of drainage, slope, erosion and pH in conjunction with slight limitations of texture, PAWC and coarse fragments. In addition to these limitations some soils (P13 and P14) pose limitations of $CaCO_3$, erosion and/or PAWC and thus become marginally suitable (S3) as reflected in the 27–43% yield (Table 5.5). The shallow soils (P1) are found to be unsuitable (N1) owing to severe limitations of depth associated with very severe limitations of PAWC and moderate limitations like erosion and coarse fragments, which are reflected in very low yield level (2 q/ha).

Most of the climatic conditions prevailing in the area corroborate with the requirements of sorghum except slight limitations of temperature and total rainfall. The annual rainfall about 1060 mm may prove to be harmful particularly when the excess rain is received during grain filling stage. Since sorghum needs sufficient soil moisture to tide over the drought stress the higher PAWC of the soils is desirable for its proper growth. This fact is also evident from the significant positive correlation (r = 0.62) of PAWC with yield of sorghum (Table 5.6).

The higher clay content and cation exchange capacity is desirable in order to enhance the nutrient and water holding capacity. The clay content and CEC also show significant positive relationship with yield. Although most of the soils are deep, clayey, fertile (more than 80% base saturation) with high nutrient and water holding capacity, but they show some problems due to drainage and erosion. Therefore, sorghum is found to be moderately suitable in most of the soils in the watershed. It has been also found by many workers that a variety of soils with clay-to-clay loam textures, moderate depth (>75 cm) and high AWC (200 mm) are suitable for successful sorghum cultivation (Singh, 1988).

Among the moderately suitable soils some soils (P5 and P9) have recorded more than 80% yield. As per the criteria suggested by FAO (1983) these soils with >80% yield level should qualify highly suitable class. But because of moderate limitations of drainage and pH they have been classified moderately suitable (S2). However, the soils may prove to be highly suitable (S1) in view of their high yield potential suggesting that the yield of sorghum on these soils can be improved by modifying the drainage and soil reaction.

TABLE 5.4 Soil-Site Suitability Criteria for Sorghum

Soil-site characteristics	Degree of limitation					
	0 (None)	1 (Slight)	2 (Moderate)	3 (Severe)	4 (Very Severe)	
		S1	S2	S3	N1	N2
• **Climatic Characteristics**						
Total rainfall (mm)	750–850	650–750, > 850	550–650	450–550	< 450	—
Rainfall in growing season (mm)	600–700	500–600	400–500	300–400	< 300	—
Length of growing period (days)	150–120	120–105	105–90	< 90	—	—
Mean temp. in growing season (°C)	32–26	26–24	24–22	22–20	< 20	—
Mean max. temp. in growing season (°C)	31	31–33	33–35	> 35	—	—
Mean min. temp. in growing season (°C)	> 22	22–18	18–15	< 15	—	—
Mean R.H. in growing season	70–60	60–50	50–40	< 40	—	—
• **Site Characteristics**						
Slope (%)	< 1	1–3	3–5	5–8	>8	—
Erosion	e1	e1	e2	e3	—	—
Saturated Hydraulic Conductivity (mmh^{-1})	20–50	10–20	5–10	1–5 , >50	< 1	—
Drainage	Well	Mod. Well	Imperfect	Poor and Excessive	V. Poor	—
PAWC (mm)	> 200	150–200	100–150	50–100	< 50	—
Stoniness (Surface)	—	< 15	15–40	40–60	>60	—
• **Soil Characteristics** *						
Texture	sic, sicl, cl	c, l, scl	sil, scl	sl	s, ls	—
Coarse fragments (Vol.%)						—
within 50 cm	< 5	5–15	15–35	35–50	>50	—
below 50 cm	5–15	15–40	40–75	> 75	—	—

TABLE 5.4 (Continued)

	Degree of limitation					
	0 (None)	**1 (Slight)**	**2 (Moderate)**	**3 (Severe)**	**4 (Very Severe)**	
Soil-site characteristics		**S1**	**S2**	**S3**	**N1**	**N2**
Depth (cm)	> 100	75–100	50–75	< 50	—	—
$CaCO_3$ (%)	< 5	5–10	10–25	>25	—	—
Exch. Ca/Mg ratio	> 2.5	1.5–2.5	0.5–1.5	<3.5	—	—
BS (%)	> 80	80–50	50–35	<35	—	—
OM (g kg^{-1}) (0–25 cm)	> 7.5	7.5–5.0	5.0–2.0	<2.0	—	—
EC (dS m^{-1})	< 2	2–4	4–8	8–10	>10	—
ESP	<< 5	< 5	≥ 5	10–15	>15	—
pH (1:2)	6.5–7.5	6.5–5.5, 7.5–8.0	5.5–5.0, 8.0–8.5	< 5.0, 8.5–9.0	>9.0	—

TABLE 5.5 Degree and Kind of Major Constraints, Suitability and Yield of Sorghum

Soils	Rainfall in growing season	Erosion	sHC (Drainage)	PAWC	Depth	Exch. Ca/Mg ratio	ESP	pH	Suitability class	Yield Q/ha	% yield to optimum #	Suitability based on actual yield
Pedon 1	*	**	--	****	***	*	--	--	N1	2	7	N1
Pedon 2	*	--	*	**	*	--	--	*	S2	12	40	S2
Pedon 3	*	**	****	--	--	**	****	*	N1	9	30	S3
Pedon 4	*	**	*	--	--	*	--	**	S2	20	67	S2
Pedon 5	*	--	**	--	--	*	*	**	S2	25	83	S1
Pedon 6	*	--	**	--	--	*	*	**	S2	18	60	S2
Pedon 7	*	*	*	*	--	*	--	**	S2	15	50	S2
Pedon 8	*	--	**	--	--	*	**	**	S2	20	67	S2
Pedon 9	*	--	**	--	--	*	--	**	S2	26	87	S1
Pedon 10	*	*	*	--	--	--	--	*	S2	20	67	S2
Pedon 11	*	*	**	*	--	*	--	*	S2	17	57	S2
Pedon 12	*	*	*	*	--	**	--	*	S2	15	50	S2
Pedon 13	*	**	***	**	--	--	--	*	S3	8	27	S3
Pedon 14	*	***	--	**	--	**	--	*	S3	14	47	S2

Based on maximum observed yield as optimum, i.e., 20 q/ha. Limitations -- No; * Slight; ** Moderate; *** Severe; **** Very Severe.

TABLE 5.6 Correlation Between Soil Attributes and Yield of Different Crops

S.No.	Yield (y)	Soil attribute (x)	r
1	Sorghum	PAWC	0.62*
2	Sorghum	Clay	0.71**
3	Sorghum	CEC	0.61*
4	Cotton	PAWC	0.77**
5	Cotton	Clay	0.82**
6	Cotton	CEC	0.73**
7	Soybean	PAWC	0.58*
8	Soybean	Clay	0.70**
9	Soybean	CEC	0.63*

Y = Dependent variable; X = Independent variable; r = Correlation co-efficient;
* = Significant at 5 percent; ** = Significant at 1 percent.

Up to a certain extent the clay content is desirable for enhancing the water and nutrient holding capacity of the soil but too high a clay content (>60), especially in the sub-soils, may impair the internal drainage of the soil. Among the climatic parameters the low relative humidity prevailing in the area has shown slight limitation because soybean needs high humidity (>80%) (Singh, 1988). It can be suggested that a short duration crop like soybean, which requires a soil depth within 1 m can be successfully grown in the area provided the soil drainage is good.

Most of the soils under study are alkaline and thus show moderate limitations of pH for the growth of soybean. This is also evident from low yield of soybean in most of the soils as compared to the optimum yield. High exchangeable sodium is also equally harmful for soybean, making such soils (P3 and P8) unsuitable.

5.4.5.2 Soil-Site Suitability Evaluation for Cotton

Cotton is the main cash crop of the area. Majority of the soils of the watershed are found to be moderately suitable (S2), because of moderate limitations of slope, erosion and drainage (Tables 5.7 and 5.8). These soils being clayey and rich in smectite minerals have restricted drainage. Some soils have severe limitation of either shallowness and associated erosion hazard or high amount of exchangeable sodium, which makes them marginally suitable for cotton. The PAWC has been considerably low in some shallow soils, hence such soils are unsuitable for cotton owing to the fact that cotton, a long duration crop, needs a deep rooting zone for sufficient soil moisture storage for its proper growth and development. A growing period of 160 to 180 days is necessary to tide over the drought stress. The LGP has been considerably reduced (128 days) on the soils that are invariably shallow with consequently lower AWC. The suitability class obtained finds good correlation with the actual performance of cotton. The marginally suitable soils have recorded very low yield (5 to 7 q/ha) showing 62–75% yield reduction (Table 5.8) whereas, moderately suitable soils have yield higher than this (10 to 18 q/ha), which justify the validity of the criteria used in the study.

It becomes apparent from the above discussion that most of the soils from the Adasa watershed provide favorable conditions for the growth of cotton such as adequate soil depth, clayey texture with high cation exchange capacity and high AWC. The properties such as PAWC, clay content and CEC show significant positive correlation with greater influence in governing the yield (Table 5.6). However, the only constraint for cotton cultivation in the area seems to be the drainage condition of the soils. Thus, the soils may prove to be potentially suitable for cotton by improving the internal soil drainage. For cotton among the soils found to be moderately suitable, some soils (P5, P6, P8 and P9) have shown only the dominant limitation of low HC. However, this limitation leads to water stagnation for some time in the rainy season, which affects the crop growth temporarily. But the other favorable conditions such as high clay, CEC, base saturation and PAWC of the soils sustain this crop for a longer period, giving more production. This is reflected in the better yields (85 to 90%) obtained in some soils, which are found to be moderately suitable as per suitability evaluation (Kadu et al., 2003).

5.4.5.3 Soil-Site Suitability Evaluation for Pigeonpea

Pigeonpea (tur) is generally grown as an intercrop in cotton and sorghum. The soil-site requirements for pigeonpea (NBSS & LUP, 1994) are more or less similar to that for cotton. Since pigeonpea is a long duration crop, it needs the availability of moisture for a longer period (LGP – 210 days). Most of the soils fulfill these requirements and are, therefore, moderately suitable (Table 5.12) with some limitations of erosion and drainage. Because of high ESP some soils are presently unsuitable, but can be potentially suitable after they are amended.

5.4.5.4 Soil-Site Suitability Evaluation for Soybean

Soybean is the most important pulse and oilseed crop of this region. It is a cash crop with a low cost-benefit ratio. Although the conditions like soil depth and high PAWC are favorable to soybean, most of the soils are found to be moderately suitable (Tables 5.9 and 5.10) because of high clay content associated with drainage problems and highly alkaline pH (yield 10–17 q/ha).

The clay content in the soils has shown significantly positive relationship with yield (Table 5.6). However, the high clay content in soils aids in improving the PAWC as well as CEC of the soils providing sufficient soil moisture and nutrients for crop production. This is also evident from the significant positive relationship of PAWC and CEC with the yield of soybean (Table 5.10). Thus, the positive and negative influence of clay content on soybean production can be justified as stated further.

5.4.5.5 Soil-Site Suitability Evaluation for Groundnut

Groundnut is an important oilseed crop, which grows well in tropical and subtropical climate with long and warm growing season of 150–180 days. High rainfall (>1000), drought conditions with less than 400 mm rainfall or cold climatic conditions are detrimental for crop growth. It grows well in well drained, friable, fine-loamy to coarse-loamy soils supplied with nutrients especially calcium, and organic matter. The fine textured soils with 60% or more clay content are not suitable for groundnut cultivation because such soils turn very hard on drying thus restricting pod formation and their development. Most of the soils of the Adasa watershed have high clay content and imperfect to moderate drainage and thus are found to be marginally suitable. These soils also pose moderate to severe limitations of organic matter. Due to severe limitations of low sHC, texture and organic matter and high pH most of the soils are unsuitable or relatively less suitable for groundnut as compared to other crops. However, some soils with relatively low clay content do not pose the problem of internal drainage and qualify as moderately suitable provided they are rich in organic matter and nearly neutral in

TABLE 5.7 Soil-Site Suitability Criteria for Cotton

	Degree of limitation				
	0 (None)	**1 (Slight)**	**2 (Moderate)**	**3 (Severe)**	**4 (Very Severe)**
Soil-site characteristics	**S1**	**S2**	**S3**	**N1**	**N2**
• Climatic Characteristics					
Total rainfall (mm)	850–1050	700–850	550–700	< 550	–
Rainfall in growing season (mm)	750–950	600–750	450–600	< 450	–
Rainfall during critical period (boll devp.)	100–120	–	–	–	–
Length of growing period (days)	160–180	135–160	120–135	< 120	–
Mean temp. in growing season (°C)	22–28	28–32	>32, 19–22	< 19	–
Mean max. temp. in growing season (°C)	–	–	–	> 36	–
Mean min. temp. in growing season (°C)	–	–	–	< 19	–
Mean R.H. in growing season	60–80	80–90	–	< 50	–
• Site Characteristics					–
Slope (%)	< 1	1–3	3–5	> 5	–
Erosion	e0	e1	e2	e3	–
Saturated Hydraulic Conductivity (mmh^{-1})	20–50	10–20	5–10	1–5, >50	< 1
Drainage	Well	Mod. Well	Imperfect	Poor and Excessive	Very poor
Water Stagnation (days)	< 1	1–2	2–3	3–5	> 5
PAWC (mm)	> 200	150–200	100–150	50–100	–
Stoniness (Surface)	< 15	< 15	15–40	> 40	–
Soil Characteristics *					–

TABLE 5.7 (Continued)

	Degree of limitation				
	0 (None)	**1 (Slight)**	**2 (Moderate)**	**3 (Severe)**	**4 (Very Severe)**
Soil-site characteristics	**S1**	**S2**	**S3**	**N1**	**N2**
Texture	sic, sicl, c	cl	sil, scl, l, sc	sl	s, ls
Coarse fragments (Vol%)					–
within 50 cm	< 5	5–15	15–35	> 35	–
below 50 cm	5–15	15–35	35–50	> 50	–
Depth (cm)	> 125	100–125	75–100	40–75	< 40
$CaCO_3$ (%)	< 5	5–10	20-Oct	>20	–
Exch. Ca/Mg ratio	> 2.5	1.5–2.5	0.5–1.5	<0.5	–
BS (%)	> 80	50–80	35–50	<35	–
OM (g kg^{-1}) (0–25 cm)	> 10	7.5–10	5.0–7.5	< 5	–
EC (dS m^{-1})	< 2	<2	2–4	> 4	–
ESP	< 5	< 5	≥ 5, 5–10	10–15	> 15
pH (1:2)	7–8	8–8.5	8.5–9, 6.5–7	> 9, <6.5	–

TABLE 5.8 Degree and Kind of Major Constraints, Suitability and Yield of Cotton

Soils	LGP	Slope	Erosion	sHC (Drainage)	Water Stagnation	PAWC	Depth	Exch. Ca/Mg ratio	OM	ESP	pH	Suitability class	Yield q/ha	% yield to optimum #	Suitability based on actual yield
Pedon 1	**	*	**	–	–	****	****	*	**	–	–	N2	–	–	–
Pedon 2	–	–	–	*	–	*	***	–	–	–	*	S3	7	35	S3
Pedon 3	–	**	**	****	**	–	*	**	–	****	*	N1	7.5	38	S3
Pedon 4	–	**	**	*	–	–	–	*	*	–	*	S2	13	65	S2
Pedon 5	–	*	–	**	–	–	–	*	*	*	*	S2	17	85	S1
Pedon 6	–	*	–	**	*	–	–	*	*	*	*	S2	15	75	S2
Pedon 7	–	*	*	*	–	–	–	*	–	–	*	S2	13	65	S2
Pedon 8	–	*	–	**	*	–	–	*	–	**	*	S2	15	75	S2
Pedon 9	–	*	–	**	*	–	–	*	–	–	*	S2	18	90	S1
Pedon 10	–	**	*	*	–	–	–	–	–	–	*	S2	14	70	S2
Pedon 11	–	*	*	**	–	*	–	*	**	–	*	S2	9.5	48	S2
Pedon 12	–	*	*	*	–	*	–	**	**	–	–	S2	10	50	S2
Pedon 13	*	**	**	***	–	**	–	–	–	–	*	S3	5	25	S3
Pedon 14	*	**	***	–	–	**	–	**	–	–	–	S3	6	30	S3

Based on maximum observed yield as optimum (i.e., 20 q/ha). Limitations – No; *Slight; **Moderate; ***Severe;**** Very Severe.

TABLE 5.9 Soil-Site Suitability Criteria for Soybean

Soil-site characteristics	Degree of limitation					
	0 (None)	1 (Slight)	2 (Moderate)	3 (Severe)	4 (Very Severe)	
	S1		S2	S3	N1	N2
• **Climatic Characteristics**						
Total rainfall (mm)	> 850	750–850	650–750	550–650	< 550	–
Rainfall in growing season (mm)	> 700	600–700	500–600	400–500	< 400	–
Length of growing period (days)	> 120	110–120	100–110	90–100	< 90	–
Mean temp. in growing season (°C)	25–28	25–28	28–30	30–34	> 34	–
Mean max. temp. in growing season (°C)	–	–	–	–	–	–
Mean min. temp. in growing season (°C)	–	–	–	< 20	–	–
Mean R.H. in growing season	> 80	70–80	60–70	50–60	< 50	–
• **Site Characteristics**						
Slope (%)	< 3	< 3	3–5	5–8	> 8	–
Erosion	–	–	–	–	–	–
Hydraulic Conductivity (mmh^{-1})	20–50	10–20	5–10	1–5, >50	< 1	–
Drainage	Well	Mod. Well	Imperfect	Poor and Excessive	V.Poor	–
PAWC (mm)	> 200	150–200	100–200	50–100	< 50	–
Stoniness (Surface)	5–10	5–10	10–15	15–25	> 25	–

• **Soil Characteristics** *						
Texture	cl, sicl	l, sil, scl	sl, c	ls	s	–
Course fragments (Vol%)						–
within 50 cm	< 5	5–15	15–25	25–35	> 35	–
Depth (cm)	> 75	60–75	50–60	40–50	< 40	–
$CaCO_3$ (%)	< 5	5–10	10–20	>20	–	–
Exch. Ca/Mg ratio	> 2.5	1.5–2.5	0.5 – 1.5	<0.5	–	–
BS (%)	> 80	80–50	50–35	<35	–	–
OM (g kg^{-1}) (0–25 cm)	> 7.5	5–7.5	2–5	<2.0	–	–
EC (dS m^{-1})	< 1	1–2	2–3	3–4	> 4	–
ESP	< 2.5	2.5–5	5–7.5	7.5–10	> 10	–
pH (1:2)	6.5 -7.5	6.5 -7.5	7.5–8.5	>8.5	–	–

TABLE 5.10 Degree and Kind of Major Constraints, Suitability and Yield of Soybean

Soils	Mean temperature in growing season	Mean RH in Growing season	Slope	HC (Drainage)	PAWC	Texture	Depth	Exch. Ca/Mg ratio	ESP	pH	Suitability class	Yield Q/ha	% yield to optimum #	Suitability based on actual yield
Pedon 1	*	*	–	–	****	–	****	*	–	–	N1	1.5	8	N1
Pedon 2	*	*	**	*	**	**	*	–	–	–	S2	7	35	S3
Pedon 3	*	*	**	***	–	**	–	**	****	***	N1	4	20	N1
Pedon 4	*	*	–	*	–	**	–	*	–	***	S2	7	35	S3
Pedon 5	*	*	–	**	–	**	–	*	–	***	S2	17	85	S1
Pedon 6	*	*	–	**	–	**	–	*	–	**	S2	15	75	S2
Pedon 7	*	*	–	*	*	**	–	*	–	**	S2	11	55	S2
Pedon 8	*	*	–	**	–	**	–	*	**	**	S3	10	50	S2
Pedon 9	*	*	–	**	–	**	–	*	–	**	S2	12	60	S2
Pedon 10	*	*	**	*	–	**	–	–	–	**	S2	14	70	S2
Pedon 11	*	*	–	**	*	**	–	*	–	**	S2	10	50	S2
Pedon 12	*	*	–	*	*	**	–	**	–	**	S2	10	50	S2
Pedon 13	*	*	**	***	**	**	–	–	–	**	S3	5	25	S3
Pedon 14	*	*	**	*	**	–	–	**	–	**	S3	7	35	S3

Based on maximum observed yield as optimum (i.e., 30 q/ha). Limitations – No; * Slight; ** Moderate; ***Severe; ****Very Severe.

reaction. The soils with very severe limitation of high ESP and alkaline pH are unsuitable (Table 5.12).

5.4.5.6 Soil-Site Suitability Evaluation for Wheat and Gram

Wheat and gram are important crops grown in the *rabi* season in this region. Their suitability class was determined on the basis of actual yield data on farmers' field. The suitability class was assigned to each soil on the basis of percent yield obtained as compared to maximum observed yield (FAO, 1983). It has been observed that (Table 5.11) most of the soils are moderately to highly suitable for wheat, which may be due to the high clay content and associated higher PAWC in these soils. This may help for residual soil moisture storage in the soils in the *rabi* season. However, some soils are marginally suitable, because of 20–30% yield due to their lower PAWC. Similarly, the residual soil moisture storage in the soils is being exploited for growing gram by the farmers of the area. The soil P3 is found to be moderately suitable for wheat because of its semi tolerance to ESP, whereas it is marginally suitable for gram because of its sensitivity to sodium.

From the above discussion it becomes clear that all the soils are not equally suitable for a particular crop. Like-wise, the suitability of each crop obtained through the soil-site suitability evaluation varies from soil to soil (Table 5.12). Most the soils have high clay content, soil depth, PAWC, CEC and BS, which is favorable for the optimum growth of these crops. However, in view of the common limitations of drainage and/or erosion it can be suggested that these crops can be grown in the area with the consideration to the better management especially to improve the drainage and organic matter status and also by checking soil loss through erosion. The soils may prove to be highly suitable for most of the crops if supplementary irrigation is provided to tide over the soil moisture stress especially at the critical growth stages of the crop. In addition to these agricultural crops, these soils can also be cultivated for citrus, which is commercially grown in this region as Nagpur Orange. It is well established that citrus is successfully grown in this area. However, it was noted from the experience of farmers that some soils have poor drainage and water stagnation problem inhibit the plants to flower properly due to lack of moisture stress. The calcium carbonate content in excess amount affects the fruit quality and its life in some of the soils.

TABLE 5.11 Suitability of soils for Wheat and Gram on the basis of actual yield*

Particulars	Pedon 1	Pedon 2	Pedon 3	Pedon 4	Pedon 5	Pedon 6	Pedon 7	Pedon 8	Pedon 9	Pedon 10	Pedon 11	Pedon 12	Pedon 13	Pedon 14
Wheat														
Actual yield of wheat (q/ha) Farmer's level of management	-	-	15	20	25	30	16	25	30	24	9	8	-	6
% yield to optimum *	-	-	50	66	83	100	53	83	100	80	30	26	-	20
Suitability class	-	-	S2	S2	S1	S1	S2	S1	S1	S1	S3	S3	-	S3
Gram														
Actual yield of gram (q/ha) Farmer's level of management	-	-	6	10	12	12	10	14	16	12	6	5	-	4
% yield to optimum *	-	-	37	62	75	75	62	87	100	75	37	31	-	25
Suitability class	-	-	S3	S2	S2	S2	S2	S1	S1	S2	S3	S3	-	S3

* Maximum observed yield in the watershed at farmer's level of management taken as optimum yield.

TABLE 5.12 Suitability of Soils for Different Crops in Adasa Watershed (Based on Soil Suitability Criteria)

Pedon No.	Crop						
	Sorghum	Cotton	Pigeonpea	Soybean	Groundnut	Wheat*	Gram*
1	N1	N2	N2	N1	N1	-	-
2	S2	S3	S2	S2	S2	-	-
3	N1	N1	N1	N1	N1	S2	S3
4	S2	S2	S2	S2	S2	S2	S2
5	S2	S2	S2	S2	S3	S1	S2
6	S2	S2	S2	S2	S3	S1	S2
7	S2	S2	S2	S2	S3	S2	S2
8	S2	S2	S2	S3	S3	S1	S1
9	S2	S2	S2	S2	S3	S1	S1
10	S2	S2	S2	S2	S3	S1	S2
11	S2	S2	S3	S2	S3	S3	S3
12	S2	S2	S2	S2	S3	S3	S3
13	S3	S3	S3	S3	S3	-	-
14	S3	S3	S2	S3	S3	S3	S3

*Suitability class for Wheat and Gram is based on actual yield obtained.

However, most of the soils of the watershed with relatively good drainage support the cultivation of citrus with higher yields.

5.4.6 SUGGESTED LAND USE AND MANAGEMENT CONSIDERATIONS

The agricultural crops can be grown well in these soil with due consideration to soil and water management. The soils developed on beveled crust and scarp slopes can be used for pastures as well as dryland horticultural crops such as ber (zizyphus spp.), aonla (Gooseberry), Custard apple (Annona), etc., with proper soil and water conservation practices such as contour bunding and graded trenching. Soils developed near the sandstone mounds with relatively lower PAWC can be used for monocropping of sorghum, soybean, groundnut, dolichus, etc. The soils developed on foot slope (head slope) have high ESP. It is desirable to keep this land fallow in *kharif* and to grow wheat in *rabi* followed by dhaincha in summer. Some of the soils with PAWC about 175 mm (P2, P11 and P12) can be used for monocropping /intercropping (i.e., cotton (early variety), soybean, sorghum + pigeonpea in *kharif* (Randhawa and Venkateswarlu, 1980). Soil P2 on sinuous spur of denuded plateau may also be used for growing groundnut and short duration crops like green gram and black gram as this soil has good drainage and clay loam texture. Double cropping is suggested for most of the soils with high PAWC (about 225 mm). High yielding varieties of cotton can be grown well with some improvement in drainage. The double cropping systems such as sorghum-safflower, soybean-chickpea, and sorghum-chickpea are suggested under rainfed cultivation (Randhawa and Venkateswarlu, 1980). Under provision of supplementary irrigation sorghum-wheat and soybean-wheat are suggested. The suitable crops for soils with PAWC about 300 mm can be wheat and gram in *rabi* on residual moisture. For most of the soils citrus is suggested as an alternative land use where good quality water is available for irrigation.

In order to obtain the sustainable production from the cropping systems, suggested above following management consideration should be followed. The poor infiltration rate of these soils can be improved by deep plowing as it opens the soil thereby improving the infiltration opportunity time of water. An opportunity time for water in the entire profile can also be increased by employing vertical mulches, wherein the trenches are opened for regular interval to depth of 30 to 45 cm and they are filled with stubbles (Umrani, 1994). The land treatments for water conservation such as contour bunding are not suitable as it leads to water stagnation in the field for long period adversely affecting the crop yields. The graded bunding can, however, be effective as the rain water is disposed off with non-erosive velocity by providing 0.2–0.3% grade to the bunds (Verma, 1988). In addition, the land treatments such as broad bed and furrow system, ridge furrow system, and raised and sunken bed system are effective for soil and water conservation here. The problems of narrow workable soil water range and consequently that of difficulties of cultivation in rainy season can be greatly overcome through this kind of land treatment. The evaporation loss through shrinkage cracks can be effectively controlled by practices such as shallow cultivation and sorghum stubble mulch. Erosion can be controlled by carefully selected land treatments and water disposal systems such as loose boulder structure, diversion bund, graded furrows and vegetative barriers prepared by live hedges with vetiver grass as well as local grasses. As the soils have heavy texture and high shrink-swell potential with consistency limitations, the timely tillage at optimum moisture condition is essential for obtaining favorable structure for the growth of arable crops.

5.5 CONCLUSIONS

The prime importance must be must be given to the improvement of internal drainage of these soils for better storage of moisture in the profile. As the sHC of these soils is adversely affected by exchangeable sodium as well as exchangeable magnesium the maintenance of proper Ca/Mg ratio on the exchange complex and in soil solution is essential for obtaining favorbale hydraulic conditions. The regular addition of FYM will enhance chemical action of CO_2 on calcite (Gupta and Abrol, 1990). The stable linkages formed between particles due to organic matter may control the slaking by reducing the rate of wetting of aggregates by water. The organic linkages involving Na^+ are generally weak and ineffective in the stabilization of structure. The Na^+ ions must therefore be replaced by multivalent cations resulting in the formation of stable linkages between particles by organic matter (Rangasamy and Olsen, 1991). As discussed earlier the sHC of these soils is considerably reduced due to ESP (≥ 5) and low Ca/Mg ratio. In addition, significant amount of $CaCO_3$ in these soils reduces the concentration of calcium on exchange complex. However, for improving the internal drainage of these soils the Ca/Mg ratio in the exchange complex should be increased to minimum value of 2 and keeping exchangeable sodium percentage low to a value of <5 for which field application of gypsum is recommended (Balpande et al., 1996; Cockroft and Tisdall, 1978). For rapid dissolution of gypsum and avoiding reduction by formation of $CaCO_3$ coatings it is necessary to have a mix of gypsum particles sizes less than 2 mm. The interaction of gypsum and deep tillage improves infiltration and consequently increases subsoil water storage (McGarity et al., 1988). In order to improve the structure of these soils and reduce the ESP level well below 5, it would be beneficial to spread gypsum on the soil surface before the onset of rainy season. Through dissolution of gypsum enough electrolyte will be released which would prevent clay dispersion and hydraulic conductivity decline both at the surface and within the soil profile. Hence, application (once a year before rainy season) of small amount of gypsum may be attempted for stabilization of green manure that helps in improvement of soil structure. It can be inferred that these soils have high potential for crop production, but in order to optimize the production on a sustainable basis the intensive cropping involving the use of high yielding varieties, proper use of fertilizers, plant protection and irrigation water is required.

KEYWORDS

- **Land Evaluation**
- **Saturated Hydraulic Conductivity**
- **Semi Arid Tropics**
- **Soil Quality**
- **Soil Suitability**
- **Vertisols**
- **Watershed**

REFERENCES

Balpande, S. S., Deshpande, S. B., & Pal, D. K. (1996). Factors and processes of soil degradation in Vertisols of the Purna valley, Maharashtra, India. *Land Degradation and Development, 7*, 313.

Bharambe, P. R., & Ghonsikar, C. P. 1985. Physico-chemical characteristics in soils in Jayakwadi Command. *Journal of Maharashtra Agricultural Universities, 10*(3), 247–249.

Bhargava, G. P., & Abrol, I. P. (1990). Nature and extent of salinity and waterlogging problems in India. *In: Technologies for Wasteland Development,* I. P. Abrol and V. V. Dhruvanarayana (Eds.), ICAR, New Delhi, pp. 307–316.

Cockroft, B., & Tisdall, J. M. (1978). Soil management, Soil structure and root activity. *In: Modification of Soil Structure*. W. W. Emerson, R. D. Bond and R. R. Dexter (Eds.). Wiley, Chichester, 387–391.

Coughlan, K. J., McGarry, D., & Smith, G. D. (1986). The physical and mechanical characterization of Vertisols. In First Regional Seminar on Management of Vertisols under Semi-arid Condition. IBSRAM Proc. No. 6, Nairobi, Kenya, 89–106.

Coulombe, C. E., Wilding, L. P., & Dixon, J. B. (1996). *Overview of Vertisols: characteristics and impacts on society. In: Advances in Agronomy,* 57. Sparks, D. L. (Ed.), Academic, Press, New York, pp. 289–375.

Dudal, R., & Eswaran, E. (1988). Distribution, properties and classification of Vertisols. In: *Vertisols: Their Distribution, Properties, Classification and Management*. Wilding, L. P., Puentes, R. (Eds.), Texas A&M University Printing Centre, College Station, Texas, pp. 1–22.

Emerson, W. W., & Bakker, A. C. (1973). The comparative effects of exchangeable calcium, magnesium and sodium on some physical properties of Red-Brown Earths-II. The spontaneous dispersion of aggregrates in water. *Australian Journal of Soil Research. 11*, 151–157.

FAO. (1976). A framework for land evaluation, Soils Bull. 32, FAO, Rome.

FAO. (1983). Guidelines land evaluation for rainfed agriculture. FAO Soils Bull. 52, FAO, Rome, 237 p.

Gardner, E. A., Shaw, R. J., Smith, G. D., & Coughlan, K. J. (1984). Plant available water capacity: concept, measurement, prediction. *In: Properties and Utilization of Cracking Clay Soils,* J. W. Mcgarity, E. H. Hoult and H. B. Co. (Eds.), Univ. of New Englad, Armidale, 164–175.

Gjems, O. (1967). *Studies on Clay Mineral Formation on Soil Profiles in Scandinavia*. Medd. Nor. Skogforsoksves. *21*, 303–415.

Gupta, R. K., & Abrol, I. P. (1990). Salt-affected soils: Their reclamation and management for crop production. *Advances in Soil Science, 11*, 223–288.

Jackson, M. L. (1979). *Soil Chemical Analysis – Advanced Course.* 2nd Edition. Published by the author, University of Wisconsin, Madison.

Kadu, P. R., Pal, D. K., & Deshpande, S. B. (1993). Effect of low exchangeable sodium on hydraulic conductivity and drainage in shrink-swell soils of Purna valley, Maharashtra. *Clay Res. 12*, 65–70.

Kadu P. R., Vaidya, P. H., Balpande, S. S., Satyavathi, P. L. A., & Pal, D. K. (2003). Use of hydraulic conductivity to evaluate the suitability of Vertisols for deep-rooted crops in semiarid parts of central India. *Soil Use and Management, 19*, 208–216.

Mermut, A. R., Padmanabham, E., Eswaran, H., & Dasog, G. S. (1996). Pedogenesis. In: *Vertisols and Technologies for Their Management.* Ahmad, N., Mermut, A. R. (Eds.), Elsevier, Amsterdam, pp. 43–61.

McGarity, J. W., Mazhoumi, H., & Hoult, E. H. (1988). Effect of soil amelioration on the yield of dryland crops on Vertisols in northern N. S. W., Australia. *In: Transactions International Workshop – Classification, Management and Use Potential of Shrink-Swell Soils.* L. R. Hirekerur, D. K. Pal, J. L. Sehgal and S. B. Deshpande (Eds.). Oxford and IBH, New Delhi, 194–198.

Murthy, R. S. Bhattacharjee, J. C., Landey, R. J., & Pofali, R. M. (1982). Distribution, characteristics and classification of Vertisols. In Vertisols and Rice Soils of the Tropics. Symposia Papers II. 12th International Conference of Soil Science, New Delhi, 3–22.

Naidu L. G. K., Ramamurthy, V., Challa, O., Hegde, R., & Krishanan, P. (2006). Manual soil-site suitability criteria for major crops. NBSS Publ. No.129, NBSS & LUP, Nagpur, 118pp.

NBSS & LUP, (1994). Proceedings – National Meet on Soil-Site Suitability Criteria for different crops, National Bureau of Soil Survey and Land Use Planning, Nagpur, NBSS & LUP Publ Feb. 7–8, 1994, 32p.

Northcote, K. H., & Skene, J. K. M. (1972). Australian Soils with saline and sodic properties. Soil Pub.27, CSIRO, Melbourne, Australia.

Palaveyev, T. D., & Penkov, M. D. (1990). Properties of surface waterlogged clay soils containing exchangeable Mg. *Soviet Soil Science, 22*, 87–96.

Pal, D. K., Deshpande, S. B., Sarma, V. A. K., & Velayutham, M. (2000). Significance of minerals in soil environment of India. NBSS review series,1. NBSS & LUP, Nagpur 68 pp.

Pal, D. K., Bhattacharyya, T., Chandran, P., Ray, S. K., Srivastav, P., Durge, S. L., & Bhuse, S. R. (2006). Significance of soil modifiers (Ca Zeolites and gypsum) in naturally degraded vertisols of peninsular India in redefining the sodic soils. *Geoderma, 136*, 210–228.

Pal, D. K., Bhattacharyya, T., & Wani, S. P. (2011). Formation and management of cracking clay soils (vertisols) to enhance crop productivity: Indian experience. *In*: Lal, R., Stewart, B. A. (Eds.), *World Soil Resources and Food Security*. CRC Press, Francis and Taylor, Boca Raton, Fl, pp. 317–343.

Randhawa, N. S., & Venkateswarlu, J. (1980). Indian Experience in the semi-arid tropics-prospects and retrospects. ICRISAT. Proceedings of the International Symposium on Development and Transfer of Technology for Rainfed Agriculture and SAT farmers, August 28–Sept., 1, 1979, Patancheru, A.P. 207–220.

Rangasamy, P., & Olsson, K. A. (1991). Sodicity and soil structure. *Aust. J. Soil. Res. 29*, 935–952.

Richards, L. A. (Ed.). (1954). *Diagnosis and Improvement of Saline and Alkali Soils.* USDA Agric. Hanb. 60, U. S. Govt. Printing Office, Washington, D.C. 160p.

Singh, S. S. (1988). *Crop Management Under Irrigated and Rainfed Conditions.* Kalyani Publ., New Delhi-Ludhiana, p.450.

Soil Survey Staff, (1999). *Soil Taxonomy; a basic system of soil classification of making and interpreting soil surveys.* U. S. Dept. Agric., Handbook No. 436, Washington, D.C. 869p.

Swindale, L. D., & Miranda, S. M. (1984). The distribution and management in dryland agriculture of vertisols in the semi-arid tropics. *In*: *The Properties and Utilization of Cracking Clay Soils: Proceedings of a Symposium, 24–28 Aug., 1981, Australia,* McGarity, J. W., Hoult, E. H., & So, H. B. (Eds.). Reviews in Rural Science No.5, University of New England, Armidale, New South Wales, Australia.

Sys, C. (1985). *Land Evaluation Part I, II, III* State Univ., Ghent Publ., Belgium, 343p.

Sys, C., Van Ranst, E., & Debaveye, J. (1991). *Land Evaluation Part I, II.* Re-edited volumes of publication No.7 of the General Administration of Cooperation Development, Brussels, Belgium, 274p.

Umrani, N. K. (1994). Soil management and rainwater conservation and use-II. Vertisols under medium rainfall. In Soil management for Sustainable Agriculture in dryland areas. Bulletin No.16. *Journal of the Indian Society of Soil Science,* pp. 41–47.

USDA-SCS, (1994). State Soil Geographic Data base (STATSGO) Data User Guide. United Department of Agriculture Soil Conservation Service. National Soil Survey Center, Lincoln, Nebraska.

Vaidya, P. H., & Pal, D. K. (2002). Microtopography as a factor in the degradation of Vertisols in central India. *Land Degradation and Development, 13*, 429–445

Verma, G. P. (1988). Land and Water management for crop production on rainfed vertisols and associated soils. In: *Transactions International Workshop–Classification, Management and Use Potential of Shrink-Swell Soils.* L. R. Hirekerur, D. K. Pal, J. L. Sehgal, & S. B. Deshpande (Eds.). Oxford & IBH, New Delhi, 153–163.

Wilding, L. P., & Coulombe, C. E. (1996). Expansive soils: distribution, morphology and genesis. In: Proceedings NATO-ARW on Clay Swelling and Expansive Soils. Baveye, P., McBride, M. B. (Eds.), Kluwer Academic, Dordrecht, The Netherlands.

CHAPTER 6

MAPPING AND CHARACTERIZATION OF SALT AFFECTED SOILS FOR RECLAMATION AND MANAGEMENT: A CASE STUDY FROM THE TRANS-GANGETIC PLAINS OF INDIA

A. K. MANDAL,[1] RANBIR SINGH,[1] P. K. JOSHI,[1] and D. K. SHARMA[2]

[1]*Principal Scientist, ICAR-Central Soil Salinity Research Institute, Zarifa Farm, Karnal, Haryana – 132 001, India*

[2]*Director, ICAR-Central Soil Salinity Research Institute, Zarifa Farm, Karnal, Haryana – 132 001, India*

CONTENTS

ABSTRACT

Soil salinity/alkalinity and poor quality groundwater are major constraints causing reduced productivity in the arid and semiarid region of the Trans-Gangetic plain of Central Haryana covering Kaithal district. Visual interpretation of Resourcesat LISS-III data for March, May and October (2009) seasons revealed the presence of salt affected soils spatially distributed as barren patches and intermixed with cropped areas in the irrigated zones. The white to yellowish white tones with high reflectance from salt crusts on barren soil surface prompted the detection of strongly sodic/saline soils. The mixed signatures (yellowish white and red mottling) of salt stress and patchy crop stand indicated the presence of slightly to moderate salt affected soils, which is authenticated by ground truth study. In absence of natural drainage, the waterlogging for stagnated water bodies in irrigated areas was easily identified by the higher energy absorption showing dark blue/black to gray shades in March and November data. Salt affected soils showing low permeability, infiltration and hydraulic conductivity, also showed similar signatures amidst the cropped areas. The prolonged use of poor quality groundwater favored salt build-up in irrigated areas and showed mixed spectral signatures (grayish red to reddish white tones) in the satellite imageries. The sodic (alkali soil dominated by sodium and carbonate salts) and saline (neutral salts dominated by chloride and sulfate) soils are common in the study area. Sodic soils were distributed in northern and central parts of Kaithal district covering Pundri (2.1%), Kaithal (3%), Guhla (1.1%) and Siwan (1.1%) blocks while saline soils were distributed in southern part covering Kalayat (2.6%) and Rajaund (1.3%) blocks. The fine texture (clay to clay loam) sub-surface soils in Ghaggar plain impaired natural drainage thus favored waterlogging and sodicity development. The precipitated calcium carbonates concretions (calcareous layer) are common in strongly sodic soils that impair salt and nutrient movements and root penetration. An area of 26301 ha (11.3%) is salt affected in Kaithal district, of which sodic and saline soils covered 17570 ha (7.3%) and 9388 ha (4%), respectively. The groundwater quality was sodic in north of Kaithal district (Guhla block), at places, showing high Residual Sodium Corbonate (RSC) (12.7 me L^{-1}), saline in the east at Kalayat block (Sodium Adsorption Ratio (SAR) 33.6) and sodic (pH 9.2) in central part of Kaithal block, and also showed high RSC (6.5 me L^{-1}) in selected samples. Based on the soil physico-chemical characteristics and the quality of groundwater, suitable reclamation and management options were also suggested.

6.1 INTRODUCTION

Overexploitation of land, water and other natural resources during last four decades has set in the process of degradation in soil, water, climate and biodiversity resources. In India unscientific and over use of soil resources led to physical, chemical and biological degradation causing irreversible loss to soil quality. The intensive agriculture, which ushered during green revolution, is now becoming a serious threat to sustainable agriculture due to deteriorating soil quality. Salt affected soils are important degraded soils in the world, which contain excess salts (soluble/exchangeable) that adversely affect plant growth and crop yield. The main causes of salt accumulation include capillary rise from subsoil salt beds or from shallow brackish groundwater, besides other factors include the indiscriminate use of irrigation water of variable qualities, weathering of rocks and the salts brought down from the upstream to the plains by rivers and subsequent deposition along with alluvial materials, ingress of sea water along the coast, salt laden sand blown by sea winds, lack of natural leaching due to topographic situation in arid and semiarid regions. Soil characteristics *viz.,* soluble salts content and soil reaction (pH) largely influence the changes in physical and chemical properties in a salt affected soil. Besides, soluble salts influence changes in proportion of exchangeable cations and the osmotic and specific ion toxicity for crop production. The presence of excess neutral salt essentially influences solute transport and non-availability of some essential nutrients required for plant growth.

Globally, five categories of salt affected soils were identified based on the nature and composition of salts (Szabolcs, 1989). These soils are saline dominated by natural salts, alkali enriched by salts capable of alkaline hydrolysis (Na_2CO_3 and $NaHCO_3$), gypsiferous with excess gypsum salts, acid sulfate soils with ferric and aluminum sulfates and others including strongly degraded subsoil and potential salinity in irrigated region. In Eastern Europe, these soils were known as Solonchak, Solonetz and Solod. The alkali soils of Europe and erstwhile United Soviet Socialist Republic (USSR) showed a good A horizon and a *natric* (sodic) subsurface B horizon. In India, these soils are known as *Kallar* or *Thur* in Punjab and Haryana, *Usar* or *Reh* in

Uttar Pradesh, *Luni* in Rajasthan, *khar* or *Kshar* in Gujarat and Maharashtra, *Chhouddu* or *Uppu* in Andhra Pradesh, *Choppan* in Karnataka, etc. The United States Soil Salinity Laboratory (Richards, 1954) proposed criteria for distinguishing sodic with saline and saline-sodic soils based on the critical limits of electrical conductivity (EC) of saturation extract, Exchangeable Sodium Percentage (ESP) and soil reaction (pH) of saturated soil paste. Soil Science Society of America (Soil Science Society of America, 1987) used EC and SAR as criteria for classification of salt affected soils. Extensive research conducted at Central Soil Salinity Research Institute (CSSRI) (Agarwal and Yadav, 1956) revealed pH 8.2 as more appropriate limit than pH 8.5 for characterizing sodic soils of the Indo-Gangetic Alluvial Plain (IGP). The adverse effect of sodicity on crop growth (Abrol et al., 1980) showed pH 8.2 as critical limit for alkalinity/sodicity in soils. Similarly, the EC values of 2 dS m^{-1} and ESP value of 5 were found as critical limits for characterizing black sodic vertisols dominated by higher content of clay with smectite mineralogy (Balpande et al., 1996). Australian workers preferred parameters of soil physical conditions and its harmful effects on plant growth for diagnosing commonly occurring sodic soils. It was found that ESP 6 was a limiting value to impairment of physical condition in a swell shrink soil in Australia (Northcote, 1979). To facilitate soil management and the influence of salts on soil properties and plant growth, two broad categories of soils *viz.,* saline and sodic were identified in India (Abrol and Bhumbla, 1978; Abrol et al., 1988). The detailed diagnostic characteristics are as follows:

Saline soils are occurring as patches with white salt encrustation on the soil surface located in the lower topographic position. Due to high salt content, high osmotic pressure and ion toxicity these soils do not support normal agriculture. Often, these soils are located in the irrigated region associated with waterlogging condition. The soil salinity is also developed by the presence of saline groundwater used for irrigation purpose in the arid and semiarid region. In the sea coast, large areas are subjected to saline water inundation that is affecting the root zone with high to very high soil salinity. The analytical data of laboratory analysis showed the presence of neutral salts such as chlorides and sulfates of sodium, calcium and magnesium higher than the prescribed limit. The pH of the saturation paste normally lies below 8.2 and the electrical conductivity usually exceeds 4 dS m^{-1} at 25°C. These soils normally have higher (>15) SAR. In the field, alkali soil are usually associated with bleached color, strong blocky structure, presence of mottles of iron and manganese, concretions/nodules of calcium and magnesium, presence of an illuvial fine textures layer dominated by silt/clay at a depth below the surface, and shows moderate to strong/violent effervescence following addition of dilute hydrochloric acid and changes to pink color following application of phenolphthalein indicator. Due to alkaline soil reaction (pH>8.2), these soils adversely affect plant growth and crop yield. The primary limitations are poor (dispersed) soil physical condition and inadequate internal drainage that cause the imbalance of nutrient availability. The laboratory investigations showed high (>15) ESP and dominance of soluble ions such as carbonate, bicarbonate of sodium, calcium and magnesium. Due to high sodium content these soils also showed high SAR.

6.1.1 STATUS OF SALT AFFECTED SOILS: GLOBAL PERSPECTIVE

Salinity related land degradation is important problem in arid and semiarid regions threatening food and nutritional security of the world. According to the FAO/UNESCO soil map of the world, 953 m ha of land (8%) is affected due to soil salinization and alkalization (Szabolcs, 1989). The worst affected areas include Africa, Asia, Australia, Europe, Latin America, Near East and North America (Koohafkan, 2012). A recent survey in Australia indicated about 2 m ha and 20,000 farms across the country is affected as a result of climate changes and 10% of the Western Australia is seriously endangered by soil salinity/alkalinity of primary and secondary origin (Australian Bureau of Statistics, 2002; McFarlane, 2004), Globally, 20% of the irrigated land (450,000 km^2) is salt affected and about 2000–5000 km^2 land lost production every year as a result of salinity (UNEP, 2009). In South Asia, annual economic loss is estimated as US$1,500 million due to salinization (UNEP, 2009). In the tropical countries, soil salinity is a serious problem affecting crop production, plant growth and soil and water quality leading to soil and environmental degradations. In Africa, a serious impact on economy has been reflected on average loss on agriculture from 28% to 76% as a result of land degradation. In India, the average production loss from such degradation varies from 40% in wheat, 45% in rice, 63% in cotton and 48% in sugarcane (Joshi and Agnihotri, 1984). The alkali land reclamation schemes in Haryana, Punjab and Uttar Pradesh (Trans-Gangetic plain) reported changes in cropping intensity by 25% in paddy, 10% in wheat, 21% in pulses and 10% in cotton (CSSRI, 2000). The economic benefit accrued from the sodic land reclamation programs showed the improvements of B:C (benefit-cost) ratio to 1.89 and 1.80 for rice and wheat, respectively (Thimmappa et al., 2013). Soil analytical data of reclaimed sodic soils showed prominent changes in soil pH and concurrent increase of organic carbon that showed improvement in fertility status (CSSRI, 2000; CSSRI, 2007).

6.1.2 STATUS OF SALT AFFECTED SOILS: INDIAN PERSPECTIVE

Current assessment of salt affected soils in India involves spatial measurements, field and laboratory studies and reconciliation to arrive at an estimated area of 6.73 m ha, which is distributed in 15 states and 13 agro-climatic regions (CSSRI, 2007; NBSS & LUP, 2006; NRSA, 1997, 2008). The saline (295 mha) and sodic (377 mha) soils are distributed in 7 physiographic regions, (Mandal et al., 2010). The occurrence of salt affected areas is primarily influenced by rainfall, 29.4% (very high), 26.2% and 19.4% exist in the ranges between 500 to 800 mm, 1000 to 1500 mm and 300 to 500 mm, respectively. These soils are associated with a wide range of parent materials, most importantly, in Pleistocene and Recent origin (38.9%), Archean Schists and Gneisses (9.5%) and Deccan and Rajmahal Traps (7.8%), respectively (Mandal et al., 2011). The irrigation through groundwater is a common practice for growing arable crops in the arid and semiarid regions, the quality of which is primary concern (average poor quality is 25%) for crop productivity. The distribution of saline groundwater, in Rajasthan (41.2%), Haryana (25.9%) and Gujarat (12.4%) states warrants necessary inputs for management (Gupta, 2010, HSMITC, 2001, Manchanda, 1976), while the alkaline groundwater in Punjab (54%), Rajasthan (35%), Haryana (30%) and Gujarat (28%) states showed necessity for promotion for reclamation. The complex saline-alkali water is also prevalent in Gujarat (52%), Rajasthan (49%), Haryana (46%) and Punjab (24%) and need special care for crop production. The expansion of canal irrigation network in arid and semiarid regions caused development of salinity and waterlogging problems in undrained areas.

In India, systematic mapping of soil salinity was originated at the Central Soil Salinity Research Institute at Karnal under the Indian Council of Agricultural Research (ICAR) New Delhi collaborating with National Bureau of Soil Survey and Land Use Planning (NBSS & LUP) Nagpur and National Remote Sensing Centre (NRSC) Hyderabad. A methodology for mapping soil salinity on reconnaissance scale (1:250,000 scale) was developed using remote sensing data in conjunction with ground truth survey, soil and laboratory studies (NRSA, 2008; Mandal and Sharma 2010; Sharma et al., 2011; Sharma and Mandal, 2006). The salt affected soils in coastal areas of Gujarat state (India) were mapped and the integrated approach of image interpretation (Landsat TM) with ground truth was followed (Joshi and Sahai, 1993). The inland soil salinity in Uttar Pradesh state (India) was mapped based on the remote sensing and soil profile studies (Saxena et al., 2004). Digital image classification of soil degradation features was combined with field observations and laboratory determinations (Mitternicht and Zinck, 1997). Singh et al. (2010) found uncertainty in the estimates of complex salt affected soils and suggested reconciliation and harmonization. Currently, Ali et al. (2015), Wu et al. (2014), Scudiero et al. (2014) and Albel and Kumar (2013) studied salt affected soils using high resolution remote sensing data and used ground truth (land use), soil studies and laboratory analysis data for quantification of soil salinity/sodicity. Keeping in view the large scale variability of salt affected soils and scale of mapping FAO (2008) reported harmonized database of salt affected lands for further planning and management.

6.1.3 SIGNIFICANCE OF THE STUDY

The first survey of salt affected soils was initiated by identifying soil alkalinity and salinity problems in the Gangetic alluvium located at Etah district in Uttar Pradesh (Leather, 1914). Further salt affected soils were detected as patches in the lower Ganges canal areas of Uttar Pradesh (Agarwal et al., 1957). Investigations revealed that salts are drained from the Himalayas and Siwalik through rivers/streams and are accumulated at the alluvial plains (Bhargava et al., 1980; Sidhu et al., 1995). The lack of adequate internal drainage in lower topographic regions prompted soil salinization (Bhargava et al., 1980). High evaporation during the dry season and lack of good quality water for leaching caused salt accumulation in soil profiles. The coexistence of salt affected soils and poor quality groundwater in central Haryana are primary constraints for agriculture (Yadav, 2003). The use of poor quality groundwater for irrigation increased salt buildup in soil profiles, which caused reduced productivity. Canal irrigation in undrained areas has also accentuated waterlogging, formation of high water table and secondary salinization in soils. The erratic rainfall and temperature patterns not only threaten agriculture but cause redistribution of salinity affected areas. In the dry regions of central Haryana, the primary dependence on groundwater for irrigation has degraded soils through the deteriorating physical and chemical properties of soils coupled with fine textured Ghaggar alluvium that caused a congenial environment (waterlogging) unfavorable for sustainable agriculture. A thorough investigation of soil and water and spatial distribution of salt affected soils using remote sensing data is required for precise assessment. The variable extents of salt affected soils (4.54 to 2.32 lakh ha) in Haryana (Abrol and Bhumbla, 1971; NRSA 2008) and the complex nature for salts, soil physical properties and drainage (Sharma et al., 2011), have been reported periodically. The complex pedogenic processes due to the anthropogenic activities using poor quality groundwater for irrigation (Jain and Kumar, 2007; Bhalla et al., 2011) were also reported in arid and semiarid areas. For precise assessment of reclamation and management, the physico-chemical characteristics and quality appraisal of irrigation water are of primary importance. This chapter addresses mapping and characterization of salt affected soils and quality appraisal of groundwater in Kaithal district of Central Haryana, India, for reclamation and management.

6.2 STUDY AREA

The Kaithal district (2317 sq.km) lies between 29°31′27.43″ N to 30°13′07.45″N and 76°09′02.99″E to 76°47′59.44″ E under the old alluvial plain covering D3.3 agro-ecological zone (Yamuna alluvial plain, hot and semiarid region with length of growing period of 90–120 days). It consists of two administrative sub-divisions *viz.,* Kaithal & Guhla and six blocks *viz.,* Kaithal, Pundri, Rajaund, Guhla, Kalayat and Siwan (Figure 6.1). The climate varies from arid to semiarid. The average rainfall of the district is 500–600 mm. The net cultivable area is 2.02 lakh ha, the area under forest is 3000 ha and barren and uncultivable land covers 2000 ha. The net irrigated area is 1.98 lakh ha mainly by canals and groundwater. The primary source of irrigation is Western Yamuna (Narwana branch and Rajaund distributary) and Bhakra (Saraswati distributary) canals. High (182%) cropping intensity is reported in Kaithal district. The primary crops include paddy in summer (*Kharif*) and wheat in winter (*Rabi*) seasons while cotton, pearl millet, sugarcane, sunflower and pulses (moong) are also practiced. Growing Dhaincha (*Sesbania aculeate*) is common practice in areas under salt affected (alkaline)

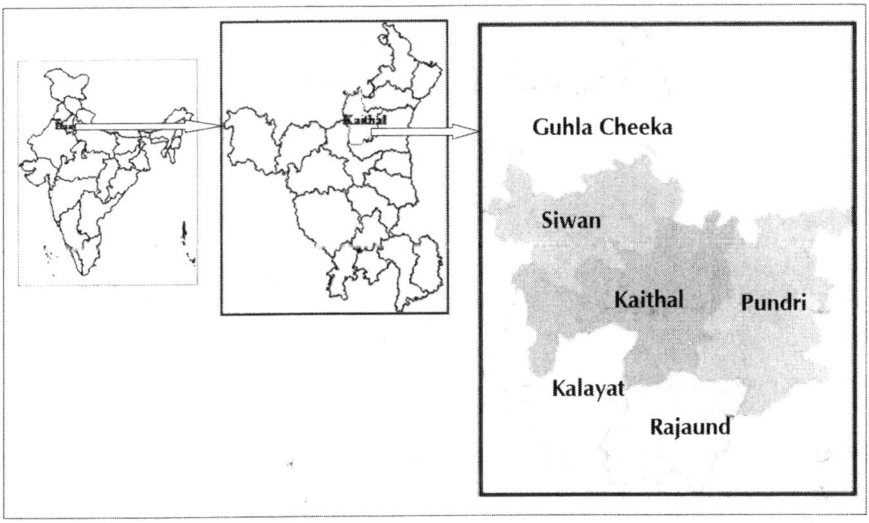

FIGURE 6.1 Location map of the study area.

soils and water to reduce salt injuries and improve soil health (physical properties). The landform is alluvial (Ghaggar alluvium) in general.

6.3 METHODOLOGY

6.3.1 DATA USED IN THE STUDY

Indian Remote Sensing (IRS) (Resourcesat) LISS III data with spatial resolution of 23.5 m and spectral resolution (Green: 0.52–0.59 μm, Red: 0.62–0.68 μm, Near Infra Red: 0.77–0.86 μm and Short-wave Infra Red: 1.55–1.70 μm) for March, May and October 2009 was used in the study (Table 6.1). *The Survey of India Topographical Maps* on 1:50,000 scale (No. 53C/1, 2,

TABLE 6.1 Particulars of Satellite Imageries

Sensor	Spectral Resolution (μm)	Spatial Resolution	Period
IRS - P6 LISS III Resourcesat I	B1 0.52–0.59 (Green)	23.5 m	March 2009
	B2 0.62–0.68 (Red)	Swath 140 km	May 2009
	B3 0.77–0.86 (NIR)		October 2009
	B4 1.55–1.70 (SWIR)		

5, 6, 9 and 10; 53B/4, 8 and 12) were used for preparing the basemap comprising of administrative and political boundaries, irrigation/drainage, infrastructure and settlements. *Software* ERDAS IMAGINE software (*ver 3.3*) was used for digital and spatial data analysis. The software Arc GIS (*ver* 9.3) was used for generation of thematic layers. A Cal Comp (A_0) digitizer, a scanner, a printer attached to a Pentium (PIV) computer equipped with Microsoft Windows XP and Office (2000), was used for entry, editing and analysis of map and attribute data. The ancillary data (State Department of Agriculture Haryana and NBSS & LUP, RC Delhi), water quality data (Gupta, 2010; Manchanda, 1976; HSMITC, 2001), crops and associated land characteristics (District Gazettes Haryana) were also collected for the study area. The salt affected soils map of Haryana (NRSA, 1997) on 1:250,000 scale were also used as legacy data. *Soil Sampling Tools* like color chart, auger, spade and knife, etc., were used. Global Positioning System (GPS) was used for collecting data related to the location of field data.

6.3.2 GEO-REFERENCING AND PREPARATION OF THE BASE MAP

The Survey of India (Government of India) topographical maps on 1:50,000 scale were used for geo-referencing and related with real world coordinates using Universal Transverse Mercator (UTM) projection with projection (ellipsoid WGS 84) and datum (WGS 1984) information. The spatial features for state, district and blocks boundaries, roads and railways, canal and river and state, district, block HQ and villages were digitized and the thematic layers were overlaid to develop a base map of the study area.

6.3.3 PROCESSING AND INTERPRETATION OF IRS DATA

The IRS imageries were processed for radiometric and geometric corrections and were geo-referenced. The False Color Composites (FCC) were prepared using combination of bands such as NIR, R and G (B321) and SWIR, NIR and R (B432) (Figure 6.2). The seasonal data were also analyzed to study the dynamics of soil salinity and waterlogging. Visual analysis was done and interpreted units were digitized following standard guidelines (Colwell, 1996). The principal component analysis was done to segregate homogenous data for visual analysis. The ratio indices Normalized Difference Vegetation Index (NDVI) and Vegetation Index (VI) were also carried out to distinguish

FIGURE 6.2 IRS-LISS-III imagery (March 2009) for Kaithal district, Central Haryana.

normal and stressed crops. The seasonal data were analyzed to study and understand the dynamics of salt affected soils and waterlogging. Overlaying thematic layers of base map and ancillary/legacy data, the interpreted units were delineated using on-screen digitization technique.

6.3.4 FIELD SURVEY FOR SOIL PROFILE STUDIES

Ground truth studies were conducted to verify and authenticate interpreted units and establish relationship between image interpretation and field conditions. During field survey important characteristics such as surface salinity and waterlogging status, seasonal behavior, topography, irrigation/drainage and crops/cropping practices were studied. GPS was used to locate sites for ground truth observations, soil profiles/soil and water sampling. The groundwater samples were collected from tube wells (used for irrigation in agriculture). The soil profiles (10 Nos.) were studied to a depth of 1.5 m at representative locations and depth-wise soil samples were collected for laboratory analysis for soil physico-chemical properties (Soil Survey Staff, 1998). These were classified using taxonomic system of soil classification (Soil Survey Division Staff, 2004). The soil samples were also characterized (Richards, 1954) for salinity/alkalinity appraisal (Table 6.2).

6.3.5 PHYSICO-CHEMICAL ANALYSIS, SOIL CLASSIFICATION AND APPRAISAL OF WATER QUALITY

Soil samples were analyzed for physico-chemical properties *viz.*, pH, EC (dS m^{-1}), soluble Na^+, K^+, Ca^{2+}, Mg^{2+}, CO_3^{2-}, HCO_3^- and Cl^- (me L^{-1}), $CaCO_3$ (<2 mm size, %) and organic carbon (%); CEC (cmol (p^+) kg^{-1}) and ESP (%),

TABLE 6.2 Keys to the Degree of Salinity/Sodicity in Salt Affected Soils (Richards, 1954)

	Saline Soil	Sodic Soil	
Degree	**EC (dS m^{-1})**	**pH**	**ESP**
Slight	4.0–8.0	8.5–9.0	<15
Moderate	8.1–30.0	9.1–9.8	15–40
Strong	>30	>9.8	>40

sand, silt and clay (%) and available N, P and K (Jackson, 1986; Singh et al., 1999). Soils are classified as saline, sodic and saline-sodic and the degrees of classes as slight, moderate and strong (Richards, 1954). The waterlogged areas were classified as permanent waterlogged (surface ponding) and sub-surface waterlogging (water table depth <1.5 m) based on NRSA (2007). Thirteen groundwater samples were collected from different locations to study the water quality for agricultural applications. These were analyzed for pHiw, EC_{iw} (dSm^{-1}), soluble Na^+, K^+, Ca^{2+}, Mg^{2+}, CO_3^{2-}, HCO_3^- and Cl^-, SAR = $[Na^+/ \{(Ca^{2+} + Mg^{2+})/2\}^{1/2}]$ and RSC = $[(CO_3^{2-} + HCO_3^-) - (Ca^{2+} + Mg^{2+})]$.

6.3.6 *MAPPING OF SALT AFFECTED AND WATERLOGGED SOILS*

An integrated approach of image interpretation, ground truth survey and laboratory analysis data for soil physico-chemical properties was used for mapping salt affected and waterlogged soils (Dwivedi, 2001; Mandal and Sharma, 2012, 2013; NRSA, 2007). The thematic layers of salt affected and waterlogged soils were linked with the physico-chemical properties of soils to develop a relational database (Mandal and Sharma, 2011). A flow chart showing mapping methodology is presented in Figure 6.3. The area statistics of salt affected and waterlogged soils were generated (Mandal and Sharma, 2011). The thematic maps of salt affected soils and waterlogged areas for Guhla and Kaithal sub-divisions were prepared.

6.4 RESULTS AND DISCUSSION

6.4.1 *INTERPRETATION OF IRS DATA*

The visual interpretation of IRS LISS III FCC (B321) for March, June and October 2009 seasons identified salt affected soils as white to yellowish

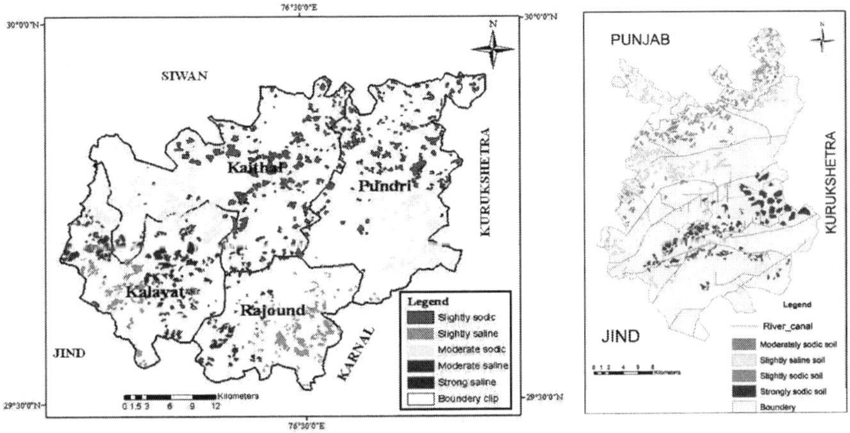

FIGURE 6.3 Methodology for mapping salt affected soils.

white patches in the old alluvial plains of Ghaggar and Saraswati covering Kaithal district. These are associated with waterlogged soils and the cropped areas that appeared as dark blue to black patches and red to dark red tones, respectively. The FCC of bands B432 (SWIR, NIR and R) showed better results due to higher contrast and clear boundaries of waterlogged areas than the FCC of bands B321 (NIR, R and G). Prominent waterlogging and salt infestation were located in the old alluvial plains of Ghaggar covering Guhla block and in the Paleo-channel of Saraswati located in Siwan block. The mixed spectral signatures for light to dark grayish tones with red mottles showed the presence of salt affected and waterlogged soils with scattered cropped areas. The ground truth data indicated prolonged irrigation of arable crops with salty groundwater that caused low permeability, infiltration and poor drainage in moderate to fine texture soils of Ghaggar plain. The post monsoon (October) data showed higher extent of waterlogged and salt affected soils in Guhla block.

Located in the west of Kaithal district, the salt affected and waterlogged soils in Siwan block were distributed as barren patches interspersed with cropped areas showing poor vegetative growth. The ground truth data confirmed the presence of sodic soil and the practice of poor quality groundwater for irrigation in wheat and rice crops. The salt affected soils were also located in the forest-covered areas and are used for growing medicinal and aromatic plants. A large area of salt affected soils were found along the Saraswati drain that flushes excess salts and water during the post-monsoon season. Waterlogging is prominent in the severely salt affected soils showing no natural drainage. Ground truth studies indicated the presence of fine texture layers at sub-surface depth inhibiting percolation of salts, water, nutrient and restricted root growth. At places, salt crusts were detected in the forest-covered areas where the groundwater is sodic in general.

In the canal-irrigated areas, waterlogging and soil salinization were identified in Kalayat block, located in the western part of Kaithal district. The rice, cotton and wheat are major summer and winter crops, respectively. Continuous irrigation in poorly drained areas caused rise of water table resulting in waterlogging. Topographically it is distributed in the low-lying flats/depressions with no or imperfect internal drainage. The waterlogging was higher during the post-monsoon season apparently due to higher irrigation for rice crop. The salt concentration was higher in post-monsoon season due to salt transport with rising water table and repeated wet and dry cycles favored salt precipitation and enriched salt concentration leading to the salt efflorescence during the dry period. Soil profile studies identified impermeable layer of calcium carbonate at a depth below the surface.

The post monsoon (October) imageries showed the patches of waterlogged and salt affected soils in the irrigated areas of Kaithal block located at the southern part of Kaithal district. The cereal crops such as rice and wheat are irrigated with salty groundwater. The ground truth studies indicated good crop growth in coarse texture soils. Waterlogging is more pronounced in the fine texture soils with low permeability and poor internal drainage. The use of salty groundwater caused salt enrichment in soil profiles resulting in low infiltration and movement of water and salts.

The March 2009 data indicated waterlogged and salt affected soils in Rajaund and Pundri blocks located in the south and east of Kaithal district. In the absence of canal irrigation and fresh water supply, the saline groundwater is used for agriculture in Rajaund block. The periodic use of salty groundwater increased salt concentration in soil profile and deteriorated soil physical properties and favored waterlogging. The salt affected soils are found in the localized patches at the low-lying areas of Rajaund block (village Mandwal). The sodic soils were identified along the Paleo-channel of river Chautang (Villages Sakra, Kheri, Sangrauli, Dosain, Buchi) in Pundri block. The IRS data (2009) showed mixed spectral signatures of dry salts and dark tones of residual moisture at the soil surface showing imprints of waterlogging. Besides, the use of sodic groundwater for irrigation in arable crops also favored sodic soil formation.

6.4.2 PHYSICO-CHEMICAL CHARACTERISTICS OF SOILS AND SUGGESTED RECOMMENDATIONS

The physico-chemical properties of soils are presented in Table 6.3. The seasonal imageries indicated barren surface and higher moisture accumulation (seasonal waterlogging) in the pre- and post monsoon seasons. Soil profile studies (March) of Pedon 1 showed salt accumulation, sodic condition (pink color develops on application of phenolphthalein indicator), moist soil strata, deep, fine soil texture, massive to moderate, fine to medium, angular to sub-angular blocky structure and the significant presence of iron and manganese mottles (1–2 mm, 10–20%) at 24–105 cm depth and calcium carbonate concretions (2–4 cm, 20–40%) at 70–135 cm depths, respectively. Based on the data of physico-chemical properties (pH 10.3 to 10.4 and ESP 75.4 to 83.2), the Pedon is classified as strongly sodic (Richards, 1954). The higher contents of $CO_3^{2-} + HCO_3^-$ (14.5 to 16.9 me L^{-1}), Na^+ (15.6 to 44.9 me L^{-1}), Cl^- (8.0 to 14.0 me L^{-1}) and SO_4^{2-} (2.8 to 7.0 me L^{-1}) ions were noted. The soil texture ranges from silty clay loam to loam showing higher clay content (12.2 to 36.6%). CEC varies from 28.7 (at surface) to 13.6 (at 187 cm) cmol (p^+) kg^{-1}. The *Calcretes* ($CaCO_3$) are present (10.2 to 11.4%) at 1 m depth. The organic carbon, available nitrogen and phosphorus contents are low throughout. These soils can be used for rice-wheat cropping following reclamation with gypsum (4–6 t ha^{-1}).

Located at the old alluvial plain of Ghaggar (Sachdev et al., 1995; Mandal, 2014) and irrigated by salty groundwater (Block Kaithal), Pedon 2 showed water stagnation and patchy crop stand in March and November satellite data. The soil is characterized by deep, massive to medium sub angular blocky structure, clayey texture and pale yellow to dark yellowish brown color. The moist to wet sub-surface layers and the presence of iron and manganese nodules showed anaerobic condition in soil profile. The physico-chemical analysis data indicated soil sodicity (pH 8.7), soil salinity (EC 6.5 to 8.6 dS m^{-1}) at sub-surface depths. The Na^+ (1.5 to 63.5 me L^{-1}), $Ca^{2+} + Mg^{2+}$ (4 to 38 me L^{-1}), Cl^- (128 to 725 me L^{-1}) and SO_4^{2-} (5.9 to 54.0 me L^{-1}) ions are prevalent. The high clay content (53 to 57%) caused poor internal drainage (Mandal and Sharma, 1997; Mandal, 2014) and higher CEC values (42.6 to 44.1 cmol (p^+) kg^{-1}). The soil is low in organic carbon (0.2 to 0.3 %), available nitrogen (11 to 78 Kg ha^{-1}) and phosphorus (16 to 53 Kg ha^{-1}) contents and showed moderate to high available potassium (466 to 509 Kg ha^{-1}). The treatment with Farm Yard Manure (FYM) or compost is required

TABLE 6.3 Physico-Chemical Properties of Soils from Kaithal District

Horizon	Depth (cm)	pH	EC (dS m^{-1})	Na^+	Ca^{2+}+ Mg^{2+}	CO_3^{2-}+ HCO_3^-	Cl^-	SO_4^{2-}	OC	ESP	Ca-CO_3	CEC cmol (p^+) kg^{-1}	Texture	Clay	Silt	Sand	Av N	Av P	Av. K
				me L^{-1}					(%)					(%)			Kg ha^{-1}		
P1: 29°47′38.9″N to 76°39′52.3″E Fine-Loamy Sodic Haplustepts, sodic soil in paleo-channel of the river Chautang, calcareous parent material																			
A1	0–24	10.3	1.6	15.6	4	15.6	8.0	2.8	0.2	76.6	1.9	24.6	sicl	36.6	38.9	24.4	31	17	208
Bw1	24–47	10.4	2.6	27.4	4	15.1	10.0	5.4	0.2	75.4	1.9	28.7	sicl	34.6	37.4	27.9	27	14	208
Bw2	47–72	10.3	2.3	22.9	4	14.5	10.0	4.2	0.2	83.2	1.2	25.4	sicl	28.0	32.1	39.8	23	10	188
B21k	72–105	10.4	3.4	28.7	4	16.7	14.0	5.7	0.2	74.8	2.3	21.1	l	18.3	24.1	57.5	23	10	151
B22k	105–135	10.4	4.0	44.9	4	16.9	13.0	7.0	0.1	80.1	11.4	14.6	l	12.3	17.6	70.0	19	10	88
BC k	135–187	10.4	3.6	39.4	4	15.2	14.0	6.4	0.1	82.4	10.2	13.6	l	12.2	16.8	71.0	19	10	93
P2: 29°46′38.9″N to 76°29′34.9″E Fine Typic Ustochrepts, waterlogged soil, rice basmati (CSR 30) irrigated by sodic water produce moderate yield																			
Ap	0–24	8.7	1.0	1.5	4	4.0	6.0	5.9	0.3	22.5	1.3	42.6	c	53.2	21.5	25.2	11	21	509
Bw1	24–61	8.0	6.5	49.5	20	4.5	9.0	41.8	0.3	27.2	0.7	44.1	c	55.6	23.2	21.1	78	16	488
Bw2	61–92	8.0	8.6	63.5	38	3.5	12.0	54.0	0.2	31.3	1.2	44.1	c	57.3	22.3	20.4	74	53	477
Bw3	92–121	7.9	8.0	44.2	24	4.5	19.0	44.2	0.2	21.8	1.3	42.6	c	57.6	23.6	18.8	70	39	466
P3: 29°46′46.4″N to 76°30′1.9″E Fine-Loamy Typic Ustochrepts, sodic soil & sodic GW, calcareous parent material, rice CSR 30 grown, low yield																			
Ap	0–21	8.6	2.5	22.3	6	4.5	10.0	13.6	0.3	44.6	3.2	13.2	sil	21.4	28.0	50.5	10	38	456
Bw1	21–62	9.2	1.2	12.4	6	4.0	8.0	5.9	0.1	52.3	3.8	19.7	sil	24.8	32.7	42.3	51	34	429
Bw2	62–94	9.2	1.7	19.4	4	4.5	8.0	8.8	0.1	45.8	1.5	22.6	sil	23.6	38.1	38.2	43	23	413
Ck	94–122	9.3	1.9	21.2	6	4.5	8.0	10.9	0.1	54.4	9.9	26.9	sil	19.6	29.3	51.0	31	13	360

P4: 29°49′6.7″N to 76°28′10.6″E, Fine Typic Ustochrepts, sodic soil & sodic GW in the old Ghaggar plain, severely waterlogged & partially barren																		
A1	0–18	8.3	2.3	18.8	8	1.5	10.0	13.7	0.4	32.3	2.3	21.5	c	40.0	24.1	35.8	10	33	392
Bw1	18–56	8.6	2.4	17.7	8	3.0	6.0	11.9	0.2	40.1	1.4	26.8	c	47.0	24.2	28.7	70	22	356
Bw2	56–89	8.4	2.5	19.9	8	4.0	8.0	16.5	0.2	37.3	1.7	31.8	c	49.8	25.4	24.7	70	12	339
Bw3	89–127	8.0	4.1	35.5	10	5.0	10.0	28.3	0.1	29.7	2.1	36.0	c	54.1	25.6	20.3	51	14	332
P5: 30°3′39″N to 76°18′33.9″E, Loamy Sodic Haplustepts, severely sodic soil & GW in the recent Ghaggar plain, barren with sparse vegetation																		
A1	0–22	9.9	6.1	64.5	1.5	6.5	23.0	33.9	0.2	56.0	2.1	16.4	l	14	17.8	68.6	88	28	377
B1	22–56	10.4	8.9	108	1.5	9.0	29.0	52.8	0.2	61.0	1.8	19.5	l	14	22.0	65.8	19	28	470
B2	56–84	10.6	5.8	56.7	1.5	8.5	16.0	38.4	0.1	60.0	3.0	20.0	l	18	22.1	59.3	15	19	203
Ck	84–121	**10.7**	5.0	42.3	2.0	**17.5**	14.0	16.5	0.1	64.0	2.9	19.5	l	21	19.6	59.1	15	25	203
P6: 30°8′56.3″N to 76°24′37.2″E, Fine-Loamy Typic Natrustalf, reclaimed sodic soil, calcareous, rice-wheat grown and produce moderate yield																		
A1	0–20	9.1	1.2	13.3	2	5.0	5.0	4.4	0.6	53.1	4.4	19.2	sil	25	28.1	47.1	86	14	321
Bt1	20–49	9.6	1.4	15.7	1	6.0	4.0	9.6	0.2	66.7	8.1	16.8	sicl	37	24.8	38.5	27	24	349
Bt2	49–88	9.7	1.6	18.5	1	7.0	4.5	9.2	0.1	69.0	5.2	20.0	sicl	29	33.3	38.0	14	27	488
Ck	88–125	9.6	1.7	19.5	3	6.0	3.0	8.4	0.2	58.5	9.5	21.2	sil	25	39.7	35.9	14	26	396
P7: 30°2′36.3″N to 76°14′49.5″E Loamy Typic Ustochrepts, sodic soil in recent alluvial plain, calcareous, rice-wheat crops showed moderate yield																		
Ap	0–22	8.9	4.4	39.2	4	5.3	12.5	22.4	0.4	40.0	2.3	24.5	l	23	21.6	55.0	82	10	243
Bw1	22–61	9.0	4.8	42.3	2	4.0	17.5	28.4	0.2	38.1	2.6	21.1	l	15	17.4	67.4	23	9	286
Bw2	61–93	9.1	3.6	34.1	4	4.8	15.0	17.9	0.2	46.6	2.3	16.3	l	15	18.1	66.6	20	14	340
Bw3	93–121	9.4	2.4	19.2	2	3.5	10.0	8.8	0.1	44.2	4.1	19.0	sil	19	31.6	49.9	16	18	303

TABLE 6.3 (Continued)

Horizon	Depth (cm)	pH	EC (dS m^{-1})	Na^+	Ca^{2+}+ Mg^{2+}	CO_3^{2-}+ HCO_3^-	Cl^-	SO_4^{2-}	OC	ESP	$CaCO_3$	CEC cmol (p^+) kg^{-1}	Texture	Clay	Silt	Sand	Av N	Av P	Av. K
				me L^{-1}					(%)					(%)			Kg ha^{-1}		
P8: 29°59′54.3″N to 76°25′41.3″E Fine-Loamy Typic Natrustalf, strongly sodic soil in old alluvial plain, natural vegetation and forestry plantations																			
A1	0–22	10.6	7.4	98.7	4	15.0	31.0	10.9	0.1	99.5	2.4	14.3	l	15.2	14.7	69.9	79	56	543
Bt1	22–56	11.0	5.5	80.0	4	15.0	18.0	12.8	0.1	91.8	2.4	20.6	cl	24.9	19.4	55.5	73	43	490
Bt2	56–91	11.1	4.6	70.4	3	20.0	15.0	9.1	0.1	96.9	0.9	25.7	cl	25.8	18.8	55.2	67	24	463
BC	91–118	11.2	5.3	83.9	3	27.0	12.0	6.2	0.1	94.1	1.4	24.6	cl	23.3	18.5	58.1	56	15	390
P9: 29°40′35.9″N to 76°13′8.5″E Coarse-Loamy Typic Ustochrepts (Saline phase), waterlogged (WT <0.5m) soil under canal irrigation, cotton crop																			
A1	0–18	7.9	21.7	168	78	3.0	112	131	0.3	51.5	0.6	14.6	sl	14.7	16.5	68.6	121	25.	437
Bw1	18–53	7.8	19.3	144	74	3.5	104	111	0.2	32.6	0.2	13.6	sl	15.8	15.6	68.5	109	15	389
Bw2	53–94	7.9	16.7	123	60	3.0	90	90	0.1	43.8	0.5	17.1	sl	19.1	18.7	62.0	90	13	320
B k	94–128	7.9	12.5	88	48	2.5	80	53	0.1	37.3	10.0	14.1	sl	11.8	23.3	64.8	82	10	277
P10: 29°52′25.3″N to 76°19′49.9″E Loamy Sodic Haplustepts, reclaimed sodic soil in the old alluvial plain, medicinal & aromatic plants growing																			
A1	0–26	9.6	9.2	117	4.0	20.0	33.0	0.6	0.1	75.9	0.1	13.6	l	13.8	21.8	64.7	115	10	425
B1k	26–58	10.4	13.7	193	4.0	44.5	32.0	4.6	0.1	91.2	0.6	14.3	l	14.1	24.7	61.1	105	6	358
B2k	58–94	10.8	13.4	193	3.0	33.0	35.0	8.1	0.1	92.3	0.7	16.6	sil	12.2	26.5	61.2	25	7	268
B3k	94–129	10.8	11.5	151	3.0	30.0	31.0	7.1	0.1	96.7	0.8	15.0	sil	11.7	25.1	63.1	20	15	186

P = Pedon, The mixed parent materials & hyperthermic temperature regimes are common in all Pedon, GW = Groundwater. AV = Available.

to improve soil physical properties and internal drainage. The alkali groundwater may be treated suitably with gypsum while the saline groundwater may be used in mixing or cyclic mode for irrigation purpose (Gupta, 2010; Singh, 2009).

The sub-angular blocky structure, silty loam texture and pale yellow to yellowish red color and sodic groundwater were characteristic features of Pedon 3 (Jain and Kumar, 2007). The scattered crop and higher water absorption (dark gray tone) for waterlogging are typical surface features in the satellite imageries. The pH (8.6 to 9.3) and ESP (44.6 to 54.4) values indicated slight to moderate sodicity and is used for rice-wheat cropping following reclamation (0–30 cm). Prominent waterlogging appeared due to fine soil texture (silt loam) and stratification caused due to irrigation by poor quality groundwater. The Na^+ (12.4 to 22.3 me L^{-1}), CO_3^{2-} and HCO_3^- (4.0 to 4.5 me L^{-1}) ions are dominating and caused sodicity development. The presence of Cl^- (8.0 to 10.0 me L^{-1}) and SO_4^{2-} (5.9 to 13.6 me L^{-1}) ions showed the presence of mixed parent materials (Sachdev et al., 1995). The $CaCO_3$ content (1.5 to 9.9%) indicated calcareous parent materials. The high CEC values (13.2 to 26.9 cmol (p^+) kg^{-1}) appeared due to higher clay contents (19.6 to 24.8%). The organic carbon (0.1 to 0.3%), available nitrogen (10 to 51 Kg ha^{-1}) and phosphorus (13 to 38 Kg ha^{-1}) contents are low. The gypsum and FYM application is suggested prior to arable cropping.

The pH (8.0 to 8.6) and ESP values (29.7 to 40.1) of Pedon 4 indicated the initiation of sodicity development. The $CaCO_3$ (<2 mm) is evenly distributed (1.4 to 2.3%) in the soil profile and higher values may be attributed to the carbonate containing irrigation water. The increase of EC values (2.3 to 4.1 dS m^{-1}) at sub-surface depths indicated the presence of salty groundwater. The Na^+ (17.7 to 35.5 me L^{-1}), Ca^{2+} + Mg^{2+} (8.0 to 10.0 me L^{-1}), CO_3^{2-} + HCO_3^- (1.5 to 5.0 me L^{-1}), Cl^- (6.0 to 10.0 me L^{-1}) and SO_4^{2-} (11.9 to 28.3 me L^{-1}) ions indicated dominance of alkaline parent materials. The higher CEC values (21.5 to 36.0 cmol (p^+) kg^{-1}) are attributed to high clay content (40.0 to 54.1%). The increasing ESP values (29.7 to 40.1) indicated the influence of alkali water irrigation. The sprinkler and drip methods of irrigation are suggested for irrigation with salty groundwater. The soil is suitable for growing low water requiring crops, fruits (horticulture) and forestry plantations to sustain in waterlogged soil affected soils and poor quality waters.

The pH values (9.9 to 10.6) of Pedon 5 indicated strongly sodic soil characterized by sparse vegetation in the Ghaggar plain (Block Guhla). The EC values (8.9 to 5.0 dS m^{-1}) indicated salt accumulation due to irrigation with salty groundwater (Qureshi et al., 1996). The high Na^+ (42.3 to 108.0 me L^{-1}) and CO_3^{2-} and HCO_3^- (6.5 to 17.5 me L^{-1}) contents cause increased soil pH and low contents of Ca^{2+}+ Mg^{2+} (1.5 to 2.0 me L^{-1}), Cl^- (14.0 to 29.0 me L^{-1}) and SO_4^{2-} (16.5 to 52.8 me L^{-1}) result in high SAR. The higher ESP (56.0 to 64.0) appeared due to the saturation with Na^+ ions. The low organic carbon (0.1 to 0.2%), available nitrogen (15 to 88 Kg ha^{-1}) and phosphorous (19 to 28 Kg ha^{-1}) contents indicated low fertility status. For reclamation, gypsum (@ 8–10 t ha^{-1}) application is necessary prior to growing arable crop such as rice and wheat.

The Pedon 6 soil is characterized by silt loam to silty clay loam texture, moderate medium to massive sub-angular blocky structure, fine consistency and yellowish brown color. Poor to imperfect drainage and the presence of lime, iron, manganese nodules and concretions of $CaCO_3$ (calcareous layer) were found at a depth below the surface. The silty clay loam soil texture at sub-surface depth with clay content ranging from 29 to 37% and ESP (66.7 to 69.0%) favored the formation of natric horizon. The pH (9.1 to 9.7) and EC (1.2 to 1.7 dS m^{-1}) values indicated strong sodicity except the surface soil which is reclaimed. Higher contents of Na^+ (13.3 to 19.5 me L^{-1}), CO_3^{2-} and HCO_3^- (5.0 to 7.0 me L^{-1}), as compared to Ca^{2+}+ Mg^{2+} (1.0 to 3.0 me L^{-1}), Cl^- (3.0 to 5.0 me L^{-1}) and SO_4^{2-} (4.4 to 9.6 me L^{-1}) indicated strong alkaline ions causing sodicity development. The higher ESP values (53.1 to 69.0) supplemented soil alkalization. The $CaCO_3$ (<2 mm) content (4.4 to 9.5%) showed precipitated calcareous parent materials at high soil pH and ESP. The Pedon can be used for arable crops following reclamation using gypsum (@6–8 t ha^{-1}). Due to the presence of *natric* horizon and poor quality groundwater, plantations of forestry, medicinal and aromatic plants and low water requiring food crops are more suitable.

The range of pH and ESP values (8.9 to 9.4 and 38.1 to 44.2) of Pedon 7 indicated moderately sodic soil and higher salt concentration (2.4 to 4.8 dS m^{-1}) at surface resulting from the irrigation with salty groundwater. The dominance of Na^+ (19.2 to 42.3 me L^{-1}) and CO_3^{2-} + HCO_3^- (3.5 to 5.3 me L^{-1}) increased soil pH. The CEC values (16.3 to 24.5 cmol (p^+) kg^{-1}) are related to clay content (15 to 23%). The $CaCO_3$ content is increased with depth (2.3 to 4.1%). The available nitrogen (16 to 83 Kg ha^{-1}) and phosphorus (9 to 18 Kg ha^{-1}) contents are low. The use of salt resistant varieties of rice and wheat is suggested to improve productivity.

In the old alluvial plain of Ghaggar (Block Siwan), the soils (Pedon 8) showed finer soil texture (loam to clay loam), massive to moderate medium sub-angular to angular blocky structure, yellowish brown to pale yellow color and poor to imperfect drainage. The pH values (10.6 to 11.2) showed strongly sodic soil. The high ESP values indicated saturation with sodium favoring hydrolysis and soil alkalinity. The Na^+ (70.4 to 98.7 me L^{-1}) and CO_3^{2-}+HCO_3^- (15.0 to 27.0 me L^{-1}) ions are dominating and favored sodicity development (Sharma et al., 2011) The CEC values (14.3 to 25.7 cmol (p^+) kg^{-1}) showed dominance of mixed clay minerals. The increasing clay content (24.9 to 25.8%) and higher ESP (91.8 to 96.9) showed *natric* horizon formation. The gypsum application @ 8 to 10 t ha^{-1} is required for reclamation and the soil can be used for arable (rice-wheat) cropping.

Located in the arid sandy alluvial plain of Ghaggar (Block Kalayat) and irrigated with Narwana branch of Western Yamuna canal Pedon 9 showed waterlogging (water table depth 0.5 m) and poor productivity. A thick layer of $CaCO_3$ concretions (2–5 mm, 50 to 60%) is found at a depth (120 cm) below the surface. The neutral soil pH (7.8 to 7.9) and higher salinity (12.5 to 21.7 dS m^{-1}) is found throughout the soil profile. The higher contents of Na^+ (88 to 168.0 me L^{-1}), Ca^{2+}+Mg^{2+} (48 to 78 me L^{-1}), Cl^- (80 to 112 me L^{-1}) and SO_4^{2-} (53 to 131 me L^{-1}) are noted. The $CaCO_3$ (<2 mm) content increased from 0.6 at surface to 10.0% below. The CEC values are low (13.6 to 17.1 cmol (p^+) kg^{-1}) due to coarse soil texture (sand 62.0 to 68.8%). The ESP values and the presence of carbonate and bicarbonate salts showed complex saline-alkaline nature. The soil needs installation of sub-surface drainage (SSD) to lower the water table depth below the root zone and reduce soil salinity. Alternately it can also be used for aquaculture

Strongly sodic soil (Pedon 10) is located at the old alluvial plain of Ghaggar in Central Haryana (Block Siwan) and is currently used for forestry, medicinal and aromatic plantations. The soil texture varies from loam to silty loam while strong to medium soil structure and pale yellow to dark yellowish brown color at surface and sub-surface depths resulted due to high soil sodicity and impaired drainage. The pH and ESP values range from 9.6 to 10.8 and 75.9 to 96.7 at surface and subsurface depths, respectively. The high Na^+ (117.0 to 193.0 me L^{-1}) and CO_3^{2-}+HCO_3^- (20.0 to 44.5 me L^{-1}) contents favored strong sodicity development. The higher soil salinity (9.2 to 13.7 dS m^{-1}) has resulted from irrigation of salty (high RSC) groundwater. The CEC values are low due to loamy soil texture (sand 61.1 to 64.7%). The soil needs suitable dosage @10 to 12 t ha^{-1} of gypsum application to neutralize CO_3^{2-}+HCO_3^- and reduce ESP. The arable cropping (rice-wheat) is suggested after reclamation.

6.5 WATER QUALITY STUDY

The physico-chemical properties of groundwater samples are presented in Table 6.4. The depth of groundwater table ranges from 76–83 m in Kaithal block, 19–21 m in Kalayat, 76–91 m in Guhla and Siwan blocks, respectively. The water samples of Kaithal block (PW 1 to 4) showed neutral to sodic pHiw (7.6 to 8.3) and dominance of carbonate and bicarbonate (8.5 to 10.5 me L^{-1}) of sodium (9.9 to 17.7 me L^{-1}), calcium and magnesium (8.0 to 10.0 me L^{-1}). PW 5 is saline (EC 12.7 dS m^{-1}), high SAR (33.6) and showed the dominance of Na^+ (164.8 me L^{-1}), $Ca^{2+} + Mg^{2+}$ (48.0 me L^{-1}), Cl^- (80.0 me L^{-1}) and SO_4^{2-} (54.8 me L^{-1}). PW 6 is sodic (pH 9.3) and high RSC (6.5) appeared due to higher $CO_3^{2-} + HCO_3^-$ (8.5 me L^{-1}) and Na^+ (9.9 me L^{-1}). The PW 7 to 13 showed sodic pH (8.6 to 9.3), high SAR (5.5 to 23.4), at places high RSC (1.0 to 12.7 me L^{-1}) and is dominated by the Na^+ (6.8 to 14.1 me L-1) and $CO_3^{2-} + HCO_3^-$ (2.5 to 15.7 me L^{-1}), Cl^- (1.7 to 20.0 me L^{-1}) and SO_4^{2-} (0.4 to 8.8 me L^{-1}) ions. The water samples (PW 7–12) showing low contents of $CO_3^{2-} + HCO_3^-$ and high SAR, can be used mixing with good quality water. PW 6 and 13 should be treated with gypsum for neutralizing residual $NaHCO_3$. PW 5 showing high SAR may be used alternately with good quality water. PW 1–4 is suitable for irrigation of salt resistant varieties.

6.6 DISTRIBUTION OF SALT AFFECTED SOILS

The area under different categories of salt affected soils was computed based on the statistics derived in Geographic Information System (GIS) (Table 6.5). Six categories of salt affected soils were identified and these are distributed in four blocks (Kaithal, Kalayat, Pundri and Rajound) of Kaithal subdivision. In Kaithal block, sodic soils (6122 ha) are dominating where sodic groundwater is used for irrigation. These are slight (4313 ha) and moderately (1809 ha) sodic in nature. Saline soils (804 ha, 0.3%) are also located at selected places only in the lower topographic zone (Figure 6.4). Saline soil is dominant (4620 ha, 2%) in the irrigated areas of Kalayat block. The patches of sodic soils (1452 ha, 0.6%) are also found in the adjoining areas. Saline soils (3063 ha, 1.3%) were dominant in irrigated areas of Rajaund block. Salt affected soils covered 4891 ha (2.1%) in Pundri block, these are primarily sodic (4723 ha, 2%) and saline (168 ha, 0.07%). Saline (1415 ha, 0.61%) and

TABLE 6.4 Quality of Groundwater Samples from Kaithal District

S. No	Location, source and depth of groundwater	pHiw	ECiw dS m^{-1}	Na^+	K^+	Ca^{2+}+ Mg^{2+}	CO_3^{2-} + HCO_3^-	Cl^-	SO_4^{2-}	RSC	SAR
				(me L^{-1})							
Water samples from Kaithal sub-division											
1	Vill.Mundri Block Kaithal (76 m)	8.3	1.1	10.3	0.06	10.0	9.0	10.0	tr	tr	4.8
2	Vill. Mundri Block Kaithal (83 m)	8.3	1.1	10.3	0.05	10.0	9.5	15.0	tr	tr	4.7
3	Vill.Sampli Kheri Block Kaithal (76 m)	7.6	1.1	9.9	0.08	8.0	10.5	12.0	tr	tr	4.9
4	Devigarh farm (HAU) Kaithal Seepage water	7.2	2.3	17.7	0.01	10.0	8.5	16.0	7.1	tr	7.9
5	Vill. Kolekha Block Kalayat (19–21 m)	7.6	12.7	164.8	0.05	48.0	5.0	80.0	54.8	tr	33.6
6.	Vill. Bhaini Majra Block Kaithal (83 m)	9.3	1.2	9.9	0.07	2.0	8.5	3.0	tr	6.5	9.9
Water samples from Guhla and Siwan blocks											
7	Vill. Sehun Majra, Block Guhla, 76 m	8.6	1.3	10.6	0.1	3.0	2.5	10.0	8.8	tr	8.6
8	Vill. Kheri Daban, Block Guhla, 76 m	9.1	1.3	12.6	0.1	1.5	2.5	10.0	6.4	1.0	14.5
9	Vill. Hansu Majra, Block Guhla, 76 m	9.3	1.4	14.1	0.1	1.0	3.0	6.0	4.7	2.0	19.9
10	Vill. Majri, Block Guhla, 91 m	9.1	1.2	12.1	0.1	2.0	4.0	20.0	2.9	2.0	12.1
11	Vill. Tatiana, Block Guhla, 91 m	9.1	1.6	16.6	0.1	1.0	3.0	5.0	7.7	2.0	23.4
12	Vill. Kamheri Block Guhla, 91 m	8.8	0.8	6.8	0.1	3.0	2.5	3.0	0.4	tr	5.5
13	Vill. Bichian, Block Siwan, 83 m	8.8	1.1	13.9	0.1	2.9	15.7	1.7	tr	12.7	11.5

TABLE 6.5 Distribution and Extent (ha) of Salt Affected Soils in Kaithal District

Name of the block	Categories of Salt Affected Soils							
	Moderately saline	**Moderately sodic**	**Slightly saline**	**Slightly sodic**	**Strongly saline**	**Strongly sodic**	**Total area**	**% of TGA**
Kaithal sub-division								
Kaithal	148	1809	50	4313	606	tr	6926	3.0
Kalayat	935	1306	1335	146	2350	tr	6072	2.6
Pundri	tr	1682	168	3041	tr	tr	4891	2.1
Rajaund	tr	09	2028	51	1026	tr	3114	1.3
Sub-total	1083	4806	3581	7551	3982	tr	21002	9.0
Guhla and Siwan blocks								
Guhla	tr	73	1415	1015	tr	143	2646	1.1
Siwan	tr	tr	255	143	tr	2255	2653	1.1
Sub-total	tr	73	1670	1158	tr	2398	5299	2.3
Grand total	1083	4879	5251	8709	3982	2398	26301	11.3

TGA = Total Geographical Area of Kaithal district.

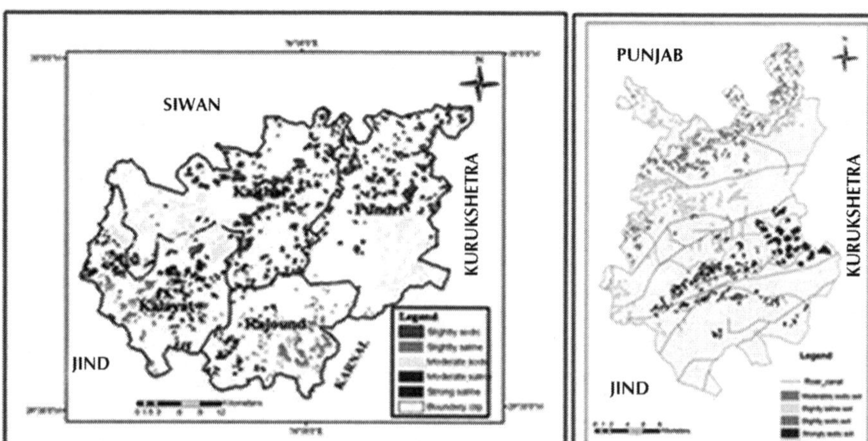

FIGURE 6.4 Distribution of salt affected soils in Guhla and Kaithal sub-divisions (Kaithal district).

sodic (1015 ha, 0.44%), soils are distributed in Guhla sub-division (Figure 6.4) covering Guhla and Siwan blocks of Kaithal district. Slightly saline and sodic soils are distributed in the irrigated areas and covered 1415 ha (0.61%) and 1015 ha, (0.44%), respectively. Strongly (143 ha, 0.06%) and moderately sodic (73 ha, 0.03%) are located along the Ghaggar plain. Salt affected soils covered 2653 ha (1.1%) in Siwan block, strongly sodic (2255 ha, 0.97%) soils are distributed along the Paleo-channel of the Saraswati river while slightly saline (255 ha, 0.1%) and slightly sodic (143 ha, 0.06%) soils are also found in the irrigated areas. Thus, 26301 ha (11.3%) is salt affected in Kaithal district, in which 6.8% (15,986 ha) and 4.4% (10,315 ha) areas are sodic and saline in nature.

6.7 CONCLUSIONS

The integrated approach of Indian Remote Sensing data interpretation with ground truth and laboratory studies facilitated the mapping and characterization of salt affected soils in Kaithal district of Central Haryana. Strongly salt affected soils were identified due to high reflectance of salty surfaces while moderate and slightly salt affected soils with mixed spectral signatures were detected in combination with ground truth and soil characteristics. The physico-chemical characteristics of salt affected soils showed complex saline and sodic nature. The dominance of carbonates and bicarbonates of sodium and higher ESP favored alkalization of soils in the Ghaggar plain. The concretionary calcium carbonate layer at sub-surface depths indicated precipitation and deposition under alkaline condition. The sodic soils are detected in the Ghaggar plain and along the paleo-channel of river Saraswati, while the saline soils are found in the irrigated region. Coarse soil texture and poor internal drainage in Kalayat block caused waterlogging and soil salinization in the irrigated region. The use of poor quality groundwater for irrigation caused salt enrichment in Kaithal block. The water samples of Kaithal and Siwan blocks are sodic and at places with high RSC, the quality of water samples in Guhla block is sodic with high SAR. Suitable management and land uses options are suggested for growing salt resistant crops, horticulture and forestry plantations with proper water management practices. The poor quality groundwater needs prior treatment for removal of salts injurious to plants.

KEYWORDS

- **Alkalinity**
- **GIS**
- **Reclamation**
- **Remote Sensing**
- **Salt Affected Soils**
- **Soil Salinity**
- **Trans-Gangetic Plain**
- **Water Quality**

REFERENCES

Abrol, I. P., & Bhumbla, D. R. (1971). Saline and Alkali Soils in India-Their occurrence and management. *World Soil Resources Report No. 41, Food and Agricultural Organization of United Nations*, pp. 42–51.

Abrol, I. P., & Bhumbla, D. R. (1978). Some comments on the terminology of salt affected soils. In Proc. Dry land saline seep control, *6*, 19–6.27, Edmonton, Canada.

Abrol, I. P., Chhabra, R., & Gupta, R. K. (1980). A fresh look at the diagnostic criteria for sodic soils. *Proceedings of the International Symposium on Salt Affected Soils*, CSSRI, Karnal, pp. 142–146.

Abrol, I. P., Yadav, J. S. P., & Massoud, F. J. (1988). *Salt Affected Soils and Their Management*, FAO Soils Bulletin 39, Rome, pp. 131.

ABS. (2002). Australian Bureau of Statistics "Salinity," Canberra 2002 (http://www.abs.gov.au/abs@.nsf/Lookup/by%20Subject/1370.0–2010-Chapter-Salinity%20 (6.2.4.4).

Agarwal, R. R., & Yadav, J. S. P. (1956). Diagnostic technique for the saline and alkali soils of the Indo-Gangetic alluvium in Uttar Pradesh. *Journal of Soil Science*, *7*, 109–121.

Agarwal, R. R., Mehrotra, C. L., & Gupta, C. P. (1957). Spread and intensity of soil alkalinity with canal irrigation in Gangetic alluvium of Uttar Pradesh. *Indian Journal of Agricultural Sciences*, *27*, 363–373.

Ali, A. A. A., Weindorf, D. C., Chakraborty, S., Sharma, A., and Li, B. (2015). Combination of proximal and remote sensing methods for rapid soil salinity quantification. *Geoderma*, 34–46, 239–240.

Allbel, A., & Kumar, L. (2013). Soil salinity mapping and monitoring in arid and semiarid regions using remote sensing technology: A review. *Advances in Remote Sensing, 2*, 373–385.

Balpande, S. S., Deshpande, S. B., & Pal, D. K. (1996). Factors and processes of soil degradation in vertisols of Purna valley, Maharashtra, India. *Land degradation and Development, 17*, 313–324.

Bhalla, A., Singh, G., Kumar, S., Shahi, J. S., & Mehta, D. (2011). Elemental analysis of groundwater from different regions of Punjab State (India) using EDXRF technique and the sources of water contamination. In *Proceedings of the International Conference of Environmental and Computer Science*, IPCBEE, IACSIT Press, Singapore, 156–164.

Bhargava, G. P., Sharma, R. C., Pal, D. K., & Abrol, I. P. (1980). A case study of distribution and formation of salt affected soils in Haryana state. In *Proceedings of International Symposium on Salt Affected Soils*, Karnal, pp. 83–90.

Colwell, R. N. (ed.) (1996). *Manual of Remote Sensing*. Vol. II, Interpretation and Applications. American Society of Photogrammetry, 1983, 2nd Ed. April 1996.

CSSRI. (2000). Impact of the Centrally Sponsored Scheme on Reclamation of Alkali Soils during VIII plan in Haryana, Punjab and Uttar Pradesh. Report Central Soil Salinity Research Institute, Karnal, India.

CSSRI. (2007). Annual Report 2006–07. Central Soil Salinity Research Institute, Karnal, India.

Dwivedi, R. S. (2001). Soil Resource Mapping: A Remote Sensing Perspective. *Remote Sensing Reviews*, *20*, 59–122.

FAO/IIASA/ISRIC/ISS-CAS/JRC. (2008). Harmonized World Soil Database (*ver.* 1.0). FAO, Rome, Italy and IIASA, Luxemburg, Austria.

Gupta, S. K. (2010). Management of alkali water. Central Soil Salinity Research Institute, Karnal India, Technical Bulletin: CSSRI/Karnal/2010/01, p. 62.

HSMITC. (2001). *Haryana Groundwater Quality Map,* Haryana State Minor irrigation and Tube Well Corporation Limited, Government of Haryana, Chandigarh, India

Jackson, M. L. (1986). *Advanced Soil Chemical Analysis,* Prentice Hall of India, New Delhi.

Jain, A. K., & Kumar, R. (2007). *Water Management Issues*. In Proceedings of the Indo-US Workshop on "*Innovative E-technologies for Distance Education and Extension Outreach for Efficient Water Management*," March 5–9, 2007, ICRISAT, Patancheru, Hyderabad, India.

Joshi, M. D., & Sahai, B. (1993). Mapping salt affected land in Saurashtra coast using Landsat satellite data. *International Journal of Remote Sensing, 14*, 1919–1929.

Joshi, P. K., & Agnihotri, A. K. (1984). An assessment of adverse effects of canal irrigation in India. *Indian Journal of Agricultural Economics, 39*, 528–536.

Koohafkan, P. (2012). "Water and Cereals in Dry lands." The Food and Agriculture Organization of the United Nations and Earth scan, Rome.

Leather, J. W. (1914). *Investigations on Usar Land in United Provinces*. Govt. Press Allahabad, pp. 88.

Manchanda, H. R. (1976). *Quality of Ground Waters in Haryana*. Haryana Agriculture University, Hisar, p. 160.

Mandal, A. K., Reddy, G. P. O., & Ravisankar, T. (2011). Digital database of salt affected soils in India using Geographic Information System. *Journal of Soil Salinity and Water Quality, 3*, 16–29.

Mandal, A. K. (2014). Characterization of some salt affected soils of Punjab for reclamation and management. *Journal of the Indian Society of Soil Science, 62*(2), 161–167.

Mandal, A. K., & Sharma, R. C. (1997). Characterization of some salt affected soils of Indira Gandhi Nahar Pariyojona command area, Rajasthan. *Agropedology*, *7*, 84–89.

Mandal, A. K., & Sharma, R. C. (2010). Delineation and characterization of waterlogged and salt affected areas in IGNP command, Rajasthan. *Journal of the Indian Society of Soil Science*, *58*(4), 449–454.

Mandal, A. K., & Sharma, R. C. (2011). Delineation and characterization of waterlogged salt affected soils in IGNP using remote sensing and GIS. *Journal of the Indian Society of Remote Sensing, 39*, 39–50.

Mandal, A. K., & Sharma, R. C. (2012). Description and characterization of typical soil monoliths from salt affected areas in Rajasthan. *Journal of the Indian Society of Soil Science*, *60*, 299–303.

Mandal, A. K., & Sharma, R. C. (2013). Mapping and characterization of waterlogged and salt affected soils in Loonkaransar area of Indira Gandhi Nahar Pariyojona for reclamation and management. *Journal of the Indian Society of Soil Science, 61*(1), 29–33.

Mandal, A. K., Sharma, R. C., Singh, G., & Dagar, J. C. (2010). Computerized database on salt affected soils in India. Technical Bulletin: CSSRI/ Karnal/02/2010, pp. 28.

McFarlane, D. J. (2004). "The extent and Potential Areas of Salt-Affected Land in Western Australia Estimated using Remote Sensing and Digital Terrain Models," *Engineering Salinity Solutions: 1st National Salinity Engineering Conference,* Barton, pp. 55–60.

Metternicht, G., & Zinck, J. A. (1997). Spatial discrimination of salt- and sodium-affected soil surfaces. *International Journal of Remote Sensing, 18*, 2571–2586.

NBSSLUP. (2006). *Soils of India*. National Bureau of Soil Survey and Land Use Planning, Nagpur India NBSS Publ. 94.

Northkote, K. H. (1979). *A Factual key for the Recognition of Australian Soils*. 4th ed. Relim Technical Publications, Adelaide, Australia.

NRSA. (1997). *Salt Affected Soils*. National Remote Sensing Agency, Department of Space, Government of India, Hyderabad.

NRSA. (2007). Manual on Nationwide Mapping of Land Degradation using Multi-temporal Satellite Data. National Remote Sensing Agency, Department of Space, Government of India, Hyderabad.

NRSA. (2008). *Mapping Salt Affected Soils of India*. National Remote Sensing Centre, Department of Space, Government of India, Hyderabad, 60 p.

Qureshi, F. M., Singh, S. K., Chaudhari, S. K., & Das, K. (1996). Genesis and Taxonomy of some saline and sodic soils of Bharatpur (Rajasthan). *Journal of the Indian Society of Soil Science, 44*, 130–135.

Richards, L. A. (Ed.) (1954). *Diagnosis and Improvement of Saline and Alkali Soils*. Agriculture Handbook No. 60, United States Department of Agriculture, Washington D.C.

Sachdev, C. B., Lal, T., Rana, K. P. C., & Sehgal, J. (1995). "*Soil of Haryana: their kinds, distribution and interpretation for Optimizing Land Use*" NBSS Publ. 44 (Soils of India Series 4), National Bureau of Soil Survey and Land Use Planning, Nagpur, 59 p.

Saxena, R. K., Sharma, R. C., Verma, K. S., Pal, D. K., & Mandal, A. K. (2004). *Salt affected soils, Etah district (U. P.)*. NBSS-CSSRI Publ. No. 108, NBSS & LUP, Nagpur, pp. 85.

Scudiero, E., Skaggs, T. H., & Cowin, D. L. (2014). Regional scale soil salinity evaluation using Landsat 7, Western San Joaquin Valley, California, USA. *Geoderma Regional, 2–3*, 82–90.

Sharma, R. C., & Mandal, A. K. (2006). Mapping of soil salinity and sodicity using digital image analysis and GIS in irrigated lands of the Indo-Gangetic Plain. *Agropedology, 16*(2), 71–76.

Sharma, R. C., Mandal, A. K., Singh, R., & Singh, Y. P. (2011). Characteristics and use potential of sodic and associated soils in CSSRI experimental farm, Lucknow, Uttar Pradesh. *Journal of the Indian Society of Soil Science, 59*, 381–387.

Sidhu, G. S., Walia, C. S., Lal, T., Rana, K. P. C., & Sehgal, J. (1995). "*Soil of Punjab for Optimizing Land Use*" NBSS Publ. 45 (Soils of India Series 4), National Bureau of Soil Survey and Land Use Planning, Nagpur, 75 p.

Singh, D., Chhonkar, P. K., & Pandey, R. N. (1999). Soil Plant Analysis A Methods manual. Indian Agricultural Research Institute, New Delhi, 146 p.

Singh, G. B. (2009). Salinity Related Desertification and Management Strategies: Indian Experience *Land Degradation and Development*, *20*, 367–385.

Singh, G., Bundela, D. S., Sethi, M., Lal, K., & Kamra, S. K. (2010). Remote Sensing and Geographic Information System for appraisal of salt-affected soils in India. *Journal of Environmental Quality*, *39*, 5–15.

Soil Science Society America. (1987). *Glossary of Soil Science Terms*. Technology committee, Soil Science Society of America, Wisconsin, USA.

Soil Survey Division Staff. (2004). *Soil Survey Manual*. United States Department of Agriculture Handbook No. 18. Scientific Publishers (India): Jodhpur.

Soil Survey Staff. (1998). '*Keys to Soil Taxonomy*,' Eighth Edition, U.S. Soil Conservation Service, Washington, D.C.

Szabolcs, I. (1989). *Salt Affected Soils*. CRC Press, New York, 274 pp.

Thimmappa, K., Tripathi, R. S., Raju, R., and Singh, Y. P. (2013). Livelihood security of resource poor farmers through alkali land reclamation: An impact analysis. *Agricultural Economics Research Review, 26*, 139–147.

United Nations Environmental Program. (2009). The Environmental Food Crisis. The environment's role of averting future food crisis. A UNEP Rapid Response Assessment, Christian Nellemann (Ed.). pp. 101.

Wu, W., Mhaimeed, A. S., Al-Shafie, W. M., Ziadat, F., Dehhibi, B., Nangia, V., & Pauw, E. D. (2014). Mapping soil salinity changes using remote sensing in Central Iraq. *Geoderma Regional, 2–3*, 21–312.

Yadav, J. S. P. (2003). Managing soil health for sustained high productivity. *Journal of the Indian Society of Soil Science, 51*, 448–465.

CHAPTER 7

CHARACTERIZATION OF COASTAL SOILS FOR ENHANCEMENT OF PRODUCTIVITY AND LIVELIHOOD SECURITY: A CASE STUDY FROM THE COASTAL PLAINS OF WEST BENGAL, INDIA

S. K. GANGOPADHYAY,[1] S. K. REZA,[2] S. MUKHOPADHYAY,[2] D. C. NAYAK,[3] and S. K. SINGH[4]

[1]*Principal Scientist, ICAR-National Bureau of Soil Survey and Land Use Planning, Regional Centre, Kolkata – 700 091, India*

[2]*Senior Scientist, ICAR-National Bureau of Soil Survey and Land Use Planning, Regional Centre, Kolkata – 700 091, India*

[3]*Principal Scientist & Head, ICAR-National Bureau of Soil Survey and Land Use Planning, Regional Centre, Kolkata – 700 091, India*

[4]*Director, ICAR-National Bureau of Soil Survey and Land Use Planning, Amravati Road, Nagpur – 440 033, India*

CONTENTS

ABSTRACT

A study was conducted in the rainfed areas of Deshapran block, Purba Medinipur district, West Bengal under coastal agro-eco sub-region 18.5 constrained by various soil and water related problems besides climatic hazards. Soil resource inventory was conducted at 1:10,000 scale using latest technologies of Remote Sensing and Geographic Information System (GIS), to identify the problems and potentials of the soils and to suggest management practices to develop appropriate and sustainable production systems that deal with the problems of food security and livelihood improvement of the resource-poor farmers. The farming community is mostly poor. Marginal and landless farmers are poorer and have no livelihood security. Soils were imperfect to poorly drained, neutral to slightly alkaline in reaction with medium to low in organic carbon (OC), low to medium in available N, low in available P and low to high in available K. Salinity in dry season and waterlogging during wet season restricts the crop choices and therefore it is a major impediment to enhance the agricultural productivity in the region. The area is mostly mono-cropped, growing rainfed traditional rice with poor yields during the monsoon (*kharif*) season. The study reveals that besides rice-fallow/rice-rice cropping sequences, paddy-vegetables cropping sequence can be introduced in the block with proper soil and water conservation measures to improve the food and livelihood security.

7.1 INTRODUCTION

The life-supporting system and socio-economic development of any country depends on the proper use of soils which are considered the most valuable natural resource. Developing countries like India are very sensitive to land use pressures and unscientific land management practices imposed by increasing population. A good understanding of soils is possible only through a scientifically conducted inventory followed by land use planning with considerations of sustainability. Comprehensive information on soil resources is therefore essential for development. Coastal area is the transitional zone between land and sea and is having influence of both land and sea. Over the years the coastal region of India are traditionally backward and disadvantaged areas with low agricultural productivity. India has a large area of land (10.78 m ha) under coastal agro-ecosystem (Velayutham et al., 1999) along its long coast line of about 8,129 km and is distributed within nine coastal states (West Bengal, Odisha, Andhra Pradesh, Tamil Nadu, Kerala, Karnataka, Maharashtra, Goa, Gujarat), two union territories (Pondicherry, Daman & Diu) and two groups of islands (Andaman & Nicobar and Lakshadweep) and encompass 4 out of 15 agro-climatic zones identified by the Planning Commission and 16 out of 127 agro-ecological zones described by the National Agricultural Research Project (NARP) of ICAR. A large area of land (3.1 m ha) under coastal agro-ecosystem is salt affected. West Bengal occupies 1.42 m ha of land under coastal agro-ecosystem out of which 0.8 m ha is salt affected. Salt affected soils in West Bengal are distributed over the four districts: North 24 Parganas, South 24 Parganas, Purba Medinipur and Howrah and lies between 87° 25′ and 89° 05′ E latitude and 21° 30′ and 23° 15′ N longitude. These areas are endowed with a variety of soils and are constrained by various soil and water related problems besides the climatic hazards. The coastal ecology is extremely fragile and vulnerable to serious degradation. The farming community of the coastal regions of the country is dominated by very poor, marginal and landless farmers (Mandal et al., 2011). High unemployment and extremely poor livelihood security are the characteristic features of the rural life of the coastal region. The area is mostly mono-cropped growing rain-fed traditional rice with poor yields during monsoon (*kharif*) season.

7.1.1 CHARACTERIZATION OF COASTAL SOILS: GLOBAL PERSPECTIVE

Global food production will have to increase by 38% by 2025 and by 57% by 2050 (Wild, 2003) if food supply to the growing world population is to be maintained at current levels. Most of the suitable land has been cultivated and expansion into new areas to increase food production is rarely possible or desirable. The aim, therefore, should be an increase in yield per unit of land rather than in the area cultivated. More efforts are needed to improve productivity as more lands are becoming degraded. It is estimated that about 15% of the total land area of the world has been degraded by soil erosion and physical and chemical degradation, including soil salinization (Wild, 2003). Current knowledge on soils and existing soil maps impart largely information that is regional in nature than local and specific. Soil mapping at larger scale (1:10,000 scale) is essential for management of soil variability at village and micro-watershed level. Therefore, need is being felt for soil mapping at larger scale with fine level of detail (Dobos and Hengle, 2009; Ravisankar and Thamappa, 2004; Rao et al., 2004) to use them for developmental planning in agriculture. The developmental planning envisages preparation of soil resource inventory, understanding of resource constraints and suggesting feasible alternatives for agricultural development.

Coastal zones are home to 40% of the world's population and support much of the world's food production. Though agriculture is the primary occupation of majority of the people in coastal region, current practices are inadequate to provide decent livelihoods to the resource poor farmers. Salinity in dry season and waterlogged situation during wet season restricts the crop choices and therefore it is a major impediment to enhance the agricultural productivity in the region. However, there are several opportunities available to enhance the farm income through better management of soil and water resources. Sustainable management of land and water resources along with integrated farming system can provide fairly better income and secured livelihoods to the farmers in the region.

7.1.2 CHARACTERIZATION OF COASTAL SOILS: INDIAN PERSPECTIVE

Several workers *viz.*, Bandopadhyay and Bandopadhyay (1982); Bandopadhyay et al. (1996), Bandopadhyay et al. (2003); Maji et al. (2001), Das et al. (1996), Bandopadhyay et al. (2014) worked on the salt affected soils specially in Sundarban delta especially on 1:50,000 scale. However, no such information is available in the farm level (1:10,000 scale) of Purba Medinipur district, West Bengal and coastal agro-eco system in particular. Realizing all this, the present study was undertaken in Deshapran block, Purba Medinipur district, West Bengal under coastal agro-eco system to demonstrate the potential of high resolution Remote Sensing data in generating soil resource database at village level on 1:10,000 scale that would facilitate natural resource management and developmental planning to enhance the livelihood security and food security of the resource poor people of the area.

7.2 STUDY AREA

Deshapran block in Kanthi subdivision of Purba Medinipur district in West Bengal is situated at 87° 45' 34″ to 87° 53'10″ E longitude and 21° 43'23″ to 21° 54'53″ N Latitude covering an area of 17,790 ha. This block is bounded by Khejuri-I block in the north, Khejuri-II block in the north-east, Kathi-I block in the west and south west, Kanthi-III block in the north-east, Bay of Bengal in the east and the Hugli river is demarcating the block in the south, east and north (partly) (Figure 7.1). The soils of the study area have been developed from the recent tertiary deposits of alluvium parent materials. Geomorphologically, the study area can be described as coastal alluvial plain with micro topographical variations. On the basis of slope, two landforms were identified *viz.*, nearly level plain (0–1% slope) and very gently sloping plain (1–3% slope). The study area falls under agro-eco sub-region 18.5 (Gangetic delta, hot, moist-subhumid agro-eco sub-region) and constitutes the eastern coastal plains of West Bengal, characterized by moist sub humid eco system with hot summers and mild winters. The area experiences seasonal water deficit from last half of January extending to the first half of May.

The climate of the area is characterized by moist sub humid ecosystems with hot summers and mild winters. The mean annual temperature varies from 26–27°C rising to a maximum of 36–37°C in the summer months of April and May and dropping to a minimum of 14–16°C in the winter months of December and January. The mean annual rainfall ranges from 1600 to 1700 mm. The zone experiences seasonal water deficit from last half of January extending to first half of May. The soil moisture control section (SMCS), remains partly or fully dry from last half of January till May accounting for more than 90 cumulative days in a year. The area thus has the Ustic soil moisture regime. Hence, the length of growing period (LGP) accounts for 240–270 days in a year. The mean annual soil temperature (MAST) is greater than 22°C and the difference between mean summer soil temperature (MSST) and mean winter soil temperature (MWST) at 50 cm depth is more than 6°C qualifying for the Hyperthermic soil temperature regime in the area. The existing land use of Deshapran block, Purba Medinipur, West Bengal follows rice -based cropping pattern. Rice is cultivated mainly in *kharif* season. However, *boro* rice is also cultivated in patches with the help of surface irrigation from ponds, wells, etc., in some areas. After *kharif* season, vegetables are cultivated with higher yield in the eastern part of the block. Brackish water pisiculture is gaining popularity in this area by impounding the coastal inlets as a promising income generating source.

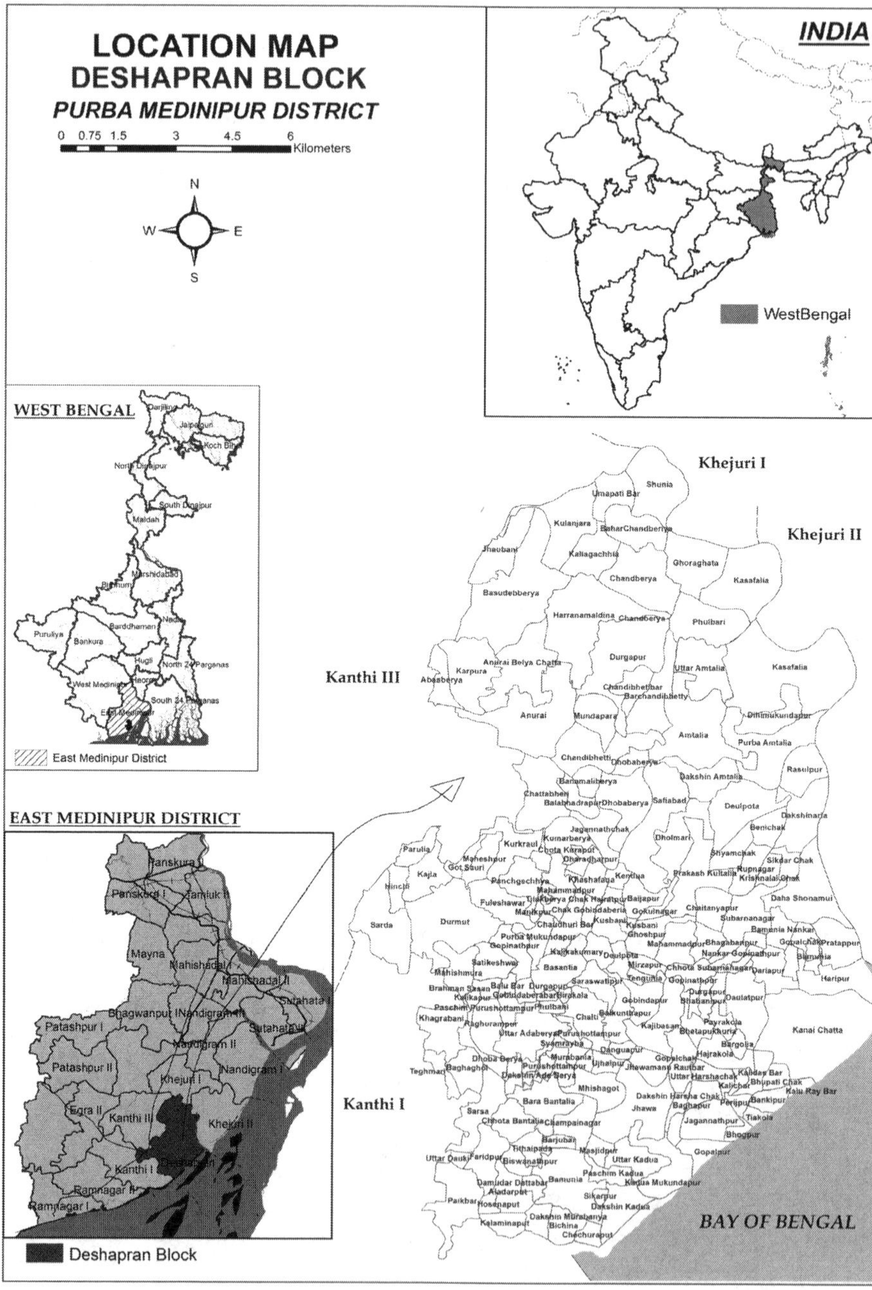

FIGURE 7.1 Location Map of Deshapran block, Purba Medinipur district, West Bengal.

7.3 METHODOLOGY

The soil resource map of the area under coastal agro-eco system was prepared on 1:10,000 scale through field survey, laboratory investigation, GIS and printing (Figure 7.2). The digital data of Indian Remote Sensing Satellite IRS-P6 LISS IV of November, 2009 were used. Ancillary data referred include Survey of India Toposheet (1:50,000 scale), climatic data, village map and geological data. The methodology adopted for this study consisted of geo-referencing and merging of Resourcesat-1 LISS IV data, on screen interpretation, ground truth collection and generation of land use/land cover map and landform map. This satellite image was superimposed on the cadastral map of the villages of the block to prepare the base map. The landform of the block was delineated along with the land use information through ground truth checking and thus the base map was finalized for soil resource

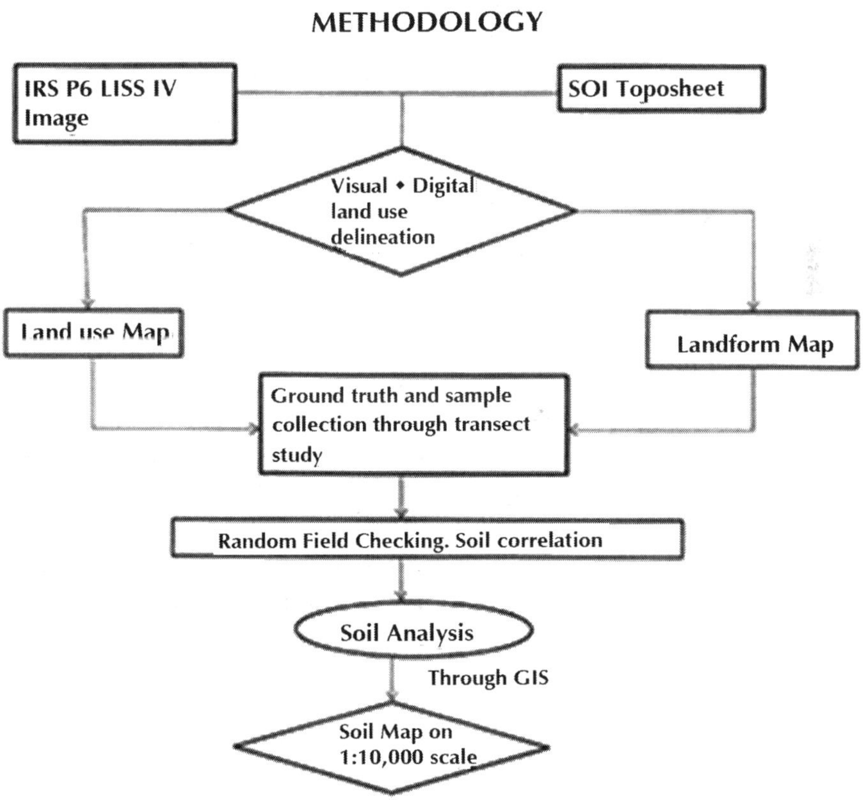

FIGURE 7.2 Flow chart of the methodology.

inventory. On the basis of preliminary interpretation, sample areas covering all the mapping units were selected in the form of sample strips of suitable size. Selected sample strips were transferred on to the base map for ground truth collection. After these, the survey work was undertaken with the objective to study soils under natural conditions and to prepare a mapping legend based on soil properties.

Detailed study of soils was carried out (1:10,000 scale) depending on the variation in land use, slope, salinity, etc. The auguring and minipits were done at shorter intervals to demarcate the variability of soils in the coastal alluvial plain. Depending on the variability of soil under different landforms supported by particular land use, the profiles were exposed up to a depth of 150 cm and the soil morphological features were studied following the standard procedure of All India Soil and Land Use Survey (1970). The soil profiles collected from different landforms under different land use were correlated to identify the soil series within the block. The horizon wise collected soil samples were analyzed for various parameters *viz*., particle size distribution, pH, OC, electrical conductivity (ECe), cation exchange capacity (CEC), exchangeable bases, base saturation, available macro and micro nutrients following standard methodology (Black, 1965; Jackson, 1973). The correlated soils were classified according to Soil Taxonomy (Soil Survey Staff, 2010) and were mapped in 1:10,000 scale using soil series with phases as the mapping unit. Using the available profile wise soil information (soil morphological and physico-chemical properties), attributes/non-spatial data were compiled for all the 18 mapping units in the block. This attribute data was linked to the spatial data. These were subsequently added to the Polygon Attribute Table (PAT) of soil coverage by following map generalization procedures. These were subsequently made into individual covers using 'Dissolve Operation' thus forming different derivative layers of the study area. Thus the thematic maps were prepared based on the soils of the mapped unit. Thematic maps of soil depth, erosion, drainage, land capability, land irrigability, soil suitability for different crops, etc., were prepared.

Soil-site Suitability Evaluation of Crops: The Food and Agriculture Organization (FAO) panel for land evaluation suggested the classification of land for specific use. The classification itself is presented in different categories; orders (S for suitable and N for unsuitable land). There are three classes (S1 to S3) under the orders S and two classes (N1 to N2) under the order N. The appraisal of the classes within an order is done according to the evaluation of land limitations or the main kinds of improvement measures required within classes. The limitations are climate (c), topography (t), wetness (w), salinity (n), soil fertility (f) and physical soil limitations (s). They are indicated by the symbol for example S2w; where w indicates wetness limitation (subclass). The limitation approach has successfully been used to provide a qualitative land evaluation based on general characteristics which are made available after a quality soil survey and a general study of other resources. Sys et al. (1993) proposed a scheme for evaluating the degree of limitation rating with the following general observations.

0 (No limitation)	Optimum crop growth
1 (Slight limitation)	Nearly optimum for crop growth
2 (Moderate limitation)	Moderate influence on crop growth decline
3 (Severe limitation)	Uneconomical for the suggested land use
4 (Very severe limitation)	Yield below profitable level not fit for suggested land use.

The soil-site parameters considered for the purpose of evaluating land for general agricultural crops, horticultural crops, other plantation crops and forestry in Deshapran block are as follows. These are considered for defining suitable classes.

Soil-site characteristics	Related Land, Soil and Crop Quality
• Climate (c)	• Available moisture
– Rainfall	
– Temperature	
• Topography and landscape (t)	• Resistance to erosion
– Slope	• Landscape position
• Wetness (w)	• Available moisture/Soil aeration
– Drainage	• Landscape position
– Flooding	• Deficiency/Toxicity of nutrients
• Physical condition (s) of soil	
– Texture	• Water availability/Soil aeration/Soil structure
– Depth	• Available space for root development
– Ground water table	• Landscape position
	• Available moisture/Soil aeration.
• Soil fertility (f)	
– pH (Soil reaction)	• Availability of plant nutrients
– Silt and clay content	• Availability of moisture nutrients
– Cation exchange capacity and base saturation	• Availability of plant nutrients
– Organic matter	• Soil structure/Availability of plant nutrients

7.4 RESULTS AND DISCUSSION

Detailed soil resource inventorization using high resolution satellite imagery helped in identifying six soil series in this block and was mapped into 18 soil mapping units using soil series with phases as the mapping unit (Figure 7.3).

(a) Soils of very gently sloping coastal alluvial plain under *kharif* paddy cultivation

Very deep, dark yellowish brown (10YR 4/6) to brown (10 YR 4/3) at the surface and very dark gray (10 YR 3/2) to dark yellowish brown (10 YR 4/3) at the sub-surface to subsoil (Table 7.1), imperfectly to moderately well drained, silt loam to silty clay in texture, neutral in reaction, low in OC, low in CEC and medium to high in base status (Tables 7.2 and 7.3) except in Sarda series which is sandy loam to sandy clay loam in texture and very strongly acidic to neutral in reaction. These soils contain higher exchangeable Ca^{+2} followed by Mg^{+2}, Na^{+} and K^{+}. The soil solution contains higher exchangeable Na^{+} in comparison to exchangeable Ca^{+2}, Mg^{+2} and K^{+} and are present mostly in the form of chlorides. These soils have ECe values ranging from <2 to 8 dS m^{-1}, ESP < 15% and pH < 8.5 indicating non-saline to moderately saline soils. The soils are mostly cultivated for *kharif* paddy only. The scarcity of ground water restricts the cultivation in the post monsoon period. The Shunia series (Fine, mixed, hyperthermic, *Typic Halaquepts*) and Mahammadpur series (Fine, mixed, hyperthermic, *Aeric Halaquepts*) are saline and Sarda series (Coarse-loamy, mixed, hyperthermic, *Aquic Haplustepts*) mostly represents non-saline soils.

(b) Soils of very gently sloping coastal alluvial plain under *kharif* paddy and vegetables

Very deep, dark yellowish brown (10YR 3/4) at the surface and dark brown (10 YR 3/3) to dark yellowish brown (10 YR 4/3) at the sub-surface to subsoil, moderately well drained, silt loam to silty clay loam in texture (Table 7.1), strongly acidic to slightly alkaline, low in OC, low in CEC and high base status (Tables 7.2 and 7.3). The soil solution contains higher exchangeable Na^{+} in comparison to exchangeable Ca^{+2}, Mg^{+2} and K^{+} mostly in the form of chlorides. These are moderately saline soils having pH less than 8.5 and exchangeable sodium percentage (ESP) more than 15 with ECe values 4 to 8 dS m^{-1}. The soils are mostly cultivated for *kharif* paddy. After paddy, vegetables are cultivated in these soils. The Kanai Chatta series

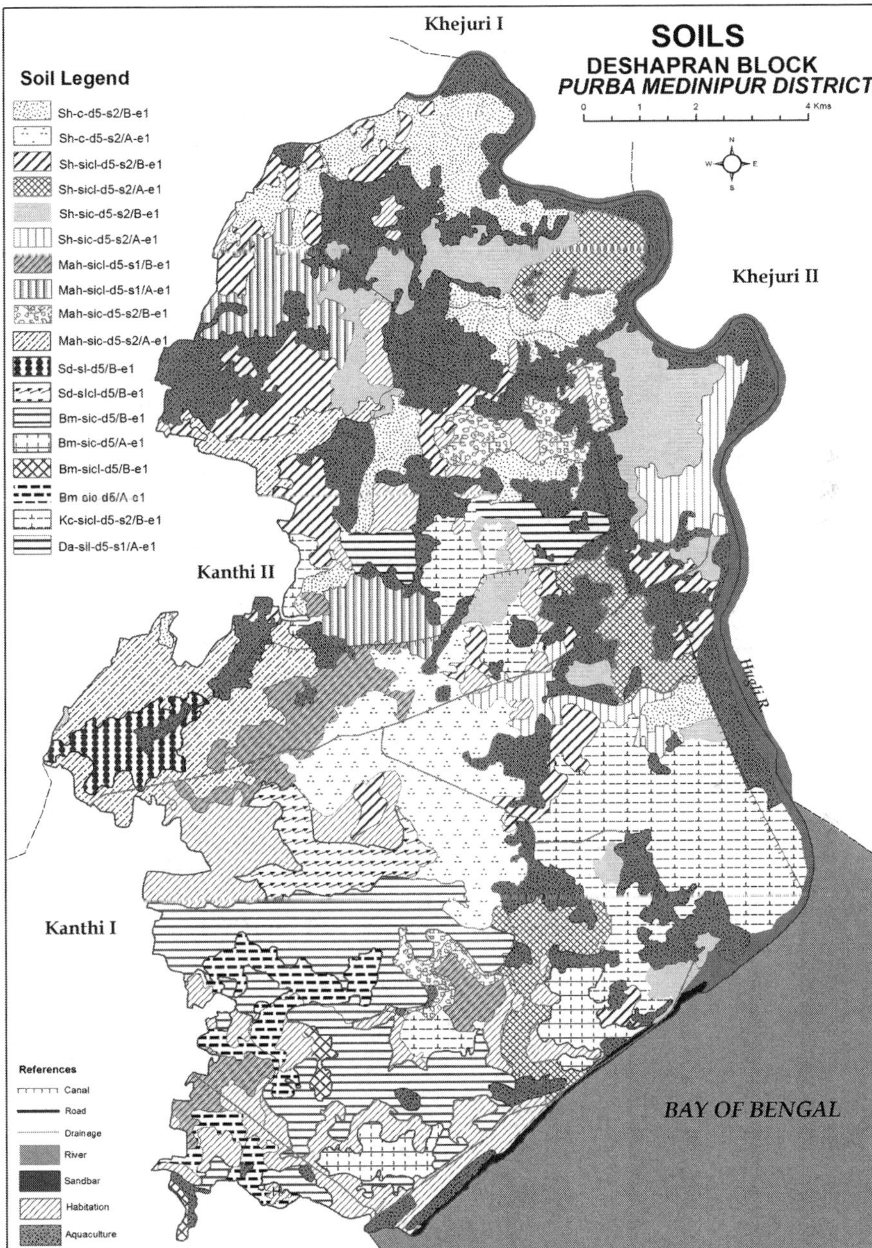

FIGURE 7.3 Soil map of the study area.

TABLE 7.1 Morphological Properties of the Soils

Horizon	Depth (m)	Matrix color (m)	Mottle color (m)	Texture	Structure	Consistency (m) (w)	Boundary	Existing Land use
Pedon-1. Shunia Series								
Ap	0 –0.14	10 YR 4/6	-	c	Puddled	h fi s p	g s	Paddy-fallow
Bw1	0.14–0.50	10 YR 4/3	-	c	m2sbk	- fi s p	g s	
Bw2	0.50–0.86	10 YR 4/2	10 YR 5/6	sic	m2sbk	- fi s p	g s	
2BCg1	0.86–1.20	10 YR 4/2	10 YR 6/8	sil	m2sbk	- fi s p	c s	
2BCg2	1.20–1.50	10 YR 4/3	10 YR 5/6	sic	m2sbk	- fi s p		
Pedon 2. Mahammadpur Series								
Ap	0–0.13	10 YR 3/3	-	sic	Puddled	h fi s p	g s	Paddy-fallow
Bw	0.13–40	10 YR 3/3	-	sic	c3sbk	- fi s p	g s	
Bg1	0.40–0.85	10 YR 3/3	10 YR 4/6	sic	m2sbk	- fi s p	c s	
Bg2	0.85–1.10	2.5 Y 3/2	10 YR 4/6	sic	m2sbk	- fi v s v p	g s	
BCg	1.10–1.50	2.5 Y 3/2	10 YR 4/6	sic	m2sbk	- fi vs v p		
Pedon 3. Sarda Series								
Ap	0–0.11	10 YR 4/3	-	sl	Puddled	h fr s sps	g s	Paddy-Paddy
Bw1	0.11–0.35	2.5 Y 3/2	10 YR 4/6	sl	m2sbk	- fr s sps	g s	
Bw2	0.35–0.75	2.5 Y 4/4	10 YR 4/6	sl	m2sbk	- frssps	c s	
Bw3	0.75–1.10	2.5 Y 4/4	10 YR 4/6	scl	m2sbk	- fr s p	g s	
Bw4	1.10–1.50	2.5 Y 4/2	10 YR 4/6	scl	m2sbk	- fr s p		
Pedon 4. Bamunia Series								
Ap	0–0.08	10 YR 5/2	-	sic	Puddled	h fi s p	c s	Paddy-Paddy

Bg1	0.08–0.25	10 YR 3/1	10 YR 4/6	sic	c3sbk	- fi s p	g s	
Bg2	0.25–0.65	10 YR 4/1	10 YR 4/6	sic	m2sbk	- fi s p	g s	
Bg3	0.65–1.00	10 YR 5/1	10 YR 4/6	sic	m2sbk	- fi v s v p	g s	
Bg4	1.00–1.25	10 YR 5/1	10 YR 4/6	sic	m2sbk	- fi vs v p		
Pedon 5. Kanai Chatta Series								
Ap	0–0.09	10 YR 3/4	-	sicl	Puddled	shfrssps	g s	Paddy-Vegetables
Bw1	0.09 –0. 35	10 YR 3/3	-	sicl	m2sbk	- fr s p	g s	
Bw2	0.35–0.70	10 YR 3/3	-	sicl	m2sbk	- fr s p	g s	
BC1	0.70–1.10	10 YR 3/3	10 YR 4/6	sicl	m2sbk	- fi s p	g s	
BC2	1.10–1.50	10 YR 3/3	10 YR 4/6	sicl	m2sbk	- fi s p		
Pedon 6. Dakshin Amtalia Series								
Ap	0–0.18	2.5 Y 5/4	10 YR 5/6	sil	Puddled	shfrssps	g s	Paddy-Paddy
Bw1	0.18–0.40	10 YR 4/3	10 YR 5/6	sicl	m2sbk	- fr s p	g s	
Bw2g	0.40–0.70	10 YR 4/2	10 YR 5/6	sicl	m2sbk	- fr s p	c s	
Bw3g	0.70–0.98	10 YR 5/1	10 YR 5/8	sicl	m2sbk	- fi s p	g s	
BC1g	0.98–1.25	10 YR 5/1	10 YR 5/8	sicl	m2sbk	- fi s p		
BC2g	1.25–1.50	10 YR 4/2	-	sil	m2sbk	- fr s p		

Note: Texture: sl – sandy loam, scl – sandy clay loam, sicl – silty clay loam, sic – silty clay, c – clay; Structure: m2 sbk – moderate medium sub angular blocky, c3sbk – strong coarse sub angular blocky; Consistency (moist): fr – friable, fi – firm, Consistency (wet):, ssps – slightly sticky and slightly plastic, s p – sticky and plastic; vsvp – very sticky very plastic; Boundary: cs – clear smooth, gs – gradual smooth.

TABLE 7.2 Physical and Chemical Characteristics of Soils

Horizon	Depth (m)	Sand (%)	Silt (%)	Clay (%)	pH H_2O	ECe (dSm^{-1})	OC (%)
Pedon-1. Shunia Series. Fine, mixed, hyperthermic, Typic Halaquepts							
Ap	0–0. 14	0.9	37.8	61.3	6.6	0.88	0.63
Bw	0.14–0.50	0.9	34.1	65.0	7.2	2.13	0.36
Bg1	0.50–0.86	1.1	46.4	52.5	7.2	4.66	0.33
2BCg1	0.86–1.20	1.9	71.1	26.4	7.2	5.25	0.28
2BCg2	1.20–1.50	1.4	52.6	46.0	7.2	7.37	0.20
Pedon 2. Mahammadpur Series. Fine, mixed, hyperthermic, Aeric Halaquepts							
Ap	0–0.13	2.3	53.0	44.7	6.3	0.63	0.53
Bw	0.13–40	0.7	54.1	45.2	7.3	0.64	0.27
Bg1	0.40–0.85	0.3	46.9	52.8	7.2	1.91	0.27
Bg2	0.85–1.10	0.4	46.4	53.2	6.9	2.98	0.32
BCg	1.10–1.50	0.9	58.2	40.9	6.7	4.33	0.30
Pedon 3. Sarda Series. Coarse-loamy, mixed, hyperthermic, Aquic Haplustepts							
Ap	0–0.11	73.9	16.9	9.2	4.8	0.44	0.27
Bw1	0.11–0.35	61.3	23.5	15.2	5.9	0.11	0.26
Bw2	0.35–0.75	61.7	23.1	15..2	6.6	0.08	0.03
Bw3	0.75–1.10	55.0	24.6	20.4	7.0	0.09	0.01
Bw4	1.10–1.50	55.1	24.5	20.4	7.6	0.13	0.01
Pedon 4. Bamunia Series., Fine, mixed, hyperthermic, Fluvaquentic Endoaquepts							
Ap	0–0.08	0.8	53.8	45.4	7.5	0.63	1.33
Bg1	0.08–0.25	0.4	49.2	50.4	7.3	0.48	0.43
Bg2	0.25–0.65	1.2	53.5	45.3	7.4	0.90	0.22
Bg3	0.65–1.00	0.9	55.6	43.5	7.2	1.34	0.36
Bg4	1.00–1.25	0.3	57.5	42.2	7.3	1.43	0.40
Pedon 5. Kanai Chatta Series. Fine-silty, mixed, hyperthermic, Typic Halaquepts							
Ap	0–0.09	3.6	66.1	30.3	5.0	1.47	0.93
Bw1	0.09–0. 35	1.5	60.2	38.3	7.7	1.57	0.51
Bw2	0.35–0.70	0.8	63.6	35.6	7.9	2.56	0.26
BC1	0.70–1.10	0.8	67.1	32.1	8.1	3.23	0.25
BC2	1.10–1.50	0.6	68.3	31.1	8.2	2.84	0.19
Pedon 6. Dakshin Amtalia Series. Fine-silty, mixed, hyperthermic, Aeric Halaquepts							
Ap	0–0.18	0.1	73.6	26.3	4.7	0.68	0.83
Bw1	0.18–0.40	0.5	63.0	36.5	7.8	0.42	0.32
Bw2g	0.40–0.70	0.2	65.9	33.9	7.9	0.74	0.30

TABLE 7.2 (Continued)

Horizon	Depth (m)	Sand (%)	Silt (%)	Clay (%)	pH H_2O	ECe (dSm^{-1})	OC (%)
Bw3g	0.70–0.98	0.4	65.9	33.7	7.8	1.26	0.25
BC1g	0.98–1.25	0.2	71.7	28.1	7.6	1.75	0.16
BC2g	1.25–1.50	0.4	73.8	25.8	7.1	2.42	0.13

(Fine-silty, mixed, hyperthermic, *Typic Halaquepts*) mostly represents these soils. Brackish water aquaculture is dominant in the study area.

(c) Soils of nearly level to very gently sloping coastal alluvial plain under *kharif* paddy and boro paddy

Very deep, grayish brown (10YR 5/2) to olive brown (2.5 Y3/4) at the surface and very dark gray (10 YR 3/1) to gray (10 YR 5/1) at the sub-surface to subsoil, imperfectly to poorly drained, silt loam to silty clay in texture at the surface and silty clay loam to silty clay at the subsurface to subsoil with characteristic redoximorphic features (Table 7.1), very strongly acidic to neutral, high in OC, low in CEC and medium to high base status (Tables 7.2 and 7.3). The soil solution contains higher exchangeable Na^+ in comparison to exchangeable Ca^{+2}, Mg^{+2} and K^+ mostly in the form of chlorides. These are slightly saline to non-saline soils having pH less than 8.5 and exchangeable sodium percentage (ESP) less than 15 with ECe values ranging from < 2 to 4 dS m^{-1}. The soils are mostly cultivated for *kharif* paddy as well as boro paddy due to the comparatively low lying situation and also the availability of pond/well water for irrigation. The Dakshin Amtalia (Fine-silty, mixed, hyperthermic, *Aeric Halaquepts*) and Bamunia (Fine, mixed, hyperthermic, *Fluvaquentic Endoaquepts*) soil series dominantly represent these soils.

The soils are mostly Inceptisols. They belong to 1 order, 2 suborders, 3 great groups, 4 sub groups and 6 soil families and series mapped into 18 mapping units. The brief description of soil series with phases is given in Table 7.4.

7.3.1 SLOPE

The slope map of Deshapran block area indicates that very gently sloping slope class covered maximum area of 7602 ha representing 42.7% of the

TABLE 7.3 Exchangeable Characteristics of Soils

Horizon	Depth (m)	Exchangeable bases [c mol (p^+) kg^{-1}]					CEC [cmol (p^+) kg^{-1}]	Base Saturation (%)	CEC/ Clay
		Ca^{+2}	Mg^{+2}	K^+	Na^+	Sum			
Pedon-1. Shunia Series. Fine, mixed, hyperthermic, Typic Halaquepts									
Ap	0 –0. 14	8.7	6.7	0.4	0.2	16.0	20.5	78	33
Bw	0.14–0.50	9.2	7.1	0.9	0.2	17.4	21.2	82	33
Bg1	0.50–0.86	6.1	3.0	1.6	0.7	11.4	14.2	80	27
2BCg1	0.86–1.20	3.2	2.6	1.3	0.7	7.8	10.4	75	39
2BCg2	1.20–1.50	6.3	4.6	0.6	0.4	11.9	15.2	78	33
Pedon 2. Mahammadpur Series. Fine, mixed, hyperthermic, Aeric Halaquepts									
Ap	0–0.13	5.4	3.2	0.9	0.4		9.9	14.1	70
Bw	0.13–40	6.3	3.9	1.1	0.6		11.9	15.8	75
Bg1	0.40–0.85	8.8	3.6	0.9	0.5		13.8	18.7	74
Bg2	0.85–1.10	8.1	4.4	1.0	0.8		14.3	19.2	74
BCg	1.10–1.50	7.0	3.6	1.1	0.6		12.3	17.1	72
Pedon 3. Sharada series Coarse-loamy, mixed, hyperthermic, Aquic Haplustepts									
Ap	0–0.11	1.1	0.5	0.2	0.1		1.9	3.4	56
Bw1	0.11–0.35	1.8	0.9	0.3	0.4		3.4	5.5	62
Bw2	0.35–0.75	1.6	0.6	0.8	0.6		3.6	5.3	68
Bw3	0.75–1.10	2.6	1.1	0.7	0.4		4.8	7.0	68
Bw4	1.10–1.50	1.9	0.9	1.0	0.8		4.6	6.6	70
Pedon 4. Bamunia Series. Fine, mixed, hyperthermic, Fluvaquentic Endoaquepts									

Ap	0–0.08	8.5	5.1	0.4	0.2	14.2	17.5	81
Bg1	0.08–0.25	9.7	5.4	0.4	0.2	15.7	19.1	82
Bg2	0.25–0.65	9.1	4.8	0.7	0.4	15.0	18.7	80
Bg3	0.65–1.00	7.8	4.9	0.6	0.2	13.5	16.3	80
Bg4	1.00–1.25	8.2	5.0	0.4	0.1	13.7	17.3	79
Pedon 5. Kanai Chatta Series. Fine-silty, mixed, hyperthermic, Typic Halaquepts								
Ap	0 – 0.09	6.1	3.8	0.3	0.2	10.4	13.1	79
Bw1	0.09 –0. 35	7.0	4.3	0.4	0.2	11.9	14.5	82
Bw2	0.35 – 0.70	6.2	3.8	0.4	0.3	10.7	13.1	82
BC1	0.70–1.10	5.7	3.5	0.5	0.2	9.9	12.2	81
BC2	1.10–1.50	5.5	4.0	0.4	0.3	10.2	12.7	80
Pedon 6. Dakshin Amtalia Series. Fine-silty, mixed, hyperthermic, Aeric Halaquepts								
Ap	0–0.18	4.1	3.9	0.4	0.2	8.6	14.8	58
Bw1	0.18–0.40	7.3	5.9	0.3	0.1	13.6	16.9	80
Bw2g	0.40–0.0	6.9	5.6	0.3	0.2	13.0	16.0	81
Bw3g	0.70–0.98	7.8	4.2	0.4	0.1	12.5	15.8	79
BC1g	0.98–1.25	4.8	3.9	1.1	0.1	9.9	13.[illegible]	75
BC2g	1.25–1.50	5.1	2.7	0.4	0.2	8.4	11.7	72

TABLE 7.4 Brief Description of Soils

Mapping unit	Soil series	Mapping symbol/ mapping Unit	Brief description	Area (ha)	Percent of TGA
1	Shunia	Sh-c–d5-s2/ B – e1	Soils of Shunia series are very deep, imperfectly drained, fine silty soils in control section with clay surface texture on very gently sloping alluvial plain with slight erosion and moderate soil salinity.	1,112	6.3
2		Sh–c–d5-s2/ A – e1	Same as Shunia series developed on nearly level alluvial plain.	996	5.6
3		Sh-sicl-d5-s2/ B – e1	Same as Shunia series with silty clay loam surface texture.	983	5.5
4		Sh-sicl-d5-s2/ A – e1	Same as Shunia series with silty clay loam surface texture and developed on nearly level alluvial plain.	723	4.1
5		Sh-sic–d5-s2/ B – e1	Same as Shunia series with silty clay surface texture.	1,035	5.8
6		Sh-sic-d5-s2/ A – e1	Same as Shunia series with silty clay surface texture and developed on nearly level alluvial plain.	434	2.5
7	Mahammadpur	Mah-sicl-d5-s1/ B – e1	Soils of Mahammadpur series are very deep, imperfectly drained, fine soils in control section with silty clay loam surface texture on very gently sloping alluvial plain with slight erosion and slight to moderate soil salinity.	579	3.3
8		Mah-sicl–d5-s1/ A – e1	Same as Mahammadpur series developed on nearly level alluvial plain.	488	2.7
9		Mah-sic-d5-s2/ B – e1	Same as Mahammadpur series with silty clay surface texture.	295	1.7

10		Mah-sic-d5-s2/ A – e1	Same as Mahammadpur series with silty clay surface texture and developed on nearly level alluvial plain.	820	4.6
11	Sarda	Sd – sl – d5/ B – e1	Soils of Sarda series are very deep, moderately well drained, coarse loamy soils in control section with sandy loam surface texture on very gently sloping alluvial plain with slight erosion.	220	1.2
12		Sd – scl – d5/ B – e1	Same as Sarda series with sandy clay loam surface texture.	303	1.7
13	Bamunia	Bm-sic-d5/ B – e1	Soils of Bamunia series are very deep, poorly drained, fine soils in control section with silty clay surface texture on nearly level alluvial plain with slight erosion.	291	1.6
14		Bm-sic-d5/ A – e1	Same as Bamunia series developed on nearly level alluvial plain.	50	0.3
15		Bm-sicl-d5-/ B – e1	Same as Bamunia series with silty clay loam surface texture.	416	2.3
16		Bm-sicl-d5-/ A – e1	Same as Bamunia series with silty clay loam surface texture and developed on nearly level alluvial plain.	1605	9.0
17	Kanai Chatta	Kc-sicl-d5-s2/ B – e1	Soils of Kanai Chatta series are very deep, imperfectly drained, fine loamy soils in control section with silty clay loam surface texture on very gently sloping alluvial plain with slight erosion and moderate salinity.	1419	8.0
18	Dakshin Amtalia	Da-sil-d5-s1/ A – e1	Soils of Dakshin Amtalia series are very deep, poorly drained, fine silty soils in control section with silt loam surface texture on nearly level alluvial plain with slight erosion and slight salinity.	233	1.3
	Misc.			5788	32.5
Total				17790	100.00

total geographical area under study. The nearly level slope class covered 4400 ha accounting for 24.7% area of the block.

7.3.2 SURFACE TEXTURE

The soils of the block have been grouped into six textural classes. Silty clay loam texture occupied maximum area (37.4%) of the block followed by silty clay (14.0%), clay (11.8%), sandy clay loam (1.7%), silt loam (1.3%) and sandy loam (1.2%). Due to dominance of silty clay loam, silty clay and clay, soils are mostly imperfectly to poorly drained having very slow permeability.

7.3.3 DRAINAGE

Soils of Deshapran block, Purba Medinipur district, West Bengal were grouped into three drainage classes *i.e.*, moderately well, imperfect and poor. The study indicates that imperfect drainage covers maximum area (49.9%) followed by poor drainage (14.6%) and moderately well (2.9%). The drainage class is supported by the textural class of the soils (Ray et al., 1997).

7.3.4 SOIL REACTION (PH)

The data on soils of Deshapran block (Table 7.2) indicates that neutral soil reaction class occupies major area (55.2%) followed by strongly acidic (10.9%) and moderately acidic (1.3%) (Figure 7.4). The reducing environment due to the saturation of soil with water for quite a long time supports the neutral pH of the soils. However, the occurrence of strongly acidic soils may be due to the ferrolysis, which replaces bases from the exchange complex with iron under reducing condition (Bandopadhyay and Maji, 1995).

7.3.5 ORGANIC CARBON

OC status in the soils of Deshapran block (Table 7.2) revealed that the soils with medium OC (0.5 to 0.75%) occupied 39.0% area of the block followed by the soils with low OC (<0.5%) covering 28.5% area (Figure 7.5).

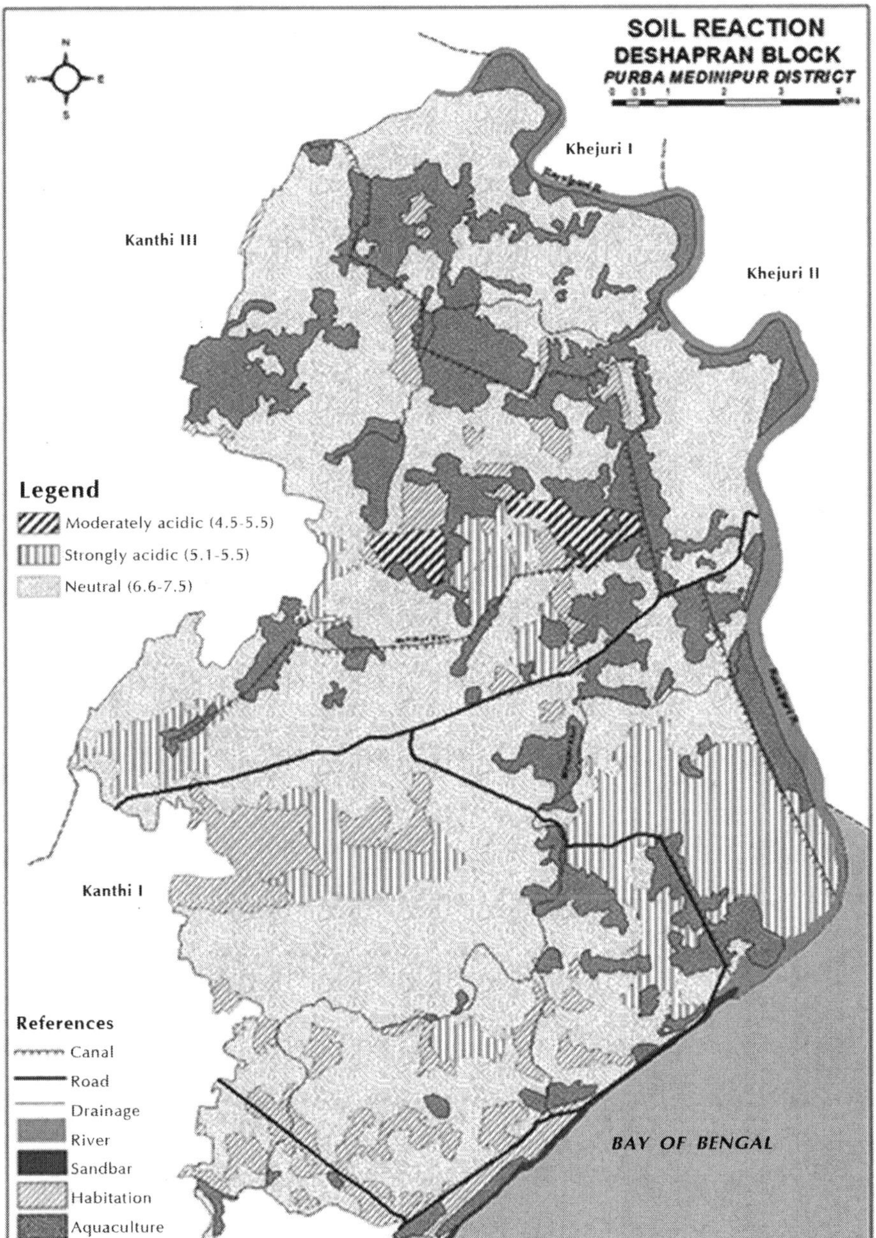

FIGURE 7.4 Soil reaction of the soils in study area.

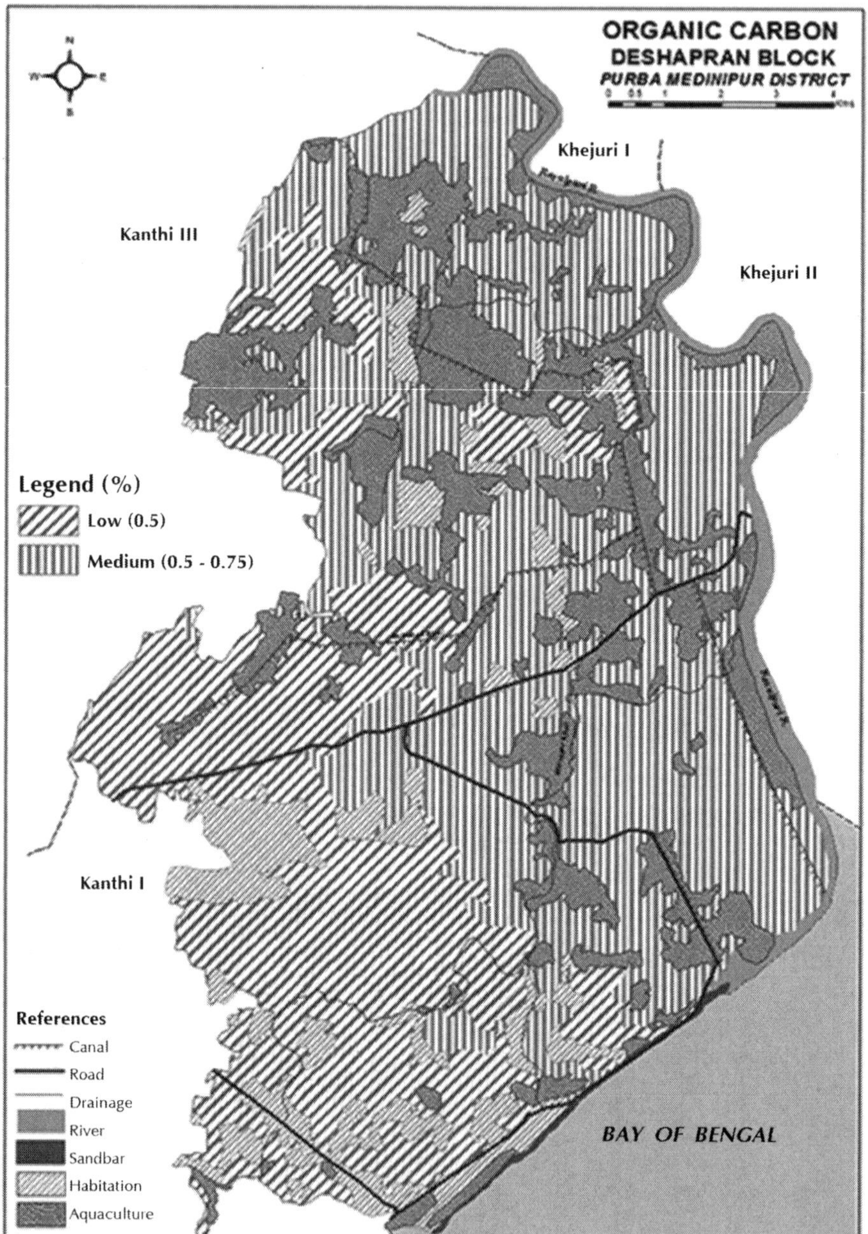

FIGURE 7.5 Organic carbon status of the soils.

7.3.6 SOIL SALINITY

Soil-water salinity along with high scarcity of good quality of irrigation water is the main constraint for intensification of crop in coastal regions besides the climatic constraints. Soil salinity is caused primarily by the tidal effects and ingress of sea water mainly through estuaries, creeks, deltas and flat lands (Bandopadhyay et al., 2003). Majority of salt-affected soils in the humid and sub-humid coastal regions are saline and are formed due to capillary rise of brackish water from ground water table present at shallow depth and sea water intrusion (Bandopadhyay et al., 2014). Brackish water aquaculture is widely practiced in this region through intrusion of sea water and after the completion of the process, the whole water is thrown away to make it dry for the next season which is the cause for increasing soil salinity affecting most of the fertile soils of this area. In Deshapran block, soils vary from non-saline to moderately saline. The result showed that only 2.9% area under Sarda and Bamunia series are non-saline in nature. About 37.7% area under Shunia, Kanai Chatta and Mahammadpur series (part) is moderately saline (ECe 4–8 dSm^{-1}) and 26.9% area under, Dakshin Amtalia is slightly saline (ECe 2–4 dSm^{-1}) (Figure 7.6).

7.3.7 SOIL FERTILITY

7.3.7.1 Macro Nutrients

Soils are generally low in fertility status. Available N content varies from 192 kg ha^{-1} (low) to 361kg ha^{-1} (medium) in these soils. Soils of Mahammadpur series, Sarda series and Kanai Chatta series are low in available nitrogen while Shunia, Bamunia and Dakshin Amtalia series are medium in available nitrogen. Available phosphorus content of these soils are low to medium and it varies from 1.6 to 27.4 kg ha^{-1}. Available potassium content of these soils varies from low to high. Sarda series is low in available K (90 kg ha^{-1}) and all the other soils are high in Potassium.

7.3.7.2 Micro Nutrients

The soils are high in available iron (84 to 208 mg kg^{-1}) and manganese (1.8 to 71.3mg kg^{-1}). Available zinc content of these soils ranges from 0.93 to 2.06

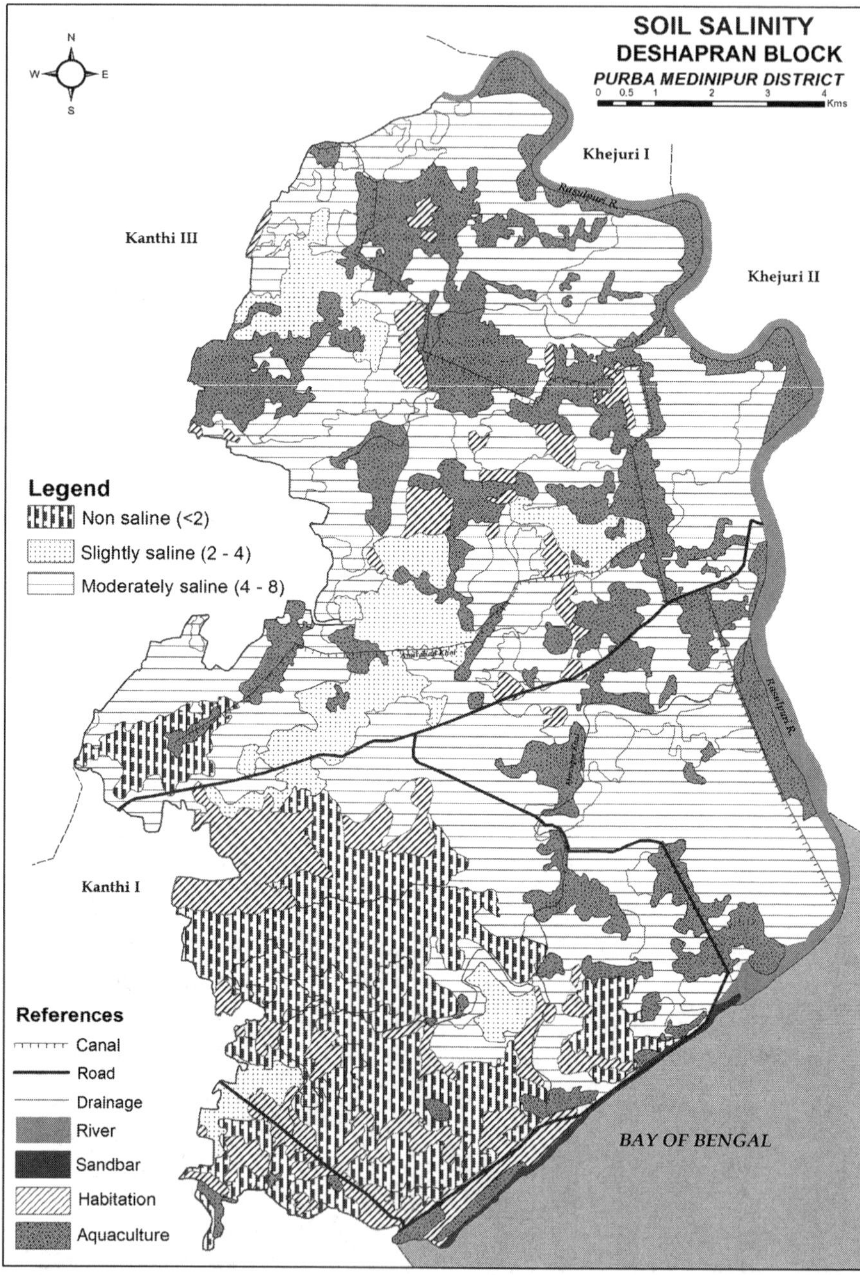

FIGURE 7.6 Soil salinity in soils.

mg kg^{-1} and it ranges from adequate to high. Shunia, Sarda and Bamunia soils are adequate in available zinc while Mahammadpur, Kanai Chatta and Dakshin Amtalia soils are high in available zinc. Available copper content of these soils varies from 1.8 (adequate) to 5.6 (high). Besides Sarda series which is adequate in copper (1.8 mg kg^{-1}), all the other soils are high in available copper.

7.3.8 LAND CAPABILITY CLASSIFICATION

The evaluation of land capability groupings, in these soils indicated that good cultivated soils with moderate soil limitation (IIs) occupies 2.9% area of the block while good cultivated soils with moderate limitation of soil and water logging (IIsw) covers 49.9% area. Moderately good cultivated soils with severe limitation of water logging (IIIw) occupy 14.6% area of the block.

7.3.9 LAND IRRIGABILITY CLASSIFICATION

The land irrigability groupings of the soils inferred that about 52.9% area of the block (2d) may be used safely under irrigation. About 14.6% area of the block has nearly level slope (3d) and it can be used for irrigated agriculture provided proper soil and water conservation measures are adopted.

7.4 SOIL-SITE SUITABILITY OF CROPS

Soils of the block were evaluated for suitability of different crops grown in that area. Paddy is the only *kharif* crop grown extensively in this area as it is the staple food of this place. Along with paddy, vegetables can be grown in the *kharif* season on the bunds with minimum management practices. In the *rabi* season crops like beans, cowpea, sunflower, potato, etc., can be grown successfully along with life saving irrigation as it can tolerate droughts and somewhat soil salinity. The soils were interpreted for evaluation of growing suitable crops as per the criteria of Sys (1976) and Sys et al. (1993).

7.4.1 SOIL SITE SUITABILITY FOR BEANS (PHASEOLUS VULGARIS)

It is found that about 56.5% area of the block is marginally suitable while 10.9% area is moderately suitable for beans due to the limitation of soil texture, soil fertility and drainage (Figure 7.7).

7.4.2 SOIL SITE SUITABILITY FOR COWPEA (VIGNA UNGUICULATA)

About 49.9% area of the block is moderately suitable to grow cowpea while 2.9% area is marginally suitable due to the limitation of soil texture, soil fertility and drainage. About 14.6% area is permanently not suitable for cowpea due to water logging and heavy texture (Figure 7.8).

7.4.3 SOIL SITE SUITABILITY FOR SUNFLOWER (HELIANTHUS ANNUUS)

About 18.1% area of the block is marginally suitable to grow sunflower while 31.8% area is moderately suitable due to the limitation of soil texture, soil fertility and drainage. About 17.5% area is permanently not suitable for sunflower due to poor drainage and heavy texture (Figure 7.9).

7.4.4 SOIL SITE SUITABILITY FOR POTATO (SOLANUM TUBEROSUM)

About 23.8% area of the block is marginally suitable for potato while 10.9% area is moderately suitable due to the limitation of soil texture, soil fertility and drainage. About 32.7% area is permanently not suitable for potato due to poor drainage and heavy texture (Figure 7.10).

7.5 MAJOR CONSTRAINTS AND MANAGEMENT OF SOILS

The agricultural productivity of land in coastal areas is constrained by various soil and water related problems besides the climatic hazards. Natural disasters like cyclone, sea water intrusion, drought, etc., that visit the coastal

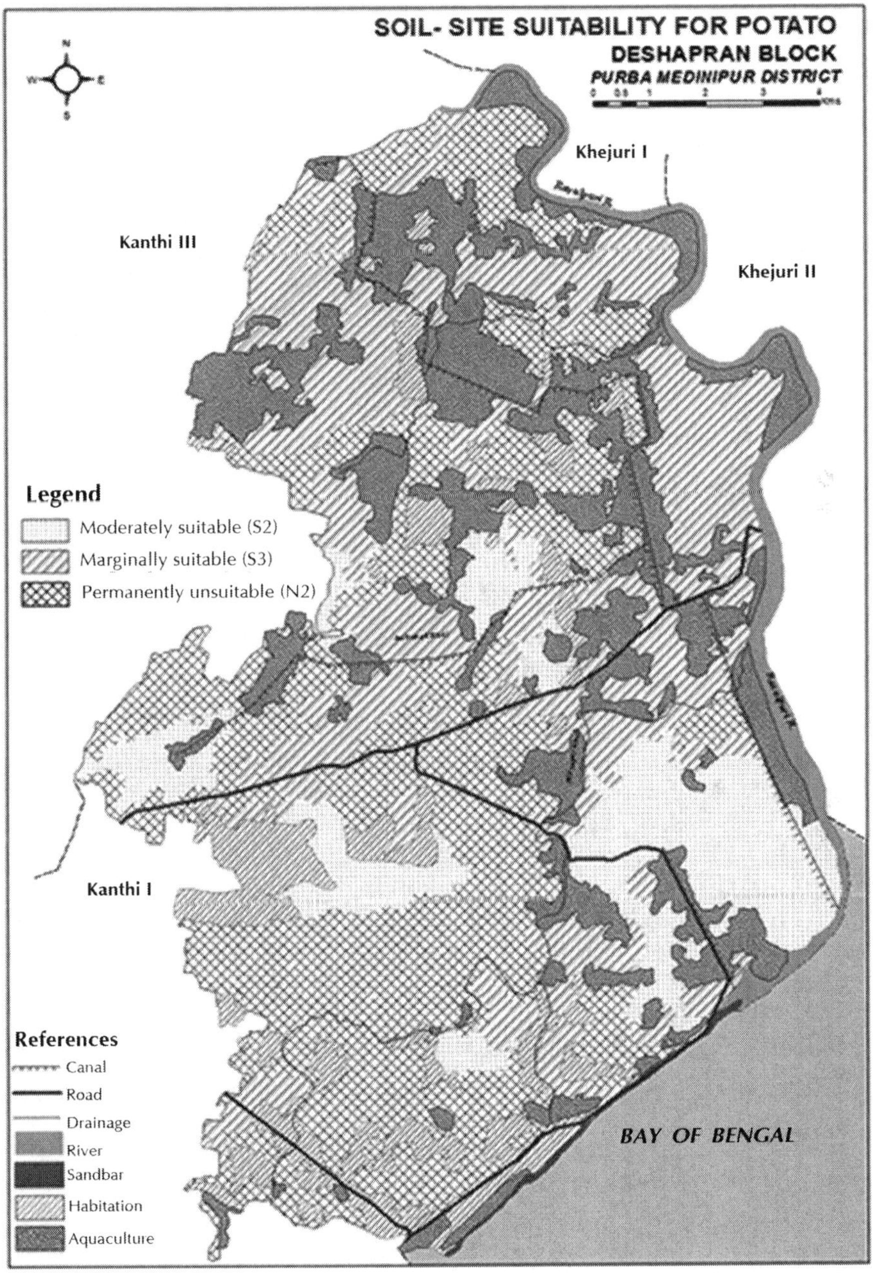

FIGURE 7.7 Soil site suitability for beans.

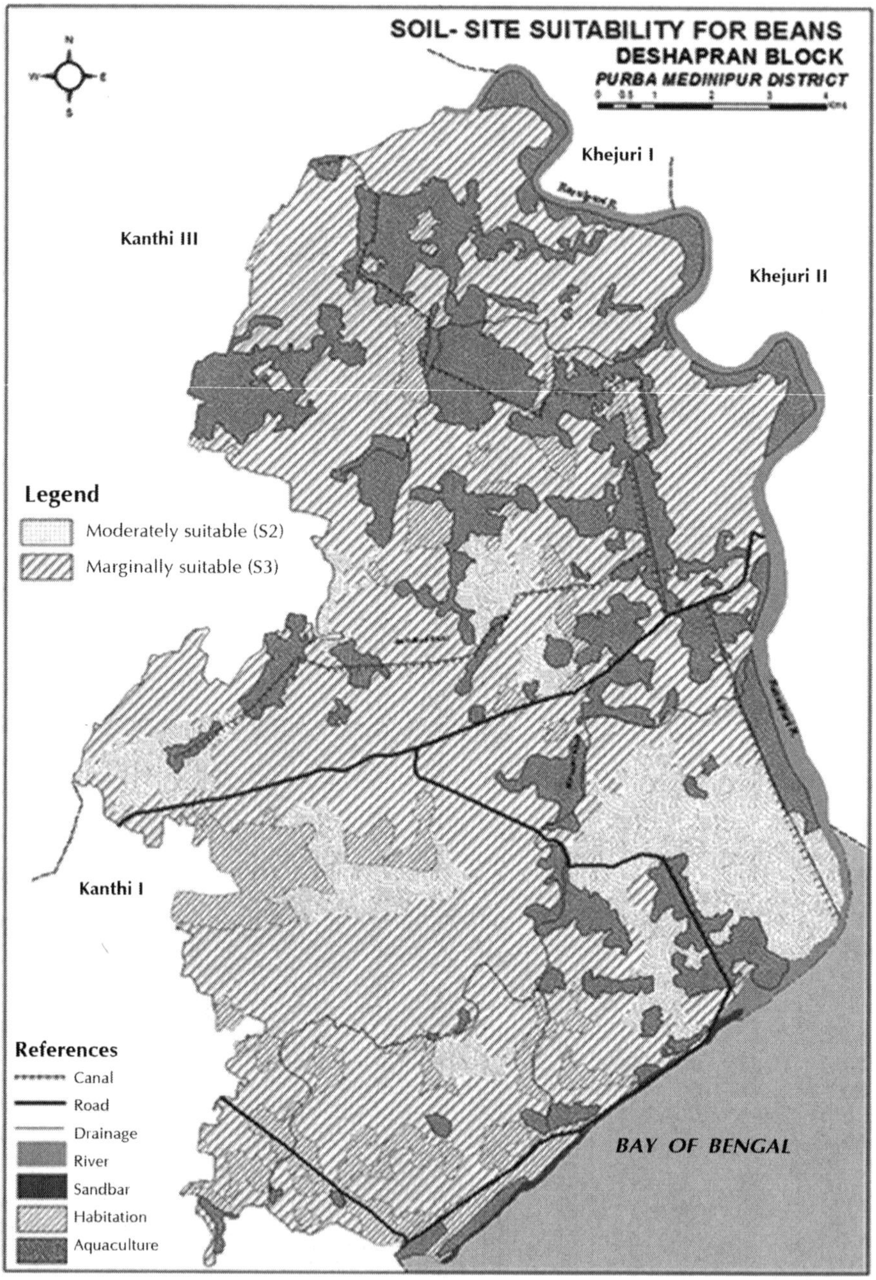

FIGURE 7.8 Soil site suitability for cowpea.

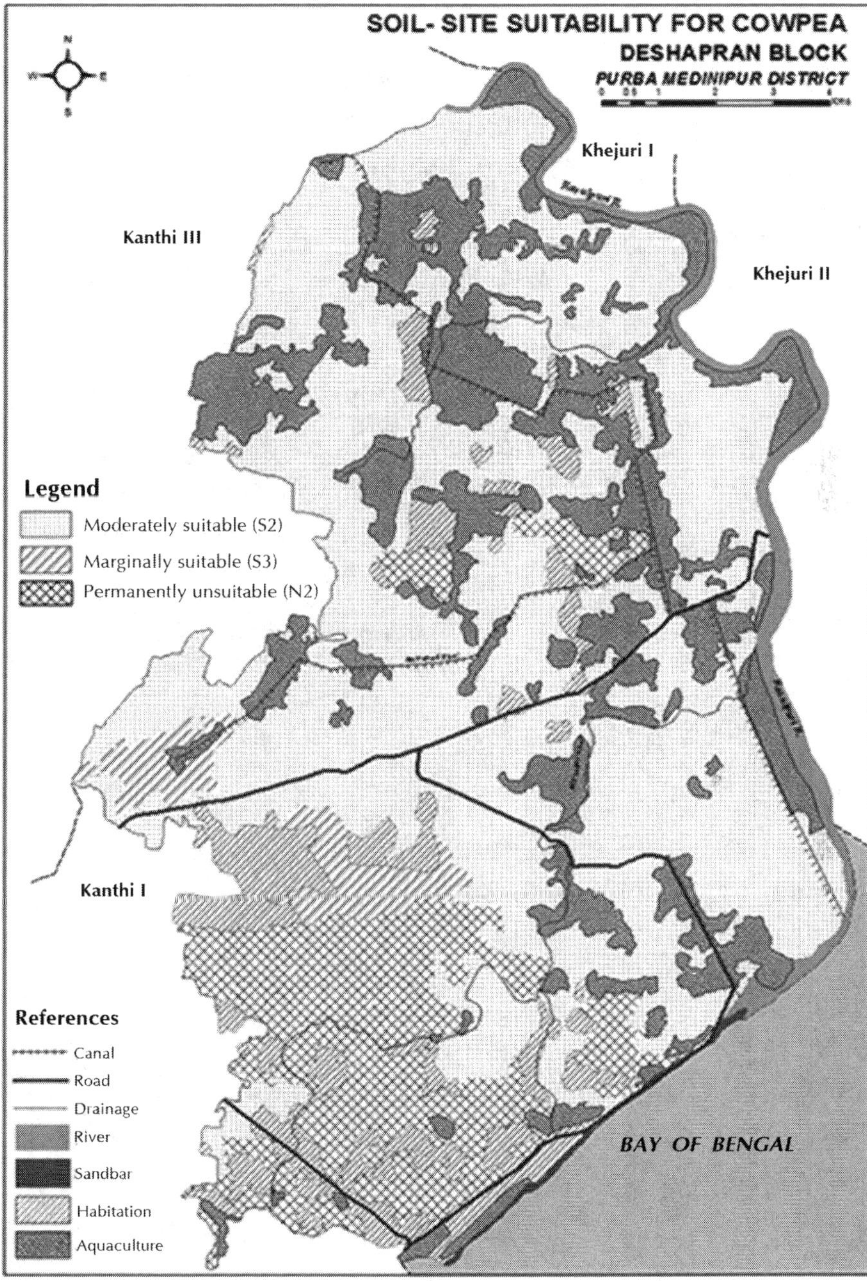

FIGURE 7.9 Soil site suitability for sunflower.

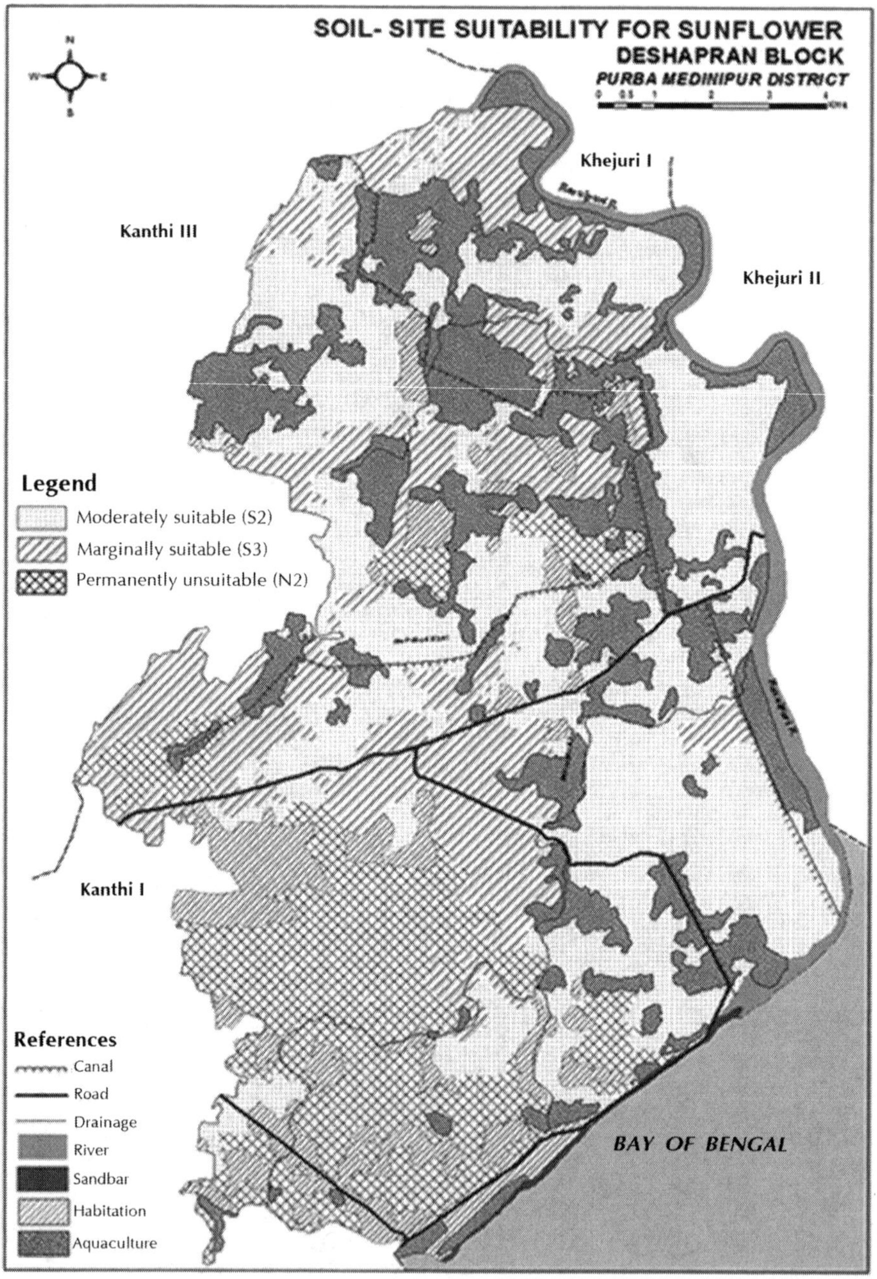

FIGURE 7.10 Soil site suitability for potato.

areas almost regularly cause colossal losses of crops, lives and properties. The area is entirely rainfed. The soil texture, drainage, flooding, scarcity of irrigation water and the brackish water aquaculture practices prevailing in this area lead to low productivity of crops.

Soil texture affects practically all the factors governing plant growth. Soil texture influences the movement and availability of soil moisture, aeration, nutrient availability and the resistance of the soil to root penetration (Bandopadhyay et al., 2014). The texture of most of the soils in this block varies from silty clay loam to silty clay. The heavy texture of soil of this block poses serious problem for tillage as they have an inappropriate consistency, not only when they are dry but also when they are wet. Soil compaction and poor structure reduces root penetration and good plant growth. Deep tillage to loosen any compacted or hardened massive layers and addition of organic manures improve the conditions.

The coastal cultivated lands are heavy textured and low lying. Besides this, presence of ground water table near the surface, flat topography, presence of compact plow sub-surface layer responsible for very poor drainage with poor permeability and infiltration capacity are the main hindrances for proper plant growth. High rainfall in the limited period causes medium to deep water logging of most of the cultivated fields of coastal region during rainy season. Due to poor internal drainage condition of land there is occasional total crop failure in *rabi* and summer season (except rice) following slightly excess rain or irrigation.

Drainage condition of the soil can be improved by installing surface and sub-surface drainage channels. The construction of alternate raised beds and furrows is another practice that will increase the drainage of condition of soil.

Low ground water level is one of the main constraints of this area. Salty irrigated water makes soils saline and thus reduces productivity. Acute scarcity of irrigation water during post monsoon periods is one of the great hindrances for improving cropping intensity and productivity of coastal lands. The rain water harvesting is the alternate source, which can alleviate the constraints to some extent. The rain water harvesting through land shaping may enhance the cropping intensity and diversity and productivity of monocropped lands to a great extent.

The introduction of salty brackish water for the profitable aquaculture practice in this area induces soil salinity affecting the highly fertile and productive agricultural land causing salinity in the long run with reduced productivity. Therefore, this profitable source of income should not be encouraged in the intensively cropped area.

7.6 STRATEGIES FOR IMPROVEMENT OF LIVELIHOOD SECURITY

Sustainable management of land and water resources along with integrated farming systems provide fairly better income and secured livelihoods to the farmers in the region. Some of the management practices that can be implemented are briefly described below (Table 7.5):

Leaching of the soil: The salinity level of salt-affected coastal soils can be reduced by leaching the soils with good quality water. This can be a better option to reclaim the cyclone affected soils of the Deshapran block. In the low-lying areas where water table remains shallow for most part of the year and the quality of ground water is poor, installation of sub-soil drainage system is more useful.

Green cover during summer: Most of the coastal areas suffer from excess water in monsoon season with problem of prolonged deep water submergence and adverse effect on crop growth. In winter and summer months, the capillary rise of the saline ground water compel the farmers to take only one rice crop in a year during the monsoon season. Introduction of second rice crop during the fallow periods, if good quality water for irrigation is available, can reduce the salinity level and increase the cropping intensity. The high salinity is due to the high evaporation rate from soil during winter and summer months, as the ground water is at shallow depth and rich in salt content. The availability of good quality irrigation water is one of the major problems in the area. However, if sufficient irrigation water of good quality is not available, a crop like chili (*Capsicum annuum*), barley (*Hordeum vulgare*), linseed (*Linum usitatissimum)*, sugar beet (*Beta vulgaris)* can be grown whose crop canopy will reduce evaporation and thus there will be reduction in soil salinity.

Application of soil amendments: Field experiments conducted at Coastal Saline Research Centre, Tamil Nadu Agricultural University, Ramanathapuram, Tamil Nadu (Ray et al., 2014) revealed that application of organic waste materials such as press mud @ 12.5 t ha^{-1} is beneficial for improving soil quality due to increased soil OC and nutrient availability and substantial reduction in the ECe of coastal saline soils. Amending coastal sandy soils with polyacrylamide @ 100–120 mg kg^{-1} is useful for increasing the aggregation of soil, which in turn increases the water holding capacity of coastal sandy soil. This plays important role in highly permeable coastal sandy soils during dry summer months.

TABLE 7.5 Livelihood Security Options Against Each Soil Mapping Unit of the Study Area

Soil Map units	Existing Land Use	Constraints	Suggested Land Use
1, 2, 3, 4, 5, 6, 7, 8, 9, 10	*Kharif* rice-fallow	– Heavy texture – Imperfect drainage – Slow permeability Low ground water level – Soil salinity	• Drought and salinity resistant short duration *Kharif* paddy may be cultivated. • Vegetables like pumpkin, cowpea, bitter gourd, snake gourd, etc., can be grown on the bunds of the water harvesting structures during *kharif* season. • Horticultural crops *viz.,* Banana, Papaya, etc., can be cultivated in the upland situation along the bunds. • Vegetables can be promoted in this area with the help of the life saving irrigation from the water harvesting structures during *rabi* season. • Bee rearing may be advocated.
11, 12	*Kharif* rice-fallow	– Coarse texture – Well to moderately well drainage – Moderate – Permeability – Low nutrient and water holding capacity	• Short duration high yielding paddy can be cultivated as *kharif* crop. • Vegetables like onion, potato, chili, pumpkin, etc., can be suitably cultivated. With the help of life saving irrigation. • Among the horticultural crops, water melon, papaya can be profitably cultivated • Sugarcane can also be cultivated as cash crop. • Groundnut can be grown profitably.
13, 14, 15, 16, 18	Paddy-paddy	– Heavy texture – Poor drainage – Slow permeability – Frequent flooding – High ground water level	• Short duration drought tolerant high yielding variety of paddy can be grown in the *kharif* season • Boro paddy may be replaced with vegetables, pulses and oil seeds. • Paddy cum pisciculture can be adopted.

TABLE 7.5 (Continued)

Soil Map units	Existing Land Use	Constraints	Suggested Land Use
17	Paddy-vegetables	– Fine-loamy textural class – Moderately well drainage – Slow permeability – Poor aeration – Occasional flooding – Soil salinity	• Drought and salinity tolerant paddy and vegetables in *kharif* season • Cultivation of pulses like green gram, bengal gram, lentil and oil seeds *viz.*, mustard, sesame with life saving irrigation during *rabi* season. • Vegetables like brinjal, chili, tomato, bitter gourd mustard, sesame, etc., can be cultivated in the *rabi* season. • Sugarcane can also be cultivated as cash crop. • Horticultural crops like banana, mango, papaya, guava, etc., can be successfully cultivated with the help of life saving irrigation. Floriculture may be taken up as a profitable cash crop in this area

Growing of suitable crops*:* Rice is the most preferable crop in the block since it is highly salt tolerant and can be grown under submerged conditions. Rice cultivation promotes the leaching of salts from soils. Selection of suitable rice variety depending upon the salinity level and depth of water regime is highly appreciable. Other than rice (*Oryza sativa),* chili (*Capsicum annuam*), guava (*Psidium guajava*) and sapota (*Manilkara zapota*) have been identified as salt tolerant vegetable and fruit crops for coastal saline soils. Growing of cashew in the coastal belt with proper irrigation and management practices may be beneficial.

Nutrient management*:* Most of the coastal soils are deficient in nitrogen due to heavy loss through volatilization, leaching and run-off. Phosphorus deficiency is also a common phenomenon in coastal saline soils. Use of nitrogenous fertilizers is very much essential to obtain higher yield of crop in coastal saline soils. Application of rock phosphate as phosphorus source is highly beneficial for coastal saline soils. Long-term fertilizer experiment (Ray et al., 2014) showed significant response of rice crop to application of nitrogenous fertilizers on coastal saline soils of West Bengal under rice-fallow cropping system. Integrated use of chemical fertilizers and farmyard manure (FYM) @15 t ha^{-1} is also a recommended practice for better use of fertilizer nutrients in coastal soils (Ray et al., 2014).

Paddy-cum-fish culture: For effective utilization of land resources and betterment of livelihood of the local people in this block, paddy-cum-fish culture can be adopted in lowland areas of without affecting the productivity of the soils. This may facilitate additional income generation to the farmers struggling for survival in the coastal regions and also uplift their socio-economic condition (Ray et al., 2014; Bandopadhyay et al., 2014).

7.7 CONCLUSIONS

This study indicates that the variability of rainfall, scarcity of irrigation water, droughts, poor drainage conditions, soil salinity, etc., are the major constraints which are responsible for low productivity in Deshapran block, Purba Medinipur district, West Bengal under coastal agro-ecosystem. Remote Sensing and GIS technology has been found to be immensely helpful for bringing out the detailed information of soil (1:10,000 scale) in respect of their extent, distribution, characteristics, problems and potential not only for management of soils for optimizing the land use but also to develop appropriate and sustainable production systems that deal with the problems of food security and livelihood improvement of the resource-poor farmers, however, institutional, financial and technical supports along with the implementation of the latest available technologies are essential.

KEYWORDS

- **Agro-Ecological Sub-Region**
- **Coastal Soils**
- **Deshapran**
- **GIS**
- **Land Use/land cover**
- **Livelihood Security**
- **Productivity**
- **Purba Medinipur**
- **Remote Sensing**
- **Soil and Water Conservation**
- **West Bengal**

REFERENCES

AIS & LUS. (1970). Soil Survey Manual. All India Survey Organization, I.A.R.I. New Delhi.

Bandopadhyay, A. K., & Bandhopadhyay, B. K. (1982). Effect of different crop rotations on salinity of coastal soils of Sundarbans (West Bengal). *Journal of the Indian Society of Soil Science, 30*, 242–244.

Bandopadhyay, B. K., & Maji, B. (1995). Nature of acid soils of Sundarban delta and suitability of classifying them as acid sulfate or potential sulfate soils. *Journal of the Indian Society of Soil Science, 43*, 251–255.

Bandopadhyay, B. K., Patra, P., & Sengupta, T. (2014). Management of degraded coastal lands for sustainable enhancement of productivity and livelihood security of farmers with special reference to coastal soils of Sundarbans, West Bengal. Presented in the one day seminar on 'Soil Management Options for Integrated Farming Towards Better Livelihood Security' Organized by Kolkata Chapter of the Indian Society of Soil Science, NBSS & LUP, Regional Centre, Kolkata at the Regional Centre, NBSS & LUP, Kolkata. pp. 50–57.

Bandopadhyay, B. K., Sen, H., & Maji, B. (1996). Transformation of sulfate salts and Black color formation in Coastal saline Soils of Sundarbans under paddy Cultivation. *Journal of the Indian Society of Coastal Agricultural Research, 14* (1&2), 39–45.

Bandopadhyay, B. K., Maji, B., Sen, H., & Tyagi, N. K. (2003). Coastal soils of West Bengal Their nature, distribution and characteristics. Bulletin No. 1/2003. Regional Research Station, Central Soil Salinity research Institute, Canning Town, south 24 Parganas, W.B., India, 62 p.

Black, C. A. (1965). '*Methods of Soil Analysis' Part 2. Chemical and Micro-biological properties* (*American Society of Agronomy*: Madison, Wisconsin, USA.

Das, T. H., Sarkar, D., Chatterjee, S, Dutta, D., & Sehgal, J. (1996). Evaluation of Soil-site suitability for different crops in Coastal saline areas of West Bengal. *Journal of the Indian Society of Coastal Agricultural Research, 14*(1&2), 31–34.

Dobos, E., & Hengle, T. (2009). Soil mapping applications developments in Soil Science. *33*, 431–179.

Jackson, M. L. (1973). *Soil Chemical Analysis*, Prentice Hall of India Pvt. Ltd., New Delhi.

Maji, B., Bandopadhyay, B. K., & Sen, H. (2001). Characterization of Coastal soils of North 24 Parganas District of West Bengal. *Journal of the Indian Society of Coastal Agricultural Research, 19* (1&2), 84–89.

Mandal, S., Bandhopadhyay, B. K., Burman, D., Sarangi, S. K., & Mahanta, K. K. (2011). Baseline report of the NAIP Project on 'Strategies for sustainable Management of Degraded Coastal Land and Water for Enhancing Livelihood Security of Farming Communities.' Central Soil Salinity Research Station, West Bengal, India, pp. 74.

Ray, P., Meena, B. L., & Nath, C. P. (2014). Management of coastal soils for improving soil quality and productivity. *Popular kheti, 2*(1), 95–99.

Ray, S. K., Reddy, R. S., & Budihal, S. L. (1997). Vertisols and associated soils development and lithologic discontinuity in coastal Godavari delta region. *Journal of the Indian Society of Coastal Agricultural Research, 15*(1), 1–14.

Ravisankar, T., & Thamappa, S. S. (2004). Satellite data interpretation and analysis for soil mapping. In: Soils and Crops. L. Venkataratnam, T. Ravisankar, & R. Sudarshana (Eds.), Hyderabad: NRSA Publications.

Rao, B. R. M., Fyzee, M. A., & Wadodkar, M. R. (2004). Utility of Remote sensing data for mapping soils at various scales and levels. In: Soils and Crops. L. Venkataratnam, T. Ravisankar and R. Sudarshana (Eds.), Hyderabad: NRSA Publications.

Soil Survey Staff. (2010). Keys to Soil Taxonomy. Eleventh Edition, USDA, Washington DC.

Sys, C. (1976). Land Evaluation (Part I, II and III). ITC, State University, Ghent, Belgium.

Sys, C., Vanranst, E., Debaveye, J., & Beernaert, F. (1993). Land Evaluation, Part III. Crop Requirements, Inter Training Centre for Post Graduate Soil Scientists, University, Ghent.

Velayutham, M., Sarkar, D., Reddy, R. S., Natarajan, A., Krishnan, P., Shiva Prasad, C. R., Challa, O., Harindranath, C. S., Shyampura, R. L., Sharma, J. P., & Bhattacharyya, T. (1999). Soil Resource and their potentialities in coastal areas of India. *Journal of the Indian Society of Coastal Agricultural Research, 17*, 29–47.

Wild, A. (2003). Soils, Land and Food: Managing the Land During the Twenty-First Century. Cambridge, U.K.: Cambridge University Press.

CHAPTER 8

ASSESSMENT OF LAND DEGRADATION VULNERABILITY: A CASE STUDY FROM PART OF WESTERN GHATS AND THE WEST COAST OF INDIA

K. S. ANIL KUMAR,[1] M. LALITHA,[2] SIDHARAM PATIL,[3] B. KALAISELVI,[2] K. M. NAIR,[1] and S. K. SINGH[4]

[1]*Principal Scientist, ICAR-National Bureau of Soil Survey and Land Use Planning, Regional Centre, Bangalore – 560 024, India*

[2]*Scientist, ICAR-National Bureau of Soil Survey and Land Use Planning, Regional Centre, Bangalore – 560 024, India*

[3]*Research Scholar, Department of Soil Science and Agricultural Chemistry, University of Agricultural Sciences, Bangalore – 560 065, India*

[4]*Director, ICAR-National Bureau of Soil Survey and Land Use Planning, Amravati Road, Nagpur – 440 033, India*

CONTENTS

ABSTRACT

A study was conducted in parts of Western Ghats and West Coast of Southern Karnataka to assess the land degradation status and vulnerability covering different physiographic units. Seven soil profiles representing different physiographic divisions *viz.*, steep hill ranges, steep low hill ranges, isolated hills and dissected hills and valleys, elongated ridges and foot hill slopes, undulating uplands, lateritic plateau and lateritic mounds, coastal plateau summits, valleys, bars and ridges and beaches and marshes (lower laterite terrace) were identified based on soil morphological, physical and chemical properties. Based on climate, terrain and soil characteristics, the land degradation status and vulnerability was evaluated and categorized into different vulnerability grades *viz.*, very low, low, medium, moderate and high. The status of degradation and vulnerability status assessed in the study area ranged from 0.45 to 0.83. Considering the scale of land degradation, undulating uplands have been placed at high level of land degradation (Molahalli) followed by coastal plateau summits and lower lateritic terrace with moderate land degradation (Murdeswar and Ullal), while steep high hills and low hills and their side slopes showed medium category of land degradation (Sullya and Belthangadi), whereas, elongated ridges and foot hill slopes were recorded as low (Brahmavar) and very low (Kollur), respectively. The simple model used to arrive at the degradation vulnerability can be replicable to hot high rainfall areas of humid tropics.

8.1 INTRODUCTION

Land degradation is one of the most obvious factors influencing the agriculture, economics, health and well being, ecosystem and hence livelihood of world population. The processes of land degradation have posed a worrisome threat to food security and it is linked with population growth and inappropriate land use options (Conacher, 2009) and results frequently from a mismatch between land quality and land use (Beinroth et al., 1994). The millennium ecosystem assessment refers land degradation as the reduction in the capacity of the land to perform ecosystem goods, functions and services that support society and development (Safriel and Adeel, 2005). The causes of land degradation are made up of natural hazards (steep slopes, impermeable soil and high intensities of rainfall), direct causes (unsuitable management practices) and underlying causes (cultivation on slopes by landless poor, non-adoption of conservation practices because of lack of security tenure) (Hegde et al., 2011). The United Nations Convention to Combat Desertification identified land degradation and desertification as one of the most pressing environmental concerns and called for a target of zero net land degradation whereby the rate of deteriorating lands would be counterbalanced by the rate of land improvement (UNCCD, 2002). Assessment of the nature and extent of land degradation using scientifically sound criteria, indicators and techniques will help to plan appropriate reclamation measures (CPC, 2004) and it will also help in determining the possible consequences (Ballayan, 2000). To arrest or prevent land degradation process one should improve the knowledge on causes and consequences of the interest phenomena and identify efficient monitoring tools (Grainger, 2014). In short, one must identify efficient tools for the detection of land degradation vulnerable areas by classifying them in different levels of land degradation vulnerability. At this aim many different methodologies have been used to study land degradation (field measurements, visual interpretation, social enquiries, mathematical models, remote sensing, environmental indicators, etc.), including the use of simple models based on indicators that synthesize information on the state and tendency of complex land degradation status and vulnerability processes (Vito et al., 2013). There are many methods used to assess land degradation *viz.*, expert opinions, land users opinions, field monitoring, observations and measurement, modeling, estimates of productivity changes and remote sensing (Kapalanga, 2008). But often these estimates of its extent and severity are highly unreliable and spurious because its results are not quantitatively replicable due to lack of baseline-measured data (Nicholson et al., 1998). The quoted statistics of 15% of the Earth's surface and 60% of dry lands are degraded (Oldeman, 1994) are acknowledged as qualitative and unsubstantiated data (Thomas and Middleton, 1994). Hence, it is hardly useful for policy making or for scientific investigations to remediate the degraded lands (Glenn et al., 1998). So there is a pressing need for accessible and accurate measurements on the extent of degradation and desertification for policy making, natural resource management and scientific research needs (Veron et al., 2006). The biophysical indicators particularly soil properties have very significant influence on the degradation rate and vulnerability potential (Onwudike, 2015). Land with better organic carbon, total nitrogen, available phosphorus, exchangeable Ca and Mg and base saturation are neither degraded nor vulnerable to degradation and therefore better soil quality indicators, while a land with low exchangeable K is extremely degraded and highly vulnerable to degradation and therefore a poor soil quality indicator (Mensah, 2015). Land with favorable texture, soil pH, exchangeable Na and effective *cation exchange capacity (*CEC) showed moderate rate of degradation and vulnerability and these might be good soil quality indicators (Amara and Momoh, 2014). This will help stakeholders in developing an effective land use plan and plan conservation measures according to the vulnerability class.

8.1.1 ASSESSMENT OF LAND DEGRADATION: GLOBAL PERSPECTIVE

Land degradation is an important global issue for the 21st century because of its adverse impact on agricultural productivity, the environment, and its effect on food security and the quality of life. Overall at present, a quarter of world population is threatened by the effects of degradation phenomena (Eswaran et al., 2001), which affect nearly 84% of agricultural lands (FAO, 2008a). In that case it is obvious that land degradation is listed among the most important socio-economic, environmental and ecological issues and cultural problems. There are different estimates on the extent and rate of land degradation based on different definitions and methodology hence there is a large variation in the available statistics. It varied from 3.6 billion ha (Dregne and Chou, 1994) to 1.9 billion ha (Oldeman, 1994) (Table 8.1). According to the European Commission, six soil degradation processes (water, wind and tillage erosion, loss of soil organic carbon, compaction, salinization and alkalinization, contamination, and decline in biodiversity) were identified as induced or worsened by bad agricultural practices (Gay et al., 2009). Many methods have been applied to assess degradation in different approaches, which use either qualitative or quantitative measures or both. Global Assessment of Land Degradation (GLASOD) is the only approach that has been applied on a worldwide scale, which is based

TABLE 8.1 Global Estimates of Soil Degradation (Scherr, 1999, courtesy of the International Food Policy Research Institute)

Region	Agricultural land			Permanent pasture			Forests		
	Total	Degraded	Percentage to total	Total	Degraded	Percentage to total	Total	Degraded	Percentage to total
	(million hectares)			(million hectares)			(million hectares)		
Africa	187	121	65	793	243	31	683	130	19
Asia	536	206	38	978	197	20	1273	344	27
South America	142	64	45	478	68	14	896	112	13
Central America	38	28	74	94	10	11	66	25	38
North America	236	63	26	274	29	11	621	4	1
Europe	287	72	25	156	54	35	353	92	26
Oceania	49	8	16	439	4	19	156	12	8
World	1475	562	38	3212	685	21	4048	719	18

on responses to a questionnaire, which was sent to recognized experts in countries around the world (Bridges and Oldeman, 1999). Soil Degradation in South and Southeast Asia (ASSOD) is another approach in which, the degree of soil degradation is expressed by degradation subtypes using qualitative terms such as impact on productivity (Lynden and Oldeman, 1997). Land Degradation Assessment in Dry lands (LADA) considers both biophysical factors and socio-economic driving forces for assessing the land degradation (FAO, 2008b). Pyke et al. (2002) developed a rapid, qualitative method for assessing degradation status of rangelands in the US using 17 indicators to assess 3 ecosystems attributes (soil and site stability, hydrological function, and biotic integrity) for a given location. Eswaran and Reich (1998) attempted to evaluate vulnerability to land degradation and desertification based on coefficient of variability of rainfall, depth of soil, extreme levels of chemical and physical conditions, resilience of soil and using the information incorporated in soil classification term. They found that about 43.3 million km^2 in arid, semi-arid, and sub-humid areas were vulnerable to land degradation and desertification in which 7.8 million km^2 was very highly vulnerable and 7.1, 13.6, 14.6 million km^2 belongs to high, moderate and low vulnerable class, respectively. The FAO and UNEP (1984) proposed a system of criteria for the evaluation of land degradation/desertification status, which contains data on plant cover, water and wind erosion and salinization. Veron et al. (2006) criticized the matrix from several perspectives, particularly the subjective nature of the data.

8.1.2 ASSESSMENT OF LAND DEGRADATION: INDIAN PERSPECTIVE

Land degradation reduces the ability of the land to perform many biophysical and chemical functions (Rashid et al., 2011). In India, initially aerial photographs were used for deriving information on degraded lands (Iyer et al., 1975). Subsequently, the application of remote sensing data gained importance in mapping degraded lands with the launch of ERTS-1/Landsat-1, Landsat-TM, SPOT and IRS Satellites (Dwivedi and Sreenivas, 1998). The estimates of land degradation by different agencies vary widely from about 53.3 m ha to 187.7 m ha, mainly due to different approaches adopted in defining degraded lands and differentiating criteria used (SAARC, 2011). Department of Land Resources (DOLR) in collaboration with Ministry of Rural Development carried out wasteland mapping using remote sensing technique and estimated 53.3 m ha of waste land in 1:1 million scale satellite imagery during 1985. Subsequently an estimate of 63.85 m ha (2000) and 55.27 m ha (2005) of wasteland in the country has been reported based on 1:50000 scale mapping during 2000 and 2005 (Bhattacharyya et al., 2015). According to National Bureau of Soil Survey and Land Use Planning (NBSS & LUP) about 120.72 m ha area is suffering from various kinds of land degradation (ICAR and NAAS, 2010). It includes area affected by water erosion 82.5 m ha, wind erosion 12.4 m ha, salinity/alkalinity 6.7 m ha, soil acidity 17.9 m ha and 1.0 m ha is under other complex problems. This was reported after the realization of the need to harmonize the area statistics on land degradation in the country, the National Academy of Agricultural Sciences (NAAS) took a major initiative to evolve a consensus among concerned organizations, *viz.*, NBSS & LUP, Nagpur, Central Soil Water Conservation and Training Institute (CSWCR&TI), Dehradun, Central Arid Zone Research Institute (CAZRI), Jodhpur, Central Soil Salinity Research Institute (CSSRI), Karnal, Forest Survey of India (FSI), Dehradun and National Remote Sensing Agency (NRSA), Hyderabad by adopting a common methodology and procedure for synthesizing the datasets on land degradation. The causes of land degradation include drought, population pressure, failure to implement appropriate technologies, poverty, constraints imposed by recent international trading agreements, and local agricultural and land use policies (Virmani et al., 1994). In Karnataka, 7.7 m ha (40% of TGA) out of the 19.1 m ha of total geographic area, is facing soil degradation problems in which water erosion is the major problem in 5.9 m ha (30.9%) of land area (Shivaprasad et al., 1998). Considerable area has been reported in Southern Karnataka with severely disturbed soil physical qualities by virtue of soil erosion. Rashid et al. (2011) used remote sensing data in conjunction with indicators such as vegetation, slope and land use and land cover for assessing the land degradation status of Kashmir region and found that 13.2% of the area has undergone moderate to high degradation, whereas about 44.1% of the area has undergone slight degradation.

8.2 STUDY AREA

Karnataka has 320 km long, 48–64 km wide coastal land, bordered by the Western Ghats on the east and the Arabian sea on the west. The coastal tract mainly consists of three districts *viz.*, the Uttara Kannada, Udupi and the Dakshina Kannada (Figure 8.1). Seven sites representing major physiographic

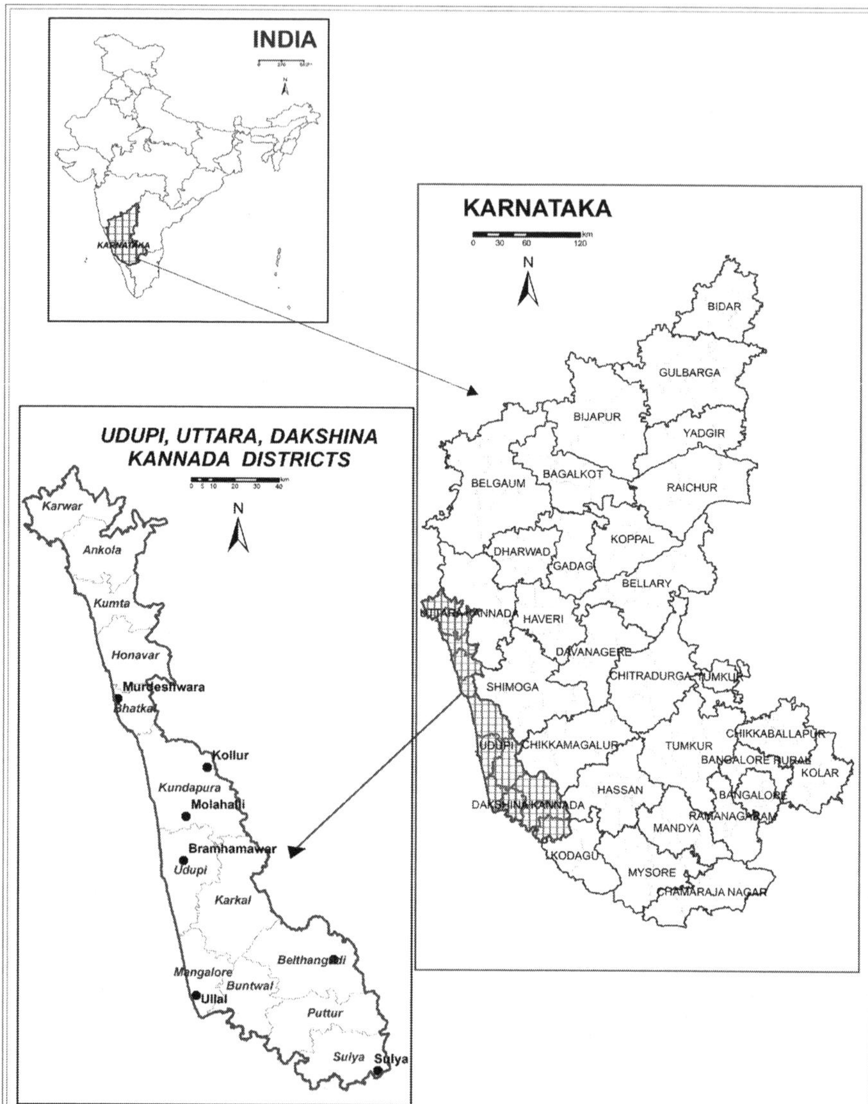

FIGURE 8.1 Location map showing study area and pedons.

units of the coastal tract were selected from these three districts for profile study in *viz.,* Sullya, Beltangadi, Kollur, Molahalli, Brahmavar, Murdeshwar and Ullal. According to delineation of National Agricultural Research Project (NARP) zones in the state, it comes under zone 10 (Coastal zone). The location details are given in Table 8.2. The western coast of Southern Karnataka

TABLE 8.2 Details About the Study Area of West Coast of Southern Karnataka

Profile no.	Broad Physiographic unit	Location	District	Latitude & longitude	Land use	Elevation (in m)	Rainfall (mm)	Soil classification	Erosion	Slope (%)*
1	Steep hill ranges	Sullya (Humid tropics)	Dakshina Kannada	12°31'11.4" N 75°31'0.4" E	Rubber plantation	252	3738	Clayey, kaolinitic, isohyperthermic, Ustic Kandihumults	Severe	5–10
2	Steep low hill ranges, isolated hills and dissected hills and valleys	Beltangadi (Humid tropics)	Dakshina Kannada	13°0'11.1" N 75°26'10.3" E	Rubber plantation	148	4485	Loamy, kaolinitic, isohyperthermic, Ustic Kanhaplohumults	Slight	1–3
3	Elongated ridges and foot hill slopes	Kollur (Humid tropics)	Udupi	13°47'59.1" N 74°53'27.5" E	Rubber plantation	141	3844	Clayey, mixed, iso hyperthermic, Kandic Palehumults	Slight	1–3
4	Undula-ting uplands	Molahalli (hot humid tropics)	Udupi	13°35' 9.4" N 74°48'06.5" E	Forest	25	3844	Laomy, kaolinitic, isohyperthermic, Kanhaplic Haplustults	Moderate	3–5
5	Lateritic plateau and lateritic monds	Brahmavar (Hot humid tropics)	Udupi	13°24'23" N 74°46'04.2" E	Rubber plantation	38	3887	Loamy, kaolinitic, isohyperthermic, Typic Kanhaplustults	slight	3–5

TABLE 8.2 (Continued)

Profile no.	Broad Physiographic unit	Location	District	Latitude & longitude	Land use	Elevation (in m)	Rainfall (mm)	Soil classification	Erosion	Slope (%)*
6	Coastal plateau summits	Murdeshwar (Hot humid tropics)	Uttar Kannada	14° 4'57.3" N 74°30'1.6" E	Cashew plantation	8	3200	Clayey-skeletal, kaolinitic, isohyperthermic, TypicKandiustults	Moderate	3–5
7	Valleys, bars and ridges and beaches and marshes (lower laterite terrace)	Ullal (Hot humid tropics)	Dakshina Kannada	12°51'28.8" N 74°51'47.5" E	Paddy	7	3769	Clayey, kaolinitic, isohyperthermic, TypicKanhaplustults	Slight	3–5

* Slope of specific pedon location.

comes under humid tropical region with mean annual rainfall of 3000–4000 mm. The length of dry season ranged between 4–6 months. Beltangadi and Sullya have experience 4 months of dry period from December to March in a year. Other pedons experience 6 months of dry period from December to May.

The study area is west coast of Southern Karnataka, which covers west facing slopes of Western Ghats including high and low hill ranges, ridges, dissected hills and narrow valleys, isolated hills, flat hill slopes, undulating uplands and lateritic plateaus as well as mounds and coastal landforms. Humid tropical climate with heavy rainfall and high temperature experienced in this region induces intense leaching of bases. Hence, these areas are predominant with deep well drained acidic soils. The major soils of these region are very deep well drained gravelly soils, deep well-drained clayey soils, moderately shallow, well drained, clayey soils and deep imperfectly drained sandy over loamy in valley region with shallow water table. The majority of the west coastal region of Karnataka is under forest plantation followed by agricultural uses. West coast of southern Karnataka is very important and significant agro-climatic zone where number of commercially important crops like rubber, cashew, coconut and paddy.

8.3 MATERIALS AND METHODS

8.3.1 VISUAL INTERPRETATION OF SATELLITE DATA

Proper interpretation of False Color Composite (FCC) imageries based on tonal variation, pattern, texture and spectral reflectance properties of soils, helps in accurate identification of degraded lands. The Survey of India toposheets (48J, 48N, 48O & 48L) of 1:250,000 scales, which cover the west of southern Karnataka, were used to prepare base maps in conjunction with satellite imageries of IRS LISS-III P6 FCC during the year 1986, 1998, 2003, 2011 and Google Earth images wherever necessary. A tracing film was overlaid on the toposheets covering the study area. Boundary of the west coast and important land features like rivers, tanks, roads, etc., were extracted. Thus a map having the above common land features was used as a base map for preparing different thematic maps. Imageries of west coast of southern Karnataka were procured from Karnataka State Remote Sensing Application Centre, Bangalore. The satellite imageries of the study area were visually interpreted in conjunction with respective toposheets, based on tonal variation, texture and pattern. Permanent structures like roads, railway lines, and water bodies were first traced, digitized and super-imposed on the interpreted satellite imageries and different types of soil/land degradation were demarcated on the imageries.

8.3.2 STUDY OF REPRESENTATIVE SOIL PROFILES

The pre-field map showing different land degradation units was overlaid on physiography map and soil map (1:50,000 scale) of Karnataka (Shivaprasad et al., 1998). The major physiographic units and its corresponding soil and its area were demarcated in ArcGIS platform. Each unit with major soil was marked as a pedon in the pre-field map, and seven such representative pedons were identified. In addition sites representing different areas for sampling were chosen on the basis of physiography, geology, vegetation, micro-climate, degree of erosion, away from field boundaries, roads and rivers. A profile of dimension of $1.5 \times 1.5 \times 2$ m^3 was dug. The landform characters such as slope, erosion, drainage, land cover, etc., and morphological properties of the pedons were recorded.

8.3.3 LABORATORY ANALYSIS OF SOIL SAMPLES

The soil parameters *viz.,* pH, EC, organic carbon (OC), soil OC (SOC) stocks, CEC, exchangeable bases, extractable acidity by barium chloride-tri-ethanol amine method and 1 *N* KCl method, effective CEC, available N, P, K, exchangeable Ca and Mg, available S and micronutrients were determined through standard procedures (Jackson et al., 1973). Degradation status map was generated by considering the climatic, soil physical and chemical variables assigning grades for each parameter depending on its impact on making soil degradation in the study area. The parameters considered are total rainfall, deviation from normal spell, thickness of surface horizon, surface texture, BD of surface horizon, OC per cent, etc. To know the status of land degradation, values have been assigned to the related soil parameters. These values of corresponding profiles were divided by the total value of all parameters. In the present study, the total value of all parameters was assigned as 25 (Table 8.3).

TABLE 8.3 Assigning of Ratings/Grades to Soil Parameters for Land Degradation in West Coast of Southern Karnataka

Rainfall (cm)	Dry months	Thickness of surface horizon	Surface texture	B.D. of surface horizon	OC % of surface horizon	SOC (kg m^{-2})	CEC/ clay ratio of control section	B.S./sum of cations ratio of surface horizon	Status of land degradation
3500–4000 = (1)	<5.0 = (1)	>20 = (1)	scl = (1)	<1.2 = (1)	>2.5 = (1)	>12 = (1)	>16 = (1)	>22.5 = (1)	<0.50 = very low
>4000 = (2)	5.0–5.5 = (2)	10–20 = (2)	gscl = (2)	>1.2 = (2)	1.5–2.5 = (2)	9–12 = (2)	12–16 = (2)	15.0–22.5 = (2)	0.50–0.60 = low
<3500 = (2)	5.5–6.0 = (3)	<10 = (3)	gc = (2)		0.75–1.5 = (3)	6–9 = (3)	8–12 = (3)	7.5–15 = (3)	0.60–0.70 = medium
			gsl = (3)		>0.75 = (4)	3–6 = (4)		<7.5 = (4)	0.70–0.80 = moderate
						0–3 = (5)			>0.80 = high

8.4 RESULTS AND DISCUSSION

8.4.1 SOIL PHYSICAL PROPERTIES

8.4.1.1 Soil Depth

Soil depth indicates the depth of the solum occurring above the parent material or hard rock and determines the effective rooting depth of plants and the capacity of the soil to hold water and nutrients. Solum depth reflects the balance between soil formation and soil loss by erosion in any area. Soil depth ranged between 109 to 155 cm or more. Except Molahalli all the pedons recorded a solum depth of more than 150 cm. The lower depth recorded in Molahalli soil profile might be due to washing out the top soil because of lack of proper soil conservation measures. This study area has the slope of 5–10%, which is responsible for formation of very deep soils in the coastal plateau summits and valleys regions. Variation in depth is due to the variation in topography, physiography and slope gradient (Sitanggang et al., 2006). Reduction in depth of solum owing to loss of top soil by sheet erosion, selective removal of finer particles and assorting of coarse grains and gravels on surface gives an indication of the degradation process operating there. This is aggravated by heavy rainfall and high temperature on steep slopes and soil surface devoid of proper vegetation cover. Less thickness of surface horizon is another indication of loss of top soil due to erosion as indicated by the morphological characteristics of soil profiles. Soil of Murdeshwar has maximum depth of 20 cm followed by soil profiles 2 and 6 with 18 cm depth, whereas, other soil profiles are only 10 cm deep.

8.4.1.2 Soil Texture

With regard to land degradation, soil texture of surface horizon was mainly considered, which includes gravelliness and it was influenced by the rainfall in removing the finer particles from the surface horizon, rainwater infiltration and percolation. Soil profile in Brahmavar has the surface texture of clay, Murdeshwar has clay loam surface texture, Kollur and Molahalli have sandy loam texture. Other soil profiles representing Sullya, Beltangadi and Ullal have sandy clay loam surface texture. Total sand percentage is much higher than the silt and comparatively higher than clay fractions in 2, 4, 5 and 7 pedons. The dominance of coarser fractions largely of siliceous nature may be due to granite gneiss parent material (Dutta et al., 2001). Appearance of hard weathered ferruginous schist and hard weathered laterite rock in pedon 3 and 4, respectively in the subsoil horizons might be due to the removal of clay, silt and fine sand particles by sheet erosion, which results in exposing rock fragments and accumulation of heavy soil particles. In the case of gravelliness, the highest gravelliness (31%) among 7 soil profiles was recorded in Molahalli soil profile which was followed by Murdeshwar soil profile (28%). The lowest gravelliness (3%) was recorded in Ullal soil profile with surface texture of clay loam. The removal of finer particles by erosion is responsible for the occurrence of coarse textured soil in the uplands (Dutta et al., 2001).

8.4.1.3 Bulk Density

The bulk density of surface horizon was lower than subsequent lower horizons. The lowest bulk density was observed in Sullya soils (1.08 Mg m^{-3}). The lower bulk density in the surface horizon is due to the high organic carbon content in the surface horizons. Higher surface bulk density was observed in pedons 1 and 4 – Murdeshwar (1.24 Mg m^{-3}) owing to the coarse texture, as a result of washing away of the clay and silt particles from the surface layer leaving well drained dense sand particles and in some cases coupled with low organic carbon (Sitanggang et al., 2006). The bulk density of all pedons has increased with increasing depth owing to dominance of illuviated compacted clay mineral in the lower horizons and low OC as compared to surface layer. In pedons 4 and 6, the bulk density increases with depth. Due to the severity of erosion, most of the soil material was removed leaving only the exposed compact layer below (Bhaskar and Subbaiah, 1995).

8.4.1.4 Soil Drainage and Erosion

Soil drainage is mainly influenced by surface soil texture. If the surface soil texture is heavy, it will be difficult for water to penetrate. Soil drainage affects the erosivity. All the pedons are well drained except pedon 7 (Ullal), which is moderately well drained. Severe soil erosion has been observed in Sullya soils due to steep hill ranges (slope 5–10%). In the study area, sheet erosion is the dominant type of erosion because of undulating slope particularly in the granite area, where large quantities of finer silt and clay particles get washed away from the top soil. In the absence of proper soil conservation measures due to continuous loss of fine particles and nutrients, the coarser particles in surface soils may result in unproductive soil over a period of time. In the upland, the sheet erosion is very active and may result in development of coarser textured surface soils (Balak Ram and Chauhan, 1992).

8.4.2 SOIL CHEMICAL PROPERTIES

8.4.2.1 Soil Reaction (pH)

In the study area, the soil reaction in soil profiles is strongly acidic to moderately acidic (Table 8.4). The reason for development of soil acidity might be high rainfall and leaching of bases. Soil acidification and consequent deficiencies of calcium and magnesium along with micronutrients B and Zn is very common in highly leached lateritic soils of coastal districts of Kerala (Kerala State Planning Board, 2013). Soil acidification is also commonly reported in high input plantation crops soils and banana and vegetable growing soils of Kerala in the same study.

8.4.2.2 Soil Organic Carbon

The pedon located at Kollur area registered highest OC content (5.5%), which was followed by the soil profile located at Brahmavar (2.5%). The lowest OC content was recorded at Molahalli (0.66%). Even though the OC varies, all soil profiles surfaces have the OC content at the high status (>0.5%) due to more foliage cover of dense forest in western Ghats, rubber cultivation and deposition of plant litter along with the alluvium (Badrinath et al., 1986).

8.4.2.3 Soil Organic Carbon Stocks

The pedon located at Kollur area registered highest SOC stocks (14.2 kg m^{-2}), which was followed by the soil profile located at Belthangadi (8.56 kg m^{-2}), Ullal (8.14 kg m^{-2}) and Brahmavar (8.01 kg m^{-2}). The lowest SOC stock was recorded at Molahalli (1.71 kg m^{-2}) and Murdeshwar (5.54 kg m^{-2}). The variations in OC stocks in all soil profiles depend on the thickness of horizon with higher OC content at the surface, soil depth, gravelliness and bulk density apart from climate and vegetation. Rajan et al. (2010) termed SOC stocks as the most reliable indicator for monitoring land degradation by soil erosion and sodicity through Principal Component Analysis after soil characterization by field studies in Kolar and Chamarajnagar districts of Karnataka, India.

8.4.2.4 Cation Exchange Capacity

CEC ranged between 8 to 28 cmol (+) kg^{-1}. The highest CEC was recorded in Kollur and Molahalli soil profiles (>16 cmol (+) kg^{-1}) and the lowest was in Murdeshwar and Ullal soil profiles (8–12 cmol (+) kg^{-1}). The CEC /clay ratio of more than 0.25 was observed in surface horizons of pedons 1, 2, 3 and 4 due to high organic carbon content, in subsurface horizon it was less than 0.25 and in rest of the pedons it was less than 0.25 throughout, which might be due to presence of low activity clay 1:1 (Kaolinitic) throughout the solum (Pujari and Moharana, 1993).

8.4.2.5 Base Saturation of Surface Horizon

Base saturation of surface horizon by sum of cations gives a rough indirect indication of the potassium, calcium and magnesium supplying capacity of soil to plant system apart from dominance of sesquioxides over bases. Highest base saturation was noted in Kollur and Brahmavar (23.5–23.9%), followed by Molahalli (20.5%), Sullya, Murdeswar and Ullal (8.6–12.3%) while least was found in Belthangadi (3.8%) owing to very high rainfall and consequent leaching away of bases. Similar observations were recorded in coffee growing areas of per-humid zones of Karnataka in a study conducted by Anil Kumar et al. (2014).

8.5 LAND DEGRADATION AND VULNERABILITY

The undulating upland physiographic unit was found highly vulnerable (0.83) to land degradation due to fragile soils, low fertility and organic carbon status, prolonged dry months and a general low input form of agriculture (Table 8.5; Figure 8.2). The physiographic units *viz.*, coastal plateau

TABLE 8.4 Horizon Wise Soil Parameters of Different Soil Profiles

Pedons and locations	Horizons	Horizon width (cm)	Vol. % Gravelliness	Texture	pH	OC (%)	CEC/ Clay	CEC [cmol (p+) kg^{-1}]	BS (%)	BD (Mg m^{-3})
Sullya	Ap	0–10	15.9	gscl	5.4	0.94	0.30	18.9	10.4	1.08
	Bt1	10–33	11.9	scl	5.2	1.22	0.20	16.1	10.3	1.04
	Bt2	33–54	7.9	c	5.1	0.59	0.12	12.0	13.0	1.05
	Bt3	54–90	15.9	gc	4.8	0.09	0.10	13.2	9.9	1.14
	Bt4	90–121	23.8	gc	5.2	0.09	0.09	8.5	17.7	1.23
	Bt5	121–150	15.9	gc	5.1	0.09	0.07	10.1	16.5	1.20
Beltangadi	Ap	0–18	27.4	gscl	5.2	2.3	0.37	24.4	3.8	1.22
	AB	18–41	21.4	gsl	5.0	0.9	0.44	15.1	7.3	1.19
	Bt1	41–62	23.0	gsc	5.3	0.46	0.13	12.3	10.6	1.22
	Bt2	62–82	25.4	gsc	5.4	0.13	0.15	9.5	21.4	1.28
	Bt3C	82–119	15.2	gscl	5.4	0.16	0.16	9.8	19.0	1.24
	BC1	119–139	15.1	gcl	5.3	0.16	0.09	6.7	18.2	1.42
	BC2	139–153	13.5	sc	5.4	0.09	0.05	6.5	16.4	1.33
Kollur	Ap	0–10	27.8	gsl	5.3	5.49	0.73	28.2	23.9	1.13
	Bt1	10–27	28.8	gcl	5.5	2.8	0.32	22.5	14.8	1.12
	Bt2	27–53	31.0	gsl	5.3	1.22	0.20	22.1	20.8	1.22
	Bt3	53–88	32.4	gl	5.4	0.99	0.21	14.6	14.7	1.21
	Bt4	88–130	38.6	vgcl	5.6	0.03	0.08	11.4	10.8	1.33
Molahalli	Ap	0–10	31.0	gsl	5.7	0.66	0.27	9.4	20.5	1.21
	Bt1	10–28	32.2	gcl	5.6	0.13	0.09	9.4	20.3	1.21

	Bt2C	28–50	32.6	gsl	5.5	0.13	0.21	8.4	17.5	1.26
	BC1	50–78	34.2	gl	5.7	0.10	0.13	6.0	25.7	1.26
	BC2	78–109	37.0	vgcl	5.5	0.06	0.12	8.5	11.7	1.34
Brahmavar	Ap	0–10	21.4	gc	5.8	2.48	0.20	18.2	23.2	1.17
	BA	10–38	19.5	gcl	5.4	0.46	0.14	13.7	20.0	1.23
	Bt1	38–59	19.1	gc	5.2	0.49	0.08	13.9	4.9	1.22
	Bt2C	59–79	13.9	cl	5.2	0.46	0.14	15.4	9.3	1.19
	BC1	79–113	22.2	gcl	5.3	0.39	0.15	9.7	7.8	1.25
	BC2	113–148	24.2	gcl	5.2	0.13	0.11	13.2	16.9	1.29
Murdeshwar	Ap	0–20	28	gcl	5.7	1.45	0.18	18.0	8.6	1.24
	Bt1	20–47	35	gsc	5.6	0.46	0.10	13.7	12.9	1.23
	Bt2	47–82	60	vgsc	5.5	0.62	0.11	14.0	14.7	1.26
	Bt3	82–112	60	vgc	5.7	0.13	0.12	11.9	22.8	1.28
	Bt4C	112–155	70	egc	5.5	0.06	0.11	12.4	17.8	1.32
Ullal	Ap	0–10	3	scl	5.2	15.9	0.21	22.3	12.3	1.12
	AB	10–20	5	Scl	5.5	11.7	0.17	20.8	13.5	1.52
	Bt1C	20–47	14	c	5.4	3.0	0.12	19.0	14.8	1.32
	Bt2C	47–78	27	gcl	5.4	2.6	0.09	16.9	14.4	1.28
	Bt3C	78–103	14	c	5.4	3.0	0.09	18.2	14.7	1.43
	BC	103–133	15	gcl	5.4	2.4	0.10	19.6	10.5	1.45
	CB	133–151	15	gcl	5.7	2.6	0.14	14.7	15.1	1.42

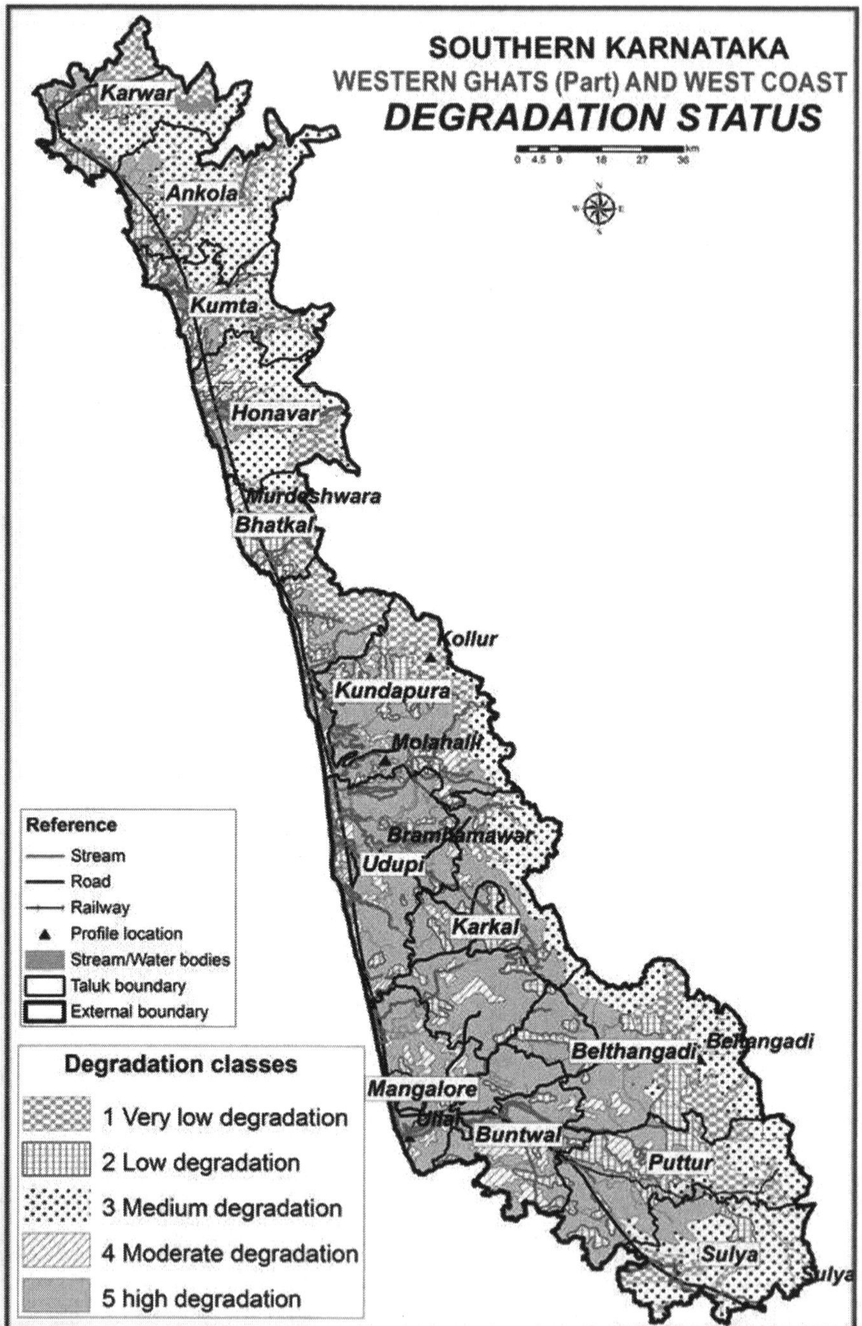

FIGURE 8.2 Land degradation status of parts of Western Ghats and West coast of Karnataka.

TABLE 8.5 Assigned Grades of Soil Parameters for Different Soil Profiles and Land Degradation

Pedon Locations	Latitude & longitude	Rainfall (cm)	Dry months	Surface soils parameters						CEC/clay ratio of control section	Status of land degradation and vulnerability
				Thickness (cm)	Texture	BD (Mg m^{-1})	OC (%)	SOC (kg m^{-2})	BS/Sum ratio		
Sullya	12°31'11.4" N 75°31' 0.4" E	3738 (1)	4.2(1)	10 (3)	gscl (2)	1.08 (1)	0.94 (3)	6.14 (3)	10.5 (3)	18–12 ((2)	19/29=0.66 (Medium)
Beltangadi	13°0'11.1" N 75°26'10.3" E	4485 (2)	4.5(1)	41 (1)	gscl (2)	1.22 (2)	2.30 (2)	8.56 (3)	3.8 (4)	12–16 (2)	19/29=0.66 (Medium)
Kollur	13°47'59.1" N 74°53'27.5" E	3860(1)	5.4(2)	10 (3)	gscl (2)	1.13 (1)	5.40 (1)	14.20 (1)	23.9 (1)	>16 (1)	13/29=0.45 (Very low)
Molahalli	13°35'9.4" N 74°48'06.5" E	3887 (1)	5.6(3)	10 (3)	gsl (3)	1.24 (2)	0.66 (4)	1.71 (5)	20.5 (2)	>16 (1)	24/29=0.83 (High)
Brahmavar	13°24'23" N 74°46'04.2" E	3887(1)	5.2(2)	10 (3)	gc (2)	1.07 (1)	2.40 (2)	8.01 (3)	23.5 (1)	12–16 (2)	17/29=0.58 (Low)
Murdeshwar	14° 4'57.3" N 74°30'1.6" E	3237(2)	5.2(2)	20 (2)	gscl (2)	1.24 (2)	1.45 (3)	5.54 (4)	8.6 (3)	8–12 (3)	23/29=0.79 (Moderate)
Ullal	12°51'28.8" N 74°51'47.5" E	3769(1)	5.2(2)	10 (3)	scl (1)	1.12 (1)	0.52 (4)	8.14 (3)	12.3 (3)	8–12 (3)	21/29=0.72 (Moderate)

summits and beaches and marshes were moderately (0.79 and 0.72, respectively) vulnerable to land degradation because of poor organic carbon status and low surface horizon thickness in lower lateritic terrace. The steep hill ranges, dissected hills and valleys were observed to be affected by medium vulnerability (0.66) to land degradation due to poor surface horizon thickness, SOC in the steep hill ranges and low base saturation to total cations ratio of surface horizon in dissected hills and valleys. The lateritic plateau has shown low vulnerability and foot hill slopes were subjected to very low vulnerability towards land degradation because of better soil parameters which reduces land degradation. The land degradation problem is not only a resource research and management issue, but also a human and social issue. The reasons and means to combat the process should be site specific and vary from region to region. A policy which supports mutual goals of optimum soil quality, clean water and sustainable farming should be adopted (Eswaran and Reich, 1998).

8.6 SUGGESTED MANAGEMENT MEASURES

- **Development of Land Resource Information Systems for Land Management**
  A scientific community has to be identified and mobilized to initiate and mount an integrated program for methods, standards, data collection and research networks for assessment and monitoring of soil and land degradation. Based on the information, hot spots can be identified for monitoring the extent of degradation, the factors causing land degradation and its impact assessment.
- **Enhance the Research and Development in Degraded Lands**
  High priorities are to be given to promote public investment in research and development aimed at identifying the root cause of land degradation and developing soil resources conserving, yield enhancing low cost technology for problematic lands.
- **Developing Suitable Land Use Planning and Policies**
  The land use plan has to be developed by considering or identifying the models which incorporate the factors (natural and human induced) that contribute to land degradation.
  Strong land use polices have to be identified, which encourage sustainable land use and management and should arrest the conversion of prime agricultural land into non-agricultural purposes.

- **Encourage Participatory Land Use Planning Involving Local Organization**
  Arrangements have to be made for collaboration between public research institutions, NGOs and local organizations for developing land use plan using locally available inputs and training should be given for effective adaptation of resource conserving and yield enhancing technologies.

8.7 CONCLUSIONS

Soil degradation status and vulnerability have been assessed by assigning scores to land quality parameters, which favor land degradation. Molahalli soils recorded high and Murdeshwar and Ullal soils showed moderate vulnerability to land degradation. The Kollur soil showed very low vulnerability to land degradation, followed by low vulnerability in Brahmavar, while Sullya and Belthangadi were assigned medium vulnerability. Lands with favorable rainfall, bulk density, base saturation, cation exchange capacity and soil reaction were neither degraded nor vulnerable to degradation in most of the places and while those lands with poor organic carbon status were tagged as highly degraded as in Molahalli sites and highly vulnerable to degradation because of high dry months. The coarse texture and low thickness of the surface horizon showed moderate rate of degradation and vulnerability and these might be good soil quality indicators in the long term if the recommended soil management strategies are adopted. Proper conservation measures like growing cover crops and *in situ* moisture conservation have to be followed. Landform specific soil and water conservation measures need to be followed in the fragile ecosystems like undulating lateritic terrains, involving mechanical measures as well as vegetative barriers.

KEYWORDS

- **Land Degradation Status and Vulnerability**
- **Land Qualities**
- **Land Resources Management**
- **Physiographic Units**

- **Red and Lateritic Soils**
- **Soil Organic Carbon Stocks**
- **Soil Parameters**
- **Southern Karnataka**
- **West Coast**
- **Western Ghats**

REFERENCES

Amara Denis, M. K., & Momoh Edwin, J. J. (2014). Fertility status, Degradation Rate and Vulnerability Potential of Soils of Sowa Chiefdom in Southern Sierra Leone. *International Journal of Interdisciplinary and Multidisciplinary Studies*, *2*(1), 151–162.

Anil Kumar, K. S., Nair, K. M., Krishnan, P., Naidu, L. G. K., & Sarkar, D. (2014). Climatic and terrain influence on acidity, exchangeable bases and cation exchange capacity in soils of coffee plantations of Karnataka, *Journal of Plantation Crops, 42*(1), 41–50.

Badrinath, M. S., Krishnan, A. M., Patil, B. N., Kenchaiah, K., & Balakrishna Rao, K. (1986). Fertility status of some typical soils of coastal Karnataka. *Journal of Indian Society of Soil Science, 34*, 436–438.

Balak Ram & Chauhan, J. S. (1992). Remote sensing in mapping the vulnerability and dynamic of the gullied land in Sikar district, Rajasthan. *Indian Cartographer*, *12*, 47–50.

Ballayan, D. (2000). Soil Degradation. Food and Agriculture Organization, Rome.

Beinroth, F. H., Eswaran, H., Reich, P. F., & Van den Berg, E. (1994). Land related stresses in agroecosystems. In: Virmani, S. M., J. C. Katyal, H. Eswaran, & I. P. Abrol. (Eds.). Stressed Ecosystems and Sustainable Agriculture. Oxford and IBH Publ. Co., New Delhi, India. 441 pp.

Bhaskar, B. P., & Subbaiah, G. V. (1995). Genesis, Characterization and classification of laterites and associated soils along the East Coast of Andhra Pradesh. *Journal of Indian Society of Soil Science, 43*, 107–112.

Bhattacharyya, R., Ghosh, B. N., Mishra, P. K., Mandal, B., Srinivasa Rao, Ch., Sarkar, D., Das, K., Anil Kumar, K. S., Lalitha, M., Hati, K., & Franzluebbers, A. (2015). Overcoming soil degradation in India: Sustainability, challenges and potential solutions. *Sustainability, 7*, 3528–3570.

Bridges, E. M., & Oldeman, L. R. (1999). Global assessment of human-induced soil degradation. *Journal of Arid Soil and Rehabilitation*, *13*(4), 319 – 325.

CPC. (2004). India Vision 2020, The Report: Report of the Committee on India Vision 2020. Central Planning Commission, Govt. of India. New Delhi.

Conacher, A. (2009). Land degradation: A global perspective. *New Zealand Geographer*, *65*(2), 91–94.

Dregne, H. E., & Chou, N. T. (1994). Global desertification dimensions and costs. *Degradation and Restoration of Arid Lands*. In: H. E. Dregne (ed.). Lubbock: Texas Technical University.

Dutta, D., Ray, S. K., Reddy, R. S., & Budhihal, S. L. (2001). Characterization and classification of paleosols in part of South India. *Journal of Indian Society of Soil Science,* 49 (4), 726–734.

Dwivedi, R. S., & Sreenivas, K. (1998). Delineation of salt affected soils and waterlogged areas in Indo-Gangetic plains using IRS 1C LISS III data. *International Journal of Remote Sensing,* 19(14), 2739–2751.

Eswaran, H., & Reich, P. F. (1998). Desertification: a global assessment and risks to sustainability. Proceedings of the 16th International Congress of Soil Science, Montpellier, France.

Eswaran, H., Lal, R., & Reich, P. F. (2001). *Land Degradation: An Overview.* In: Bridges, E. M., Hannam, I. D., Oldeman, L. R., DeVries, W. T. P., Scherr, S. J., Sombatpanit, S. (eds.) Response to Land Degradation. Enfield: Science Publishers Inc., pp. 20–35.

FAO. (2008a). Land Degradation on the Rise – One Fourth of the World's Population Affected Says New Study. Report by Food and Agricultural Organization. http://www.fao.org/newsroom/en/news/2008/1000874/index.html.

FAO. (2008b). LADA project documents FAO, internet website: http://www.fao.org/ag/agl/agll/lada/ladaprojectdoc.pdf (accessed July 20, 2008).

FAO and UNEP. (1984). Provisional Methodology for Assessment and Mapping of Desertification. Food and Agriculture Organization of the United Nations, United Nations Environmental Program, Rome. p 73.

Gay, S. H., Louwagie, G., Sammeth, F., Ratinger, T., Maréchal, B., Prosperi, P., Rusco, E., Terres, J., van der Velde, M., Baldock, D., Bowyer, C., Cooper, T., Fenn, I., Hagemann, N., Prager, K., Heyn, N., & Schuler, J. (2009). Final Report on the project 'Sustainable Agriculture and Soil Conservation.' EUR 23820 EN, 150 pp. Office for Official Publications of the European Communities, Luxembourg.

Glenn, E., Smith, S. M., & Squires, V. (1998). On our failure to control desertification: Implications for global change issues, and a research agenda for the future. *Environmental Science and Policy,* 1, 71–78.

Grainger, A. (2014). Is Land Degradation Neutrality feasible in dry areas? *Journal of Arid Environments. 7,* doi: 10.1016/j.jaridenv.2014.05.014.

ICAR & NAAS. (2010). Degraded and Wastelands of India: Status and Spatial Distribution, Indian Council of Agricultural Research and National Academy of Agricultural Sciences, PUSA, New Delhi. p. 158.

Iyer, H. S., Singh, A. N., & Kumar, R. (1975). Problem area inventory of parts of Hoshiarpur district through photo-interpretation. *Journal of Indian Photointerpretations, 3*(2), 79.

Jackson, M. L. (1973). *Soil Chemical Analysis.* Oxford IBH Publishing Co., Bombay.

Kapalanga, T. S. (2008). A Review of Land Degradation Assessment Methods. Final Report, Land Restoration Training Program, Iceland, pp. 17–68.

Kerala State Planning Board. (2013). Soil Fertility Assessment and Information Management for Enhancing Crop Productivity in Kerala, (eds.) P. Rajasekharan, K. M. Nair, G. Rajasree, P. Sureshkumar, & M. C. Narayanan Kutty, Kerala State Planning Board, Thiruvananthapuram, p. 514.

Lynden Van, G. W. J., & Oldeman, L. R. (1997). The Assessment of the Status of Human-induced Soil Degradation in South and South East Asia. UNEP, FAO and ISRIC, Wageningen.

Mensah, A. K. (2015). Role of re-vegetation in restoring fertility of degraded mined soils in Ghana: A review. *International Journal of Biodiversity Conservation, 7,* 57–80.

Nicholson, S. E., Tucker, C. J., & Ba, M. B. (1998). Desertification, drought, and surface vegetation: An example from the West African Sahel. *Bulletin of American Meteorological Society, 79,* 815–830.

Oldeman, L. R. (1994). The global extent of land degradation. In: D. J. Greenland and I. Szabolcs, (eds). *Land Resilience and Sustainable Land Use*, 99–118. Wallingford: CABI.

Onwudike, S. U. (2015). Effect of Land Use Types on Vulnerability Potential and Degradation Rate of Soils of Similar Lithology in a Tropical Soil of Owerri, Southeastern Nigeria. *International Journal of Soil Science, 10*, 177–185.

Pujari, K. L., & Moharana, G. (1993). Characterization and classification of some lateritic soils of Orissa. In red and lateritic soils of India resource appraisal and management, pp. 46–52. NBSS publ. 37, NBSS & LUP, Nagpur, India.

Pyke, D. A., Herrick, J. E., Shaver, P., & Pellant, M. (2002). Rangeland health attributes and indicators for qualitative assessment. *Journal of Range Management, 55*, 584–297.

Rajan, K., Natarajan, A., Anil Kumar, K. S., Badrinath, M. S., & Gowda, R. C. (2010). Soil organic carbon-the most reliable indicator for monitoring land degradation by soil erosion, *Current Science, 99*(6), 823–827.

Hegde, R., Natarajan, A. Naidu, L. G. K., & Dipak Sarkar (2011). Soil Degradation. In book *Soil Erosion Issues in Agriculture* (Ed.) Danilo Godone and Silvia Stanchi. In Tech Publication, Italy.

Rashid, M., Lone, M. A., & Romshoo, S. A. (2011). Geospatial tools for assessing land degradation in Budgam district, Kashmir Himalaya, India, *Journal of Earth System Science, 120*(3), 423–433.

Safriel, U., & Adeel, Z. (2005). *Dry land systems*. In: Hassan, R., Scholes, R. J., Ash, N., (Eds.). *Ecosystems and Human Well-Being: Current State and Trends*. Island Press: London, UK. p. 917.

SAARC. (2011). *Strategies for Arresting Land Degradation in South Asian Countries*. SAARC Agriculture Centre, BARC Complex, Farmgate Dhaka, Bangladesh. p. 271.

Scherr, S. J. (1999). Soil Degradation A Threat to Developing-Country Food Security by 2020? Washington: International Food Policy Research Institute.

Shivaprasad, C. R., Reddy, R. S., Seghal, J., & Velayutham, M. (1998). Soils of Karnataka for optimizing land use. NBSS & LUP, Nagpur, publ. 47, p. 15.

Sitanggang, M., Rao, V. S., Ahmed, N., & Mahapatra, S. K. (2006). Characterization and classification of soils in watershed area of Shikolpur, Gurgaon district, Haryana. *Journal of Indian Society of Soil Science, 54*, 106–110.

Thomas, D. S., & Middleton, N. J. (1994). *Desertification: Exploding the Myth*; John Wiley & Sons Ltd.: Hoboken, NJ, USA.

UNCCD. (2002). Recommendations and Conclusions of the African Regional Conference Preparatory to the First Session of the Committee for the Review of the Implementation of the United Nations Convention to Combat Desertification (UNCCD–CRIC1); Secretariat of the Convention to Combat Desertification: Windhoek, Namibia.

Veron, S. R., Paruelo, J. M., & Oesterheld, M. (2006). Assessing desertification. *Journal of Arid Environments, 66*, 751–763.

Virmani, S. M., Katyal, J. C., Eswaran, H., & Abrol. I. P. (Eds.) (1994). *Stressed Agroecosystems and Sustainable Agriculture*. Oxford & IBH Publishing Co., New Delhi, India.

Vito, I., Mariagrazia, D., Maria, L., Tiziana, S., Maria, R., & Maria, M. (2013). Integrated Indicators for the Estimation of Vulnerability to Land Degradation, Soil Processes and Current Trends in Quality Assessment, Dr. Maria C. Hernandez Soriano (Ed.), ISBN: 978-953-51-1029-3, In Tech, doi: 10.5772/52870.

CHAPTER 9

MAJOR AND TRACE ELEMENT GEOCHEMISTRY IN FERRUGINOUS SOILS DEVELOPED UNDER THE HOT HUMID MALABAR REGION, INDIA

P. N. DUBEY,[1] B. P. BHASKAR,[2] P. CHANDRAN,[2] B. SINGH,[3] and B. K. MISHRA[1]

[1]*Senior Scientist, ICAR-National Research Centre on Seed Spices, Ajmer – 305 206, India*

[2]*Principal Scientist, ICAR-National Bureau of Soil Survey and Land Use Planning, Amravati Road, Nagpur – 440 033, India*

[3]*Director, ICAR-National Research Centre on Seed Spices, Ajmer – 305 206, India*

CONTENTS

ABSTRACT

The lateritic soils of Malabar region of Kerala have a strong geographical association with varying degree of chemical weathering in relation to rainfall and lithology. In the study, four representative ferruginous soils on lateritic landscape developed over Precambrian rocks in parts of Kottayam of Kerala were studied to understand the weathering pattern and genesis through geochemistry. These soils are strongly to moderately acidic, reddish brown with low Cation Exchange Capacity (CEC) and base saturation having SiO_2 – 33 to 57%, Al_2O_3 – 16–31%, Fe_2O_3 – 8 to 15% and TiO_2 – 0.7 to 1.4%. The Kanjirapalli (P3) and Athirampuzha (P4) soil series were more intensely weathered as compared to the Kinalur (P1) and Chingavanam series (P2) with silica to alumina-iron ratio less than 2 and had a significant negative relationship with CIA (R = –0.75**), Harnois index (R = –0.678**), Richie index (R = –0.953**) and Plagioclase Index of Weathering (R = –0.705**). The Si-Al-Fe ternary diagram showed strong kaolinization process associated with dark red subsoils indicating well drained conditions. The Kanjirapalli and Athirampuzha soils have silica to sesiquioxide ratio less than 2 indicating lateritic nature with low bases to Al ratio in control section. The trace elemental concentration ranges were above the values of world soils having an enrichment index more than 1 in Kanjirapalli series (P3) and Ni contamination in genetic horizons (Ni>200 μgg-1). The cluster analysis showed similar major oxide concentration pattern in Group – 1 and Group – 2 but varied in trace elemental pattern with Cr>Ba>Cu in Group – 1 and Cu>Cr>Ba in Group – 2 soils whereas Zr>Ni>Mn in Group – 3 to Ni>Mn>Zr in Group – 4 soils. The study further showed that differential rates of weathering in soils under tropical climate were further accelerated due to anthropogenic activities such as improper land use practices and deforestation on steep slopes.

9.1 INTRODUCTION

India has long been an important area for study of laterites ever since Buchanan (1807) made the first historic description and suggested the term 'laterite.' Laterites are widespread in India and well investigated, more recently by Valeton (1994), Sahasrabudhe and Deshmukh (1981), Bruckner (1989), Bruhn (1990), Achyuthan (1996) and Widdowson and Cox (1996). Studies of widespread coastal laterite in Western India demonstrate the ubiquity of lateritization process in producing a residuum of consistently similar physical and chemical character despite major variations in the underlying protolith (Widdowson and Gunnell, 1999). The range of alteration products in this autochthonous profile display a number of important elemental enrichment and depletion patterns relative to the parent rock protolith. Such patterns may have characteristic laterite profiles in general, and consistent with the qualitative laterite definition given by Sivaranjasingham et al. (1962), and the quantitative chemical definitions suggested by Schellman (1986).

9.1.1 STUDY OF LATERITE SOILS: GLOBAL PERSPECTIVE

Lateritic soils (Ferralsols *sensu stricto*) are considered to be the product of weathering of basement rock under tropical climate conditions. They are rich in Fe and Al, low in silica, chemically acidic and exhibit a soil profile different from that of other soils (Foos, 1991; Tardy, 1997; Tardy and Nahon, 1985). Studies regarding the mineralogy, major element composition, isotopes, fluids and weathering rates have been carried out on soils developed around the Equator (Wimpenny et al., 2007). Chemical weathering is one of the most important processes that change the chemical composition of soils and distribution of elements in weathering products that differ from parent rocks (Nesbitt and Markovics, 1997). Chemical compositions of soils have been used effectively to evaluate weathering and soil formation conditions, to trace the provenance of soils (Nesbitt et al., 1996) and to reconstruct paleoclimate records (Wei et al., 2004; Zabel et al., 2001). Thus, quite a number of studies have been carried out in the past several decades to investigate chemical weathering (Little and Aeolus Lee, 2006; Price et al., 1991). Previous studies show elements that are conserved in temperate zone, such as Ti and Zr are mobile during extreme chemical weathering in tropical regions (Cornu et al., 1999). Probing into element behavior during weathering is pivotal to understanding element mobilization and redistribution during chemical weathering. In addition, laterites, the products of extreme weathering, account for over 85% of the present world soil cover (Nahon, 2003).

Quantitative characterization of weathering in soils is made through development of weathering indices (Abbaslou et al., 2013; Price and Velbel, 2003; Yousefifard et al., 2014). The chemical index of alteration (CIA, Nesbitt and Young, 1982), chemical index of alteration (CIW, Harnois,

1988) and plagioclase index of alteration, (PIA, Fedo et al., 1995) serve as examples of the decomposition of unstable minerals. According to the principles of soil genesis, alkali and alkaline earth elements move through soil horizons prior to silicon as weathering progresses (Souri et al., 2006). The genetic relationship between major elements and some trace elements of the major constituent materials in the lateritic cover of the Southern Cameroon forest plateau region was studied. The application of chemical indices for weathering in lateritic profile over chlorite schists in South Cameroon were reported by Kamang Beyala et al. (2009) and in lacustraine sediments of tecocomulic lake, Central Mexico by Ray et al. (2008). The degree of chemical weathering in 126 North America soils was assessed and established relationship of precipitation with chemical index of weathering, molar ratio of bases to alumina and potash or soda to alumina (Sheldon et al., 2002).

9.1.2 STUDY OF LATERITE SOILS: INDIAN PERSPECTIVE

Laterite and lateritic soils are formations peculiar to India and some other tropical countries with intermittently moist climate. In India, they cover a total area of about 24.8 m ha in the states of Karnataka, Kerala, Madhya Pradesh, the Eastern Ghat regions of Orissa, Maharashtra and parts of Assam (Raychaudhuri, 1980). The general characteristics of laterite associated soils in Kerala were reported by Satyanarayana and Thomas (1962) and of Nellore district of Andhra Pradesh by Bhaskar and Subbaih (1995). Silica as invariant was used to work out geochemical mass balance sheet of laterite associated soils of Somasila project (Bhaskar et al., 1999). The distinguishing feature of these laterite soils is development of strong chroma and redder hue due to accumulation of clay and relatively minor accumulation of Fe and Al sesquioxides in the B horizon with silica to sesequioxide ratio less than 2 (Chandran et al., 2005). It was reported that the dark red appearance of subsoils have dominance of iron oxide minerals *viz.*, lepidocrocite, haematite and goethite in the clay fraction and high proportion of dithionite soluble iron oxides over oxalate soluble (Basavaraj and Sarma, 1993). The variability of lateritic soils in Malabar region was due to variations in length of dry period and of duration of soil moisture deficit (Nair et al., 2011). While compiling the soil data on 1:250,000 scale, it was reported that the lateritic soils cover an area of 18.09 m ha (5.5%) in association with red soils covering 87.98 m ha, 26.8%) by Bhattacharryya et al. (2009). He also reported that in Kerala, the lateritic soils classified under Ultisols occupy an area of 2.29 m ha in association with Inceptisols (0.91 m ha). In India, therefore, comprehensive understanding of the behavior of elements during extreme weathering in the tropical laterite profile may aid in understanding of the mechanisms of weathering and help to explain the chemical records in soils. In this paper, we compare the composition of weathering products in terms of elements mobilization and redistribution during the processes of weathering in tropical climates.

9.1.3 IMPORTANCE OF THE STUDY

Laterites cover one-fourth of the total geographical area of 329 m ha on the basis of soil resource inventory (Bhattacharyya et al., 2009; Sehgal, 1998). The geochemical trends in lateritic profiles of Kerala were reported by Narayana Swamy (1992) indicating the depletion of silica and enrichment of sesquioxides and TiO_2 during the process of weathering. Later, Ramahashay et al. (1987) reported the occurrence of halloysite in association with kaolinite, goethite, gibbsite and quartz in these soils possessing high plasticity and cation exchange capacity. The recent review on pedology of red ferruginous soils of India was made by Pal et al. (2014) describing several pedological and edaphological aspects of Alfisols, Mollisols and Ultisols mainly of humid tropical (HT) climate but little emphasis made on geochemical records of these soils. The lateritic soils are dominant in Southern mid land zone (SMZ) mostly used for rubber (Hevea *brasiliensis*) under terraced hill slopes from Tiruvananthapuram to Kottayam. The intensive monsoon climate in the region is strongly interlinked with strong chemical weathering of lateritic soils and occurrence of devastating landslides. Among various geoenvironmental factors, slope plays vital role for many land use associations and cautions towards necessity of adopting land use control (Chattopadhyay, 2015). In India, therefore, comprehensive understanding of the behavior of elements during extreme weathering in the tropical laterite profile may aid in our understanding of the mechanisms of weathering for drawing meaningful geotechnical measures in the region. There is scanty literature on geochemical interpretations of lateritic soils dealing with elemental organizations as controlled by geology in Kerala. Hence, in the present study, an attempt was made for geochemical characterization of lateritic soils so as to decipher the pedogenic association and interrelationships of these soils with geology and bioclimatic conditions.

9.2 STUDY AREA

The Kerala region constitutes an important segment of the western passive continental margin of the Indian shield, and is bound by Western Ghats on the east and the Arabian Sea on the west. The rock types can be grouped into: (i) the Precambrian rocks (ii) the Tertiary formations and (iii) the Recent to sub-Recent sediments. The study is confined to the parts of Kottayam, Ernakulam and Kozhikode in Kerala, where wide spread Precambrian rocks like charnockites with narrow bands of pyroxene granulites and magnetite-quartz rocks are dominant. These rocks also constitute the high ranges of Western Ghats and extend to the Nilgiri massif (Damodaran, 1954). In Nilambur Valley, Kozhikode district, magnetite-quartzite, pyroxene and hornblende granulites and charnockite gneiss constitute the charnockite group (Sarwarkar, 1980). Kerala has tropical monsoon climate with mean annual rainfall of ~3265 mm. Most of the rainfall is received through the monsoon that usually arrives in early June and lasts till late September. May is the hottest month, seeing day temperatures of over 35° C with short winter between mid-December and February as marked by nights of around 21°C and days of around 28°C. The soil moisture regime is *ustic* with *isohyperthermic* soil temperature regime (Eswaran et al., 1990). Kerala forms part of agro-ecological sub regions 19.1 and 19.2 (Western Ghats and coastal plains-hot humid and per humid eco-region with red and lateritic and alluvium derived soils) having a growing period of 210–270 days and more (Sehgal et al., 1992). The dominant natural vegetation in midlands of Kerala comprises of Canarium *strictum*, *Mesa ferrea*, Dipterocarpus *spps.*, Callophylum *elatum*, Cullinia *excels*, Palaquium *ellipticum*, Tectonia *grandis*, Terminalia *tomentosa*, Dalbergia *latifolia*, Xylia *xylocarpa*, Pterocarpus *marsupium*, Santalum *album*, Avecinnia *spps.and tropical mangrove tree-* Rhizophora *spp*s., etc., and patches of scrub vegetation with tropical fruit trees like jack fruit (Artocarpus *heterophyllus*) and cashew (Anacardium *occidentale*).

9.3 METHODOLOGY

9.3.1 SOILS

Broadly, the soils of this region have been categorized as red sandy or sandy clay loams (Kamath, 1985) or more specifically by Bourgeon (1989) as mainly *ferrallitics* (French soil taxonomy) or subgroups of Ultisols, Alfisols and Inceptisols as per USDA Soil Taxonomy (USDA, NRCS, 2008). Kerala has substantial portion of laterites around 60% of the exposed surface area (Krishnan et al., 1996; Harindranath et al., 1999) developed over rocks of different composition and age. The soil profiles are located in the midlands (≤ 300 m) of Kerala having very gentle to moderate slope laterites to gently sloping to moderately steeply sloping laterite mounds and narrow valleys. Four lateritic soil profiles developed over charnockite, granite – gneiss, laterite, ferruginous quartzite and quartz-mica-schist (Narayana Swamy, 1992) were selected from Kerala. Brief description of four representative soil series under study in Kerala (Figure 9.1) are described below.

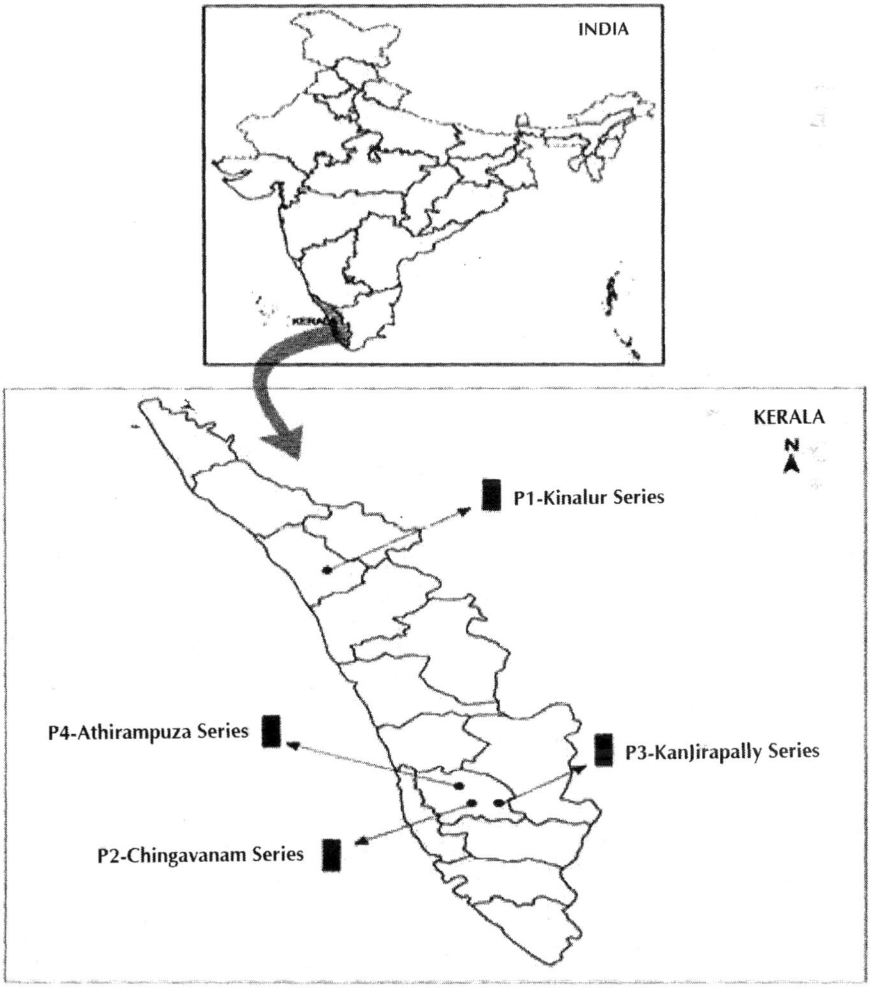

FIGURE 9.1 Location map of soil profiles under the study in Kerala.

P1: Kinalur series is a member of clayey, kaolinitic, *Ustic Kanhaplohumults*. These soils are very deep, well drained, dark red, gravelly sandy clay loam surface and red to reddish brown gravelly sandy clay sub soils with argillic horizon occurring on uplands and mounds. P2: Chingavanam series is a member of loamy skeletal, kaolinitic, family of *Typic Kandiustults* (Laterite). These soils are deep, well drained dark red surface horizons and red argillic Bt horizons with sandy clay loam texture. These soils occur on midlands having >15% slopes. P3: Kanjirapalli series is a member of clayey-skeletal, kaolinitic, *Ustic Kandihumults*. These soils are very deep, well drained formed on charnokite with dark reddish brown gravelly sandy clay loam surface and red to reddish brown gravelly sandy clay sub soils with argillic horizon occurring on uplands mounds and hills having more than 30% slopes. P4: Athirampuza series is a member of clayey-skeletal, kaolinitic, *Ustic Kanhaplohumults*. These soils are deep, well drained formed on laterized rocks (charnokites) with yellowish red gravelly clay surface and red to yellowish red gravelly argillic clay subsoils. This soil occurs on uplands, mounds and hills having slopes more than 30%. This is a competing series of Arur series (Harindranath et al., 1999).

9.3.2 LABORATORY ANALYSIS

Horizon wise soil samples for each soil series were collected and passed through 2 mm sieve after air drying. The fine earth fraction was used for laboratory analysis for particle size distribution (International pipette method), pH (1:2.5), organic carbon (OC, Walkley Black), Cation exchange capacity (CEC) and exchangeable bases (ammonium acteate) and percent base saturation was estimated as sum of [bases/CEC]100. The ECEC (effective cation exchange capacity) was derived as summation of exchangeable aluminum and exchangeable bases. 1M KCl extractable aluminum was determined titrimetrically, apparent CEC and ECEC was estimated as [CEC/clay]×100 and [ECEC/clay]×100 (Jackson, 1975; Sarma et al., 1987). Elemental analysis was carried out using 1 mm soil fraction by acid digestion (HF) for all elements except silica. Silica was estimated separately by sodium carbonate fusion using platinum crucibles (Page et al., 1982). Molar concentrations were estimated by dividing the elemental concentrations with atomic weight of the elements. The weathering indices were calculated by various methods as listed below.

SUMMARY OF WEATHERING INDICES (INDEX FORMULA)

Index	Formula	Reference
Chemical index of alteration (CIA)	CIA= $[Al_2O_3/Al_2O_3+Na_2O+CaO=K_2O]100$	Nesbitt and Young (1989)
WIP	$[(2Na_2O/(0.35) + (MgO/0.90) + (2K_2O/0.25) + (CaO/0.70)]100$	Parker (1970)
Si/Al		Birkeland (1999)
Bases/Alumina		
Bases/R_2O_3		
Reiche product index (RPI)	$100[(SiO_2)/(SiO_2+R_2O_3)]$	Reiche (1943)
Vogt ratio	$[Al_2O_3 + K_2O]/[MgO + CaO + Na_2O]$	Vogt (1927)
Ignition loss index	H_2O*	Jayawardena (1993)

Weathering indices used in the present study represent widely the nature and impact of chemical processes in ferruginous soils of Kerala. The approach is to identify relative ratio of mobile to immobile components with the advancement of pedogenesis. Some of the weathering indices were evaluated considering the following points as: Weathering index of Parker (WIP) considers susceptibility of Na^+, K^+, Ca^{2+} and Mg^{2+} during transformation of feldspars to clay minerals by including in the denominator Nicholls' values of bond strength as a measure of the energy necessary to break the cation-to-oxygen bonds of the respective oxides. These different values are considered to reflect the probability of an element to be mobilized during the weathering process. Values of WIP are commonly between ≥ 100 and 0 with the least weathered rocks having the highest values. The WIP implicitly assumes that Ca^{2+} in a silicate rock is contained in silicate minerals. This index does not consider immobile reference like Al_2O_3 in the formula, which would help to monitor relative changes of composition of the relevant mineral components. To overcome this problem, Chemical Index of Alteration (CIA) using major element oxides (Nesbitt and Young, 1982) was proposed to monitor the hydrolysis of feldspar and changes in major cations as a measure of chemical weathering. The Ignition Loss Index or H_2O^+ used by Jayawardena (1993) indicated the amount of crystalline water within the weathered material. An increasing H_2O^+ content is caused by hydration and clay formation during weathering. The Vogt Ratio (VR) attempted (incorrectly) to determine the ratio of immobile to mobile cations, but assumed that potassium remained stable within the weathering system. The Silica-Alumina Ratio provides a measure of the total element loss indicating the degree of weathering in humid climates on acidic rocks. The Reiche product index and weathering ratio were used to track movement of less mobile elements with respect to silica.

9.4 RESULTS AND DISCUSSION

9.4.1 SOIL CHARACTERISTICS

The characteristic red and latertic soils of Kerala developed over granite (P1), laterite (P2&P3) and charnokite (P4) are listed in Table 9.1. Kinalur series (P1) has very strongly acid, dark brown A horizon. This soil has dark reddish (5YR3/4) to reddish brown (5YR4/4) or red (2.5YR/6), strongly acid kandic B horizon with more than 30% clay and base saturation less than 35%. The kandic horizon has an apparent CEC less than 16 cmol/kg clay and an apparent ECEC less than 12 cmol/kg within 100 cm to further classify as *Kanhaplic Haplustults*. The very strongly acid Chingavanam series (P2) have 15 cm thick, dark red (2.5YR3/6) Ap horizons and red (2.5YR4/6) to reddish brown (2.5YR4/4) clay rich kandic horizons (30 to 36% clay) with base saturation less than 35%, to be classified as *Typic Kandiustults*.

The very strongly acid Kanjiraplli series (P3) and Athiram puzha series (P4) are classified as *Ustic Kandihumults* because these soils have OC more than 0.9% in upper 15 cm of kandic horizon with low base saturation. These soils have reddish brown Ap horizons (5YR3/3 to 3/4) and dark reddish brown (5YR3/4, P5) to reddish brown (5YR4/6, P4) but changed to red (2.5YR4/6) in lower kandic horizons. The apparent CEC and ECEC is similar to that of kandic horizons reported elsewhere in India (Bhaskar et al., 2005; Bhattacharyya et al., 2006). The occurrence of kandic gentic horizons in ferruginous soils of rubber growing areas in Kerala and Tamil nadu were reported (Krishnan et al., 1996). In XRD study conducted in a similar soil, Basavaraj and Sarma (1993) reported dominance of iron oxide minerals *viz.*, lepidocrocite, haematite and goethite in the clay fraction. High proportion of dithionite soluble iron oxides over oxalate soluble indicated good drainage in soil profile and presence of Fe as a substitute ion for octahedral Al in kaolinite, apart from Fe being carried as mineral component by amorphous ferri-aluminosilicates.

TABLE 9.1 Selected Morphological, Physical and Chemical Properties of Soils

Depth (cm)	Horizon	Matrix color	Texture	Particle size distribution (%) Sand	Silt	Clay	pH	OC g/kg	CEC	ECEC	CEC/C*100	ECEC/C*100	Base saturation (%)
									cmol(P+)kg^{-1}				
P1 – Kinalur series, Kozhikode, Kerala: Fine loamy, isohyperthermic family of Kaolinitic, *Kanhaplic Haplustults* (Granite Gneiss)													
0–10	Ap1	7.5YR 3/4	SCl	47.2	21.2	31.6	4.5	31.9	7.4	2.3	23.41	7.27	12.5
10–36	AB	5 YR 3/4	SCl	48.7	18.7	32.6	5.1	22.2	5.4	1.8	16.56	5.52	9.0
36–48	Bt1	5 YR 4/4	SCl	48.5	19.3	32.2	5.3	14.8	6.1	1.6	18.94	4.96	9.2
48–66	Bt2	5 YR 4/4	SCl	47.1	19.5	33.4	5.2	12.0	5.4	1.4	16.16	4.19	10.8
66–80	Bt3	2.5YR 4/6	SCl	50.7	19.2	30.1	5.1	6.1	5.1	1.6	16.94	5.31	13.7
P2 – Chingavanam series, Kottayam, Kerala:Loamy skeletal, Kaolinitic family of *Typic Kandiustults* (Laterite)													
0–15	Ap	2.5YR 3/6	SCl	50.2	16.3	33.5	4.5	8.7	4.5	2.4	13.43	7.16	44.2
15–41	AB	2.5 YR 4/6	SCl	52.2	17.9	29.9	4.7	3.5	3.1	0.9	10.36	3.01	19.7
41–66	Bt1	2.5 YR 4/6	SCl	47.4	22.1	30.5	4.7	2.3	3.0	1.3	9.83	4.26	23.3
66–99	Bt2	2.5 YR 4/4	SCl	46.9	20.4	32.7	4.9	3.1	3.6	1.3	11.00	3.97	19.2
99–140	Bt3	2.5 YR 4/4	SCl	39.5	24.3	36.2	4.8	3.9	3.5	1.9	9.66	5.24	37.1
140–176	Bt4	2.5 YR 4/4	SCl	45.2	22.8	32.0	5.0	3.1	3.5	1.7	10.93	5.31	
P3 – Kanjirapally series, Kottayam, Kerala: Loamy skeletal, Kaolinitic family of *Ustic Kandihumults* (Charnokite)													
0–13	Ap	5YR 3/3	SCl	69.9	9.0	21.2	4.8	23.5	4.5	1.6	21.22	7.54	21.7
13–32	Bt1	5 YR 3/4	SCl	56.1	12.5	31.3	4.4	18.6	3.5	1.1	11.18	3.51	17.1
32–56	Bt2	2.5 YR 3/6	SCl	59.9	11.1	28.9	4.5	15.0	3.7	0.8	12.80	2.76	14.0
56–83	Bt3	2.5 YR 3/6	SCl	62.8	11.4	25.7	4.5	9.0	4.1	0.7	15.95	2.72	13.2

TABLE 9.1 (Continued)

Depth (cm)	Horizon	Matrix color	Texture	Particle size distribution (%) Sand	Silt	Clay	pH	OC g/kg	CEC	ECEC	CEC/C*100	ECEC/C*100	Base saturation (%)
									cmol(P+)kg^{-1}				
83–112	Bt4	2.5YR 4/6	SCl	57.1	11.3	28.5	4.4	11.1	3.9	0.6	13.68	2.10	13.2
122–150	Bt5	2.5YR 4/6	SCl	62.6	13.4	24.0	4.7	12.2	4.0	0.7	16.66	2.91	16.8
P4 – Athirampuza series, Ernakulam, Kerala:Clayey skeletal, Kaolinitic family of *Ustic Kandihumults* (Laterite)													
0–16	Ap	5 YR 4/4	SC	51.7	11.0	37.3	4.0	23.6	9.0	2.6	24.12	6.97	6.3
16–42	Bt1	5 YR 4/6	C	39.7	8.0	52.3	3.8	12.6	8.1	3.4	15.48	6.50	5.7
42–66	Bt2	5 YR 4/6	C	33.3	6.7	60.0	4.1	9.9	7.3	3.3	12.16	5.5	5.1
66–89	Bt3	2.5 YR 4/6	C	34.6	11.8	53.6	4.3	6.7	6.4	3.1	11.94	5.78	6.7
89–105	BC	2.5 YR 4/6	C	30.3	16.3	53.4	4.1	6.3	7.0	2.4	13.10	4.49	14.0

9.4.2 ELEMENTAL COMPOSITION

The elemental composition of four lateritic and associated soils of Kerala is presented in Table 9.2. Among major elemental oxide, SiO_2 is dominant with its gradational decrease in concentrations to less than 40% in kandic horizons of Kanjirapalli series (P3) but 40 to 50% with irregular trends in Chingavanam series (P2) and Athirampuzha series (P4) and exceeding 50% in Kinalur series (P2). Next to SiO_2, Al_2O_3 concentration is 25 to 31% with irregular depth trends in Kanjirapalli (P3) and of increasing trends in Athirampuzha series (P4). The Chingavanam series (P2) have Al_2O_3 concentration of 20 to 22% with fairly uniform depth distribution but less than 20% in Kinalur (P1, irregular trends). The soils with more than 20% Al_2O_3 with high values of chemical index of alteration and a positive relation with TiO_2 (R^2=0.473*, significant at 5% level) may represent Paleosols in Peninsular India (Sreenivas and Srinivasan, 1994). He further used ternary diagram of Al_2O_3-K_2O-Fe_2O_3 to distinguish transported clay sediments from Paleosols. The Al_2O_3-K_2O-Fe_2O_3 ternary diagram (Figure 9.2) constructed for these soils showed that A-K-F ratio is similar to that of residual clays with low potassic nature. The Fe_2O_3 content is 3.7% in kandic B horizons (P1) and reached to a maximum of 16.3% in Bt3 horizon of P4. These soils have 1.1 to 2.2% of TiO_2 with high concentration of 1.9% in P4 but decrease its concentrations in Bt horizons. The concentration of alkali and alkaline earth elements are less than 1.0% in majority of soil horizons with the exceptions in soils where CaO and Na_2O exceed 1% (P1/P2).

9.4.3 TRACE ELEMENTS

Trace element concentration in the lateritic profiles is mainly driven by weathering of bedrocks. Trace elements are defined as those elements that are present at concentrations below 1000 $mgkg^{-1}$ or 0.1% (Rollinson, 1993). The distribution of trace elements with depth varies in different soil profiles due to variations in mobility characteristics, oxidation or intense leaching, effects of OC contents and iron/ manganese co – precipitation under humid climate (Table 9.3). The concentration range of these elements is as follows: Cu – 9 – 450 μgg^{-1}, Zn – 109–327 μgg^{-1}, Mn – 117–1571 μgg^{-1}, Cr – 48–281 μgg^{-1}, Ni – 5 – 990 μgg^{-1}, Ba – 19.7 -94 μgg^{-1}, Sr – 11.2 – 40 μgg^{-1}, and Zr – 13–513 μgg^{-1}. The Cu content is 304 μgg^{-1} in Ap horizon but decreases to

TABLE 9.2 Elemental Composition of Selected Soil

Depth (cm)	Horizon	Elemental composition																	
		SiO_2	Al_2O_3	Fe_2O_3	TiO_2	CaO	MgO	Na_2O	K_2O	Cu	Zn	Mn	Ni	Cr	Cd	Zr	Ba	Sr	LOI
		(%)								µg/g									(%)
P1 – Kinalur series																			
0–10	Ap1	55.80	17.40	1.20	0.80	1.59	0.68	1.97	1.08	304	142	324	60	48	8	14	33	40	16.91
10–36	Ap2	54.40	16.40	8.10	0.70	2.13	0.71	2.46	1.00	47	156	311	56	82	9	13	68	34	14.02
36–48	Bt1	56.30	17.80	3.70	1.00	1.59	1.15	3.09	1.17	43	234	359	33	89	8	15	54	36	14.17
48–66	Bt2	57.20	17.00	3.90	0.80	1.39	0.65	2.74	0.98	42	204	275	49	94	9	15	40	33	13.26
66–80	Bt3	50.30	17.80	13.20	0.70	1.05	0.77	1.87	1.05	43	327	250	49	72	8	13	60	31	13.05
P2 – Chingavanam series																			
0–15	Ap	45.10	22.00	8.20	1.70	1.46	0.41	2.67	0.42	64	172	221	38	105	10	17	90	28	16.1
15–41	Bt1	43.70	21.20	13.50	1.80	1.1	0.41	2.79	0.45	67	274	192	18	74	11	16	76	23	14.52
41–66	Bt2	44.30	20.60	15.90	1.40	1.43	0.33	3.33	0.42	450	184	180	22	107	11	19	59	23	12.37
66–99	Bt3	40.10	20.40	16.30	1.30	1.05	0.4	4.13	0.48	55	213	199	18	81	14	16	64	24	16.53
99–140	Bt4	45.50	20.60	13.80	1.20	1.19	0.44	3.44	0.49	70	151	160	26	110	11	16	69	25	12.97
140–176	Bt5	46.70	22.00	11.80	1.30	1.23	0.39	3.05	0.46	74	145	172	5	122	10	17	78	25	11.9

P3 – Kanjirapally series																			
0–13	Ap	45.40	26.10	6.40	1.10	0.74	0.54	0.2	0.52	17	151	535	624	127	0	371	38.64	17.27	18.18
13–32	Bt1	37.90	27.30	12.70	1.10	0.46	0.49	0.17	0.51	34	140	707	384	99	0	350	36.21	15.79	19.59
32–56	Bt2	37.30	30.90	11.60	1.00	0.54	0.67	0.17	0.53	20	126	784	990	167	26	261	30.45	14.08	18.93
56–83	Bt3	36.10	27.60	14.90	1.10	0.57	0.48	0.19	0.63	251	137	484	416	121	67	513	32.3	13.88	18.52
83–112	Bt4	33.70	30.00	14.90	1.20	0.52	0.31	0.13	0.65	59	117	299	175	62	0	221	52.4	11.2	18.96
112–150	Bt5	33.00	31.30	12.70	1.20	0.42	0.29	0.15	0.5	316	109	357	507	69	0	256	19.74	11.94	20.39
P4 – Athirampuza series																			
0–16	Ap	46.80	24.60	8.30	1.90	0.44	0.43	0.22	0.79	15	185	360	296	116	0	220	25.91	20.8	16.56
16–42	Bt1	43.70	29.20	8.10	1.50	0.33	0.43	0.17	0.62	9	164	268	233	81	0	360	24.5	24.68	16.22
42–66	Bt2	39.20	30.40	12.50	1.40	0.55	0.41	0.17	0.69	162	148	169	156	91	65	473	42.81	24.78	15.12
66–89	Bt3	40.70	30.60	10.20	1.40	0.36	0.41	0.2	0.69	18	200	182	286	104	13	376	31.91	21.62	15.94
89–105	BC	41.60	31.80	7.80	1.40	0.33	0.45	0.17	0.71	18	145	188	415	67	13	269	26.67	21.69	15.41

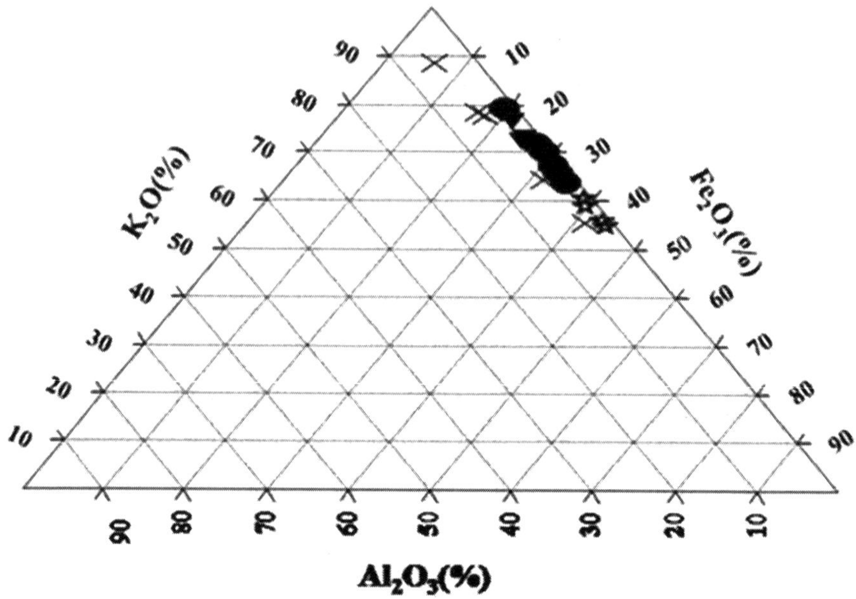

FIGURE 9.2 $Al_2O_3 - K_2O - Fe_2O_3$ ternary diagram for ferruginous soils of Kerala.

42 to 43 μgg^{-1} in genetic Bt horizons(P1) while in Chingavanam series (P2), the Cu content is 64 μgg^{-1} in Ap horizon but reached to 450 μgg^{-1} in Bt2 layer with its concentrations varying from 55 to 74 μgg^{-1}. In Kanjirapalli series (P3), the Cu content is 17 μgg^{-1} in Ap horizon but varied from 20 to 59 μgg^{-1} in genetic layers with maximum value of 251 (Bt3) to 316 μgg^{-1} (Bt5). The Cu content in Athirampuza (P4) is 15 μgg^{-1} in Ap layer, 9 to 18 μgg^{-1} in B horizon with 162 μgg^{-1} in Bt2 layer (Table 9.2).

The Cu anamolies are well expressed due to greater dispersion of Cu in iron rich lateritic soils. The Zn contents are more than 200 μgg^{-1} in genetic horizons of P1 and P2 as compared to P3 and P4. The mean Zn concentration is 180 μgg^{-1}, which is three times more than world average for soils (90 mg/kg, Bowen, 1979). The surface enrichment of Mn contents was noted (268 to 535 μgg^{-1}) in A horizons of all soils with decreasing values in Bt horizons except in P3 where its concentration reached to more than 700 μgg^{-1} in Bt1 and Bt2 layers. The Ni contents are less than 100 μgg^{-1} in P1 and P2 but reached to a maxima of 990 μgg^{-1} in Bt2 horizon of P3 and 415 μgg^{-1} in BC horizon of P4. The mean Ni content is 220 μgg^{-1}, which is almost four times higher than the world average for soils (60 mg/kg, given by Bowen, 1979). On comparing the Ni contents (> 200 μgg^{-1}) with the Great London Council (2001), these soils are registered as contaminated.

TABLE 9.3 Summary Statistics (min, max, mean, median, and standard deviation) of Major/Minor Elements, LOI, and Weathering Indices of Bulk Soil Samples ($n = 21$)

Summary statistics	SiO_2	Al_2O_3	Fe_2O_3	TiO_2	CaO	MgO	Na_2O	K_2O	Cu	Zn	Mn	Ni	Cr	Zr	Ba	Sr
	%								µg/g							
Maximum	57.20	31.80	16.30	1.90	2.13	1.15	4.13	1.17	450.0	327.0	784.00	990.00	167.00	513.	90.00	40.0
Minimum	33.00	16.40	1.20	0.70	0.33	0.29	0.13	0.42	9.00	109.0	160.00	5.00	48.00	13.00	19.74	1.20
Mean	44.31	24.23	10.44	1.23	0.93	0.51	1.52	0.67	99.00	173.8	317.09	220.73	94.91	174.5	47.84	23.6
SD	7.10	5.42	4.18	0.33	0.52	0.20	1.45	0.24	120.3	52.33	171.80	252.79	26.69	176.3	20.14	7.84
CV (%)	16.02	22.35	40.06	26.8	55.8	38.2	95.2	35.0	121.6	30.10	54.18	114.53	28.12	101.0	42.09	33.2
World soil mean*	59.90	15.10	5.00	0.67	1.96	1.49	1.35	1.69	30.00	70.00	700.00	20.00	80.00	230.0	-	-

The strong negative relation of Ni with SiO_2 (R = –0.51**, significant at 1% level) indicates its decrease with increase of SiO_2 but has a strong relation with Al_2O_3 (0.67**) suggesting the occurrence of Al rich phases such as clay minerals excert significant control on abundance of Ni (Al Chalabi, 2004). The Cr contents varied from 48 to 167 μgg^{-1} with mean of 94.9 μgg^{-1}, which is more than average of world soil (68 μgg^{-1}, Callender, 2004). The mean strontium content is 23.62 μgg^{-1} with strong positive relation with SiO_2 (r = 0.48**) and MgO (r = 0.78**) suggesting its strong association with plagioclase feldspars. The source of Sr is andesine and presence of Zr in these soils can be explained by presence of plagioclase feldspars. It was reported by Brantley et al. (1999) that Sr releases from feldspar at pH3.0 where bytownite releases Sr by factor of 60 to 400 times than anorthite and microcline, respectively. The Ba content is 47.84 μgg^{-1} with strong positive correlation with CaO (r = 0.67**) and K_2O (r = 0.77**). Alkalis and alkaline earth elements, such as K, Na and Sr, are the most active elements and easily removed from the profile during chemical weathering (Nesbitt et al., 1980). It can be inferred from the major and trace element records that these soils are intensively weathered under tropical monsoon climate with an annual precipitation more than 2000 mm and the consecutive trapping of Sr by clay minerals.

9.4.4 WEATHERING INDICES

Chemical weathering indices were used for evaluating degree of weathering in four lateritic profiles of Kerala by incorporating bulk major elemental oxide chemistry in a single metric (Table 9.4). The indices are based on the principle that the ratio between concentrations of mobile (SiO_2, CaO, MgO, and Na_2O) and immobile elements (Al_2O_3, Fe_2O_3, TiO_2) should decrease over time as leaching progresses (Souri et al., 2006) (Figure 9.2). The ratios of more active elements to that of more stable elements, like the molecular ratio of Si to Al or Si to (Al+Fe), have been used as a criteria in soil genetic classification systems.

The molar ratio of Si to Al (Ruxton, 1968) was used to evaluate loss of SiO_2 with weathering that show value less than 2 in kandic horizons of Kanjirapalli series (P3) indicating presence of kaolinite and Al chlorite with more weathering. The wider ratios (> 3) in P1 is an indication of persistence of primary minerals such as quartz and muscovite and the values in between

TABLE 9.4 Weathering Indices and Ratios in Soils

Depth	Horizon	mSi/ mAl	mSi/ mR_2O_3	VR	CIA	RPI	WIP	CIW	PIA
P1 – Kinalur series									
0–10	Ap1	5.44	4.94	1.11	68.15	18.82	39.77	73.94	71.52
10–36	Ap2	5.63	4.11	1.37	62.67	22.03	44.59	67.44	64.75
36–48	Bt1	5.37	4.46	1.20	63.71	21.03	52.70	69.06	66.24
48–66	Bt2	5.71	4.73	1.21	65.77	20.12	44.82	70.73	68.34
66–80	Bt3	4.80	3.15	1.52	71.98	26.60	37.28	78.12	76.08
ώ (weighted mean)		5.43	4.23	1.30	65.83	21.84	43.97	–	–
P2 – Chingavanam series									
0–15	Ap	3.48	2.60	1.34	73.77	28.84	35.57	75.74	75.07
15–41	Bt1	3.50	2.31	1.52	74.07	31.50	36.19	76.29	75.55
41–66	Bt2	3.65	2.31	1.58	69.94	31.91	41.36	71.83	71.05
66–99	Bt3	3.34	2.10	1.59	68.02	31.84	48.83	70.09	69.15
99–140	Bt4	3.75	2.50	1.50	70.24	30.35	43.08	72.48	71.57
140–176	Bt5	3.61	2.54	1.43	73.08	30.60	39.00	75.20	74.46
ώ		3.57	2.38	1.50	71.23	30.94	41.42	–	–
P3 – Kanjirapally series									
0–13	Ap	2.95	2.44	1.21	90.82	30.98	12.77	93.97	93.75
13–32	Bt1	2.36	1.75	1.35	92.98	36.11	11.49	96.07	95.94
32–56	Bt2	2.05	1.60	1.28	93.23	38.82	12.49	96.07	95.95
56–83	Bt3	2.22	1.60	1.40	91.65	37.78	13.67	95.35	95.14
83–112	Bt4	1.91	1.40	1.38	92.70	40.26	12.81	96.28	96.12
122–150	Bt5	1.79	1.37	1.31	94.18	40.15	10.51	96.87	96.78
ώ		2.10	1.59	1.33	92.84	38.23	12.16	–	–
P4 – Athirampuzha series									
0–16	Ap	3.23	2.46	1.31	90.36	31.70	15.81	95.49	95.21
16–42	Bt1	2.54	2.04	1.24	93.51	35.59	12.60	97.08	96.95
42–66	Bt2	2.19	1.66	1.32	92.24	39.40	14.12	95.96	95.79
66–89	Bt3	2.26	1.78	1.27	93.12	38.15	13.91	96.88	96.75
89–105	BC	2.23	1.83	1.21	93.55	37.83	13.96	97.31	97.19
ώ		2.67	2.10	1.39	92.57	40.37	15.26	-	–

ω = weighted mean.

2 and 3 in other soils is an indication of moderate weathering. This observation is further confirmed with silica to sesquioxide ratio (SiO_2/R_2O_3). The classification is possible on the basis of geochemistry of this ratio (Gidigasu, 1976). The ratio is in between 1.33 to 2.0 in kandic horizons of P3 and P4 to define as lateritic (Raychaudhuri, 1980) whereas in other soils this ratio is above 2 to define as non-lateritic tropical soils. The Kanjirapalli (P3) and Athirampuzha (P4) soils are subjected to intensive continuous weathering with hydrolysis of silicates, processes of argillation (mobilization and accumulation of clay particles due to sudden wetting of dry soils under extreme acid conditions) and rubification (transformation of ferric oxide to bright red hematite). It was recognized that Fe and Al are retained preferentially because of the precipitation of poorly crystalline soild phases (e.g., ferrihydrite) from supersaturated solutions (Chadwick et al., 2003). These ferrallitic soils are characterized by low silica to sesiquioxide ratio and base saturation. It is interesting to note that high degree of weathering in P3/P4 profiles is reflected by low SiO_2/R_2O_3 ratios and high CIA, PIA and CIW (Ranger, 2002). The triangular plots of SiO_2-Al_2O_3-Fe_2O_3 (Figure 9.3) clearly show that these soils are subjected to weak laterization in P1, moderate in case of P2 to strong laterization process in P3/P4 (Schellmann, 1981).

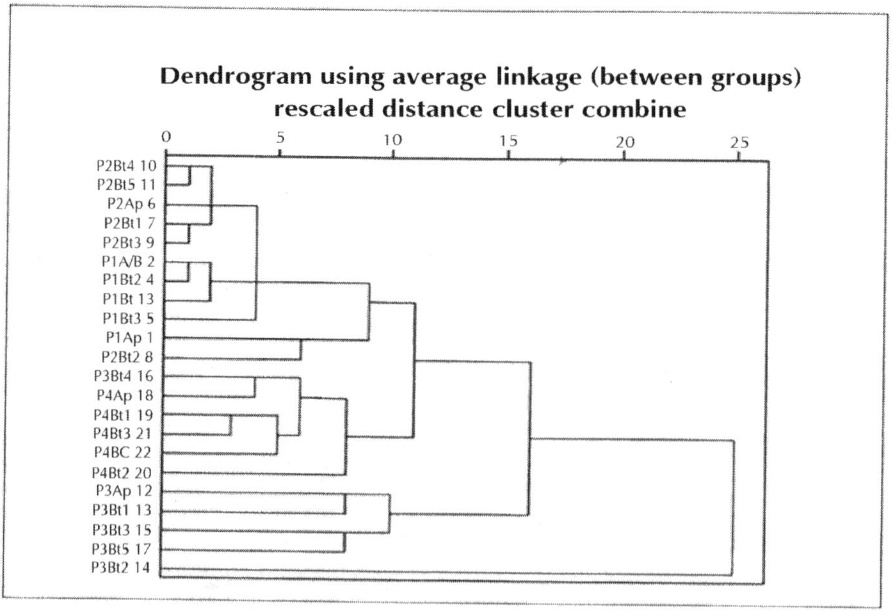

FIGURE 9.3 Soil grouping based on chemical affinity.

The Vogt residual index values in these soils vary from 1.21 to 2.07 showing slight variations with depth and are less than 1.4 in majority of kandic horizons. The reported vogt index values are in agreement with the values reported in humid tropical soils of Cameroon (Che et al., 2012). The chemical index of alteration (CIA), Harnois chemical weathering index (CIW) and plagioclase index of weathering (PIA) are more than 90 in P3 and P4 indicating high degree of weathering as reported in lateritic profiles of Kerala (Sanjinkumar et al., 2011). The parker index further shows decreasing values signifying greater weathering in P3 and P4 as compared to the other soils. The parker index values are less than 15 in P3/P4 and more than 40 in P1/P2. Such kind of variations in parker index were reported in highly weathered soils by Eswaran and Raghu Mohan (1973).

9.4.5 SOIL GROUPING

Cluster analysis of data on seventeen elements in the four types of ferruginous soils was carried out to estimate Euclidean distance to measure similarity using single linkage method (Table 9.5). The horizon wise major and trace elemental concentrations in these soils are grouped into four clusters (Table 9.6; Figure 9.4) that displayed different patterns of dominance of these elements as shown below. The major element concentration pattern in Group-1 and 2 is similar with an order of Si > Al > Fe > Na > Ti > Ca > K > Mg with Cr > Ba > Cu in Group-1 and Cu > Cr > Ba in Group-2 soils while it is Si > Al > Fe > Ti > K > Ca > Mg >Na with Zr > > Mn in Group-3 and Ni > Mn > Zr in Group-4 soils. In all soils ferric iron oxide is abundant in the

TABLE 9.5 Cluster Analysis of Data on Seventeen Elements in the Four Types of Ferruginous Soils

Clusters	Horizon sequences	Major elements	Trace elements
Group -1	P2Bt4- P2Bt5- P2Ap-P2Bt3	Si>Al>Fe>Na>Ti>Ca>K>Mg	Mn>Zn>Cr>Ba>Cu>Sr>Ni>Zr
Group-2	P1Bt3-P1Bt4-P1A/B-P1Bt2-P1Bt1-P1Ap-P2Bt2		Mn>Zn>Cu>Cr>Ba>Ni>Sr>Zr
Group-3	P3Bt4-P4Ap-P4Bt1-P4Bt3-P4BC-P4Bt2	Si>Al>Fe>Ti>K>Ca>Mg>Na	Zr>Ni>Mn>Zn>Cr>Cu>Ba>Sr
Gropu-4	P3Ap-P3Bt1-P3Bt3-P3Bt5-P3Bt2		Ni>Mn>Zr>Zn>Cr>Cu>Ba>Sr

TABLE 9.6 Group Wise Summary of Statistics for Major and Trace Elements in Soils

Statistics	SiO_2	Al_2O_3	Fe_2O_3	TiO_2	CaO	MgO	Na_2O	K_2O	Cu	Zn	Mn	Ni	Cr	Zr	Ba	Sr	LOI
	%								µg/g								%
Group-I																	
Mean	44.35	21.25	12.53	1.38	1.23	0.41	3.32	0.46	65.75	170.25	188.00	21.75	104.50	16.50	75.25	25.50	14.38
SD	2.91	0.87	3.42	0.22	0.17	0.02	0.62	0.03	8.26	30.76	27.39	13.87	17.21	0.58	11.41	1.73	2.29
CV (%)	6.57	4.09	27.31	16.13	13.81	5.27	18.76	6.69	12.56	18.07	14.57	63.75	16.47	3.50	15.17	6.79	15.92
Group-II																	
Mean	51.71	18.31	8.50	1.03	1.47	0.67	2.61	0.88	142.29	217.29	270.14	41.00	80.86	15.00	55.71	31.43	14.04
SD	5.72	1.84	5.77	0.42	0.36	0.27	0.54	0.31	166.01	66.13	67.24	16.67	18.84	2.08	15.04	6.40	1.46
CV (%)	11.06	10.05	67.84	40.75	24.64	39.67	20.88	35.20	116.67	30.44	24.89	40.67	23.30	13.88	27.00	20.36	10.43
Group-III																	
Mean	40.95	29.43	10.30	1.47	0.42	0.41	0.18	0.69	46.83	159.83	244.33	260.17	86.83	319.83	34.03	20.80	16.37
SD	4.42	2.52	2.87	0.23	0.10	0.05	0.03	0.06	59.19	29.86	77.02	94.62	21.01	100.55	11.22	4.99	1.37
CV(%)	10.81	8.55	27.84	15.94	23.01	12.21	17.42	8.40	126.39	18.68	31.52	36.37	24.19	31.44	32.98	24.00	8.39
Group-IV																	
Mean	37.94	28.64	11.66	1.10	0.55	0.49	0.18	0.54	127.60	132.60	573.40	584.20	116.60	350.20	31.47	14.59	19.12
SD	4.58	2.32	3.18	0.07	0.12	0.14	0.02	0.05	144.30	15.92	172.12	245.25	36.20	104.60	7.30	2.03	0.88
CV(%)	12.07	8.10	27.23	6.43	22.71	27.70	11.08	9.78	113.09	12.00	30.02	41.98	31.05	29.87	23.20	13.88	4.61

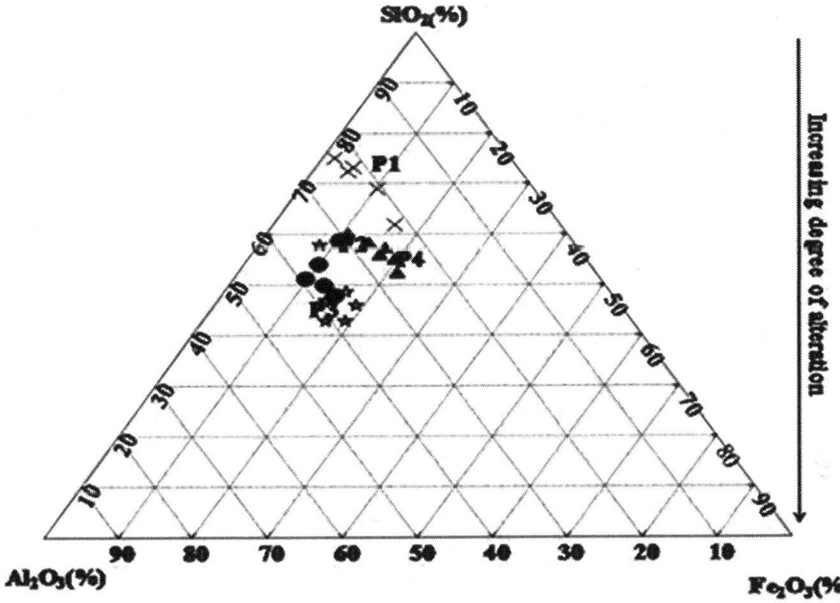

FIGURE 9.4 Si-Al-Fe ternary diagram modified by Hill et al. (2000) for laterite profiles.

lower and intermediate horizons while abundance of titanium is observed in Group-3 and Group-4, which is controlled by the original mafic lithology that bears remarkable high percentages of titaniferous pyroxenes and Fe-Ti oxide minerals (Moufti, 2010).

9.4.6 PEDOGENIC THRESHOLDS

The pedogenic characteristics of ferruginous soils at study sites considered to determine thresholds are as: (i) the ratio of total Al to total Fe; (ii) pH; (iii) base saturation; (iv) the weathering indices of CIA, CIW, PIA and VR. The ratio of total Al to total Fe in soils is more than unit value indicating consistent increase with considerable decrease in base saturation (<20%) in argillic B horizons. The loss of base cations is due to drop in pH towards acidity and the increased Al that competes with base cations for the exchange sites. These soils have very strongly acidic (<pH of 4.0) to strongly (pH of 4–5) to moderately acidic reaction (pH <5.5) leading to rise of exchangeable acidity with dominance of monomeric inorganic Al and dramatic increase in the ratio of exchangeable acidity to total CEC (>60% of Al saturation (Bhaskar et al., 2009; Thomas and Hargrove, 1984). This pH range under study area within a relatively narrow range of climate window, represents a threshold transition from an exchange complex dominated by Al, hydroxyl Al species and H^+ (Sparks, 1995). The area comes under hot humid to perhumid climate with length of growing period 210–270 days and greater leaching intensity of low base saturation (less than 30%) and acid environment even under small changes of rainfall at study sites. The soil profiles show only minor differences in base saturation and pH because they all are being buffered by Al hydrolysis. It is pertinent to say that current weathering rates in soil profiles (high values CIA, CIW and PIA) at the high rainfall may actually be operating at slower rate because of the depletion of primary minerals (Stewart et al., 2001). It is interesting to mention here that the top soils of these profiles meet the criteria to define them as modic that display characteristics like incompletely mineralized OM (intensive rubber growing area and frequent additions of leaf) and an OC content of >0.6%, blackish brownish in the hue of 5YR with base saturation <10% and pH <5.0 (FAO, 1998). The occurrence of modic topsoils in ferruginous soils of Shillong plateau were reported and proposed to use modic at subgroup level for Sombrihumults in Shillong plateau (Bhaskar et al., 2009). The difficulties in identifying the clay skins in argillic horizons in soils of Kerala and in Northern states was very well expressed in review of red ferruginous soils of India by Pal et al. (2014) and made them to place some soils in the subgroups of Dystrudepts. In the study sites, Al:Fe ratio ranges from 10.97 (P1) to 0.94 (P2) with maximum values in between 2.0 to 3.0 in genetic horizons. This threshold reflects the onset of anaerobiosis leading to iron reduction (Thompson et al., 2006). The CIA values more than 90 in P3 and P4 indicates intense chemical weathering corresponding to pH values less than 4.5 whereas CIA values less than 75 in P1and P2 indicate less intense weathering with a corresponding pH more than 4.7. The assumption that Ca, Na and K decrease as weathering intensity increases and that Al stays mostly immobile is valid here. These results are in accordance with observations in other weathering studies and theoretical considerations of the element behavior, suggesting that K release is low as compared to the Na release. This is due to stronger weathering resistance of K phases such as K-feldspar and due to the fixation of K on clay minerals (Yang et al., 2004; Reeder et al., 2006). Moreover, high CIA values, suggest derivation from a stable terrain (autochthonous in nature) of SMZ of Kerala. The high PIA values (P3 and P4) further indicate that the plagioclases in parent rock displayed increasing chemical weathering with steadily decreasing contents of plagioclases and enrichment in secondary aluminous clay minerals (Roy et al., 2008).

9.5 RECOMMENDATIONS

The lateritic soils are dominant in Southern mid land zone (SMZ) mostly used for rubber (*Hevea brasiliensis*) under terraced hill slopes from Tiruvananthapuram to Kottayam. The soil domains along southern mid land zone, where pedogenic thresholds are bound to be relatively straightforward with erosion and deposition on slopes yielding patches of fertile soil well supplied with rock-derived nutrients within what would otherwise (on the basis of rainfall alone) represent a domain of infertile, iron-enriched soils. This midland zone could represent a gradual accumulation of the influences of weathering and leaching leading to changing process domain boundaries over the long history of these sites. The geochemical thresholds such as molar ratio of l Al to Fe ratio >1 with Al_2O_3 more than 20% and high values of chemical index of alteration (>90%) may represent Paleosols with low potassic nature and sluggish weathering rates in soil profiles under the high rainfall conditions. It is possible that the distribution of pedogenic thresholds and soil process domains reflect the legacy of a climate shift and potential for indigenous agricultural systems to be tied up with the weathering/biological uplift.

The soil productive capacity is determined with soil OC status being a structural and functional component in providing the critical linkage between management and productivity. However, no quantitative means (deductive or inductive) in defining the critical threshold values for soil OC is established because of critical threshold values for ferruginous soils may vary with climatic regions and land use. It is logical that pH should be a key chemical indicator measured in soil surveys and soil data bases as it influences so many biological and chemical relationships simultaneously. It is interesting to note that in the typically more acid tropical subgroups of Ultisols, Al saturation (inverse of percent base saturation) is found to be sensitive and meaningful indicator of crop response than soil pH. This underscores the importance of base saturation rather than CEC itself, as an index of base cation availability in soils. The inclusion of CEC as a critical attribute for agricultural soils is important to the soil's nutrient supplying power. The underlying assumption of 'good' management (i.e., alleviation of nutrient deficiencies through routine fertilizer amendments) also explains why productivity indices for agricultural soils did not include any reference to nutrient chemistry unless particular exchangeable cations are suspected to limit productivity (mostly K in agricultural soils). In acid ferruginous soils, CEC per se is far less important to the soil's nutrient supplying power than percent base saturation (BS). It is therefore base saturation that determines the influence of the exchange complex on soil solution chemistry and acidity and elevated Al concentrations leading to possible toxicity (low saturation soils) with threshold of percentage base saturation <15% of effective CEC. Among trace elements, these soils have mean Ni contents 220 $\mu g g^{-1}$, which is almost four times higher than world average for soils and are registered as contaminated. Our challenge is to expand our knowledge of these ferruginous soil properties to predict their dynamic behavior soil processes and the impact of management practices on those processes. Ability to meet this challenge will play a key role in determining the sustainability of soil management activities and its management effects on soil productivity. Further studies are essential to understand chemical characteristics of soils at landscape level along rainfall gradient in Kerala so as to create a sound baseline geochemical data and for understanding the influence of exogenic processes on changes in soil properties in relation to land use history.

9.6 CONCLUSIONS

The four representative ferruginous soils classified under the subgroups of Ultisols in Kerala were examined to assess intensity of weathering through indices in relation to rainfall and parent material. The triangular plots of SiO_2-Al_2O_3-Fe_2O_3 clearly showed that these soils are subjected to weak laterization in Kinalur (P1), moderate in case of Chingavanam series (P2) to strong laterization process in Kanjirapalli (P3) and Athirampuzha (P4). These soils with more than 20% Al_2O_3 and high values of CIA, CIW, PIA in argillic B horizons may represent Paleosols with residual clays of low potassic nature as shown in ternary diagram of A-K-F system. The top soils of these profiles meet the criteria to define as modic with an OC content of >0.6%, blackish brownish in the hue of 5YR, base saturation <10% and pH <5.0. The clustering of element chemistry have yielded two distinct groups in relation to lithology with high levels of Ni content. The pedon level information was used to throw some light on pedogenic thresholds and soil process domains to link with the legacy of a climate shift as varied in elemental concentrations and its potential for enhancing agricultural systems in the region.

ACKNOWLEDGEMENT

This research was carried out under a project at the ICAR-National Bureau of Soil Survey and Land Use Planning, Nagpur.

KEYWORDS

- **Chemical Index of Weathering**
- **Enrichment Index**
- **Index of Laterization**
- **Kerala**
- **Trace Elements**

REFERENCES

Achyuthan, H. (1996). Geomorphic evolution and genesis of laterites around the east coast of Madras, Tamil Nadu, India, *Geomorphology, 16*(1), 71–76.

Abbaslou, H., Abtahi, A., & Baghernejad, M. (2013). Effect of weathering and mineralogy on the distribution of major and trace elements (Hormozgan province, Southern Iran). *International Journal of Forest, Soil and Erosion, 3*(1), 15–25

Al Chalabi, S. N. A. (2004). *Minerology and origin of chromate in qalander area, Northern Iraq*. MSc. Thesis. Salahaddin University-Erbil., Iraq, 104 pp.

Basavaraj, B., & Sarma, V. A. K. (1993). *Iron Bearing Clay Mineralogy of Some Selected Soils of Peninsular India.* In: Sehgal, J., Sarma, V. A. K., Batta, R. K., Gajbhiye, K. S., Nagabhushana, S. R., & Venugopal, K. R. (Eds.). Red and Lateritic Soils of India–Resource Appraisal and Management, NBSS Publ. 37, NBSS & LUP, Nagpur, India, p. 143–152.

Bhaskar, B. P., & Subbaiah, G. V. (1995). Genesis, characterization and classification of laterites and associated soils along the east coast of Andhra Pradesh, *Journal of the Indian Society of Soil Science, 43*(1), 107–112.

Bhaskar, B. P., Raja, P., & Srinivas, C. V. (1999). Genetic interpretation of pedological characteristics of laterite associated soils of somasila project, Andhra Pradesh. *Gondwana Geological Magazine, 14*(2), 39–47.

Bhaskar, B. P., Saxena, R. K., Vadivelu, S., Baruah, U., Butte, P. S., & Dutta, D. P. (2005). Pedogenesis in high altitude soils of Meghalaya plateau. *Agropedology, 14*(1), 9–23.

Bhaskar, B. P., Saxena, R. K., Vadivelu, S., Baruah, U., Sarkar, D., Raja, P., & Butte, P. S. (2009). Intricacy in classification of pine-growing soils in Shillong Plateau, Meghalaya, India. *Soil Survey Horizons*, *50*(1), 11–16.

Bhattacharyya, T., Pal, D. K., Lal, S., Chandran, P., & Ray, S. K. (2006). Formation and persistence of Mollisols on Zeolitic Deccan basalt of humid tropical India. *Geoderma,* 136, 609–620.

Bhattacharyya, T., Sarkar, D., Sehgal, J. L., Velayutham, M., Gajbhiye, K. S., Nagar, A. P., & Nimkhedkar, S. S. (2009). *Soil Taxonomic Database of India and the States (1:250,000 scale*). NBSS Publ.143. NBSS & LUP, Nagpur.

Birkeland, P. W. (1999). Soils and Geomorphology. Oxford Univ. Press, New York. 430 pp.

Bourgeon, G. (1989). Explanatory booklet on the reconnaissance soil map of forest area Western Karnataka and Goa. Institut Francais de Pondichery, 96 pp. + 2 annexes dont 1 carte, 1989, Travauxdela Section Scientique et Technique. Hors Serie N 20, Head of Ecology Department, Institut Francais de Pondichery.

Bowen, H. J. M. (1979). *Environmental Chemistry of the Elements*. Academic Press, London, UK, p. 333.

Brantley, S., Cesley, L. J., & Stillings, L. L. (1999). Isotopic Ratios and Release Rates of Strontium Measured from Weathering Feldspars. *Geochimica et Cosmochimica Acta, 62*(9), 1493–1500.

Bruckner, H. (1989). Kiistennahe Tieflander in Indien – ein Beitrag zur Geomorphologie der Tropen. – Diisseldorfer Geogr. Schr, 28.

Bruhn, N. (1990). Substratgenese – Rumpfflachendynamik. Bodenbildung unci Tiefenverwitterung in saprolitisch zersetzten granitischen Gneisen aus Siidindien. – Kieler Geogr. Schr, 74.

Buchanan, F. (1807). A journey from Madras through the countries of Mysore, Kanara, and Malabar, East India Company, London, *2*, 436–461.

Callender, E. (2004). *Heavy Metals in the Environment-Historical Trends*. In H. D. Holland, K. K. Turekian (Executive Eds.), Treatise on Geochemistry (pp. 67–105), B. S. Lollar (Ed.). *Environmental Geochemistry,* (Vol. 9). Oxford: Elsevier-Pergamon.

Chadwick, O. A., Gavenda, R. T., Kelly, E. F., Ziegler, K., Olson, C. G., Elliott, W. C., & Hendricks, D. M. (2003). The impact of climate on the biogeochemical functioning of volcanic soils. *Chemical Geology, 202*, 195–223.

Chandran, P., Ray, S., Bhatacharyya, T., Srivastava, P., Krishnan, P., & Pal, D. K. (2005). Lateritic soils (Ultisols) of Kerala, India: their genesis and taxonomy. *Australian Journal of Soil Research, 43*, 839–852.

Chattopadhyay, S. (2015). Environmental consequences of rubber plantations in Kerala. *National Research Program on Plantation Development*. Discussion paper No. 44, 1–54.

Che, V. B., Fontijn, K., Ernst, G. G. J., Kervyn, M., Elburg, M., Van Ranst, E., & Suh, C. E. (2012). Evaluating the degree of weathering in landslide-prone soils in the humid tropics: The case of Limbe, SW Cameroon. *Geoderma, 170*, 378–389.

Cornu, S., Lucas, Y., Lebon, E., Ambrosid, J. P., Luizãoe, F., Rouillerf, J., Bonnayg, M., & Nealh, C. (1999). Evidence of titanium mobility in soil profiles, Manaus, central Amazonia. *Geoderma, 91*, 281–295.

Damodaran, A. (1954). Progress Report for the field season 1952–53. A note on the traverses in Malnad areas of partsod Malabar and Nilagiri district. Geological Survey of India (Unpublished).

Eswaran, H., & Raghu Mohan, N. G. (1973). The microfabric of petroplinthite. *Soil Science Society* of *America Proceedings, 37*, 79–82.

Eswaran, H., De Coninck, F., & Varghese, T. (1990). Role of plinthite and related forms in soil degradation, In *Advance in Soil Science, 11*, 109–127.

FAO. (1998). Soil map of the world. Revised legend, by FAO–UNESCO–ISRIC. World Soil Resources Report No. 60, Rome.

Fedo, C. M., Nesbitt, H. W., & Young, G. M. (1995). Unraveling the effects of potassium meta somatism in sedimentary rocks and paleosols, with implications for paleo weathering conditions and provenance. *Geology, 23*(10), 921–924.

Foos, A. M. (1991). Aluminous lateritic soils, Eleuthera Bahamas: A modern analog to carbonate Palesols. *Journal of Sedimentary Petrology, 61*, 340–348.

Gidigasu, M. D. (1976). Laterite soil engineering. Development in geotechnical engineering. Elsevier Scientific Publishing Company.

Great London Council. (2001). Guidelines for Contaminated Soils (on-line) http://www. contaminatedland.co.uk/std-guid/kelly-l.htm. Society of the Chemical Industry [Retrieved 11 May 2009].

Harindranath, C. S., Venugopal, K. R., Raghu Mohan, N. G., Sehgal, J. L., & Velayuthm, M. (1999). Soils of Goa for optimizing land use. *NBSSPubl.*74. NBSS & LUP, Nagpur, p. 131.

Harnois, L. (1998). The CIW index: A new chemical index of weathering. *Sedimentary Geology, 55*(3–4), 319–322.

Jayawardane, U. (1993). Des. Use of H_2 O+ for classification of residual soils. In: *Geotechnical Engineering of Hard Soils – Soft Rocks*. Anagnostopoulos *et al.* (eds.), Balkema, Rotterdam, 169–171.

Kamath, S. U. (1985). Karnataka State Gazetteer: Uttara Kannada District, Gazetteer of India, Government of Karnataka, Bangalore.

Kamanag Beyala, V., Onana, V. L., Priso, E. N. E., Parisot, J. C., & Ekodeek, G. E. (2009). Behaviour of REE and mass balance calculations in a lateritic profiles over chlorite schists in South Cameroon. *Chemie der Erde: Geochemistry, 69*, 61–73.

Krishnan, P., Venugopal, K. R., & Sehgal, J. L. (1996). Soils of Kerala for optimizing land use. *NBSSPubl.* 48b. NBSS & LUP, Nagpur. p. 54, Soil Series of India, p. 10.

Little, M. G., & Aeolus Lee, C. T. (2006). On the formation of an inverted weathering profile on Mount Kilimanjaro, Tanzania: buried paleosol or groundwater weathering. *Chemical Geology, 235*, 205–221.

Jackson, M. L. (1975). Soil *Chemical Analysis–Advanced Course*. Univ. of Wisconsin, College of Agric., Dept. Soil Sci., Madison, WI Prentice-Hall, Inc., Englewood Cliffs, NJ.

Moufti, A. M. B. (2010). Field, Mineralogical and Geochemical Characteristics of As-Sarat Laterite Profiles, SW Saudi Arabia. *JAKU: Earth Science, 21*(2), 47–75.

Nahon, D. (2003). Weathering in tropical zone. Significance through ancient and still active mechanisms. *Comptes Rendus Geoscience, 335*, 1109–1119.

Nair, K. M., Anil Kumar, K. S., Krishnan, P., Naidu L. G. K., & Sarkar D. (2011). Variability of Lateritic Soil Development in Humid Tropical Environment, *Clay Research, 30*(2), 12–20.

Narayana Swamy. (1992). *Geochemistry and Genesis of Laterite in Parts of Cannore District*, North Kerala. PhD thesis. The Cochin University of Science and Technology, pp. 1–111.

Nesbitt, H. W., Markovics G., & Price R. C. (1980). Chemical processes affecting alkalis and alkaline earths during continental weathering. *Geochimica et Cosmochimica Acta. 44*, 1659–1666.

Nesbitt, H. W., & Young, G. M. (1989). Formation and Diagnosis of Weathering Profiles. *Journal of Geology, 97*, 129–147.

Nesbitt, H. W., & Young, G. M. (1982). Early Proterozoic climates and plate motion inferred from major element chemistry of lutites. *Nature, 299*, 715–717.

Nesbitt, H. W., & Markovics, G. (1997).Weathering of granodioritic crust, long-term storage of elements in weathering profiles, and petrogenesis of siliciclastic sediments. *Geochimica et Cosmochimica Acta, 61*, 1653–1670.

Nesbitt, H. W., Young, G. M., McLennan, S. M., & Keays, R. R. (1996). Effects of chemical weathering and sorting on the petrogenesis of siliciclastic sediments, with implications for provenance studies. *Journal of Geology*, 104, 525–542.

Page, A. L., Miller, R. H., & Keeney, D. R. (Eds.) (1982). Methods of Soil Analysis, Part-2, Chemical and microbiological properties. *American Society of Agronomy and Soil Science Society of America.* Madison, Wisconsin, 1159 p.

Pal, D. K., Wani, S. P., Sharawat, K. L., & Srivastava, P. (2014). Red ferruginous soils of tropical Indian environments: A review of the pedogenic processes and its implications for edaphology. *Catena,* 121, 260–278.

Parker, A. (1970). An Index of Weathering for Silicate Rocks, *Geological Magazine,* 107, 501–504.

Price, J. R., & Velbel, M. A. (2003). Chemical weathering indices applied to weathering profiles developed on heterogeneous felsic metamorphic parent rocks. *Chemical Geology,* 202, 397–416.

Price, R. C., Gray, C. M., Wilson, R. E., Frey, F. A., & Taylor, S. R. (1991). The effects of weathering on rare-earth element, Y and Ba abundances in tertiary basalts from south-eastern Australia. *Journal of Chemical Geology,* 93, 245–265.

Ramahashay, B. C., Rao, K. S., Mehta, V. K., & Bhavana, P. R. (1987). Minerology and Geochemistry of lateritic soil profiles in Kerala, India. *Chemical Geology*, 60, 327–330.

Ranger, J., Allie, S., Gelhaye, D., Pollier, B., Turpault, M. P., & Granier, A. (2002). Nutrient budgets for a rotation of a Douglas-fir plantation in the Beaujolais (France) based on a chronosequence study. *Forest Ecology* and *Management*, 171, 3–16.

Raychaudhuri, S. P. (1980). The occurrence, distribution, classification and management of laterite and lateritic soils. *O. R. S. T. O. M., sér. Pédol., uol.*, XVIII, noa, 3–4, 249–252.

Roy, P. D., Caballero, M., Lozano, R., & Smykatz kloss, W. (2008). Geochemistry of late Quaternary sediments from tecocomulco lake, Central Mexico: implications to chemical weathering and provenance. *Chemie der Erde:Geochemistry, 68*(4), 383–393.

Reeder, S., Taylor, H., Shaw, R. A., & Demetriades, A. (2006). Introduction to the chem-istry and geochemistry of the elements. In: Tarvainen, T., de Vos, M. (Eds.), Geochemical Atlas of Europe. Part 2. *Interpretation of Geochemical Maps*, Additional Tables, Figures, Maps, and Related Publications. Geological Survey of Finland, Espoo, pp. 48–429.

Reiche, P. (1943). Graphic representation of chemical weathering. *Journal of Sedimentary Petrology, 13*, 53–68.

Rollison, H. R. (1993). *Using Geochemical Data: Evaluation, Presentation, Interpretation.* Pearson Education Limited, NY, USA, pp. 352.

Ruxton, B. P. (1968). Measure of the degree of chemical weathering of rocks, *Journal of Geolology, 76*, 518–527.

Sahashrabudhe, Y. S., & Deshmukh, S. S. (1981). The Laterites of the Maharashtra State – Proceedings of the International Seminar on Lateritisation Processes (Trivandrum, India 11–14 December, 1979), 209–220.

Sajinkumar, K. S., Anbazhagan, S., Pradeep kumar, A. P., & Rani, V. R. (2011). Weathering and Landslide Occurrences in Partsof Western Ghats, Kerala. *Journal of Geological Society of India, 78*, 249–257.

Sarma, V. A. K., Krishnan, P., & Budhihal, S. L. (1987). Laboratory Methods. NBSS Publ. Technical Bulletin 14. NBSS & LUP. Nagpur.

Satyanarayana, K. V. S., & Thomas, P. K. (1962). Studies on laterite associated soils. II. Chemical composition of lateritic profiles. *Journal of the Indian Society of Soil Science, 10*, 213–222.

Sawarakar A. R. (1980). Geological and Geomorphological features in part of Nilambur valley, Kozhikkode district, Kerala State with special reference to the Alluvial Gold deposits in the area, GSI special publication, *5*, 29–37.

Schellmann, W. (1981). Consideration on the definition and classification of laterites. In Leterization process. *Proceedings of International Seminar on Laterization process*. Trivandrum, India, pp. 1–10. *Geological Society of India*, Kolkata.

Schellman, W. (1986). On the geochemistry of laterites. *Chemie der Erde, 45*, 39–52.

Sehgal, J., Mandal, D. K., Mandal, C., & Vadivelu, S. (1992). Agroecological regions of India. Second edition. Technical bulletin. NBSS & LUP, Publ. 24. p. 130. NBSS & LUP, Nagpur.

Sehgal, J. L. (1998). Red and lateritic soils: an overview. In: Sehgal, J., Blum, W. E., & Gajbhiye, K. S. (eds.) Managing Red and Lateritic Soils for Sustainable Agriculture, Vol. I. Oxford & IBH Publishing Co. Pvt. Ltd, New Delhi, India.

Sheldon, N. D., Retallack, G. J., & Tanaka, S. (2002). Geochemical climofunctions from North American soils and application to Paleosols across the Eocene – Oligocene Boundary in Oregon. *The Journal of Geology, 110*(6), 687–696.

Sivaranjasingham, S., Alexander, L. T., Cady, J. G., & Cline, M. G. (1962). Laterite. *Advances in Agronomy, 14*, 1–60.

Sparks, D. L. (1995). *Environmental Soil Chemistry*. Academic Press, San Diego.

Sreenivas, B., & Srinivasan, R. (1994). Identification of Paleosols in the Precambrian metapeltic assemblages of Peninsular India – A major element geochemical approach. *Current Science, 67*(2), 89–94.

Stewart, B. W., Capo, R. C., & Chadwick, O. A. (2001). Effects of rainfall onweathering rate, base cation provenance and Sr isotope composition of Hawaiian soils. *Geochim Cosmochim Acta, 65*, 1087–99.

Souri, B., Watanabe, M., & Sakagami, K. (2006). Contribution of Parker and Product indexes to evaluate weathering condition of yellow brown forest soils in Japan. *Geoderma, 130*, 346–355.

Tardy, Y. (1997). Petrology of Laterites and Tropical Soils. ISBN 90-5410-678-6. http://www.books.google.com/books.

Tardy, Y., & Nahon, D. (1985). Geochemistry of laterites, stability of Al-geothite, Al-hematite and Fe3+ Kaolinite in Bauxites and Ferricretes:an approach to the mechanism of concretion formation. *American Journal of Science, 285*, 865–903.

Thomas, G. W., & Hargrove, W. L. (1984). The chemistry of soil acidity. In: Adams, F.(ed)., Soil Acidity and Liming. *American Society of Agronomy*, Madison, WI.

Thompson, A., Chadwick, O. A., Rancourt, D. G., & Chorover J. (2006). Iron-oxide crystallinity increases during soil redox oscillations.*Geochim Cosmochim Acta.* 2006. *70*, 1710–1727.

USDA, NRCS. (2008). Official Soil Series Descriptions. Available online at (verified 7 May 2008). a. Official Soil Series Descriptions. Available online at http://soils.usda. Gov/technical/classification/osd/index.html (verified 7 May 2008).

Valeton, I. (1994). Element concentration and formation of ore deposits by weathering. In: Schwarz, Germann, K. (eds.): Lateritization processes and supergene ore formation. *Catena, 21*(2–3), 99–129.

Vogt, T. (1927). Sulitjelmefeltets geologiog petrografi. *Norsk Geologisk Tidsskrift, 121*, 1–560.

Wei G., Ying, L., Xianhua, L., Lei, S., & Dianyong, F. (2004). Major and trace element variations of the sediments at ODP Site 1144, South China Sea, during the last 230 ka and their paleoclimate implications. *Palaeogeography and Palaeoecology, 212*, 331–342.

Widdowson, M., & Cox, K. G. (1996). Uplift and erosional history of the Deccan Traps, India: Evidence from laterites and drainage patterns of the Western Ghats and Konkan Coast. *Earth and Planetary Science Letters, 137*, 57–69.

Widdowson, M., & Gunnell, Y. (1999). Lateritization, geomorphology and geodynamics of a passive continental margin: the Konkan and Kanara lowlands of western peninsular India. *Special Publications: International Association of Sedimentologists, 27*, 245–274.

Wimpenny, J., Gannoun, A., Burton, K. W., Widdowson, M., R. H. James, R. H., & R. Gíslason, S. R. (2007). Rhenium and osmium isotope and elemental behavior accompanying laterite formation in the Deccan region of India. *Earth and Planetary Science Letters, 261*, 239–258.

Yang, S. Y., Li, C. X., Yang, D. Y., & Li, X. S. (2004). Chemical weathering of the loess deposits in the lower Changjiang Valley, China, and paleoclimatic implications. *Quaternary International, 117*, 27–34.

Yousefifard, M., Youbi, S. A., Jalallian, A., Khadun, H., & Makkizaddi, M. A. (2014). Mass balance of major elements in relation to weathering in soils developed on Igneous rocks in semi arid region, North Western Iran. *Journal of Mountain Science, 9*, 41–58.

Zabel, M., Schneider, R. R., Wagner, T., Adegbie, A. T., Vries, U. D., & Kolonic, S. (2001). Late Quaternary climate changes in central Africa as inferred from terrigenous input to the Niger fan. *Quaternary Research, 56*, 207–217.

PART II

GEOSPATIAL TECHNOLOGIES IN LAND RESOURCE MAPPING AND MANAGEMENT

CHAPTER 10

GEOSPATIAL TOOLS AND TECHNIQUES IN LAND RESOURCE INVENTORY

SURESH KUMAR

Scientist–SG and Head, Agriculture and Soils Department, Indian Institute of Remote Sensing, 4 Kalidas Road, Dehradun – 248 001, India

CONTENTS

ABSTRACT

Soil, an important component of land resource inventory is necessary to deal with the issues of food security, environmental problem, sustainability of natural resources and climate change impact on soil resource degradation and their sustainability. Satellite remote sensing data and integrated use of Geographic Information Systems (GIS) and Global Positioning System (GPS) technologies offers unique potential in soil resource inventory. Optical remote sensing data are commonly being used for retrieving thematic information. Several Global freely available Digital Elevation Models (DEMs) are very useful sources to derive topographic parameters facilitating in digital soil mapping. The growth of open source softwares has a particular impact on the potential to analyze geospatial data. They are providing large opportunities for predictive soil mapping to generate accurate spatially explicit soil maps. Soil surveyors follows soil–landscape model to capture the relationship between soil and their formative environment. The spatial extents of these soil formative environments are interpreted using satellite data to delineate soil–landscape units known as physiographic units for soil mapping. In digital soil mapping, laboratory analyzed soil data and remote sensing data are integrated with statistical methods to infer spatial pattern and distribution of soil properties. It follows SCORPAN equation in spatial prediction of soil types. Various approaches such as multivariate statistical analysis, geostatistics, artificial neural network (ANN), expert systems, etc., have emerged as advanced tools in digital soil mapping. Statistical regression models, generalized linear models (GLMs) and general additive models (GAMs) are used in predictive soil mapping. Geostatistics in association with GIS serves as an advanced method in quantifying the spatial pattern of soil properties and environmental variables. Hyperspectral remote sensing data are being used to derive spectral information of soil properties and to map them quantitatively. However, to retrieve soil hydrological properties, microwave remote sensing data are used.

10.1 INTRODUCTION

Land Resource Inventory plays a vital role in natural resource management of a country. It is essential for appraisal of land resource potential, land use planning and policy-formulation. Existing soil information of many countries including India is often at inappropriate scales to meet the users need for optimal planning and sustainable use of land resources. Available soil and terrain information is generally old, and has been derived from surveys following inconsistent approaches and methods. These data do not adequately represent soil variability at an appropriate resolution in space or time. Soil resource information is available at smaller scales and is not up-to-date. It poses a major bottleneck in modeling environmental impact assessment as well as potential assessment at larger scale to meet the need at operational level. A new, comprehensive survey of soil resources is unlikely to happen in the near future. Therefore, creation of a new, more reliable and comprehensive soil database requires alternative, easy and inexpensive methods for generating soil information.

Geospatial tools such as Remote Sensing (RS), GIS, GPS, geostatistics and 3D visualization have emerged as most powerful tools in assessing and monitoring the land resources for sustainable development, utilization and conservation planning. Spatially explicit data of soil properties and nutrients are needed for better assessment of soil resources and precision agriculture. The advent of GIS, the greater accessibility of high-resolution multispectral and hyper-spectral remote sensing data and the development of spatial statistical techniques greatly enhanced the ability to predict spatial distribution of soils. In recent years, availability of high-resolution satellite data and terrain data has provided large opportunities for predictive soil mapping to generate accurate spatially explicit soil maps (McBratney et al., 2003).

RS data available in variety of spatial, spectral and temporal resolutions are being widely used in land resource mapping, monitoring and evaluation of land resources. High-resolution stereo satellite data and microwave SAR (synthetic Aperture Radar) data have been used to generate DEMs to study topography of earth surface. Today, several Global DEMs such as Shuttle Radar Topographic Mission (SRTM) (http://earthexplorer.usgs.gov; http://glovis.usgs.gov; http://glcf.umiacs.umd.edu); and Advanced Spaceborne Thermal Emission and Reflection radiometer (ASTER) DEMs http://ws.csiss.gmu.edu/DEMExplorer) are freely available for the users. India also generated CartoDEM (30 m) from stereo Cartosat satellite data and it is freely available in public domain. (http://www.bhuvan.nrsc.gov.in). These DEMs are very useful sources to derive topographic parameters comprising of a set of primary and secondary terrain attributes useful in digital soil mapping (DSM) applications. It provides advanced method for improving soil information and spatial analysis of soil data. Knowledge of spatial variation of soil properties is important in precision farming and environmental modeling. Future studies will focus on the improved integration of proximal and remote sensing using scaling based approaches to make optimal use of all data sources available (Mulder et al., 2011).

In recent years, the rapid development of geospatial technologies (RS, GIS, GPS and computer software) have made available new tools and capabilities to land resource planner and users for inventory and mapping of soil resources at various scales. The growth of open source softwares have had a particular impact on the potential to analyze geospatial data. This chapter deals with the potential of geospatial tools and techniques being used by soil scientists and researchers in soil resource mapping.

10.1.1 LAND RESOURCE INVENTORY: GLOBAL PERSPECTIVE

In recent decades, the soil science community has made tremendous efforts in generating regional and global soil databases. Digitized soil maps are available for most of the world at smaller scale of 1:1 million scale or coarser (Grunwald et al., 2011). These digital maps are basically compilation of soil maps prepared by various organizations through digitization of hard copies. Currently, there are several soil databases available at map scales smaller than 1:250,000; namely the Harmonized World Soil Database at a map scale of 1:5 million scale developed by the FAO-UNESCO (FAO et al., 2008); Soil Survey Geographic Data Base (SSURGO) (Soil Survey Staff, 2010b); the Australian Soil Resource Information System (ASRIS) (CSIRO, 2010); Canadian Soil Information System (CANSIS) and the National Soil Database (NSDB) of Canada (AAFC, 2010) and the Russian soil map at 1:2.5 million scale (Stolbovoi and McCallum, 2002). The European Soil Database available at 1:1 million scale (Le Bas et al., 1998). The Soil and Terrain Digital Database (SOTER), which incorporates quantitative information on soils and terrain at map scales 1:1 million and 1:5 million scale (Oldeman and van Engelen, 1993) is available of world soil.

10.1.2 LAND RESOURCE INVENTORY: INDIAN PERSPECTIVE

Under FAO/UNESCO's scheme of World Soil Map project, soil map of India was generated with 25 broad soil classes represented on a 1:7 million scale map (Govinda Rajan, 1971). This map was refined following USDA soil classification at the group level and published on 1:6.3 million scale (Murthy and Pandey, 1983). During the past 30 years, the National Bureau of Soil Survey and Land Use Planning (NBSS & LUP), mapped the soil resource of the country on 1:7 million scale at sub-order association level.

Later, NBSS & LUP surveyed soils of the country and published soil maps for all the states and union territories of the country on 1: 250,000 scale providing soil information at the level of family association (Bhattacharyya et al., 2009). Soil map at 1:50,000 scale has been prepared for a part of the country covering nearly 36% of total geographical area under various operational projects (DOS, 1999). Soil information of very limited area of the country is available at larger than 1:50,000 scale.

10.2 GEOSPATIAL TOOLS

Geospatial tools include earth observing RS, GIS, GPS, Digital Terrain Models (DTMs) and terrain analysis (TA) softwares. These tools are used for creating, viewing, managing, analyzing, and utilizing geospatial data in land resource inventory. These tools are discussed as below:

10.2.1 SATELLITE REMOTE SENSING DATA

Satellite remote sensing provides information on various earth surface features. It collects emitted and reflected energy from the objects, which help in inventorying, mapping and monitoring of land resources, including soils. Most of the earth observing remote sensing satellite collects data in one or several bands of the visible, infrared (IR) or microwave portions of the spectrum. Each part of the spectrum provides typical information about various earth resources. Optical remote sensing data such as Indian Remote Sensing (IRS), IKONOS, QUICKBIRD, Landsat Thematic Mapper (TM), SPOT, ASTER and Moderate Resolution Imaging Spectroradiometer (MODIS) are being commonly used as representatives for retrieving thematic information of soil types, vegetation, topography and parent material (geology and geomorphology) information. Higher resolution satellite data provides more spatial variation information of earth surfaces. Medium resolution satellite data (Landsat MSS, Landsat TM, and IRS IC/1 D) with 80–180 m spatial resolution are suitable for soil resource mapping at 1:50,000 and 1:100,000 scale. Very high-resolution satellite data (Resourcesat, Cartosat, IKONOS and Quickbird) with spatial resolution of 1 to 4 meters are suitable for mapping of soils at 1:25,000 to 1:10,000 scale. A list of various remote sensing data available to derive earth surface features including soils is described in Table 10.1.

TABLE 10.1 Remote Sensing Satellite Data, Their Resolutions and Level of Mapping

RS data	Sensor*	Spectral bands**	Spatial resolution (meter)	Frequency of revisit (days)	Scale of mapping
Fine resolution					
Cartosat-2	PAN	1	1	24	1:8,000 or larger
Cartosat-1	PAN	1	2.5	24	1:10,000
Resourcesat-1,2	MS	3	5.8	24	1:25,000
IKONOS	MS	4	1–4	5	1:10,000, 1:25,000
QUICKBIRD	PAN	1	0.61	5	1:5,000
	MS	4	2.44		1:10,000
WORLDVIEW	PAN	1	0.55	2–6	1:5,000
SPOT	PAN	1	2.5	3–5	1:10,000
	MS	4	20		
Medium resolution					
IRS – 1C/1D	MS	4	23.5	24	
Landsat-7 ETM+	MS	8	30	16 days	1:50,000
ASTER	MS	15	15, 30, 90	16	
SPOT	MS	4	20	5	
Coarse resolution					
MODIS	MS	7	250, 500, 1000	daily	1:2,50,000
AVHRR	MS	5	100	daily	
IRS-1C/1D	MS	3	56	5	

*PAN: Panchromatic band; MS: Multispectral bands.

**4 bands refers to blue (0.52–0.59 μm), green (0.62–0.69 μm), red (0.77–0.86 μm) and near infra-red (0.77–0.86 μm) bands whereas 4 bands refers to green, red, infrared and short wave infrared (1.55–1.70 μm). 3 bands refer to green, red and near infra-red bands and. 2 bands refers to red and near infrared bands. 1 band for panchromatic band (0.5–0.85 μm) only.

10.2.1.1 Hyperspectral Remote Sensing (HRS)

Spectrometry (IS) or hyperspectral technology, as an advanced tool that provides high spectral resolution data significantly broadens the utility for mapping of the soil surface from a more precise chemical and physical point of view. Hyperspectral remote sensing data sets are composed of about 100 to 200 spectral bands of relatively narrow bandwidths (5–10 nm), whereas, multispectral broad band remote sensing data sets are normally composed of about 5 to 10 bands of relatively large bandwidths (50–100 nm). Hyperspectral imaging capacity in remote sensing was a major breakthrough that opened the avenues of research in various fields like mineralogical mapping, environmental geology, vegetation sciences, soils, and hydrology. The launch of NASA's EO-1 Hyperion sensor in November 2000 marked the establishment of spaceborne hyperspectral capabilities. HySPIRI is a new NASA initiative to be placed in orbit in years to come to cover the globe on periodical basis. Ben-Dor et al. (2009) have given a detailed review of application potentials of imaging spectroscopy to study various soil properties.

10.2.1.2 Microwave Remote Sensing Data

Microwave remote sensing satellites operate in frequency bands in the range of 3 GHz to 30 GHz. There are both passive and active sensors used in microwave remote sensing. Spaceborne active microwave sensors, synthetic aperture radar (SAR), are able to provide high spatial resolution (up to 10 m) and are more sensitive to surface characteristics than passive microwave sensors. However, passive microwave sensors (radiometers) provide low spatial resolutions (20 to 50 km). Active microwave SAR sensors at C-band (4.0–8.0 GHz) and L-band (1.0–2.0 GHz) frequencies are commonly used for soil moisture studies.

European Remote Sensing (ERS) satellites have been providing active microwave remote sensing data in C-band (5.3 GHz) for 1991–1996 (ERS-1) and as ERS-2 since 1996. India launched its first civilian microwave satellite – Radar Imaging Satellite-1 (RISAT-1) on April 26, 2012. It operates in C-band (5.35 GHz) and provides data with resolution from 1–50 m with multi-polarization capabilities. The Advanced Microwave Scanning Radiometer on Earth Observing System (AMSR-E) on the AQUA satellite is a passive microwave sensor. It provides brightness temperature at five frequencies. AMSR-E C-band (6.9 GHz) and X-band (10.7 GHz) channels are suitable for soil moisture assessment (Njoku et al., 2003). TRMM Microwave Imager (TMI) is measuring microwave emission at five frequencies from 10.7 GHz to 85.5 GHz since 1997.

10.2.1.3 Geographic Information System (GIS)

GIS is a computer based information system capable of capturing, storing, analyzing, and displaying geographically referenced information. It stores spatial as well as non-spatial data and work with both the vector and the raster geographic models. The vector model is generally used for describing the discrete features, while the raster model does it for the continuous features. Various commercial and open source GIS softwares are available for the users. Open source softwares have emerged as most promising tools in analyzing natural resources.

Free and Open source GIS softwares such as ILWIS, GRASS, PCRaster, QGIS, PostGIS, OSSIM, GVSIG, Mapguide, SAGA and uDig are available (Table 10.2) to the user for geospatial analysis. These softwares generally allow the user to create modifications of the software, port it to new operating systems and processor architectures and share it with others. Lists of some freely available GIS software are given below:

10.2.1.4 Global Positioning System (GPS)

GPS is a satellite-based positioning system, which operates using L-band radio signals to provide highly accurate positional data. GPS accuracies typically range between 15–25 meters (standard GPS positioning); in differential mode (differential GPS positioning or DGPS) the manufacture-specified accuracies range from 1–5 meters. The GPS, which consists of 24 satellites in near circular orbits at about 20,200 km altitude, which provides full coverage with signals from minimum 4 satellites available to the user, at any place on the earth. The GPS signals are made freely available to users and its global importance is unlikely to diminish in spite of other options becoming available in near future. Indian Regional Navigational Satellite System (IRNSS) consists of constellation of seven satellites and it is designed to offer accurate position information within India and up to 1,500 kilometers outside of the country's boundaries. This system would provide a position accuracy of better than 20 meters in the primary service area.

10.2.2 DIGITAL ELEVATION MODELS (DEMS) AND TERRAIN ANALYSIS SOFTWARES

Topography is one of the most important characteristics of the earth surface. It can be characterized with the use of DEM. DEM refers to the digital representation of the continuous variation of relief (elevation) over space. The term digital terrain model (DTM) is also commonly used. Because the

TABLE 10.2 Open Source GIS Software

GIS Software	Salient features	Website
GRASS	Geographic Resources Analysis Support System, commonly referred to as GRASS, is a Geographic Information System (GIS) used for geospatial data management and analysis, image processing, graphics/maps production, spatial modeling and visualization.	http://grass.itc.it/
PCRaster	The PCRaster Environmental Modeling language is a computer language for construction of iterative spatiotemporal environmental models. It is raster-based analysis with strong hydrological modeling.	http://pcraster.geo.uu.nl/downloads.htm
QGIS	Quantum GIS (QGIS) is a user friendly Open Source Geographic Information System (GIS) that runs on Linux, Unix, Mac OSX, and Windows. QGIS supports vector, raster, and database formats.	http://www.qgis.org/
PostGIS	PostGIS adds support for geographic objects to the PostgreSQL object-relational database. Used as a backend spatial database for geographic information systems (GIS)	http://postgis.refractions.net/
GVSIG	gvSIG is a desktop tool designed to manage geographic information. It is characterized by a user-friendly interface that can easily access the most common raster and vector formats.	http://www.gvsig.gva.es/
ILWIS	The Integrated Land and Water Information System (ILWIS) is a PC-based GIS & Remote Sensing software, developed by ITC. ILWIS comprises a complete package of image processing, spatial analysis and digital mapping.	http://www.itc.nl/ilwis
SAGA	GIS designed with geosciences in mind, especially terrain and hydrographic analysis. Powerful raster analysis and programmability.	http://www.saga-gis.org/
uDig	User-friendly Desktop Internet GIS (uDig) both a Geospatial application and a platform through which developers can create new derived applications.	http://udig.refractions.net/

term "terrain" often implies attributes of landscape other than altitudes of the landscape. The term DEM is preferred for models containing only elevation data. Digital elevation data sets are generated from stereo optical satellite data, SAR Interferometry, digitizing the contour lines on topographic maps or conducting ground surveys (Wilson and Gallant, 2000).

Most commonly used Global DEM are SRTM and ASTER. SRTM global elevation data is available between 60°N and 56°S. SRTM DEM data are available in three different spatial resolution of 30 arc seconds (1 km), 3 arc-seconds (90 m) resolutions for the world and 1 arc-second (30 m) resolution. The ASTER global DEM with spatial resolution of 30 m is available between 83°N and 83°S of the earth. It has absolute accuracies of 20 m vertically and 30 m horizontally. CartoDEM of Indian region is available at 30 m spatial resolution. CartoDEM derived from Cartosat-1 stereo satellite data has planimetric accuracy of 15 m and vertical accuracy of 8 m (Krishnamurthy et al., 2008). ALOS/PRISM is a panchromatic remote-sensing instrument specially designed for stereo mapping and can be used to generate fine resolution DEM with absolute vertical accuracy of about 6 m and the absolute horizontal accuracy is 2.5 m. LIDAR data sets provide very high resolution DEM but are very costly. Many different airborne LIDAR sensors, such as Airborne Laser Terrain Mapper (ALTM) of Optech are available that provide DEM with vertical accuracy of 8–11 cm and a horizontal accuracy of 2–3cm on the ground. These DEMs are used to derive various terrain characteristics using terrain analysis software inbuilt with GIS software. Some of the terrain analysis softwares available freely on websites are listed below (Table 10.3).

10.3 SOIL RESOURCE INVENTORY

The spectral reflectance of soils is a cumulative property, which derives from the inherent spectral behavior of heterogeneous combinations of minerals, organic matter and soil water (Baumgardner et al., 1985). Spectral absorptions and reflectance changes in the Vis-NIR (400–2500 nm) range of the reflected electromagnetic radiation provides diagnostic features, which can be used to identify directly important mineralogical constituents of soils. Soil properties such as soil organic matter, texture, moisture content and aggregation have direct influence on Vis-NIR spectra. The Vis-NIR spectra are influenced not only by the chemistry of a material,

TABLE 10.3 Terrain Analysis Software

Terrain Software	Salient features	Website
Landserf	Surface analysis package, Java based, cross-platform with excellent analysis and visualization facilities.	http://www.landserf.org
ILWIS	PC-based GIS & Remote Sensing software developed by ITC. Support terrain analysis/hydrological analysis.	http://www.itc.nl/ilwis
TAS (WHITEBOX)	Terrain Analysis System – Compact, stand-alone program. Provides wide range of terrain analysis/hydrological analysis functions and index computations. Now superseded by WHITEBOX	http://www.uoguelph.ca/~hydrogeo/Whitebox/index.html
TAUDEM	Terrain Analysis Using Digital Elevation Models – ArcGIS Add-in/toolbar. Provides wide range of terrain analysis/hydrological analysis.	http://hydrology.usu.edu/taudem/taudem5.0/
TOPAZ	Digital Topographic and Watershed Analysis. Support most geographic information systems (GIS) for display, registration to other data layers and spatial analysis.	http://gcmd.nasa.gov/records/TOPAZ.html
Quantum GIS (QGIS)	Desktop GIS software support raster and vector analysis, terrain analysis and hydrological analysis.	http://www.qgis.org/

but also by its physical structure (Ben-Dor et al., 2006). Satellite imageries (Standard False Color Composites – FCCs) are commonly used to interpret soil-forming factors governing variability of soils represented by soil–landscape units. Satellite imageries provide synoptic coverage of large area that facilitates in interpreting soil forming factors by visual approach. Soil forming factors affect the soil pedogenic processes that determine soil development and its soil properties. Soil development variations are further explained by catena concept that attributes soil variation. It follows conventional soil–landscape model (CSM) based on the Clorpt model which define the state of a soil system governed by soil forming factors (Jenny, 1941). The state factor model is expressed by the following equation:

$$S = f(cl, o, r, p, t)$$

where, soil (S) is considered to be a function of climate (cl), organisms (o), relief (r), and parent material (p) acting through time (t).

10.3.1 VISUAL ANALYSIS APPROACH AS CONVENTIONAL METHOD IN SOIL RESOURCE INVENTORY

Conventional soil survey is based on the soil–landscape equation or concept. The soil–landscape model captures the relationships between the soils in the area and the different landscape units. Soil surveyor detects different soil formative environments through visual interpretation of geological maps, topographical maps and satellite images. The spatial extents of the soil formative environments are then used to delineate soil–landscape units known as physiographic units. These physiographic units are often initially delineated on satellite images, then field checked and compiled onto a base map. Soil profiles are then studied in these physiographic units to characterize and classify the soils to prepare physiographic-soil map (Biswas, 1987; Karale et al., 1981; Mulders and Epema, 1986).

Using the satellite imageries of an area, a soil scientist learns many relationships between the image characteristics and soil and landscape features, but many uncertainties inevitably remain. Some landforms are less easily identified, but most images contain clues that narrow the choices of the kinds of landforms represented. Experience in interpreting tone patterns, configuration of relief, and patterns of drainage ways commonly permits correlation of these patterns with kinds of geologic deposits and geomorphic features in an area. Topography (relief) can be interpreted with drainage feature or constructing 3D of the area using DEM. Topography features help to locate many soil boundaries on the map. It also identifies many kinds of landforms, which are commonly related to kinds of soil. Many landform terraces, flood plains, sand dunes, etc., can be identified and delineated reliably from their shapes, relative heights, and slopes. Further divisions within these units can be made based on land use/land cover and erosion hazards. Thus, soils are surveyed and mapped though visual interpretation following a 3-tier approach comprising interpretation of remotely sensed imagery, field survey (including laboratory analysis of soil samples) and cartography (Sehgal et al., 1988).

10.3.2 DIGITAL APPROACH IN SOIL RESOURCE INVENTORY

In traditional soil mapping the surveyor translates soil–landscape models to lines on maps (e.g., soil classes), whereas in digital approach environmental variables are translated in a digital framework. The mapping of landscape properties can be used to distinguish areas of soils having similar causal factors (such as climate, topography, and parent materials) and are involved in the formation of both landscapes and soils. Soil landscape facets are a way of dividing a soil landscape into discrete sub-units, each containing a distinct soil type or suite of soil types. A common way of spatially segmenting the landscape is to divide it into internally more or less homogeneous and mutually contrasting landform units. Soil delineation depicted on a soil map is more closely linked to geomorphic processes. Landform maps are typically suitable predictors of soil types because soil development often occurs in response to the underlying lithology and water movement in the landscape. Digital soil–landscape modeling can be achieved by generating landform map and integrating with topography (slope/aspect) and vegetation cover types. In a small area or at large scale in area of uniform geology, landform elements/types govern the parent material of soils.

Digital approach of soil mapping requires incorporation of landforms, topography (slope/aspect) and vegetation (land use/land cover) information to derive soil–landscape unit. The DEM derived information are integrated in GIS with landform and geomorphologic units for their further subdivision in order to derive soil–landscape map. It helps to delineate precise soil boundaries. Soil patterns are intimately related to slope, aspect and elevation. Klingebiel et al. (1987) demonstrated by their results, major benefit of using DEM derived data to improve the quality of soil maps. The important merit of this approach is that soil landscape patterns are much more evident than with conventional soil map. It requires generating digital landform types and terrain parameters information to derive landscape unit which can be discussed as below:

10.3.2.1 Landform Analysis

Terrain attributes (TAs) play an important role in quantitatively describing the characteristics of landform and its structure. Using DEM/3D modeling, landscape can be better visualized leading to a better understanding of certain relation in the landscape. Terrain plays a fundamental role in modulating the earth surface and landform configuration frequently governs the movement of materials and water on the landscape (Moore et al., 1993; Park and Vlek, 2002). Many recent efforts have outlined techniques for processing DEM data to automatically classify landforms (Burrough et al., 2001; MacMillan et al., 2000).

10.3.2.2 Terrain Analysis

Digital terrain characterization is the quantitative analysis of topographic surfaces. The automation of terrain analysis and the use of DEMs facilitate easy quantification of the topographic attributes. There are primary terrain parameters and some more complex parameters which are derived with the combined use of the primary terrain parameters (Table 10.4). The primary

TABLE 10.4 Major Primary and Secondary Terrain Attributes from DTM for Soil Mapping

Terrain parameters	Description	Relation with soil type
Primary		
Slope	Steepness of the area	Soil varies along the hill slopes described as catena concept.
Aspect	Slope azimuth	Soil varies with change in aspect on same parent material. In general soil of northern aspect are more developed than southern aspect.
Specific Catchment Area (As)	Total upslope catchment plan area draining across a unit length of contour.	Governs erosion and deposition processes in the landscape.
Secondary parameters		
Terrain Wetness Index (TWI)	TWI = ln (As/tanβ) Where, As is specific catchment area and β is slope in radians.	It is well studied indicator of soil property and soil moisture distribution at different landscape positions.
Stream Power index	SPI = As * tanβ	Erosive power of overland flow and could be used to identify suitable locations for soil conservation measures to reduce the effect of concentrated surface runoff.
Sediment Transport Index	STI = (As/22.13)0.6* (sinβ/0.0896)1.3	Used to characterize the water flows in landscapes.

(Source: Wilson J. P., & Gallant J. C. 2000).

attributes are direct descriptors of the terrain features, like the slope, curvature or aspect, while secondary attributes describe more complex characteristics of the landform, which are linked to certain terrain-regulated processes, like stream power index (SPI) or the compound topographic index (CTI). These terrain attributes quantify hydrological, environmental, vegetation, climatic and other aspects of a terrain. These terrain parameters can be used for calculating "terrain-adjusted" climatic variables, like temperature, solar irradiation, long wave surface radiation, reflected radiation, etc., which are important factors in the energy balance of the surface and thus in the soil formation. Among major five factors of soil formation, the influence of topography is so dominant that it can be used to predict the spatial distribution of soils in many landscapes.

10.3.2.3 Deriving Soil–Landscape Units: An Example

The soil–landscape units were prepared for a watershed lying in Tehri Garhwal district of Uttarakhand state, India. It was generated by integration of landform, aspect map and land use/ land cover map. ASTER DEM of 30 m spatial resolution was used to generate topographic position index (TPI) using ArcGIS 10.1 software. The landscape was delineated for the landform elements namely ridgetop, shoulder, backslope, toe slope and valleys based on Topographic Position Index (TPI). Using TPI and slope, landscape was delineated into slope position (i.e., ridgetop, valley, mid slope, etc.). The landform with the negative value, depicted a valley, positive value was interpreted as hill top while near to zero was classified as mid slope or flat surface. According to the TPI values, landform map was divided into five classes *viz.,* hill top, upper slope, middle slope, lower slope and valley. Aspect map was also generated using DEM. Major aspect classes (north and south aspects) were considered for generating soil landscape units. Land use/land cover map was prepared by visual analysis of satellite data. All the three maps were integrated in GIS software to derive soil–landscape units in the watershed. After intersecting and generalization, 15 different soil–landscape units namely, H1 – Hill top, H2 – Upper hillslope, H3 – Mid hillslope, H4 – Lower hillslope and H5 – valley where A and F represent agriculture and forest cover types and 1 and 2 as north and south aspects of the hillslopes, respectively were identified to map soil types in the watershed (Figure 10.1).

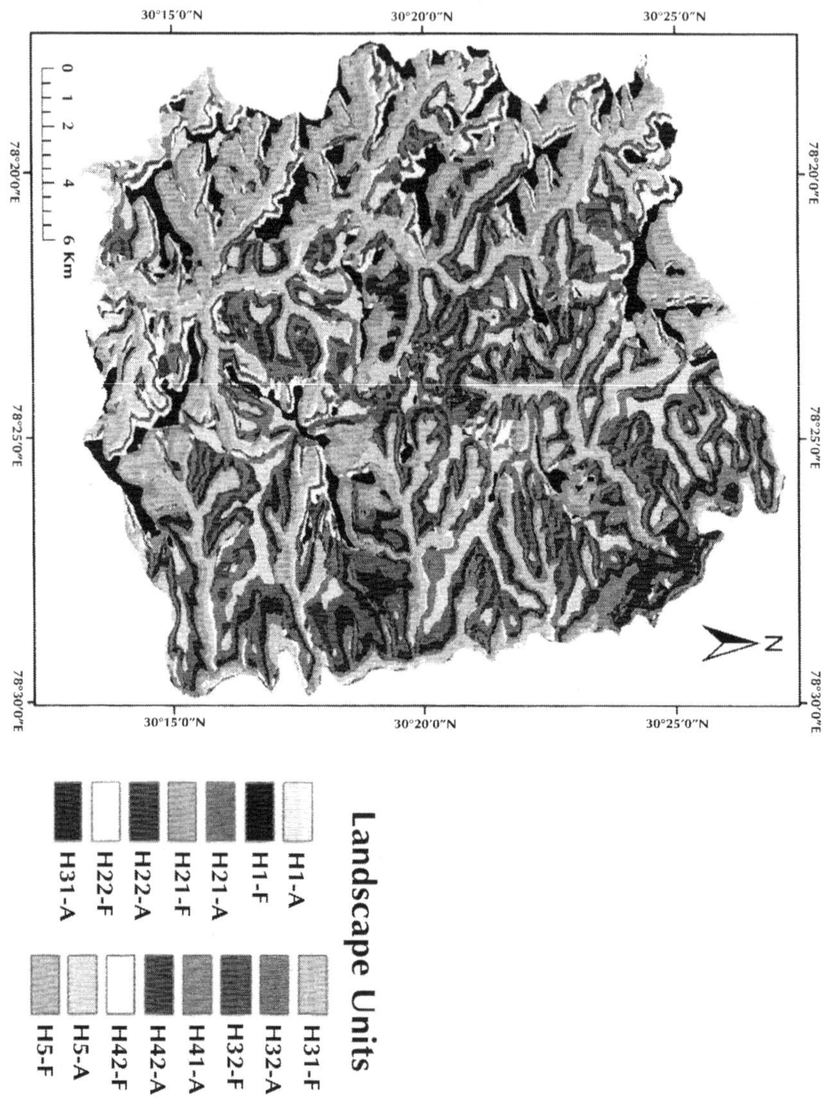

FIGURE 10.1 Landscape units in the watershed.

10.3.3 DIGITAL SOIL MAPPING (DSM)

There is considerable increase in demand of digital soil maps showing continuous varying soil properties. It refers to the computer-aided analyzes and prediction of soil properties and/or soil classes. DSM uses measured soil data with remote sensing and terrain attributes and a range of (geo) statistical modeling to predict the characteristics of soil across landscapes. DSM uses this concept to develop empirical models that relate observations of soil properties with environmental variables governed by soil forming factors (i.e., CLORPT). McBratney et al. (2003) identified 7 factors (C: Climate; O: Organism; R: Relief; P: Parent material; A: Age, time; and N: Geographic position) for soil spatial prediction and formulated the so called SCORPAN equation:

$$Sa = f(S, C, O, R, P, A, N) \text{ and } Scl = f(S, C, O, R, P, A, N)$$

where, Sa is the estimated soil attribute value and *Scl* is the estimated soil property class.

The generation of digital soil maps using these input environmental variables representing the six factors in the SCORPAN model is integral part of digital soil resource inventory. Satellite Remote sensing data provide indirect information of some of these environmental factors. These factors are described as:

Climate factor is represented by mean annual temperature (T) and mean annual rainfall (P) which is mainly used to derive potential evapotranspiration (PET). PET and available soil water capacity (AWC) are used in characterizing soil moisture regimes (SMR). Karimi and Bastiaanssen (2014) reviewed the reliability of remote sensing algorithms to accurately determine the spatial distribution of actual evapotranspiration, rainfall and land use. The main soil forming factor is organism (O), which can be described as macro-organism (i.e., land use/land cover. Remote sensing data are widely used to obtain land use/ land cover information as well as biomass estimation various land use and covers. Remote sensing provides information concerning the spatial distribution and the temporal variability of dynamic land cover characteristics. Bishop and McBratney (2001) used yield-monitored wheat yield to aid in the prediction of soil clay content.

Topography or relief (R) has been recognized as one of the soil forming factors (Jenny, 1941). It denotes the configuration of the land surface. Locally it influences the climate and vegetation in the area. It defines the way how the water moves through the landscape and transport soil materials into the soil. DEMs are used to derive various terrain parameters such as altitude, slope, aspect, different curvatures, upslope area and compound topographic index, in predicting soil attributes and soil classes (McKenzie et al., 2000). Parent material (P) provides the bulk and has the distinct effect on soils. Parent material information can be obtained from digitized geological maps. It provides information of lithology that determines the physico-chemical properties of soils. Elapsed time (A) represents age depicting how long pedogenesis has been occurring and could differentiate soil classes and properties. Spatial location (N) information is now much easier due to the advent of GPS. Nowadays, several GPS are available that offer spatial accuracy of 1-m meter. It is not a factor of soil formation but provide coordinate of sample location to study spatial pattern and its association with environmental variables. It assists in integrating environmental parameters and analyzes their relation.

10.3.4 VARIOUS METHODS OF DIGITAL SOIL MAPPING (DSM)

There are various approaches described in DSM (Hengl and MacMillan, 2009). DSM is indeed flexible, quantitative and accurate. The choice of the approach is determined by a number of factors such as expert knowledge; soil–landscape complexity; the availability of datasets (predictor, environmental variables), and the competency of expert. Some of the important approaches in DSM are described below:

10.3.4.1 Statistical Based Regression Model: Prediction of Continuous Soil Attributes

Regression models fall into two categories: Generalized Linear Models (GLMs) and General Additive Models (GAMs). GLMs assume a linear relationship between predictor and target variables, with regression coefficients as output. A statistically-based system develops quantitative models generated from a model fitting process. It defines the relationship between predictor and environmental variables and co-variables for spatial predictions. McKenzie and Austin (1993) used GLMs to predict soil attributes (clay content, CEC, EC, pH and bulk density) using environmental variables as predictors. Minasny and McBratney (2002) predicted soil hydraulic properties in the form of pedotransfer functions. Chang and Islam (2000) predicted soil texture from multi-temporal remotely sensed brightness temperature and soil moisture maps. Linear models with independent variables such as slope, curvature, wetness index and soil profile information are used to derive soil attribute maps. GAMs are suitable for non-linear relationships between predictor and target variables. The most suitable of non-linear models is the logit model, which is built using logistic regression. Campling et al. (2002) used (multiple) logistic regression to model soil drainage classes from terrain attributes and vegetation indices as calculated from a Landsat TM image.

10.3.4.2 Tree Models-Classification and Regression

Classification and regression tree analysis is a data mining technique. Tree-based models are fitted by successively splitting a database of continuous and/or discrete environmental covariates into increasingly homogenous groupings. The result of model fitting is a decision tree that can be employed to predict individual soil property values or class types over an area of interest. It can handle both categorical and continuous data, for prediction of discrete soil classes or continuous soil attributes. The advantage of regression trees over linear model is the ability to deal with nonlinearity. Regression trees are widely used for the prediction of soil attributes (Pachepsky et al., 2001; Shatar and McBratney, 1999).

10.3.4.3 Geostatistics

Geostatistics are extensively used for quantifying the spatial pattern of soil properties and environmental variables (Webster and Oliver, 2001; Zhang and McGrath 2004). It provides advanced tools to carry out spatial interpolation. Geostatistical methods for quantitative soil survey have proven useful at large scales but their utility at intermediate and smaller scales is less clear (Webster, 1997). It allows for more in-depth analysis of prediction uncertainties and spatial processes compared to the 'CLORPT' approach.

Geostatistical models, in general, require higher sampling densities and spatial dependence of the observations (Hengl et al., 2003). It is data driven rather than knowledge driven, and deal with geographical space rather than feature space. Geostatistical interpolation techniques such as kriging and co-kriging are widely used to predict soil properties at unsampled locations and to better understand their spatial variation in the landscape. Kriging uses a mathematical model of the semi-variogram to calculate estimates of the surface at the grid nodes. It calculates the value at the interpolation point using a weighted average of the values at the data points. Simple kriging may not be the method of choice in heterogeneous areas because the mean is deemed constant and known throughout the area. It provides satisfactory results in homogenous landscapes (Goovaerts, 1999). In Co-kriging, spatial dependency of soil properties can be improved by incorporating secondary variables (Zhang and McGrath, 2004). Studies have showed that co-kriging improves over kriging in predicting soil properties (Goovaerts, 1997; Ping and Dobermann 2006).

In this chapter, a study carried out for spatial prediction of total soil carbon (TC) at watershed level using geo-statistical analysis was presented. Soil samples in the watershed were collected in grid sampling approach. Total 113 soil samples were collected spaced at 250 m (Figure 10.1). Soil carbon in the watershed ranged from 0.22 to 2.66%. CartoDEM having spatial resolution of 10 m was used to derive terrain attributes such as slope and terrain wetness index (TWI). A significant correlation between TWI and total soil carbon was found. Geostatistical method such as Kriging and Cogriging was used to map total carbon in the watershed. Cokriging was attempted with TWI as associated terrain parameter. The study revealed that Cokriging with TWI has improved the spatial prediction of soil nutrients. Figure 10.2 (a1 and a2) shows the spatial predicted maps with both the methods.

10.3.4.4 Fuzzy Logic Approach

Soils appear to be more of continuous variables than discrete objects (Qi et al., 2006). Many studies have shown that within soil map unit variance is often

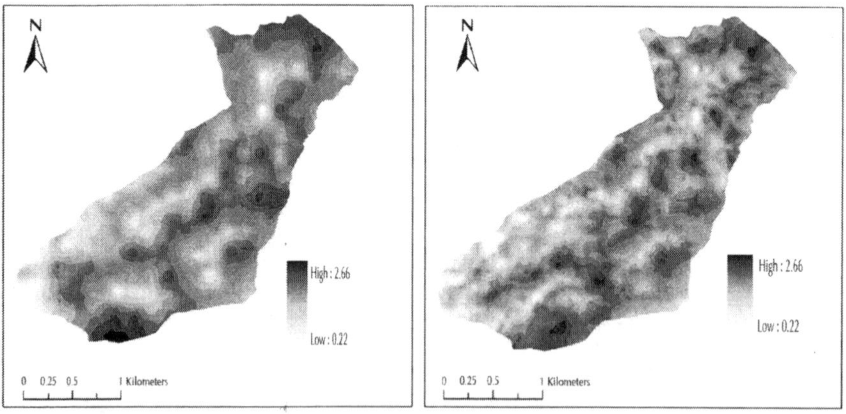

FIGURE 10.2 Spatial prediction of soil nutrient (total soil carbon %) by ordinary (a1) and cokriging with TWI (a2). (Reprinted from Kumar, S., & Singh, R. P. (2016). Spatial distribution of soil nutrients in a watershed of Himalayan landscape using terrain attributes and geostatistical methods. Environmental Earth Sciences, 75 : 473. © 2016 Springer. With permission.)

unacceptably high. Such variables are better predicted through fuzzy logic approach so that the uncertainties related to conceptualization can be reduced. Fuzzy soil mapping can overcome this difficulty by producing continuous soil maps (Burrough, 1989; Zhu et al., 2001). Application of fuzzy logic requires establishment of knowledge bases or models for the fuzzy membership criteria. Zhu et al. (1997) used fuzzy logic to infer the membership of a soil to particular classes from the environmental variables, such as parent material, elevation, aspect, gradient, profile curvature and canopy coverage.

10.3.4.5 Artificial Neural Networks (ANN)

ANNs are sophisticated computer programs able to model complex functional relationships, when applied to the soil mapping problem use a set of variables related to soil forming factors and the respective soil type as training data in order to construct rules that can be extended to the unmapped areas. It attempts to build a mathematical model that supposedly works in an analogous way to the human brain. The mathematical model of a neural network comprises of a set of simple functions linked together by weights. The network consists of a set of input units, output units, and hidden units, which link the inputs to outputs. The hidden units extract useful information from inputs and use them to predict the outputs. It relies on human understanding of relationships between the predictors (e.g., location-specific soil data and observations) and the target variables. Multi-layer Perceptron (MLP) and Self-Organizing Map (SOM) are the most frequently-used ANNs for DSM, these networks utilize different classification approaches (Sarmento et al., 2010). DSM may combine GIS with advanced techniques such as ANN, which have enabled mapping the spatial distribution of soils in a cheaper, more accurate, reproducible, and flexible way.

10.3.4.6 Expert (Knowledge-Based) Systems

Expert systems are computer programs which use symbolic knowledge to simulate the behavior of human experts (Stock, 1987). Expert system comprises a knowledge base, an inference engine, and a user interface. The knowledge base contains facts and rules, which the program uses to search for a solution to the problem. The inference engine uses the knowledge base to infer logically valid conclusions, and to logically justify conclusions at the completion of the program (Davis and McDonald, 1993). The knowledge base may be represented as a probability matrix or by using a mega-rule structure (Skidmore et al., 1991). Skidmore et al. (1996) integrated Bayesian expert system with a commercially available GIS to map forest soils into five soil landscape classes by utilizing a digital terrain model and vegetation map, as well as knowledge provided by a soil scientist.

10.4 HYPERSPECTRAL REMOTE SENSING DATA ANALYSIS

Hyperspectral remote sensing data are not employed for soil–landscape analysis as in conventional soil mapping. They are used to analyze spectral reflectance for deriving soil properties. Most approaches for analyzing hyperspectral images concentrate on the spectral information in individual image cells. Recently, several soil color based hyperspectral indices such as Brightness Index (BI), Colouration Index (CI), Hue Index (HI), Redness Index (RI) and Saturation Index (SI) were successfully used for quantitative mapping using Hyperspectral satellite data (Hyperion) (Ghosh et al., 2012). Ben-Dor et al. (2002) assessed hyperspectral airborne sensor data for quantification and generation of soil properties maps of organic matter, soil field moisture, soil saturated moisture, and soil salinity. Multivariate relationships of reflectance with soil properties are often developed in remote sensing to help prediction of soil attributes. Viscarra Rossel et al. (2006) provided a review of the literature comparing quantitative prediction of various soils attributes using multivariate statistical techniques and spectral response in the visible, near infrared (NIR) and mid-infrared region of the electromagnetic spectrum.

Suresh Kumar et al. (2015) used hyperspectral remote sensing (Hyperion EO-1) data of May 15, 2007 of Indo-Gangetic plains, Mathura, India to identify sensitive spectral bands (band 9, 20, 22, 28, 29, and 46) to soil salinity parameters. These bands were used to compute spectral indexes *viz.*, salinity index (SI), brightness index (BI) and normalized differential salinity index (NDSI). The relationship between these spectral indices and electrical conductivity (EC) was established to generate maps showing severity of salt-affected soils of the area. Among these spectral indices, SI showed highest correlation coefficient (r^2) with the parameters of ECe (r^2=0.78). The SI map was categorized into classes of normal, slight, moderate and highly showing the spatial distribution of severity of salt affected soils.

10.5 MICROWAVE REMOTE SENSING IN SOIL STUDIES

Microwave remote sensing is used to retrieve soil hydrological properties. The penetration of microwave energy into the ground depends on the dielectric constant of the upper layer, frequency and radar polarization. The dielectric constant is comprised of the permittivity or real part and the loss factor or imaginary part. Research studies indicate that soil salinity has no influence on the real part of the dielectric constant whereas; the imaginary part is dependent and increases with increase in salinity for all three textured soils (Sreenivas et al., 1995; Ulaby et al., 1986). Temporal Envisat polarimetric SAR data (ASD, 2006) was used to map spatial soil salinity variation in part of Unao district of Uttar Pradesh. Chatterjee et al. (2009) mapped ravenous land using ERS-1 SAR data. Saha (2011) reviewed microwave remote sensing techniques (active and passive) for spatial assessment of soil quality parameters such as soil salinity, soil erosion, soil physical properties (soil texture and hydraulic properties; drainage condition); and soil surface roughness.

10.6 CONCLUSIONS

Development of new generation of high spatial resolution, enhanced spectral coverage, revisit capabilities and stereo viewing has opened new vistas in various application areas. GIS coupled with RS, GPS and computer technology is providing new methods for data acquisition, storage, processing analysis and modeling. Satellite imageries provide synoptic coverage of large area that facilitates in interpreting soil forming factors by visual approach. The spatial extents of the soil formative environments are then used to delineate soil–landscape units. Terrain plays a fundamental role in modulating the earth surface and landform configuration frequently governs the variability of soils on the landscape. The DEM derived information are integrated in GIS environment with landform and geomorphologic units for their further subdivision in order to derive soil–landscape map. DSM can model high-quality digital soil maps in a fast and cost-effective way. Automatic classification of landforms using DTM is highly promising in soil mapping. Digital terrain analysis allows the generation of a suite of variables that reflect geomorphic, climatic and hydrologic processes. DSM uses this concept to develop empirical models that relate observations of soil properties with environmental variables governed by soil forming factors for soil spatial prediction. Geostatistical methods such as kriging and co-kriging are widely used to predict soil properties at unsampled locations and to better understand their spatial variation in the landscape. Various multivariate predictive modeling techniques such as multivariate relationship, partial least square regression and ANN, etc., were used by several researchers to predict spatial variation of surface soil properties using hyperspectral remote sensing data. Microwave remote sensing is used to retrieve soil hydrological properties as well as to characterize the salt affected soils.

KEYWORDS

- **Artificial Neural Network**
- **Digital Soil Mapping**
- **Geostatistics**
- **GIS**
- **Hyperspectral Remote Sensing**
- **Remote Sensing**
- **Soil Mapping**
- **Terrain Analysis**

REFERENCES

AAFC. (2010). Agriculture and Agri-Food Canada. The Canadian Soil Information System (CANSIS) and the National Soil Database (NSDB). Available online at http://sis.agr.gc.ca/cansis/intro.html.

Baumgardner, M. F., Silva, L. F., Biehl, L. L., & Stoner, E. R. (1985). Reflectance properties of soils. *Advances in Agronomy*, *38*, 1–44.

Ben-Dor, E., Levin, N., Singer, A., Karnieli, A., Braun, O., & Kidron, G. J. (2006). Quantitative mapping of the soil rubification process on sand dunes using an airborne hyperspectral sensor. *Geoderma, 131*(1–2), 1–21.

Ben-Dor, E., Chabrillat, S., Dematte, J. A. M., Taylor, G. R., Hill, J., Whiting, M., & Sommers, S. (2009). Using imaging spectroscopy to study soil properties. *Remote Sensing of Environment, 113*, 538–555.

Ben-Dor, E., Patkin, K., Banin, A., & Karnieli, A. (2002). Mapping of several soil properties using DAIS-7915 hyperspectral scanner data: A Case study over clayey soils in Israel. *International Journal of Remote Sensing, 23*(6), 1043–1062.

Bhattacharyya, T., Sarkar, D., Sehgal, J. L., Velaytham, M., Gajbhiye, K. S., Nagar, A. P., & Nimkhedkar, S. S. (2009). Soil Taxonomic Database of India and the States (1:250,000 scale), *NBSSLUP Publ.* 143, National Bureau of Soil Survey and Land Use Planning, Nagpur, pp. 266.

Bishop, T. F. A., & McBratney, A. B. (2001). A comparison of prediction methods for the creation of field-extent soil property maps. *Geoderma, 103*(1), 149–160.

Biswas, R. R. (1987). A soil map through landsat satellite imagery in part of the Auranga catchment in Ranchi and Palamon district of Bihar, India. *International Journal of Remote Sensing, 4*, 541–543.

Burrough, P. A. (1989). Fuzzy mathematical methods for soil survey and land evaluation. *Journal of Soil Science, 40*, 477–492.

Burrough, P. A., Wilson, P. S., van Gaans, P. F. M., & Hansen, A. J. (2001). Fuzzy k-means classification of topo-climatic data as an aid to forest mapping in the Greater Yellowstone Area, U.S.A. *Landscape Ecology, 16*, 523–546.

Campling, P., Gobin, A., & Feyen, J. (2002). Logistic modeling to spatially predict the probability of soil drainage classes. *Soil Science Society of America Journal, 66*, 1390–1401.

Chang, D. H., & Islam, S. (2000). Estimation of soil physical properties using remote sensing and artificial neural network. *Remote Sensing of Environment, 74*, 534–544.

Chatterjee, R. S., Saha, S. K., Suresh, Kumar, Sarika, M., Lakhera, R. C., & Dadhwal, V. K. (2009). Interferometric SAR for characterization of ravines as a function of their density, depth, and surface cover, *ISPRS Journal of Photogrammetry and Remote Sensing, 64*, 472–481.

CSIRO. (2010). Australian Department of Agriculture, Fisheries and Forestry. Australian Soil Resource Information System (ASRIS). Available online at http://www.asris.csiro.au/index_ie.html

Davis, J. R., & McDonald, G. (1993). Applying a Rule-Based Decision Support System to Local Government Planning, Springer-Verlag, Berlin.

DOS. (1999). Application of remote sensing in soil resource mapping, Peer Review Document, Department of Space (DOS).

FAO, IIASA, ISRIC, ISS-CAS, JRC, (2008). Harmonized World Soil Database (version 1.0). FAO, Rome, Italy and IIASA, Luxemburg, Austria.

Ghosh, G., Kumar, S., & Saha, S. K. (2012). Hyperspectral satellite data in mapping salt-affected soils using linear spectral unmixing analysis. *Journal Indian Society of Remote Sensing, 40*(1), 129–136.

Goovaerts, P. (1997). Geostatistics for Natural Resources Evaluation. Oxford University Press, New York, 487 pp.

Goovaerts, P. (1999). Geostatistics in soil science: state of the art and perspectives, *Geoderma, 89*, 1–45.

Govinda Rajan, S. V. (1971). Soil Map of India. *In: Review of Soil Research in India* (Kanwar, J. S., & Raychaudhuri, S. P., Eds.), *1*, 7.

Grunwald, S., Thompson, J. A., & Boettinger, J. L. (2011). Digital soil mapping and modeling at continental scales – finding solutions for global issues. *Soil Science Society of America. Journal,* (SSSAJ 75th Anniversary Special Paper) *75*(4), 1201–1213.

Hengl, T., & MacMillan, R. A. (2009). Geomorphetry – A Key to Landscape Mapping and Modeling. In: *Geomorphetry: Concepts, Software, Applications. Developments in Soil Science*. T. Hengl, H. I. Reuter (Eds.), Amsterdam, Elsevier, pp. 433–460.

Hengl, T., Rossiter, D., & Stein, A. (2003). Soil sampling strategies for spatial prediction by correlation with auxiliary maps. *Australian Journal of Soil Research, 41*(8), 1403–1422.

Jenny, H. (1941). Factors of soil formation, a system of quantitative pedology. McGraw-Hill, New York.

Karale, R. L., Bali Y. P., & Rao K. V. (1981). Soil mapping using remote sensing techniques. *Proceedings Indian Academy of Science and Engineering Sciences, 3*, 197–208.

Karimi, P., & Bastiaanssen, W. G. M. (2014). Spatial evapotranspiration, rainfall and land use data in water accounting – Part 1: Review of the accuracy of the remote sensing data. *Hydrological Earth System Science, 11*, 1–51.

Klingbiel, A. A., Horvath, E. H., Moore, D. G., & Reybold, W. U. (1987). Use of slope, aspect and elevation maps derived from digital elevation model data in making soil survey. In: *Soil Survey Techniques*, Reybold, W. U., & Peterson, G. W. (Eds.). *SSSA Special Publication No. 20*, 77–90.

Krishnamurthy, Y. V. N., Rao, S. S., Rao, D. S. P., & Jayaraman, V. (2008). Analysis of DEM Generated using Cartosat-1 Stereo Data over Mausanne Les Alpiles. *The International Archives of the Photogrammetry, Remote Sensing and Spatial Information Sciencesm XXXVII*, 1343–1348.

Kumar, S., & Singh, R. P. (2016). Spatial distribution of soil nutrients in a watershed of Himalayan landscape using terrain attributes and geostatistical methods. Environmental Earth Sciences, 75 : 473.

Le Bas, C., King, D., Jamagne, M., & Daroussin, J. (1998). The European Soil Information System. European Soil Bureau Research, Luxembourg.

MacMillan, R. A., Pettapiece, W. W., Nolan, S. C., & Goddard, T. W. (2000). A generic procedure for automatically segmenting landforms into landform elements using DEMs, heuristic rules and fuzzy logic. *Fuzzy Sets and Systems, 113*(1), 81–109.

McBratney, A. B., Mendonça Santos, M. L., & Minasny, B. (2003). On Digital Soil Mapping. *Geoderma, 117*, 3–52.

McKenzie, N. J., & Austin, M. P. (1993). A quantitative Australian approach to medium and small scale surveys based on soil stratigraphy and environmental correlation. *Geoderma, 57*, 329–355.

McKenzie, N. J., Gessler, P. E., Ryan, P. J., & O'Connell, D. A. (2000). The role of terrain analysis in soil mapping. *In: Terrain Analysis Principles and Applications. Gallant,* J. C., Wilson, J. P. (Eds.), John Wiley and Sons, pp. 245–265.

Minasny, B., & McBratney, A. B. (2002). The neuro-method for fitting neural network parametric pedotransfer functions. *Soil Science Society of America Journal, 66*(2), 352–361.

Moore, I. D., Gessler, P. E., Nielsen, G. A., & Peterson, G. A. (1993). Soil attribute prediction using terrain analysis. *Soil Science Society of America Journal, 57*, 443–452.

Mulder, V. L., de Bruin, S., Schaepman, M. E., & Mayr, T. R. (2011). The use of remote sensing in soil and terrain mapping – A review. *Geoderma, 162*, 1–19.

Mulders, M. A., & Epema, G. F. (1986). The thematic mapper: A new tool for soil mapping in arid area. *ITC Journal, 1*, 24–29.

Murthy, R. S., & Pandey, S. (1983). Soil Map of India (1, 6.3 m). National Bureau of Soil Survey and Land Use Planning, Indian Council of Agricultural Research, Nagpur.

Njoku, E., Jackson, T., Lakshmi, V., Chan, T., & Nghiem, S. (2003). Soil moisture retrieval from AMSR-E, *IEEE Geosc. Remote Sensing Letters, 41*(2), 215–229.

Oldeman, L. R. and van Engelen, V. W. P. (1993). A world soils and terrain digital database (SOTER) – An improved assessment of land resources. *Geoderma, 60*(1–4), 309–325.

Pachepsky, Y. A., Timlin, D. J., & Rawls, W. J. (2001). Soil water retention as related to topographic variables. *Soil Science Society of America Journal, 65*(6), 1787–1795.

Park, S. J., & Vlek, L. G. (2002). Environmental correlation of three-dimensional soil spatial variability: a comparison of three environmental correlation techniques. *Geoderma, 109*, 117–140.

Ping, J. L., & Dobermann, A. (2006). Variation in the precision of soil organic carbon maps due to different laboratory and spatial prediction methods. *Soil Science, 171*(5), 374–387.

Qi, F., Zhu, A. X., Harrower, M., & Burt, J. E. (2006). Fuzzy soil mapping based on prototype category theory. *Geoderma,* 136(3–4), 774–787.

Saha, S. K. (2011). Microwave Remote Sensing in Soil Quality Assessment. International Archives of the Photogrammetry, *Remote Sensing and Spatial Information Sciences,* Vol. XXXVIII-8/W20, 2011 ISPRS Bhopal 2011 Workshop, 8 November 2011, Bhopal, India.

Sarmento, E. C., Giasson, E., Weber, E., Flores, C. A., & Hasenack, H. (2010). Comparison of four machine learning algorithms for digital soil mapping in the Vale dos Vinhedos, RS, Brasil. In International Workshop on Digital Soil Mapping, 4. Anais. Roma: CRA-RPS, 2010.

Sehgal, J, L., Saxena, R. K., & Verma, K. S. (1988). Soil resource inventory of India using image interpretation technique, *Proc. 5th Symposium. Remote Sensing Working Group of the International Social of Soil Science*, April 11–15, Budapest, Hungary.

Shatar, T. M., & McBratney, A. B. (1999.) Empirical modeling of relationships between sorghum yield and soil properties. *Precision Agriculture, 1*, 249– 276.

Skidmore, A. K., Ryan, P. J., Dawes, W., Short, D., & Loughlin, E. O. (1991). Use of an expert system to map forest soils from a geographical information system, *International Journal of Geographical Information Systems*, *5*(4), 431–445.

Skidmore, A. K., Watford, F., Luckananurug, P., & Ryan, P. J. (1996). An operational GIS expert system for mapping forest soil. *Photogrammetric Engineering and Remote Sensing, 62*, 501–511.

Soil Survey Staff. (2010). Natural Resources Conservation Service, United States Department of Agriculture. Soil Survey Geographic (SSURGO). Available online at http://soildatamart.nrcs.usda.gov.

Sreenivas, K., Venkataratnam, L., & Rao, P. V. N. (1995). Dielectric properties of salt-affected soils, *International Journal of Remote Sensing, 16*(4), 641–649.

Stock, M. (1987). A1 and Expert Systems: An Overview, *AI Applications*, *1*(1), 9–17.

Stolbovoi, V., & McCallum, I. (2002). *Land Resources of Russia*. Luxemburg, Austria: International Institute for Applied Systems Analysis and the Russian Academy of Science. CD-ROM.

Suresh Kumar, Ghosh, G., & Saha, S. K. (2015). Hyperspectral remote sensing data derived spectral indices in characterizing salt-affected soils: a case study of Indo-Gangetic plains of India. *Journal of Environmental Earth Sciences*, *73*(7), 3299–3308.

Ulaby, F. T., Moore, R. K., & Fung, A. K. (1986). Microwave Remote Sensing Active and Passive – From Theory to Applications, 3. Dedham, *Massachusetts: Artech House*.

Viscarra Rossel, R. A., Walvoort, D. J. J., McBratney, A. B., Janik, L. J., & Skjemstad, J. O. (2006) Visible, near infrared, mid infrared or combined diffuse reflectance spectroscopy for simultaneous assessment of various soil properties. *Geoderma, 131*, 59–75.

Webster, R. (1997). Soil resources and their assessment. *Philosophical Transactions of the Royal Society, London*, Ser. B *352*, 963–973.

Webster, R., & Oliver, M. A. (2001). Geostatistics for Environmental Scientists. Statistics in Practice. *Wiley*, Chichester, p. 265.

Wilson, J. P., & Gallant, J. C. (2000). Digital terrain analysis. In: *Terrain Analysis; Principles and Applications.* Wilson, J. P., & Gallant, J. D. (Eds). John Wiley and Sons: New York, NY, 1–27.

Zhang, G., & McGrath, J. (2004). Geostatistical and GIS analyzes on soil organic carbon concentrations in grassland of south-eastern Ireland from two different periods. *Geoderma, 119*, 261–275.

Zhu, A. X., Band, L. E., Vertessy, R., & Dutton, B. (1997). Derivation of soil properties using a soil land inference model (SoLIM). *Soil Science Society of America Journal, 61*(2), 523–533.

Zhu, A. X., Hudson, B., Burt, J., Lubich, K., & Simonson, D. (2001). Soil mapping using GIS, expert knowledge, and fuzzy logic. *Soil Science Society of America Journal, 65*(5), 1463–1472.

CHAPTER 11

MORPHOMETRIC ANALYSIS USING GIS TECHNIQUES: A CASE STUDY FROM THE BASALTIC TERRAIN OF CENTRAL INDIA

NISHA SAHU,[1] G. P. OBI REDDY,[2] NIRMAL KUMAR,[1]
M. S. S. NAGARAJU,[2] RAJEEV SRIVASTAVA,[3] and S. K. SINGH[4]

[1]*Scientist, ICAR-National Bureau of Soil Survey and Land Use Planning, Amravati Road, Nagpur – 440 033, India*

[2]*Principal Scientist, ICAR-National Bureau of Soil Survey and Land Use Planning, Amravati Road, Nagpur – 440 033, India*

[3]*Principal Scientist & Head, ICAR-National Bureau of Soil Survey and Land Use Planning, Amravati Road, Nagpur – 440 033, India*

[4]*Director, ICAR-National Bureau of Soil Survey and Land Use Planning, Amravati Road, Nagpur – 440 033, India*

CONTENTS

ABSTRACT

Morphometric analysis is important in many hydrological investigations and is inevitable in development and management of drainage basin. The present study depicts the process to evaluate the various morphometric parameters of Miniwada watershed in Nagpur district, Maharashtra, India through measurement of linear, aerial and relief aspects of watershed using Geographic Information System (GIS). The drainage network of watershed was delineated using ESRI Software ArcGIS 10.2. The drainage pattern is dendritic and the analysis reveals that the stream order varies from 1 to 4. The total number of stream segments of all orders counted as 37, out of which the maximum area (70.27%) is covered by 1st order streams and the minimum (2.70%) by 4th order. The bifurcation ratio reflecting geological and tectonic characteristics of the watershed were estimated at 3.08 and it indicates that the watershed has suffered less structural disturbance and the drainage pattern has not been distorted by its structural disturbance. The drainage density of watershed 3.63 km/sq km indicates the closeness of spacing of channels, thus providing a quantitative measure of the average length of stream channel for the whole watershed. The systematic analysis of various parameters in GIS helps to understand the soil resources distribution, watershed prioritization, planning and management at watershed level. Hence, it is concluded that GIS techniques proved to be competent tools in the morphometric analysis as these tools not only upgrade and monitor parameters, but also permit the analysis of drainage information in association with other resources and environmental parameters.

11.1 INTRODUCTION

India has diverse geographical features and varied climates. Proper planning and management of available natural resources is necessary for progress and economic development in agriculture, which are mainstay of 70% people living in the country, directly or indirectly. Keeping in view ever increasing population and need for food security, it is realized that the water and the land resources need to be developed, used and managed in an integrated and comprehensive manner. The optimal and sustainable development of these resources is prerequisite so that it is assessed rationally to avoid any future problems regarding its qualitative and quantitative availability. Water, which is one of the precious natural resource, vital for sustaining all life on the earth is becoming scarce due to various reasons including reduction in infiltration rates, runoff, uneconomical use, overexploitation of the surface water resources, etc., due to changes in land use patterns and degradation of land cover. Quantitative morphometric characterization of a drainage basin is considered to be the most appropriate method for the proper planning and management of watershed, as it enables us to understand the relationship among different aspects of drainage pattern of the basin, and also to make a comparative evaluation of different drainage basins, developed in various geologic and climatic regimes. Development of a drainage system and the flowing pattern over space and time are influenced by several variables such as geology, geomorphology, structural components, soil and vegetation of the area through which it flows. Using watershed is the most logical choice for morphometric analysis. It involves quantitative measurements of the geographic parameters and provides measurement, mathematical analysis of the configuration of the earth's surface as well as shape and dimensions of its landforms and quantitative description of the drainage system, which is an important aspect for the characterization of watersheds (Agarwal, 1998; Clarke, 1966; Strahler, 1964). A watershed is the surface area drained by a part or the totality of one or several given water courses and can be taken as a basic erosional landscape element where land and water resources interact in a perceptible manner. In fact, there are many primary determinants of running water systems like geology, relief and climate which are the functioning at the basin scale (Mesa, 2006). Hence, quantities analysis and their interrelationship are important to support decisions for various themes and to understand the influence of drainage morphometry on landforms and their characteristics. It also provides quantitative description of the basin geometry to understand initial slope or inequalities in the rock hardness, structural controls, recent diastrophism, geological and geomorphic history of drainage basin (Strahler, 1964).

11.1.1 MORPHOMETRIC ANALYSIS: GLOBAL PERSPECTIVE

A major emphasis in geomorphology over the past several decades has been on the development of quantitative physiographic methods to describe the evolution and behavior of surface drainage networks (Abrahams, 1984; Horton, 1945; Leopold and Maddock, 1953). Drainage characteristics of many river basins and sub basins in different parts of the globe have been studied using a traditional method which is laborious and cumbersome such as field observations and topographic maps (Horton, 1945; Krishnamurthy et al., 1996; Strahler, 1957, 1964) or alternatively with advanced methods using remote sensing (RS) and GIS (Macka, 2001; Sreedevi et al., 2009), the morphometric analysis of natural drain and its drainage network can be better achieved. In traditional methods, it is difficult to examine all drainage networks from field observations due to their extent throughout rough terrain and/or vast areas whereas RS and GIS techniques provide a speed, precision, fast and inexpensive way for calculating morphometric analysis (Farr and Kobrick, 2000; Grohmann, 2004; Grohmann et al., 2007; Smith and Sandwell, 2003). The method of quantitative analysis of drainage basins was developed by Horton (1945), and modified by Strahler (1964) in conventional means but recently GIS and satellite remote sensing provides a complete tool to analyze, to update and to correlate the measurements with periodic changes. In recent years, it has emerged as a powerful tool to evaluate various morphometric parameters of the watersheds for future developmental planning and to provide a flexible environment for analysis of the spatial information more precisely (Reddy et al., 2004). Evaluation of morphometric parameters necessitates the analysis of various drainage parameters such as ordering of the various streams, measurement of basin area and perimeter, length, drainage density, stream frequency, bifurcation ratio, texture ratio, circulatory ratio and ruggedness number (Nag and Chakraborty, 2003; Nooka et al., 2005). Morphometric parameters such as stream order, together with soil and land use, also play very important role in generating water resources action plan for locating recharge and discharge areas. Stream orders have been classified based on its relative position in the stream network, which help us to understand the similarities and differences amongst them. The study of stream order in drainage basin helps to identify the natural environment of a place. It has been classified based on its relative position in the stream network. Different types of stream order classification system have been developed, in that one of the earliest method and most commonly used method was developed by Strahler (1952).

11.1.2 MORPHOMETRIC ANALYSIS: INDIAN PERSPECTIVE

The morphometric characteristics at watershed scale may contain important information regarding its formation and development because all hydrologic and geomorphic processes occur within the watershed. It can be achieved through measurement of linear, aerial and relief aspects, which helps better understanding of the basin development and management (Nautiyal, 1994; Reddy et al., 2002). It involves the evaluation of stream parameters through the measurements of various stream properties. A number of morphometric studies have been carried out in many parts of the world, as well as, in different Indian watersheds and subsequently used for water resources development and management projects as well as for watershed characterization and prioritization (Ali and Singh, 2002; Anyadike and Phil-Eze, 1989; Chalam et al., 1996; Chaudhary and Sharma, 1998; Faniran and Ojo, 1980; Kumar et al., 2001; Srinivasan and Subramanian, 1999; Singh et al., 2003) and all have arrived to the conclusion that RS and GIS are powerful tools for studying basin morphometry and its continuous monitoring. The influence of drainage morphometry is very important in understanding the landform processes, soil physical properties and erosional characteristics. The source of the watershed drainage lines have been discussed since they were made predominantly by surface fluvial runoff which has very important climatic, geologic and biologic effects (Hynek and Phillips, 2003; Laity and Malin, 1985; Pareta, 2004; Sharp and Malin, 1975). Without the morphometric analysis of these catchments, the challenges of environmental management, groundwater analysis, engineering project, water pollution and health of the populace will be at risk of environmental mismanagement.

11.1.3 IMPORTANCE OF THE STUDY

The application of various morphometric techniques is a major advancement in the quantitative and qualitative description of geometry and network of drainage basins (Macka, 2001; Sreedevi et al., 2009). Several authors have studied morphometric properties of drainage basins as indicators of structural influence on drainage development and neotectonic activity (Bali et al., 2012; Das et al., 2011). In many studies, morphometric analysis has been used to assess the groundwater potentiality of the basins and to locate suitable sites for construction of check dams and artificial recharge structures (Jasmin and Mallikarjuna, 2013; Mishra et al., 2011). Watershed prioritization based on morphometric characteristics has also been carried out and aids in the mapping of high flood potential and erosion prone zones (Romshoo et al., 2012; Wakode et al., 2011). Drainage morphology along with slope map was also explored for locating and selecting the water harvesting structure like percolation tank, pond, check dams, etc. Further, the quantitative analysis of morphometric parameters is found to be of immense utility in river basin evaluation, watershed prioritization for soil and water conservation, pedology, environmental assessment and natural resources management at watershed level. Therefore, the morphometric techniques parameters were considered to describe and classify the various parameters of basin. The main objective of this study is to evaluate various morphometric parameters of Miniwada watershed, Nagpur district, basaltic terrain of Central India using GIS tools. The study area enjoys hot sub-tropical conditions. The main occupation of the people is agriculture. They depend on groundwater, because surface water resources are scarce. Due to erratic rainfall pattern and uncontrolled abstraction, groundwater levels have declined to deeper levels, which are the main source of water that meets the agricultural, industrial and household requirements. Therefore, the study of morphometric characteristics becomes important for developing the surface and groundwater resources in the study area.

11.2 STUDY AREA

The study area lies between 21° 5′ 47″ to 21° 7′ 26″ North latitudes and 78° 40′ 41″ to 78° 43′ 26″ East longitudes in Katol tehsil, western part of Nagpur district, Maharashtra with an area of 1053 ha (Figure 11.1). The elevation of the basin ranges from 407 m to 475 m above the mean sea level (MSL) and falling in Survey of India (SOI) toposheet No. 55K/12 on 1:50,000 scale. Miniwada Watershed constitutes two villages Mhasala and Miniwada. The climate is mainly hot sub tropical with general dryness throughout the year except during the south-west monsoon season. The year may be divided into four seasons. The period from the middle of November to the end of February constitutes the winter season. The hot summer season extends from March to June. This is followed by the south-west monsoon season, which extends up to the end of September. October and November constitutes the post-monsoon season. Temperature rises rapidly after February till May, which is the hottest month of the year. In May, the mean daily maximum temperature records 42.1°C and mean annual temperature of 28°C. Both the day and night temperatures decrease rapidly from October to till December, which is the coldest month in the year. During the month of December, the mean daily maximum and mean minimum temperatures are 28.3°C and 10.2°C,

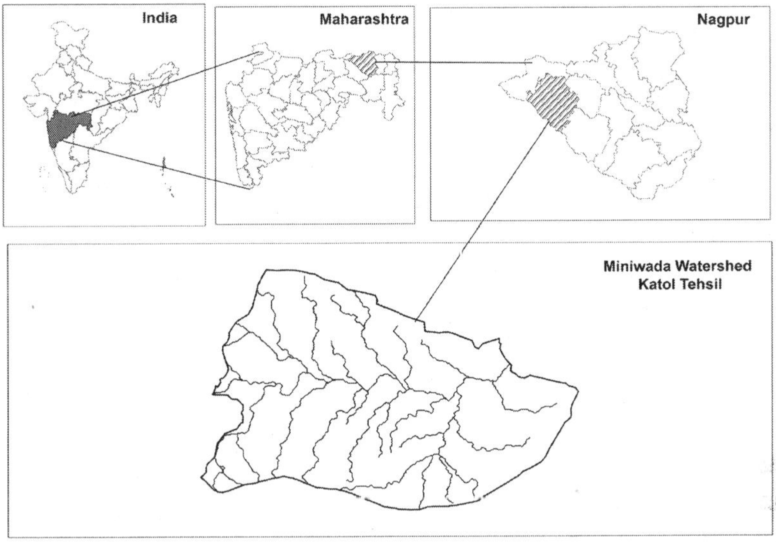

FIGURE 11.1 Location map of study area. (Reprinted from Sahu, N., Reddy, G.P.O., Kumar, N., Nagaraju, M. S. S., Srivastava, R., & S. K. Singh (2016). Morphometric analysis in basaltic Terrain of Central India using GIS techniques: a case study, Applied Water Science, 2016. Creative Commons Attribution 4.0 International License. http://creativecommons.org/licenses/by/4.0)

respectively. The area qualifies for hyperthermic soil temperature regime. The mean annual rainfall is 980 mm, which is lower than the average rainfall of 1205 mm of Nagpur district.

The Miniwada watershed is covered by spread of basaltic lava flows, commonly known as Traps. As usual, there is a considerable variation among the beds of this formation, some of the layers being dense and compact, and others amygdaoidal and more friable. Due to rapid cooling after extrusion, the resultant basaltic rocks possess an aphanitic texture, which is generally dark gray to dark greenish gray in color. Columnar joints and spheroidal weathering are vital features of these rocks. The soils are generally porous and swell considerably on addition of water and dry up developing cracks on losing the moisture. The main field crops are cotton (*Gossypium spp.*), soybean (*Glycine max*), sorghum (*Sorghum bicolor*), pigeonpea (*Cajanus cajan*), gram (*Cicer arientinum*), wheat (*Triticum aestivum*), etc. Vegetables like chili (*Capsiccum annum*), onion (*Allium cepa*) and brinjal (*Solanum melongena*) and in fruit crops mandarin (*Citrus reticulata*) is mainly growing. The natural vegetation comprises of teak (*Tectona grandis*), babul (*Acacia spp.*), palash (*Butea frandosa*), neem (*Azadirachta indica*), mahua (*Madhuca longifolia*), ber (*Ziziphus mauritiana*), anjan (*Terminalia arjuna*), etc.

11.3 METHODOLOGY

In the present study, the Miniwada watershed was delineated and drainage network was extracted using the SOI topographic map (55 K/12) on 1:50,000 scale and was georeferenced using WGS 84 datum, Universal Transverse Mercator (UTM) zone 44N projection in ArcGIS desktop *ver.* 10.2.2. The attributes were assigned to create the digital database for drainage layer of the river basin. The ridgelines in the toposheet were identified, which act as dividing lines for the runoff (Figure 11.2). The parameters computed using GIS technique includes stream order, stream length, bifurcation ratio, drainage density, stream frequency, form factor, circulatory ratio, elongation ratio,

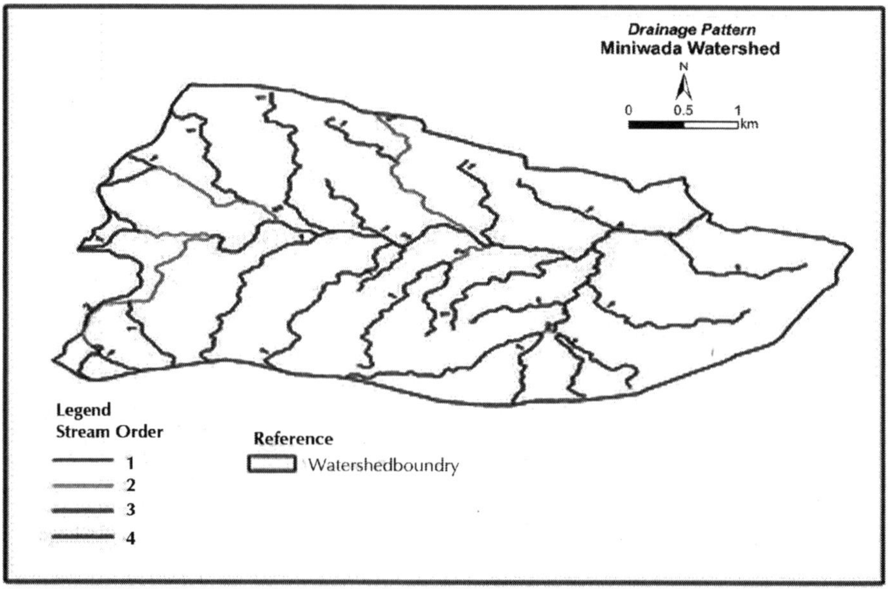

FIGURE 11.2 Digitized drainage map of Miniwada Watershed. (Reprinted from Sahu, N., Reddy, G.P.O., Kumar, N., Nagaraju, M. S. S., Srivastava, R., & S. K. Singh (2016). Morphometric analysis in basaltic Terrain of Central India using GIS techniques: a case study, Applied Water Science, 2016. Creative Commons Attribution 4.0 International License. http://creativecommons.org/licenses/by/4.0)

relief ratio and ruggedness number. The input parameters for the present study are area, perimeter, elevation, stream length, etc., which were obtained from digitized coverage of drainage network map.

The morphometric parameters analyzed were divided into three categories related to their orientation in space. They are linear aspects, aerial aspects and relief aspects. The linear aspects were studied using the methods of Horton (1945), Schumn (1956) and Strahler (1957, 1964), the areal aspects using those of Horton (1932, 1945), Miller (1953) and Schumn (1956) and the relief aspects employing the techniques of Schumn (1956). The morphometric parameters were computed using the standard formulae presented in Table 11.1.

11.4 RESULTS AND DISCUSSION

The study of basin morphometry relates basin and stream network geometries to the transmission of water and sediment through the basin. Systematic description of the geometry of a drainage basin and its stream channel requires measurement of linear, areal and relief (gradient) aspects of the channel network. In the present study, the morphometric analysis has been carried out about parameters as stream order, stream length, bifurcation ratio, stream length ratio, basin length, drainage density, stream frequency, elongation ratio, circularity ratio, form factor, basin relief, relief ratio using mathematical formulae as given in Table 11.1. The properties of the stream networks are highly important to study the landform development processes (Strahler and Strahler, 2002). Morphometric parameters, such as basin relief, basin shape and stream length also influence basin discharge pattern strongly through their varying effects on lag time. Dury (1952) and Christian (1957) applied various methods for landform analysis, which could be classified in different ways and their results presented in the form of graphs, maps or statistical indices. The natural runoff is one of the most potent geomorphic agencies in shaping the landscape of an area. The land area that contributes water to the main stream through smaller ones forms its catchment area or the drainage basin. The arrangement of streams in a drainage system constitutes the drainage pattern that, in turn, reflects mainly structural or lithologic controls of the underlying rocks. The drainage pattern of basin is dendritic in nature.

TABLE 11.1 Formulae Adopted for Computation of Morphometric Parameters

S. No.	Morphometric Parameters	Formula/Relationship	Reference
Linear Aspects			
1.	Stream order	Hierarchical rank	Strahler (1964)
2.	Stream length	Length of stream	Horton (1945)
3.	Mean Stream length	Lsm = Lu/Nu, where, Lu = Total stream length of order "u," Nu = Total no. of stream segments of order u	Strahler (1964)
4.	Stream length ratio	RL = Lu/Lu–1, where, Lu = Total stream length of order "u," Lu–1 = the total stream length of its next lower order.	Horton (1945)
5.	Bifurcation ratio	Rb = Nu/Nu+1, Nu = total number of stream segments of order "u," Nu+1 = number of stream segments of the next higher order.	Schumn (1956)
6.	Mean Bifurcation ratio	Rbm = average of the bifurcation ratio of all order	Strahler (1957)
Aerial Aspects			
7.	Drainage Texture	Dt = Nu/P, where Nu is the total number of streams of all order, P is the perimeter of the basin in km^2	Horton (1945)
8.	Texture ratio	Rt = N1/P, where N1 is the total number of first order streams	Horton (1932)
9.	Drainage density	D = Lu/A, where A is the total area of the basin (km)^2, Lu is the total stream length of all orders.	Horton (1932)
10.	Stream frequency	Fs = Nu/A, Where Nu is the total number of streams of all order, A is basin area in km^2	Horton (1932)
11.	Form factor	Ff = A/Lb^2 is the square of the basin length (km), A is the basin area in km^2	Horton (1932)
12.	Circulatory ratio	Rc = 4π A/P2, where A is the area (km)^2 and p is the perimeter (km) of the watershed	Miller (1953)
13.	Elongation ratio	Re = 2sqrt (A/π)/Lb, where A is the area (km)^2 and Lb = basin length.	Schumn (1956)
14.	Length of overland flow	Lg = 1/(D*2), where D is the drainage density	Horton (1945)
15.	Constant channel maintenance	Inverse of drainage density	Schumn (1956)
16.	Shape index	Reciprocal of form factor	Horton (1932)
Relief Aspects			
17.	Relief	(Elevation of basin mouth)-(Elevation of highest point on the basin)	Schumn (1956)
18.	Relief ratio	Rr = H/Lb, where H = total relief (relative relief) of the basin, Lb = basin length.	Schumn (1956)

(Reprinted from Sahu, N., Reddy, G.P.O., Kumar, N., Nagaraju, M. S. S., Srivastava, R., & S. K. Singh (2016). Morphometric analysis in basaltic Terrain of Central India using GIS techniques: a case study, Applied Water Science, 2016.

11.4.1 LINEAR MORPHOMETRIC PARAMETERS

Linear aspects of the basins are closely linked with the channel patterns of the drainage network wherein, the topological characteristics of the stream segments in terms of open links of the network system are analyzed. The morphometric investigation of the linear parameters of the basins includes stream order (U), stream number (Nu), bifurcation ratio (Rb), stream length (Lu), mean stream length (Lsm), stream length ratio (RL). Some of the important linear aspects have been computed as shown in Table 11.2.

11.4.1.1 Stream Order (U)

Stream Order (U) expresses the hierarchal relationship between the individual stream segments that make up a drainage network. The first step in drainage basin analysis is the designation of stream orders, following a system introduced by Horton, 1945 but later Strahler (1952) has proposed this ordering system with some modifications. The streams have been ranked following Strahler (1964) stream ordering system based on hierarchic ranking of streams. In the Strahler (1952) system, a stream segment with no tributaries is a first-order stream, a second-order stream is formed by joining of two first-order streams, where two second-order streams join, the stream is designated as third-order, and so forth. Stream order of the whole basin is of fourth order (Figure 11.2).

11.4.1.2 Stream Number (Nu)

The total order wise stream segments are known as stream number. The details of stream characteristics conform to (Horton, 1932) "law of stream

TABLE 11.2 Linear Aspects of the Drainage Network of the Study Area

Stream order (U)	No. of streams (Nu)	Total length of streams (in mts)	Mean stream length (in mts)	Bifurcation ratio (Rb)	Stream length ratio (RL)
1	26	27085.01	1041.73		
2	8	5627.75	703.47	3.25	0.68
3	2	4523.87	2261.93	4	3.22
4	1	1004.97	1004.97	2	0.44
Total	37	38241.60	5012.10	Avg. 3.08	4.34

(Reprinted from Sahu, N., Reddy, G.P.O., Kumar, N., Nagaraju, M. S. S., Srivastava, R., & S. K. Singh (2016). Morphometric analysis in basaltic Terrain of Central India using GIS techniques: a case study, Applied Water Science, 2016.

numbers" which states that the number of streams of different orders in a given drainage basin tends closely to approximate an inverse geometric ratio. According to Horton's principle the number of streams is negatively correlated with the order, (i.e., stream number decreases with increase in stream order. Stream number (Nu) here supports Horton's law (i.e., stream number decrease with increase in stream order in this basin (Figure 11.3(a)). The correlation coefficient for the watershed is 0.984, which is quite satisfactory. It is expressed in the form of negative exponential function as:

$$N\mu = Rb\ (K-\mu) \tag{1}$$

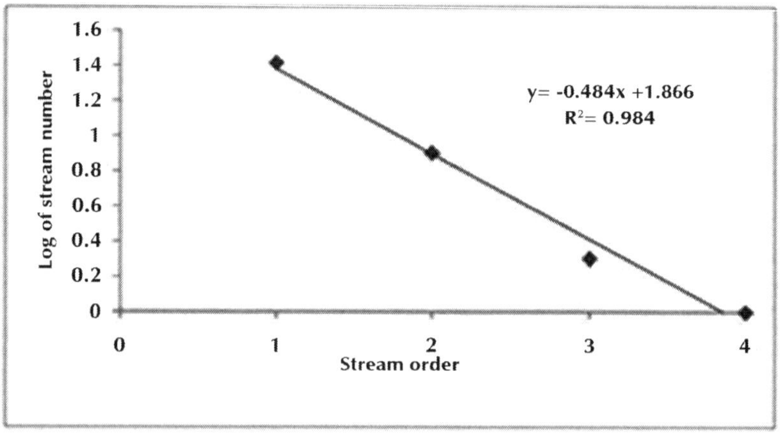

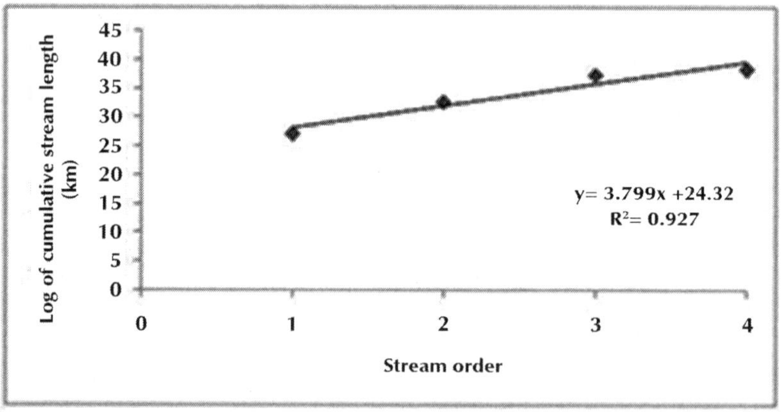

FIGURE 11.3 (a) Regression of logarithm of number of streams and stream order, (b) Regression of logarithm of cumulative stream length and stream order. (Reprinted from Sahu, N., Reddy, G.P.O., Kumar, N., Nagaraju, M. S. S., Srivastava, R., & S. K. Singh (2016). Morphometric analysis in basaltic Terrain of Central India using GIS techniques: a case study, Applied Water Science, 2016. Creative Commons Attribution 4.0 International License. http://creativecommons.org/licenses/by/4.0)

Also, the total number of stream segments of the catchment can be calculated as:

$$\Sigma\mu = \text{Rb K}-1/\text{Rb}-1 \quad (2)$$

where, $N\mu$ = Number of stream segments of a given order; Rb = Constant Bifurcation ratio; μ = Basin order; and K = Highest order of the basin.

11.4.1.3 Stream Length (Lu)

Stream length is indicative of chronological developments of the stream segments including interlude tectonic disturbances. The stream length characteristics of the basin conform (Horton, 1945) second "law of stream length," which states that the average length of streams of each order in a drainage basin tends closely to approximate a direct geometric ratio. The numbers of streams of various orders in the basin are counted and their lengths are measured. The length of various orders in drainage basin has been calculated using ArcGIS. The length of first orders is 27085.01 m, second order stream is 5627.75 m, third order stream is 4523.87 m and fourth order stream is 1004.97 m. The length of stream segments is maximum for first order stream and decreases as the stream order increases. Any deviation from its general behavior indicates that the terrain is characterized by high relief/moderately steep slopes, underlain by varying lithology and probable uplift across the basin. The plot of logarithm of stream length (ordinate) as a function of stream order (abscissa) should yield a set of points lying essentially along a straight line and it results when basin evolution follows the erosion laws acting on geological material with homogeneous weathering erosion characteristics. In general, the total length of stream segments decreases with increasing stream order. Horton suggested the following positive exponential function model of stream lengths.

$$L\mu = \text{L1 RL } (\mu-1) \quad (3)$$

where, $L\mu$ = Cumulative Mean Length of the given order; L1 = Mean Length of the first order; RL = Constant length ratio; and μ = Given order.

The regression line plotted on semi log graph (Figure 11.3b) tends to validate Horton's Law of stream lengths as the coefficient of correlation is 0.927.

11.4.1.4 Mean Stream Length (Lsm)

The mean stream length (Lsm) of a channel is a dimensional property revealing the characteristic size of components of a drainage network and its contributing basin surfaces (Strahler, 1964). It is the total length of all streams of order "u" in a given drainage basin divided by number of streams of order "u." It is generally observed that the Lsm of any given order is greater than that of the lower order and less than that of its higher order. This general observation fails in the case of second order streams and this deviation in the value of Lsm is due to the slope and topographical variations. The value varies from 0.70 km to 2.26 km.

11.4.1.5 Stream Length Ratio (RL)

Stream length ratio (RL) is defined as the ratio of the mean length of an order to the next lower order of stream segment. The stream length ratio of the basin ranges from 0.44–3.22. The RL between streams of different orders reveals variation in basin. This may be attributed to variation in slope and topography. It is generally observed that, there is an increasing trend in the length ratio from lower order to higher order indicating their mature geomorphic stages and if there is a change from one order to another order, it indicates their late youth stage of geomorphic development (Singh and Singh, 1997). Abnormal low value (0.44) of the stream length ratio between third and fourth streams in the present study revealed that the river basin was subjected to neo-tectonic adjustments, as suggested by Lahiri (1996), resulting in late youth stage of geomorphic development (Table 11.2).

11.4.1.6 Bifurcation Ratio (Rb)

Bifurcation Ratio is a dimensionless parameter that expresses the ratio of the number of streams of any given order (Nu) to the number in the next lower order (Horton, 1945; Schumn, 1956). The bifurcation ratio was introduced by Horton (1945) and modified by Strahler (1952). Horton (1945) considered bifurcation ratio as an index of relief and dissections. Strahler (1957) demonstrated that bifurcation ratio shows a small range of variation for different regions or for different environment except where the powerful geological control dominates. It is generally observed that the Rb value varies between

3 and 5, and not the same from one order to its next order, then these irregularities are attributed to geological and lithological development of a drainage basin (Strahler, 1964). Bifurcation ratio is an important parameter that expresses the degree of ramification of drainage network (Mesa, 2006) and reflecting geological and tectonic characteristics of the watershed. The present estimate as 3.08 indicates that the watershed has suffered less structural disturbance and the drainage pattern has not been distorted by structural disturbance whereas high Rb value indicates high structural complexity and low permeability of the terrain (Nag and Chakroborty, 2003). According to Kale and Gupta (2001), bifurcation ratios ranging from 3 to 5 indicate natural drainage system characteristics within a homogeneous rock. Hence, the drainage basin morphometry of the Basin may have been affected by human activities.

11.4.2 AREAL MORPHOMETRIC PARAMETERS

Area of a basin (A) and perimeter (P) are the important parameters in quantitative geo-morphology. Basin area directly affects the size of the storm hydrograph, the magnitudes of peak and mean runoff. The maximum flood discharge per unit area is inversely related to size (Smart and Surkan, 1967). The aerial aspects of the drainage basin such as basin area (A), drainage density (D), stream frequency (Fs), texture ratio (Rt), elongation ratio (Re), circularity ratio (Rc) and form factor ratio (Ff) were calculated and results have been given in Table 11.3.

11.4.2.1 Basin Area (A)

The drainage area is defined as a collecting area from which water would go to a stream or river. The boundary of the area is determined by the ridge separating water flowing in opposite directions. The area of the basin was computed by converting the merged geo-referenced and rectified SOI toposheet (55 K/12) of 1971 on scale 1:50,000 of the basin into polygon form. The total area of the drainage basin is 1053 ha.

11.4.2.2 Basin Length (Lb)

Basin length (Lb) has been given different meanings by different workers (Gardiner, 1975; Gregory and Walling, 1968; Schumn, 1956). According to

TABLE 11.3 Aerial and Relief Aspects of the Study Area

Morphometric characteristics	Estimated Values
Area (A)	1053 ha
Perimeter (P)	20485 m
Length of Basin (Lb)	4402 m
Form factor (Ff)	0.54
Shape index (Sb)	1.84
Circulatory ratio (Rc)	0.32
Elongation ratio (Re)	0.83
Drainage density (D)	3.63 km/km^2
Stream frequency (Fs)	0.04 per ha
Constant of channel maintenance (C)	0.28 km^2/km
Length of overland flow (Lg)	0.14 m
Texture ratio (Rt)	1.27
Relief (H)	0.07 km
Relief ratio (Rr)	0.02
Ruggedness number (Rn)	0.25
Drainage texture (Dt)	1.81

Gregory and Walling (1968), the Lb is the longest length of the basin, from the catchment to the point of confluence. The length of the basin is 4402 m, as shown in Table 11.3.

11.4.2.3 Basin Perimeter (P)

Basin perimeter is the outer boundary of the drainage basin that encloses its area. It is measured along the divides between basins and may be used as an indicator of basin size and shape (Schumn, 1956). The authors have computed the basin perimeter by using ArcGIS 10.2.2 software, which is 20485 m.

11.4.2.4 Drainage Texture (Dt)

Drainage texture (Dt) is one of the important drainage parameters in morphometric analysis, which indicates relative spacing of drainage lines, which are more prominent in impermeable material compared to the permeable ones. Horton (1945) defined drainage texture as the total number of stream segments of all orders divided by the perimeter of the watershed. He also recognized infiltration capacity as the dominant factor influencing drainage texture which includes drainage density and stream frequency as well. Drainage texture (Dt) depends upon a number of natural factors such as climate, rainfall, vegetation, lithology, soil type, infiltration capacity, relief and stage of development. Smith (1954) classified drainage density into five different classes of drainage texture, (i.e., less than 2, indicates very coarse, between 2 and 4 is coarse, between 4 and 6 is moderate, between 6 and 8 is fine and greater than 8 is very fine drainage texture. The soft or weak rocks devoid of vegetation generally exhibit a fine texture, whereas, in massive and resistant rocks coarse drainage texture is developed. The basin has a value of 1.81 which falls under very coarse texture (Table 11.3).

11.4.2.5 Drainage Density (D)

Drainage density is defined as the cumulative length of all streams in a basin divided by the area of the basin (Strahler, 1957). It is a measure of average length of streams per unit drainage area, and describes the spacing of drainage channels. Drainage density has been interpreted to reflect the interaction between climate and geology (Ritter and Major, 1995). Horton (1932) introduced drainage density as an expression to indicate the closeness of spacing of channels. Thus, drainage density is the ratio of total channel segment lengths cumulated for all orders within a basin to the basin area. It is considered as an important indicator of the linear scale of landform elements in stream eroded topography. Density factor is related to climate, lithology, relief, infiltration capacity, vegetative cover, surface roughness and runoff index. Out of which only surface roughness has no significant correlation with drainage density.

The amount and type of precipitation directly influences the quantity and character of surface runoff. An area with high precipitation (such as thundershowers) loses greater percentage of rainfall as runoff, resulting in more surface drainage channels. Density of vegetation and rainfall absorption capacity of soils influence the rate of surface run-off and affects the drainage texture of an area. Langbein (1947) recognized significance of drainage density as a factor determining the time of travel by water and stated that drainage density values between 0.55 and 2.09 km/km^2 correspond to humid regions. Nag (1998) found that low drainage density generally results in areas of highly resistant rocks or permeable subsoil material, dense vegetation and low relief. High drainage density results due to weak or impermeable subsurface material, sparse vegetation and mountainous relief. Low drainage density leads to coarse drainage texture whereas high drainage density leads to fine drainage texture. Basin possess high drainage density (i.e., 3.63 km/km^2, which is indicative of less permeable material, sparse vegetative cover and moderate to high relief (Table 11.3).

11.4.2.6 Stream Frequency (Fs)

Horton, 1932 introduced stream frequency (Fs) or channel frequency as the ratio of total number of stream segments of all orders to the basin area. Reddy et al., 2002 found that lower Fs values indicate permeable sub-surface material and low relief, whereas, higher values are the characteristic of resistant sub-surface material, sparse vegetation and high relief. It mainly depends on the lithology of the basin and reflects the texture of the drainage network. It is an index of the various stages of landscape evolution. The occurrence of stream segments depends on the nature and structure of rocks, vegetation cover, nature and amount of rainfall and soil permeability. The stream frequency of the basin is 0.04 per ha (Table 11.3).

11.4.2.7 Texture Ratio (Rt)

Texture ratio (Rt) is an important factor in the drainage morphometric analysis, which depends on the underlying lithology, infiltration capacity and relief aspect of the terrain. The texture ratio is expressed as the ratio between the first order streams and perimeter of the basin. In the present study the texture ratio of the watershed is 1.27.

11.4.2.8 Circularity Ratio (Rc)

Miller (1953) and Strahler (1964), defined circularity ratio (Rc), as the ratio of the area of the basin (A) to the area of a circle having the same circumference as the perimeter (P) of the basin. Circularity ratio (Rc) is influenced by the length and frequency of streams, geological structures, land use/land cover, climate, relief and slope of the basin (Chopra et al., 2005). The circularity ratio of the basin is 0.32 (Table 11.3), which is indicative of the lack of circularity. This significant ratio indicates the dendritic stage of a basin. Its low, medium and high values are indicative of the youth, mature and old stages of the life cycle of the tributary basins.

11.4.2.9 Elongation Ratio (Re)

Elongation ratio is the ratio of the diameter of a circle of the same area as the basin to maximum basin length. According to Mustafa and Yusuf (1999), values of elongation ratio range from 0.4–1.0. Strahler (1964) states that its ratio runs between 0.6 and 1.0 over a wide variety of climatic and geologic types. It is a very significant index in the analysis of basin shape which helps to give an idea about the hydrological character of a drainage basin. Values near to 1.0 are typical of regions of low relief (Strahler, 1964). The varying slopes of watershed can be classified with the help of the index of elongation ratio, (i.e., circular (0.9–0.10), oval (0.8–0.9), less elongated (0.7–0.8), elongated (0.5–0.7), and more elongated (less than 0.5). The elongation ratio of Miniwada watershed is 0.83, which indicates that the watershed is less elongated and oval in shape (Table 11.3).

11.4.2.10 Form Factor (Ff)

Quantitative expression of drainage basin outline form through a form factor ratio (Ff), which is the dimensionless ratio of basin area to the square of basin length (Horton, 1932). Basin shape may be indexed by simple dimensionless ratios of the basic measurements of area, perimeter and length (Singh, 1998). The Form factor (Ff) has direct relation to the stream flow and the shape of the watershed. In the present study the Form factor (Ff) is 0.54, which is low, showing that the watershed is elongated in nature with less side flow for shorter duration and high main flow for longer duration. The elongated nature of the basin has implication on both hydrologic and geomorphic processes. Mustafa and Yusuf (1999) have noted that the flow of water in elongated basins is distributed over a longer period than in circular ones. Flood flows of such elongated basins are easier to manage than of the circular basin (Christopher et al., 2010).

11.4.2.11 Length of Overland Flow (Lg)

Horton (1945) defined length of overland flow (Lg) as the length of water over the ground before it gets concentrated in definite stream channels. He considered it as one of the most important independent variables affecting hydrologic and physiographic development of drainage basins. This factor relates inversely to the average slope of the channel and is quite synonymous with the length of sheet flow to a large degree. The average length of overland flow is approximately half the average distance between stream channels and is therefore, approximately equals to half of reciprocal of drainage density (Horton, 1945). Overland flow is significantly affected by infiltration (exfiltration) and percolation through the soil, both varying in time and space. The value of Length of Overland Flow of the basin is 0.14 (Table 11.3) and is equals to the half of the constant of channel maintenance.

11.4.2.12 Constant of Channel Maintenance (C)

Schumn (1956) has used inverse of drainage density as another parameter called constant of channel maintenance (C), defined in units of square feet per foot. It has dimensions of length and therefore increases in magnitude as the scale of the landform units increases. In short, it measures the number of square feet of watershed surface area required to sustain one linear foot of channel. It is the area of the basin surface needed to sustain a unit length of stream channel and is expressed by the inverse of drainage density and depends on the rock type, permeability, climatic regime, vegetation cover as well as duration of erosion. The Constant of Channel Maintenance of the basin is 0.28 (Table 11.3). This low value indicates high structural disturbances, low permeability, steep to very steep slopes and high surface run off.

11.4.2.13 Shape Index (Sb)

Rate of water and sediment yield along the length and relief of the drainage basin is largely affected by the shape. It is the reciprocal of form factor. The shape index value for watershed is 1.84 as shown in Table 11.3.

11.4.3 RELIEF MORPHOMETRIC PARAMETERS

The relief aspects of the drainage basins are significantly linked with the study of three dimensional features involving area, volume and altitude of vertical dimension of landforms to analyze different geo-hydrological characteristics. Some of the important relief parameters that are related to the study have been analyzed as shown in Table 11.3.

11.4.3.1 Basin Relief (H)

Basin relief is an important factor in understanding the geomorphic processes and landform characteristics. The total basin relief of the drainage basin is 0.07 km. It has been observed that a high degree of correlation exists among relief and drainage frequency and stream channel slopes. Relief aspect of the watersheds plays an important role in drainage development, surface and sub-surface water flow, permeability, landform development and associated features of the terrain. There is a correlation between hydrological characteristics and the relief ratio (Rr) of a drainage basin.

11.4.3.2 Relief Ratio (Rr)

Relief ratio (Rr) measures the overall steepness of a drainage basin and is an indicator of the intensity of erosional process operating on slope of the basin (Schumn, 1956). The relief ratio of the basin is 0.02 that indicates moderate relief and steep to moderate slope. The relief ratio Rr, as suggested by Schumn (1956) was defined as the total relief of the catchment (elevation difference between the lowest and the highest points in the basin) and the longest dimension of the basin parallel to the principal drainage line. The low value of Rr for the basin (0.02) indicates moderate relief and the presence of basement rocks that are exposed in the form of small ridges and mounds with moderate to lower degree slopes (Table 11.3).

11.4.3.3 Ruggedness Number (Rn)

Strahler's (1964) ruggedness number is the product of the basin relief and the drainage density and usefully combines slope steepness with its length.

Calculated accordingly, the Miniwada watershed has a ruggedness number of 0.25 (Table 11.3). The low ruggedness value of watershed implies that area is less prone to soil erosion and have intrinsic structural complexity in association with relief and drainage density. It is computed using the equation:

$$Rn = H \times D \quad (4)$$

where, Rn = Ruggedness number; H = Watershed relief (km); D = Drainage density km/Km^2.

11.5 CONCLUSIONS

Remote Sensing and GIS have been proved to be efficient tools in drainage delineation and updating drainage, which is used for the morphometric analysis. The quantitative analysis of morphometric parameters is found to be of immense utility in river basin evaluation, watershed prioritization for soil and water conservation, and natural resources management at micro level. The study reveals that the Miniwada watershed of 1053 ha is approaching towards elongated basin having fourth stream order, high drainage density, low relief ratio and low infiltration with high bifurcation ratio. Drainage network of the basin exhibits as mainly dendritic type, which indicates the homogeneity in texture and lack of structural control. The study has also shown that the watershed is in conformity with the Horton's law of stream number and stream length. The low value of bifurcation ratio (3.08) indicates that the watershed has suffered less structural disturbance. High value of elongation ratio compared to circulatory ratio indicates the elongated shape of the watershed. The drainage density is evaluated to be 3.63 Km/Sq km which indicates that the area is of less permeable material, sparse vegetative cover and moderate to high relief. The circularity ratio for the watershed is 0.32, which indicates mature nature of topography. The elongation ratio is 0.83, which indicates that the watershed is elongated. The elongated shape of the basin is mainly due to the guiding effect of thrusting and faulting. Relief ratio (0.02) reflects that the watershed be treated with soil and water conservation measures. Results reported in the study will be useful in developing functional relationships between geomorphological parameters and hydrological variables and also provides the beneficial parameters for the assessment of the groundwater potential zones, identification of sites for water harvesting structures, water resource management, runoff and geographic characteristics of the drainage system. In future, drainage morphology along with slope map can also be explored for locating and selecting the water harvesting structure like percolation tank, pond, check dams, etc.

KEYWORDS

- **Aerial Parameters**
- **Drainage Basin**
- **Geographic Information System**
- **Linear Parameters**
- **Miniwada Watershed**
- **Morphometric Parameters**
- **Relief Parameters**

REFERENCES

Abrahams, A. D. (1984). Channel networks: a geomorphological perspective. *Water Resource Research, 20*, 161–168.

Agarwal, C. S. (1998). Study of drainage pattern through aerial data in Naugarh area of Varanasi district, U.P. *Journal of the Indian Society of Remote Sensing, 24*(4), 169–175.

Ali, S., & Singh, R. (2002). Morphological and Hydrological investigation in Hirakund catchment for watershed management planning. *Journal of Soil and Water Conservation (India). 1*(4), 246–256.

Anyadike, R. N. C., & Phil-Eze, P. O. (1989). Runoff response to basin parameters in southeastern Nigeria. Geografiska Annaler. Series A, *Physical Geography, 71*(1/2), 75–84.

Bali, R., Agarwal, K. K., Nawaz Ali, S., Rastogi, S. K., & Krishna, K. (2012). Drainage morphometry of Himalayan Glacio-fluvial basin, India: hydrologic and neotectonic implications. *Environmental Earth Sciences, 66*(4), 1163–1174.

Chalam, B. N. S., Krishnaveni, M., & Karmegam, M. (1996). Correlation of runoff with geomorphic parameters. *Journal of Applied Hydrology, 9*(3–4), 24–31.

Chaudhary, R. S., & Sharma, P. D. (1998). Erosion hazard assessment and treatment prioritization of Giri river catchment, North Western Himalaya. *Indian Journal of Soil Conservation, 26*(1), 6–11.

Christian, C. S. (1957). The concept of land units and land, 9th Pacific Science Congress. Department of Science, Bangkok, Thailand. *20*, 74–81.

Christopher, O., Idown, A. O., & Olugbenga, A. S. (2010). Hydrological analysis of Onitsha north east drainage basin using Geoinformatic techniques. *World Applied Science Journal, 11*, 1297–1302.

Chopra, R., Diman, R. D., & Sharma, P. K. (2005). Morphometric Analysis of Sub-Watershed in Gurdaspur District, Punjab Using Remote Sensing and GIS Techniques. *Journal of the Indian Society of Remote Sensing*, *33*(4), 531–539.

Clarke, J. I. (1966). Morphometry from Maps, Essays in Geomorphology. Elsevier Publ. Co., New York, 235–274.

Das, J. D., Shujat, Y., & Saraf, A. K. (2011). Spatial technologies in deriving the morphotectonic characteristics of tectonically active Western Tripura Region, Northeast India. *Journal of the Indian Society of Remote Sensing*, *39*(2), 249–258.

Dury, G. H. (1952). Methods of Cartographical Analysis in Geomorphological Research, Silver Jubilee Volume, Indian Geographical Society, Madras, 136–139.

Faniran, A., & Ojo, O. (1980). Man's Physical Environment: An Intermediate Physical Geography. London: Heinemann (Chapter 8).

Farr, T. G., & Kobrick, M. (2000). Shuttle radar topography mission produces a wealth of data. American Geophys. Union, EOS. *81*, 583–585.

Gardiner, V. (1975). Drainage basin morphometry, British Geomorphological Group, Technical Bulletin. *14*, 48.

Gregory, K. J., & Walling, D. E. (1968). The variation of drainage density within a catchment, International Association of Scientific Hydrology – Bulletin. *13*, 61–68.

Grohmann, C. H. (2004). Morphometric analysis in geographic information systems: applications of free software GRASS and R. *Computers and Geosciences*, *30*, 1055–1067.

Grohmann, C. H., Riccomini, C., & Alves, F. M. (2007). SRTM-based morphotectonic analysis of the Pocos de caldas alkaline massif Southeastern Brazil. *Computers and Geosciences*, *3*, 10–19.

Horton, R. E. (1932). Drainage basin characteristics, *Transactions of American Geophysical Union*, *13*, 350–361.

Horton, R. E. (1945). Erosional development of streams and their drainage basins; Hydrophysical approach to quantitative morphology. *Geological Society of America Bulletin*, *56*, 275–370.

Hynek, B. M., & Phillips, R. J. (2003). New data reveal mature, Integrated Drainage Systems on Mars Indicative of Past Precipitation. *Geology*, *31*, 757–760.

Jasmin, I., & Mallikarjuna, P. (2013). Morphometric analysis of Araniar river basin using remote sensing and geographical information system in the assessment of groundwater potential. *Arab Journal of Geosciences*, *6*(10), 3683–3692.

Kale, V. S., & Gupta, A. (2001). Introduction to Geomorphology. Orient Blackswan Private Limited, Calcutta.

Krishnamurthy, J., Srinivas, G., Jayaram, V., & Chandrasekhar, M. G. (1996). Influence of rock type and structure in the development of drainage networks in typical hard rock terrain. *ITC Journal, 3/4*, 252–259.

Kumar, R., Lohani, A. K., Kumar, S., Chatterjee, C. & Nema, R. K. (2001). GIS based morphometric analysis of Ajay river basin upto Sarath gauging site of South Bihar. *Journal of Applied Hydrology, 15*(4), 45–54.

Lahiri, S. (1996). Channel Pattern as Signature of Neotectonic Movements- A Case Study from Brahmaputra Valley in Assam. *Journal of the Indian Society of Remote Sensing,* 24(4), 265–272.

Laity, J. E., & Malin, M. C. (1985). Sapping processes and the development of Theatre-Headed Valley networks on the Colorado plateau. *Bulletin. Geological Society of America*, *96*, 203–217.

Langbein, W. B. (1947). Topographic Characteristics of Drainage Basins, US Geological Survey, Water-Supply Paper, No. 956-C, 125–157.

Leopold, L. B., & Maddock, T. (1953). The hydraulic geometry of stream channels and some physiographic implications. USGS professional paper. *252*, 1–57.

Macka, Z. (2001). Determination of texture of topography from large scale contour maps. *Geografski Vestnik, 73*(2), 53–62.

Mesa, L. M. (2006). Morphometric analysis of a subtropical Andean basin (Tucumam, Argentina). *Environmental Geology, 50*, 1235–1242.

Miller, V. C. (1953). A Quantitative geomorphic study of drainage basin characteristics in the Clinch Mountain area, Virginia and Tennessee, Proj. NR 389–402, Tech Rep 3, Columbia University, Department of Geology, ONR, New York.

Mishra, A., Dubey, D. P., & Tiwari, R. N. (2011). Morphometric analysis of Tons basin, Rewa District, Madhya Pradesh, based on watershed approach. *Earth Science India, 4*(3), 171–180.

Mustafa, S., & Yusuf, M. I. (1999). A textbook of hydrology and water resources. (Ist ed.). Abuja: Jenas Prints and Publishing Company (Chapter 5).

Nag, S. K. (1998). Morphometric Analysis Using Remote Sensing Techniques in the Chaka Sub Basin, Purulia District, West Bengal. *Journal of the Indian Society of Remote Sensing, 26*(1–2), 69–76.

Nag, S. K., & Chakroborty, S. (2003). Influence of rock types and structures in the development of drainage networks in hard rock area. *Journal of the Indian Society of Remote Sensing, 31*(1), 25–35.

Nautiyal, M. D. (1994). Morphometric analysis of drainage basin using aerial photographs, a case study of Khairkuli basin, District Dehradun. U.P. *Journal of the Indian Society of Remote Sensing, 22*(4), 251–261.

Nooka, R. K., Srivastava, Y. K., Venkatesteswara Rao, V., Amminedu, E., & Murthy, K. S. R. (2005). Check dam positioning by prioritization of micro watersheds using SYI model and morphometric analysis: Remote sensing and GIS perspective. *Journal of the Indian Society of Remote Sensing, 33*(1), 25–38.

Pareta, K. (2004). Hydro-Geomorphology of Sagar District (M.P.): A Study through Remote Sensing Technique, Proceeding in XIX M. P. Young Scientist Congress, Madhya Pradesh Council of Science & Technology (MAPCOST), Bhopal.

Reddy, G. P. O., Maji, A. K., & Gajbhiye, K. S. (2002). GIS for morphometric analysis of drainage basins. *GIS India, 11*(4), 9–14.

Reddy. G. P. O., Maji, A. K., & Gajbhiye, K. S. (2004). Drainage morphometry and its influence on landform characteristics in a basaltic terrain, Central India – A remote sensing and GIS approach. *International Journal of Applied Earth Observation and Geoinformation, 6*(1), 1–16.

Ritter, F. E., & Major, N. P. (1995). Useful Mechanisms for Developing Simulations for Cognitive Models. AISB Quarterly. *91*, 7–18.

Romshoo, S. A., Bhat, S. A., & Rashid, I. (2012). Geoinformatics for assessing the morphometric control on hydrological response at watershed scale in the Upper Indus basin. *Journal of Earth System Science, 121*(3), 659–686.

Sahu, N., Reddy, G.P.O., Kumar, N., Nagaraju, M. S. S., Srivastava, R., & S. K. Singh (2016). Morphometric analysis in basaltic Terrain of Central India using GIS techniques: a case study, *Applied Water Science,* (in press), DOI 10.1007/s13201-016-0442-z.

Schumn, S. A. (1956). Evaluation of drainage systems and slopes in badlands at Perth Amboy, New Jersey. *Bulletin of Geological Society of America, 67*, 597–646.

Sharp, R. P., & Malin, M. C. (1975). Channels on mars, *Bulletin of the Geol. Society of America*, 86, 593–609.

Singh, S. (1998). Geomorphology. Prayag Pustak Bhawan, Allahabad, pp. 334–412.

Singh, S., & Singh, M. C. (1997). Morphometric Analysis of Kanhar River Basin. *National Geographical Journal of India, 43*(1), 31–43.

Singh, R. K., Bhatt, C. M., & Hari Prasad, V. (2003). Morphological study of a watershed using remote sensing and GIS techniques. *Journal of Hydrolology*, *26*(1–2), 55–66.

Smart, S., & Surkan, A. J. (1967). The relationship between mainstream length and area in drainage basins. *Water Resources Research*, *3*, 963–973.

Smith, K. G. (1954). Standards for Grading Texture of Erosional Topography. *American Journal of Science*, *248*, 655–668.

Smith, B., & Sandwell, D. (2003). Accuracy and resolution of shuttle radar topography mission data. *Geophysical Research Letters, 30*(9), 20–21.

Sreedevi, P. D., Owais, S., Khan, H. H., & Ahmed, S. (2009). Morphometric Analysis of a Watershed of South India Using SRTM Data and GIS. *Journal of the Geological Society of India*, *73*, 543–552.

Srinavasan, P. R., & Subramanian, V. (1999) Ground water targeting through morphometric analysis in Mamundiyar river basin, Tamilnadu. *The Deccan Geographer*, *37*(1), 22–23.

Strahler, A. N. (1952). Hypsometric Analysis of Erosional Topography. Bulletin *of the Geological Society of America*, *63*(11), 1117–1142.

Strahler, A. N. (1957). Quantitative analysis of watershed geomorphology, *Transactions of American Geophysics Union*, *38*, 913–920.

Strahler, A. N. (1964). Quantitative geomorphology of drainage basin and Channel networks. *Handbook of Applied Hydrology*, McGraw Hill, New York, Section 4–11.

Strahler, A. N., & Strahler, A. H. (2002). *A Textbook of Physical Geography*, John Wiley and Sons, New York.

Wakode, H. B., Dutta, D., Desai, V. R., Baier, K., & Azzam, R. (2011). Morphometric analysis of the upper catchment of Kosi River using GIS techniques. *Arabian Journal of Geosciences*, *6*(2), 395–408.

CHAPTER 12

DIGITAL TERRAIN ANALYSIS AND GEOMORPHOLOGICAL MAPPING USING REMOTE SENSING AND GIS: A CASE STUDY FROM CENTRAL INDIA

B. S. MANJARE,[1] PRITI JAMBHULKAR,[2] M. A. PADHYE,[2] and S. S. GIRHE[2]

[1]*Assistant Professor, Department of Geology, RTM, Nagpur University, Nagpur – 440 001, India*

[2]*Research Scholar, Department of Geology, RTM, Nagpur University, Nagpur – 440 001, India*

CONTENTS

ABSTRACT

In the study, an attempt has been made to delineate the distinct geomorphic units in Kolar river sub basin, Nagpur distract, Maharashtra of Central India, using integrated remote sensing and Geographic Information System (GIS) techniques especially Shuttle Radar Topographic Mission (SRTM), Digital Elevation Model (DEM) of 90 m resolution by analyzing the elevation, slope and image characteristics. DEM of study area has been generated from SRTM elevation data of 90 m resolution using software ArcGIS *ver.* 10.1. The analysis of the remote sensing data included delineation of the various terrain features, such as geology, structures, geomorphological units and their characters. Geologically, study area consists of Deccan trap, Quartz muscovite schist and small patches of the sandstone and limestone. From the DEM and satellite data the structure, processes and factor are clearly observed on the landforms, which give a complex appearance to slope profile. The steep escarpment slope followed by the talus slope or debris slope with convex profile make it clear that DEM and satellite data plays very important role in the landform delineation.

12.1 INTRODUCTION

Geomorphology is the study of the nature and origin of landforms, particularly of the formative processes of weathering and erosion that occur in the atmosphere and hydrosphere. This process shapes the Earth's surface and generates the sediments that circulate in the rock cycle. Landforms are the result of the interactions among the geosphere, atmosphere and hydrosphere. In recent years, the advances in computer technologies and digital data acquisition/processing have led to the improvement of the knowledge on geomorphic processes and the increase of the use of predictive models and quantitative measurements to analyze, monitor, and understand the landform changes (Summerfield, 1997; Wood, 1996). Application of aerial photography, satellite remote sensing data in conjunction with toposheets and Digital Terrain Models (DTMs) will be of immense help in geomorphic interpretation and allow us to analyze various geomorphic processes.

Geomorphological maps can be considered as graphical inventories of a landscape depicting landforms and surface as well as subsurface materials. Sketches and maps of landscapes and landforms (Dykes, 2008) have been fundamental methods to analyze and visualize Earth surface features ever since early geomorphological research. The widespread distribution and extended graphical capabilities of GIS as well as the availability of high-resolution remote sensing data, such as aerial and satellite imagery and Digital Elevation Models (DEMs) has led to rejuvenation of various methods (Lee, 2001; Paron and Claessens, 2011; Smith et al., 2011). Geomorphological maps can act as a preliminary tool for land management and geomorphological and geological risk management, as well as, providing baseline data for other applied sectors of environmental research such as landscape ecology, forestry or soil science (Cooke and Doornkamp, 1990; Dramis et al., 2011; Paron and Claessens, 2011).

A drainage network can be extracted from a DEM with an arbitrary drainage density or resolution (Tarboton et al., 1991). The characteristics of the extracted network depend on the definition of channel sources on the digital land surface topography. Once the channel sources are defined, the essential topology and morphometric characteristics of the drainage network are implicitly defined because of their close dependence channel source definition. Thus, the proper identification of channel sources is critical for extraction of a representative drainage network from DEMs. The extracted stream network in hydrologic analysis is important because the network indirectly determines the hill slope travel distance and network link lengths. The characteristics of the extracted network depend extensively on the definition of channel initiations on the digital landscapes. Once the channel initiations are defined, the essential topology and morphometric characteristics of the corresponding downstream drainage network are implicitly predefined because of their close dependence on channel initiation definition.

DEM is simply called digital description of earth's surface or terrain condition of earth as a whole or part of it (Bolstad and Stowe, 1994). DEMs are being used widely to extract important elevation and topographic information. DEMs are used for visual analysis of topography, landscapes and landforms other than modeling of surface processes (Welch, 1990). DEM, Digital Elevation Data (DED), Digital Terrain Data (DTD) (Campbell, 2002) or Digital Terrain Model (DTM) all consists of different arrangements of individual points of x and y coordinates of horizontal geographic positions. Z is the vertical elevation value that is relative to a given datum for a set of x, y points (Bernhardsen, 1999; Bolstad and Stowe, 1994; Welch, 1990). Relatively large DEM grid cells make up the model domain in order to reduce the computation time. This allows quick model calibrations and model sensitivity analysis. A major disadvantage of the use of low-resolution input data is the loss of important small-scale features that can seriously affect the modeling results. If the input DEM is at a higher grid resolution, during the transformation or re-sampling of the original DEM data to a lower model resolution, important topographic details are lost mainly as a result of averaging. If the input DEM is already at a low resolution, it does not represent the actual on-ground topographic features, which might significantly affect the accuracy and reliability of the results from the modeling exercise. Applications of GIS, provide opportunities for modeling, analyzing and displaying geographical phenomena connected with topography and relief. The utility of the DEM is evidenced by the widespread availability of digital topographic data and ever- increasing applications of DEM. They have been found useful in many fields of study such as geomorphometry, as these are primarily related to surface processes such as terrain analysis (Reddy et al., 2013), soil–landscape modeling, landslides zonation mapping, (Hengl and Evans, 2009), archaeology (Menze et al., 2006), forestry (Simard et al., 2006). The derivation of topographic attributes relies on digital elevation data sets that may be acquired from satellite imagery, digitizing the contour lines on topographic maps, or conducting ground surveys (Wilson and Gallant, 2000). In the present study, the SRTM DEM (90 m) data were used for elevation and slope analysis.

12.1.2 DIGITAL TERRAIN ANALYSIS: GLOBAL PERSPECTIVE

Traditionally, DTM have been used to build elevation models of the Earth's surface. They are 3D in nature and thus lend themselves to several important uses. Terrain and slope analysis, globally, are used for demarcating sites of sedimentation, stable zones, etc., which ultimately provide important data to develop final remedy structures and to deliver solutions for various environmental problems. DEMs are being used for a wide variety of scientific and commercial purposes. Geological land observations are also dependent on accurate elevation to determine the things such as land subsidence, landslides or avalanches, and extracting geomorphological information. Some DTM will be raster based (grids) while others will be vector based (point derived-interpolated).The relief model allows a quick overview of the given area.

12.1.3 DIGITAL TERRAIN ANALYSIS: INDIAN PERSPECTIVE

In India, DTM is used in various sectors for several purposes. One of the most important uses of DTM is for planning and management of water resources. About 25 watershed basin areas were generated using Survey of India (SOI) 20-meter contour interval data for the State of Meghalaya under the state's "Integrated Basin Development and Livelihood Program (IBDLH)" implemented by State Basin Development Authority. Then all basins were superimposed on 3D Terrain Model and were analyzed for possible location of small and multiple reservoirs based on suitable terrain conditions like topography, locality and ecological impact. About 40 reservoir locations across the state were identified along with attributes like water storage capacity at different raised heights of the reservoir, list of benefitted habitations, etc., provided to the state Government for further ground level planning and implementation. Watershed development and preparation of a virtual 3D model was done after event of Flash Floods during 2006 in Barmer district in Rajasthan (Singh, 2012). Planning of rural roads under Pradhanmantri gramin sadak yojna (PMGSY), cadastral mapping in Haryana state, preparation of city base maps (Kodge and Hiremath, 2011) planning of infrastructural development in difficult terrains in Kohima of Nagaland state, 3D Digital Terrain Models for internal security in several states of India are some of the important applications of DTM in India.

12.1.4 IMPORTANCE OF THE STUDY

The availability of near-global terrain information from the SRTM mission provides a new source to those interested in terrain analysis. A traditional terrain analysis product is slope configuration. Slope configuration analysis has been used in several terrain analysis products, which includes various DEM. SRTM DEM data has unique characteristics that can cause differences in slope configuration analysis when compared to traditionally collected terrain data sets. The study was carried out with the objectives to extract information about geomorphic features in the study area from the IRS LISS III satellite imagery of 23.5m spatial resolutions and to categories those features and use of SRTM DEM for determining the slope factor in the study area.

12.2 STUDY AREA

The Kolar river sub basin is situated North-western part of Nagpur district, Maharashtra, Central India and lies between 21° 10' 00" N to 21° 30' 00" N latitudes and 78° 40' 00" E to 78° 10' 00" E longitudes (Figure 12.1). Kolar river flowing southeast from above the town of Saoner to its juncture with the Kanhan River, which is the tributary of Godavari river. The Kolar river forms the boundary between Saoner taluka and Ramtek taluka of the Nagpur district.

12.3 METHODOLOGY

12.3.1 DATA USED IN THE STUDY

The Indian remote sensing satellite IRS-1C and 1D of (April, 2012) multi resolution sensor as panchromatic (PAN) and linear imaging self scanner (LISS-III-2008) image with ground resolution of 5.8 and 23.5 m, respectively. The satellite data was rectified with the help of SOI toposheets. The Geological map was prepared from district resource map of Nagpur district (GSI, 2000). SRTM DEM (90 m) resolution was used in the study.

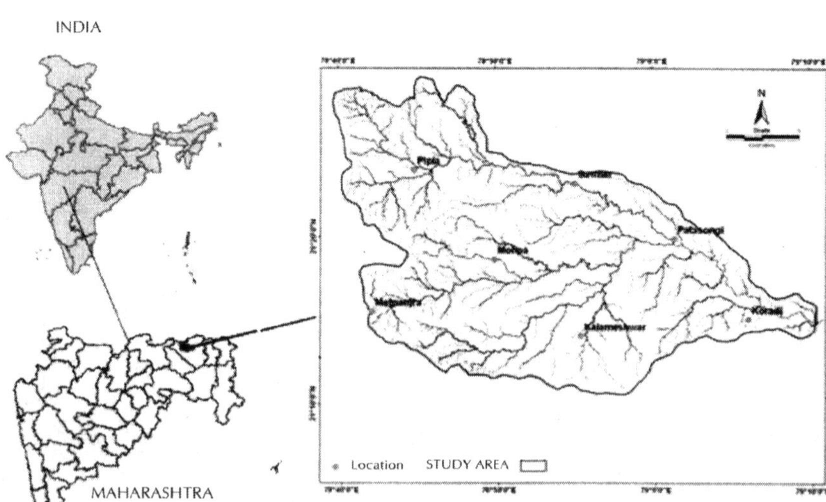

FIGURE 12.1 Location and drainage map of the study area.

12.3.2 PROCESSING OF SATELLITE DATA

In order to analyze morphometric properties of the river basin, Indian Remote Sensing Satellite (IRS-1C) LISS III and SOI toposheets of the study area were processed. The SOI toposheets and digital satellite data were geometrically rectified and geo referenced by collecting required ground control points (GCPs) using UTM projection and WGS 84 datum. Further, all geocoded images were mosaic using ERDAS Imagine *ver.* 8.6 image processing software. Digitization work was carried out using Arc GIS ver. 10.1 software for analysis of basin morphometry.

12.3.3 PROCESSING OF SRTM DEM DATA

In the study, a 30×40 km SRTM DEM was used and resampled at a grid resolution of 90 m. Few data gaps present in the DEM of the study area, were filled using the 'fill' option of GIS by averaging the elevation values of the surrounding pixels. The positional accuracy is within the sub pixel range when compared with 1:50,000 topographic maps. Random tests of the vertical error of the SRTM data with reference to the 1:50,000 maps showed a similar absolute accuracy, in accordance with technical sensor specifications (Rabus et al., 2003). The shaded terrain view of the SRTM DEM of the study area is shown in Figure 12.2. Stream-flow channels are well represented and without any significant errors. The visual appeal of the DEMs can be greatly improved by applying a stretched color ramp to highlight regions of elevation differences and using the "Hill shade" function in GIS to visualize the changes in relief patterns.

12.3.4 VISUAL INTERPRETATION OF SATELLITE DATA

In order to delineate the distinct geomorphic units, the geo-coded IRS-LISS-3 satellite imagery was used. Basic image characteristics like tone, texture, shape, color, associations, etc., were used along with field parameters, such as topography, relief, slope factor, surface cover, soil and vegetation cover in delineating the distinct geomorphic units of the study area. The original SRTM data has been projected in a Geographic (Lat/Long) projection, with the WGS84 horizontal datum and the EGM96 vertical datum.

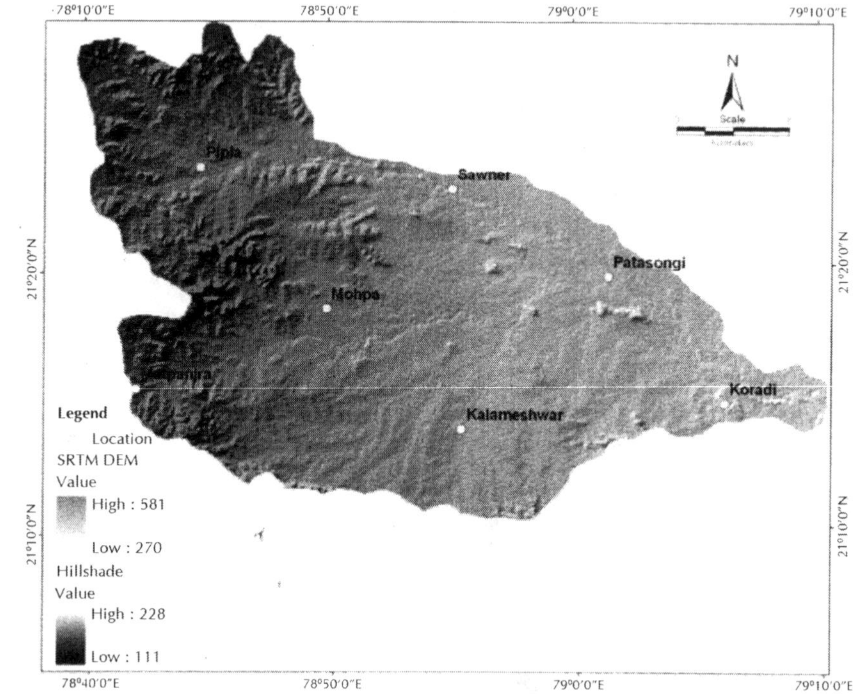

FIGURE 12.2 SRTM DEM (90 m) showing elevation of the study area.

12.3.5 DELINEATION OF DISTINCT GEOMORPHIC UNITS

The geomorphic mapping has been carried out using IRS -LISS III data of the study area. The recognition elements such as color, texture, pattern, drainage pattern, size, shape, land use/land cover, erosion, etc., have been utilized for the convergence of evidence. The geomorphic units identified and delineated in the study area have been categorized under: (i) Units of depositional origin, 2) Units of denudational origin, 3) Unit of structural origin, and 4) Unit of fluvial origin.

12.4 RESULTS AND DISCUSSION

12.4.1 GEOLOGY

Geologically, the study area mainly consists of Deccan Trap basalt (641.70 sq.km), Intertrappean beds (5.78 sq.km), Conglomerate/Sandstone/Limestone

(6.82 sq.km), Sandstone/Ferruginous Sandstone (42.45 sq.km), Quartzite/ Quartzite Muscovite Schist (24.43 sq.km). Deccan basalts are mainly vesicular and amygdaloidal type, which are encountered at top portion of the Geological sequence in the study area followed by fractured and jointed but comparatively less weathered basalt. Hard variety of Deccan basalt in the area shows fine to medium grained texture. Deccan trap is classified into three stratigraphic groups (i.e., upper, middle and lower traps. This classification is based on the distribution and relative proportion of Intertrappean sedimentary beds. The lava flows at some places are separated by Intertrappean sedimentary beds. The thickness of Intertrappean beds ranges from a few centimeters up to two meters and are composed of cherts, impure limestones and pyroclastic material including trap detritus. Deccan traps are followed by Kamthi group of rocks predominantly composed of soft and coarse grained sandstones, Kamthi sandstone also contains fine grained micaceous sandstones, hard and gritty sandstones and homogeneous and compact shales. Conglomerates are composed of white quartz pebbles set in a matrix of grit. Interstratified with this conglomerate is fine red argillaceous sandstone. Massive Quartzite and Mica-Schist belonging to Sausar series forms the base unit in the study area. The geological formations of Kolar river sub basin are shown in Figure 12.3.

12.4.2 DRAINAGE PATTERN

The study area has dendritic to sub dendritic drainage pattern. It is characterized by a tree like branching system in which tributaries join the gently curving main stream at acute angles. The occurrence of this drainage system indicates homogeneous, uniform soil and rock material. They are governed by the topography of the land, whether a particular region is dominated by hard or soft rocks, and the gradient of the land. A basin represents all of the stream tributaries that flow to some location along the stream channel. The number, size, and shape of the drainage basins found in the area vary and the larger the topographic map, the more information on the drainage basin is available.

12.4.3 SLOPE ANALYSIS

On a grid DEM, slope calculation was performed using 3x3 moving window to derive finite differential. In the study, second order finite difference was

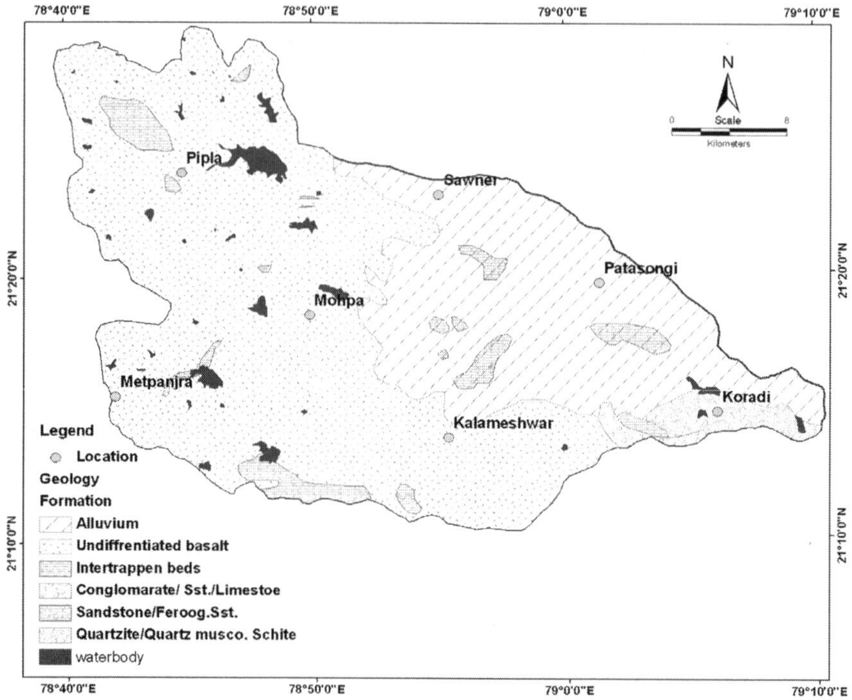

FIGURE 12.3 Geological formations of the study area.

used. Four Closest Neighbors (FCN) algorithms (Guth, 1995; Raaflaub and Colloins, 2006) were used for computing the slope. It takes into account two orthogonal components of slope, slope in x direction and slope in y direction. In other words, the algorithm used the four cardinal neighbors (i.e., North, South, East and West) represents a second order finite difference relationship. This defines the steepness and downhill direction. The slope information is useful in understanding the topography, geomorphology, soil types and their erodability, surface drainage (Manjare, 2013).

The slope map was prepared for study area using 1:50,000 scale SOI topographical maps with 20m contour interval and support from the SRTM (DEM) 90 m data. The raster layer was georeferenced using ARC GIS software (10.1). The isotangent lines were digitized and final isopleth map (a map displaying the distribution of an attribute in terms of lines connecting points of equal value) was prepared. The slope in the study area were calculated in the percent and categorized in to six slope classes namely <1%

(Nearly level), 1–3% (Very gentle slope), 3–5% (Gentle slope), 5–10% (Moderate slope), 10–15% (Strong slope) and 15–30% (Moderately steep slope) (Figure 12.4).

Nearly level slope to Very gentle slope is the lowest category of slope in this region and is associated with extremely flat part of the area. This class covers almost all part of the study area except in western part of Matpanjra village and right to the Mohpa town, which is drained by the river and streams and represents an erosional surface of the current cycle. It also covers the flat land around Patsaongi, Mohpa, Pipla and Saoner village (Figure 12.4). Gentle slope was observed in the study area towards the south west to the Metpanjra and Mohpa village.

Moderately to strong slope can be seen in patches in different parts of the study area. It mainly covers around Metpanjra and along the Pipla and Mohpa village. This slope zone is prominent along the contacts of hill ridges and flat grounds. Pediments are included in this category (Figure 12.4).

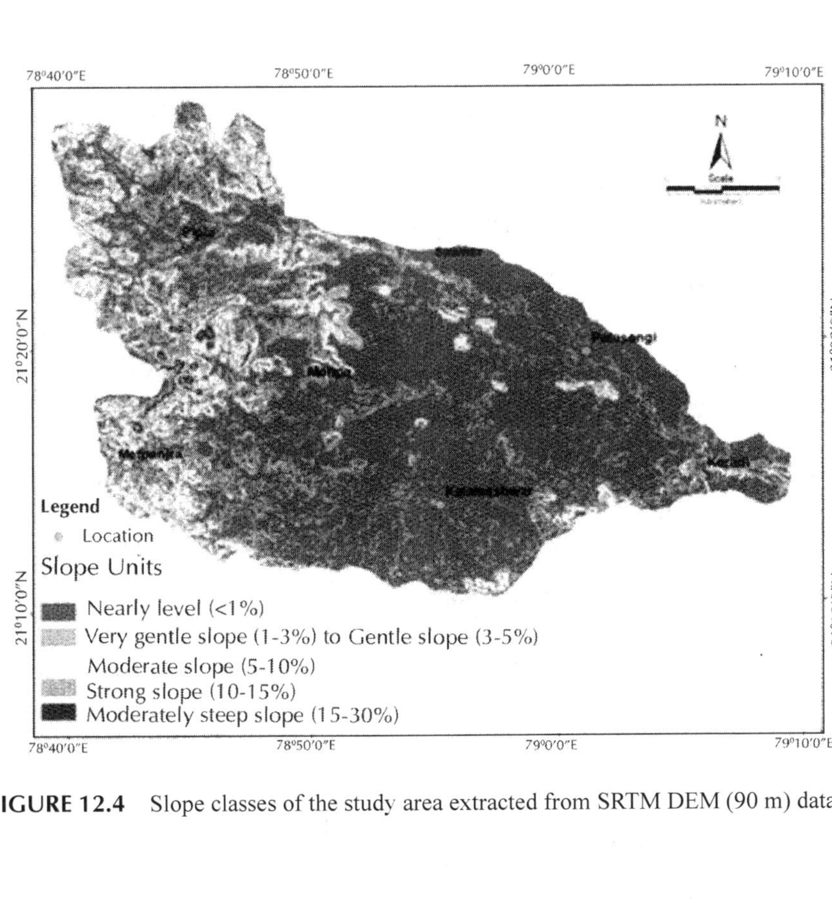

FIGURE 12.4 Slope classes of the study area extracted from SRTM DEM (90 m) data.

Strong slope can be seen more in Pipla village area of the study area. The area of this slope angle ranges occur in narrow parallel continuous stretch along with the next higher slope. Moderately steep slope is observed near North West and south west part of the study area. It is also found along the south west part of the study area. It is mostly from the upper part of debris slope. It is seen more in the south west near Metpanjra village and near right to the Pipla small town and in the south west part of the study area with steep escarpment slope and it shows more erosional features (Figure 12.4).

12.5 GEOMORPHIC UNITS AND THEIR CHARACTERISTICS

In the study area, distinct geomorphic units were delineated based on the analysis of SRTM DEM (90 m), SOI toposheets and satellite imageries. Based on the genesis and processes of geomorphic units, the delineated geomorphological units were grouped under four categories namely structural, denudational, depositional and fluvial.

12.5.1 UNITS OF STRUCTURAL ORIGIN

The effect of geologic structure is neither obvious nor striking and its influence is lacking. The apparent lack of structural control of topography may simply indicate homogeneity of structure with resultant homogeneity of topography or may result from the large scale of the structural units. The entire basaltic terrain can be classified as Deccan trap plateau. However, on the basis of dissection the unit is divided in following three units.

The landform of *highly dissected plateau* occurs in the north-western and the central part of the district. It is represented by high hills with severe dissection. The thin soil cover and rock outcrops mark these landforms. The forest is the dominant land use of this landform (Table 12.1 and Figure 12.5). The highly dissected plateaus are fractured and weathered surfaces and occurring in marginal part of the sub-basin. This unit is severely dissected by the streams of Kolar river giving rise to a terrain consisting of flat topped ridges and steep scarps. The dissected plateau is characterized by the shallow soil cover, moderately high relief, moderately steep slope, rocky and rugged terrain and hard and compact basalt bed rock. This unit is structurally controlled by numerous joints, fractures and lineaments which facilitate some infiltration and mostly act as runoff zones.

TABLE 12.1 Area Wise with Percentage of Geomorphic Units of the Study Area

S. No.	Geomorphic Units	Area (in sq. km)	Area (in %)
1.	Upper plateau on Deccan trap	290.61	45.73
2.	Plateau on Deccan trap	400.81	38.77
3.	Structural Hills	19.32	3.04
4.	Denudational Hills	7.81	1.22
5.	Pediment	10.70	1.68
6.	Pediplain Weathered	252.63	39.75
7.	Flood Plain	38.91	6.12
8.	Residual Hills	10.68	1.68
	Total Area	1031.47	

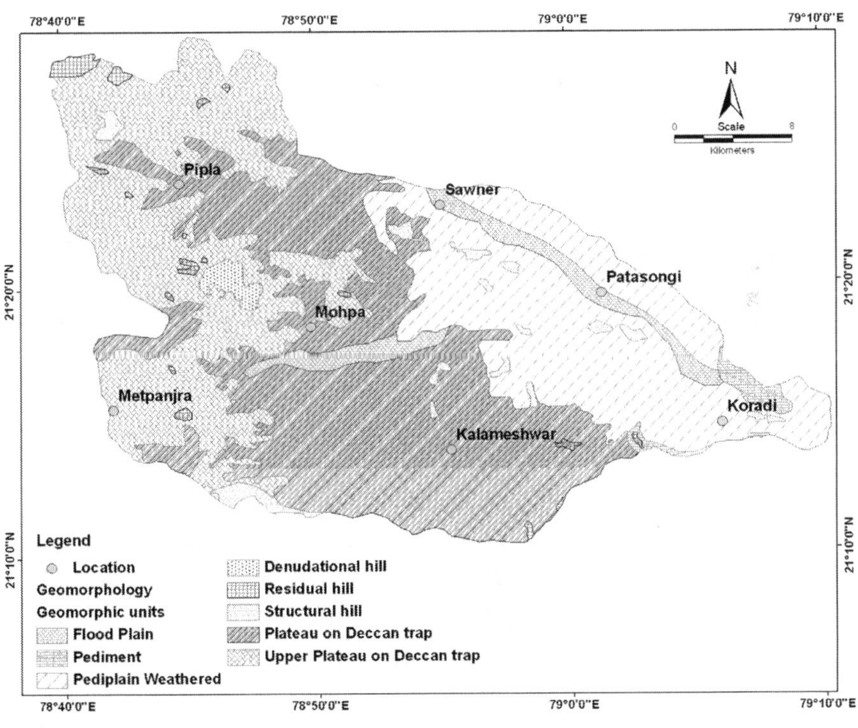

FIGURE 12.5 Geomorphic units of the study area.

The *moderately dissected basaltic plateau* landforms are seen around the highly dissected basaltic plateau units in the south eastern part of the study area. The moderately high hills and medium dissection are the characteristics of this landform and this serves as the recharge belt. The major land covers of this unit are the scrubs (Figure 12.5). The slightly dissected plateau landform is mostly undulating landform occupying the valley areas and the plains. Moderate to thick soil cover, appreciable zone of weathering and less dissection are the main characteristics of this landform (Figure 12.5). Most of the study area is covered by Deccan trap, which is in southern part of the Kalmeshwar village. Some part of Mohpa and Pipla villages is also occupied by the plateau on Deccan trap in West- North direction (Figure 12.5).

Structural hills are noticed as a broad zone of dissected hilly topography in the western part of the area. This unit represents a typical stream caved morphology comprising crested radiation eroded ridges along the down slopes. The structural complexity, presence of impermeable sub-surface and high relief are the cause for existence of trellis and sub-parallel to parallel drainage pattern with high density and stream frequency. The cluster of hills is characterized by rill and gully erosion, smooth and rounded hilltop. Free faces are common on the precipitous sidewalls. The height of the ridges is found to decrease progressively from west to east. It signifies intense linear head ward stream erosion and slope retreat under high hydraulic gradients of stream wok and river piracy activities. At places, where the sparse vegetation exists, this geomorphic unit is susceptible to severe erosion hazards. In the present study, most of the western parts have hill and ridge topography (Figure 12.5). The area covered by the structural hills and occupies about the area of 19.32 sq.km (Table 12.1).

12.5.2 UNITS OF DENUDATIONAL ORIGIN

The fluvio-denudational geomorphologic processes are actively involved landscape reduction process. The physico-chemical and biological weathering and multiple slope dissections under the influence of steep slopes, high drainage density pattern and precipitation conditions lead to the development of ridge-valley land systems in the western part of the area. Mass wasting in the form of creep and debris fall are common processes of land erosion along the scarp slopes and escarpments. Denudational hills, Pediments, have been identified and mapped under denudational landforms.

Denudational hills are the remnants of the natural dynamic process of denudation and weathering. The geomorphic forms of denudational hills occur as exfoliation domes, linear ridges, mesas, low mounds and tors with partial scree or debris covered at the foot slopes. The geomorphic expression and shape of the denudational hills are controlled by lithology and spacing of structural features like joints and fractures occurring in them. The denudational hills in basic intrusive occur as narrow linear ridges within the pediplain where dykes are seen. Denudational hills are the massive hills with resistant rock bodies that are formed due to differential erosional and weathering processes. These hills are composed of Vindhyan sediments, which are fractured, jointed having no soil cover moderate to steep slope. on the satellite image, these landforms were identified by green color due to thick forest cover. Denudational hills occupy western portions of the study area (Figure 12.5). These landforms represent low lying isolated hillocks and occupy west of Mohpa village with an area of 7.81 sq.km (Table 12.1).

A *pediment* is a very gently sloping (1–3%) inclined bedrock surface. It typically slopes down from the base of a steeper retreating desert cliff or escarpment, but may continue to exist after the mountain has eroded away. It is caused by erosion. It develops when sheets of running water (laminar sheet flows) wash over it in intense rainfall events. It may be thinly covered with fluvial gravel that has washed over it from the foot of mountains produced by cliff retreat erosion. It is typically a concave surface gently sloping away from mountainous desert areas. In the study area, sediments are gently sloping areas or erosional surface of bed rock. Pediments may or may not be covered by a thin layer of alluvium and are mostly developed at the foot of the hills occurring along the eastern margin and of the study area. These landforms show light brown color and fine texture (Figure 12.5) and cover an area of 10.70 sq.km.

12.5.3 UNITS OF DEPOSITIONAL ORIGIN

The fluvio-depositional landforms are formed under the influence of corrosive and erosive dynamics and slope retreat processes. Active physico-chemical weathering accompanied by sheet wash and strip removal of debris are responsible for development of various depositional landforms. *Pedipalins* have been identified and mapped under these landforms. This pediplains are formed due to intensive weathering under semi arid climatic conditions, representing final stage of the cyclic erosion. These are identified in the imageries due to gray tone on false color composite. Pediplains have developed in the north eastern part of study area as a result of continuous erosional process. The pediplains are low in relief and consist of deposited sediment that is regularly carried out from upland catchments, subdued hill slopes and pediments. The deposited sediments are admixed with sandy loam and clay fragments. The pediplains predominantly occupy the north-east parts of the study area from Koradi, Patsaongi, Saoner villages with an area of 252.63 sq.km.

12.5.4 UNITS OF FLUVIAL ORIGIN

A flood plain is an area of land adjacent to a stream or river that stretches from the banks of its channel to the base of the enclosing valley walls and experiences flooding during periods of high discharge. It includes the flood way, which consists of the stream channel and adjacent areas that actively carry flood flows downstream, and the flood fringe, which are areas inundated by the flood, but which do not experience a strong current. In other words, a flood plain is an area near a river or a stream, which floods when the water level reaches flood stage. In the study, along the Kolar river it occupies the Saoner, Patsaongi and Koradi villages. The flood plains have a slope towards the eastern part of the area and this unit covers an area of 38.91 sq. km.

Residual hills are the end products of the process of pediplaning, which reduces the original mountain masses into a series of scattered knolls standing on the pediplains. Residual hills occur as small hills comprise of more resistant formations formed due to differential erosion process and these geomorphic units were noticed in the western and south western parts of the study area (Figure 12.5). These features were noticed as dark pink patches with forest cover on the satellite imageries. The residual hills covered an area of 10.68 sq.km (Table 12.1).

12.6 CONCLUSIONS

In the study area, distinct geomorphological units such as Denudational hills (7.81 sq.km) and Pediments (10.70 sq.km) under Denudational landforms, Structural hills (19.32 sq.km), Plateau on Deccan trap (400.81 sq.km),

Upper plateau on Deccan trap (260.61 sq.km) under Structural landforms, Residual hills (10.68 sq.km); Flood plain (38.91 sq.km) underFluvial landforms have been delineated. The analysis of the land degradation status in the study area reveals that the deciduous forests in the hilly catchment areas are being indiscriminately cleared for cultivation, which led to rill and gully formations in uplands. The information generated at geomorphic sub unit could be effectively used for prioritizing the areas for adoption of appropriate soil and water conservation measures, river basins management, hazards mitigation, environmental planning. The impact of structure, processes and factor are clearly observed on the landforms, which give a complex appearance to slope profile. The summit plateaus are bordered by the steep escarpment slope, which are followed by the talus slope or debris slope with convex profile. A pediment particularly in the foot hills region and the debris slope, whereas, in the other parts the pediments slope is absent. The hilly terrain with steep slope is more extensive than the low angled gentle pediment slope. The presence of steep slope of the Deccan trap is unique feature observed in the study area. The scrap face is conspicuous in most parts of the study area and found bordering to the plateaus. The integrated analysis of IRS-1C and 1D LISS-III, PAN and SRTM DEM (90 m) data will be of immense help in delineation of distinct geomorphic units and analysis of their process for adoption of appropriate soil and water conservation measures, river basins management, hazards mitigation and environmental planning in the study area.

KEYWORDS

- **Central India**
- **Digital Terrain Analysis**
- **Geomorphology**
- **GIS**
- **Kolar River Sub Basin**
- **Remote Sensing**
- **Slope**
- **SRTM DEM**

REFERENCES

Bernhardsen, T., & Stove, M. (1999). Geographic Information Systems: An Introduction, 2nd ed.; John Wiley and Sons, Inc: New York.

Bolstad, P. V., & T. Stowe. (1994). An evaluation of DEM accuracy: Elevation, slope, and aspect, *Photogrammetric Engineering & Remote Sensing, 60*(11), 1327–1332.

Campbell, L. B. (2002). Introduction to Remote Sensing. 3rd ed.; The Guilford Press, New York.

CGIAR-CSI – http://srtm.csi.cgiar.org, Consultative Group on International Agricultural Research Consortium, Accessed on 15th January, 2016.

Cooke, R. U., & Doornkamp, J. C. (1990). Geomorphology in environmental management, A New Introduction. Clarendon Press: Oxford.

Dramis, S., Guida, D., & Cestari, A. (2011). Nature and Aims of Geomorphological Mapping. In: Geomorphological Mapping: methods and applications. Smith, M. J., Paron, P., Griffiths, J. (Eds.). Elsevier: London, pp. 39–74.

Dykes, A. P. (2008). Geomorphological maps of Irish peat landslides created using hand-held GPS. *Journal of Maps* 258–276.

GSI. (2000). *District Resource Map of Nagpur District*, Central Region, Geological Survey of India.

Guth, P. L. (1995). Slope and aspect calculations on gridded digital elevation models: examples from a geomorphometric toolbox for personal computers. *Zeitschrift für Geomorphologie N. F. Supplement Band, 101*, 31–52.

Hengl, T., & Evans, I. S. (2009), Mathematical and Digital models of the Land Surface. *In: Geomorphometry: Concepts, Software, Applications*, T. Hengl, and H. I. Reuter (Eds.), pp. 31–63 (The Netherlands: Elsevier).

Kodge, B. G., & Hiremath, P. S. (2011) Computer Modeling of 3D Geological Surface, International *Journal of Computer Science and Information Security, 9*(2), 175–179.

Lee, E. M. (2001). Geomorphological mapping. *Geological Society Special Publication 18*, 53–56.

Manjare, B. S. (2013). Morphometric Slope Analysis of Part of Salbardi and Adjoining Region of Amravati District, Maharashtra and Betul District of Madhya Pradesh, India. *Indian Journal of Geomorphology, 18*(2), 59–67.

Menze, B. H., Ur, J. A., & Sherratt, A. G. (2006). Detection of Ancient Settlement Mounds: Archaeological Survey Based on the SRTM Terrain Model. *Photogrammetric Engineering & Remote Sensing, 72*, 321–327.

Paron, P., & Claessens, L. (2011). Makers and users of geomorphological maps. In: Geomorphological Mapping: methods and applications. Smith, M. J., Paron, P, Griffiths, J. (Eds.). Elsevier: London, pp. 75–106.

Raaflaub, L. D., & Collins, M. J. (2006). The effect of error in gridded digital elevation models on the estimation of topographic parameters. *Environmental Modeling and Software, 21*, 710–732.

Rabus, B., Eineder, M., Roth, A., & Bamler, R. (2003). The shuttle radar topography mission a new class of digital elevation models acquired by space borne radar. ISPRS *Journal of Photogrammetry and Remote Sensing, 57*(4), 241–262.

Reddy, G. P. O., Nagaraju, M. S. S., Ramteke, I. K., & Sarkar, D. (2013). Terrain Characterization for soil resource mapping in part of semi-tract of Central India using high resolution satellite data and GIS, *Journal of the Indian Society of Remote Sensing, 41*(2), 331–343.

Simard, M., Zhang, K., Rivera-Monroy, V. H., Ross, M. S., Ruiz, P. L., Castañeda-Moya, E., Twilley, R. R., & Rodriquez, E. (2006). Mapping height and biomass of mangrove forests in everglades national Park with SRTM Elevation Data. *Photogrammetric Engineering & Remote Sensing, 72*, 299–311.

Singh, C. G., Prusty, G., & Srivastava, H. (2012). Terrain Analysis of Desert Area Using Geo-Information Techniques, *International Journal of Remote Sensing & Geoscience*, *2*(3), 33–38.

Smith, M. J., Griffiths, J., & Paron, P. (eds.). (2011). Geomorphological Mapping: methods and applications. Elsevier: London. pp. 75–106

Summerfield, M. A. (1997). Global Geomorphology: An introduction to the study of landforms. Essex, England: Longman.

Tarboton, D. G., Bras, R. F., & Rodriguez-Iturbe, I. (1991). On the extraction of channel networks from digital elevation data. *Hydrological Processes, 5*, 81–100.

Welch, R. (1990). 3-D Terrain Modeling for GIS Applications. GIS World, *3*(5), 26–30.

Wilson, J. P., & Gallant, J. C. (2000). Terrain Analysis Principles and Applications, John Wiley and Sons, New York, 479 p.

Wood, J. D. (1996). The geomorphological characterization of digital elevation models. PhD Thesis, University of Leicester, United Kingdom.

CHAPTER 13

TIME SERIES SATELLITE DATA AND GIS FOR CROP ACREAGE ESTIMATION: A CASE STUDY FROM CENTRAL INDIA

A. R. PIMPALE,[1] P. B. RAJANKAR,[2] S. B. WADATKAR,[3] and I. K. RAMTEKE[4]

[1]*Assistant Professor, Section of Agricultural Engineering, College of Agriculture, Nagpur – 440 001, India*

[2]*Associate Scientist, Maharashtra Remote Sensing Applications Centre, Nagpur – 440 010, India*

[3]*Head, Department of Irrigation and Drainage Engineering, Dr. PDKV, Akola – 444 104, India*

[4]*Scientific Associate, Maharashtra Remote Sensing Applications Centre, Nagpur – 440 010, India*

CONTENTS

ABSTRACT

Estimation of crop production in advance plays a key role in the economy of agrarian countries, particularly in developing countries like India in order to obtain cropped area and the average yields. Reliable and timely estimates of crop area coverage are of great importance to planners and policy makers and researchers throughout the world. Sorghum and wheat are dominant crops grown in *rabi* season in Maharashtra, Central India, and it assumes great importance in estimation of acreage. Traditionally, the estimates of crop acreage are obtained through complete enumeration by revenue agencies (Giradwari system), sample surveys and personal assessment by village *patwari*. Application of remote sensing and Geographic Information System (GIS) can provide more accurate and timely estimates of crop acreages. The present chapter deals with the methodology adopted for classifying *rabi* sorghum and wheat crops using time series Normalized Difference Vegetation Index (NDVI) derived from multi-date Indian Remote Sensing (IRS)-P6, Advanced Wide Field Sensor (AWiFS) 56 m resolution data for *rabi* season of 2012–13. Study area covers the dominant *rabi* sorghum and wheat growing districts of Maharashtra state namely Pune, Solapur, Ahmednagar, Beed and Osmanabad. Two-stage classification of the datasets has been carried out through unsupervised Iterative Self Organizing Data Analysis Technique (ISODATA) to label classes based on temporal spectral profiles of wheat and *rabi* sorghum as well as other competing crops, vectorisation of classified image, manual editing and labeling of mixed class polygons. The decision rule based integration was followed to generate final classified image. This hybrid classification technique takes advantage of inherent clustering tendency of land use/land cover classes in feature space with temporal dimension added to it in terms of NDVI time series data. It also makes use of signatures of known crop classes for labeling the clusters. Analysis of *rabi* sorghum and wheat acreage estimation in the study area shows 16,69,599 ha and 1,89,481 ha, respectively, which is deviating by 8.81% and 9.78%, respectively from the reference data as reported by Government of Maharashtra. It is observed that this technique is simple, time saving, less subjective and requires less expertise compared to hierarchical classification technique.

13.1 INTRODUCTION

The economy of most developing countries is agrarian and agriculture is principal source of income and source of raw material for a large number of industries in these countries. Accelerating the growth of agriculture production is therefore necessary not only to meet the rising demand for food, but also to increase income of those dependent on agriculture to ensure inclusiveness. Globally, climate change is a new challenge for agriculture. Agriculture has to face both the types of droughts conditions (i.e., excess rainfall and scanty rainfall as well as extreme weather situations. Food grain production is the major concern in order to meet the demand of food for rapidly increasing world population. The crop growth is dependent on multiple factors like judicious use of land and water resources, use of high yielding variety, optimal use of fertilizers and suitable management practices. Knowledge of food grain production well in advance of harvest period helps the planners, policy makers, administrators and researchers to make important and wise decisions with respect to adapt suitable measures to meet shortages if any, buffer stock, fixing minimum support prizes, procurement, processing, storage, public distribution, export, import and proper planning of other related issues according to market situations. Therefore, it becomes necessary for stakeholders to have knowledge of the scenario of food production in order to frame strategy for inadequate or very high food grain production for import or export of a particular food grain. Crop production estimates have two components-estimating the area under the crop and forecasting yield per hectare. This chapter focuses on crop acreage estimation using temporal satellite data and GIS. In many countries, conventional techniques of data collection are based on field visits and reports are used for crop inventory. These methods are subjective, very costly and time consuming. For example in India traditionally, the estimates of crop acreage are obtained through complete enumeration by revenue agencies, sample surveys and personal assessment by village officials, which takes quite long time. Since these officials have multiple responsibilities, these records may not be accurate and reliable. The advent of remote sensing technology and spectral indices based classification techniques have opened new vistas of enhancing systems, revolving around crop discrimination, identification, classification and mapping. Multi-date, multi-spectral data in the form of several vegetation indices have been utilized for crop identification and area estimation worldwide.

13.1.1 CROP ACREAGE ESTIMATION: GLOBAL PERSPECTIVE

The first global experiment to demonstrate the operational capability of remote sensing for crop inventory and forecasting was carried out in USA in 1980. It was called as Large Area Crop Inventory Experiment (LACIE) (McDonald and Hall, 1980). The major thrust was to develop, test and evaluate the system for predicting wheat production through the use of LANDSAT data for area estimation and weather based empirical regression yield models to predict yield. Since then, many other countries have carried out similar work. Another major work was taken up by European Union under the special task called Monitoring Agriculture through Remote Sensing (MARS) for crop production estimates across the Europe (Sherman, 1993) using a sampling approach, multi-date crop identification and field survey information. Guruge (1996) monitored the development of sugarcane in Buttala area of Monaragala district, Sri Lanka from 1983 to 1994 using aerial photographs and IRS images. Sugarcane areas were computed using digital images for four years (1983, 1988, 1992 and 1994). In Greece, Tsiligirides (1998) estimated acreage, productivity and production of soft durum wheat for the Hellenic regions of Macedonia and Thrace, by using area frame of square segments methodology in combination with Landsat TM images. Liu et al. (1999) predicted winter wheat production in North China plain with an accuracy of over 96% compared with on the spot measurement by using Landsat TM data and Advanced Very High Resolution Radiometer (AVHRR) time series data. In Australia, Potgieter et al. (2005) used multi temporal Moderate Resolution Imaging Spectroradiometer (MODIS) data for determining winter wheat acreage in the Darling Downs region by generating Enhanced Vegetation Index (EVI) imagery and adopting multivariate approaches such as Harmonic Analysis of Time Series (HANTS) and Principal Component Analysis (PCA). The results were encouraging. An operational remote sensing based Crop Monitoring and Production Forecast Program (CROPMON) was reported from Hungary by Csornoi et al. (2008). In this program the crop area estimation (including winter wheat) was based on the quantitative analysis of multispectral high resolution images (Landsat TM and IRS –1C/1D LISS III).

13.1.2 CROP ACREAGE ESTIMATION: INDIAN PERSPECTIVE

In India, experiments of crop identification and area coverage estimation by using space/remote sensing satellite data started since 1980. Ayyangar et al. (1980) conducted experiment in agriculture area around Mandya town of Karnataka state. They used the spectral temporal profiles with red and near infrared bands with spectral information obtained through modular multispectral scanner from a height of 1000 m above ground level, for crop identification and condition assessment of rice and sugarcane crops. They analyzed and interpreted the spectral data from ground information. It was possible to distinguish between wet and dry paddy fields. The spectral information in the form of ratio also enabled to distinguish between rice and sugarcane. Munshi (1982), in his study used space borne data and employed visual mapping for wheat crop in four districts of Punjab state. Dadhwal and Parihar (1985) made attempt towards the use of satellite data in wheat crop acreage estimation in the Karnal district of Haryana using Landsat MSS data. The work emphasized the use of single date data and supervised Maximum Likelihood (MXL) classification approach. Based on the encouraging results of this study a project called Large Area Crop Acreage (LACA) was initiated. Later on the objective of yield forecasting was also added and was named as Crop Acreage and Production Estimation (CAPE) since 1988. In this project, single date high-resolution RS data were used. Further improvements in this project gave rise to project called "Forecasting Agricultural Output Using Space, Agro-meteorology and Land–based Observations" (FASAL) for forecasting of multiple crops. Apart from these national level experiments many other attempts are being made to improve accuracy on regional scale for particular crops (Dua et al., 2007 and Gontia and Tiwari, 2010; Kimothi et al., 1997; Panigrahy and Chakraborty, 1996; Rao and Rao, 1987; Sharma, 1995). The stacked NDVI of multi date remote sensing satellite data found additional accurate in crop identification like sugarcane acreage estimation and crop production forecast (Krishna Rao et al., 2002).

For crop discrimination and area estimation by remote sensing, initially single date satellite images with coarse resolution were used. Gradually with the improvement in spatial and temporal resolution multidate images are utilized for crop classification. In addition, high resolution satellite imageries at much cheaper and affordable cost are becoming available. Thus, more accuracy is possible on regional scale classification of crops for precision management of crops. The most commonly used classification methodologies used in remote sensing are unsupervised procedures such as Iterative Self Organizing Data Analysis Technique (ISODATA) and in supervised methods, the most popular is the Maximum Likelihood (MXL) algorithm, where precise filtering of the information registered on imageries is carried out with image analysis algorithms. Researchers have used various classifiers for classifying the RS data for different applications (Tateishi et al., 1991;

Friedl and Brodley, 1997; Hastings, 1997; Friedl et al., 1999). Multi-date IRS WiFS and AWiFS data have been used for various agricultural applications including cropping pattern monitoring, crop yield modeling and crop classification (Oza et al., 2008; Rajak et al., 2002, 2005).

13.1.3 IMPORTANCE OF THE STUDY

The present study was conducted in five districts of Maharashtra, located in central region of the India with an aim to estimate area under *rabi* sorghum and wheat using hybrid classification technique using multi temporal remote sensing data. During *rabi* season, sorghum and wheat are the major crops and are cultivated in large contiguous areas in the study area. Monitoring of crop condition, development of crop growth profile and prediction of yield based on remotely sensed multi-spectral data of these crops holds great promise. In this study, a hybrid technique of multi-date RS data classification based on two-stage ISODATA clustering and visual classification has been followed. It takes advantage of the natural groupings of pixels based on their spectral properties by way of ISODATA clustering and discrimination of spectrally similar classes by way of visual vector polygon analysis.

13.2 STUDY AREA

The study area comprises of five districts of Maharashtra state of India (i.e., Pune, Solapur, Ahmednagar, Beed and Osmanabad (Figure 13.1) wherein spatially extensive and contiguous patches contribute to sorghum and wheat production. It covers an area of approximately 65,716 Km^2. It is located between 73° 17′ 19″ E to 76° 47′ 42″ E longitudes and 19° 58′ 57″ N to 17° 03′ 56″ N latitudes. Most parts of the study area are water scarcity zones with average annual rainfall between 500–700 mm with uncertainty and erratic distribution. Rainfall pattern is bimodal and droughts occur very often. The climate of the study area is hot and dry. Soils of the study area are generally of trap origin. Study area is the part of Deccan Trap, composed of Precambrian Crystalline Rocks. Considering varying characteristics, soils of the region are classified as lateritic soils, medium black soils, coarse shallow gray soils and alluvial soils. (Challa, et al., 1995). A continuous belt of lateritic soils stretches in north-south direction occupying mostly Western Ghat in Pune and Ahmednagar districts. Medium black soils are very fertile and occupy large area of the study region in Pune, Ahmednagar, Beed and

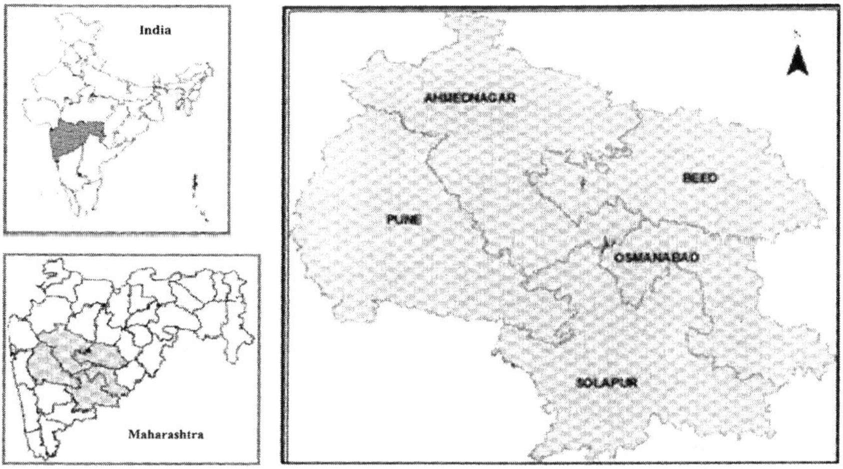

FIGURE 13.1 Location of the study area.

Solapur district. These soils are also known as black cotton soils, regur, vertisols or gurmosols, covering major portion of the region. Coarse shallow gray soils are found in Central part of Beed district running east west direction and north-west to south-east middle part of the Osmanabad district. River valleys and terraces are dominated by alluvial soils. Sorghum and wheat are dominant crops in the *rabi* season, hence, these two crops were selected for present investigation. The other *rabi* crops of the area are chickpea, pigeon pea, maize, onion, garlic and lucern. Horticultural crops like pomegranate, grapes, banana and *ber* are also noticed. Biannual cash crop sugarcane is also one of the competing crops present in the field during the *rabi* season. Most of the *rabi* crops in the region are grown on the residual soil moisture, except some irrigated areas in the command. Wheat is grown on assured irrigation sources. The major sources of irrigation are canals in the command areas. Most parts of the districts under study are falling in water scarcity zone with average annual rainfall between 500–700 mm with uncertainty and ill distribution.

13.3 MATERIALS AND METHODS

13.3.1 REMOTELY SENSED DATA USED

Multi-date, multispectral satellite images of IRS-P6, AWiFS for five consecutive months (contained in eight dates) of *rabi* season (October, November, December, January and February) of the year 2012–13 were obtained from

National Remote Sensing Centre (NRSC), Hyderabad. The details of satellite data used are given in Table 13.1. AWiFS sensor have four bands, band- 1 is Green (0.5–0.59 μm), band-2 is Red (0.62–0.68 μm), band-3 is Near Infra Red (NIR) (0.77–0.86 μm) and band-4 is Middle infra red (MIR) (1.55–1.70 μm) with resolution of 56 m near Nadir and 70 m near edge. It covers swath of 740 km with radiometric resolution of 10 bit. The projection and datum of the data products are Lambert Conformal Conic and WGS 84, respectively. These bands were downloaded and 4 band raw images of the given path and rows were obtained by stacking the layer. Rectangular subsets covering the study area were obtained using data preparation module of ERDAS Imagine. The subset images are shown in Figure 13.2.

13.3.2 ANCILLARY DATA

The Survey of India (SOI) toposheets on 1:50,000 scale covering the study area were used in the present study as a reference to digitize the administrative boundaries of the districts. The non-crop mask (NCM) of the study area was prepared by utilizing the data of water bodies, urban sprawl, forests, wastelands and other non crop features from visual interpretation of satellite data. The vector layer indicating base features such as roads (national highways, express highways, district roads, etc.), railways, dams canal, etc., in the study area were also digitized. The district-wise crop data (Area, Production and Yield data) published by the Department of Agriculture, Government of Maharashtra, on the website www.mahaagri.co.in (Crop statistics, 2012) related to study area was collected as a reference for analyzing the results of crop acreage estimation.

TABLE 13.1 Multi-Date IRS-P6 AWiFS Data Used for the Study

S No	Satellite	Sensor	Path	Row	Date of Pass
1	IRS-P6	AWiFS	097	058	19–10–2012
2	IRS-P6	AWiFS	098	059	11–11–2012
3	IRS-P6	AWiFS	098	059	29–11–2012
4	IRS-P6	AWiFS	098	059	11–12–2012
5	IRS-P6	AWiFS	098	062	27–12–2012
6	IRS-P6	AWiFS	097	059	11–01–2013
7	IRS-P6	AWiFS	097	058	23–01–2013
8	IRS-P6	AWiFS	097	059	04–02–2013

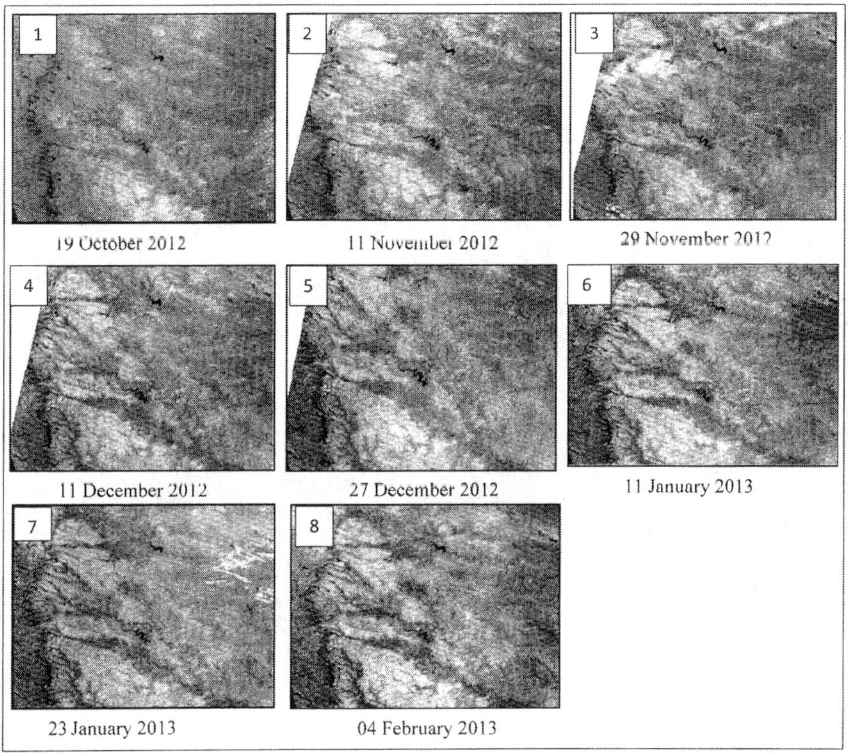

FIGURE 13.2 IRS-P6 AWiFS subset images used for study.

13.3.3 GROUND TRUTH DATA

Ground truth (GT) work was carried out in December 2012 coinciding with the season of *rabi* sorghum and wheat crop in the study area. Data were collected from 82 sites of *rabi* sorghum, wheat and other crops. (Figure 13.3). The handheld Global Positioning System (GPS), geotagged camera and a mobile with Android Apps for collecting formatted field information were used to obtain crop and field information, photographs, locations and elevations of the sites. At the different locations of fields visited, *rabi* sorghum and wheat crops were at different stages. The information regarding the age was obtained by interviewing the farmers and by using criteria suggested by Kansas State University USA (Vanderlip, 1993) and Feekes scale of wheat development (Herbek and Lee, 2009) for sorghum and wheat, respectively. The information obtained from each site was recorded in GT proforma sheets. The GT of other coexisting crops like *rabi* sorghum, chickpea,

FIGURE 13.3 Observations of *rabi* sorghum, wheat and other crops during Ground Truth.

sugarcane, etc., was also collected. The GT work in operation and location points are depicted in Figures 13.3 and 13.4, respectively.

13.4 DATA PROCESSING

Detailed methodology followed in data processing to obtain the acreage of *rabi* sorghum and wheat crops in the study area using remotely sensed data and GT observations data is described in the following subsections.

13.4.1 PREPARATION OF REFERENCE TEMPORAL SPECTRAL PROFILES (RTSP)

NDVI (Rouse et al., 1973) was extracted from the subset images for all the dates of pass to get 8 NDVI images and were stacked together in the form of single stack layer image. The GT locations were digitized to get point vector layer (Figure 13.4). Based on the details collected during field visits, crop polygons (training windows) were marked on the image having acquisition date close to field observation date and crop polygon vector layer was prepared using ArcGIS. The geographical locations of GT field sites were transferred on the images and pure sorghum and wheat multi-date NDVIs were extracted by overlaying the crop polygon vector layer on the stack layer images using signature editor function of ERDAS Imagine. The temporal NDVI of other coexisting crops were also obtained. These NDVI values were exported to Microsoft Excel and arranged weekwise considering the age of crops in terms of week at different locations on dates of images (dates of pass). The date of pass close to the dates of GT and the interval between the different dates of pass was also considered while arranging the NDVI values as per age (in weeks). These NDVI values on the different dates of acquisition were plotted against the days after sowing represent the general signature of the crops. Accordingly, the signatures of *rabi* sorghum and wheat crop were developed. These signatures provide Referencc Tcmporal Spectral Profiles (RTSPs) of *rabi* sorghum and wheat. The sugarcane being important competing crop, its RTSP was also prepared.

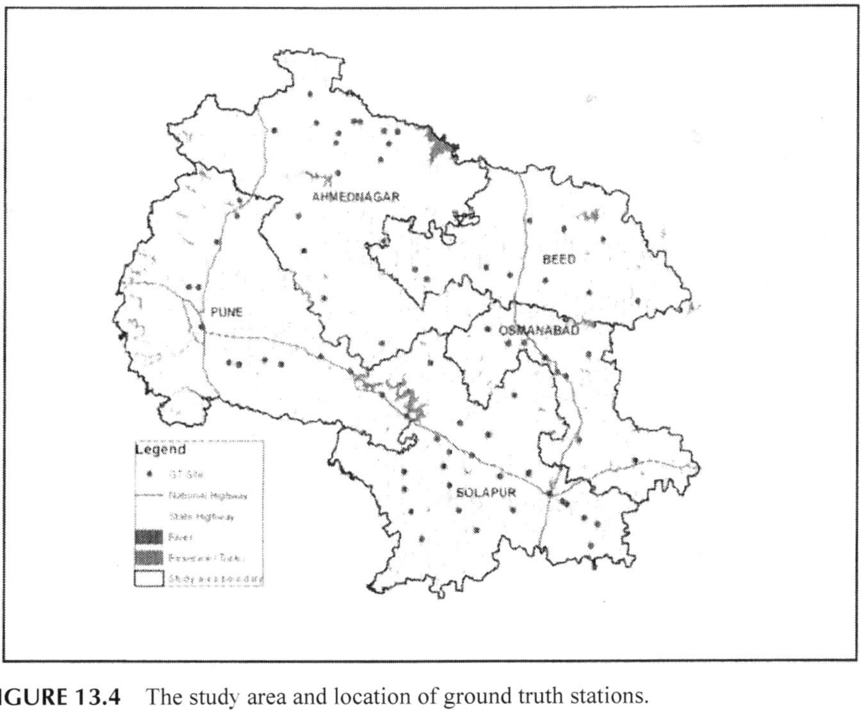

FIGURE 13.4 The study area and location of ground truth stations.

13.4.2 DATA LOAD REDUCTION OF MULTI-DATE REMOTE SENSING DATASET

Non Crop Mask (NCM) was prepared using visual interpretation, exhibiting information on wasteland, forest, slope, built-up areas, etc. To reduce the number of data pixels to be subjected for ISODATA clustering, all pixels belonging to non-agricultural areas were masked out. It was carried out by multiplying NDVI stacked data with NCM image. The NCM image contains 1 and 0 for all pixels belonging to crops and non-agriculture, respectively. The output multi-layer dataset is then used for further processing in ISODATA clustering stage-I and stage-II.

13.4.3 ISODATA CLUSTERING AND LABELING (STAGE-I AND STAGE-II)

The NDVI stack image containing NDVI of all the 8 dates were used for classification. Unsupervised classification was carried out in ERDAS Imagine software. Total 100 classes were opted with 20 iterations and 0.950 coverage threshold with skip factor X=1 and Y=1. A signature file containing signatures of all the 100 classes was also generated. The signature file so generated was used to generate Temporal Spectral Profile (TSPs) for each class. Each of these TSPs of the 100 classes was then compared visually with available RTSPs of *rabi* sorghum and wheat crops. The TSP of cluster class is assigned to the RTSP, which matches most in temporal domain within threshold limit of RTSP. Matching of TSPs with RTSPs was carried out by visual observation and with analyst's expert knowledge of the region.

TSP of some clusters was not assigned to any of the RTSPs and such clusters were put under "Mix crop" class. Also due to lack of RTSPs some cluster classes remained "unclassified" in the classified image. The pixels belonging to "Mix crop" class and "unclassified" class were further fed to ISODATA clustering (stage-II) and same procedure was repeated as in the case of ISODATA clustering (stage-I). Final output of ISODATA stage-I and Stage-II were stored separately.

13.4.4 VISUAL INTERPRETATION OF RESIDUAL PIXELS

Unclassified pixels and/or pixels, which could not be assigned to any crop class with confidence using available RTSPs even after ISODATA Stage-I and Stage-II, were smoothened using median filter and were converted to vector coverage. This vector coverage was then overlaid on multi-dates RS images. Polygons were visually analyzed and labeled accordingly.

13.4.5 INTEGRATION OF CLASSIFIED IMAGES/VECTOR AND ACREAGE ESTIMATION

The output of ISODATA stage-I and stage-II analysis and visually interpreted polygons are integrated to create the final classified image. For extracting *rabi* sorghum and wheat acreages, a stratified random sampling approach was adopted. Using the statistical sampling technique number of *rabi* sorghum and wheat pixels were estimated and were converted into acreage by multiplying with spatial resolution of AWiFS sensor. District-wise *rabi* sorghum and wheat acreages were obtained by subjecting the classified image and study area vector for GIS analysis majority count tool in ERDAS Imagine. The district-wise area statistics was then extrapolated for total acreage estimation of the study area. The flow of methodology adopted in this study is depicted in Figure 13.5.

13.6 RESULTS AND DISCUSSION

The classified image was obtained after analysis is shown in Figure 13.6. After comparing with known RTSPs of *rabi* sorgum, wheat, sugarcane, chickpea, etc., the cluster classes were assigned to a particular crop with which it matches the most. It was found that TSP of 24 and 15 classes were matching with RTSP of *rabi* sorghum and wheat, respectively. The analysis of *rabi* sorghum and wheat acreage estimation for the study area comprising five districts of Maharashtra state shows that about sorghum and wheat crops occupy about 16.69 lakh ha and 1.89 lakh ha, respectively. More than 95% pixels classified as *rabi* sorghum and wheat were correctly classified using the hybrid technique. It is observed that many linear features like canals and roads were correctly classified as non-crop class in the present analysis. The results of acreage estimation and its comparison with the reference figures received from crop statistics by Department of Agriculture, Government of Maharashtra are given in Table 13.2. Relative deviations in terms of % variation are estimated district-wise and crop-wise.

Remote sensing based *rabi* sorghum acreage for the year 2012–13 when compared with acreage data collected from Department of Agriculture showed an average deviation of 8.81% on study area basis. When district-wise results

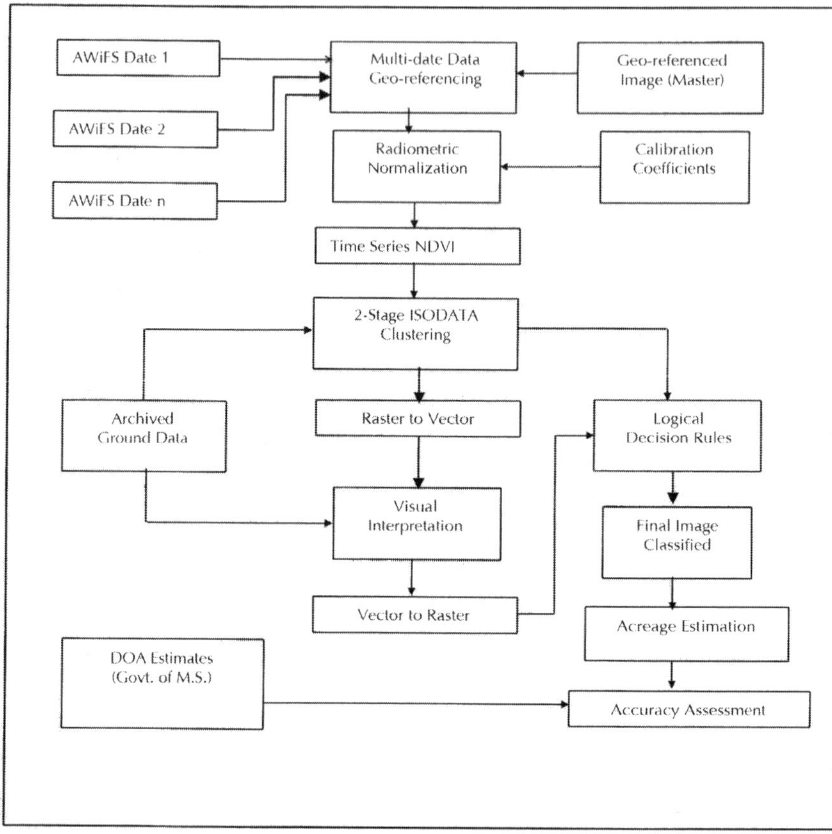

FIGURE 13.5 Flow diagram of methodology adopted for acreage estimation.

are studied lowest variation of 5.72% was observed in Osmanabad district while Pune district showed highest variation of 13.54%. The overall results indicated high overall accuracy of 91.91%, whereas district-wise accuracy of acreage estimation is lower. This difference may be because of mixing of crop pixels due to comparatively smaller size of plots and other competing crops such as maize, lucern, chickpea, pigeonpea, etc., in the same season in the fields.

Area under wheat crop in the study area for the year 2012–13 was estimated by remote sensing as 1.89 lakh ha against actual area of 1.72 lakh ha reported by Department of Agriculture showing 9.78% over estimation. This may be because of similar spectral profiles of coexisting crops like oats. Wheat being irrigated crop, soil moisture also affects the reflectance of crop. Lowest variation of 6.31% was observed in Ahmednagar district, whereas, highest variation 12.66% was observed in Pune district.

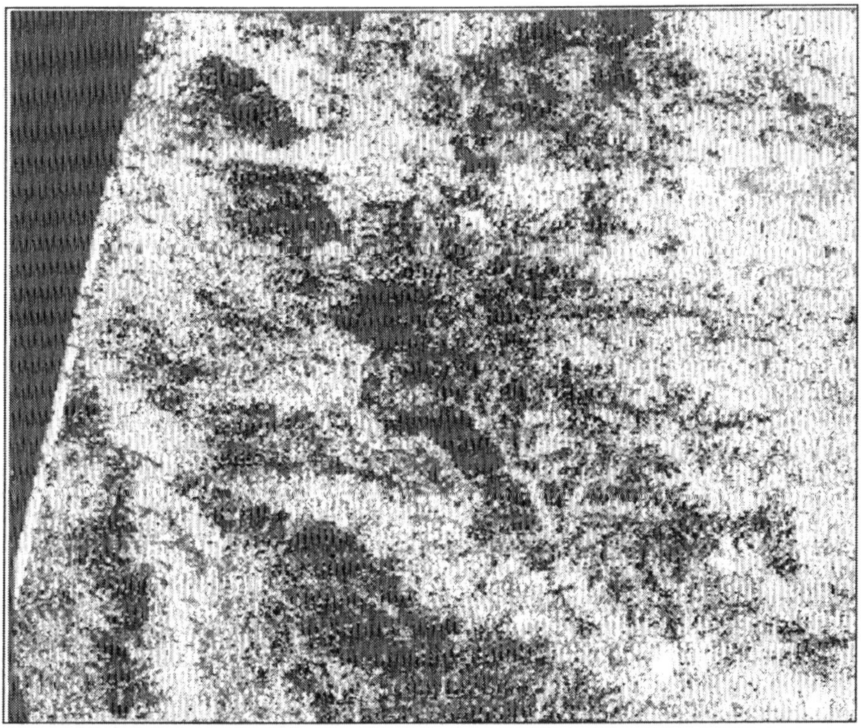

FIGURE 13.6 Classified image obtained by unsupervised classification.

TABLE 13.2 Estimated District-Wise Acreage of *Rabi* Sorghum and Wheat

Crop	District	RS Estimates (ha)	DOA estimates (ha)	Deviation (%)
Rabi sorghum	Ahmednagar	485792	444800	9.22
	Pune	201073	177100	13.54
	Solapur	523202	480700	8.84
	Beed	200195	186500	7.34
	Osmanabad	259337	245300	5.72
	Total	1669599	1534400	8.81
Wheat	Ahmednagar	41357	38900	6.31
	Pune	59708	53000	12.66
	Solapur	30861	28800	7.16
	Beed	31805	27700	11.21
	Osmanabad	26750	24200	10.53
	Total	189481	172600	9.78

Different researchers across the world have used single date or the multidate satellite data of different satellites in order to classify the crops and find its coverage. Similar results related to accuracy of acreage estimation were found by Dua et al. (2007) who used IRS-P6 AWiFS data for the potato acreage estimation in dominant potato growing areas in the Indo-Gangetic region of India and fond accuracy of estimation as 90%. Pakhale et al. (2010) determined area under wheat cultivation in Karnal district of Hariyana, India using Landsat ETM+ image with accuracy of 96.8%. Goswami et al. (2012) carried out wheat acreage estimation in Indore district of Madhya Pradesh using single date, cloud free Resourcesat-1 LISS-III and found relative deviation of 19% as compared to data from Land Commissioner. Many other works across the world have similar results.

13.7 CONCLUSIONS

The hybrid technique used in this study demonstrates the utility of multidate remote sensing data classification, integrated two-stage ISODATA clustering and visual classification of unknown vector polygons, which resulted accuracy of 91.19 and 90.22% for *rabi* sorghum and wheat, respectively. The two-stage ISODATA clustering yielded in a classified image having 93 to 98% pure-class pixels and 2 to 7% mixed-class pixels. It is observed that many linear features like canals and roads were correctly classified as non-crop classes in the present study. Such a holistic approach where multidate RS data is fed to rigorous and repetitive analysis, makes it conceptually sounder than those methods, involving single-date spectral indexes which rely solely on spectral data within a narrow critical period of maximum vegetative growth phase. Therefore, it is concluded that this method could be used for quick and more accurate estimation of crop acreages. The accuracy can be improved by using remote sensing data of high spatial resolution.

KEYWORDS

- **Acreage Estimation**
- **Central India**
- **Crop Inventory**
- **GIS**

- **Ground Truth**
- **Hybrid Classification**
- **Isodata Clustering**
- **NDVI**
- **Remote Sensing**

REFERENCES

Ayyangar, R., Rao, P. P. N., & Rao, K. R. (1980). Crop covers and crop phenological information from red and infrared spectral responses, *Journal of the Indian Society of Remote Sensing,* 8, 23–29.

Challa, O., Vadivelu, S., & Sehgal, J. (1995). Soils of Maharashtra: Their Kinds, Distribution, Characterization, and Interpretations for Optimizing Land Use Volume 54 of NBSS publ, Volume 5 of NBSS publ: Soils of India Series.

Crop statistics. (2012). District-wise Area, Production and Productivity. http://www.mahaagri.gov.in/statistics/crop statistics/year 2012 (Accessed August 12, 2014).

Csornoi, G., Wirnhardt, C., Suba Z., Nador, G. L., Tikasz L., Martinovich L., Kocsis A., Zelei G., Lazlo, I., & Bognar, E. (2008). CROPMON: Hungarian crop production forecast by remote sensing. *In: Proc. Workshop Remote Sensing Support Crop Yield Forecast Area Estimates.* ISPRS Archives XXXVI 8/W48, pp. 25–30.

Dadhwal, V, K., & Parihar, J. S. (1985). Estimation of 1983–84 wheat acreage of Karnal district (Haryana) using Landsat MSS digital data. Technical Note, IRS-UP/SAC/CPF/TN/09/85. Space Applications Centre, Ahmedabad, India.

Dua, V. K., Panigrahy, S., Mehta, R. L., Lal, S. S., Govindakrishnan, P. M., Singh, J. P., Arora, R. K., Kumar, M., Kumar, P., Rawat, S., Bhatt, N., & Ahmad, I. (2007). Estimation of potato acreage in the Indo-Gangetic region using IRS AWiFS data. *Journal of Potato, 34*(1–2), 123–124.

Friedl, M. A., & Brodley, C. E. (1997). Decision tree classification of land cover from remotely sensed data. *Remote Sensing of Environment, 61*, 399–409.

Friedl, M. A., Brodley, C. E., & Strahler, A. H. (1999). Maximizing land cover classification accuracies produced by decision tree at continental to global scales. *IEEE Transactions on Geoscience and Remote Sensing, 37*, 969–977.

Gontia, N. K., & Tiwari, K. N. (2010). Estimation of crop coefficient and evapotranspiration of wheat (*Triticuum aerstivum*) in a irrigation command using remote sensing and GIS. *Water Resources Management, 24*, 1399–1414.

Goswami, S. B., Saxena, A., & Bairagi, G. D. (2012). Remote sensing and GIS based wheat crop acreage estimation of Indore district, M. P. *International Journal of Emerging Technology and Advanced Engineering, 2*(3), 200–203.

Guruge, P. A. P. (1996). The use of remote sensing data for identifying development of sugarcane in Buttala area. Proc. of XVII Asian Conference on Remote Sensing, November 4–8, 1996, Sri Lanka. Online. http://www.gisdevelopment.net/acrs (accessed August 4, 2013).

Hastings, D. A. (1997). Land cover classification: some new techniques, new source data. Proceedings, 18th Asian Conference on Remote Sensing. Tokyo, Asian Association of Remote Sensing. pp. JS-2-1 to JS-2-6.

Herbek, J., & Lee, C. (2009). "Section 2. Growth and Development." *A Comprehensive Guide to Wheat Management in Kentucky*. University of Kentucky College of Agriculture.

Kimothi, M. M., Kalubarm, M. H., Dutta, S., Thapa, R., & Sood, R. K. (1997). Remote sensing of horticultural plantations in Kumarsain tehsil in Shimla district, Himachal Pradesh. *Journal of the Indian Society of Remote Sensing, 25*, 19–26.

Krishna Rao, P. V., Venkateswara, V., & Venkataratnam, L. (2002). Remote Sensing: A Technology for assessment of sugarcane crop acreage and yield. *International Journal of Sugar Crops and Related Industries, 4*(3&4), 97–101.

Liu, H., Yang, X., & Wang, N. (1999). Remote sensing based estimation for winter wheat yield in North China plain. *Chinese Geographical Science, 9*(1), 40–48.

MacDonald, R. B., & Hall F. G. (1980). Global crop forecasting. *Science*, *208*, 670–679.

Munshi, M. K. (1982). A Study of the Determination of Wheat Crop Statistics in India through the Utilisation of Landsat Data. PhD Thesis. IIT Delhi.

Oza, M. P., Pandya, M. R., & Rajak, D. R. (2008). Evaluation and use of Resources at-I data for agricultural applications. *International Journal of Applied Earth Observation and Geoinformation, 10*, 194–205.

Pakhale, G., Gupta, P., & Jyoti, N. (2010). Crop and Irrigation Water Requirement Estimation By Remote Sensing and GIS: A Case Study of Karnal District, Haryana, India, *International Journal of Engineering and Technology, 2*(4), 207–211.

Panigrahy, S., & Chakraborty, M. (1996). A case study for Burdhaman district of West Bengal. Indo-U.S Workshop on Remote Sensing and its Applications, Mumbai, India.

Potgieter, A. B., Apan, A., Hammer, G., & Dunn P. (2005). Spying on the winter wheat crop-Generating objective planted area and crop production estimates using MODIS imagery. *In: Proc. Nation. Bienniel Conf. Spatial Intelligence*, Innovation Praxis, 2005, Spatial Sci. Inst., Melbourne, Australia.

Rajak, D. R., Oza, M. P., Bhagia, N., & Dadhwal, V. K. (2002). Monitoring Cropping Pattern Changes Using Multi-temporal WiFS Data. *Geocarto International*, *17*(2), 51–56.

Rajak, D. R., Oza, M. P., Bhagia, N., & Dadhwal, V. K. (2005). Spectral Wheat Growth Profile in Punjab Using IRS WiFS Data. *Journal of the Indian Society of Remote Sensing, 33*(2), 345–52.

Rao, P. P. N., & Rao, V. R. (1987). Rice crop identification and area estimation using remotely sensed data from Indian cropping patterns. *International Journal of Remote Sensing, 8*, 639–650.

Sharma, R. (1995). Coconut inventory using remote sensing data, SAC Courier. 20. Space Applications Centre, Ahmedabad, India.

Tateishi, R., Kajiwara, K., & Odajima, T. (1991). Global land cover classification by phenological methods using NOAA GVI data. *Asian-Pacific Remote Sensing Journal*, *4*(1), 41–50.

Tsiligirides, T. A. (1998). Remote sensing as a tool for agricultural statistics: a case study of area frame sampling methodology in Hellas. *Computers and Electronics in Agriculture, 20*(1), 45–77.

Vanderlip, R. L. (1993). How a Sorghum Plant Develops. Contribution No. 1203, Agronomy Department, Kansas Agricultural Experiment Station, Kansas State University, Manhattan, *66506*, 1–19.

CHAPTER 14

ASSESSMENT OF GROSS PRIMARY PRODUCTIVITY IN THE SEMI ARID AGRICULTURAL REGION OF CENTRAL INDIA USING TEMPORAL MODIS DATA

JUGAL KISHORE MANI,[1] A. O. VARGHESE,[1] and A. K. JOSHI[2]

[1]*Scientist, Regional Remote Sensing Centre-Central, NRSC, ISRO, Amravati Road, Nagpur, Maharashtra – 440 033, India*

[2]*General Manager, Regional Remote Sensing Centre-Central, NRSC, ISRO, Amravati Road, Nagpur, Maharashtra – 440 033, India*

CONTENTS

ABSTRACT

Earth observation driven ecosystem modeling have played a major role in the estimation of carbon budget component as gross primary productivity (GPP) and net primary production (NPP) over terrestrial ecosystem, including agriculture. The quantification of regional contribution from different agro-ecosystems is necessary for the estimation of total carbon budget. The present study aim to assess seasonal and interannual dynamics of gross primary productivity in semi arid agricultural region of central India using Moderate Resolution Imaging Spectroradiometer (MODIS) 8-day data. Vidarbha lies between 17°57' - 21°46' N latitude and 75°57' - 80°59' E longitude. It covers an area of 98 lakh ha, accounting for 32% area of Maharashtra State. MODIS satellite data during the period of 2009 to 2013 have been used in the present study. Result shows that the annual GPP was found highest during 2011 (1.64 kg C m^{-2} $year^{-1}$) followed by 2013 (1.57 kg C m^{-2} $year^{-1}$), 2010 (1.52 kg C m^{-2} $year^{-1}$), 2012 (1.42 kg C m^{-2} $year^{-1}$) and 2009 (1.34 kg C m^{-2} $year^{-1}$) in Vidarbha region. The average GPP of five years ranged from 0 to 1.47 kg C m^{-2}, which is at par with the range reported for Indian region (0.2 to 1.5 kg C m^{-2}). Average highest GPP was observed in the month of September during 2009–2013. Achalpur and Chandur Bazar taluka of Amravati, Ralegaon taluka of Yavatmal and Selu taluka of Wardha district are more productive in Vidarbha region in terms of GPP. The distribution of GPP was between 0.015–0.029, 0.010–0.023 and 0.000–0.005 kg C m^{-2} during *kharif* (July–October), *rabi* (November–February) and *zaid* (March–May), respectively in most of the districts of Vidarbha.

14.1 INTRODUCTION

Temporal and spatial changes in terrestrial biological productivity have a large impact on humankind because terrestrial ecosystems not only create environments suitable for human habitation, but also provide materials essential for survival, such as food, fiber and fuel. Crop productivity is a result of the interception of solar radiation by the vegetation canopy. Thus, it is possible to use remotely sensed data from satellites, including reflected and emitted radiation from the earth in various wavelengths of the electromagnetic spectrum, as a powerful and expedient tool for assessing crop gross primary productivity at regional and global scales. Since 1999, the

National Aeronautics and Space Administration (NASA) has provided GPP estimates for the entire globe based on the MODIS 1 km products (Running et al., 2000, 2004). Recognizing the growing need for improved earth observations, group on earth observations (GEO) plans to produce globally harmonized data sets on global, national and local scales, using common algorithms, variables and units, as well as, to develop an integrated model that stitches all carbon observations together (Ciais et al., 2010).

Quantification of the magnitude of net terrestrial carbon (C) uptake and how it varies inter-annually, is an important question with future potential sequestration influenced by both increased atmospheric CO_2 and changing climate (Nemani et al., 2003). Therefore, estimation of the net ecosystem exchange (NEE) and the gross primary production (GPP) of terrestrial ecosystems for regions, continents, or the globe can improve our understanding of the feedbacks between the terrestrial biosphere and the atmosphere in the context of global change and facilitate climate policy-making (Canadell et al., 2000; Xiao et al., 2008). Terrestrial GPP also provides important societal services through provision of food, fiber and energy. Methods for quantifying dynamics in terrestrial GPP are therefore required to improve climate forecasts and ensure long-term security in services provided by terrestrial ecosystems (Bunn and Goetz, 2006; Schimel, 2007). Sur and Choi (2013) studied the interactions between carbon and water circulation in terrestrial ecosystems and how their ecological procedures generated by the photosynthesis of vegetation influence in climatological perspectives. They observed that accurate assessments of spatio-temporal variations in GPP play a crucial role in the evaluation of carbon balance as well as have considerable effects on climate change.

14.1.1 ASSESSMENT OF GROSS PRIMARY PRODUCTIVITY: GLOBAL PERSPECTIVE

Global land ecosystems in recent times appeared as major sink of global carbon cycle and curtailing the increasing CO_2 concentration in the atmosphere (Zhao and Running, 2010). As of now most studies related to primary productivity and carbon budget in India and abroad have focused on forest ecosystem due to their potential to sequester high magnitude of atmospheric CO_2. Agro-ecosystems have received less attention by carbon science groups worldwide and in India, in-spite covering around 15 m km^2 all over the world and about more than half of land area in India. Indian terrestrial ecosystem acts as spatially and temporally variant carbon source and sink due to monsoon based climate system, diverse land use and land cover distribution and cultural practices. Various agriculture practices and agroecological regions in India necessitiated the assessment of GPP at regional level to understand the contribution of carbon budget of that particular region to the global carbon budget. The cropping systems in semi-arid areas however have comparatively less magnitude of GPP throughout the growing season. The present study envisaged to estimate the contribution of semi-arid agriculture ecosystem of central India in global carbon cycle and to understand its sessional and annual dynamics.

14.1.2 ASSESSMENT OF GROSS PRIMARY PRODUCTIVITY: INDIAN PERSPECTIVE

Earth observation driven ecosystem modeling has played a major role in estimation of carbon budget component as GPP and NPP over terrestrial ecosystem, including agriculture. The quantification of regional contribution from different agro-ecosystems is necessary for total carbon budget estimation. There has been an increasing trend in assimilating EO data towards understanding of land surface processes, climate measurements and carbon pools and fluxes (IPCC, 1995). Phenological shifts, vegetation cover and type changes (Chaturvedi et al., 2011; Gopalakrishnan et al., 2011), fire monitoring and biomass burning (NRSA, 2006), feedbacks to regional climate and CO_2-fertilization (Bala et al., 2011), estimation of net carbon flux caused by deforestation and afforestation in India (Kaul et al., 2009; Sharma et al., 2011), assessment of potential carbon sequentration of different ecosystems in Nagpur district (Wanjari et al., 2009), carbon sequestration potential in above ground biomass of natural reserve forest of central India (Juwarkar et al., 2011) are a few recent studies reported from India using EO data.

The Indian subcontinent has diverse vegetation with the climate varying from monsoonal in the south to temperate in the north. The biological productivity of vegetation cover is, therefore, largely controlled by water and temperature stresses. The monsoon based climate system, diverse land use and land cover distribution and cultural practices pose complex issues in monitoring, assessment and simulation of Indian carbon cycle. Several studies reported lack of spatially and temporally consistent databases, need for calibration and validation of models, and development of national framework to maintain consistency and completeness in efforts and reduction of uncertainty.

Indian terrestrial ecosystem acts as spatially and temporally variant carbon source and sink due to monsoon based climate system, diverse land use and land cover distribution and cultural practices. The agriculture covering around 180 m ha and forests covering 68 m ha contributes largely to terrestrial carbon dynamics in India (Chhabra et al., 2004; Kaul et al., 2009). The diverse cropping patterns, wetland rice ecosystems and temporally variant dry land agriculture makes the assessment of Indian agriculture carbon fluxes more complex. The diverse forest structure and composition, phenology, fire regimes, biotic disturbances, extractions and large reservoir of trees outside forests plays larger role in carbon sequestration and shaping the carbon budget (Bhat and Ravindranath, 2011). Unlike the regular forest inventories, soil inventories are rarely repeated on a regular basis and require chrono-sequenced sampling to understand the impacts of different management regimes and land use changes on carbon dynamics.

A large number of studies have compared results derived from remote sensing-based models with in situ measurements (Turner et al., 2006; Heinsch et al., 2006; Yuan et al., 2007; Sims et al., 2008; Mahadevan et al., 2008; Xiao et al., 2010). However, all of these studies are based on relatively small in situ data sets and none have explicitly examined both spatial and temporal variations in remotely sensed proxies (Hashimoto et al., 2012) and modeled estimates with corresponding variations in situ measurements of GPP. Many current models of ecosystem carbon exchange based on remote sensing, such as the MODIS product termed MOD17, still require considerable input from ground based meteorological measurements and look up tables based on vegetation type. Since these data are often not available at the same spatial scale as the remote sensing imagery, they can introduce substantial errors into the carbon exchange estimates (Sims et al., 2008).

Many remote-sensing-based methods are used to estimate the GPP on a variety of scales ranging from local to regional and even global. However, it is very difficult to find a model capable of estimating the GPP of an entire vegetation ecosystem. Studying the suitability of models with regard to specific areas and objectives is therefore very important (Yuan et al., 2007). One of the most frequently used methods for estimating GPP with remotely sensed data is the light use efficiency (LUE) approach (Sjöström et al., 2011), which links GPP to a linear combination of remotely sensed variables like enhanced vegetation index (EVI) and normalized difference vegetation index (NDVI) and climatic variables like photosynthetically active radiation (PAR) and temperature.

Although many of the studies revealed that global terrestrial ecosystem was a net carbon sink, but carbon uptakes and releases had large spatial and temporal variability due to the impacts of environmental and biological factors, such as climatic conditions, vegetation distribution, land use, etc. The studies conducted over India have strongly envisaged a need for national level initiative for understanding of terrestrial carbon cycle considering the diversity of terrestrial ecosystems, monsoon climate. The present study was aimed to assess of gross primary productivity in semi arid agricultural region of central India at regional level using temporal MODIS data.

14.1.3 IMPORTANCE OF THE STUDY

The climatic patterns of the world are changing and spatial distribution of global terrestrial carbon; the food and fiber of the world is an important factor in the changing climate. Knowledge of how the terrestrial carbon stock is changing, its distribution and quantity, is important in understanding how the patterns of the world are changing and large-scale models using remotely sensed data have emerged for this purpose. There is a need for a comprehensive assessment and management to improve primary productivity of regional scale using remote sensing data. Keeping this in view, the present study was aimed with the objectives to quantify the GPP of Vidarbha region during 2009–2013, to assess seasonal dynamics of GPP and prioritize the areas of higher productivity in Vidarbha region of Central India.

14.2 STUDY AREA

India is a large country, situated in the tropics between 8–38°N latitudes in the northern hemisphere and 68–98°E longitude eastward from prime meridian. The country has diverse vegetation cover with climate varying from monsoonal in the south to temperate in the north. The climate over India has undergone large changes during past two decades with the clear warming trends with the background of global warming. The variation of temperature and precipitation had distinct regional differences (Dash and Hunt, 2007).

The area selected for the present study is Vidarbha region of central India, which is mostly characterized by dryland farming (Challa and Wadodkar, 1996). The climatic condition of this region can be broadly described as semi-arid. It is richly endowed with natural resources of varied types. The region comprises of eleven districts *viz.*, Buldana, Akola, Washim, Amravati, Yavatmal, Wardha, Nagpur, Bhandara, Gondia, Chandrapur and Gadchiroli (Figure 14.1). It lies in between 17°57' - 21°46' N Latitude and 75°57' - 75°57' E Longitude and covers an area of 98 lakh ha, which is 32% area of Maharashtra. It has Satpura hill ranges in north bound the region, Ajanta hill ranges at the south west and a series of high hill ranges along the eastern border. It is divided into Godavari and Tapi river basins. The mean annual rainfall ranges from 700 mm at the west to 1700 mm at the east. This region mostly receives adequate rainfall in aggregate during monsoon period but suffers from vagaries of distribution and consequently the scarcity and semi-scarcity conditions. Some areas are even drought prone. Vidarbha region has been divided into three agro-climatic zones based on rainfall, soil types and vegetation *viz.*, western Vidarbha Zone (Rainfall: 700 to 950 mm), Central Vidarbha Zone (Rainfall: 950 to 1250 mm) and Eastern Vidarbha Zone (Rainfall: <1250 mm).

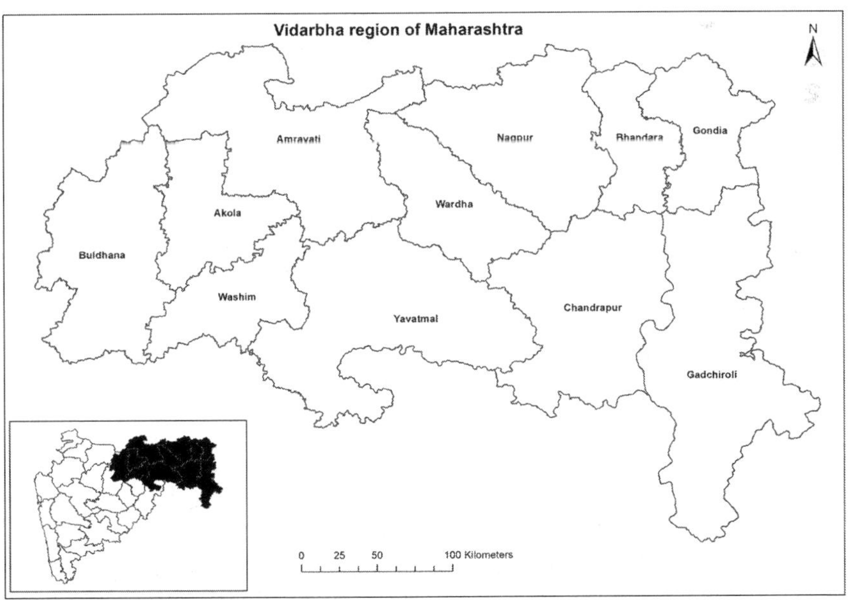

FIGURE 14.1 Vidarbha region of Maharashtra.

Western Vidarbha soils are derived from trap rock and have varying depth depending upon their physiography. Most of the soils are calcareous, highly base saturated, fairly well drained, well supplied with potash, moderate to low in phosphate, but low in organic matter content and slightly alkaline. Soils of the Central Vidarbha are derived from basalt, black in color and having varying depth depending upon their physiography. Vertisols, Inceptisols and Entisols are the dominant soil orders occurring here. Eastern Vidarbha soils are chiefly derived from rocks like granite-gneiss and schist. They are yellowish brown to red in color and coarse in texture. These soils predominantly contain Kaolinite (1:1), non-expanding clay mineral. Paddy is predominant crop of eastern Vidarbha zone. Bhandara district is known as paddy growing area of Vidarbha region. Main cropping system of eastern Vidarbha is paddy followed by *rabi* jowar, pulses (gram, lathyrus) and oilseeds. Moderate rainfall is received by central Vidarbha zone and main crops of this zone are cotton, soybean, sorghum, pigeon pea in *kharif* and wheat and gram in *rabi* season. Sorghum is a predominant crop of western Vidarbha zone followed by cotton, oilseeds (groundnut, sesamum, safflower and niger) and pulses (pigeon pea, green gram, Bengal gram and lentil). Area under paddy is increasing. Sugarcane and summer crops are grown on availability of irrigation.

14.3 MATERIALS AND METHODS

14.3.1 DATA USED

The MODIS multispectral sensor of the NASA Earth Observing System (EOS) is a key instrument aboard the Terra and Aqua satellites, acquiring data in 36 spectral bands from 450 nm to 2100 nm. These data provide important insights for global dynamics research both on the land and in the oceans products. The overpass time of the Terra satellite is approximately 10:30 AM when descending and 10:30 PM when ascending. The Aqua satellite overpass time is approximately 1:30 AM when descending and 1:30 PM when ascending. All MODIS land products are projected on the integerized sinusoidal (ISIN) 10° grid, where the globe is tiled for production and distribution purposes with 36 tiles along the east-west axis, and 18 tiles along the north-south axis, each about 1200 x 1200 km^2. MODIS is meeting the stated geolocation requirement of 0.1 pixels at 2 standard deviations for the 1 km bands (Wolfe et al., 2002). The Collection 4 projection is sinusoidal (SIN), while Collections 1–3 use an integerized sinusoidal projection (ISIN). At a 1 km spatial resolution, the difference between the SIN and ISIN projections is negligible. The decision to switch from the ISIN to the SIN projection was made to make the data more compatible with current image processing software.

For many applications it may be convenient to reproject MODIS data from the ISIN or SIN projection to a different projection that is more suited to the area of interest. Few proprietary image processing or Geographic Information System (GIS) software have the capability to reproject MODIS data from an ISIN projection. Fortunately, however, there are good tools, which are simple to download and are freely available. The primary tool currently used to reproject MODIS data in both formats is the MODIS reprojection tool (MRT). This tool can be downloaded at http://edcdaac.usgs.gov/tools/modis.

MOD17 is the standard product on primary production and is derived from MODIS data with 1 km spatial resolution. MOD17A2 8 days data provided by NASA were used to study the GPP of Vidarbha region from 2009–2013. The gross primary production product (MOD17A2) is designed to provide an accurate regular measure of the growth of the terrestrial vegetation using daily MODIS land cover, fAPAR/LAI and surface meteorology at 1 km for the global vegetated land surface (Field et al., 1998).

GPP is calculated on a daily basis, 8-day summations of GPP are created and these summations are available to the public. The summations are named for the first day included in the 8-day period. Each summation consists of 8 consecutive days of data, and there are 46 such summations created for each calendar year of data collection. To obtain an estimate of daily GPP for this 8- day period, it is necessary to divide the value obtained during a data download by eight for the first 45 values/year and by five (or six in a leap year) for the final period.

14.3.2 DESCRIPTION OF MODIS17 ALGORITHM

MOD17 consists of two products, MOD17A2 and MOD17A3. MOD17A2 contains both 8-day GPP and 8-day PSNnet, whereas MOD17A3 contains annual sums of NPP (Heinsch et al., 2003). The GPP is a function of sunlight and active radiation-related parameters. The GPP is one of the parameters representing the material circulation of the terrestrial ecosystem and is used for detecting and measuring the change in carbon circulation (Hwang et al., 2008; Running et al., 2000; Yu et al., 2008; Yuan et al., 2010; Zhao et al., 2005). Primary production can be divided into the GPP, which represents the total amount of organic matter produced by vegetation through photosynthesis, and the NPP, which excludes the breathing quantity of the vegetation from total organic matter (Tian et al., 2010). These two indices are expressed as the amount of carbon per unit area. In this study, the 8-day GPP output with spatial resolution of 1 km from the MODIS primary product provided by the NASA was used. The MODIS 17 GPP algorithm is associated with carbon circulation. The GPP is a function of sunlight and active radiation-related parameters. The MODIS GPP algorithm is based on the radiation use efficiency of vegetation (Running et al., 2000; Tian et al., 2010) as expressed in the following equations:

$$GPP = \varepsilon \times FPAR \times PAR \quad (1)$$

$$\varepsilon = \varepsilon_{max} \times f(T_{min}) \times f(VPD) \quad (2)$$

$$PAR = 0.45\ X\ R_s \quad (3)$$

where ε is the radiation use efficiency of vegetation, PAR is the photosynthetically available radiation (MJ m^{-2} d^{-1}), ε_{max} is the maximum radiation use efficiency, T_{min} is the daily minimum temperature (°C), VPD is the vapor pressure deficit (the difference between the vapor pressure and actual pressure) (Pa), R_s is the short wave radiant energy (MJ m^{-2} d^{-1}), and FPAR is the fraction of the PAR absorbed by the plant with a value ranging from 0 to 1. When the PAR absorbed by the plant is large or the value of ε is high, then the GPP value increases. Equation (1) shows that the MODIS GPP algorithm assumes that the value of ε is determined according to the vegetation type (Heinsch et al., 2003). The PAR is assumed to be 45% of Rs [Eq. (3)], which expresses the change in use efficiency determined by the opening and closing of the stomata according to T_{min} and VPD, which have upper and lower limits ranging from ε max to 0. The MODIS 15 FPAR product is used due to its similarity with normalized difference vegetation index (NDVI) (Running et al., 2000). The meteorological data of the MODIS GPP algorithm used the data assimilation office (DAO) provided by the NASA, having a spatial resolution of 1.00° × 1.25° as input data. The upper and lower thresholds of

ε_{max}, T_{min} and VPD determined the opening and closing of the stomata use constant, which were determined according to the vegetation type (Heinsch et al., 2003). Analyses were carried out using MODIS Tool, ERDAS Imagine and Arc GIS software with spatial analyst extensions.

14.4 RESULTS AND DISCUSSION

14.4.1 SEASONAL DYNAMICS

GPP was varied among seasons throughout each year. The accumulation of 8 days GPP was more during *kharif* season than other seasons among five years (Figure 14.2). This might be due to plenty of monsoon rainfall during *kharif* season, which leads to accumulation of GPP uniformly. The *rabi* season GPP was also found to be good but in this season sown crops depend upon irrigation. The GPP was much lower during *zaid* season because of scarcity of irrigated water. So, the rate of metabolic activities of crops is lower under limited availability of water. The 8 days GPP on 29th August, 2013 was increased sharply because of favorable weather conditions like plenty of water, bright sunshine and optimum temperature prevailed during that time, which leads to carbon synthesis.

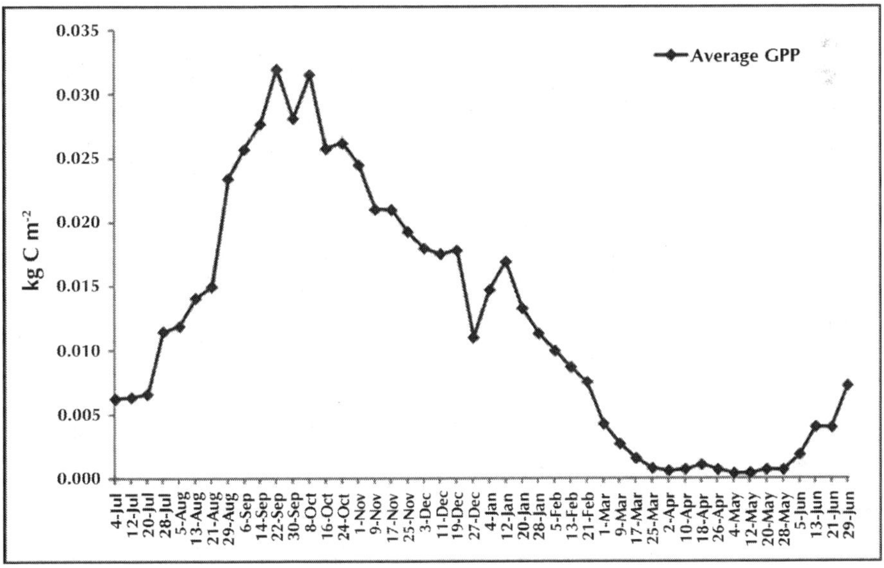

FIGURE 14.2 Average gross primary productivity of Vidarbha region from 2009 to 2013.

Five-year average of *kharif*, *rabi* and *zaid* season GPP is presented in Figure 14.3a–c. The highest value of GPP observed during *kharif* was 0.042 kg C m^{-2}, for *rabi* 0.041 kg C m^{-2} and *zaid* (0.0147 kg C m^{-2}) in Vidarbha region. The distribution of GPP was between 0.015–0.029, 0.010–0.023 and 0.000–0.005 kg C m^{-2} during *kharif*, *rabi* and *zaid*, respectively in most of the districts of Vidarbha. The GPP was found more during *kharif* as compared to *rabi* and *zaid*. The uniform distribution of GPP was observed during *kharif* season because of more amount of rainfall received during this season as compared to other season, which accelerates carbon assimilation through photosynthesis by vegetation. Yavatmal and Wasim districts and part of Buldhana, Amravati and Wardha district were found to be more productive during *kharif* season.

14.4.2 INTERANNUAL VARIATION

Intensity of GPP was varied interannualy within the study area (Figure 14.4a–e). The highest GPP was found during 2011 (1.64 kg C m^{-2} $year^{-1}$) followed by 2013 (1.57 kg C m^{-2} $year^{-1}$), 2010 (1.52 kg C m^{-2} $year^{-1}$), 2012

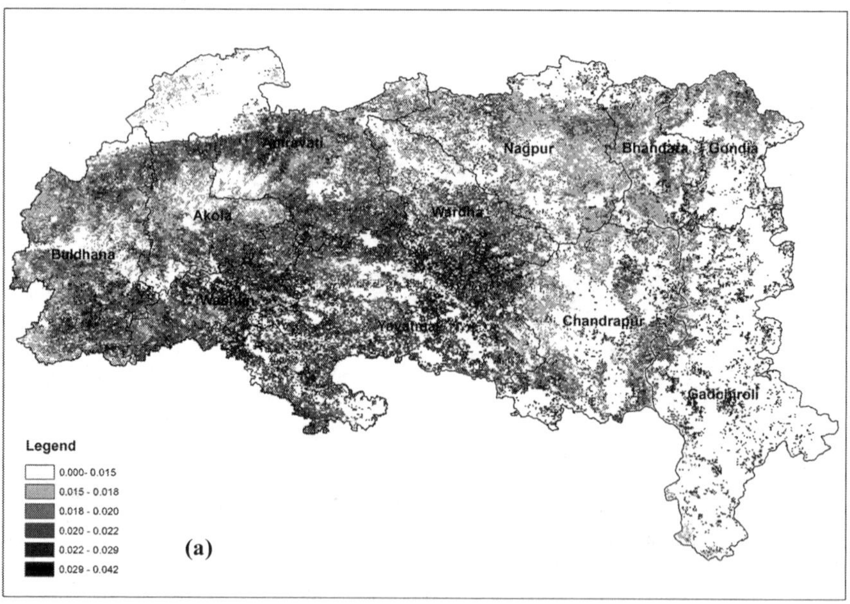

FIGURE 14.3 Gross primary productivity of Vidarbha region (a) *kharif* season (b) *rabi* season and (c) *zaid* season.

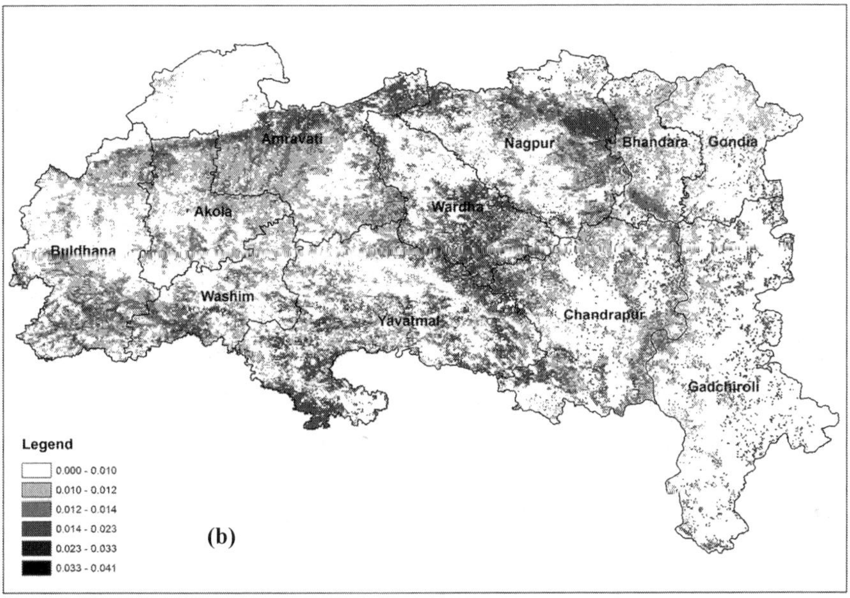

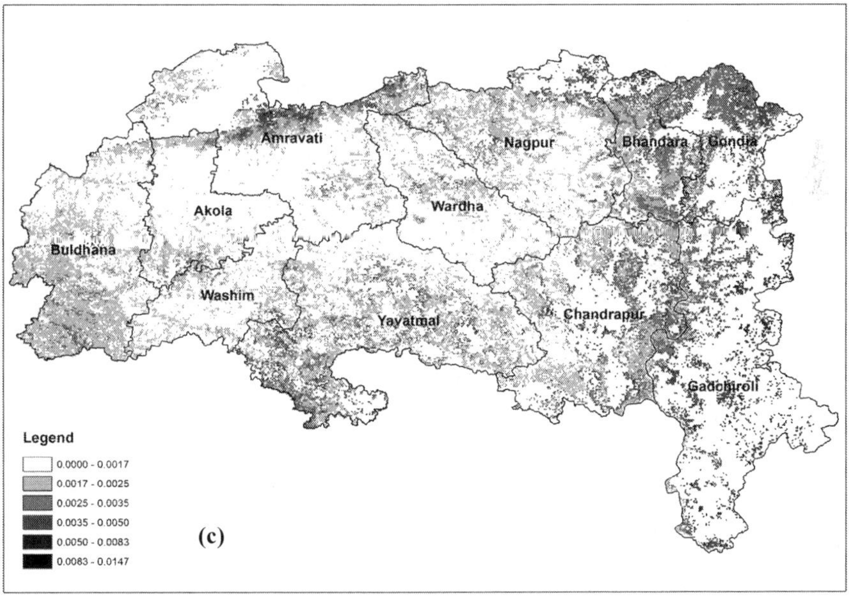

FIGURE 14.3 (Continued)

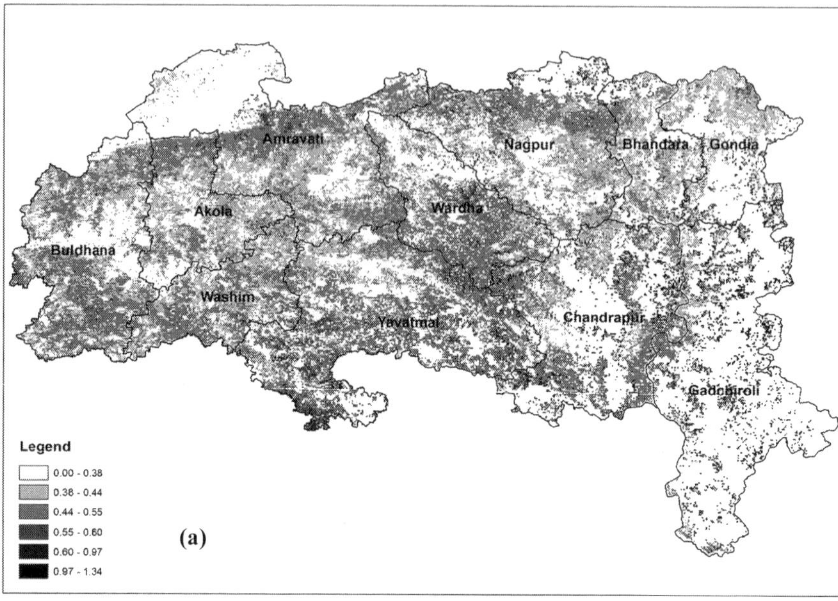

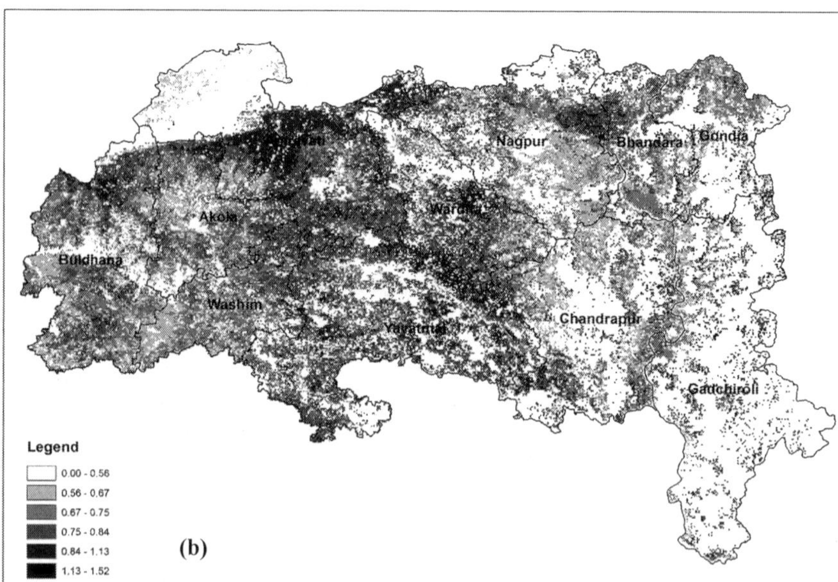

FIGURE 14.4 Gross primary productivity of Vidarbha region (a) 2009 (b) 2010 (c) 2011 (d) 2012 (e) 2013 and (f) average of five years (2009 to 2013).

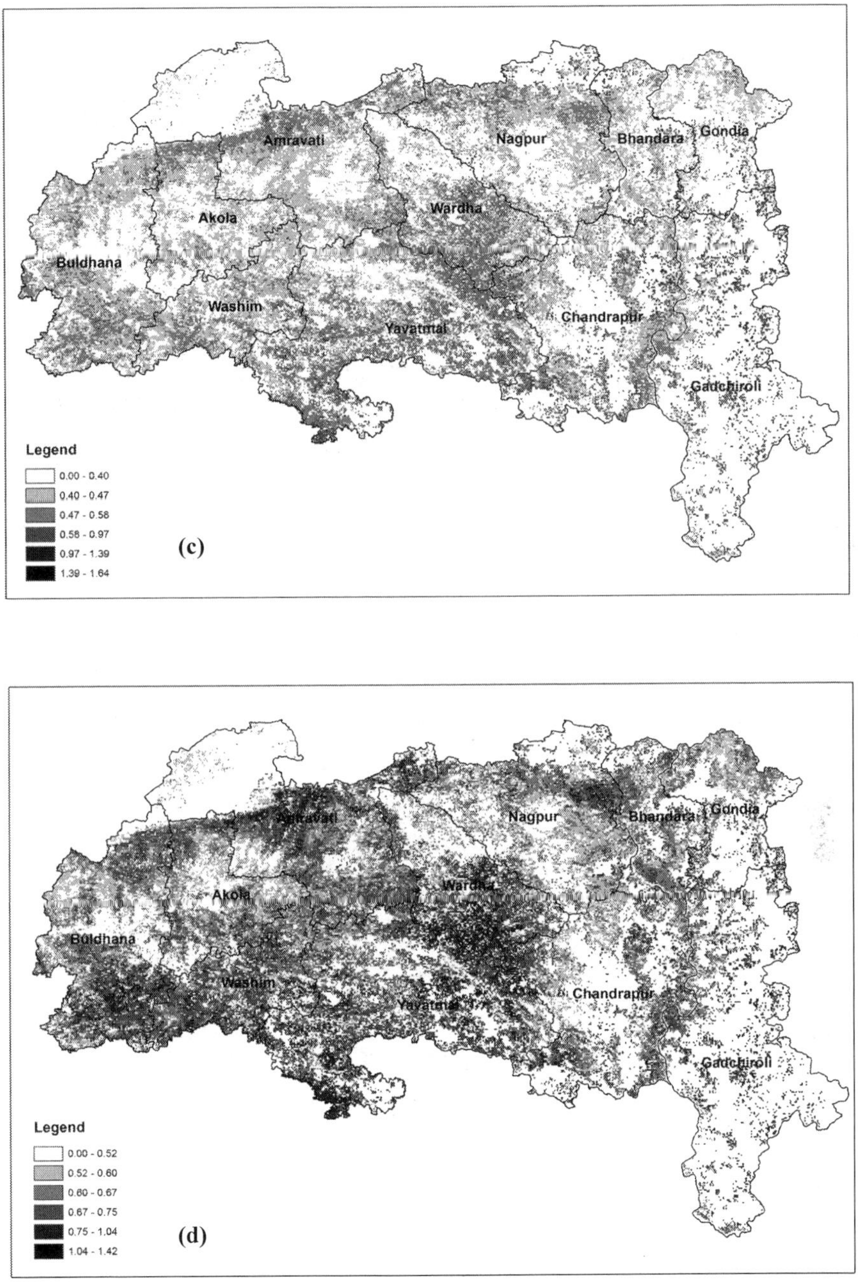

FIGURE 14.4 (Continued)

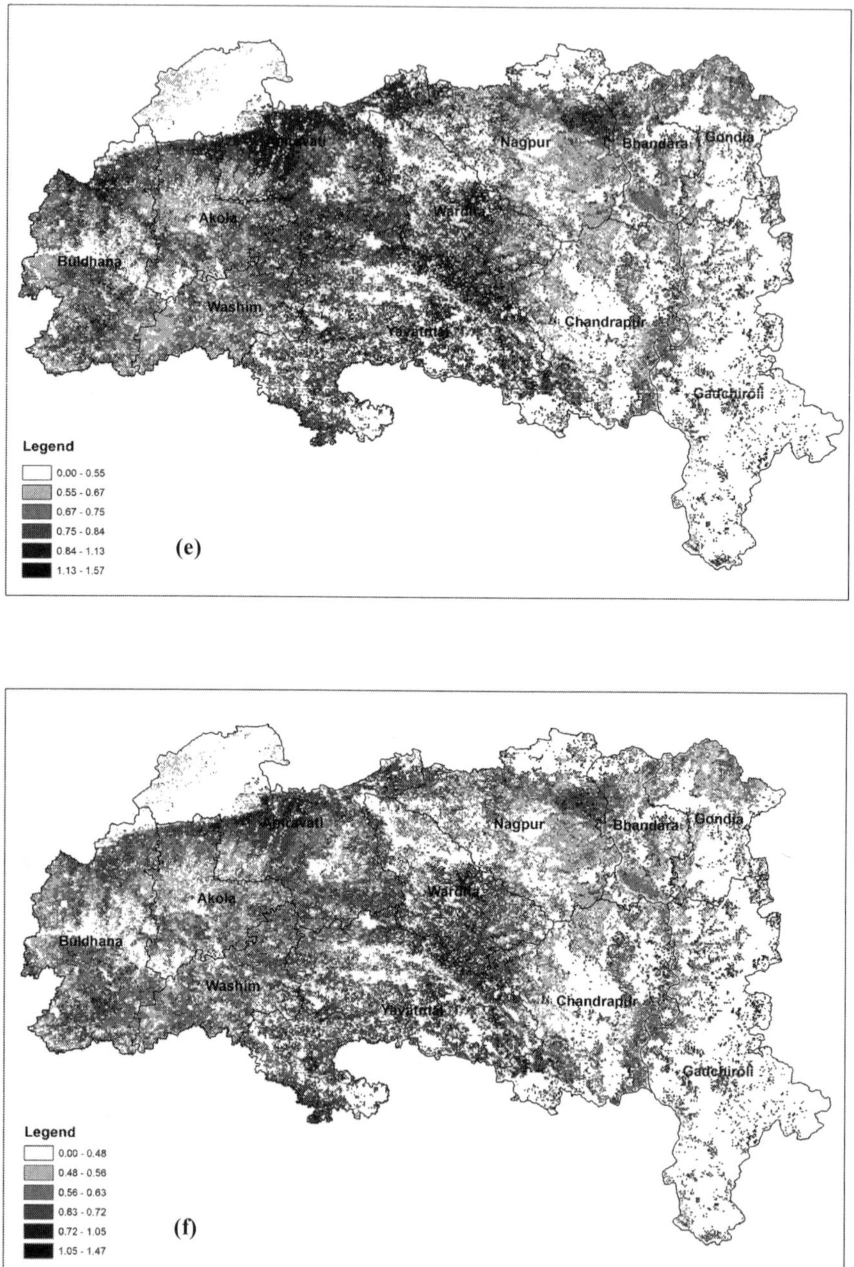

FIGURE 14.4 (Continued)

(1.42 kg C m^{-2} $year^{-1}$) and 2009 (1.34 kg C m^{-2} $year^{-1}$). The distribution of GPP was between 0.38–0.55, 0.56–0.84, 0.40–0.58, 0.52–0.75 and 0.55–0.84 kg C m^{-2} $year^{-1}$ during 2009, 2010, 2011, 2012 and 2013, respectively in most of the areas of Vidarbha. The uniform distribution of GPP was more during 2013 as compared to other years. The less distribution of GPP was observed during 2009. This might be due to scarcity of rainfall as influenced by El Niño effect over India in 2009.

In addition to the reasons mentioned above, the remote sensing methods tested here did not effectively explain spatial variation in annual GPP in crops for every year, probably because agricultural practices that are not captured by remote sensing exert significant control on GPP in croplands. Specifically, application of fertilizers (Eugster et al., 2010), variation in crop varieties (Moors et al., 2010), irrigation, and harvest practices significantly modify productivity in croplands (Suyker et al., 2004; Verma et al., 2005). These practices are not directly observable from remote sensing, and as a result, variation in productivity arising from these practices are not well reproduced by remote sensing-based models (Chen et al., 2011; Zhang et al., 2008).

14.4.3 SPATIAL DISTRIBUTION

The average GPP from 2009 to 2013 are represented in Figure 14.4f. The distribution of GPP was observed between 0.56 to 0.72 kg C m^{-2} in most of the districts of Vidarbha region. The average GPP of five years ranged from 0.0 to 1.47 kg C m^{-2}. This finding is at par with the GPP range reported for Indian subcontinent (0.2 to 1.5 kg C m^{-2}) by Patel et al. (2011). From the Figure 14.4f, it is clearly indicates that Achalpur and Chandur Bazar taluka of Amravati, Ralegaon taluka of Yavatmal and Selu taluka of Wardha district are more productive in Vidarbha region. Different land use for each district of Vidarbha region could affect the value of GPP, because the different land use has different vegetation type, distribution, and different photosynthetic pathway type (As-syakur et al., 2010).

14.5 CONCLUSIONS

The study assessed gross primary productivity in semi arid agricultural region of Central India. The results show that maximum annual GPP in Vidarbha region is 1.64 kg C m^{-2} $year^{-1}$ recorded in 2011. Average highest

GPP was observed in the month of September during 2009–2013. Achalpur and Chandur Bazar taluka of Amravati, Ralegaon taluka of Yavatmal and Selu taluka of Wardha district are more productive. The GPP distribution was less during the *zaid* season and higher during the *kharif* season. *Kharif* season has significant contribution to carbon storage by terrestrial biosphere over Vidarbha region. Further research is needed to evaluate the GPP estimation models, either on the basis of remote sensing observations or on a combination of climate variables in other ecosystems. The underlying mechanism and the effects of climate variables on the temporal patterns of GPP should be better understood to allow incorporation into future models.

ACKNOWLEDGEMENT

Authors extend their sincere gratitude to Dr. J. R. Sharma, CGM, RC's and Dr. V. K. Dadhwal, Director, NRSC for their support. The technical support provided by the RRSC-C staff especially Dr. Deemat C Mathew is duly acknowledged. We are also thankful to the USGS Earth Explorer data center for online accessibility of MODIS data.

KEYWORDS

- **Assessment**
- **Central India**
- **Gross Primary Productivity**
- **MODIS Data**
- **Semi Arid Agricultural Region**

REFERENCES

As-syakur, A. R., Osawa, T., & Adnyana, I. W. S. (2010). Estimation of gross primary production using satellite data and GIS in urban area, Denpasar. *International Journal of Earth Sciences, 7*, 84–95.

Bala, G., Gopalakrishnan, R., Jayaraman, M., Nemani, R., & Ravindranath, N. H. (2011). CO_2-fertilization and potential future terrestrial carbon uptake in India. *Mitigation and Adaptation Strategies for Global Change, 16*(2), 143–160.

Bhat, D. M., & Ravindranath, N. H. (2011). Above-ground standing biomass and carbon stock dynamics under a varied degree of anthropogenic pressure in tropical rain forests of Uttara Kannada district, Western Ghats, India. *Taiwania, 56*(2), 85–96.

Bunn, A. G., & Goetz, S. J. (2006). Trends in satellite-observed circumpolar photosynthetic activity from 1982 to 2003: The influence of seasonality, cover type, and vegetation density. *Earth Interact, 10*(12), 1–19.

Canadell, J. G., Mooney, H. A., Baldocchi, D. D., Berry, J. A., Ehleringer, B., & Field, C. B. (2000). Carbon metabolism of the terrestrial biosphere: A multi-technique approach for improved understanding. *Ecosystems, 3*(2), 115–130.

Challa, O., & Wadodkar, M. R. (1996). Delineation of drought prone areas in Maharashtra: A soil-climate based approach. *Agropedology, 6*(2), 21–27.

Chaturvedi, R. K., Gopalakrishnan, R., Jayaraman, M., Bala, G., Joshi, N. V., Sukumar, R., & Ravindranath, N. H. (2011). Impact of climate change on Indian forests: a dynamic vegetation modeling approach. *Mitigation and Adaptation Strategies for Global Change, 16*(2), 119–142.

Chen, T. X., van der Werf, G. R., Dolman, A. J., & Groenendijk, M. (2011). Evaluation of cropland maximum light use efficiency using eddy flux measurements in North America and Europe. *Geophysical Research Letters,* 38: L14707, doi: 10.1029/2011GL047533.

Chhabra, A., & Dadhwal, V. K. (2004). Estimating terrestrial net primary productivity over India using satellite data. *Current Science, 86*(2), 269–271.

Ciais, P., Dolman, A. J., Dargaville, R., Barrie, L., Bombelli, A., Butler, J., Canadell, P., & Moriyama, T. (2010). Geo Carbon Strategy, Geo Secretariat Geneva, FAO, Rome.

Dash, S. K., & Hunt, J. C. R. (2007). Variability of climate change in India. *Current Science, 93*(6), 782–788.

Eugster, W., Moffat, A. M., Ceschia, E., Aubinet, M., Ammann, C., Osborne, B., Davis, P. A., Smith, P., Jacobs, C., Moors, E., Le Dantec, V., Beziat, P., Saunders, M., Jans, W., Grunwald, T., Rebmann, C., Kutsch, W. L., Czerny, R., Janous, D., Moureaux, C., Dufranne, D., Carrara, A., Magliulo, V., Di Tommasi, P., Olesen, J. E., Schelde, K., Olioso, A., Bernhofer, C., Cellier, P., Larmanou, E., Loubet, B., Wattenbach, M., Marloie, O., Sanz, M. J., Sogaard, H., & Buchmann, N. (2010). Management effects on European cropland respiration. *Agriculture, Ecosystems and Environment, 139*(3), 346–362.

Field, C. B., Behrenfeld, M. J., Randerson, J. T., & Falkowski, P. (1998). Primary production of the biosphere: Integrating terrestrial and oceanic components. *Science, 28*(5374), 237–240.

Gopalakrishnan, R., Jayaraman, M., Bala, G., & Ravindranath, N. H. (2011). Climate change and Indian forests. *Current Science, 101*(3), 348–355.

Hashimoto, H., Wang, W., Milesi, C., White, M. A., Ganguly, S., Gamo, M., Hirata, R., Myneni, R. B., & Nemani, R. R. (2012). Exploring simple algorithms for estimating gross primary production in forested areas from satellite data. *Remote Sensing, 4*(1), 303–326.

Heinsch, F. A., Reeeves, M., Votava, P., Kang, S., Milesi, C., & Zhao, M. (2003). Users guide GPP and NPP (MOD17A2/A3) products. NASA MODIS land algorithm (version 2.0.), University of Montana, NTSG.

Heinsch, F. A., Zhao, M., Running, S. W., Kimball, J. S., Nemani, R. R., & Davis, K. J. (2006). Evaluation of remote sensing based terrestrial productivity from MODIS using regional tower eddy flux network observations. *IEEE Transactions* on *Geoscience* and *Remote Sensing, 44*(7), 1908–1925.

Hwang, T., Kang, S., Kim, J., Kim, Y., Lee, D., & Band, L. (2008). Evaluating drought effect on MODIS gross primary production (GPP) with an eco-hydrological model in the mountainous forest, East Asia. *Global Change Biology, 14*(5), 1037–1056.

IPCC. (1995). Greenhouse gas inventory reporting instructions-IPCC guidelines for national greenhouse gas inventories, IPCC, IEA, OECD, Volumes 1–3.

Juwarkar, A. A., Varghese, A. O., Sigh, S. K., Aher, V. V., & Thawale, P. R. (2011). Carbon sequestration potential in above ground biomass of natural reserve forest of Central India. *International Journal of Agricultural Research and Reviews*, *1*(2), 80–86.

Kaul, M., Dadhwal, V. K., & Mohren, G. M. J. (2009). Land use change and net C flux in Indian forests. *Forest Ecology and Management*, *258*(2), 100–108.

Mahadevan, P., Wofsy, S. C., Matross, D. M., Xiao, X. M., Dunn, A. L., Lin, J. C., Gerbig, C., Munger, J. W., Chow, V. Y., & Gottlieb, E. W. (2008). A satellite-based biosphere parameterization for net ecosystem CO_2 exchange: Vegetation Photosynthesis and Respiration Model (VPRM). *Global Biogeochemical Cycles,* 22(2), B2005, doi: 10.1029/2006GB002735.

Moors, E. J., Jacobs, C., Jans, W., Supit, I., Kutsch, W. L., Bernhofer, C., Beziat, P., Buchmann, N., Carrara, A., Ceschia, E., Elbers, J., Eugster, W., Kruijt, B., Loubet, B., Magliulo, E., Moureaux, C., Olioso, A., Saunders, M., & Soegaard, H. (2010). Variability in carbon exchange of European croplands. *Agriculture, Ecosystems and Environment, 139*(3), 325–335.

Nemani, R. R., Keeling, C. D., Hashimoto, H., Jolly, W. M., Piper, S. C., & Tucker, C. J. (2003). Climate-driven increases in global terrestrial net primary production from1982 to 1999. *Science, 300*, 1560–1563.

NRSA. (2006). Perspectives of geoinformatics in forest fire management. Indian Forest Fire Response and Assessment System), Technical Report, National Remote Sensing Agency, Hyderabad, India.

Patel, N. R., Dadhwal, V. K., Agrawal, S., & Saha, S. K. (2011). Satellite driven estimation of primary productivity of agro-ecosystems in India. International Archive of the photogrammetry, *Remote Sensing and Spatial Information Science.* XXXVIII-8/W20, 134–139.

Running, S. W., Nemani, R. R., Heinsch, F. A., Zhao, M. S., Reeves, M., & Hashimoto, H. (2004). A continuous satellite-derived measure of global terrestrial primary production. *Bioscience*, *54*(6), 547–560.

Running, S. W., Thornton, P. E., Nemani, R., & Glassy, J. M. (2000). Global terrestrial gross and net primary productivity from the earth observing system. *In: Methods in Ecosystem Science.* Sala, O. E.; Jackson, R. B.; Mooney, H. A.; Howarth, R. W. (Ed.). Springer-Verlag, New York, 44–57.

Schimel, D. (2007). Carbon cycle conundrums. *Proceedings of the National Acadamey of Sciences.* USA, *104*(47), 18353–18354.

Sharma, S. K., Choudhury, A., Sarkar, P., Biswas, S., Singh, A., Dadhich, P. K., Singh, A. K., Majumdar, S., Bhatia, A., Mohini, M., Kumar, R., Jha, C. S., Murthy, M. S. R., Ravindranath, N. H., Bhattacharya, J. K., Karthik, M., Bhattacharya, S., & Chauhan, R. (2011). Greenhouse gas inventory estimates for India. *Current Science, 101*(3), 405–415.

Sims, D. A., Rahman, A. F., Cordova, V. D., El-Masri, B. Z., Baldocchi, D. D., & Bolstad, P. V. (2008). A new model of gross primary productivity for North American ecosystems based solely on the enhanced vegetation index and land surface temperature from MODIS. *Remote Sensing of Environment, 112*(4), 1633–1646.

Sjöström, M., Ardö, J., Arneth, A., Boulain, N., Cappelaere, B., Eklundh, L., de Grandcourt, A., Kutsch, W. L., Merbold, L., Nouvellon, Y., Scholes, R. J., & Schubert, P. (2011). Exploring the potential of MODIS EVI for modeling gross primary production across African ecosystems. *Remote Sensing of Environment, 115*(4), 1081–1089.

Sur, C. and; Choi, M. (2013). Evaluating ecohydrological impacts of vegetation activities on climatological perspectives using MODIS gross primary productivity and evapotranspiration products at Korean regional flux network site. *Remote Sensing, 5*(5), 2534–2553.

Suyker, A. E., Verma, S. B., Burba, G. G., Arkebauer, T. J., Walters, D. T., & Hubbard, K. G. (2004). Growing season carbon dioxide exchange in irrigated and rainfed maize. *Agricultural and Forest Meteorology, 124*(1), 1–13.

Tian, H., Chen, G., Liu, M., Zhang, C., Sun, G., Lu, C., Xu, X., Ren, W., Pan, S., & Chappelka, A. (2010). Model estimates of net primary productivity, evapotranspiration, and water use efficiency in the terrestrial ecosystems of the southern United States during 1895–2007. *Forest Ecology and. Management, 259*(7), 1311–1327.

Turner, D. P., Ritts, W. D., Zhao, M. S., Kurc, S. A., Dunn, A. L., Wofsy, S. C., Small, E. E., & Running, S. W. (2006). Assessing interannual variation in MODIS-based estimates of gross primary production. *IEEE Transactions on Geoscience and Remote Sensing, 44*(7), 1899–1907.

Verma, S. B., Dobermann, A., Cassman, K. G., Walters, D. T., Knops, J. M., Arkebauer, T. J., Suyker, A. E., Burba, G. G., Amos, B., Yang, H. S., Ginting, D., Hubbard, K. G., Gitelson, A. A., & Walter-Shea, E. A. (2005). Annual carbon dioxide exchange in irrigated and rainfed maize-based agroecosystems. *Agricultural and Forest Meteorology, 131*(1), 77–96.

Wanjari, T., Varghese, A. O., Tawale, P., Nair, R., Singh, S. K., & Juwarkar, A. A. (2009). Assessment of the potential for carbon sequestration through afforestation, forestry and agriculture models using remote sensing and GIS. In Proc. of the ISRS Symposium on Advances in Geo-spatial technologies with special emphasis on sustainable rainfed Agriculture, Nagpur, India, September 17–19.

Wolfe, R. E., Nishiham, M., Fleig, A. J., Kuyper, J. A., Roy, D. P., & Storey, J. C. (2002). Achieving sub-pixel geolocation accuracy in support of MODIS land science. *Remote Sensing of Environment, 83*(1), 31–49.

Xiao, J. F., Zhuang, Q. L., Law, B. E., Chen, J. Q., Baldocchi, D. D., Cook, D. R., Oren, R., Richardson, A. D., Wharton, S., Ma, S. Y., Martin, T. A., Verma, S. B., Suyker, A. E., Scott, R. L., Monson, R. K., Litvak, M., Hollinger, D. Y., Sun, G., Davis, K. J., Bolstad, P. V., Burns, S. P., Curtis, P. S., Drake, B. G., Falk, M., Fischer, M. L., Foster, D. R., Gu, L. H., Hadley, J. L., Katul, G. G., Roser, Y., McNulty, S., Meyers, T. P., Munger, J. W., Noormets, A., Oechel, W. C., Paw, K. T., Schmid, H. P., Starr, G., Torn, M. S., & Wofsy, S. C. (2010). A continuous measure of gross primary production for the conterminous United States derived from MODIS and AmeriFlux data. *Remote Sensing of Environment, 114*(3), 576–591.

Xiao, J., Zhuang, Q., Baldocchi, D. D., Law, B. E., Richardson, A. D., & Chen, J. (2008). Estimation of net ecosystem carbon exchange for the conterminous United States by combining MODIS and AmeriFlux data. *Agricultural and Forest Meteorology*, 148(11), 1827–1847.

Yu, G., Song, X., Wang, Q., Liu, Y., Guan, D., Yan, J., Sun, X., Zhang, L., & Wen, X. (2008). Water-use efficiency of forest ecosystems in eastern China and its relations to climatic variables. *New Phytologist, 177*(4), 927–937.

Yuan, W., Liu, S., Yu, G., Bonnefond, J. M., Chen, J., Davis, K., Desai, A. R., Goldstein, A. H., Gianelle, D., & Rossi, F. (2010). Global estimates of evapotranspiration and gross primary production based on MODIS and global meteorology data. *Remote Sensing of Environment, 114*(7), 1416–1431.

Yuan, W., Liu, S., Zhou, G., Tieszen, L. L., Baldocchi, D., Bernhofer, C., Gholz, H., Goldstein, A. H., Goulden, M. L., Hollinger, D. Y., Hu, Y. M., Law, B. E., Stoy, P. C., Vesala, T., & Wofsy, S. C. (2007). Deriving a light use efficiency model from eddy covariance flux data for predicting daily gross primary production across biomes. *Agricultural and Forest Meteorology, 143*(3), 189–207.

Zhang, Y. Q., Yu, Q., Jiang, J., & Tang, Y. H. (2008). Calibration of Terra/MODIS gross primary production over an irrigated cropland on the North China Plain and an alpine meadow on the Tibetan Plateau. *Global Change Biology, 14*(4), 757–767.

Zhao, M., & Running, S. (2010). Drought-induced reduction in global terrestrial net primary production from 2000 through 2009. *Science, 329*(5994), 940–942.

Zhao, M., Heinsch, F. A., Nemani, R. R., & Running, S. W. (2005). Improvement of the MODIS terrestrial gross and net primary production global dataset. *Remote Sensing of Environment, 95*(2), 164–176.

CHAPTER 15

LAND SUITABILITY EVALUATION FOR SOYBEAN USING TEMPORAL SATELLITE DATA AND GIS: A CASE STUDY FROM CENTRAL INDIA

NIRMAL KUMAR,[1] G. P. OBI REDDY,[2] S. CHATTERJI,[2] RAJEEV SRIVASTAVA,[3] and S. K. SINGH[4]

[1]*Scientist, ICAR – National Bureau of Soil Survey and Land Use Planning, Amravati Road, Nagpur – 440 033, India*

[2]*Pr. Scientist, ICAR – National Bureau of Soil Survey and Land Use Planning, Amravati Road, Nagpur – 440 033, India*

[3]*Pr. Scientist & Head, ICAR – National Bureau of Soil Survey and Land Use Planning, Amravati Road, Nagpur – 440 033, India*

[4]*Director, ICAR – National Bureau of Soil Survey and Land Use Planning, Amravati Road, Nagpur – 440 033, India*

CONTENTS

ABSTRACT

In the present study, an attempt was made to disaggregate the soil suitability classes for soybean in to spatially finite grid cells (as opposed to map units) with some indicator of land suitability such as multi-temporal Normalized Difference Vegetation Index (NDVI) data over 4 years (2008–11) and terrain parameters (*viz.,* slope and topographic wetness index) derived from Advanced Space borne Thermal Emission and Reflection Radiometer (ASTER) digital elevation models of 30 m in combination with the ancillary soil resource inventory datasets. Based on suitability criteria for soybean, four main factors, i.e., texture, drainage, depth and soil reaction (pH) available in soil survey report of Wardha district and terrain parameters have been selected. The soil productivity factors have been replaced with principal component (PC) of Moderate Resolution Imaging Spectroradiometer (MODIS) 16 days composite time series NDVI data (250 m) for 4 years (from January 2008 to December, 2011). Multi-year time series NDVI reflects growth of vegetation for different seasons and thus indirectly reflects soil productivity. Instead of using multi-year average monthly NDVI images over the growing seasons, a few standardized principal components of the original NDVI images are used to screen out anomalies related to inter-annual climate variability and different farming practices and images of undesirable dates (e.g., peak NDVI images) effectively. The PC 3 was found to be strongly associated with the NDVI of the period of peak growth of soybean crop in the district. Based on these parameters, the culturable areas of the district were categorized as very good, good, moderate, poor, and very poor by adopting the logical criteria. These categories were arrived at by integrating the various layers with corresponding weights in Geographic Information System (GIS) using multi-criteria analysis. The analysis shows that majority area of the district is suitable for soybean. Majority of culturable area (84%) comes under moderate, good and very good categories. Tehsil wise analysis shows that Arvi tehsil is least suitable with 44% of culturable areas under poor and very poor categories. Less than 25% of culturable area – least among the tehsils-comes under good and very good categories for Arvi tehsil. Hinganghat, Samudrapur and Seloo are the most favorable tehsils with more than 90% of culturable area above poor category. The significantly different seasonal NDVI for all the classes validates the findings. The model provides a better insight to the suitability of land parcels for specific crop production as the suitability classes are quantified and evaluated on pixel basis.

15.1 INTRODUCTION

Appropriate land use decisions are vital to achieve optimum productivity of the land and to ensure environmental sustainability. A Land Suitability Analysis (LSA) is a preliminary step when assessing whether land is likely to be practical and successful for sustainable use of an intended crop or not. Soil resource inventory datasets have been used extensively in environmental and crop suitability studies and have proven to be highly useful for land evaluation (Mandal et al., 2005). However, these dataset are mapping units based polygons, which are soil series associations containing up to several different components or soil series in case of 1:50,000 scale. With this type of data the variations within the polygons cannot be detected. Increasing model sophistication and computing resources and the availability of high-resolution terrain and land use/land cover data enable to model and generate spatially more precise land suitability information. Lack of locational specificity relating to the soil properties, particularly related to soil productivity is becoming increasingly problematic in productivity and predictive modeling (Li et al., 2012). NDVI derived from Earth Observation (EO) data have increasingly being applied to the identification of soil conditions and soil productivity (Khaldoune et al., 2011; Kumar et al., 2014; Zribi et al., 2011). It is closely associated with vegetative reflectance and this relationship could be used as a proxy for soil productivity (Kumar et al., 2014; Li et al., 2012). Topography, including elevation, slope and aspect is a critical factor in LSA, as it affects soil properties such as hydrological processes and soil development (Beven et al., 1995; Pennock et al., 1987; Moore et al., 1993; Seibert et al., 1997), as well as soil movement and land management practices.

Multi-criterion evaluation (MCE) technique is widely used for LSA. This involves multiple criterions like soil properties, topography, vegetation, and climatic conditions to find solutions of different problems related to land with multiple alternatives (Jankowski, 1995). GIS is useful to analyze the multiple geo-spatial data with higher flexibility and precision in LSA (Mokarram and Aminzadeh, 2010). Therefore, Multi-criterion Decision Making (MCDM) technique has been integrated with GIS techniques in different studies for land use decision support (Mendas and Delali, 2012) in complex problems of land management with prioritized alternatives (Malczewski, 2006).

15.1.1 LAND SUITABILITY EVALUATION FOR SOYBEAN: GLOBAL PERSPECTIVE

Soybean (*Glycine max* L.; Fabaceae) is grown world-wide as an important staple and commercial crop. The area grown to soybean around the world is 99.5 m ha (FAO/STAT, 2009). Soybean accounted for 56% of production of the main world oilseed crops in 2011 with a total production of 251.5 m tons (ASA, 2012). Method based on FAO guidelines (FAO, 1976) is most popular qualitative LSA technique for soybean (Moshia et al., 2008; Sudaryono et al., 2011). A qualitative LSA for soybean based on FAO guidelines was conducted over five consecutive years from 1999 at Syferkuil experimental farm, Limpopo Province, South Africa (Moshia et al., 2008). The farm was divided into suitability classes of highly suitable (S1), suitable (S2), and unsuitable (N1), and permanently unsuitable (N2) classes based on A 1:10 000 corrected scale aerial photograph and laboratory analysis. Sundaryono et al. (2011) applied FAO method to assess the potential fertility and land suitability for extensification and development of soybean crop in Bumi Nabung and Rumbia districts in Indonesia.

MCE technique is widely used for LSA to detect the potential lands for agriculture (Bandyopadhyay et al., 2009; Kumar et al., 2014; Mustafa et al., 2011; Prakash, 2003; Olayeye et al., 2008; Shalaby et al., 2006; Zolekar and Bhagat, 2015), plantation (Bhagat, 2009; Zolekar and Bhagat, 2014), watershed management (Steiner et al., 2000), settlements (Soltani et al., 2012), industries (Kauko, 2006), etc. LSA for individual crops like, soybean (He et al., 2011; Kamkar et al., 2014), cotton (Walke et al., 2012) and citrus (Zabihi et al., 2015) have also been reported. He et al. (2011) applied analytical Hierarchical Process (AHP) – a type of MCE – to evaluate and map the potentially suitable planting areas for Chinese soybeans. Kamkar et al. (2014) used weighted overlay in GIS to assess the possibility and performance of a canola (*Brassica napus* L.) – soybean (*Glycine max* L.) rotation in Golestan province with raster layers, including climatic (precipitation, temperature), topographic (aspects and slope) and soil (texture, pH, EC). The consistency of results adopted from final overlaid maps with real statistics in the study region show that GIS as a systemic approach that can play a vital role in saving time and reducing research costs.

15.1.2 LAND SUITABILITY EVALUATION FOR SOYBEAN-INDIAN PERSPECTIVE

Soybean ranks first among the major oil seed crops of the world and has now found a prominent place in India. Soybean cultivation in India started long ago but its successful cultivation has increased over last two decades (Mahna, 2005). Soybean contributes significantly to the Indian edible oil pool. Presently soybean contributes 43% to the total oilseeds and 25% to the total oil production in the country. Currently, India ranks fourth in respect to production of soybean in the world. The crop helps earn valuable foreign exchange (Rs. 62,000 millions in 2012–13) by way of soya meal exports (FICCI, 2015). This increased cultivation has revolutionized the rural economy and improved the socio-economic status of farmers.

Qualitative method for LSA for soybean is attempted based on methodology given by FAO (1976) and modified by Sys et al., (1993) (Shukla et al., 2009; Patel et al., 2012; Karthikeyan et al., 2013; Meena et al., 2014). The method provides five suitability classes *viz*., S1 (highly suitable), S2 (moderately suitable), S3 (marginally suitable), N1 (presently unsuitable but potentially suitable) and N2 (permanently unsuitable) based on the soil-site requirements of the crop. Karthikeyan et al. (2013) applied LSA methods like Productivity Index (Riquier et al., 1970) and Storie Index (Storie, 1978) to assess potential productivity of soybean in Dhar district, Madhya Pradesh, India. These indices can be characterized as a multiplicative parametric method to express the joint influence of soil factors on suitability for productivity of agricultural crops. These do not include management factors. MCE methods applied to assess LSA for soybean in India has not been reported.

15.1.3 IMPORTANCE OF THE STUDY

Production of soybean in India is dominated by Maharashtra and Madhya Pradesh which contribute to 89% of the total Soybean production in country (FICCI, 2015). The gap between demand and production of edible oils in India has increased sharply in recent years making India world's top vegetable oil importer. Among the major oilseeds for edible oil, soybean has emerged as one of the important crops. The district Wardha, along with Jabalpur, Vidisha, Patancheru, Bhopal, and Betul are identified as sites with high and stable yield potential with CV <30% (Singh et al., 2006). However, the yield gap was found to be highest for sites of Wardha district with mean simulated yield and mean observed yield of 3108 and 907 kg ha^{-1}, respectively and CV of 20% (Singh et al., 2006). The gap between demand and production and the yield gap between simulated and observed yields certainly highlights the need of quantitative LSA of the area for identifying strategy/investment for improving crop production. Quantitative LSA for soybean has not been reported in India.

These studies are generally practiced following traditional land evaluation methods using soil resource inventory datasets. This involves map units which consist of more than one component, the proportion of the total area covered by each component is provided, but the location within the map unit is unknown. These data may not have the soil fertility status of the units, at least, not the present fertility status. In the present study an attempt was made to segment soil mapping unit polygons into finer resolution cells using some indicator of land suitability such as multi-temporal MODIS NDVI time series data in combination with terrain and legacy soil data to create more accurate and detailed identical soil units and to generate land suitability map for soybean crop in the area at a regional scale.

15.2 STUDY AREA

Wardha District is located in the Vidarbha region of Maharashtra state (Figure 15.1) and is named after its most important river, the Wardha. Wardha district lies between 20° 18′ North and 21° 21′ North latitudes and 78° 04′ East to 79° 15′ East longitudes. The district covers an area of 6309 sq. km with three sub-divisions, which are further sub-divided into eight tehsils namely, Wardha, Seloo and Deoli (Wardha subdivision); Hinganghat and Samudrapur (Hinganghat subdivision); and Arvi, Ashti and Karanja (Arvi subdivision). The geology of Wardha district basically consists of Deccan Trap lava flows with some patches of Gondwana formations, Lametas and the alluvium along the major river courses (http://wardha.nic.in/htmldocs/geo.asp). The physiographic features of the district can be organized into three distinct geographical units: the uplands of the north and north east with Talegaon plateau, the narrow Arvi plains to the west of the first unit, and the Wardha – Hinganghat plains. The district is characterized by hot

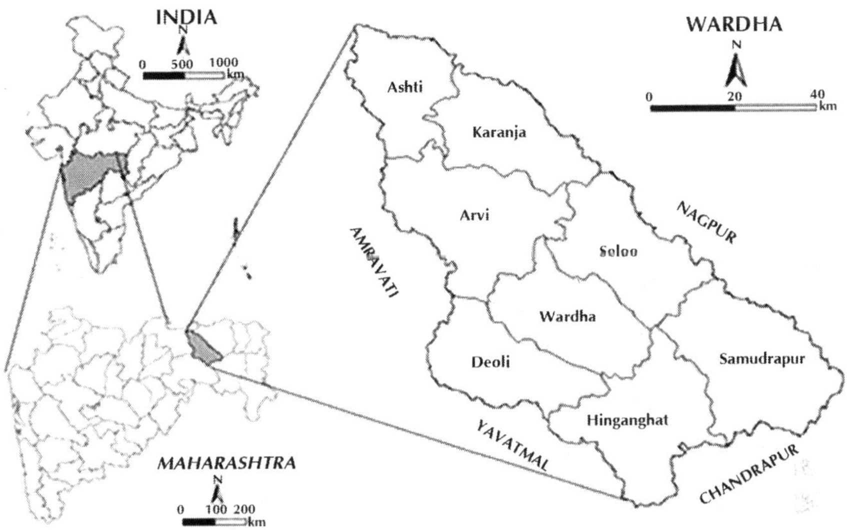

FIGURE 15.1 Location map of study area.

summers and a general dryness throughout the year except during the south-west monsoon when the humidity is above 60%. The average annual rainfall in the district is 1090.3 mm, out of which 87% is received during June to September. In general, the rainfall increases from west to east. Temperatures increase steadily from about the beginning of March. May is the hottest month of the year with the mean daily maximum temperature at about 43°C and the mean daily minimum at about 27°C. The major land use category in the district is agriculture. *Kharif* crops are widespread in the southern part comprising the tehsils of Samudrapur, Hinganghat, Wardha and Deoli. In the northern tehsils of Ashti, Arvi, Karanja, and Seloo, agriculture and forest coverage occupy more or less equal areas. A significantly large area under orchards is found in Hinganghat tehsil, with smaller patches in Samudrapur and Arvi tehsils. Deciduous forest is spread noticeably in Selu, Karnaja and Arvi tehsils with degraded forest around the fringes. The important *kharif* crops are cotton, sorghum, and pulses, and since the mid-1980s, soybean has become an extremely important *kharif* crop in Wardha district. In the *rabi s*eason, wheat and gram are cultivated. The cropping pattern of the district has been changed from once a cereal (mainly Jowar) – cotton in 1960s to oilseed (mainly Soybean) – cotton at present (Figure 15.2).

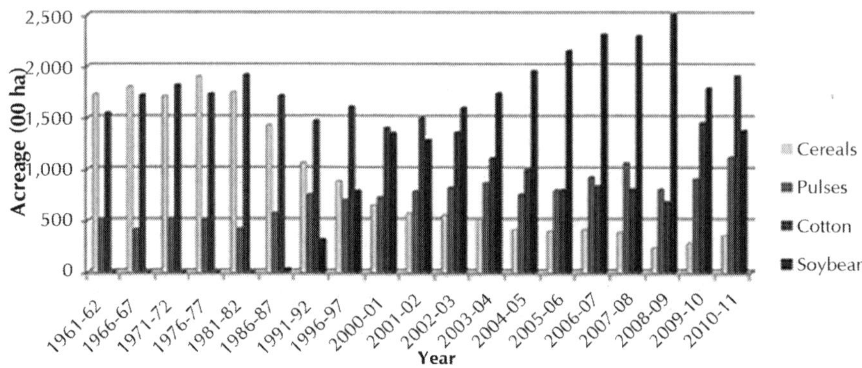

FIGURE 15.2 Change of cropping pattern in Wardha district over a period of five decades (1961–62 to 2010–11).

15.3 METHODOLOGY

15.3.1 DATA SOURCES

In the study, the time-series MODIS NDVI for the period from January, 2008 to December, 2011 in conjunction with other ancillary data were used to interpret LSA for soybean. The MODIS product used in this study was the Terra 16-day composite, 250 m NDVI (MOD13Q1) acquired from NASA LPDAC collections (http://earthexplorer.usgs.gov/). A total of 91 images were acquired. The NDVI bands then extracted and reprojected from sinusoidal projection system to UTM projection system. All the 91 layers then stacked in a single file and the area of interest (Wardha district) was masked out for further analysis. Three tiles of Advanced Space borne Thermal Emission and Reflection Radiometer (ASTER) GDEM (ASTGTM_N20E078, ASTGTM_N20E079 and ASTGTM_N21E078) covering the district were downloaded (http://earthexplorer.usgs.gov/) and mosaicked in a single file. The area under Wardha district was masked out for generating primary (slope, aspect) and secondary terrain (curvature, TWI) variables. Soil property information, such as soil types, soil texture, soil pH, soil depth, and soil fertility were extracted from reconnaissance survey report of Wardha district prepared by the ICAR-NBSS & LUP, Nagpur (Sharma et al., 2008). Three scenes of IRS P-6 LISS III data covering the district were utilized for generating land use land cover of the district. The raw data were preprocessed (corrected geometrically and mosaicked) and study area was clipped from the entire scene.

15.3.2 LAND USE/LAND COVER

Maximum likelihood (MLH) classification method was applied on IRS P6 LISS 3 image of October 2010 after preprocessing. Accuracy was compared for the classified images. To increase the accuracy of the classification in hilly areas, elevation was included as additional information to the classifier. From the classified image, the area under agriculture was extracted for LSA of soybean crop.

15.3.3 LAND SUITABILITY ANALYSIS

For the objective, we conducted multi-criteria analysis using ArcGIS® Desktop 10 (ESRI). The variables used for classification were divided into three groups; (1) soil productivity, represented by the first two principal components of the NDVI images, (2) terrain parameters, represented by slope and TWI, and (3) soil attributes like texture, depth, pH and drainage – important for the crops grown in the region were obtained from the soil legacy data. All these rasters were resampled to 250 m spatial resolution to be complementary with 250 m MODIS NDVI data. Only the culturable areas were used in the modeling process hence, agricultural areas were masked out from all the rasters before using them in the model. These inputs are discussed in more detail in the following sections.

15.3.3.1 Soil Productivity (Principal Component Analysis)

Multi-year average monthly NDVI can reflect general information regarding growth of vegetation for different seasons and thus indirectly reflect soil productivity (Li et al., 2012). However, NDVI for the same month of different years may show inconsistent patterns due to inter-annual climate variability and different farming practices (Li et al., 2012). Further, NDVI is expected to saturate for high biomass values (Santin-Janin et al., 2009), especially at the peak of the growing season. Therefore, instead of using multi-year average monthly NDVI images over the whole growing seasons for analysis, we chose to elicit a few standardized principal components (Eastman and Fulk, 1993) of the original NDVI images so that anomalies and images of undesirable dates (e.g., peak NDVI images) can be effectively screened out.

Table 15.1 shows variance explained by each component image. Figure 15.3 shows Principal Component Analysis (PCA) loadings against different dates of NDVI images for the first four components. The loadings of Component 1 are high and are strongly associated with the spring and late summer NDVI as it has high loadings (over 0.8) with the NDVI of January to June (except some anomalies). Component 1 has a negative loading factor for *kharif* season. Component 2 is with positive loadings for *kharif* season covering a period from August to January. Component 3 is positively correlated to *kharif* NDVI images, thus capturing the characteristics of NDVI of August to October. Component 4 accounts for a small amount of variance and primarily captures summer (April, May) NDVI. Principal components after the 4th component represent patterns related to other environmental factors and noise (Eastman and Fulk, 1993).

15.3.3.2 Terrain Variables

Slope can influence water flow patterns and drainage capacity and subsequently soil moisture (Beven et al., 1995). The topographic wetness

TABLE 15.1 Variance Explained by Each PCA Component

Component	C1	C2	C3	C4	C5	C6	C7	C8	C9	C10
(%)	39.68	28.80	15.30	7.05	2.75	2.62	1.37	0.75	0.50	0.31

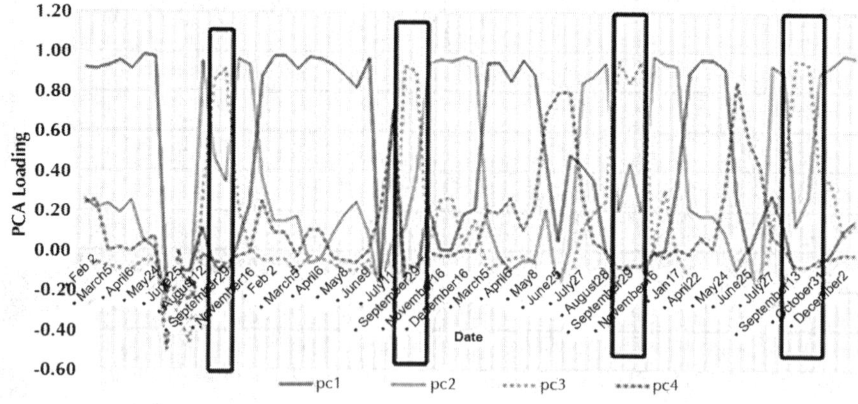

FIGURE 15.3 PCA loadings against different dates of NDVI images for the first four components.

index (TWI) is highly correlated with several soil attributes, such as, horizon depth, texture, organic matter content, phosphorus, and soil moisture (Moore et al., 1993). Setiyono et al. (2010) used a location at 190–357 m above sea level to study soybean growth and yield at near-optimal growth conditions. Soybean yield was negatively related to altitude and slope indicating that lower, flatter areas of a field had higher yields (Cox et al., 2003; Kravchenko and Bullock, 2000; McConkey and Ulrich, 1997). ASTER GDEM with 30 m spatial resolution has been utilized to generate slope and TWI in ArcGIS Desktop 10. Slope was generated as per cent rise and classified in standard ranges before being resampled to 250 m. TWI was developed by Beven and Kirkby (1979) within the runoff model TOPMODEL. It is defined as ln(a/tanβ) where *a* is the local upslope area draining through a certain point per unit contour length and tanβ is the local slope. TWI was calculated in ArcGIS and reclassified according to expert knowledge before resampled to 250 m.

15.3.3.3 Soil Attributes

Soil resource database has been generated from reconnaissance survey maps of Wardha district (Shrama et al., 2008). The Wardha district has high diversity of soil types located over different landforms. About 38 soil series with 33 soil mapping units (soil series association) have been identified in the district by reconnaissance soil survey of the district (Shrama et al., 2008) at 1:50,000 scale. Taxonomically the soils in Wardha come mainly under Entisols (120,175 ha), Vertisols (110,307 ha) and Inceptisols (400,913 ha). Individual themes of soil depth, soil pH, soil drainage and soil texture were reclassified and rasterized to make it assimilable in the GIS model.

15.3.3.4 Multi-Criteria Analysis

The methodology flow chart for generation of land suitability units is shown in the Figure 15.4. The model was used by Linear Combination Method of ranking and weighting score in order to create land suitability map for soybean crop in the district. The modeling was done using the AcrGIS desktop 10 because of its ability to accommodate various files formats in common use. The different rasters were given influence percent

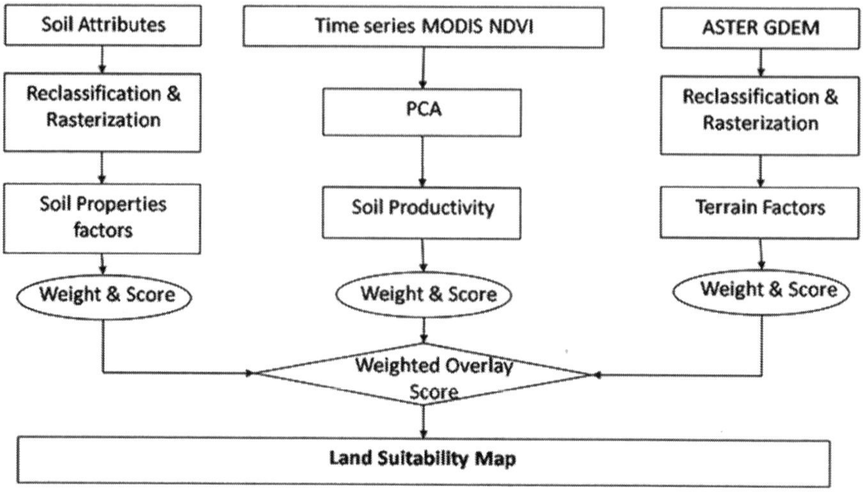

FIGURE 15.4 Methodology flow chart for LSA. (Adapted from Kumar et al., 2014)

based on their effect on the productivity of the crops in the area and expert knowledge (Table 15.3). The classes within each rasters were assigned a scale value ranging from 1 to 10 based on the crop suitability criteria and expert suggestions. Once the land suitability maps were generated, classes were redefined with merging small classes to the nearest one.

15.4 RESULTS AND DISCUSSION

15.4.1 LAND USE/LAND COVER

Eight classes namely, dense forest, open forest, croplands, fallows scrubs, wastelands, settlements, and water bodies were identified in the imagery and well distributed training sites with ground truth were provided to the classifier. Apart from the four bands of LISS III data, ASTER elevation data (resampled to 24 m) were also included for better accuracy. The accuracies of the classifier with and without inclusion of elevation data were 88.51 (kappa 0.866) and 92.7 (kappa 0.915), respectively. The percent of areas under different classes are given in the Table 15.2. The classification results of MLH with elevation data were selected for further analysis. For further analysis only the area under agriculture (culturable lands) were considered and were masked out of the total area (Figure 15.5).

TABLE 15.2 Percentage of Area Under Different Classes by MLH Classifier

Class	% TGA
Dense forest	16.92
Croplands	17.94
Fallows	21.63
Scrubs	20.49
Wastelands	4.35
Settlements	2.12
Water bodies	1.94
Open forest	14.62

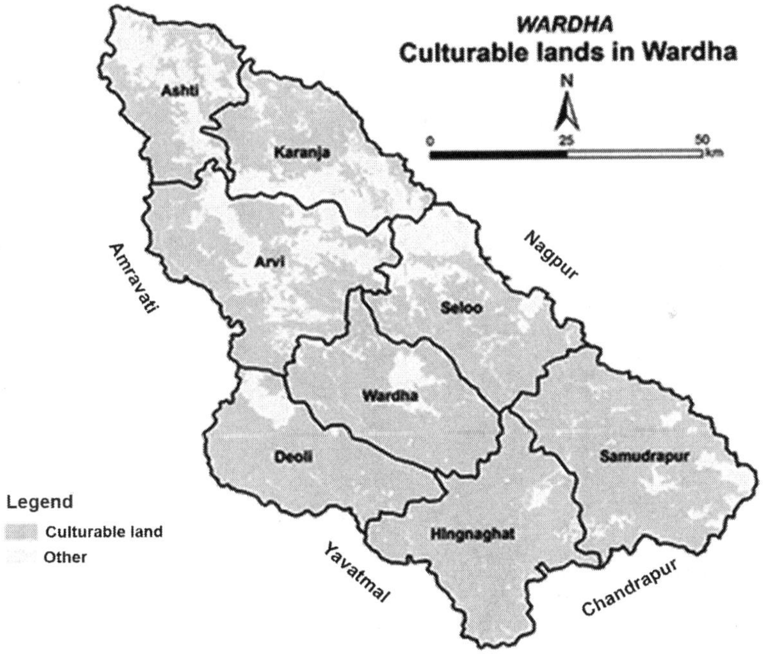

FIGURE 15.5 Distribution of Culturable lands in Wardha district

15.4.2 TERRAIN ANALYSIS

The slope of the district ranges from nearly level-to-level to very steep slopes (Table 15.3). The four northern tehsils (Arvi, Ashti, Karanja and Seloo) are having more than 50% area under slope more than 3%, i.e., undulating and hilly terrain. The southern tehsils (Hinganghat, Samudrapur and Deoli) are having maximum area under level to nearly level and very gently sloping lands. The TWI value for the district varies from –1 to 26.9. The TWI raster was reclassified in six classes for further analysis (Table 15.3). Areas with

TABLE 15.3 Parameters Identified, Their Weights and Scales for LPP of Soybean

Layers	Influence %	Field	Scale Value	Layers	Influence %	Field	Scale Value
PC 3 (4 yrs. 16 days MODIS NDVI composite)	25	<0	1	Soil Depth	10	Rock outcrop	1
		0–0.1	1			1	4
		0.1–0.3	2			2	6
		0.3–0.5	4			3	8
		0.5–0.7	8			5	10
		0.7–0.9	9			6	10
		>0.9	10			7	10
TWI	15	5–10	2	Texture	10	Rock outcrop	1
		10–15	4			C	9
		15–20	6			Cl	10
		20–25	8			Gc	6
		>25	10			Gl	6
Slope	20	0–1%	10			Sc	7
		1–3%	10			Scl	8
		3–8%	7	Drainage	10	Rock outcrop	1
		8–15%	4			2	3
		15–30%	2			4	6
		30–50 %	1			5	10
		>50%	0			6	4
						7	3
				pH	10	Rock outcrop	1
						7–7.5	10
						7.5–8	9
						8–8.5	7
						>8.5	6

low TWI values represent places with small catchments, and steep slopes or hills. Whereas, areas with high TWI values represent places with large catchments, and gentle slopes are depressions or plains. The tehsils with steep slopes and small catchments (Arvi, Ashti, Karajna and Seloo tehsils) are having more than 15% area under lower TWI. Concave, low gradient areas gather water (low TWI values), whereas steep, convex areas shed water (high TWI values). TWI values typically range from less than 1 (dry cells) to greater than 20 (wet cells). The northern hilly and undulating areas of tehsils Arvi, Ashti, Karajna and Seloo are having larger percentage of areas under lower TWI indicating drier cells and lesser area under higher TWI indicating wet cells.

15.4.3 SOIL ATTRIBUTES

15.4.3.1 Soil Depth

Soil depth of the district varies from extremely shallow (0 to 10 cm) to very deep (>150 cm) (Figure 15.6a). Maximum area of the district (36%) comes under deep soils followed by very shallow soils (30%). Very deep soils are limited to only Hinganghat tehsil. More than 60% area of Wardha and Arvi tehsils comes under soils less than 25 cm, i.e., extremely shallow and very shallow. The extremely shallow soils coincide with hilly areas and rock outcrops. The plain areas of Seloo, Hinganghat and Samudrapur tehsils show deep and very deep soils.

15.4.3.2 Soil Drainage

The soil drainage of Wardha district ranges from *poor* to *excessive* (Figure 15.6b). Majority of the area in Wardha district comes under moderately well (53%) and well-drained (18%) soils. The excessively and somewhat excessively drained soils are mainly concentrated to Arvi, Ashti, Seloo, and Wardha tehsils with more than 20% area in each tehsil. Poor drainage conditions prevail mainly in Arvi (6.7%), Samudrapur (8%) and Ashti (22%) soils. The soils of plain areas of Hinganghat and Samudrapur tehsils are mostly moderately well drained. The hilly tehsils' soils are somewhat excessive to excessively drained.

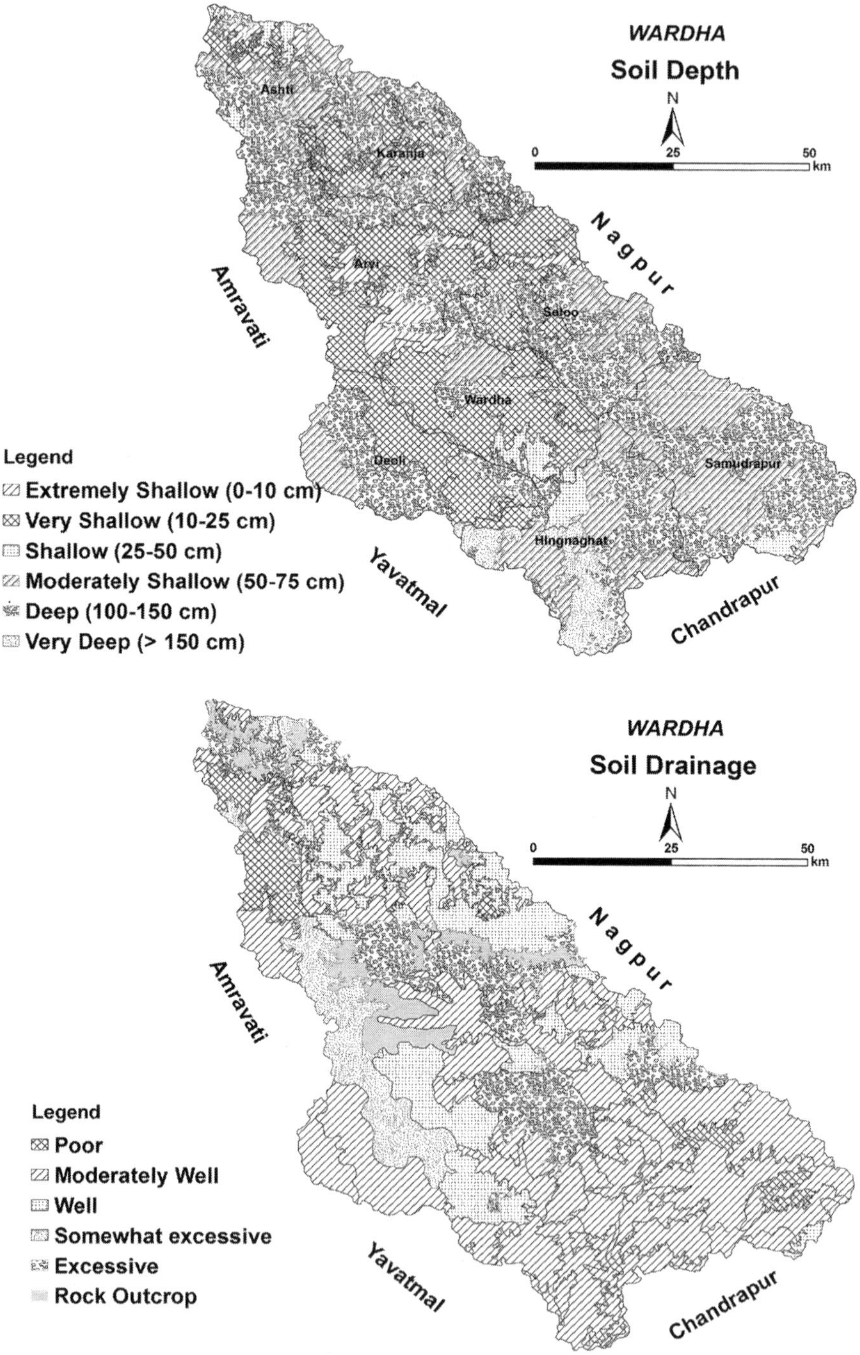

FIGURE 15.6 Soil depth (a), soil drainage (b), soil reaction (c), and soil texture (d) maps of Wardha district.

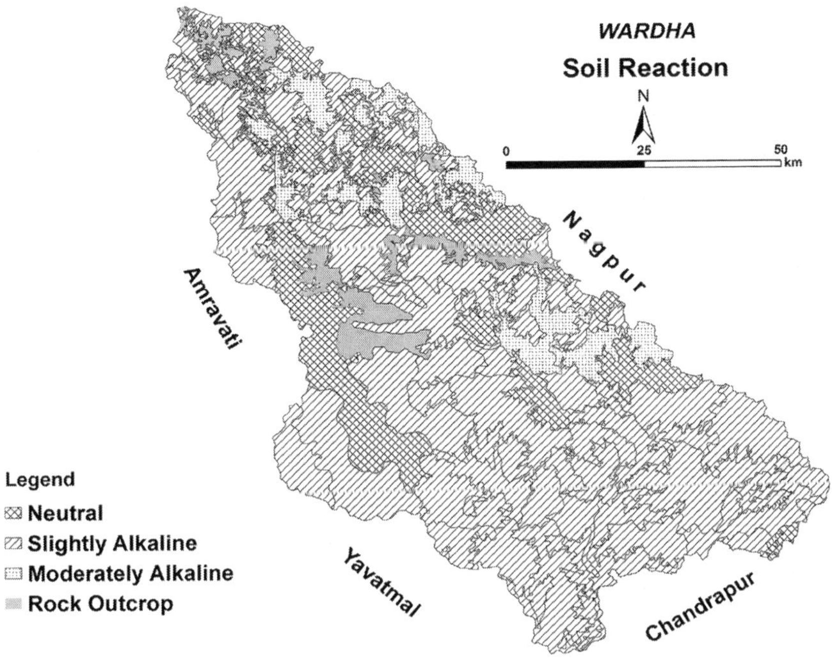

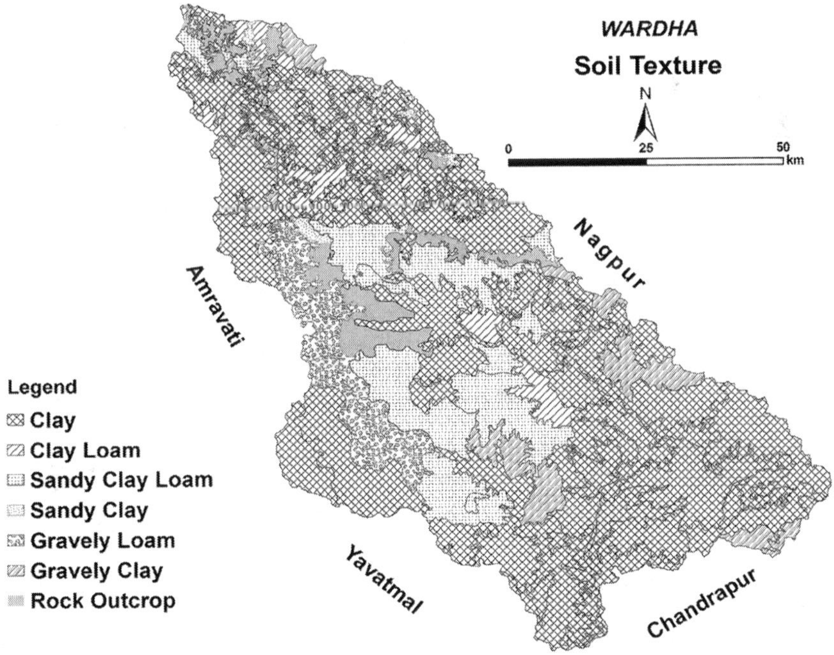

FIGURE 15.6 (Continued)

15.4.3.3 Soil Reaction

The pH of the Wardha district soils varies from 7 (neutral) to 8.7 (moderately alkaline). Majority of the area (68%) comes under slightly alkaline soils (Figure 15.6c). Moderately alkaline soils are mainly concentrated in Arvi, Ashti, Karanja, and Seloo tehsils. Wardha, Hinganghat, and Samudrapur soils are mostly slightly alkaline (>85%).

15.4.3.4 Soil Texture

The texture of Wardha soils varies from clay to sandy and gravely clay (Figure 15.6d). Overall more than 60% area of the district comes under clayey soils with Hinganghat, Karanja and Samudrapur tehsils having more than 80% area under clay soils. Next to clay are sandy clay loam soils in area (14%) with maximum proportion in Wardha tehsil. Gravely clay soils are restricted to 6% area with occurrence in Ashti, Karanja, Seloo and Wardha tehsils mainly. Gravely loam soils are found mainly in Arvi and Deoli tehsils.

15.4.4 LAND SUITABILITY ANALYSIS FOR SOYBEAN

The cropping pattern of the study area suggests that the cultivation of soybean starts on the onset of monsoon *i.e.,* mid July to first week of August and harvested during August-October. From the PC loading factors in Figure 15.4 we can observe that the PC 3 is strongly associated with the NDVI of the period of peak growth of soybean crop in the district. Hence, the PC 3 is considered as a proxy for soil productivity in LSA for soybean. The percent influence and scale values of classes within each factor including PC3, terrain variables and soil properties were decided with expert suggestions and crop suitability criteria (NBSS & LUP, 1994) and are shown in Table 15.3. A scale value of 10 indicates the most favorable conditions and a value of one indicates the least favorable conditions in relation to that particular parameter. After weighted overlay of all the factors the output classes were defined from very poor to very good suitability class for soybean (Figure 15.7). The tehsil wise distribution of these classes is shown in Table 15.4.

The analysis shows that overall for soybean the district's lands are suitable. As visible from the Table 15.4, majority of culturable area (84%) comes under moderate, good and very good class for soybean. Tehsil wise,

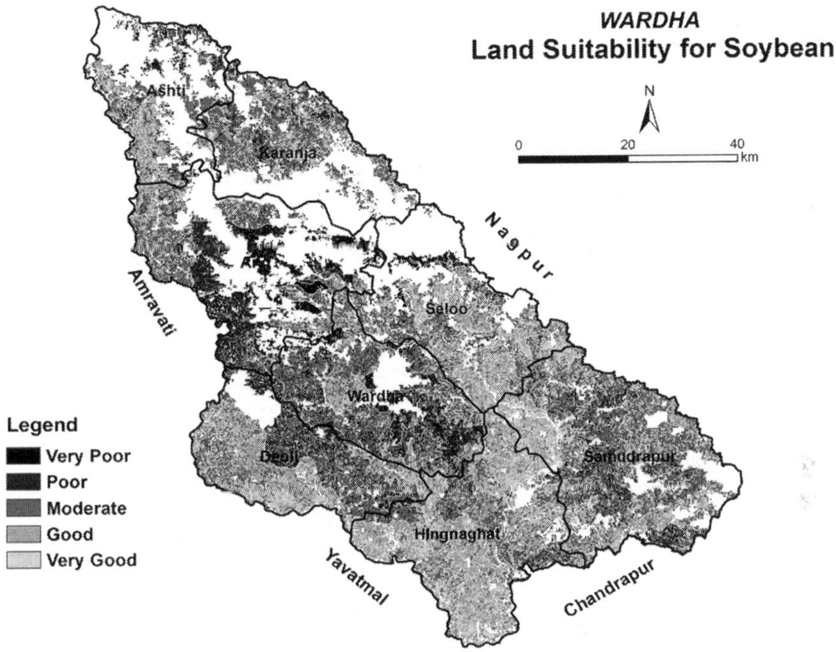

FIGURE 15.7 Land suitability for soybean in Wardha district.

Arvi is least suitable with 44% of culturable areas under poor and very poor categories. Less than 25% of culturable area – least among the tehsils—comes under good and very good categories for Arvi tehsil. Hinganghat, Samudrapur and Seloo are the most favorable tehsils with more than 90% of culturable area above poor category. The disaggregated land suitability map is shown as Figure 15.5. The polygons of the original soil map were shown in solid red lines. It can be observed that a single soil mapping unit now covers various suitability classes, which would have been a single class by traditional method.

15.5 CONCLUSIONS

The study attempted quantitative land suitability analysis for soybean for Wardha district using temporal NDVI data, terrain attributes, and legacy soil inventory data. The analysis shows that majority of the culturable area (84%) comes under moderate, good and very good suitability class for soybean.

TABLE 15.4 Tehsil Wise Distribution of Land Suitability Classes for Soybean

	Very Poor		Poor		Moderate		Good		Very Good		
Tehsil	Area	%*	Area	%*	Area	%*	Area	%*	Area	%*	Total
Arvi	8392.6	15.0	16266.3	29.0	18776.9	33.4	10218.0	18.2	2504.6	4.5	56158.4
Ashti	1159.0	4.3	3912.5	15.1	10243.4	39.5	7886.0	30.4	2809.9	10.8	25960.9
Deoli	70.6	0.1	7148.9	12.5	20684.8	36.2	20875.2	36.5	8343.9	14.6	57123.4
Hinganghat	30.3	0.0	1726.9	2.2	18263.6	23.3	38771.3	49.5	19571.5	25.0	78363.5
Karanja	110.9	0.3	3863.1	10.6	21324.2	58.6	8765.0	24.1	2308.7	6.3	36371.9
Samudrapur	79.3	0.1	7559.7	8.9	35764.2	42.2	28194.0	33.3	13191.4	15.6	84788.6
Seloo	1727.4	3.7	2722.1	5.7	13283.4	27.8	20817.4	43.5	9291.5	19.4	47841.8
Wardha	2128.8	3.5	13545.6	22.6	26734.5	44.6	14010.3	23.4	3468.5	5.8	59877.7
Total	13438.5	3.1	56745.0	12.7	165075.1	37.0	149537.2	33.5	61490.1	13.8	446486.2

* Percentage to total area.

The developed model provides a pixel level land suitability map to provide a better insight to the suitability of land for soybean production in the area. The suitability classes are more precise and quantified as the parameters are judged and evaluated on pixel basis *i.e.,* the soil units are having finite assessment.

KEYWORDS

- **ASTER**
- **Central India**
- **MODIS**
- **Multi-Criteria Analysis**
- **NDVI**
- **PCA**
- **Soil Suitability**
- **Soybean**

REFERENCES

ASA (American Soybean Association), (2012). Soy stats: a reference guide to important soybean facts and figures, St. Louis, MO, USA. http://www.soystats.com/ (last accessed on 08.12.15).

Bandyopadhyay, S., Jaiswal, R. K., Hegde, V. S., & Jayaraman, V. (2009). Assessment of land suitability potentials for agriculture using a remote sensing and GIS based approach. *International Journal of Remote Sensing, 30*(4), 879–895.

Beven, K. J., & Kirkby, M. J. (1979). A physically based, variable contributing area model of basin hydrology, *Hydrolological Sciences Bulletin, 24*(1), 43–69.

Beven, K. J., Lamb, R., Quinn, P., Romanowicz, R., & Freer, J. (1995). Top model. *In: Computer Models of Watershed Hydrology.* Singh V. P. (Ed.), Water Resources Publications, Colorado.

Bhagat, V. S. (2009). Use of Landsat ETM+ data for detection of potential areas for afforestation. *International Journal of Remote Sensing, 30*(10), 2607–2617.

Cox, M. S., Gerard, P. D., Wardlaw, M. C., & Abshire, M. J. (2003). Variability of selected soil properties and their relationships with soybean yield. *Soil Science Society of America Journal, 67*(4), 1296–1302.

Eastman, J. R., & Fulk, M. (1993). Long sequence time series evaluation using standardized principal components. *Photogrammetric Engineering and Remote Sensing, 59*(8), 1307–1312.

FAO. (1976). A framework for land evaluation, Soil Bulletin 32. Food and Agriculture Organization of the United Nations, Rome, Italy.

FAO/STAT. (2009). Food and Agriculture Organization of the United Nations. http://faostat.fao.org/site/339/default.aspx (last accessed 08.12.15).

FICCI. (2015). Evaluation of the PPPIAD Project on soybean. Documentation of the project on Improving productivity of Soybean in Maharashtra by Archer Daniels Midland Company (ADM) and Department of Agriculture, Government of Maharashtra.

He, W., Yang, S., Guo, R., Chen, Y., Zhou, W., Jia, C., & Sun, G. (2010). CCTA (3), volume 346 of IFIP Advances in Information and Communication Technology, 357–366. Springer.

http://earthexplorer.usgs.gov/. Accessed in July, 2015.

http://wardha.nic.in/htmldocs/geo.asp. Accessed in July, 2015.

Jankowski, P. (1995). Integrating geographical information systems and multiple criteria decision-making methods. *International Journal of Geographical Information Systems*, *9*(3), 251–273.

Kamkar, B., Dorri, M. A., & daSilva, J. A. T. (2014). Assessment of land suitability and the possibility and performance of a canola (*Brassica napus* L.) – soybean (*Glycine max* L.) rotation in four basins of Golestan province, Iran. *The Egyptian Journal of Remote Sensing and Space Sciences*, *17*(1), 95–104.

Karthikeyan, K., Pushpanjali, Prasad, J., & Sarkar, D. (2013). Suitability and productivity assessment of soybean (*Glycine max* L.) – growing soils of Dhar district, Madhya Pradesh, India. *Legume Research, 36*(5), 442–447.

Kauko, T. (2006). What makes a location attractive for the housing consumer? Preliminary findings from metropolitan Helsinki and Randstad Holland using the analytical hierarchy process. *Journal Housing and the Built Environment*, *21*(2), 159–176.

Khaldoune, J., Van Bochove, E., Bernier, M., & Nolin, M. C. (2011). Mapping agricultural frozen soil on the watershed scale using remote sensing data. *Applied and Environmental Soil Science*, 1–16.

Kravchenko, A. N., & Bullock, D. G. (2000). Correlation of corn and soybean grain yield with topography and soil properties. *Agronomy Journal, 92*(1), 75–83.

Kumar, N., Reddy, G. P. O., & Chatterji, S. (2014). GIS modeling to assess land productivity potential for agriculture in sub-humid (dry) ecosystem of Wardha district, Maharashtra. *In*: *Applied Geoinformatics for Society and Environment.* Vyas, A., Behr, F. J., Schröder, D. (Eds.), Stuttgart University of Applied Sciences Publications, Hochschule für Technik Stuttgart No. 137, Stuttgart, Germany.

Li, Z., Huffman, T., Aining, Z., Fuqun, Z., & Brian, Mc. C. (2012). Spatially locating soil classes within complex soil polygons – Mapping soil capability for agriculture in Saskatchewan Canada. *Agriculture, Ecosystems and Environment, 152*, 59–67.

Mahna, S. K. (2005). Production, regional distribution of cultivars, and agricultural aspects of soybean in India. *In: Nitrogen Fixation in Agriculture, Forestry, Ecology, and the Environment,* D. Werner, & W. E. Newton (Eds.), Springer, 43–66.

Malczewski, J. (2006). GIS-based multi-criteria decision analysis: a survey of the literature. *International Journal of Geographical Information Science*, *20*(7), 703–726.

Mandal, D. K., Mandal, C., & Venugopalan, M. V. (2005). Suitability of cotton cultivation in shrink–swell soils in central India. *Agricultural Systems, 84*, 55–75.

McConkey, B. G., Ulric, D. J., & Dyck. F. B. (1997). Slope position and subsoiling effects on soil water and spring wheat yield. *Canadian Journal of Soil Science*, *77*(1), 83–90.

Meena, G. L., Singh, R. S., Meena, S., Meena, R. H., & Meena, R. S. (2014). Assessment of land suitability for soybean (*Glycine max*) in Bundi district, Rajasthan. *Agropedology, 24*(2), 146–156.

Mendas, A., & Delali, A. (2012). Integration of Multi-Criteria Decision Analysis in GIS to develop land suitability for agriculture: application to durum wheat cultivation in the region of Mleta in Algeria. *Computers and Electronics in Agriculture, 83*, 117–126.

Mokarram, M., & Aminzadeh, F. (2010). GIS-based multicriteria land suitability evaluation using ordered weight averaging with fuzzy quantifier: a case study in Shavur Plain, Iran. The International Archives of the Photogrammetry, *Remote Sensing and Spatial Information Sciences, 38*(II), 508–512.

Moore, I. D., Gessler, P. E., Nielsen, G. A., & Peterson, G. A. (1993). Soil attributes prediction using terrain analysis. *Soil Science Society of America Journal, 57*(2), 443–452.

Moshia, M. E., Mashatola, M. B., Shaker, P., Fouché, P. S., & Boshomane, M. A. W. (2008). Land suitability assessment and precision farming prospects for irrigated maize-soybean intercropping in syferkuil experimental farm using geospatial information technology. *Journal of Agriculture and Social Research, 8*(2), 138–147.

Mustafa, A. A., Singh, M., Sahoo, R. N., Ahmed, N., Khanna, M., Sarangi, A., & Mishra, A. K. (2011). Land suitability analysis for different crops: a multi criteria decision making approach using remote sensing and GIS. *Researcher, 3*(12), 61–84.

NBSS & LUP. (1994). Proceedings of the National meet on soil-site suitability criteria for different crops (Feb, 7–8), NBSS & LUP, Nagpur, 30 pp.

Olaleye, A. O., Akinbola, G. E., Marake, V. M., Molet, S. F., Mapheshoane, B. (2008). Soil in suitability evaluation for irrigated lowland rice culture in southwestern Nigeria: management implications for sustainability. *Communications in Soil Science and Plant Analysis, 39*(19–20), 2920–2938.

Parakash, T. N. (2003). Land Suitability Analysis for Agricultural Crops: A Fuzzy Multicriteria Decision Making Approach. MSc Thesis, The International Institute for Geoinformation Science and Earth Observation (ITC), Enschede, The Netherlands.

Patel, H. P., Savalia, S. G., & Chopada, M. C. (2012). Soil-site suitability evaluation for soybean in meghal irrigation command area of southern saurashtra region of Gujarat. *An Asian Journal of Soil Science, 7*(1), 117–123.

Pennock, D. J., Zebarth, B. J., & De Jong, E. (1987). Landform classification and soil distribution in hummocky terrain, Saskatchewan, Canada. *Geoderma, 40*(3–4), 297–315.

Riquier, J., Bramao, D. L., & Comet, J. P. (1970). A new system of appraisal in terms of actual and potential productivity. FAO Soil Resources, Development and Conservation Services, Land and Water Development Division, FAO, Rome, Italy.

Santin-Janin, H., Gare, M., Chapuis, J. L., & Pontier, D. (2009). Assessing the performance of NDVI as a proxy for plant biomass using non-linear models: a case study on the Kerguelen archipelago. *Polar Biology 32*(6), 861–871.

Seibert, J., Bishop, K. H., & Nyberg, L. (1997). A test of TOPMODEL's ability to predict spatially distributed groundwater levels. *Hydrological Processes, 11*(9), 1131–1144.

Setiyono, T. D., Cassman, K. G., Specht, J. E., Dobermann, A., Weiss, A., Yang, H., Conley, S. P., Robinson, A. P., Pedersen, P., & De Bruinh, J. L. (2010). Simulation of soybean growth and yield in near-optimal growth conditions. *Field Crops Research, 119*(1), 161–174.

Shalaby, A., Ouma, Y. O., & Tateishi, R. (2006). Land suitability assessment for perennial crops using remote sensing and geographic information systems: a case study in Northwestern Egypt. *Archives in Agronomy and Soil Science, 52*(3), 243–261.

Sharma, J. P., Mandal, C., Raja, P., Nair, K. M., Bhaskar, B. P., & Sarkar, D. (2008). Reconnaissance soil survey, mapping, correlation and classification of Wardha district, Maharashtra. NBSS Publication 595, NBSS & LUP, Nagpur.

Shukla, E. A., Prasad, J., Nagaraju, M. S. S., Srivastava, R., & Kauraw, D. L. (2009). Use of remote sensing in characterization and management of Dhami micro-watershed of Chandrapur district of Maharashtra. *Journal of Indian Society of Remote Sensing, 37*(1), 129–137.

Singh, P., Vijaya, D., Srinivas, K., & Wan, S. P. (2006). Potential productivity, yield gap, and water balance of soybean-chickpea sequential system at selected benchmark sites in India. Global Theme 3: Water, Soil, and Agrobiodiversity Management for Ecosystem Health. Report no. 1. Patancheru 502 324, Andhra Pradesh, India: International Crops Research Institute for the Semi-Arid Tropics. 52 pp.

Soltani, S. R., Mahiny, A. S., Monavari, S. M., & Alesheik, A. A. (2012). A soft approach to conflict resolution in multi-criteria evaluation of urban land use suitability. *World Applied Science Journal, 19*(7), 1066–1077.

Steiner, F., McSherry, L., & Cohen, J. (2000). Land suitability analysis for the upper Gila River watershed. *Landscape and Urban Planning, 50*(4), 199–214.

Storie, R., & Earl, R. (1978). *Storie Index Soil Rating, Special Publication, 3203*, Division of Agricultural Sciences, University of California, USA.

Sudaryono, P., & Wijanarko, A. (2011). Land Suitability for Developing Soybean Crops in Bumi Nabung and Rumbia Districts, Central Lampung. *Journal of Tropical Soils, 16*(1), 85–92.

Sys, Ir. Van Ranst, C. E., Debaveye, J., & Beernaert, F. (1993). Land evaluation crop requirements. *Agricultural Publication No. 7*, Part III, General, FAO, Rome, Italy.

Walke, N., Reddy, G. P. O., Maji, A. K., & Thayalan S. (2012). GIS based multi-criteria overlay analysis in soil- suitability valuation for Cotton (Gossypium spp.): A case study in black soil region of Central India. *Computers and Geosciences, 41*, 108–118.

Zabihi, H., Ahmad, A., Vogeler, I., Said, M. N., Golmohammadi, M., Golein, B., & Nilashi, M. (2015). Land suitability procedure for sustainable citrus planning using the application of the analytical network process approach and GIS. *Computers and Electronics in Agriculture, 117*, 114–126.

Zolekar, R. B., & Bhagat, V. S. (2014). Use of IRS P6 LISS-IV data for land suitability analysis for cashew plantation in hilly zone. *Asian Journal of Geoinformatics, 14*(3), 23–35.

Zolekar, R. B., & Bhagat, V. S. (2015). Multi-criteria land suitability analysis for agriculture in hilly zone: Remote sensing and GIS approach. *Computers and Electronics in Agriculture, 118*, 300–321.

Zribi, M., Baghdadi, N., & Nolin, M. (2011). Remote sensing of soil. Applied and Environmental Soil Science, 1–2.

CHAPTER 16

GEOSPATIAL TECHNOLOGIES IN DEVELOPMENT OF SOIL INFORMATION SYSTEM AND PROTOTYPE GEOPORTAL

G. P. OBI REDDY,[1] S. K. SINGH,[2] NIRMAL KUMAR,[3] C. MANDAL,[1] RAJEEV SRIVASTAVA,[4] A. K. MAJI,[5] and DIPAK SARKAR[6]

[1]*Principal Scientist, ICAR-National Bureau of Soil Survey and Land Use Planning, Amravati Road, Nagpur – 440 033, India*

[2]*Director, ICAR-National Bureau of Soil Survey and Land Use Planning, Amravati Road, Nagpur – 440 033, India*

[3]*Scientist, ICAR-National Bureau of Soil Survey and Land Use Planning, Amravati Road, Nagpur – 440 033, India*

[4]*Principal Scientist & Head, ICAR-National Bureau of Soil Survey and Land Use Planning, Amravati Road, Nagpur – 440 033, India*

[5]*Ex-Director (Act.), ICAR-National Bureau of Soil Survey and Land Use Planning, Amravati Road, Nagpur – 440 033, India*

[6]*Ex-Director, ICAR-National Bureau of Soil Survey and Land Use Planning, Amravati Road, Nagpur – 440 033, India*

CONTENTS

ABSTRACT

The availability of accurate, reliable and timely information on natural resources enable and rationalize the decision-making process by the decision makers, planners and resource managers. In this chapter, an attempt has been made to demonstrate the capabilities of geospatial technologies in development of Soil Information System (SIS) and prototype Geoportal to store, query, update, manage and visualize the spatial and non-spatial soil information of India on 1:250,000 scale and provide robust web map services. The available state wise soil resource maps were digitized to generate seamless mosaic, standardized spatial and non-spatial data and develop SIS on 1:250,000 scale using various capabilities of Geographic Information System (GIS). The schema has been developed for input soil parameters to effectively store, update, manage and query the soil resource database from seamless SIS. Attributions of various soil parameters were standardized in seamless soil polygon layer of India for entry of database on various soil themes. The standardized soil resource database contains information about the soil site characteristics, physical and chemical properties. The prototype standalone Geoportal has been designed and developed on local host to provide Web Map Services (WMS), Web Catalogue Services (WCS), interface tools and basic Geoportal functionalities. Further, Geoportal has been enriched with raster images of Shuttle Radar Topographic Mission (SRTM) Digital Elevation Model (DEM) (90 m), Advanced Space borne Thermal Emission and Reflection Radiometer (ASTER) Global Digital Elevation Model (GDEM) (30 m) and Cartosat DEM (30 m), Indian Remote Sensing Satellite (IRS) P-6 LISS-III (23 m) and Advanced Wide Field Sensor (AWiFS) (56 m) and LANDSAT (30 m) data of India. The web map applications have been created using the ArcGIS Viewer and Flex plug-in by displaying the layers on the web map. Besides the existing soils, soil loss and climatic data, soil profile point database have also been developed. The developed WMS architecture allows the user to add new map service to the existing service, produce a map for area of interest and answer basic queries about the content of the map. The WCS provides the facility to retrieve data and find out the data relevant to the application needs. The developed SIS and Prototype Geoportal pave the way for development of next generation operational Geoportal to enforce data consistency, standards, sharable protocols and build cross-domain knowledge based applications for effective utilization and management of soil and allied resource databases to develop location specific agricultural land use planning in order to enhance the crop productivity in the country.

16.1 INTRODUCTION

In the context of growing population pressure, increasing demand of land for non-agricultural uses, declining per capita land availability, land degradation, climate change and food security, the use of geospatial technologies in development of reliable database on land resources for their optimum utilization assumes greater significance. Further, climate change is posing the greatest threat to agriculture production and food security in the 21st Century (Pedercini, 2012), particularly, in agriculture-based countries like India with a possible impact on natural resource management and crop yields. Sustainable management of land resources depends on their judicious use by adopting appropriate technology under prevailing socio-economic conditions. In order to utilize the land resources for sustained production, it is prerequisite that comprehensive information on land resources, more particularly, on soils is essential to provide systematic and location-specific information to the planners, decision makers and developmental agencies. In this direction, development of GIS based location-specific database plays a vital role in diverse planning, research and development activities of the present and future use of soil resources in agricultural land use planning in developing countries like India. Therefore, the knowledge of soils, in respect of their extent, distribution, characterization and use potential, is extremely important for optimized land resources utilization.

Across the globe, many national survey and mapping agencies produce several spatial databases and maintain at different scales to develop various spatial thematic maps (Kazemi, 2003; Lee, 2003). Soils provide the basis for food production and are increasingly under pressure by competing land uses for cropping, forestry, and pasture/rangeland but also for urbanization and infrastructure, mining, etc. Maintaining multiple databases from small scale to large scale is resource-intensive, time consuming and cumbersome (Arnold and Wight, 2003). In order to serve multiple-purpose and multiple-scale applications, management of an automated soil resource database on GIS platform is essential and it plays a key role in providing solutions to many location specific problems. Detailed knowledge of soil characteristics with respect to their extent, distribution, characterization, potentials and limitations are important for their management, conservation and soil resources based agricultural land use planning. The main drawback of the soil analog maps is difficult to integrate with most of the other forms of natural resource data that are based either on a polygon or a grid. Hence, in the last three decades there has been increasing demand to standardize and organize the soil resource information in different countries in the form of SIS and Geoportal so as to use it more effectively and efficiently in spatially explicit agricultural planning activities. The capabilities of these systems range from mere storage and retrieval of data to integrated dynamic modeling using GIS technology for evaluating current and future policy requirements at local, regional and national scale.

In recent times, web-based GIS tools are increasingly used for basic mapping, data visualization, complex web mapping and data analysis. In spite of this apparent migration of GIS to the web-platform, desktop or client-side GIS tools are likely to continue to be needed for development of variety of use cases in Geoportal. The word portal stems from the Latin word 'porta' and indicates an entrance point (Annoni et al., 2004). Geoportal may be defined as an Internet or intranet entry point with the tools for retrieving metadata, searching, visualizing, downloading and disseminating of geoinformation and providing the geoinformation based services (Fisher, 2006; Maguire and Longley, 2005; Tait, 2005). The aim of Geoportal is to serve as a one-stop shop providing access to geoinformation content and services so that they can be easily shared and reused for various applications. In order to provide this facility effectively and efficiently, an ideal Geoportal should have access to the majority of geoinformation datasets/services.

Design and development of prototype soil Geoportal at national level assumes a paramount importance as national level soils data infrastructure aid the agricultural land use planning process. Geoportal should have the standardization in data themes as per the national and international standards, have standardized metadata information, ability to search data with selecting area from map or record from table, interactive data view within the web map viewer and data download ability. Though the advanced technology and standards are generally in place for the creation of operational Geoportals to manage, support and develop services, the full potential of large scale development and implementation of effective national level Geoportals in India especially on soils is not fully explored. In recent times, the demand for dedicated thematic Geoportals, which provide solutions and solve problems of specific thematic areas are increasing. Hence, development and maintenance of effective SIS and Geoportal is essential to store, manage, analyze and disseminate information on soil and allied resources to generate integrated land use plans and meet the demand of food, fiber, fodder and other products of the growing population. This is one of the guiding principles to develop SIS and Prototype Geoportal for soil resource database management and provide applications for sustainable agricultural land use planning in India.

16.1.1 SOIL INFORMATION SYSTEMS: GLOBAL PERSPECTIVE

The advanced geospatial technologies play a key role, not only for storing and management of spatial data but also perform range of spatial analysis tasks like data integration and transformation of basic soil data into a variety of quantitative soil information (McBratney et al., 2002), devising new methods for different users to search, query and access a range of suitably formatted soil information. Globally, some of the countries have developed their own SIS. Beaudette and O'Geen (2009) developed a framework to construct a web-based interface to the Natural Resources Conservation Service (NRCS) State Soil Geographic (STATSGO) Data and Soil Survey Geographic (SSURGO) Database to support sophisticated data storage, querying, map composition, data presentation, and contextual links to related materials. Australian Soil Resource Information System (ASRIS) provides access to the best available soil and land resource information in a consistent format across the country, however, the level of details depends on the survey coverage in each region. ASRIS contains a set of spatial and temporal databases that maintain national soil and land information in a consistent and usable format (ASRIS, 2011). The Canadian Soil Information Service (CanSIS) (http://sis.agr.gc.ca/cansis/) manages and provides access to soil and land resource information of Canada. It maintains the national repository of soil information such as soil data, maps, technical reports, and standards and procedures through its National Soil Database (NSDB). Soil and Terrain Database (SOTER) program was launched by FAO, ISRIC–World Soil Information and the United Nations Environmental Program (UNEP), to create a soil and terrain digital database (Baumgardner, 1986) with a global coverage of SOTER attribute data at a scale of 1:1 million. The SOTER concept is based on the relationship between the physiography, parent materials and soils within an area. The SOTER methodology was developed as a land resources information system for the scale of 1:1 m (Van Engelen and Wen 1995). The SOTER methodology has been applied at a range of scales, from 1:50,000 to 1:5 m, using a similar standard database structure (ISRIC et al., 2003; ISSS et al., 1998).

Within the framework of the Global Earth Observation System of Systems (GEOSS), the e-SOTER (http://www.esoter.net) was developed by European countries to deliver a web-based regional pilot platform with data, methodology, and applications, using remote sensing to validate, augment and extend existing data. Detailed digital elevation models (DEMs), advances in remote sensing, and new analytical tools were extensively used in *e-SOTER* for landform analysis, parent material detection and soil pattern recognition-both to extend the legacy soil data and to build a framework for new data acquisition. ISRIC World Inventory of Soil Emission Potential (WISE) is a comprehensive repository of global primary data on soil profiles. Under the WISE, ISRIC has consolidated select attribute data for over 10,250 soil profiles, with about 47,800 horizons, from 149 countries in the world. Profiles were selected from data holdings provided by the Natural Resources Conservation Service (USDA-NRCS), the Food and Agriculture Organization (FAO-SDB), and ISRIC (ISRIC-ISIS) (Batjes, 2006; 2008). WISE data have been used for a wide range of applications, which includes the development of harmonized sets of derived soil properties of the main soil types of the world, gap-filling in the primary SOTER database, global modeling of environmental change, analysis of global ecosystems, up-scaling and down-scaling of green house gases, crop simulation and agro-ecological zoning (ISRIC-WISE, 2008). The initiative of Digital Soil Mapping (DSM) (www.globalsoilmap.net) by Working Group of the International Union of Soil Science (IUSS) with an aim to assist decisions in a range of global issues like food production, climate change, and land degradation. Many projects have been initiated aiming towards the compilation of new digital soil maps of the world. Under this initiative, a number of

Regional and National soil data collection programs are on, like the Africa Soil Information Service (AFSIS), the European Soil Information System (EUSIS) and others. In recent times, the task of the Global Soil Partnership (GSP) was initiated to build a partnership among the various soil data collection programs in order to develop synergies, cost savings by avoiding duplication of efforts and provide a common soil data and information platform responding to the various user needs at global, regional, national and local scales.

16.1.2 SOIL INFORMATION SYSTEMS: INDIAN PERSPECTIVE

Inventory and characterization of soil resources is a requisite for evaluation, proper utilization and sustainable management of land resources in the country. Realizing the importance of soils and its role in national agricultural planning, in 1980's, National Bureau of Soil Survey and Land Use Planning (NBSS & LUP) the Indian premier organization in soil resource inventory under the Indian Council of Agricultural Research (ICAR), undertook the gigantic task of soil resource inventory in the country on 1:250,000 scale with an aim to generate a country wide soil resource database and provide soil information for viable agricultural land use planning in the country. The three tier approach *viz.,* image analysis, soil survey, laboratory analysis and mapping and cartography as outlined by Sehgal et al. (1987) was followed to generate state-wise spatial database of soil resources and soil based thematic maps. This available state wise soil resource data has been used to develop the seamless soil resource database of India. Of late, GIS has been increasingly used to generate digital soil resource database and development of various derivative thematic databases for various applications. Technological development in the field of mapping and development of spatial database in GIS is very fast, following the trend from manual cartography to advanced web based applications. Realizing the importance of digital soil resource databases in GIS, NBSS & LUP has developed SIS on 1:1 m scale in GIS under Integrated National Agricultural Resource Information System (INARIS) of National Agricultural Technology Project (NATP). The soil attributes database on soil mapping units, soil site, physical and chemical properties was compiled, codified and extended legend was prepared as per the INARIS data structure (Maji et al., 2004). Attempts have been made to develop state level SIS for various states like Arunchal Pradesh (Maji et al., 2001), Nagaland (Maji et al., 2000), and Tripura (Bhattacharyya et al., 2010). In order to meet the demands of wider applications, digital soil resource databases of India in the form of SIS has been designed and developed as a GIS based spatial database management system with a set of files *viz.*, spatial and non-spatial to store and retrieve for systematic analysis and reporting (Reddy et al., 2014, 2016). GIS based soil resource database at district level for the selected districts of the country have also been developed under SIS for district level agriculture planning (Maji et al., 1998; Reddy et al., 2014a).

16.1.3 IMPORTANCE OF SOIL INFORMATION SYSTEM AND GEOPORTAL

Agriculture sector is the mainstay of the Indian economy, contributing about 15% of national Gross Domestic Product (GDP) and more importantly, about half of India's population is significantly dependent on agriculture and allied activities for their livelihood (GOI, 2011). The maintenance of spatial databases in GIS on the valuable agricultural resources is the primary necessity of the present digital era to develop sustainable agricultural land use plans in order to achieve higher crop productivity. Advances in geospatial and information technology, specifically the use of GIS based SIS and Geoportal greatly assist in accessing the information on soil resources by the planners, decision makers and land users for land evaluation and development of suitable land use plans at different levels like national, regional, district and block. GIS technology provides a powerful tool to present and analyze soil information (King et al., 1995) and integrates common database operations, such as query and statistical analysis with unique visualization offered through maps. Geoportals act as a gateway to collection of information resources, including data sets, services, cookbooks, tutorials, tools and an organized collection of links to many other sites usually through catalogs. Thus, Geoportal is a web environment that allows an organization or a community of information users and providers to aggregate and share geographic content. Further, Geoportal helps to maintain one master database in order to reduce data handling and data duplication, as well as to describe how desired products and databases should be dynamically developed "on-the-fly" from the single database using an automated generalization procedure. Realizing the importance of the availability of the accurate and reliable information on soil resources, ICAR-NBSS & LUP took an initiative to develop SIS and develop prototype Geoportal to support the national efforts of digital India to promote sustainable agricultural development and enhance food security in the country. In this chapter an attempt has been made to demonstrate the capabilities of geospatial technologies in processing the soil geospatial database, development of protocols and procedures, design and development of SIS and prototype Geoportal and its applications. Finally, this chapter will also outline a way forward for operational Geoportal and the challenges to be addressed in providing the user-friendly Geoportal services.

16.2 DEVELOPMENT OF SOIL INFORMATION SYSTEM

16.2.1 DATA SOURCES AND ASSIMILATION OF SPATIAL DATABASE

The spatial soil resource database generated under the mapping and characterization of the soil resources of the country under Soil Resource Mapping (SRM) project carried out by NBSS & LUP at 1:250,000 scale has been used to generate state-wise spatial database on soil resources. The soil maps of different states generated at toposheet level (hard copies) under the above mentioned project have been used as input data in GIS for digitization and generation of state-wise soil information. State-wise soil sheets were scanned, georeferenced and mosaic has been generated in GIS. The georeferenced soil sheets were used for onscreen digitization in GIS to digitize soil polygon features and associated layers. At the state level, the digitized soil polygon layers were cleaned for removal of geometrical errors and polygon topology was built to assign unique mapping units Ids in polygon layer. In order to develop seamless digital soil resource resources of India, the state-wise soil resource polygon layers were brought on common platform in GIS, the necessary geometric errors were corrected during the edge matching and transformed to Lambert Conformal Conic (LCC) projection system with geographic extents in latitudes/longitudes. The digitization of various associated layers like infrastructure, rivers and settlements were also carried out. The topology was built for the seamless polygon layer of soils and other administrative unit maps. This available state wise soil resource data have been used to develop the seamless soil resource database of India. Besides the soil resources data on 1:250,000 scale, soil resources data on 1:1 m scale, soil loss, climatic and soil profile point database, the raster database on digital elevation models of SRTM DEM (90 m), ASTER DEM (30 m) and Cartosat DEM (30 m), satellite image database on IRS P-6 LISS-III (23M) and Advanced Wide Field Sensor (AWiFS) (56 m) and LANDSAT (30 m) data of India were also used to develop SIS. GIS based SIS was designed to assimilate geo-referenced spatial and non-spatial database (Burrough and McDonnell, 1998). In other words, GIS worked as a database system with specific capabilities for developing spatially referenced data, as well as, perform a set of operations for working with the datasets. Assimilation process spatial and non-spatial data for SIS and Prototype Geoportal is shown in Figure 16.1.

16.2.2 STANDARDIZATION OF ATTRIBUTE DATABASE

The standardization of spatial data relates to various aspects of data preparation and management involving data formats, data dictionaries, data quality control, map layouts, and many others. Although a comprehensive set of data standards are being followed in the generation soil and allied databases, number of standards and norms needs to be adopted for enhancing

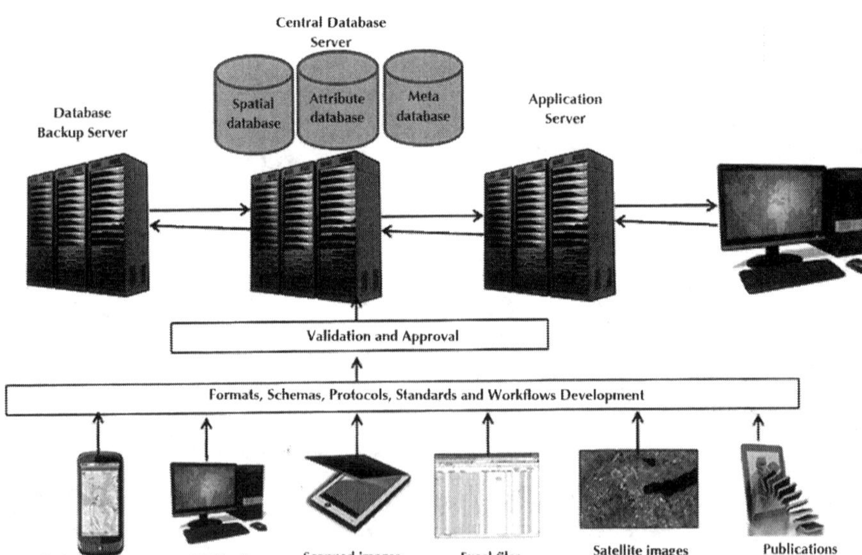

FIGURE 16.1 Assimilation of spatial and non-spatial database for Soil information system and Geoportal.

capabilities of data exchange and compatibility geographic layers and attribute databases, which are being used in Geoportal. The uniform standards have been followed in generation of soil physico-chemical properties in the dBase IV environment and linked to the soil polygon layer in GIS. Attributions of various soil parameters at the state level were standardized in a seamless soil polygon layer of India for entry of various thematic databases to master soil layer. Prior to development of SIS, the inventory and available data formats were reviewed in order to understand the level of effort that is needed in developing SIS and viable prototype Geoportal. The standardized soil resource database contains information about the soil site characteristics *viz.,* slope, landform, land use, climate, flooding, erosion, parent material, depth and morphological properties comprising of particle size, color, structure, texture, stoniness, internal drainage and chemical properties like pH, Cation Exchange Capacity (CEC), Organic Carbon (OC), mineralogy, salinity, sodicity and calcareousness. The standard codes have been generated for all the soil thematic parameters.

16.2.3 DEVELOPMENT OF DATABASE SCHEMA

Database schema refers to the diagram and document that layout the structure of the database and the relationships that exist between elements of the database. A schema is like a blueprint for a database that explains a knowledgeable builder exactly how to construct it. A schema at its simplest consists of an arrangement of tables and the relationships between them. Because organizations differ so widely in the kind of work they do and the types of data they need to handle, it is impossible to provide a cookbook schema for every application. The critical part of a database schema is actually to create a diagram that shows the relationships among the various tables in the database. Relationships have a property called cardinality that describes the type of relationship. The possibilities for relationships are one to one, one to many, many to one and many to many. Additionally, relationships may have the property of being required (mandatory) or optional. The schema has been developed for input parameters to define the data elements, sub elements, short names, data type, unit of measure, maximum and minimum value of each parameter and brief description to effectively store, update, manage and query the geospatial soil database from SIS. The standardized schema allows to store, analyze, visualize and query the soil resource and allied datasets across the administrative units and geographic regions in SIS.

16.2.4 DEVELOPMENT OF DATABASE MANAGEMENT SYSTEM

Data Management Systems are being used to manage large amount of geo-spatial data and it is necessary to use a Data Base Management System (DBMS) for the same. A spatial database offers spatial types (geometry types such as point, line and polygon), spatial functions and allows for the creation of spatial indexes to improve the performance of the spatial queries and functions. A DBMS provides many advantages to manage and work with large amount of data, such as the support for multi-users (concurrency), data consistency and integrity, persistent storage, and query capabilities through a programming interface (Garcia-Molina et al., 2002).

16.2.5 DEVELOPMENT OF METADATA STANDARDS

Metadata literally means "data about data." Meta-information communicates the structure, syntax, and semantic content of information to be processed (or processing instructions). Metadata is the term used to describe the summary information or characteristics of a set of data. Standardization and development of metadata standards are particularly beneficial for data management and integration. The challenges associated with maintaining quality soil data in digital format require a clear understanding of data content standards and effective coordination and management of the organization's data requirements. In SIS, common standards are adopted for soil data, which allows to integrate soil information with allied spatial information systems for wider applications in agricultural and land use planning. Metadata are structured, encoded data that describe characteristics of information-bearing entities to aid in the identification, discovery, assessment, and management of the described entities. It explains the database structures of SIS and its related databases. This includes the structure of each table, data element definitions, lists of codes, and relationship information. The developed metadata schema with minimum fields for SIS provides a formal structure to identify the soil resource databases of a given state and that information assist to identify, discover and use of detail soil information of that state.

16.2.6 DEVELOPMENT OF SOIL INFORMATION SYSTEM IN GIS

Consistent and compatible information on soils arranged in logical order is important to analyze and interpret for objective specific queries. Information systems are the software and hardware systems that support data-intensive applications. GIS is used to integrate, store, edit, analyze, and display geo-referenced information. The usefulness of an information system will depend on its ability to provide decision makers with the right data at the right time. In order to meet the demands of wider applications, digital soil resource databases of India in the form of SIS has been designed and developed as a GIS based spatial database management system with a set of files *viz.,* spatial and non-spatial to store and retrieve for systematic analysis and reporting. Advanced GIS techniques have been used to develop the seamless SIS to integrate with other resource databases. GIS is a tool to efficiently capture, store, update, manipulate, analyze and display all forms of geographically referenced data. It allows mapping, modeling, querying, analyzing and displaying large quantities of diverse data, all held together within a single database. The geocoded spatial data defines an object that has an orientation and relationship with other objects in two or three-dimensional space, also known as topological data and stores in topological database. On the other hand, attribute data stored in a relational database describes the objects in detail. GIS links these two databases by manipulating a one-to-one relationship between records of object location in the topological database and records of the object attribute in relational databases by using end-user defined common identification index or code. The true value of GIS can only be realized if the proper tools used to collect spatial data and integrate them with attribute data to analyze the processes or phenomena. GIS is also a potential tool for handling voluminous remotely sensed data and has the capability to support spatial statistical analysis (Balaselvakumar and Saravanan, 2006; Singh et al., 2003) using the information generated in SIS. Thus, SIS developed on GIS platform provides the ability to search, discover, visualize and update geospatial soil resource database through a customized user interface.

16.3 DEVELOPMENT OF PROTOTYPE SOIL GEOPORTAL

During the past few years, many definitions have been created to express the meaning of Geoportal (Tait, 2005; Fisher, 2006). Geoportal is defined as a website considered as an entry point to geographic content on the web or, more simply, a website where geographic content can be discovered (Tait, 2005). Geoportals organize content and services, such as directories, search tools, support resources, data and applications. They provide capabilities to query metadata records for relevant data and services, and then link directly to the on-line content services themselves. A geographic portal is a website where the discovery of geographic content is a primary focus. More specifically, a Geoportal may be defined as an internet or intranet entry point with the tools for retrieving metadata, searching, visualizing, downloading, disseminating and in some cases the ordering of geographic information services. Data and data services were cataloged systematically according to a metadata standards and schema designated for Geoportal. This data cataloging and maintenance reflect the type of data that is to be published using the Geoportal. Since a Geoportal is really about data, the data inventorying and maintenance element of Geoportal is important aspect. Though data can be maintained and associated metadata can be created and published on a Geoportal by entities other than the portal's host organization, the metadata could be generated prior to the publication to ensure its completeness and conformity to established standards and schemas.

In the study, web-mapping application was created using the ArcGIS Viewer for Flex plugin. Initially, the spatial layers to be displayed on the web map were created and saved as ".mxd" format in ArcMap and then shared as a map package. The map package was accessed through Geoportal and published to the server hosting the web mapping services. Various user-friendly utility tools, like data query, extract and legend development were developed in the prototype Geoportal. A chart widget, directions widget, pop ups and the locator for locating addresses were also added. After the necessary details added, the Geoportal was configured on the local host. The data workflow in the prototype Geoportal is shown in Figure 16.2. Web services are independent, self-describing, modular applications that can be published, located and dynamically invoked (Weerawarana et al., 2005). They are entirely based on Extensible Markup Language (XML) technologies, and are discoverable based on their descriptions and terms and conditions available based on the web services metadata. The user requires little to no knowledge about the implementation of the web service but has to be familiar with the interface of the service. The web service architecture allows a new service to be created from existing ones; this is a very valuable aspect of web services. WMS is able to produce a map of a specific selection of data,

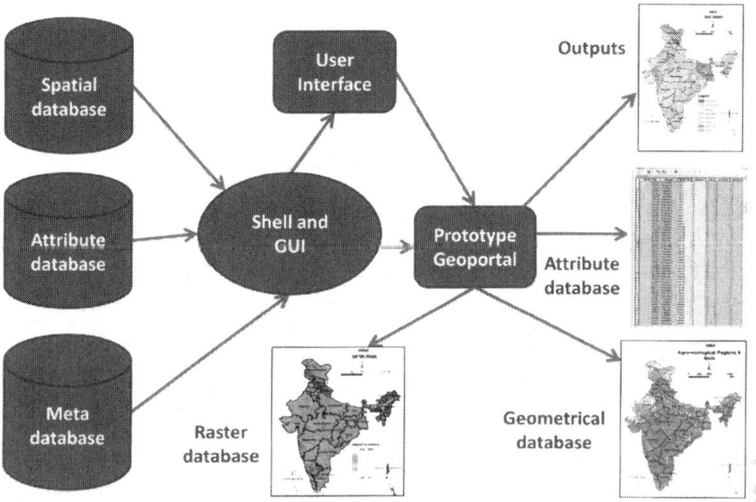

FIGURE 16.2 Conceptual design and data workflow in the prototype Geoportal.

answer basic queries about the content of the map, and provide clients with a list of other maps it can produce. The main obstacle encountered in assimilation of spatial datasets is the diversity of their formats. These problems have been overcome by abstracting all possible data formats into a single generic model. Then, any data can be combined, visualized, overlaid and analyzed together. The data catalog is able to retrieve the data of which the metadata match the search criteria. The catalog allows the user to find data relevant to application needs.

16.3.1 PUBLISHING OF DATA CONTENT

The publishing process entails addition, deletion, and modification of metadata content. Depending on the sophistication of the site, publishing can be manual, through a web page interface, or automated through a web service interface (metadata harvesting). The administration function is simply an extension of the publishing function with one additional capability: the review/approval of metadata content submitted for publishing on the portal web site. Geoportal requirements have been controlled through administrative privileges granted to certain administrators who can edit and validate the content to publish. Additionally, administrators are responsible for publishing site-level versus content-level metadata.

16.3.2 GEOGRAPHIC DATA SEARCH

Search functions are aggregations of building block tools, which are executed in sequential steps. The first step in many applications is to locate a place through one of several methods including a place name search using a gazetteer tool, an address search using a geocode tool, or simply selecting a location from a list. For example, using a gazetteer tool, user can enter a place name, execute a search for that place, and return a list of candidate locations, allowing a selection to be made. Once a place is identified, most Geoportal applications then execute a second step in the search process; they search for a particular set of features or objects that are usually the focus of the Geoportal. This search could be for particular location based on a neighborhood name or for geographic web services with coverage of a particular location. In many cases searches allow both geographic and attribute criteria for searching. The geographic search can use geographic content directly or content metadata.

16.3.3 GEOGRAPHIC DATA VISUALIZATION

In Geoportal, a map allows the user to fully explore and examine the published content. Additional functions like pan, zoom, and feature identify capabilities aid the user in more thoroughly evaluating the published content. Geoportal also support the ability to view multiple map services in a fused or single map image. Visualization interfaces in Geoportals can be a very powerful tool for spatial data, which can aid users with information visualization and data exploration. Using ArcGIS server, the protype Geoportal with WMS, WCS, interface tools and basic Geoportal functionalities have been developed, which provides the set up to design, develop and operate a Geoportal. Various thematic map services have been deployed as individual services with uniform projection standards. A thematic map is an example of information visualization; a specific type of map that is designed to communicate information about a single topic or theme (Tyner, 2010; Wade & Sommer, 2006). Thematic maps allow users to analyze data in several ways, through identifying pattern and classification of data, for example. A thematic web service allows users to specify a link to the data source, enabling users to produce thematic maps. The Catalogue Service component provides the underlying database and data management capabilities that enable users to post and discover metadata records. Geoportal server helps the user to build and administer a Geoportal website where data producers can register their geospatial resources for users to discover and consume. It also helps the user to maintain the quality and availability of registered resources by providing tools to evaluate new entries, control access to metadata and integrate the Geoportal with other enterprise systems. Prototype Geoportal provides an easy-to-use interface for users to discover and access resources registered with the Geoportal to use in their own applications. Geoportal provides distributed GIS applications with capabilities for searching, mapping, publishing and administrating geographic information.

16.3.4 DEVELOPMENT OF PROTOTYPE GEOPORTAL SERVICES

Web frameworks support the development of dynamic web applications by providing libraries, packages and templates into a single connected piece of software and, thus, alleviating the tasks of dealing with low-level details. The developed Geoportal framework allows easily to combine the visualization of geodata and to publish them without having to deal with every single detail regarding projection and interactivity functions. Geoportal also offer basic Graphical User Interface (GUI) objects that can be extended and modified with additional functions and/or plugins. They usually offer an Application Programming Interface (API) to allow an easier use and access to the libraries. The services and standards of Geoportal are important because they allow devices to communicate through services and using a common protocol set by standards, and therefore enabling the core aspects of the interoperability concept between devices. Web services of Geoportal are applications, which offer functionality and which are accessible by other applications over the Web and through Internet-protocol. They support direct and remote interactions from machine to machine, using standards-based interface (Alonso et al., 2004). Thus, they allow delivering remotely available functionality from a service provider to a service consumer. Web services use the concept of request-response mechanism: a client or service consumer sends a request to the server and in return the server sends a response, containing the results asked for, to the client. In the field of GIS, web services are used for a wide range of tasks including the display and retrieval of data, processing operations, and search functions. The concept of web framework and of the available options for web mapping and Geoportal allow presenting the geospatial data to the users in a simple and coherent manner. Some of the common services of prototype Geoportal are given below.

- ***WMS (Web Map Services):*** An implementation of the WMS standard provides the client with spatially referenced 2D maps in the form of an image, such as JPG and PNG, or SVG dynamically from geographic information that can be displayed in a browser. Figure 16.3 depicts the agro-ecological regions and agro-ecological sub region maps as WMS through prototype Geoportal.
- ***WCS (Web Catalogue Service):*** Web Catalogue Service supports the ability to publish and search collections of descriptive information (Metadata) for data, services and related information objects. Metadata in catalogs represent resources characteristics that can query and presented for evaluation and further processing by both humans and software.
- ***WFS (Web Feature Service):*** The WFS standard describes a standard that should allow the sharing of geodata by allowing a client to retrieve only the geodata they are seeking for and not a whole file. In comparison with a WMS, where the client receives the visualization for the data, this service delivers the actual data (Open Geospatial Consortium 2010).

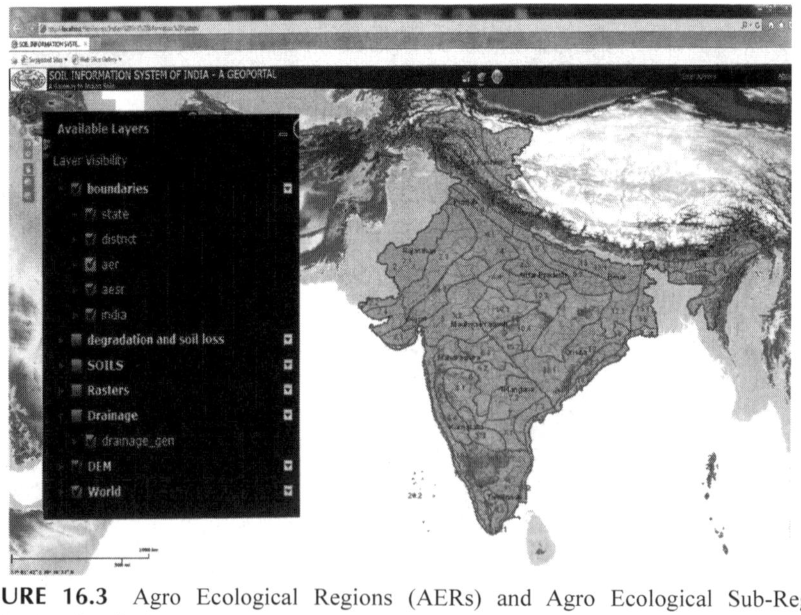

FIGURE 16.3 Agro Ecological Regions (AERs) and Agro Ecological Sub-Regions (AESR's) of India in prototype Geoportal.

16.4 APPLICATIONS OF SOIL INFORMATION SYSTEM AND PROTOTYPE GEOPORTAL

With the availability of digital soil resources and allied databases in SIS and prototype Geoportal within the GIS framework, the applications of soil survey data enhanced considerably in providing the solutions in the fields of thematic mapping and analysis, digital terrain analysis, soil–landscape modeling, assessment and monitoring of degraded lands, watershed hydrology, planning and management, spatial modeling for crop suitability evaluation, development of spatial decision support systems and input at various levels in policy formulation. Some of the important applications of SIS and Prototype Geoportal in land resources management are discussed in the following subsections.

16.4.1 THEMATIC MAPPING, ANALYSIS AND REPORTS

Thematic mapping, analysis and reports generation play an important role to understand particular soil based theme in spatial perspective and it helps to interpret theme for various applications. The standardized database in SIS offers multiple map generations in a continually varying range of scales. Potential applications of GIS based SIS and Geoportal include constructing thematic maps showing the spatial variability of particular parameter related to soil and allied theme. In combination of other GIS layers, it is also possible to obtain new layers, which are the result of the intersection or difference of thematic layers connected to the corresponding databases. By applying reclassification techniques on SIS, the soil based thematic maps like landforms, soil depth, texture, erosion, soil reaction (pH), etc., at various scales could be generated and reclassified. Spatial modeling could be followed to simultaneously analyze both the spatial and thematic characteristics of soil information. The thematic component of soil information could be analyzed through various spatial analysis techniques, such as topological map overlay, buffer generation, feature extraction, feature merging, and relational database operations like join and relate.

16.4.2 DIGITAL TERRAIN MODELING

Digital terrain modeling has many applications in hydrological and geomorphological studies. The availability of satellite based new topographic digital datasets has opened new avenues for hydrologic and geomorphologic studies including analysis of surface morphology (Frankel and Dolan, 2007). Seamless mosaic of SRTM DEM (90 m) and ASTER DEM (30 m) of India developed under the framework of prototype Geoportal enable to analyze and characterize various geomorphometric parameters namely slope, aspect, hill shade, plane curvature, profile curvature, total curvature, flow direction, flow accumulation and topographic wetness index at state and agro-ecological regions of India (Reddy et al., 2012). Delineation of terrain parameters, such as drainage network and watershed boundaries from collateral data and remotely sensed data crucially depends on generation of an accurate DEM. DEMs developed in Geoportal could be effectively used in terrain analysis and it involves both interactive and interpretative methods, requiring repeated visualization of the resulting classified maps and refinement of classification parameters. DEM based landform analysis with respect to their nature, extent, spatial distribution, potential and limitations is very useful for evaluation and optimal utilization of land resources (Reddy et al., 2013). High resolution satellite data provides reliable source of information to prepare comprehensive and detail inventory of landforms (Reddy and Maji, 2003) and land degradation mapping (Reddy et al., 2004). Digital datasets on high resolution DEMs, slope, elevation range, contours and stream network pattern available in Geoportal could be effectively used in quantitative analysis of landforms, their mapping and classification. SRTM (90 m) Digital Terrain Model (DTM) and automatically extracted drainage pattern of northeastern region of India in prototype Geoportal are shown in Figure 16.4.

16.4.3 SOIL–LANDSCAPE ANALYSIS AND MODELING

Soil and terrain information play important role in land resource management, land use planning and monitoring the environmental impacts of development. Technological advances during the last few decades have created a tremendous potential for improvement in the way that soil maps are produced (McKensie et al., 2000). Remote sensing and photogrammetric techniques provide spatially explicit, digital data representations of the Earth's surface that can be combined with digital datasets available in Geoportal to allow efficient characterization and analysis of vast amounts of data. The integration of remote sensing data and legacy soil data in SIS offers possibilities to segment the landscape into internally more homogeneous soil–landscape

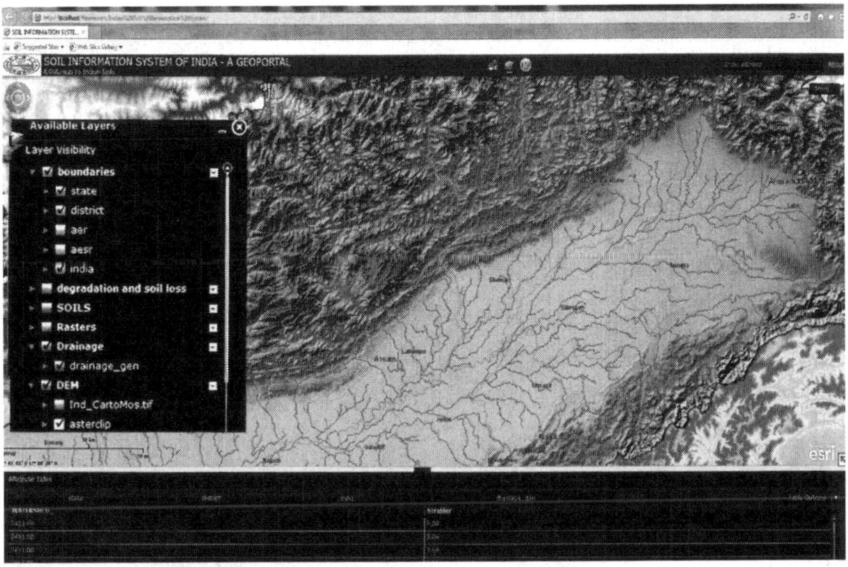

FIGURE 16.4 SRTM (90 m) Digital Terrain Model and drainage pattern in northeastern region of India in prototype Geoportal.

units for which soil composition can be assessed by sampling using classical or more advanced methods. Soil–landscape and modeling can be divided into two parts. Firstly, a detailed landform map, derived from topographical sheets and satellite data analysis through visual interpretation, can be prepared to form a base map in soil survey, which shows spatial variation in terrain features and their characteristics. Secondly, based on the interpretation, the potentials and limitations of the soils can be obtained and such information can be used to construct geospatial database in Geoportal. The future soil–landscape and modeling lies in using GIS to model spatial soil variation from more easily mapped environmental variables. In predictive models, numerical or statistical model of the relationship among environmental variables and soil properties, could be effectively applied to a geographic database to create a predictive soil map (Franklin, 1995). SIS of India developed on 1:25,000 scale in prototype Geoportal is shown in Figure 16.5.

16.4.4 ASSESSMENT AND MONITORING OF LAND DEGRADATION

Land degradation in general, implies temporary or permanent recession from a higher to a lower status of productivity through deterioration of physical,

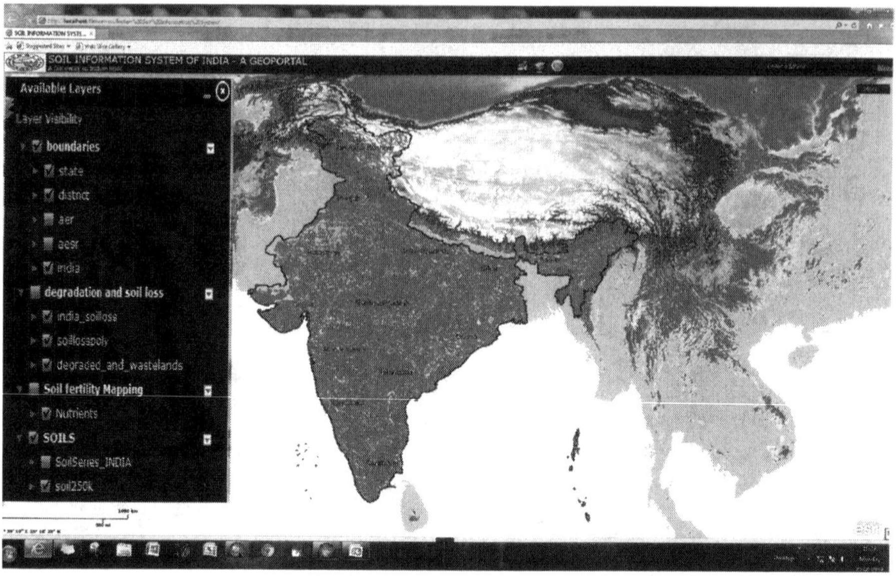

FIGURE 16.5 Soil Information System of India on 1:25,000 scale available in prototype Geoportal.

chemical and biological aspects. It is a complex ensemble of surface processes (e.g., wind erosion, water erosion, soil compaction, Salinization, and soil water logging). In India alone, out of a 329 m ha geographical area, 120.72 m ha of land are affected by various kinds of land degradation (ICAR and NAAS, 2010). Soil erosion by water is a major concern in landscape management and conservation planning. Among the new technologies, geospatial technologies are effective for detecting, assessing, mapping, and monitoring the land degradation. The potential and actual soil loss at state level has been estimated through integrated analysis of Universal Soil Loss Equation (USLE) factors in GIS (Reddy et al., 2004a, 2013; Sidhu et al., 2013). Temporal ssatellite data have become valuable tools in studying the spatial extent of degraded lands and for monitoring the changes that have taken place over a period of time due to reclamation/conservation measures. In particular, remotely sensed satellite data can effectively reveal the spatial extent, magnitude and temporal behavior of lands affected by waterlogging and subsequent salinization/alkalinization (Sujatha et al., 2000). Gao, (2008) reported that degraded lands can be mapped more than 70% accurately from

ASTER data using both per-pixel and object oriented image classification methods. Digital remote sensing data in conjunction with the soil resource database available in SIS and prototype Geoportal have immense potential to detect, map and monitor degradation problems with spatial and spectral resolution and for the detection of degraded areas including their spread effects with time (Sabins, 1987; and Sujatha et al., 2000). Gao and Liu, (2008) and Lu et al. (2007) reported that temporal remotely sensed imageries with necessary ground truth data are good at revealing the land that has been affected by degradation to various degrees and effective in identifying and mapping land degradation risks. Geoportal proved to be an effective mechanism in handling spatial data on various kinds of degraded lands in conjunction with other resources database at different scales. Spatial data query of harmonized land degradation database in prototype Geoportal is shown in Figure 16.6.

16.4.5 WATERSHED HYDROLOGY, PLANNING AND MANAGEMENT

Watershed is a geo-hydrological unit area that drains to a common point and is considered as an appropriate physical unit for land resources evaluation,

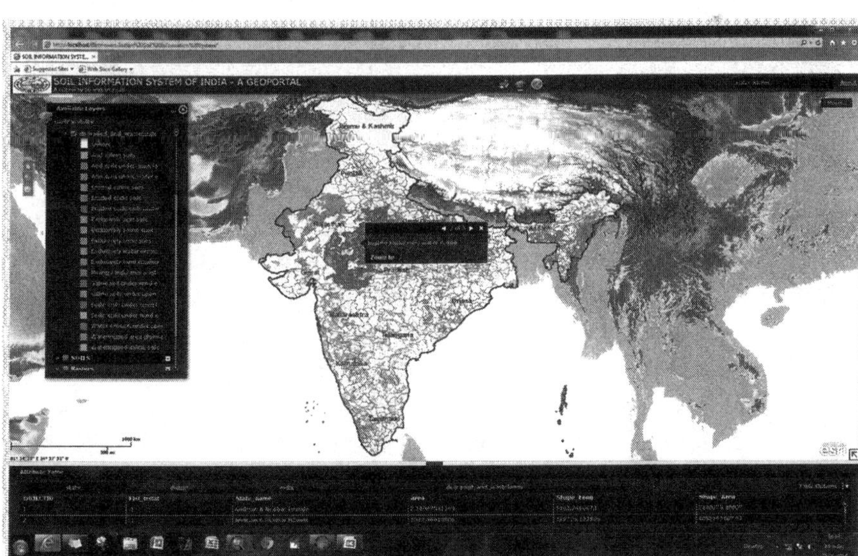

FIGURE 16.6 Spatial Data Query of harmonized land degradation database in prototype Geoportal.

planning and management. Watershed management implies the rational utilization of land and water resources for optimum production with minimum hazard to natural environment (Bhat, 1989). The remote sensing and GIS tools have opened new paths in large-scale hydrological studies. The DEMs developed in SIS and Geoportal have immense potential for generation of detailed drainage pattern, 3-D perspective views, terrain analysis and extraction of other terrain variables, which can be used as an input in land resource inventory, mapping and management. Assessment of surface and sub-surface hydrological parameters through analysis of remotely sensed data in conjunction with soils data and field surveys provide a mechanism to identify hydro-geomorphological units (Krishnamurthy and Srinivas, 1995; Reddy et al., 1999). Digital terrain datasets available in Geoportal enable us to disseminate the geographic data on watersheds more efficiently for hydrological monitoring, analyzing, planning and management. GIS technology has been integrated with these surface or subsurface hydrological models emphasizing the utility and significance of topographic attributes of the terrain for various hydrological applications. The reason of adopting GIS technology is because it allows the spatial information to be displaced in integrative ways that are readily comprehensible and visual. The use of a GIS in combination with hydrological models allows to perform multi-objective analyzes incorporating a wide range of geographical information and data in an accurate and efficient manner (McDonnell, 1996). Despite the advent of recent integrated approaches to watershed planning and development, as well as the large research effort to enhance the interface between GIS technology and hydrological models, a need remains to make the GIS-based hydrological models more reliable, effective and easy to implement for research and engineering purposes. In addition, a GIS methodology for predicting runoff and flood risks spatially within the corresponding watershed needs to be developed. These applications of GIS for flood prediction and watershed management would greatly improve the capabilities of the hydrological modeling and risk assessment, offering users a powerful capability to analyze the spatially distributed hydrological variables. Most of these topographic attributes can be calculated from the directional derivatives of a topographic surface. These attributes may affect soil characteristics, distribution and abundance of soil water, susceptibility of landscapes to erosion by water, and the distribution and abundance of flora and fauna. These topographic indices are being used frequently in many hydrological models.

16.4.6 SPATIAL MODELING FOR LAND SUITABILITY EVALUATION

Land suitability evaluation is a prerequisite to achieve optimum utilization of available land for agricultural production in a sustainable manner. In order to practice it one has to grow the crops where they suit best and for which first and the foremost requirement is to carry out land suitability analysis (Ahamed et al., 2000; Baja et al., 2001; Sys, 1985). Hence, proper evaluation of land resources is essential for sustainable agriculture land use planning. The common problem in land suitability analysis is variability and complexity of different kind of data and information. Soils and allied datasets available in SIS and Geoportal offer huge storage of data, flexible and more powerful tool than conventional data processing systems, as it provides different kinds of datasets and manipulating the datasets into new datasets that can be displayed in the form of thematic maps (Foote and Lynch, 1996). Overlay and modeling techniques in Geoportal help in land suitability evaluation for wide variety of crops. These exercises could be executed taking into consideration the factors like climate (rainfall, temperature) and soil (depth, texture, slope, drainage, CEC, pH) parameters. The GIS based SIS will be of immense help in providing support for making spatial decisions in land suitability evaluation (Malczewski, 1999). Exclusive modules could be developed in Geoportal with a set of procedures to facilitate the data input, compute to support decision-making activities and provide desirable outputs. The GIS capabilities for supporting spatial decisions can be analyzed in the context of the decision making process in land suitability analysis.

16.4.7 DEVELOPMENT OF SPATIAL DECISION SUPPORT SYSTEMS

A wide range of knowledge-based decision support tools and analytical environments would be required to support various spatial decisions in resource management and mitigation strategies. SIS and Geospatial platforms have the potential to being effectively used for developing such Spatial Decision Support Systems (SDSS). Key technologies include GPS, GIS, electronic sensors, and ruggedized computers for field data acquisition and operation control could be integrated to Geoportal to provide wide range of solutions and services. Although it is now relatively easy to collect geospatial data for various site-specific decisions, it is more difficult to know how to most effectively use those data in making resource management decisions through SDSS. The important step in these management decisions is to understand the spatial relationship of their influence in final decision while developing the SDSS.

16.5 WAY FORWARD

GIS communities are recognizing that providing access to geographic content is an important GIS activity that requires a long-term vision in order to realize the possible applications and utility that GIS offers. Web services, service-oriented architectures and distributed GIS are the foundation technologies through which users will realize the benefits of GIS, and geographic portals play a key role, guiding the way to the emergence of effective GIS. Efforts are on to migrate from protype Geoportal platform to operational Geoportal platform to provide advance services like Web Feature Services (WFS) and Web Processing Service (WPS) besides WMS and WCS to facilitate the publication of geospatial processes, their discovery and their use by clients, Web Map Tile Service (WMTS) are also planned to render maps as image tiles in days to come. The potential of Geography Markup Language (GML) model to share and store geographic information and Keyhole Markup Language (KML) to encode and transport the representations of geographic data for display in an earth browser like Google Earth could also explored to develop more stable and robust operational Geoportal in days to come. Geoportal in future should be interoperable using standardized software interfaces to connect to the many spatially related services offered by the different providers. Eventually, it should connect the different theme Geoportals to provide a single entry point to all geoinformation related datasets and services across the service providers.

16.6 CONCLUSIONS

The study demonstrates that the advanced capabilities of GIS are capable of producing sophisticated SIS, Geoportals and graphical outputs but it is important to note that the outputs are only as good as the input data. Further, the data discovery tools and spatial data mining techniques enable the users to query and obtain the information from related datasets to address the issues and challenges like climate change adaptation, mitigation and sustainable intensification of agriculture and food security, etc. Integration of SIS with remotely sensed data and SDSS in GIS provide enormous potential to address the various emerging issues in land use planning and sustainable land resource management. The development of GIS based SIS enables to catalog, store, manage, process and analyze voluminous soil resource and allied databases. The development of tools for spatial data query and mining and necessary pre and post processing tools would enhance the capabilities of SIS. The system supports the seamless integration of both spatial and non-spatial queries to support efficient search and discovery of the datasets. The availability of digital soil resource databases in GIS will be of immense help in various applications like thematic mapping and analysis, soil–landscape modeling, land suitability evaluation, assessment and monitoring of degraded lands, assessment of land productivity potential, development of SDSS and provide input at various levels in policy formulation.

The developed Geoportal helps to reduce the time and redundancy of data production. It enables the users to maintain data integrity to easily share the authoritative version of data among its users. It also enables easy search and discovery of existing geospatial data and services by allowing users to create and manage descriptions of their geospatial resources and supporting easy-to-use, sophisticated, data discovery technologies. Prototype Geoportal with geocoding capabilities was successfully developed to enable easy search, discovery of data and it helps to disseminate the information and provide the services in future. The results show that the capabilities Geoportal are enormous in database storage, management and analysis. Once it is operational, it offers viable solutions to the data interoperability among the departments/organizations. The implementation of full-fledged Geoportal can lead to effective decision-making tool and user can be able to access the data from the Geoportal. However, more research is required to improve the analysis and geoprocessing capabilities of Geoportal especially in development of services to address some of the important applications. The experience gained in the study will be of immense help to widen the scope of the study in future for the development of operational Geoportal with advanced analysis tools and capabilities to provide the various application services in management of land resources and location specific land use planning for sustainable agriculture.

KEYWORDS

- **Database Schema**
- **Digital Elevation Models (DEMs)**
- **Geographic Information System (GIS)**
- **Integrated National Agricultural Resource Information System (INARIS)**
- **Metadata**
- **Prototype Geoportal**
- **Relational Database Management System (RDBMS)**
- **Soil Information System (SIS)**
- **Web Catalogue Service (WCS)**
- **Web Map Service (WMS)**

REFERENCES

Ahamed, T. R. N., Rao, K. G., & Murthy, J. S. R. (2000). GIS based fuzzy membership model for crop land suitability analysis. *Agricultural Systems, 63*, 75–95.

Alonso, G., Casati, F., Kuno, H., & Machiraju, V. (2004). Web Services: Concepts, Architectures and Applications. Berlin: Springer Berlin-Heidelberg.

Annoni, A., Bernard, L., Fullerton, K., de Groof, H., Kanellopoulos, I., Millot, M., Peedell, S., Rase, D., Smits, P., & Vanderhaegen, M. (2004). Towards a European spatial data infrastructure: the INSPIRE initiative. *In*: *Proceedings of the 7th International Global Spatial Data Infrastructure Conference*, Bangalore, India, February 2–4, 2004, 11 pp.

Arnold, L., & G. Wright. (2003). Analyzing product-specific behavior to support process dependent updates in a dynamic spatial updating model. Proceedings of the Spatial Sciences Conference 2003, September, Canberra, Australia, pp. 1–10.

ASRIS. (2011). ASRIS-Australian Soil Resource Information System. http://www.asris.csiro.au. Accessed on 15th July, 2015.

Baja, S., Chapman, M. D., & Dragonvich, D. (2002). A conceptual model for defining and assessing land management units using a fuzzy modeling approach in GIS environment. *Environmental Management. 29*, 647–661.

Balaselvakumar, S., & Saravanan, S. (2006). 'Remote Sensing Techniques for Agriculture Survey.' http://www.gisdevelopment.net/application/agriculture/overview/agrio014.htm.

Batjes, N. H. (2006). ISRIC-WISE derived soil properties on a 5 by 5 arc-minutes global grid (*ver.* 1.1). Report 2006/02, ISRIC -World Soil Information, Wageningen.

Batjes, N. H. (2008). ISRIC-WISE Harmonized Global Soil Profile Data set (*ver* 3.1). Report 2008/02, ISRIC-World Soil Information, Wageningen (with data set).

Baumgardner, M. F. (1986). World Soils and Terrain Digital Database at a scale of 1, 1M (SOTER), ISSS, Wageningen.

Beaudette, D., & O'Geen, A. T. (2009). Soil-Web: An online soil survey for California, Arizona and Nevada. *Computers and Geosciences, 35*, 2119–2128.

Bhat, L. S. (1989). Status report on Land Systems Analysis for Evolution of resources at Micro level. Discussion paper for fifth review meeting for NRDMS project, DST Govt. of India.

Bhattacharyya, T., Sarkar, Dipak., Pal, D. K., Mandal, C., Baruah, U., Telpande, B., & Vaidya, P. H. (2010). Soil information system for resource management – Tripura as a case study, *Current Science*, *99*(9), 1208–1217.

Burrough, P. A., & McDonnell, R. A. (1998). Principles of geographical information systems, Oxford University Press, Oxford.

Fisher, F. (2006). Report about Recent Trends in the GIS sector: INTERGEO 2006, *GEOInformatics*, *9*(7), 6–8.

Foote, K. E., & Lynch, M. (1996). Geographic Information Systems as an Integrating Technology: Context, Concepts and Definition. Austin, University of Texas.

Frankel, K. L., & Dolan, J. F. (2007). Characterizing arid-region alluvial fan surface roughness with airborne laser swath mapping digital topographic data, *Journal of Geophysical Research – Earth Surface*, 112, F02025, doi: 10.1029/2006JF000644.

Franklin, J. (1995). Predictive vegetation mapping: geographic modeling of biospatial patterns in relation to environmental gradients. *Progress in Physical Geography,* 19, 474–90.

Gao, J. (2008). Mapping of land degradation from ASTER data: a comparison of object-based and pixel-based methods. *GIS Science and Remote Sensing, 45*(2), 149–166.

Gao, J., & Liu, Y. (2008). Mapping of land degradation from space: a comparative study of Landsat ETM+ and ASTER data, *International Journal of Remote Sensing, 29*(14), 4029–4043.

Garcia-Molina, H., Ullman, J. D., & Widom, J. (2002). Database Systems: The Complete Book. International Edition. New Jersey: Prentice Hall. First Edition, ISBN 978-0-13-031995-1.

Government of India. (2011). "Faster, Sustainable and More Inclusive Growth: An Approach to the 12th Five Year Plan (Draft)," Planning Commission, Government of India, New Delhi.

http://sis.agr.gc.ca/cansis/, Canadian Soil Information Service, Accessed on 19 December, 2015.

http://www.esoter.net, Accessed on 25 December, 2015.

ICAR and NAAS. (2010). Degraded and Wastelands of India Status and Spatial Distribution, Indian Council of Agricultural Research and National Academy of Agricultural Science, New Delhi p 158.

ISRIC, FAO, & UNEP. (2003). Soil and Terrain Database for Southern Africa (1:2 million scale), FAO Land and Water Digital Media Series No. 25. FAO, Rome.

ISRIC-WISE. (2008). References and examples of WISE database applications. PDF ISRIC 2007. ISRIC World Soil Database.

ISSS, ISRIC and FAO. (1998). World Reference Base for Soil Resources, Food and Agriculture Organization of the United Nations, Rome.

Kazemi, F. (2003). A generalization framework to derive multi-scale GEODATA. Proceedings of the Spatial Sciences Conference 2003, September, Canberra, Australia, pp. 1–12.

King, D., Burrill, A., Daroussin, J., Le Bas C., Tavernier R., & Van Ranst, E. (1995). *The EU Soil Geographic Database. In: Agriculture; European Land Information Systems for*

Agro-Environmental Monitoring. European Commission Joint Research Centre. Italy. pp, 43–60.

Krishnamurthy, J., & Srinivas, G. (1995), Role of geological and geomorphological factors in groundwater exploration: A study using IRS-LISS-II data. *International Journal of Remote Sensing,* 16, 2595- 2618.

Lee, D. (2003). Generalization within a geoprocessing framework. Proceedings of the GEO-PRO 2003 Workshop, Mexico City, November 2003, pp. 1–10.

Lu, D., Batistella, M., Mausel, P., & Moran, E. (2007). Mapping and monitoring land degradation risks in the western Brazilian Amazon using multi-temporal Landsat TM/ETM+ images, *Land Degradation & Development, 18*, 41–54.

Maguire, D. J., & Longley, P. A. (2005). The Emergence of Geoportals and Their Role in Spatial Data Infrastructures, Computers, *Environment and Urban Systems*, p.29.

Maji, A. K., Dubey, P. N., Verma, T. P., Chamuah, G. S., Sehgal, J., & Velayutham, M. (2000). Soils of Nagaland for optimizing land use, NBSS Publ. 47b, NBSS & LUP, Nagpur, India.

Maji, A. K., Krishna, N. D. R., & Challa, O. (1998). Geographical Information System in analysis and interpretation of soil resource data for land use planning. *Journal of Indian Society of Soil Science, 46*, 260–263.

Maji, A. K., Nayak, D. C., Krishna, N. D. R., Srinivas, C. V., Kamble, K., Reddy, G. P. O., & Velayutham, M. (2001). Soil Information System of Arunachal Pradesh in a GIS environment for land use planning. *International Journal of Applied Earth Observation and Geoinformation, 3*, 69–77.

Maji, A. K., Reddy, G. P. O., & CCPI- Regional Centers. (2004). NATP-(MM) Project on Integrated National Agricultural Resource Information System (INARIS Sub-Project-Soil Resource Database, Annual Report, 47–51.

Malczewski, J. (1999). GIS and Multicriteria Decision Analysis. USA and Canada. John Wiley & Sons.

McBratney, A. B., Minasny, B., Cattle, S. R., & Vervoort, R. W. (2002). From pedotransfer functions to soil inference systems. *Geoderma, 109*, 41–73.

McDonnell, R. A. (1996). Including the spatial dimension: using geographical information systems in hydrology. Progress in Physical Geography *20*(2), 159–177.

McKenzie, N. J., Jacquier, D. W., Ashton L. J., & Cresswell, H. P. (2000). Estimation of Soil Properties Using the Atlas of Australian Soils. CSIRO Land and Water Technical Report 11/00. http://www.clw.csiro.au/publications/technical/.

Open Geospatial Consortium. (2010). OGC WCS 2.0 interface standard – core [Internet]. Open Geospatial Consortium Inc., Available from: http://www.opengeospatial.org/standards/wcs.

Pedercini, M., Kanamaru, H., & Derwisch, S. (2012). Potential impacts of climate change on food security in Mali. Natural Resources Management and Environment Department, FAO, Rome.

Reddy, G. P. O., Mondal, C., Srivastava, R. Bhattacharyya, T. Naidu, L. G. K., Sidhu, G. S. Baruah, U, Singh, R. S., Kumar, N., Singh, S. K., & Sarkar, D. (2014). Development of Indian Soil Information System- A Geoportal, NBSS & LUP, Nagpur, Project Report, pp, 34.

Reddy, G. P. O., Singh, S. K., Mondal, C., Srivastava, R., Bhattacharyya, T., Naidu, L. G. K., Sidhu, G. S., Baruah, U., Singh R. S., Kumar, N., & Sarkar, D. (2014a). Development of District Soil Information System (DSIS) on 1, 50,000 Scale (50 Districts), NBSS & LUP, Nagpur, Project Report, 160 pp.

Reddy, G. P. O., Nagaraju, M. S. S., Ramteke, I. K., & Sarkar, D. (2013). Terrain Characterization for Soil Resource Mapping Using IRS-P6 Data and GIS – A Case Study From Basaltic Terrain of Central India, *Journal of the Indian Society of Remote Sensing, 41,* 331–343.

Reddy, G. P. O., & Maji, A. K. (2003). Delineation and Characterization of Geomorphological features in a part of Lower Maharahstra Metamorphic Plateau, using IRS-ID LISS-III data. *Journal of the Indian Society of Remote sensing, 31*(4), 241–250.

Reddy, G. P. O. Shekinah, D. E., Maurya, U. K., Thayalan, S., Jagdish Prasad, Ray, S. K., & Bhaskar, B. P. (1999). Landscape-soil relationship in part of Bazargaon plateau, Maharashtra, *The Geographical Review, 63*(3), 280–291.

Reddy, G. P. O. Maji, A. K., Chary, G. R., Srinivas, C. V., Tiwary, P., & Gajbhiye, K. S. (2004a). GIS and Remote sensing Applications in Prioritization of River sub basins using Morphometric and USLE Parameters: A Case study, *Asian Journal of Geoinformatics, 4*(4), 35–49.

Reddy, G. P. O. Sarkar, D., Mandal, C., Srivastava, R., Bhattacharyya, T., Naidu, L. G. K., Sidhu, G. S., Baruah, U., Singh, S. K., Singh, R. S., Nair, K. M., Sen, T. K., Chandran, P., Sahoo, A. K., Srinivas, S., Kumar, N., & Sapana C. (2016). Digital Soil Resource Database and Information System *In: Geospatial Technology for Integrated Natural Resources Management* (R. S. Dwivedi, & P. S. Roy, Eds.), Yes Dee Publishing Pvt Ltd, Chennai. 321–351.

Reddy, G. P. O., Maji, A. K., & Gajbhiye, K. S. (2004). Drainage morphometry and its influence on landform characteristics in a basaltic terrain, Central India – A remote sensing and GIS approach. *International Journal of Applied Earth Observation and Geoinformation, 6,* 1–16.

Reddy, G. P. O., Maji, A. K. Das, S. N., & Srivastava, R. (2012). Development of GIS Based Seamless Mosaic of SRTM Elevation Data of India to Analyze and Characterize the Selected Geomorphic Parameters, NBSS & LUP, Nagpur, Project Report, 54p.

Sabins, F. F. Jr. (1987). Remote Sensing: Principles and Interpretation, 2e édition, W. H. Freeman and Co., New York, x + 449 p.

Sehgal, J., Saxena, R. K., & Vadivelu, S. (1987). Field Manual, Soil Resource Mapping of different states in India. NBSS Publ. 13, NBSS & LUP, Nagpur, India.

Sidhu, G. S., Pal, Sharmistha, Tiwari, A. K., Sarkar, D., & Sharda, V. N. (2013). Soil erosion in Punjab, NBSS Publ. 151, NBSS & LUP, Nagpur, p 33.

Singh, R., Sahoo, P., & Rai, A. (2003). Use of Remote Sensing and GIS Technology in agricultural surveys. From http://www.gisdevelopment.net/application/ agriculture/cropping/mio368.htm

Sujatha, G., Dwivedi, R. S., Sreenivas, K., & Venkataratnam, L. (2000). Mapping and monitoring of degraded lands in part of Jaunpur district of Uttar Pradesh using temporal spaceborne multispectral data. *International Journal of Remote Sensing, 21,* 519–531.

Sys, C. (1985). Land evaluation. Algemeen Bestuur vande Ontwikkelingss, Ghent, Belgium: International Training Centre for Post-Graduate Soil Scientists. State University of Ghent.

Tait, M. (2005). Implementing Geoportals: applications of Distributed GIS, *Computers, Environment and Urban Systems, 29,* 33–47.

Tyner, J. A. (2010). Principles of Map Design, The Guilford Press: New York, 259 p.

Van Engelen, V. W. P., & Wen, T. T. (Eds.). (1995). Global and National Soils and Terrain Digital Databases (SOTER). Procedures Manual (revised edition). ISSSUNEP-FAO-ISRIC, Wageningen, 125 p.

Wade, T., & Sommer, S. (2006). A to Z GIS. An illustrated dictionary of geographic information systems. Redlands, CA: ESRI Press. 288 pp.

Weerawarana, S., Curbera, F., Leymann, F., Storey, T., & Ferguson, D. F. (2005). Web Services Platform Architecture: SOAP, WSDL, WS-Policy, WS-Addressing, WS-BPEL, WS-Reliable, Messaging and More, 4. Printing ed. Prentice Hall/PTR, Upper Saddle River, New Jersey.

PART III

SOIL NUTRIENT STATUS AND MANAGEMENT

CHAPTER 17

GEO-REFERENCED SOIL FERTILITY MONITORING FOR OPTIMIZED FERTILIZER USE: A CASE STUDY FROM THE SEMI-ARID REGION OF WESTERN INDIA

R. N. KATKAR,[1] V. K. KHARCHE,[2] S. R. LAKHE,[3] U. D. IKHE,[4] A. B. AGE,[1] and D. V. MALI[1]

[1]*Associate Professor, Department of Soil Science and Agricultural Chemistry, Dr. Panjabrao Deshmukh Krishi Vidyapeeth, Akola – 444 104, India*

[2]*Professor, Department of Soil Science and Agricultural Chemistry, Dr. Panjabrao Deshmukh Krishi Vidyapeeth, Akola – 444 104, India*

[3]*Research Associate, Department of Soil Science and Agricultural Chemistry, Dr. Panjabrao Deshmukh Krishi Vidyapeeth, Akola – 444 104, India*

[4]*Senior Research Fellow, Department of Soil Science and Agricultural Chemistry, Dr. Panjabrao Deshmukh Krishi Vidyapeeth, Akola – 444 104, India*

CONTENTS

ABSTRACT

The geo-referenced soil fertility monitoring was carried out in Jalna district, Maharashtra. The surface soil samples were collected in 96 villages from eight tehsils of Jalna district during 2012. Two soil samples were collected from fields of each of these small (< 1 ha), medium (1–3 ha) and large (> 3 ha) group of land holdings. The geo-referenced soil samples (576) were collected and coordinates of sampling sites were recorded using Global Positioning System. The soil pH ranged from 6.61 to 8.91 while electrical conductivity varied from 0.101 to 0.285 dS m^{-1}. The free $CaCO_3$ content was found to range from 1.38 to 11.42%. Mantha tehsil recorded the highest mean value of 9.31%. The calcium carbonate content was found to be distributed to 2.95, 55.21 and 41.84% in low, medium and high category. Organic carbon content in the soil ranged from 1.10 to 9.86 g kg^{-1}. Jalna tehsil recorded the highest mean organic carbon content of 4.54 g kg^{-1} followed by Ambad (4.21 g kg^{-1}) tehsil. The available nitrogen varied from 81.5 to 294.4 kg ha^{-1}. The available nitrogen status was observed to the tune of 95.14 % under low category. The available phosphorus ranged from 2.70 – 49.20 kg ha^{-1} in which 40.45% samples were found deficient while available potassium ranged between 93.0 – 619.0 kg ha^{-1} wherein 68.06% samples were found high while 27.08% samples were found medium. The available sulfur ranged from 4.53–30.41 mg kg^{-1} in which 24.65% samples were deficient and 67.36% samples were found medium. The available zinc ranged from 0.21 to 3.19 mg kg^{-1} and 54.2% samples were deficient whereas 43.9% samples were in medium category. The available iron ranged from 2.13 to 27.56 mg kg^{-1} out of which 29.5% samples were deficient. The available copper ranged from 0.30 to 5.98 mg kg^{-1} while available Mn ranged from 2.10 to 20.14. The nutrient index of nitrogen was low; phosphorus and sulfur were medium whereas the nutrient index of potassium was found high. The nutrient index of zinc was found low while that of iron was medium. The nutrient indices of Mn and Cu were recorded high. It could be inferred that, fertility of soils in Jalna district of

Maharashtra was low in nitrogen and zinc, medium in phosphorus and iron while high in potassium, copper and manganese.

17.1 INTRODUCTION

The wide spread deficiencies of micronutrients in soils of agriculturally progressive states are being observed in post green revolution. The micronutrients are important for maintaining soil health and increasing use efficiency of major nutrients and ultimately the crop productivity. These are required in very small quantity. The soil must supply micronutrients for desired growth of plants and synthesis of human food (Shukla, 2011). The deficiencies of micronutrients have become major constraints in sustainable productivity and sustainability of soils. Whenever a micronutrient is deficient, the abnormal growth of plant is observed, which sometime causes complete failure of crops. Grains and flower are not formed under severe deficiency. The main sources of these micronutrients are parent material, organic and inorganic material. The availability of micronutrients is particularly sensitive to changes in soil environment. In the present era of precision farming, the inputs such as fertilizer, crop varieties and management practices are matched precisely with the variability of soil and climatic conditions so that inputs are applied as per the location specific requirements of the crop (Singh, 2007). The introduction of information technology has provided tools *viz.,* Global Positioning System (GPS) which helps in collecting the systematic set of georeferenced samples and generating the spatial data about the distribution of nutrients with Geographic Information System (GIS). GIS based fertility mapping for different cropping system in various agro-ecologies needs to be emphasized and these maps needs to be integrated with farmenrs crop management practices for developing site specific nutrient management prescriptions (Gangwar et al., 2013).

17.1.1 SOIL FERTILITY MONITORING: GLOBAL PERSPECTIVE

The survey carried out in many countries indicated that the fertilizer practices are more skewed towards the nitrogen and phosphorus application and sulfur application is almost neglected. The fast decline in available sulfur in soil is mainly due to higher crop removal by high yielding genotypes, high cropping intensity, poor replenishment in soil due to use of sulfur free fertilizers. In spite of the proven benefits of adequate zinc nutrition, the zinc deficiency is the fifth leading risk factor for disease in developing countries with high mortality rates (WHO, 2002). Zinc shows the strongest effect on immune system. About sixty percent of the agricultural land is suffering from zinc deficiency. The area under heavy rainfall with light texture soils are mostly deficient in nitrogen, sulfur and boron.

17.1.2 SOIL FERTILITY MONITORING: INDIAN PERSPECTIVE

Maji et al. (2000) used GIS application in bringing out the different aspects of the rice wheat-based cropping system in the Indo-Gangetic Plains of India whereas, Singh et al. (2009) used remote sensing and GIS to study potential suitability for different water harvesting structures in Soankhad watershed of Kandi region situated in Talwara block in Hoshiarpur district of Punjab. The soil factors *viz.,* texture, pH, organic matter content, calcium carbonate content, type of clay minerals and interactions among the nutrients markedly regulate the availability of nutrients in soils (Malewar, 2005). Imbalanced and inadequate use of fertilizers coupled with low efficiency of other inputs, the response efficiency of chemical fertilizer nutrients has declined tremendously under intensive agriculture in recent years. About forty percent deficiency of sulfur has been observed in various states of India. Zinc is reported to be deficient of 55% and likely to increase upto 63% by 2025.

17.1.3 NEED OF SOIL FERTILITY MONITORING

The intensive cultivation of soils and use of improved high yielding crop varieties which takes up nutrients from the soil are the major causes of deficiency, continuous application of one or two macronutrients may in due course deplete the soil reserve of sulfur and micronutrients and limit the crop performance. For sustainability purpose there is need to know the nutrients status of the soil. There is an urgent need to monitor the soil fertility in semi-arid areas where agricultural productivity is expected to be higher to cater to the need of swelling population, Keeping this in view, the present investigation was undertaken with the objectives to assess the status of major and micronutrients in soils, their relationships and to identify and delineate areas of nutrient deficiencies in Jalna district of Maharashtra.

17.2 STUDY AREA

The study area is located in Western India and situated on Deccan plateau. It lies on the northern side of the Godavari river basin and southeast part of the Ajantha ranges. Geologically, the district is covered with Deccan trap of weathered basaltic material. The different landforms in a region constitute its physical set-up. The average annual rainfall of the Jalna district is 816 mm with wide variability in different tehsils. The major area falls under Agro-Ecological Sub Region (AESR) 6.3, with the length of growing period (LGP) ranging from 120–150 days. The soils in Jalna district are black with considerable variation in texture and depth. The soils are light, medium and heavy texture. The soils along the riverbanks especially in Ambad and Partur blocks are deep black and quite fertile. The soils in northern parts of the district (i.e., in Jalna, Bhokardan and Jaffrabad blocks are coarser. The most important river in Jalna district is Godavari, which flows for about 60 kilometers along the Southern boundary of the district. Its principle tributaries are Dudhana, which flows from central part of the district and Galhati, which passes through Ambad tehsil. The other tributaries of Purna and Khelna are Girija and Dudhana, respectively, are flowing through this district substantially benefiting the agricultural activity. It is characterized by semi-arid eco-system with shallow and medium to deep black soils belonging to *Ustorthents, Haplusterts, Haplustepts* (Challa et al., 1995, Challa et al., 1999). The major crops cultivated in Jalna district are cotton, soybean, pigeon pea, sorghum, pulses. The minor crops like pearl millet, maize, safflower, sunflower and seasamum are also grown. Fruit crops like sweet orange, guava, sapota, etc., are also grown. The district consists of eight tehsils *viz.*, Bhokardan, Jafrabad, Jalna, Badnapur, Ambad, Ghansawngi, Partur and Mantha.

17.3 METHODOLOGY

Stratified random sampling technique was employed to identify the villages for sampling in the district. In order to monitor the soil fertility in the area the modern tool like GPS was used to record the Latitude, Longitude and altitude of the sampling sites so that in the future the sampling sites can be located to monitor the fertility after specific period. The surface soil samples (0–20 cm) were collected in 96 villages from 8 tehsils of Jalna district during

Novmber, 2012. Six farmers from each village were selected based on land holdings. Two soil samples were collected from fields of small (less than 1 ha), medium (1–3 ha) and large (above 3 ha) land holding group. The 576 geo-referenced soil samples were collected from all the tehsils and Latitude, Longitude were recorded using GPS. Soil pH and EC were determined in soil: water suspensions (1:2.5 w/v) as described by Jackson (1973). Organic carbon was determined by dichromate wet oxidation method described by Walkley and Black (1934). Free $CaCO_3$ was determined by Rapid Titration method (Piper, 1966). The available N was estimated by alkaline permanganate method (Subbiah and Asija, 1956) and available P by Olsen's method (Olsen et al., 1954) and available K by ammonium acetate extraction method (Jackson, 1967). The available S was estimated by turbidimetric method (Chesnin and Yien, 1951). The available zinc, iron, copper and manganese were extracted with 0.005 *M* diethylene triamine penta acetic acid (DTPA) and the concentrations of nutrients were determined on Atomic Absorption Spectrophotometer (Lindsay and Norvell, 1978). The nutrient indices were calculated by using the formula given by Parker et al. (1951). The georeferenced data of nutrients determined were used to prepare the fertility maps using ArcGIS software and fertility maps were prepared.

17.4 RESULTS AND DISCUSSION

17.4.1 SOIL PROPERTIES

The pH of soils in Jalna district was neutral to alkaline (6.61–8.91). The highest pH was observed (Table 17.1) in Ghansawangi tehsil (8.30) and lowest in Jalna tehsil (8.23). The alkaline reaction of soil was probably due to presence of sufficient free lime content in soil (Kaushal et al., 1980). The EC varied from 0.101 to 0.285 dS m^{-1} indicating that all the soils are non-saline in nature and suitable for healthy plant growth. Organic carbon content in the soil ranged from 1.10 to 9.86 g kg^{-1}. Jalna tehsil recorded the highest mean organic carbon content of 4.54 g kg^{-1} followed by Ambad (4.21 g kg^{-1}) and Bhokardan (4.18 g kg^{-1}) tehsil. The organic carbon deficiency was noticed in 55.7% samples. The high temperature during summer and non-addition of organic matter regularly might be the reason for deficiency in Jalna district. The organic carbon deficiency was observed in 62.9, 56.6, and 56.2% areas in Ghansawangi, Jaffrabad and Bhokardhan tehsiles,

TABLE 17.1 Chemical Properties of Soils in Jalna District

	pH (1:2.5)		EC (dS m^{-1})		$CaCO_3$ (%)		Organic carbon (g kg^{-1})	
Tehsil	**Range**	**Mean**	**Range**	**Mean**	**Range**	**Mean**	**Range**	**Mean**
Bhokardan	6.6–7.8	7.3	0.10–0.19	0.14	2.3–8.0	5.4	1.5–8.9	4.2
Jafrabad	7.2–7.9	7.6	0.11–0.18	0.14	1.4–8.3	4.9	1.2–8.8	4.1
Jalna	6.7–7.7	7.2	0.10–0.19	0.13	1.6–8.7	5.9	1.1–9.8	4.5
Badnapur	6.8–7.9	7.3	0.11–0.18	0.15	4.4–11.4	8.9	1.1–9.6	4.0
Ambad	7.2–8.9	8.0	0.12–0.28	0.17	2.9–11.3	7.9	1.3–9.2	4.2
Ghansa-wangi	7.8–8.7	8.3	0.10–0.27	0.18	4.6–11.3	9.1	1.2–9.4	3.9
Partur	7.6–8.7	8.2	0.12–0.25	0.17	4.5–1.4	8.7	1.3–8.7	4.1
Mantha	7.7–8.9	8.2	0.12–0.19	0.16	5.4–1.3	9.3	1.3–8.7	4.2
Jalna district	6.6–8.9	7.7	0.10–0.28	0.15	1.4–11.4	7.4	1.1–9.8	4.2

respectively. The free $CaCO_3$ content was found to range from 1.38 to 11.42%. Mantha tehsil recorded the highest mean value of CaCO3 (9.31%) followed by Ghansawangi (9.13%) and Badnapur (8.99%). The calcium carbonate content belonged to medium category in 55.21% area and high category in 41.84% area. High calcium carbonate is harmful; it reduces the concentration of micronutrient cations in soils to such a level that the sensitive plant suffers from deficiency of micronutrients (Deb et al., 2009). The highest calcium carbonate content was noticed in Mantha and Ghansawangi tehsils.

17.4.2 MAJOR NUTRIENTS STATUS

The available nitrogen content ranged from 81.5 to 294.4 kg ha^{-1}, which showed 95.14% deficiency (Table 17.2). The deficiency of available nitrogen might be due to very less addition of organic manures and heavy uptake under intensive cultivation of improved high yielding varieties of different crops. The available P varied from very low (2.70 to 49.20 kg ha^{-1}) to very high indicating 40.45% deficiency. Ghansawangi, Partur and Ambad tehsils showed more than 60% deficiency. The deficiency of available P may be because of its fixation in the form of calcium

TABLE 17.2 Nutrients Status of Jalna District

	N		P		K		S	
	(kg ha^{-1}) (mg kg^{-1})							
Tehsil	**Range**	**PSD**	**Range**	**PSD**	**Range**	**PSD**	**Range**	**PSD**
Bhokardan	90.3–288.6	96.8	6.74–63.77	39.5	134–459	2.08	4.5–13.6	33.3
Jafrabad	84.5–285.9	95.0	7.83–43.20	48.3	145–448	1.66	6.5–22.1	16.7
Jalna	98.5–294.4	95.5	11.53–68.52	14.4	101–437	6.66	7.9–23.9	23.3
Badnapur	85.9–291.7	96.3	6.68–49.44	25.9	93–437	16.67	7.8–15.9	29.6
Ambad	89.5–292.9	94.1	2.70–36.50	60.7	115–617	7.14	8.1–29.6	20.2
Ghansa-wangi	94.7–284.5	91.6	6.35–38.66	63.8	129–452	2.77	8.3–30.4	18.1
Partur	87.1–290.0	94.4	6.43–28.74	62.9	152–619	0.0	8.2–17.5	25.9
Mantha	81.5–289.6	96.9	6.51–39.13	12.1	146–578	3.03	8.5–19.6	28.7
Jalna district	81.5–294.4	95.1	2.70–49.20	40.4	93–619	4.86	4.5–30.4	24.6

*PSD – Percent sample deficient.

phosphate due to alkaline nature of soil (Bhandari, 2013). The available potassium ranged between 93.0 to 619 kg ha^{-1} wherein 68.06% samples were found high while 27.08% samples were found medium (Figure 17.1). During post green revolution era, the available potassium was considered very high in black cotton soils. The recent trends indicated potassium deficiency and the crops are responding to its addition (Subba Rao and Srinivasa Rao, 1996). Moreover, the results of long-term fertilizers experiments indicate mining of K from soils under major cropping system. The available sulfur varied from low to very high (4.53 to 30.41 g kg^{-1}) with 24.65% deficiency whereas 67.36% samples were found medium (Table 17.2). The intensive cultivation of crops and application of fertilizers devoid of sulfur might be depleting the sulfur from soil. The application of balanced nutrition to the crops under intensive cultivation is essential for maintaining the soil fertility and sustainable productivity (Ghosh et al., 2004).

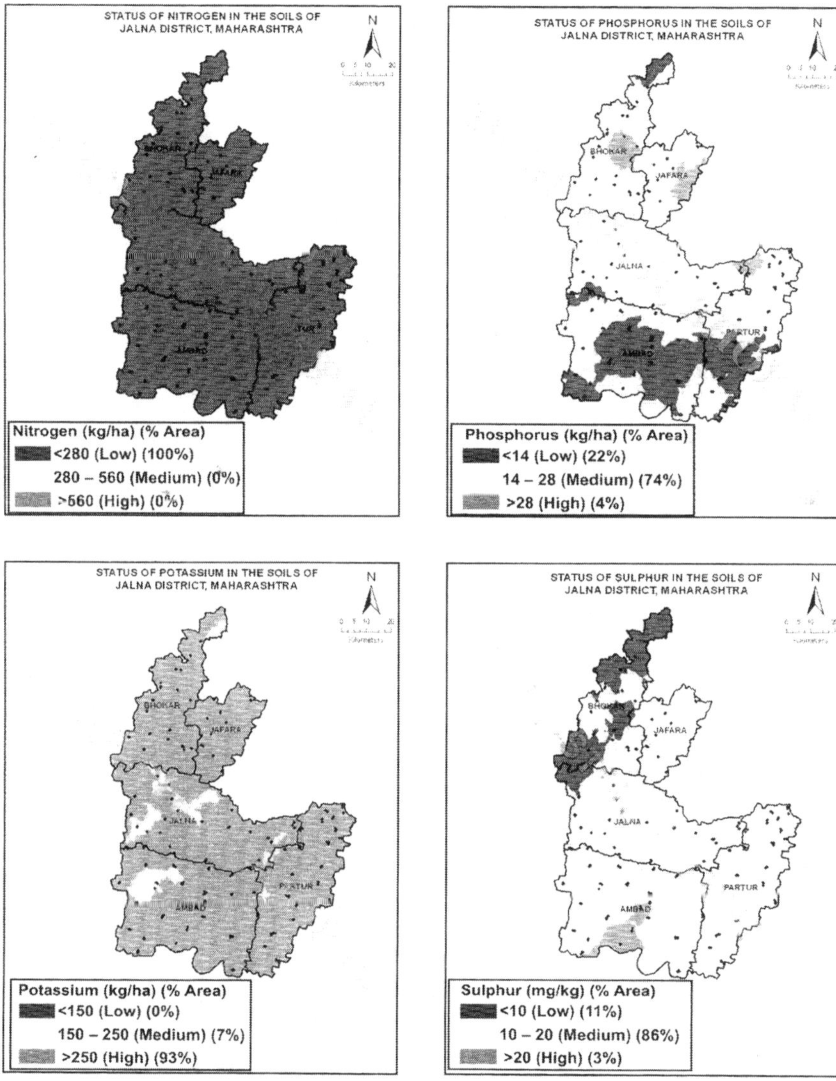

FIGURE 17.1 Fertility maps of N, P, K and S in Jalna district of Western India.

17.4.3 MICRO NUTRIENTS STATUS

The DTPA-Zn in soils of Jalna district ranged from 0.21 to 3.19 mg kg^{-1} (Table 17.3) indicating 54.2% deficiency, whereas 43.9% samples of available Zn were in medium category showing widespread deficiency of zinc. The highest deficiency of zinc was observed in Jaffrabad (68.3%) tehsil

TABLE 17.3 DTPA-Extractable Micronutrients Status in Soil

	Zn		Fe		Cu		Mn	
	(mg kg^{-1})							
Tehsil	**Range**	**PSD**	**Range**	**PSD**	**Range**	**PSD**	**Range**	**PSD**
Bhokardan	0.28–1.73	64.6	2.39–16.24	27.1	0.34–1.78	0	2.72–14.50	0
Jafrabad	0.34–1.86	68.3	2.13–19.08	35.0	0.30–1.91	0	3.52–9.70	0
Jalna	0.27–3.08	52.2	2.46–22.83	36.7	0.44–2.35	0	2.23–13.88	0
Badnapur	0.29–2.35	51.9	2.58–27.56	25.9	0.75–1.99	0	2.10–12.25	0
Ambad	0.30–3.19	47.6	2.16–15.30	36.9	0.50–2.13	0	3.95–18.76	0
Ghansawangi	0.27–1.61	45.8	2.18–15.30	27.8	0.48–3.65	0	3.48–19.82	0
Partur	0.22–1.67	55.6	2.27–24.83	24.1	0.35–5.98	0	3.57–20.14	0
Mantha	0.21–1.62	47.0	2.33–21.49	18.2	0.30–4.02	0	2.25–19.92	0
Jalna district	0.21–3.19	54.2	2.13–27.56	29.5	0.30–5.98	0	2.10–20.14	0

*PSD – Percent sample deficient.

followed by Bhokardhan and Partur. The availability of micronutrient cations are generally low in alkaline soils and crops grown on these soils suffer from hidden hunger (Malewar, 2005).

The major crops grown in these tehsils are cotton, pigeon pea, soybean, wheat, etc., and their intensive cultivation might have mined the zinc along with N, P and K. This indicated wide spread zinc deficiency in Jalna district. It might be due to high nutrient requirement of recently introduced high yielding varieties. The imbalanced use of N, P and K fertilizers, reduction in organic carbon contents of soil and decline in the level of micronutrient in soil below critical level is also leading to zinc deficiency (Singh, 2003). The less availability of organic manures and non- application of manures to soil results in micronutrient deficiency. Moreover, the farmers are not testing the soils for micronutrients which are not being added along with macronutrients. The widespread deficiency of zinc in intensively cultivated districts of Western Maharashtra having predominant alkaline, calcareous, black clayey soils was reported by Patil and Kharche (2001). It has also been reported that the soils of Maharashtra did not show response to application of zinc during seventies (Kharche et al., 2003). However, afterwards due to intensification of agriculture the soils became deficient in zinc. The deficiency of nutrients creates imbalance in soils which results into nutritional stress in plants (Malewar, 2005). Sakal (2001) reported zinc deficiency as most serious constraint to sustainable productivity in several states. Micronutrient deficiency of plants occur more frequently in calcareous soils with high pH such as those found in arid and semiarid regions (Alloway, 2006). The great variation was noticed in iron content (2.13 to 27.56 mg kg^{-1}) in Jalna district. The deficiency percentage of iron was 29.5, while 67.9% samples under medium category (Table 17.3) indicating that the soils are becoming deficient in iron followed by zinc. Increased removal of micronutrients as a consequence of adoption of high yielding varieties and intensive cropping together with a shift towards the use of high analysis NPK fertilizers have caused decline in the level of micronutrients in the soil below the critical level which are required for normal productivity of crops (Zende, 1987). Singh (2003) also noticed the heavy depletion of micronutrients with time, particularly under intensive cropping of rice-wheat, maize-wheat and rice-rice. Patil et al. (2004) reported 40.0 and 34.7% soils deficient in zinc and iron, respectively.

The DTPA extractable Cu in the Jalna soils ranged from 0.30 to 5.98 mg kg^{-1} (Table 17.3). Patil and Sonar (1994) reported that in swell-shrink soils of Maharashtra, available Cu ranges from 0.58 to 1.7 mg kg^{-1}. The data further showed that Cu deficiency was not noticed. It indicates that the majority of soils in Jalna district were found adequate in Cu content. The DTPA-Mn status of soils ranged from 2.10 to 20.14 mg kg^{-1} (Table 17.3). Gajbhe et al. (1976) reported that available Mn content in surface soils of Marathwada ranged from 13.3 to 65.20 mg kg^{-1}. The deficiencies of Mn and Cu in the soils of Jalna district were not observed.

17.4.4 NUTRIENT INDICES

In Jalna district, the nutrient indices (Table 17.4) were found low in available N (1.04) and Zn (1.48), whereas, medium in available P (1.72), S (1.83) and Fe (1.73) and high in K (2.63), Mn (2.46) and Cu (2.73) (Figure 17.2). The areas where the status of nutrients are medium, may show deficiency in near future if the due care was not taken for addition of organic manures and

TABLE 17.4 Status of Micronutrients and Nutrient Indices in Jalna District

Nutrients	Percent samples			Nutrient indices
	Low	Medium	High	
N	95.1	4.86	0	1.04
P	40.4	46.3	13.2	1.72
K	4.86	27.1	68.1	2.63
S	24.6	67.4	7.99	1.83
Zn	54.2	43.9	1.9	1.48
Fe	29.5	67.9	2.6	1.73
Cu	0	27.3	72.7	1.73
Mn	0	54.2	45.8	2.73

inorganic micronutrient fertilizers based on soil testing by the cultivators in the districts for intensive cultivation of different crops (Malewar, 2005).

The GIS based maps generated under the study will be useful for guiding the amount and kind of nutrients to be applied for soil health management. The information technology based GPS-GIS technique has been found useful for systematic mapping and delineation of micronutrient status in soils which revealed widespread deficiency of nitrogen and zinc followed by iron and sulfur. Mapping of the current status of micronutrients in soils of Jalna district will be helpful to suggest the efficient ways and methods for enhancing the yields by using organic manures and inorganic fertilizers in the areas of major and micro-nutrient deficiency.

17.5 NUTRIENT STATUS UNDER MAJOR CROPPING SYSTEMS

The 36–63% soil samples collected from the sites where the cotton was grown, showed zinc deficiency while under sorghum crop, deficiency was found in less samples (24 to 47%). Deficiency of sulfur was found in highest number of samples (21–39%) while under cotton (17–34%) and sorghum (14–25%). The nitrogen deficiency was recorded under almost all the crops. The phosphorus deficiency was noticed more pronounced under cotton and sorghum crops as compared to soybean. The sweet orange is the major fruit crop of the district; 43–68% sampling sites under this crop indicated pronounced deficiency of zinc and iron deficiency was observed in 24–36% samples.

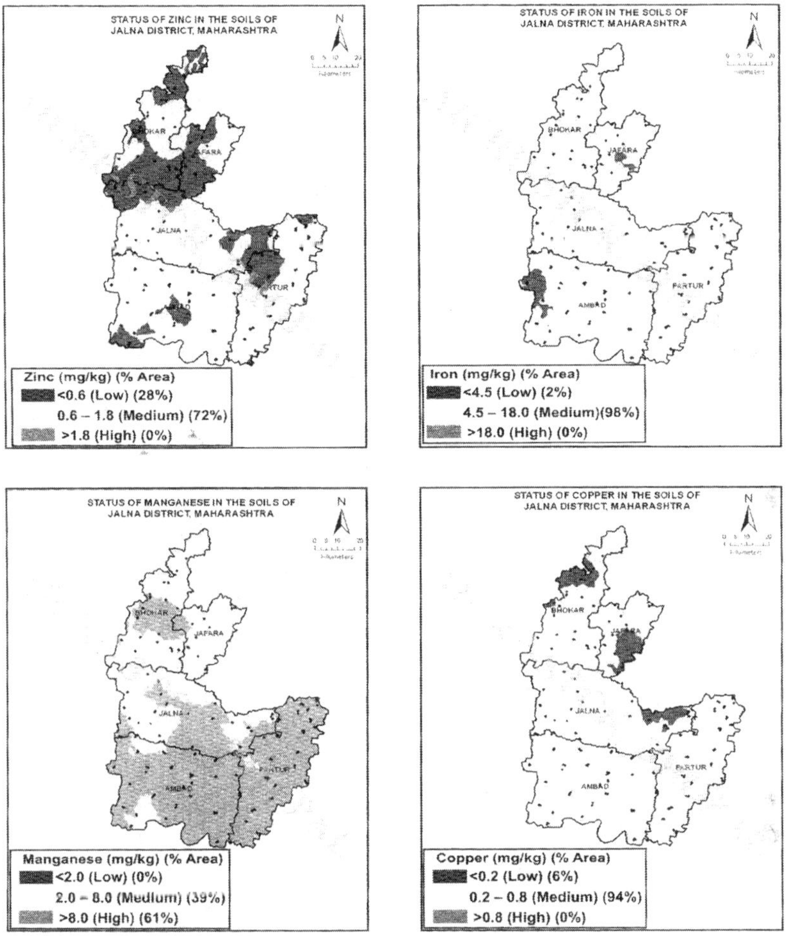

FIGURE 17.2 Micronutrient status maps of Zn, Fe, Mn and Cu in Jalna district of Western India.

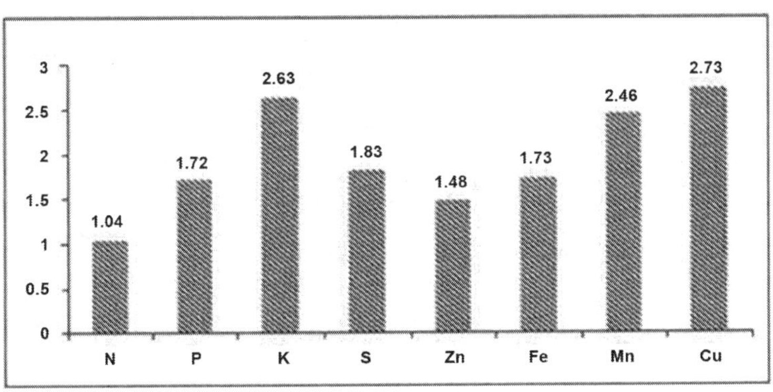

FIGURE 17.3 Nutrient indices of macro and micronutrient in Jalna district.

17.6 CONCLUSIONS

It is inferred that, soil fertility status in Jalna district of Maharashtra was low with respect to nitrogen and zinc; medium in phosphorus and iron while high in potassium, copper and manganese. The current status of plant essential nutrients in soils is of practical utility to planners, extension workers, scientists and policy makers and this type of data are pre requisite for site specific nutrient management.

KEYWORDS

- **Assessment**
- **Jalna District**
- **Micronutrients**
- **Nutrient Indices**
- **Organic Carbon Status**
- **Semi-Arid Region**
- **Soil Fertility**

REFERENCES

Alloway, B. J. (2006). Zinc in soils and crop nutrition. Online book published by the International Zinc Association, Brussels, Belgium. Available via http://www.zinc-crps.org/.

Bhandari, S. C. (2013). Soil phosphorus management and environment aspects. *Journal of Indian Society of Soil Science, 61* (Supplement), 49–58.

Challa, O., Vadivelu, S., & Sehgal, J. (1995). Soils of Maharashtra for Optimizing Land Use. NBSS Publ. 54b (Soils of India series). National Bureau of Soil Survey and Land Use Planning (ICAR), Nagpur India, 16.

Challa, O., Gajbhiye, K. S., & Velayutham, M. (1999). Soil Series of Maharashtra, NBSS Publ. *79*, 410

Chesnin, L., & Yein. C. H. (1951). Turbidimetric determination of available sulfates. *Soil Science Society of America Proceedings, 15*, 149–157.

Deb, D. L., Sakal, L. R., & Datta, S. P. (2009). Micronutrients in fundamental of Soil Science. *Journal of Indian Society of Soil Science,* 441–490.

Gajbhe, M. V., Lande, M. G., & Varade, S. B. (1976). Soils of Marathwada. *Journal of Maharashtra Agricultural Universities, 1*, 55–59.

Gangwar, V., Singh, V. K., & Ravi, S. (2013). Fertilizer best management practices in important cropping systems. *Indian Journal of Fertilizers*, *9*(4), 34–51.

Ghosh, P. K., Bandopadhyay, K. K., Misra, A. K., & Rao, A. S. (2004). Balanced fertilization for maintaining soil health sustainable agriculture. *Fertilizer News, 49*(4), 13–25, 35.

Jackson, M. L. (1967). Soil Chemical Analysis, Prentice Hall of India Pvt. Ltd New Delhi.

Jackson, M. L. (1973). Soil Chemical Analysis (Eds.) Prentice Hall of India, Pvt. Ltd, New Delhi.

Kaushal, G. S., Sinha, B. R., & Sinha, S. B. (1980). Morphology and taxonomy of black soils under Bargi irrigation project in Madhya Pradesh. *Journal of Indian Society of Soil Science, 34*, 329–333.

Kharche, V. K., Pharande, A. L., & Patil, J. D. (2003). Micronutrient in soils and plants: An overview for Western Maharashtra. Interface Session on Micronutrient, MCEAR, Pune.

Lindsay, W. L., & Norvell, W. A. (1978). Development of a DTPA soil test for zinc, iron, manganese and copper. *Soil Science Society of America Journal, 42*, 421–428.

Maji, A. K., Pande, S., Velayutham, M., & Ahmed, M. I. (2000). Changes in wheat productivity in the Indo-Gangetic plains of India. Pages 61–64, *In:* GIS application in cropping system analysis – Case studies in Asia: proceedings of the International Workshop on Harmonization of Databases for GIS Analysis of Cropping Systems in the Asia Region, 18–29 August 1997, ICRISAT, Patancheru, India.

Malewar, G. U. (2005). Micronutrient stresses in soils and crops: Serious sickness and clinical approaches for sustainable agriculture. *Journal of Indian Society of Soil Science,* 53, 484–499.

Olsen, S. R., Cole, C. V., Watanable, F. S., & Dean, L. A. (1954). Estimation of available phosphorus in soil by extraction with $NaHCO_3$. Cir. U.S. Dept. Agric. 939.

Parker, F. W., Nelson, E., Winters, E., & Miles, K. F. (1951). The broad interpretation and application of soil test information. *Agronomy Journal, 43*, 105–112.

Patil, D. B., Bharambe, P. R., Deshmukh, P. W., Rane, P. V., & Guldekar, S. V. (2004). Micronutrient status in soils of Vidharba. *Technical Bulletin* Dr. PDKV, Akola.

Patil, J. D., & Kharche, V. K. (2001). Crop response to micronutrients in soils of western Maharashtra – An overview. *Journal of Maharashtra Agricultural Universities,* 26, 127–129.

Patil, Y. M., & Sonar, K. R. (1994). Status of major and micronutrients in swell-shrink soils of Maharashtra. *Journal of Maharashtra Agricultural Universities, 19*, 169–172.

Piper, C. S. (1966). Soil and Plant Analysis, Hans. Pub. Bombay. Asian Ed. 368–374.

Rao, A. S., & Srinivasarao, Ch. (1996). Potassium status and crop response to potassium on the soils of agro-ecological regions of India. IPI Research Topic No. 20, International Potash Institute, Basel, Switzerland, 1–57.

Sakal, R. (2001). Efficient management of micronutrient for sustainable crop production. *Journal of Indian Society of Soil Science, 49*, 593–608.

Shukla, A. K. (2011). Micronutrient research in India: Current status and future strategies. *Journal of Indian Society of Soil Science, 59* (Supplement), 88–98.

Singh, M. V. (2007). Micronutrient Deficiencies in Global Crop Production, 93–125.

Singh, J. P., Singh, D., & Litoria, P. K. (2009). "Selection of Suitable sites for water harvesting structures in Soankhad Watershed, Punjab using Remote Sensing and Geographical Information System (RS & GIS) Approach: A Case Study," *Journal of the Indian Society of Remote Sensing, 37*, 21–35.

Singh, M. V. (2003). Issues and challenges in promoting safe use of micronutrients in agriculture during new millennium: Interface on micronutrients, MCEAR, Pune.

Subbiah, B. V., & Asija, G. L. (1956). A rapid procedure for the estimation of available nitrogen in soils. *Current Science, 25*, 258–260.

Walkley, A., & Black, C. A. (1934). An examination of the different method for determining soil organic matter and proposal modification of the chromic acid titration method. *Soil Science, 37*, 29–38.

World Health Organization (WHO), (2002). The Word Health Report.

Zende, G. K. (1987). Invited Paper in Proceedings on Micronutrient Stress in Crop Plants, MPKV, Rahuri.

CHAPTER 18

ASSESSMENT OF SOIL FERTILITY STATUS IN SOYBEAN GROWING SOILS: A CASE STUDY FROM THE SEMI-ARID REGION OF CENTRAL INDIA

A. S. GAJARE,[1] and A. S. DHAWAN[2]

[1]*Research Scholar, Department of Soil Science and Agricultural Chemistry, Dr. Panjabrao Deshmukh Krishi Vidyapeeth, Akola – 444 104, India*

[2]*Director of Extension Education, Vasantrao Naik Marathwada Agricultural University, Parbhani – 431 402, India*

CONTENTS

ABSTRACT

The edible oil needs of an ever increasing population of the country have been greatly supplemented by the yellow revolution, which used to be just 5 m tons at the time of India's Independence. A significant role has been played by introduction of soybean (*Glycine man* (L.) Merr.) and sunflower (*Helianthusannutrs* L.) in late sixties for commercial cultivation. At present, soybean covers an area of nearly 7 m ha in the country and can be placed neck and neck to traditionally cultivated and established crops like groundnut (*Arachis hypogaea* L.) and rapeseed-mustard group of crops. Hence, the case study of semi arid region of central India has been carried out to know the fertility status with respect to major macro and micronutrients of major soybean growing soils of Latur district in the year 2009–10. For this purpose, 140 representative soil samples (0–15 cm depth) were collected from seven tehsils (20 villages from each tehsil) of Latur district and analyzed for available Phosphorus (P), Sulfur (S) and some micronutrients *viz.*, Iron (Fe), Manganese (Mn), Zinc (Zn) and Copper (Cu) following standard procedures. The available phosphorus in these soils varied from 0.20 to 21.15 kg ha^{-1} with an average value of 7.42 kg ha^{-1}, out of 140 soil samples 66.43% were low (< 10 kg ha^{-1}) and 33.57% samples were found medium (10 To 25 kg ha^{-1}) in available P content. Available sulfur ranged from 3.62 to 53.52 mg kg^{-1} with a mean value of 21.47 mg kg^{-1}, about 12.14% samples were low (<5mg kg^{-1}), 32.14% medium (5 to 10 mg kg^{-1}) and 55.72% high (> 10 mg kg^{-1}) in sulfur content. In case of micronutrients, available Zn ranged from 0.12 to 2.96 mg kg^{-1}, the available Fe varied from 0.48 to 17.1 mg kg^{-1}, available Mn content were 0.16 to 21.12 mg kg^{-1}, Cu was very high varying from 0.25 to 12.19 mg kg^{-1}. Renapur and Chakur tehasils were found deficient in available P, all tehsils were depleted in case of Zn, available Fe, Mn, Cu and S were adequate in all tehsils. Soil Nutrient Index values of the soybean growing soils of Latur district were high in S, Fe, Mn and Cu, whereas low in available P and Zn. All the seven tehsils were high in available Mn and Cu, while all the tehsils were low in available Zn. In case of available Fe, Renapur tehsil belonged to medium category while remaining six tehsils were in high category. The soybean growing soils of Latur district were found depleted with respect to available phosphorus, whereas they were adequate in available sulfur.

18.1 INTRODUCTION

Soybean is a triple benefit crop as it contains about 20% oil, 38–42% high quality protein and being leguminous crop fixes atmospheric nitrogen in soils @65–100 kg per hectare (Quayum et al., 1985). The capacity of soybean plants to fix nitrogen through nodules can be utilized for succeeding crops in soybean-based cropping system, thus reducing the requirement of fertilizer input in cropping system. It builds up the soil fertility by fixing large amount of atmospheric nitrogen through root nodules and fall leaf. (Mendhe, et al., 2008). Venkatswralu, (1985) also reported that under hostile moisture regime and intermittent moisture stress the soybean is most efficient crop due to shorter duration and inherent genetic capability to produce yield. It is also tolerant to wide range of soil conditions and can be grown in all types of soils (Martin et al., 1976).

Soybean being oilseed legume phosphorus plays a very important role in its nutrition. In order to improve and sustain performance of the soybean crop adequate supply of phosphorus and sulfur is to be insured. Phosphorus is the structural component of plant cell and required for early root development and growth. It helps in flowering and fruiting, an adequate supply of P_2O_5 in the early stages helps in initiating its reproductive parts. It hastens maturity and improves quality of grain. During pod filling stage, phosphorus is transported from leaves, branches and main stem to grain. Out of total uptake of phosphorus by soybean plant, 60% phosphorus comes in to grain therefore application of P is must. Further, excess of phosphorus may cause in some cases trace element deficiency particularly zinc and iron (Pattanayak et al., 2009).

Besides phosphorus, sulfur is an essential secondary plant nutrient, which plays a vital role in biosynthesis of primary metabolism for improving yield and quality of oilseed crop and for achieving better yield under balance fertilization. Sulfur is constituent of amino acid namely cystein and mithionine. It is necessary for chlorophyll formation and biosynthesis of oil from enzyme and vitamins. Sulfur also plays an important role in synthesis of certain vitamins and protein in the formation of flavored compounds. Since, nitrogen and sulfur are closely linked in protein metabolism, their combined effect on crop could be synergistic. Use of high analysis, low or no sulfur containing fertilizers, decrease in atmospheric input of sulfur from burning of fossils fuels, increasing removal of soil sulfur through higher yields, higher cropping intensity, wide gap between removal and addition of sulfur, losses of sulfur through leaching and erosion leads to a marked increase in area of sulfur deficiency in soils and crops (Sundaran, 1986).

Sulfur deficiency has been recorded in India with the expansion of area under oilseed (Khamparia et al., 2009), increased use of sulfur free fertilizers and inadequate use of organic manures in intensive cropped area and this problem is likely to aggravate in the near future. Tandon (1986) indicated that the extent of sulfur deficiency is 15 to 20% in the country. Whereas, 35% soils are moderately deficient in available sulfur in Marathwada (Lande et al., 1977). The area speculated as sufficient in sulfur have now started showing sulfur deficiency in some districts of Maharashtra (Lande et al., 1977).

18.1.1 SOYBEAN GROWING COUNTRIES

Soybean is grown in 111.54 m ha in the world and produces 276.03 m tons. India, with 12.20 m ha area ranks fourth and with 11.94 m tons of production ranks fifth in the world. While the USA tops the list in soybean acreage and production. The other major soybean producing countries are Brazil, Argentina and China (Table 18.1).

Soybeans were a crucial crop in East Asia long before written records began. There is evidence for soybean domestication between 9000 and 8600 BP (before present) in China, between 7000 BP and 5000 BP in Japan and 3000 BP in Korea.

18.1.2 NUTRIENT STATUS OF SOYBEAN GROWING SOILS OF INDIA

Indian farmers cultivated soybean from ages in northern and northeastern hills as food plant and is a part of routine diet of the people (Tiwari et al., 1999). Black-seeded soybean has been grown since early times in the northern and north-eastern hills and in scattered areas in the central part

TABLE 18.1 Harvested Area of Soybean in the World (2013)

Rank	Top 5 Countries	Harvested Area (in m ha)
1	United States of America	30.70
2	Brazil	27.90
3	Argentina	19.41
4	India	12.20
5	China	6.79

Source: Indiastat (2015).

of the country. Indian agriculture during last decade is marked by distinct change of cropping pattern from cereal to Soybean *(Glycine Max L.Merill)* especially in arid, semi-arid and dry sub humid parts of Deccan Plateau. Although soybean is cultivated from prehistoric time in China and introduced in USA 1804, its journey to India is only 1880 A.D. (Rathore et al., 2011). Popularity of soybean-based cropping system has gained importance since 1970, as edible oil crisis appeared in India. The development of many short duration varieties and high price of soybean grain, maintenance free cultivations makes it more popular.

The acreage under cultivation has been steadily increasing from 300 ha in 1968 to 7.7 m ha in 2009–2010. India ranks 5th in terms of cultivated area of 7.7 m ha with production of 6.2 m tons. Madhya Pradesh, Maharashtra and Rajasthan are the major soybean-growing states in that order. Madhya Pradesh contributes to the extent of 65% with respect to area and 56% to production (2009–10). Madhya Pradesh and Maharashtra grow soybean largely on Vertisols and associated soils and constitute about 86.8% of total area. Addition of Rajasthan makes this figure to 87.67%. For immediate gains in national production, the concerted efforts to enhance the productivity are essentially required in these 3 states, followed by other potential areas (Joshi 2003). The other upcoming states are southern peninsula like Andhra Pradesh and Karnataka.

Maharashtra is one of the leading soybean growing state occupying about 2.65 m ha with an annual production of 32.37 lakh million metric tons with an average, productivity of 12.53 q/ha (2010–2011). More than 75–80% area lies in Vidharbha with cultivable area of 16.40 lakh ha and productivity of 9.59 q/ha. Soybean is the popular *kharif* crop amongst farmers of Vidarbha region (Mandal et al., 2005), Whereas during the year 2009–2010 area under soybean crop in Latur District was 1.945 lakh ha with an annual production of 4318 Mt and productivity of 2220 kg ha^{-1} (Anonymous, 2010).

Recently Pathak et al. (2010) estimated nutrient budget of Indian agriculture and showed that there were positive balances of N (1.4 Mt) and P (1.0 Mt) and a negative balance of K (3.3 Mt). Krishna Prasad et al. (2004), Krishna Prasad and Badarinath (2006) and Murugan and Dhadwal (2007) also estimated net accumulation of N in soil (positive N balance) ranging from 1.9 to 14.4 Mt. However, these estimates differed from that of Fertilizer Association of India (FAI, 2000), which projected a negative balance of N. This is because of exclusion of addition of N through irrigation, rain and crop residues in the study. Panda et al. (2007) observed positive balances of N and P and negative balance of K in the long-term experiments in rice-rice systems in the treatments with N and P application.

18.1.2 IMPORTANCE OF THE STUDY

Besides phosphorus and sulfur some of the micronutrients like Fe, Mn, Zn and Cu are required to improve and sustain performance of the soybean crop (Mali and Ismail, 2002). Zinc promotes growth hormones, starch formation, seed maturation and production, whereas, Iron helps in the absorption of other nutrients. It also plays an important role in chlorophyll formation along with Mn. Copper has some indirect effect on nodule formation. The analysis of soils from sixteen states and two union territories for micronutrient cations indicated that Zn deficiency was the most serious constrain to sustainable productivity in 11 states (Takkar, 1996). The physico-chemical characteristics such as pH, EC, Calcium carbonate and organic carbon are important as these affect availability of nutrients in the soil thereby crop growth and production.

However, there is a large decline in yield levels of the crop in recent years and farmers are shifting to other crops. Low soil fertility and inadequate manuring are the major causes for the low yield of the soybean crop. The nutrient imbalance on account of deficit supplementation of nutrients has been documented in soils of India, particularly in dry lands (Katyal and Reddy 1997). Such gap (imbalance) is going to widen further in future, if adequate care is not taken to replenish the soil fertility by fertilizers or and by organic resources. The work done under Long Term Fertilizer Experiments (LTFE) program at Jabalpur, Indore and also at the Indian Institute of Soil Science has clearly brought out that nutrient management through integrated approach can sustain the productivity of soybean-based cropping systems. The work reported by Dravid et al. (2000) on differential dependence of soybean varieties on native and applied sources and supplementation of nutrients based on crop-growth stage when needed most and application of nutrients, particularly nitrogen and sulfur can further facilitate the refinement of fertilizer management schedule to achieve optimum performance of soybean-based cropping systems. There appears a need to readjust the supplementation of sulfur (Sharma et al., 2004) and potassium to meet the need of the crop to enhance the productivity (Joshi 2004). Dispensing with the prevailing faulty nutrient-management practices and mitigating the micronutrient deficiency by integrated approach to achieve sustainable production has adequately been suggested (Joshi 2004).

Because of intensive cultivation of oilseed crops and the role and requirement of sulfur in oil seed production, the status of available S in soil cannot be ignored. In recent years, the deficiency of sulfur is becoming common due to use of high analysis S free fertilizers coupled higher sulfur removal by the crops.

The deficiency of sulfur is expected in oilseed dominated area of the region, because of heavy sulfur requirement and higher removal by the oilseed crops. Availability of sulfur is influenced by several soil factors. Establishing the relationship between various soil factors and sulfur content of soils is important. Hence, the case study has been carried out in semi arid region of the central India

The present study was undertaken with an objective to assess the fertility status in soybean growing soils of Latur district, Maharashtra, India. An attempt was made to know the chemical properties of soybean growing soils, to determine the fertility status of soybean growing soils, to find out the status of micronutrient cations, to find correlation between chemical properties and available nutrients and to prepare nutrient index value of soybean growing soils of Latur district.

18.2 STUDY AREA

Latur district of Maharashtra state lies between 18°05′ to 18°75′ North latitude and 73°25′ to 77°25′ East latitude with elevation of 613 m from mean sea level (Figure 18.1). Geographical area of Latur district is 7166 sq. km. The district may be divided into two regions the Balaghat plateau and the northeastern region consisting of Ahmadpur and Udgir tehsils. Latur district is bound by Nanded district to the northeast; the state border with Karnataka to the east and southeast; Osmanabad district to the south-west; Beed district to the west; and Parbhani district to the northwest.

Soybean can be grown in fairly wide range of climates and soils. Latur district comes under semiarid and tropical region of Maharashtra state. It receives

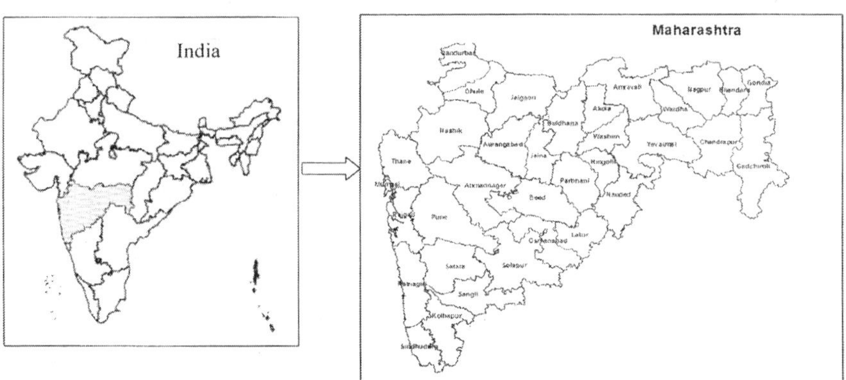

FIGURE 18.1 Location map of the study area.

about 750–800 mm precipitation. This is usually during the monsoon months from July to October. Summers begin from early March to July. Summers are dry and hot. The mean minimum and maximum temperatures of Latur district are 13.9–39.6°C, respectively, though at the peak they may reach 41°C. November to January is the winter season (around 13.9–21.8°C). January to March are the months with moderate temperature and the optimum temperature for rapid germination of the soybean is 30°C. The metrological data of Latur district during the crop growing season (2009–2012) is shown in Figure 18.2.

The district lies on the basaltic rock also known as "trap" rock. The deep soils in the district are seen to extend in the valleys of Manjra, Terna, Rena, Manar, Tawarja, Tiru and Gharni. These rivers are the branches of river Godavari, which runs from west to east. Soils of varying depth developed as a result of undulating topography exist in other parts of the district. The soils are mostly black varying in texture from clay to clay loams. Sandy loam soils are present in hilly areas. The soils of the district are heavy, medium and light. However, the major portion is under medium and heavy soils. Majority soils in these areas were categorized under the order Vertisol followed by Inceptisol and Entisol. The soils with a normal pH of 7 and fair degree of water retention are suitable for its cultivation. It grows well in sandy loam to clay soil. The methods and package of practices followed for soybean cultivation in Maharashtra are shown in Table 18.2.

18.3 MATERIALS AND METHODS

18.3.1 SITE SELECTION

The major soybean growing soils of Latur district were selected for the study after traversing the area. The major soybean growing seven tehsils (Latur,

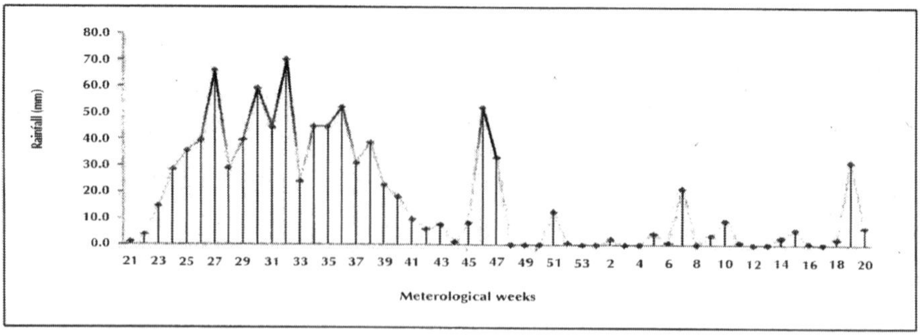

FIGURE 18.2 Weekly rainfall data (2009–2012) of Latur district, Maharashtra.

TABLE 18.2 Methods and Package of Practices Followed for Soybean Cultivation in Maharashtra

Variety	JS-335, Samrudhi, TAMS-38
Crop Maturity Days	90–115 days (seed to seed)
Seed bed	Well prepared, deep, friable, 1 Ploughing, 1 Harrowing
Sowing	Done on flat beds and by drilling
Sowing season	June 15 to July 1
Date of time of harvest	1st week of October
Seed Rate	30 kg/acre.
Spacing	15–30 cm row-to-row
Seed Treatment	Farmers rarely apply Rhizhobium treatment to seeds before sowing
Fertilizer Application	1 bag DAP/Acre
Weeding	By sickle and mostly use of weedicides
Harvesting	By sickle
Yield	8–10 Quintal/ha
Market Rate	1800–2000 Rs/q

Renapur, Nilanga, Ausa, Udgir, Chakur and Ahmedpur) were selected for collection of soil samples. 20 soil samples from each tehsils were collected at the depth of 0–15 cm from farmer's field (Figure 18.3). One hundred and forty soil samples were collected from seven tehsil of Latur district where soybean crop is being grown since last 3–4 years as a component of dominant cropping system in the district keeping in view the physiographic characteristics in different cross sections of the area.

18.3.2 PREPARATION OF SOIL SAMPLES

Soil samples collected from different villages of Latur district were brought to the laboratory, thoroughly mixed, air dried in shade, ground with porcelain mortar and pestle and passed through 2 mm sieve for general analysis. For micronutrient analysis, it was passed through muslin cloth. The sieved soil samples were stored in cloth bags with proper labeling for subsequent analysis. All the precautions outlined by Jackson (1973) were scrupulously followed in order to avoid contamination.

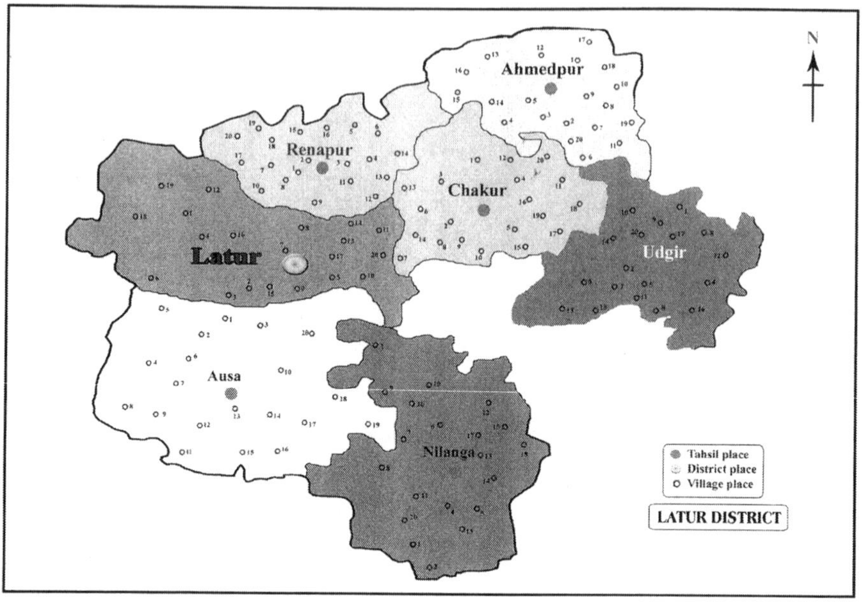

FIGURE 18.3 Sample site locations in selected villages of Soybean growing soils in Latur district.

18.3.3 ANALYSIS OF AVAILABLE MAJOR NUTRIENTS

Soil pH is one of the most decisive measurements of the chemical properties of a soil. The concept of pH is based on the ionic product of pure water. The pH and KCl pH of a soil was measured by ELICO pH meter, after equilibrating the soil with water in the ratio of 1:2 soil: water suspensions (Jackson, 1973). A fairly qualitative estimate of the salt content of the solutions extracted from soils or of natural waters can be made from their electrical conductance. The EC of a soil: water (1:2) suspension was determined by ELICO conductivity bridge (Jackson, 1973). Organic carbon was determined in soil samples ground to 100 mesh size, by modified Walkley and Black rapid titration procedure (Jackson, 1973). Calcium carbonate equivalent of the soils were determined by treating 100-mesh soil sample with excess hydrochloric acid of known volume (and a known strength of $\approx 0.5N$ HCl) to neutralize all carbonates. The samples were heated. The excess of hydrochloric acid was back titrated with standard NaOH solution using phenolphthalein as an indicator (Piper, 1966). Available phosphorus was determined by extracting the soil with 0.5 M sodium bicarbonate, pH

8.5 (Olsen's reagent) as an extracting reagent and determined using double beam UV-VIS spectrophotometer with stannous chloride indicator as described by Olsen et al. (1954). Available Sulfur was estimated by extracting the soil with 0.15% $CaCl_2$ in 1:5 ratio. The S content in the filtrate was determined with turbidity method using spectrophotometer (Williams and Steinberg, 1959).

18.3.4 ANALYSIS OF AVAILABLE MICRONUTRIENTS

Micronutrient cations *viz.*, Fe, Mn, Zn and Cu in soil were determined as per the procedure described by Lindsay and Norvell (1978). For this 10 g finely sieved soil (0.5 mm) was shaken in 20 ml of 0.05 M DTPA solution (Diethylenetriamine penta acetic acid containing 0.1 M triethanol amine and 0.01 M calcium chloride, adjusted to pH 7.3 with HCL) for two hours and filtrate through Whatman No. 42 filter paper and clear filtrate was collected. The filtrate was used for measurement of Fe, Zn, Mn and Cu using Atomic Absorption Spectrophotometer (AAS-200), at different wavelength for Fe, Zn, Mn, and Cu.

18.3.5 SOIL NUTRIENT INDEX

Nutrient index was calculated as per the formula suggested by Ramamoorthy and Bajaj (1969) and the values for low, medium and high will be taken as <1.67, 1.67–2.33, >2.33, respectively.

$$\text{NIV} = \frac{(\text{No. of samples "Low"} - 1) + (\text{No. of samples "Medium"} - 2) + (\text{No. of samples "High"} - 3)}{\text{Total number of samples}}$$

Rating of NIV

Category	Value
Low	<1.67
Medium	1.67–2.33
High	>2.33

The correlations of chemical characteristics with available nutrients were worked out as per standard method given by Panse and Sukhatme (1967).

18.4 RESULTS AND DISCUSSION

18.4.1 CHEMICAL PROPERTIES

The data presented in Table 18.3 shows that the pH of the soybean growing soils of Latur district ranged from 6.39 to 8.89 with an average value of 7.98. These soils were neutral to alkaline in reaction. The relative high value of pH of these soils might be due to high degree of base saturation (Mali and Raut, 2001).

The EC values of these soils ranged from 0.12 to 1.96 dSm^{-1} with an average value of 0.31 dSm^{-1}. The low EC value observed in these soils may be due to leaching of salts from surface layer soils. These results are in conformity with Dwivedi et al. (2005).

The organic carbon content varied from 0.15 to 1.03% with an average value of 0.56%. The variation in the organic carbon content of soil may be attributed to the factor like high temperature of Latur district (39.3°C), which was responsible to hasten the rate of oxidation and very little addition of crop residues in the soil (Malewar, 1995).The $CaCO_3$ ranged from 0.66 to 14.26% with a mean value of 4.25%. The low to medium $CaCO_3$ content in the soils might be due to the presence of $CaCO_3$ in different proportions and in powdery form with hyper thermic temperate regime of Latur district. Similar findings were reported by Pharande et al. (1996).

18.4.2 STATUS OF AVAILABLE MAJOR NUTRIENTS

The available phosphorus content of the study area ranges from 0.2 to 21.15 kg ha^{-1} with an average value of 7.42 kg ha^{-1} (Table 18.4). Among seven

TABLE 18.3 Characteristics of Soybean Growing Soils of Latur District

Soil characteristics	Range	Mean
pH	6.39–8.89	7.98
EC (dSm^{-1})	0.12–1.96	0.31
Organic carbon (%)	0.15–1.03	0.56
$CaCO_3$ (%)	0.66–14.26	4.25

TABLE 18.4 Characterization of Available Phosphorus and Sulfur in Soybean Growing Soils of Latur District

	Phosphorus (kg ha^{-1})				Sulfur (mg kg^{-1})			
Tehsil	**Range (Mean)**	**< 10 Low**	**10 to 25 Medium**	**> 25 High**	**Range (Mean)**	**< 5 Low**	**5 to 10 Medium**	**> 10 High**
Ausa	6.59–18.3 (12.77)	4	16	-	4.32–52.3 (25.99)	4	4	12
Renapur	1.02–15.93 (5.86)	18	2	-	3.68–32.08 (16.98)	3	7	10
Latur	0.23–12.48 (4.06)	17	3	-	6.95–53.28 (20.43)	-	9	11
Nilanga	4.21–21.15 (10.65)	9	11	-	4.5–49.47 (28.50)	2	3	15
Chakur	0.2–12.35 (4.05)	18	2	-	3.62–41.72 (15.11)	4	7	9
Ahmed-pur	0.5–13.12 (4.63)	17	3	-	3.92–53.52 (21.53)	4	5	11
Udgir	2.42–17.18 (9.90)	10	10	-	5.85–51.03 (21.77)	-	10	10
Total		93	47	-		17	45	78
Percentage		66.43	33.57	-		12.14	32.14	55.72

Note: Figures in parenthesis indicate mean of twenty samples.

tehsils of Latur district, the lower range of available P content was 0.2 to 12.35 kg ha^{-1} with an average value of 4.05 kg ha^{-1} in Chakur tehsil. Whereas, the higher range of available P was 6.59 to 18.3 kg ha^{-1} with an average value of 12.77 kg ha^{-1} in Ausa tehsil followed by Nilanga tehsil (4.21 to 21.15 kg ha^{-1} with a mean value of 10.65 kg ha^{-1}). Out of 140 soil samples 66.43% were low (< 10 kg ha^{-1}) and 33.57% samples were medium (10 to 25 kg ha^{-1}) in available P content. More et al. (2005) studied the oilseed dominated area of Parbhani and reported that available P ranged from 0.5 to 14.28 kg ha^{-1}. Low available P in shrink-swell soil was also noticed by Patil and Sonar (1994). Similar results were found by Gajare et al. (2013) in swell-shrink soils of Latur district. Low available phosphorus content of these soils could be attributed to their high P fixing capacity, which prevents readily available form in the soil solution (Hundal et al., 2006). Beside this the continuous growing of high P requiring crop like soybean with inadequate P fertilizers might have depleted the available P balance.

The available sulfur ranged from 3.62 to 53.52 mg kg^{-1} with an average value of 21.47 mg kg^{-1}. Among seven tehsils the lower range of sulfur content was 3.62 to 41.72 mg kg^{-1} with an average value of 15.11 mg kg^{-1} in Chakur tehsil (Table 18.4). While, higher range of 4.5 to 49.47 mg kg^{-1} with an average value of 28.50 mg kg^{-1} was observed in Nilanga tehsil. It was followed by Ausa tehsil (4.32 to 52.3 mg kg^{-1} with mean value of 25.99 mg kg^{-1}). Out of 140 samples 12.14, 32.14 and 55.72% samples were categorized as low (<5 mg kg^{-1}), medium (5 to 10 mg kg^{-1}) and high (> 10 mg kg^{-1}), respectively, in sulfur content. From the above results, it was inferred that soils of Latur district were sufficient in sulfur content. The high amount of sulfur content in soils could be attributed to high amount of clay content, which can absorb varying amounts of sulfur (Mali and Raut, 2001). Singh et al. (2006) reported that the available sulfur content in valley hill soils of Manipur ranged from 10.0 to 7.00 mg kg^{-1} with an average value of 26.6 mg kg^{-1}. Chauhan et al. (2012) reported that the available sulfur ranged from 0.9 to 50.2 mg kg^{-1} with an average value of 13.9 mg kg^{-1}.

18.4.3 STATUS OF AVAILABLE MICRONUTRIENTS

The status of available Zn content in soybean growing soils of Latur district ranges from 0.12 to 2.96 mg kg^{-1} with an average value of 0.59 mg kg^{-1} (Table 18.5). Based on the critical limit (0.6 mg kg^{-1}) for DTPA-Zn, 67.14% samples were found to be deficient in Zn, which is in conformity of the findings of Patil and Kharche (2006). This might be due to the fact that in well-drained, aerated, calcareous soils, Zinc exits in oxidized state and their availability becomes very low (Malewar, 1995). It was also observed that Vertisol and Alfisol in Maharashtra region and some arid soils of Rajasthan are low in available content of Zn (Mahesh Kumar et al., 2011; Pharande et al., 1996).

The available Fe content ranges from 0.48 to 17.1 mg kg^{-1} with an average value of 6.10 mg kg^{-1} (Table 18.5). This high Fe content (65.72%) was due to the presence of minerals like feldspar, magnetite, hematite and limonite, which together constitute bulk of trap rock in these soils. The available Fe in Vertisol of Maharashtra was in the range from 3.52 to 19.44 mg kg^{-1}. Verma et al. (2008) and Chauhan et al. (2012) showed that available Fe ranged from 4.5 to 22.5 mg kg^{-1}. The present results are more or less in accordance with the above findings.

TABLE 18.5 Characterization of Micronutrients (Zn, Fe, Mn and Cu) of Soybean Growing Soils of Latur District

Tehsil	Zn (mg kg^{-1})				Fe (mg kg^{-1})				Mn (mg kg^{-1})				Cu (mg kg^{-1})			
	Range (Mean)	<0.6 L.	0.6 to 1.2 M.	>1.2 H.	Range (Mean)	<2.5 L.	2.5 to 4.5 M.	>4.5 H.	Range (Mean)	<2.0 L.	2.0 to 5.0 M.	>5.0 H.	Range (Mean)	<0.3 L.	0.3 to 0.5 M.	>0.5 H.
Ausa	0.27–1.74 (0.62)	13	5	2	2.65–9.4 (5.59)	–	9	11	1.25–13.48 (6.62)	1	5	14	1.44–12.19 (3.44)	–	–	20
Renapur	0.22–2.84 (0.66)	16	1	3	0.48–12.12 (4.62)	6	5	9	1.16–11.38 (4.43)	2	11	7	0.98–6.1 (2.41)	–	–	20
Latur	0.18–2.96 (0.69)	11	7	2	0.98–10.41 (5.93)	3	4	13	0.6–11.3 (5.29)	2	7	11	0.82–8.74 (2.65)	–	–	20
Nilanga	0.29–1.86 (0.64)	11	8	1	3.76–17.1 (6.92)	–	2	18	2.35–14.28 (6.94)	–	7	13	1.08–7.12 (2.85)	–	–	20
Chakur	0.12–0.95 (0.44)	15	5	–	1.98–10.43 (6.06)	1	5	14	0.68–4.88 (2.44)	6	13	1	0.25–5.5 (2.35)	1	–	19
Ahmed-pur	0.23–1.75 (0.60)	12	7	1	2.66–11.32 (6.66)	–	7	13	1.18–9.15 (4.72)	2	8	10	1.38–6.5 (4.01)	–	–	20
Udgir	0.2–1.24 (0.49)	16	3	1	1.18–12.86 (6.94)	2	4	14	0.16–21.12 (6.78)	2	5	13	0.98–7.65 (3.52)	–	–	20
Total		94	36	10		12	36	92		15	56	69		1	–	139
Percentage		67.14	25.71	7.14		8.57	25.71	65.72		10.71	40	49.29		0.71	–	99.29

Note: Figures in parenthesis indicates mean of twenty samples; L. – Low; M. – Medium; H. – High.

The available Mn content in these soils varied from 0.16 to 21.12 mg kg^{-1} with an average value of 5.32 mg kg^{-1} (Table 18.5). Among 140 samples, 40.00% samples were medium and 49.29% samples were high in available Zn content. This sufficient status of Mn might be due to the fact that lower oxidation (reduced) states of Mn was more soluble than higher oxidation state at normal pH range of soils and oxidation of divalent Mn^{++} to trivalent Mn^{+++} by certain fungi and bacteria, certain organic compounds synthesized by microorganisms or released by the plants as root exudates. Dwivedi et al. (2005) reported that the available Mn ranged from 2.23 to 49.1 mg kg^{-1} and Chauhan et al. (2012) showed that available Mn ranged from 1.70 to 25.70 mg kg^{-1}. Similar results were also found by Gajare et al. (2015).

The soybean growing soils of Latur district were high in Cu content (0.25–12.19 mg kg^{-1}) with an average value of 3.03 mg kg^{-1} (Table 18.5). Out of 140 samples 99.29% of soil samples were high in available Cu content. Most of soil the samples were high in available Cu. This Cu content could be attributed to the difference in geology, physiography and degree of weathering in these soils. Ghosh et al. (2010) showed that the content of available Cu ranged from 0.5 to 4.0 mg kg^{-1} and Chauhan et al. (2012) showed that the content of available Cu ranged from 0.24 to 5.90 mg kg^{-1}.

18.4.4 CORRELATION BETWEEN CHEMICAL PROPERTIES AND AVAILABLE NUTRIENTS

It was observed from the data that the pH showed negative correlation with phosphorus (–0.376**) and Sulfur (–0.162) but it was significant in case of available Phosphorus (Table 18.6). The availability of Phosphorus in soil is determined largely by the ionic form of this element. The correlation between pH and available P_2O_5 observed in this study was also reported by Sakal et al. (1986). Soil pH showed negative correlation with available Sulfur. This might be due to effect of soil pH on availability of Sulfur. When pH of soil increased the availability of sulfur decreased. Similar findings were observed by Mahapatra and Sahu (1996). Further data revealed that the organic carbon showed positive and highly significant correlation with available Phosphorus (0.362**) and Sulfur (0.269**) due to the fact that in mineral soil one-half to two-third of the total P is organic. The organic P compounds can move in soils to a great depth up to 2 to 4 feet, than can

TABLE 18.6 Correlation Coefficient Between Chemical Properties and Available Nutrients

Chemical properties	P	S	Zn	Fe	Mn	Cu
pH	–0.376**	–0.162	–0.249*	–0.132	–0.114	–0.190
EC	0.186	0.157	0.304**	0.281**	–0.278**	0.199*
OC	0.362**	0.269**	0.171	0.217*	0.185	0.310**
$CaCO_3$	–0.253*	–0.311**	–0.198*	–0.291*	–0.179	–0.170

inorganic P in solution (Tisdale et al., 1997). The $CaCO_3$ were negative and significantly correlated with available Phosphorus (–0.253**) and Sulfur (–0.311**). The availability of P was also found to decrease with increasing $CaCO_3$ in soil. This decreasing P availability may also be associated with increased pH due to high content of $CaCO_3$ in soil. The amount and reactivity of calcium carbonate influences the P fixation. In case of $CaCO_3$, it showed negative relationship with available Sulfur because it increases the pH of the soil and decreases the availability of Sulfur. Similar results were also observed by Sharma et al. (2003).

The increase in DTPA-Zn with Decrease in soil pH was substantiated by the overall negative correlation obtained between Zn and soil pH (r = 249**). There was positive correlation between Fe and OC (r = 0.217**) and negative and significant correlation was obtained between Fe and soil pH (r = –132**). This is substantiated by the overall negative correlation of DTPA-Mn with soil pH (r = –0.114) and positive correlation obtained between DTPA-Mn and organic carbon (r = 0.185). The DTPA-Cu content increased with increase in OC (r = 0.310**) and decreased in pH (r = –0.190) of the soils (Table 18.6).

18.5 NUTRIENT INDEX VALUE

18.5.1 SOIL NUTRIENT INDEX FOR PHOSPHORUS, SULFUR AND MICRONUTRIENTS

The soil nutrient indices for phosphorus and sulfur were presented in Table 18.7. Among seven tehsils Renapur, Latur, Nilanga, Chakur, Ahmedpur and Udgir are low in phosphorus content except Ausa, which has medium phosphorus content. The values worked out for Renapur, Latur, Nilanga, Chakur, Ahmedpur, Udgir and Ausa were 1.1, 1.15, 1.55, 1.1, 1.15, 1.5, and 1.8,

TABLE 18.7 Nutrient Index Value for Phosphorus and Sulfur Content

		Phosphorus		Sulfur	
S. No.	Tehsil	NIV	Category	NIV	Category
1	Ausa	1.8	Medium	2.4	High
2	Renapur	1.1	Low	3.5	High
3	Latur	1.15	Low	2.55	High
4	Nilanga	1.55	Low	2.65	High
5	Chakur	1.1	Low	2.25	Medium
6	Ahmedpur	1.15	Low	2.35	High
7	Udgir	1.5	Low	2.5	High
8	Latur district	1.34	Low	2.44	High

respectively. While all the tehsils were categorized as High in Sulfur content, Chakur, which was in medium category. The values worked out for tehsils Ausa, Renapur, Latur, Nilanga, Ahmedpur, Udgir and Chakur were 2.4, 2.35, 2.55, 2.65, 2.35, 2.5, and 2.25, respectively.

The Nutrient Index Value (NIV) for micronutrients (Zn, Fe, Mn and Cu) content in soils are shown in Table 18.8. The NIV for available Zinc was low in all the seven tehsils. The values of Nutrient Index of Zinc were 1.45 (Ausa), 1.35 (Renapur), 1.55 (Latur), 1.5 (Nilanga), 1.25 (Chakur), 1.45 (Ahmedpur) and 1.25 (Udgir). Ausa, Latur, Nilanga, Chakur, Ahmedpur and Udgir were found under category High for available Iron (Fe) and Renapur was categorized as Medium in Fe content. The values worked out for Nutrient

TABLE 18.8 Nutrient index value for Zn, Fe, Mn and Cu content.

		Zn		Fe		Mn		Cu	
S. No.	Tehasil	NIV	Category	NIV	Category	NIV	Category	NIV	Category
1	Ausa	1.45	Low	2.55	High	2.60	High	3	High
2	Renapur	1.35	Low	2.15	Medium	1.95	High	3	High
3	Latur	1.55	Low	2.50	High	2.45	High	3	High
4	Nilanga	1.50	Low	2.90	High	2.90	High	3	High
5	Chakur	1.25	Low	2.65	High	2.90	High	2.9	High
6	Ahmedpur	1.45	Low	2.65	High	2.75	High	3	High
7	Udgir	1.25	Low	2.60	High	2.60	High	3	High
8	Latur district	1.40	Low	2.57	High	2.39	High	2.99	High

Index of Fe were 2.55 (Ausa), 2.5 (Latur), 2.9 (Nilanga), 2.65 (Chakur), 2.65 (Ahmedpur), 2.6 (Udgir), and 2.15 (Renapur). All the seven tehsils were found high for both Manganese and Copper and the values worked out for Nutrient Index of Manganese were 2.6, 1.9, 2.45, 2.9, 2.9, 2.75 and 2.6 for Ausa, Renapur, Latur, Nilanga, Chakur, Ahmedpur and Udgir, respectively. For copper, nutrient index values were 3 for Ausa, Renapur, Latur, Nilanga, Ahmedpur, Udgir and 2.9 for Chakur.

18.6 SOIL NUTRIENT INDEX OF LATUR DISTRICT

The analysis Soil Nutrient Index of Latur district shows that it is high for sulfur, Fe, Mn and Cu, while it was low for phosphorus and Zn. The value worked out for sulfur, Fe, Mn, Cu, phosphorus and Zn were 2.44, 2.57, 2.39, 2.99, 1.34, and 1.4, respectively (Figure 18.4). From the analysis of nutrient index of Latur district with respect to phosphorus, sulfur and micronutrients like Zn, Fe, Mn and Cu, it is observed that the soybean growing soils of the district were depleted with respect to available phosphorus and zinc, while adequate with respect to available sulfur and other micronutrients.

18.7 CONCLUSIONS

From the analysis of soybean growing soils of Latur district with respect to phosphorus and sulfur it is revealed that the soils were depleted with respect to available phosphorus and adequate with available sulfur. Though available sulfur was adequate, application of single super phosphate should be preferred over sulfur free DAP or other complex fertilizers as a source of

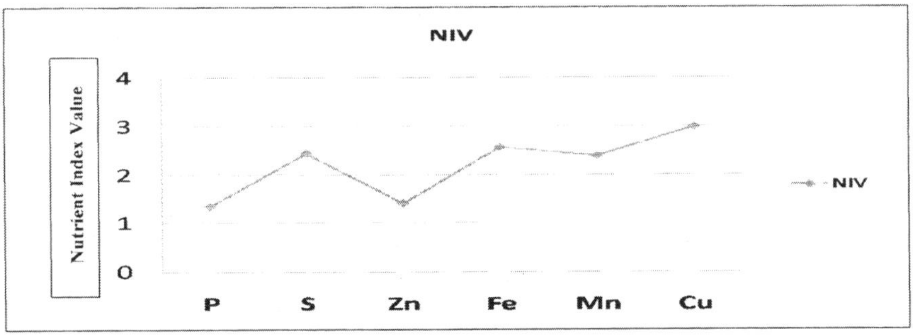

FIGURE 18.4 Nutrient index value of soybean growing soils of Latur district.

phosphorus to prevent sulfur depletion. Among the micronutrients, available Zn became a limiting factor and therefore, application of $ZnSO_4$ (25 kg $ZnSO_4$/ha) could help to sustain soybean yield.

KEYWORDS

- **Available P, S**
- **Correlation Coefficient**
- **Micronutrients**
- **Nutrient Indexing**
- **Soil Properties**
- **Soybean Growing Soils**

REFERENCES

Anonymous, (2010). Agricultural secretariat Government of Maharashtra survey number. 191/2010.

Chauhan, N., Sharma, G. D., Khamparia, R. S., & R. K. Sahu, (2012). Status of sulfur and micronutrients in medium black soils of Dewas district, Madhya Pradesh. *Agropedology, 22*(1), 66–68.

Dravid, M. S., Joshi, O. P., & Vyas, B. N. (2000). P utilization in soybean as influenced by plant genotypes and growth promoter (Vipul) under field conditions. *Journal of Nuclear Research in Agriculture and Biology, 29*(2), 83–6.

Dwivedi, S. K., Shrama, V. K., & Bhardwaj, V. (2005). Status of available nutrients soils of cold arid region of Ladakh. *Journal of the Indian Society of Soil Science, 53*(3), 421–423.

FAI. (2000). Fertilizer Statistics. Fertilizer Association of India, New Delhi.

Gajare A. S., Dhawan, A. S., Ghodke, S. K., & Bhor, S. D. (2013). Available sulfur and phosphorus status of soybean growing soils of Latur district. *An Asian Journal of Soil Science, 8*(1), 94–97

Gajare, A. S., Mandal, D. K., Mandal, C., & Prasad, J. (2015). Cationic micronutrient status in cotton growing shrink-swell soils of Jalgaon district of Maharashtra. *Indian Journal of Fertilizers, 11*(7), 66–69.

Ghosh, S. K., Swain, P. K., & Sarkar, D. (2010). Distribution of DTPA extractable micronutrient cations in some soils of Hot Dry Subhumid agro-ecological zone of West Bengal. *Agropedology, 20*(1), 80–84.

Hundal, H. S., Singh, R. D., & Machandra, J. S. (2006). Available nutrient and heavy metal status of soils of Punjab, North-west India. *Journal of the Indian Society of Soil Science, 54*(1), 50–56.

Indiastat (2015). *http://www.indiaagristat.com/agriculture/2/agriculturalarealanduse.aspx,* (Accessed 15 February, 2016).

Jackson, M. L. (1967). Soil Chemical Analysis, Prentice Hall of India Pvt. Ltd. New Delhi.

Jackson, M. L. (1973). Soil Chemical Analysis, Prentice Hall of India Pvt. Ltd. New Delhi.

Joshi, O. P. (2003). Future perspective of soybean in India. *Soybean Research*, l, 29–42.

Joshi, O. P. (2004). Soil fertility management in India. *In: Proceedings of VII World Soybean Research Conference.* VI International Soybean Processing and Utilization Conference, III Congresso Brasileiro De Soja, held at Foz du Iguvassu, Brasil Moscardi et al. (Eds.), pp. 400–406.

Katyal, J. C., & Reddy, K. C. K. (1997). *Plant nutrient supply needs*: Rainfed food crops. *In*: *Plant Nutrient Needs, Supply, Efficiency and Policy Issues, 2000–2025*, pp. 91–113. Kanwar, J. S., & Katyal, J. C. (Eds.). National Academy of Agricultural Sciences, New Delhi.

Khamparia, R. S., Sharma, B. L., Singh, M. V., & Sharma, G. D. (2009). Four decades of research in micro, secondary nutrient and pollutant element in soil and plants. Research bulletin, Department of Soil Science and Agricultural Chemistry, JNKKV, Jabalpur, M.P., 48 pp.

Krishna Prasad, V., & Badarinath, K. V. S. (2006). Soil surface nitrogen losses from agriculture in India: A regional inventory within agroecological zones (2000–2001) *International Journal of Sustainable Development, 13*, 173–182.

Krishna Prasad, V., Badarinath, K. V. S., Yonemura, S., & Tsuruta, H. (2004). Regional inventory of soil surface nitrogen balances in Indian agriculture (2000–2001). *Journal of Environment Management, 73*, 209–218.

Lande, M. G., Varade, S. B., & Badhe, N. N. (1977). Sulfur status and it's relationship with physico-chemical properties of Marathwada soils. *Journal of Maharashtra Agricultural Universities*, *2*(3), 195–201.

Lindsay, W. L., & Narvell, W. A. (1978). Development of DTPA soil test for Zn, Fe, Mn and Cu. *Soil Science Society of America Journal, 42*, 421–428.

Mahapatra, P. K., & Sahu, S. K. (1996). Relationship between some secondary and micronutrient status of soils growing groundnut and content in Groundnut plants. *Journal of the Indian Society of Soil Science*, *44*(1), 100–103.

Mahesh kumar, Raina P. R., & Sharma, B. K. (2011). Distribution of DTPA extractable micronutrients in arid soils of Churu district, Rajasthan. *Agropedology*, *21*(1), 44–48.

Malewar, G. U. (1995). Micronutrient availability as influenced by cropping pattern Maharathwada region of Maharashtra. *Journal of Maharashtra Agricultural Universities, 20*(3), 330–333.

Mali, C. V., & Ismail, S. (2002). Characterization and fertility status of Marathwada soils, sulfur and micronutrients in soil resource inventory of Marathwada, Department of Agricultural Chemistry and Soil Science, bulletin, MAU, Parbhani. pp. 72–87.

Mali, C. V., & Raut, P. D. (2001). Available sulfur and physico-chemical characteristics of oilseed dominated area of Latur district. *Journal of Maharashtra Agricultural Universities, 26*(1), 117–118.

Mandal, D. K., Mandal, C., & Venugopalan, M. V. (2005). Suitability of cotton cultivation in shrink–swell soils in central India. *Agricultural Systems*, *84*, 55–75

Martin, J. H., Leonardand, W. H., & Stamp, D. L. (1976). Soybeans: In principles of field crop production. MacMillan Publ. Co. Inc., New York, pp. 691–700.

Mendhe, S. N., Kolte, H. Kashikar, V. G., & Gadekar, S. R. (2008). Pulses Pigeonpea, *Kharif* crop production. Agromet Pub. Nagpur, India, pp. 61–67.

More, S. D., Adasule, P. B., & More, A. B. (2005). Soil water and plant analysis for quality agricultural production. Bulletin, MAU Publication. pp. 65–75.

Murugan, A. V., & Dadhwal, V. K. (2007). Indian agriculture and nitrogen cycle. In Agricultural nitrogen use & its environmental implications (Abrol, Y. P., Raghuram, N., & Sachdev, M. S., Eds.), IK International Publishing House Pvt. Ltd. New Delhi, 2007, pp. 9–28.

Olsen, S. R., Cole, C. V., Watanabe, F. S., & Dean, L. A. (1954). Estimation of available phosphorus in soil by extraction with sodium bicarbonate. USDA Circular 939, 19.

Panda, D, Samantaray, R.N, Misra, A. K., & Senapati, H. K. (2007). Nutrient balance in rice. *Indian Journal of Fertilizer 3*, 33–38

Panse, V. G., & Sukhatme, P. N. (1967). Statistical methods for agricultural workers. IARI, New Delhi, 145–156.

Pathak, H, Mohanty, S., Jain, N., & Bhatia, A. (2010). Nitrogen, phosphorus, and potassium budgets in Indian agriculture. *Nutrient Cycling in Agro-Ecosystems, 86*, 287–299.

Patil, J. D., & Kharche, V. K. (2006). Micronutrients deficiencies in soils of western Maharashtra. *Indian Journal of Fertilizers, 2*, 55–58.

Patil, Y. M., & Sonar, K. R. (1994). Status of major and micronutrients of swell-shrink soils of Maharashtra. *Journal of Maharashtra Agricultural University, 19*(2), 169–172.

Pattanayak, S. K., Sureshkumar, P., & Tarafdar, J. C. (2009). New Vista in Phosphorus Research. *Journal of the Indian Society of Soil Science, 57*(4), 536–545

Pharande, A. L., Raskar, B. N., & Nipunage, M. V. (1996). Micronutrients status of important Vertisol and Alfisol soils series of western Maharashtra. *Journal of Maharashtra Agricultural Universities 21*(2), 182–185.

Piper, C. S. (1966). Soil and plant analysis. Hans publication, Bombay.

Quayum, A., Rao, M. S. S., & Kerkatta, V. (1985). Soybean in Bihar plateau. In oilseed production constraints and opportunities. Oxford and IBH Publ. Co. New Delhi, 219–232.

Rathore, P. S. (2011). Techniques and management of field crop production. 264–271.

Ramamurthy, B., & Bajaj, J. C. (1969). Available nitrogen, phosphorus and potassium status of Indian soils. *Fertilizer News, 14*, 25–36.

Sakal, R., Singh, B. P., Singh, A. P., Singh, R. B., Singh, S. P., & Jha, S. N. (1986). Availability of zinc, copper, iron and manganese in sub-Himalayan hill and forest soils as influenced by certain soil properties. *Journal of the Indian Society of Soil Science, 34*, 191–193.

Sharma, O. P., Tiwari S. C., & Raghuwanshi, R. K. (2004). Effect of doses and sources of sulfur on nodulation, yield, oil and protein content of soybean and soil properties. *Soybean Research, 2*, 35–40.

Sharma, R. P., Singh, M., & Sharma, J. P. (2003). Correlation studies on micronutrients vis-à-vis soil properties in some soils of Nagaur district in Semi-arid region of Rajasthan. *Journal of the Indian Society of Soil Science, 51*(4), 522–526.

Singh, A. H., Singh, K., Singh, L. N., Singh, N. G., Chongtham, N., & Singh, A. K. (2006). Status of forms of sulfur in acidic soil of Mainpur. *Journal of the Indian Society of Soil Science, 54*(3), 351–353.

Sundaran, K. P. (1986). Frank notes: Importance of Sulfur in Indian Agriculture. *Fertilizer news, 31*(9), 8.

Takkar, P. N. (1996). Micronutrient research and sustainable Agricultural Productivity in India. *Journal of the Indian Society of Soil Science, 44*(4), 562–581.

Tandon, H. L. S. (1986). Sulfur research and development in Indian Agriculture. *Fertilizer news,* 31(9), 9–16.

Tisdale, S. L., Nelson, W. L. Beaton, J. D., & Halvin, J. L. (1997). Soil fertility and fertilizers, 5th Edition, Macmillan Publishing Co., New Delhi.

Tiwari, S. P., & Joshi, O. P. (1999). Strategy for enhancing the production and productivity of soybean in India. *In*: Group meeting on strategic issues in doubling the productivity of oilseeds Based Production Systems, Gujarat Agricultural University Campus, Junagadh held during 10–11th September, 1999.

Venkatswralu, J. (1985). Crops and cropping system in dyrland agriculture. In efficient management of dryland crops CRIDA, Hyderabad.

Verma, V. K., Setia, R. K., & Sharma, P. K. (2008). Distribution of micronutrient cations in different physiographic units of semi-arid region of Punjab. *Agropedology, 18*(1), 58–65.

Williams, C. H., & Steinberg, A. (1959). Soil sulfur fraction as chemical indices of available sulfur in some Australian soils. *Australian Journal of Agricultural Research, 10*, 340–350.

CHAPTER 19

SOIL ACIDITY AND POOR NUTRIENT STATUS: EMERGING ISSUES FOR AGRICULTURAL LAND USE PLANNING IN JAMTARA DISTRICT OF JHARKHAND

A. K. SAHOO,[1] S. K. SINGH,[2] D. C. NAYAK,[3] S. MUKHOPADHYAY,[4] T. BANERJEE,[1] D. SARKAR,[5] A. K. SARKAR,[6] B. K. AGARWAL,[7] D. K. SHAHI[8] and ARVIND KUMAR[9]

[1]*Principal Scientist, ICAR-National Bureau of Soil Survey and Land Use Planning, Regional Centre, Kolkata – 700 091, India*

[2]*Director, ICAR-National Bureau of Soil Survey and Land Use Planning, Amravati Road, Nagpur – 440 033, India*

[3]*Principal Scientist & Head, ICAR-National Bureau of Soil Survey and Land Use Planning, Regional Centre, Kolkata – 700 091, India*

[4]*Senior Scientist, ICAR-National Bureau of Soil Survey and Land Use Planning, Regional Centre, Kolkata – 700 091, India*

[5]*Ex-Director, ICAR-National Bureau of Soil Survey and Land Use Planning, Amravati Road, Nagpur – 440 033, India*

[6]*Ex-Dean (Agriculture), Birsa Agricultural University, Ranchi – 834 006, India*

[7]*Chief Scientist cum Professor, Birsa Agricultural University, Ranchi – 834 006, India*

[8]*Chairman, Department of Soil Science and Agricultural Chemistry, Birsa Agricultural University, Ranchi – 834 006, India*

[9]*Assistant Professor, Birsa Agricultural University, Rachi-834 006, India*

CONTENTS

ABSTRACT

Soil fertility status of Jamtara district of Jharkhand was evaluated and mapped using Geographic Information System (GIS) environment based on the geo-referenced database generated through the analysis of seven thousand two hundred nine (7209) composite surface (0–20 cm) soil samples collected at 500 m grid interval from six blocks (Fatehpur, Jamtara, Karmatarn, Kundahit, Nala and Narayanpur) of the district using Global Positioning System (GPS). Soils were analyzed for pH, organic carbon, available phosphorous, potassium and sulfur, DTPA extractable zinc, copper, iron and manganese and hot water soluble boron with standard procedures. Data reveal that soils of majority area (52.8 to 85.6%) of all the blocks are strongly to extremely acidic (pH < 5.5). The organic carbon and available phosphorus content of the soils are low in majority area of all the blocks in the district. Aerial extension of low content of organic carbon ($<0.50\%$) and available phosphorus (<10 kg ha^{-1}) varied from 56.6 to 80.1% and 50.8 to 84.6% area in different blocks, respectively. Available contents of potassium and sulfur are found to be low in 28.1 and 24.4% area of the district, respectively. Deficiency in available contents of zinc (DTPA extractable) and boron (hot water soluble) are also observed in 44.6 and 29.1% area of the district, respectively. It is evident that soil acidity is an "inherent/core problem" of the soils which may be responsible for soil degradation and therefore warrants for amelioration of the same on priority. The low content of organic matter and available phosphorus along with moderate contents of available potassium and sulfur as well as deficiency of zinc and boron certainly reflects in general poor fertility status of the district and thus requires adequate soil management practices on priority towards optimizing agricultural production on sustainable basis. Land resource inventorying on 1:10,000 scale (detailed soil survey) will be of immense importance towards identification of site and situation specific problem areas for agricultural land use planning for the district.

19.1 INTRODUCTION

Soil is the most precious basic natural resource, on which depends the security of food, nutrition, income, biodiversity, livelihood and environment. But, its indiscriminate and unscientific management is a major threat to sustainability of agriculture. Over the years, there is a loss in the productive potential of soils, which is attributed to imbalance in nutrient use, mining of nutrients from soil, poor replenishment, loss of carbon, pollution, decrease in biological activities, waterlogging, soil salinity, etc. Thus, land use is not commensurate with land capability. With increased cropping intensity in major groups of soils, disproportionate increase in the rate of nutrient application, the inherent soil fertility is showing a decline with reduction in crop uptake of essential plant nutrients. This has led to lowering of crop yields with deterioration in crop quality with lower value of economic produce. Increased extraction of several nutrients from soil has resulted in multi-nutrient deficiencies, poor efficiency of applied nutrients and lower returns from money spent on fertilizers and other inputs. This process of degradation has to be reversed for the benefit of masses, for whom agriculture is the prime source of livelihood.

Spatial variability in soil parameters including nutrients has been attributed to the parent material, topography, landforms, cropping pattern and fertilization history. Blanket nutrient recommendations further widen the variability and enhance the risk of soil degradation in terms of soil organic carbon and nutrient depletion, acidity, green house gas emission, water and environmental pollution. Unfavorable economics on account of blanket recommendation are bound to adversely influence the enthusiasm of farmers to enhance investments in new technologies. This results in low farm productivity and poor soil health that jeopardize food security and agricultural sustainability. Site Specific Nutrient Management (SSNM) is one of the logical steps to overcome the problems but inadequate infrastructure for soil testing laboratory in the country limits its applicability. Soil parameters, mapping including nutrients involving geo-referenced soil sampling, laboratory analysis, database structuring in GIS and subsequent interpolation using appropriate geo-statistics provides valid ground for SSNM. The possibility of using such map as a fertilizer decision support tool to guide nutrient application in a site-specific mode is explored. Nutrient sufficiency/deficiency zones of single and multiple nutrients prepared using logical algorithms in GIS are one step forward for SSNM (NBSS & LUP 2006, Sen et al., 2012, Sahoo et al., 2013). Repetitive survey using these maps can give a clear visual indication of the changing fertility scenario at block/village/district level on real time scale. This kind of information is imperative for the area affected with wide spread acidification, single and multiple nutrient deficiency, water stress and intermittent drought spell during monsoon. The major objective of the study was to prepare GIS aided block wise soil acidity (soil pH) and soil nutrient maps (organic carbon, available P, K, S and available Fe, Mn, Zn, Cu, B) based on analytical data of geo-referenced surface soil samples for identification of site specific problems in different blocks towards formulating optimum land use plan of Jamtara district.

19.2 GEOGRAPHICAL SETTINGS

Jamtara district of Jharkhand state of India, a part of Chotanagpur plateau, is located between 23°47'36" to 24°09'54" N and 86°28'5" to 87°17'53" E covering an area of 1802 sq.km. The district comprises one subdivision namely Jamtara and six blocks *viz.*, Fatehpur, Jamtara, Karmatarn, Kundahit, Nala and Narayanpur (Figure19.1). It consists of rolling open land surface with long ridges and intervening depressions. Average elevation of the district ranges between 150 and 300 m. Geologically, the area is comprised with granite-gneiss and basaltic trap with sedimentary beds. Ajay is the major river flowing through the district. Owing to its position near West Bengal and hilly landscape of the region climatic condition is slightly different from the rest of the state. The district receives an annual rainfall of 1500 mm and most of the rainfall occurs during the rainy season. During winter season, the temperature varies between 8 and 21°C and during summer season it varies between 28 and 45°C. Since the area consists of hilly landscape and receives more rainfall, it has considerable vegetative cover. But due to ruthless exploitation, most of the forest has turned into bushes devoid of big trees. As per the land utilization statistics of Jharkhand (FAI, 2011–12), majority area of the district is under fallow land (39.6% of TGA) and net sown area of the district is 336.4 sq.km (18.6% of TGA). Important crops of the district are paddy, maize and wheat. Only 6.42% area of agricultural use is irrigated and major source of irrigation is well and tanks. The soils occurring on different landforms have been characterized during soil resource mapping of the state on 1:250,000 scale (Haldar et al., 1996) and three soil orders namely Entisols, Inceptisols and Alfisols were observed in Jamtara district. Inceptisols were the dominant soils covering 50.8% of TGA followed by Alfisols (39.3%) and Entisols (6.6%).

19.3 METHODOLOGY

The base map of all the blocks of Jamtara district was prepared on 1:50,000 scale using Survey of India toposheets (72L/8, 12, 16, 72P/4, 73I/9, 13 and 73M/1,5) and PS/Block maps of all the blocks and maps were demarcated with grid points at 500 m interval. Seven thousand two hundred nine (7209) composite surface soil samples (0–20 cm) from demarcated grid points in six blocks of Jamtara district (Table 19.1a) and other related information were collected using GPS through field survey.

Soil samples were air dried, processed and analyzed for pH, organic carbon, available phosphorous and potassium (Page et al., 1982), available sulfur by using 0.15% $CaCl_2$ as the extractant (William and Steinbergs, 1959), available (DTPA extractable) Fe, Mn, Zn and Cu (Lindsay and Norvell, 1978) and available B (hot water soluble) by Carmine method (Hatcher and Wilcox, 1950). The rating of soils for different parameters is given in Table 19.1b.

TABLE 19.1A Number of Collected Surface Soil Samples in Different Blocks

Blocks	Area (km^2)	Surface soil samples
Fatehpur	263.7	1064
Jamtara	347.1	1277
Karmatarn	177.1	712
Kundahit	296.5	1231
Nala	434.0	1740
Narayanpur	283.6	1185
Total	1802	7209

TABLE 19.1B Grouping of Soils for Different Parameters

Parameters	Rating	Reference
pH		Soil Survey Manual (IARI, 1970)
Extremely acidic	<4.5	
Very strongly acidic	4.5 to 5.0	
Strongly acidic	5.1 to 5.5	
Moderately acidic	5.6 to 6.0	
Slightly acidic	6.1 to 6.5	
Neutral	6.6 to 7.3	
Slightly alkaline	7.4 to 7.8	
Moderately alkaline	7.9 to 8.4	
Strongly alkaline	8.5 to 9.0	
Organic Carbon		Singh et al. (2005)
Low	<0.50%	
Medium	0.50 to 0.75%	
High	>0.75%	
Available Phosphorous		Singh et al. (2005)
Low	<10 kg ha^{-1}	
Medium	10 to 25 kg ha^{-1}	
High	>25 kg ha^{-1}	
Available Potassium		Singh et al. (2005)
Low	<108 kg ha^{-1}	
Medium	108 to 280 kg ha^{-1}	
High	>280 kg ha^{-1}	
Available Sulfur		Mehta et al. (1988)
Low	<10 mg kg^{-1}	
Medium	10 to 20 mg kg^{-1}	
High	>20 mg kg^{-1}	
Available Micronutrients	**Critical Limits**	Lindsay and Norvel (1978)
DTPA extractable Fe	<4.5 mg kg^{-1}	
DTPA extractable Mn	<2.0 mg kg^{-1}	
DTPA extractable Cu	<0.2 mg kg^{-1}	
DTPA extractable Zn	<0.5 mg kg^{-1}	Follet and Lindsay (1970)
Hot water soluble B	<0.5 mg kg^{-1}	Berger and Truog (1940)

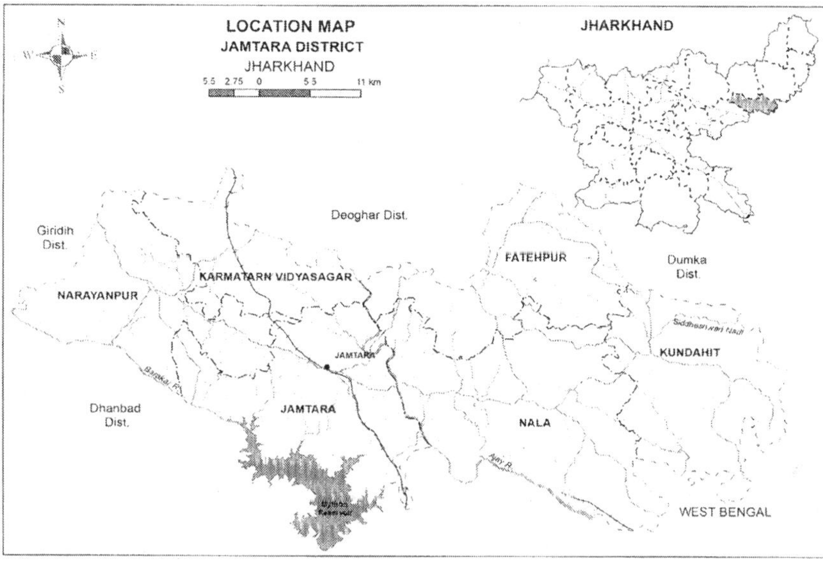

FIGURE 19.1 Location map of the study area.

The maps listed in Table 19.1b parameters have been prepared using GIS from data generated by analysis of grid soil samples.

19.4 RESULTS AND DISCUSSION

19.4.1 SOIL REACTION (pH)

Soil pH is an important soil property, which affects the availability of several plant nutrients. It is a measure of acidity and alkalinity and reflects the status of base saturation. The soils of the block have been grouped under six soil reaction classes according to Soil Survey Manual (IARI, 1970). The soil pH in different blocks ranges from 4.0 to 7.6. The soil reaction classes with area are given in Table 19.2. The data reveal that strong to extremely acidic (pH<5.5) soils cover 81.3, 53.5, 54.7, 52.8, 85.6 and 76.8% area of Fatehpur, Jamtara, Karmatarn, Kundahit, Nala and Narayanpur blocks, respectively. Spatial distribution of soil pH in Karmatarn block is shown in Figure 19.2.

19.4.2 ORGANIC CARBON

The effect of soil organic matter on soil properties is well recognized. Soil organic matter plays a vital role in supplying plant nutrients, cation

TABLE 19.2 Soils Under Different Reaction Classes

Soil reaction (pH)	Fatehpur	Jamtara	Karmatarn	Kundahit	Nala	Narayanpur
	Area in km^2 (% TGA)					
Extremely acidic (<4.5)	37.2 (14.1)	23.2 (6.7)	13.0 (7.3)	11.9 (4.0)	94.6 (21.8)	103.4 (36.5)
Very strongly acidic (4.5 to 5.0)	88.1 (33.4)	71.2 (20.5)	30.4 (17.2)	53.4 (18.0)	169.7 (39.1)	73.5 (25.9)
Strongly acidic (5.1 to 5.5)	89.2 (33.8)	91.3 (26.3)	53.6 (30.2)	91.3 (30.8)	107.2 (24.7)	40.7 (14.4)
Moderately acidic (5.6 to 6.0)	34.7 (13.2)	65.2 (18.8)	42.9 (24.3)	67.0 (22.6)	38.2 (8.8)	25.9 (9.1)
Slightly acidic (6.1 to 6.5)	7.0 (2.6)	36.1 (10.4)	22.8 (12.9)	47.4 (16.0)	5.2 (1.2)	18.0 (6.3)
Neutral (6.6 to 7.3)	2.0 (0.8)	21.2 (6.1)	10.3 (5.8)	21.6 (7.3)	2.2 (0.5)	13.0 (4.6)
Slightly alkaline (pH 7.4 to 7.8)	-	6.6 (1.9)	-	-	2.6 (0.6)	-
Miscellaneous	5.5 (2.1)	32.3 (9.3)	4.1 (2.3)	3.9 (1.3)	14.3 **(3.3)**	9.1 (3.2)
Total	263.7 (100.0)	347.1 (100.0)	177.1 (100.0)	296.5 (100.0)	434.0 (100.0)	283.6 (100.0)

exchange capacity, improving soil aggregation and hence, water retention and soil biological activity. The organic carbon content in different blocks varies from 0.03 to 1.31%. They are mapped into three classes. i.e., low (< 0.5%), medium (0.5 to 0.75%) and high (> 0.75%) (Table 19.3). It is observed that the organic carbon content in the soils is low in majority area (56.6 to 80.1%) of all the blocks. The spatial distribution of organic carbon in the soils of Nala block is shown in Figure 19.3.

19.4.3 MACRONUTRIENTS

Nutrients like phosphorus (P) and potassium (K) are considered as primary nutrients and sulfur (S) as secondary nutrient. These nutrients are essential for proper growth, development and yield differentiation of plants and are generally required by plants in large quantity.

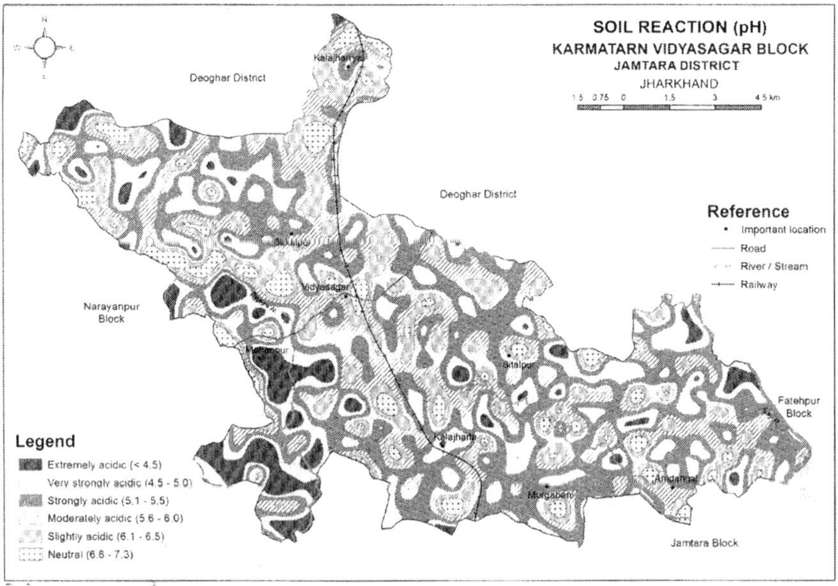

FIGURE 19.2 Soil pH map of Karmatarn Vidyasagar block of Jamtara district.

TABLE 19.3 Organic Carbon Status

Available OC (%)	Fatehpur	Jamtara	Karmatarn	Kundahit	Nala	Narayanpur
			Area in km² (% TGA)			
Low (<0.50)	149.4 (56.6)	240.5 (69.3)	139.8 (79.0)	215.7 (72.8)	296.4 (68.3)	227.1 (80.1)
Medium (0.50–0.75)	79.6 (30.2)	74.3 (21.4)	33.2 (18.7)	76.9 (25.9)	102.4 (23.6)	47.4 (16.7)
High (>75)	29.2 (11.1)	-	-	-	20.9 (4.8)	-
Miscellaneous	5.5 (2.1)	32.3 (9.3)	4.1 (2.3)	3.9 (1.3)	14.3 **(3.3)**	9.1 (3.2)
Total	263.7 (100)	347.1 (100.0)	177.1 (100.0)	296.5 (100.0)	434.0 (100.0)	283.6 (100.0)

19.4.3.1 Available Phosphorus

Phosphorus is important component of adenosine di-phosphate (ADP) and adenosine tri-phosphate (ATP), which are involved in energy transformation in plant. It is essential component of deoxyribonucleic acid (DNA), the seat

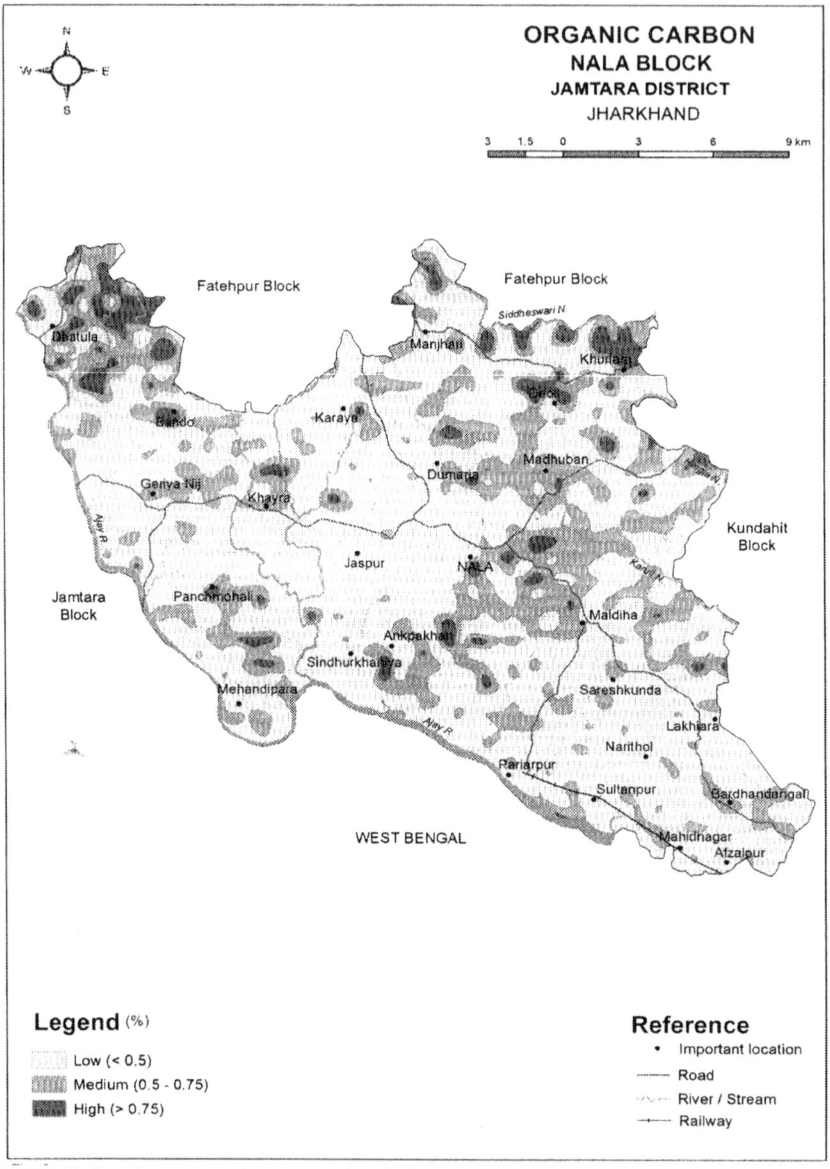

FIGURE 19.3 Organic carbon map of Nala block of Jamtara district.

of genetic inheritance in plant and animal. The availability of phosphorous is restricted under acidic and alkaline soil reaction mainly due to P-fixation. In acidic conditions it gets fixed with aluminum and iron and in alkaline conditions with calcium. Available phosphorus content in these soils ranges between 0.81 and 53.0 kg ha^{-1} and their distribution in different blocks is shown in Table 19.4. Data reveals that the soils of the major area of all the blocks are low in available phosphorous and the aerial extension of low content of available phosphorous varies from 50.8 to 84.6% area. The spatial distribution of available phosphorous in the soils of Narayanpur block is shown in Figure 19.4.

TABLE 19.4 Available Phosphorous Status

Available P (kg ha^{-1})	Fatehpur	Jamtara	Karmatarn	Kundahit	Nala	Narayanpur
	Area in km^2 (% TGA)					
Low (< 10)	191.7 (72.7)	213.1 (61.4)	105.3 (59.5)	150.7 (50.8)	367.2 (84.6)	165.3 (58.3)
Medium (10–25)	51.9 (19.7)	90.9 (26.2)	55.5 (31.3)	128.0 (43.2)	41.2 (9.5)	86.9 (30.6)
High (> 25)	14.6 (5.5)	10.8 (3.1)	12.2 (6.9)	13.9 (4.7)	11.3 (2.6)	22.3 (7.9)
Miscellaneous	5.5 (2.1)	32.3 (9.3)	4.1 (2.3)	3.9 (1.3)	14.3 **(3.3)**	9.1 (3.2)
Total	263.7 (100)	347.1 (100.0)	177.1 (100.0)	296.5 (100.0)	434.0 (100.0)	283.6 (100.0)

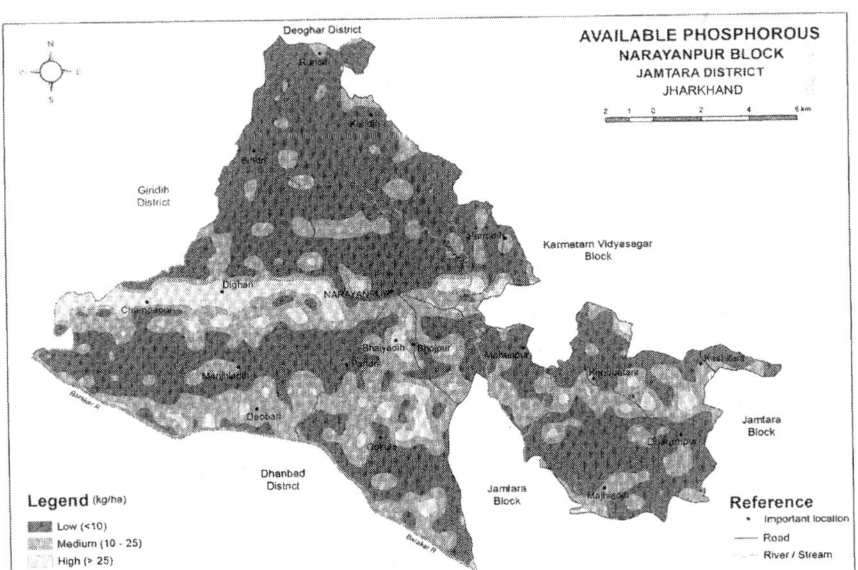

FIGURE 19.4 Available phosphorous status in Narayanpur block of Jamtara district.

19.4.3.2 Available Potassium

Potassium is an activator of various enzymes responsible for plant processes like energy metabolism, starch synthesis, nitrate reduction and sugar degradation. It is extremely mobile in plants and helps to regulate opening and closing of stomata in the leaves and uptake of water by root cells. It is important in grain formation and tuber development and encourages crop resistance for certain fungal and bacterial diseases. Available potassium content in these soils ranges between 29.2 and 612.3 kg ha^{-1} and details about area and distribution is shown in Table 19.5. The data revealed that the major area (47.6–60.1%) of Jamtara, Karmatanr, Kundahit, Nala and Narayanpur blocks have medium content (108 to 280 kg ha^{-1}) of available potassium and 16.9–41.7% area of different blocks are low (<108 kg ha^{-1}) in available potassium content. The spatial distribution of available potassium in the soils of Fatehpur block is shown in Figure 19.5.

19.4.3.3 Available Sulfur

Sulfur is essential in synthesis of sulfur containing amino acids (cystine, cysteine and methionine), chlorophyll and metabolites including co-enzyme A, biotin, thiamine, or vitamin B1 and glutathione. It activates many proteolytic enzymes, increases root growth and nodule formation and stimulates seed formation. The available sulfur content in the soils varies from 0.97 to

TABLE 19.5 Available Potassium Status

Available K (kg ha^{-1})	Fatehpur	Jamtara	Karmatarn	Kundahit	Nala	Narayanpur
	Area in km^2 (% TGA)					
Low (<108)	110.1 (41.7)	80.5 (23.2)	29.9 (16.9)	72.2 (24.4)	149.7 (34.5)	64.7 (22.8)
Medium (108–280)	98.4 (37.3)	177.7 (51.2)	106.5 (60.1)	141.2 (47.6)	236.1 (54.4)	156.6 (55.2)
High (>280)	49.7 (18.9)	56.6 (16.3)	36.6 (20.7)	79.2 (26.7)	33.9 (7.8)	53.2 (18.8)
Miscellaneous	5.5 (2.1)	32.3 (9.3)	4.1 (2.3)	3.9 (1.3)	14.3 **(3.3)**	9.1 (3.2)
Total	263.7 (100)	347.1 (100.0)	177.1 (100.0)	296.5 (100.0)	434.0 (100.0)	283.6 (100.0)

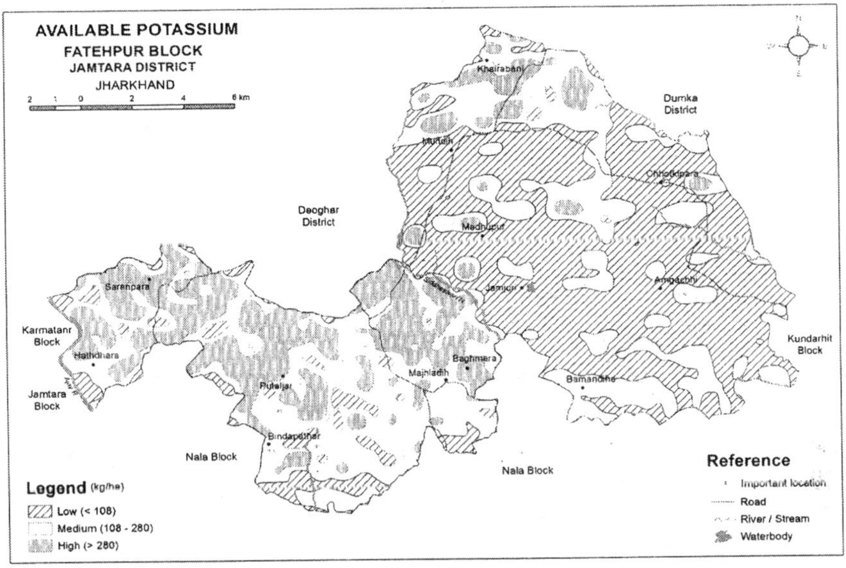

FIGURE 19.5 Available potassium status in Fatehpur block of Jamtara district.

TABLE 19.6 Available Sulfur Status

Available S (mg kg^{-1})	Fatehpur	Jamtara	Karmatarn	Kundahit	Nala	Narayanpur
	Area in km^2 (% TGA)					
Low (<10)	60.4 (22.9)	83.3 (24.0)	48.9 (27.6)	40.6 (13.7)	118.5 (27.3)	87.7 (30.9)
Medium (10 20)	168.3 (63.8)	178.4 (51.4)	109.7 (62.0)	204.9 (69.1)	250.9 (57.8)	177.9 (62.7)
High (>20)	29.5 (11.2)	53.1 (15.3)	14.4 (8.1)	47.1 (15.9)	50.3 (11.6)	8.9 (3.2)
Miscellaneous	5.5 (2.1)	32.3 (9.3)	4.1 (2.3)	3.9 (1.3)	14.3 **(3.3)**	9.1 (3.2)
Total	263.7 (100)	347.1 (100.0)	177.1 (100.0)	296.5 (100.0)	434.0 (100.0)	283.6 (100.0)

52.30 mg kg^{-1} and details about area and distribution is given in Table 19.6. Majority soils (51.4 to 69.1% of TGA) of different blocks have medium content (10 to 20 mg kg^{-1}) of available sulfur. Soils of 13.7 to 30.9% area of these blocks are low (<10 mg kg^{-1}) in available sulfur content. The spatial distribution of available sulfur in the soils of Kundahit block is shown in Figure 19.6.

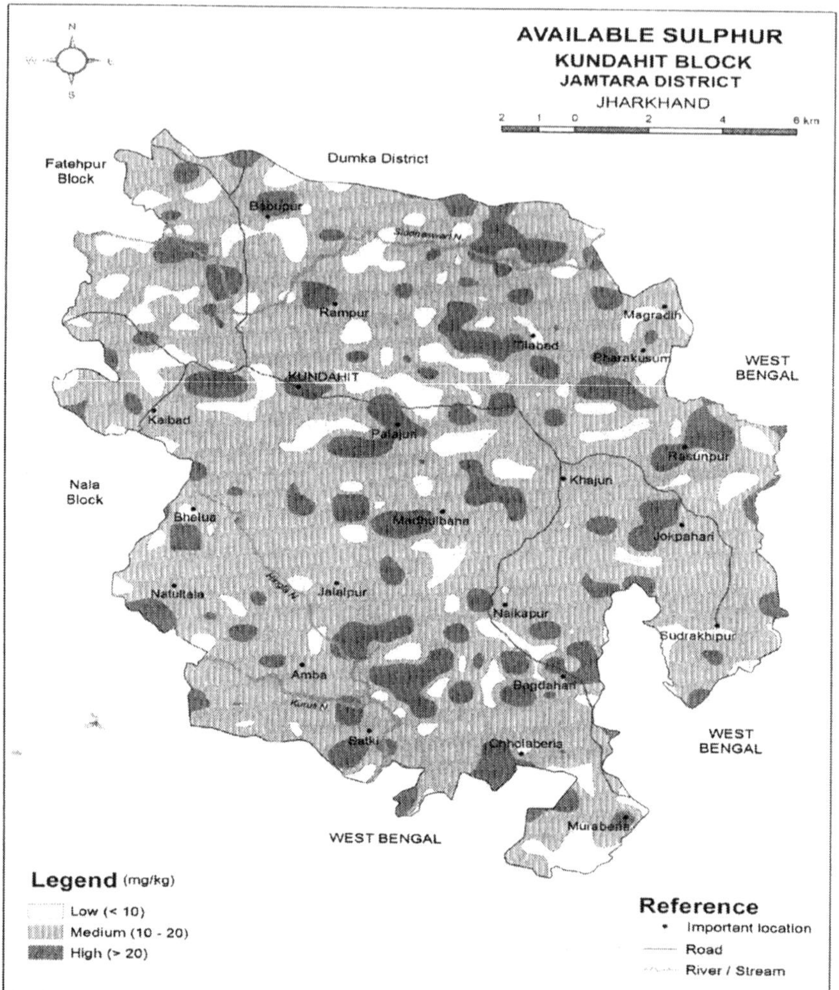

FIGURE 19.6 Available sulfur status in Kundahit block of Jamtara district.

19.4.4 MICRONUTRIENTS

Proper understanding of micronutrients availability in soils and extent of their deficiencies is the pre-requisite for efficient management of micronutrient fertilizer to sustain crop productivity. Therefore, it is essential to know the micronutrients status of soil before introducing any type of land use.

19.4.4.1 Available Iron

Iron is constituent of cytochromes, haems and nonhaem enzymes. It is capable of acting as electron carrier in many enzyme systems that bring about oxidation-reduction reactions in plants. It promotes starch formation and seed maturation. The available iron content in the surface soils ranges between 0.9 and 92.0 mg kg^{-1}. As per the critical limit of available iron (<4.5 mg kg^{-1}), soils of 1.4, 27.9, 19.4, 3.1 and 9.8% area of Fatehpur, Jamtara, Karmatarn, Nala and Narayanpur blocks are deficient in available iron, respectively. Majority of the soils (31.9 to 49.0% of TGA) have available iron content between the ranges of 25 to 50 mg kg^{-1}. The details of area and distribution of iron in different blocks are presented in Table 19.7. The spatial distribution of available iron in the soils of Nala block is shown in Figure 19.7.

19.4.4.2 Available Manganese

Manganese is essential in photosynthesis and nitrogen transformations in plants. It activates decarboxylase, dehydrogenase, and oxidase enzymes.

TABLE 19.7 Available Iron Status

Available Fe (mg kg^{-1})	Fatehpur	Jamtara	Karmatarn	Kundahit	Nala	Narayanpur
	Area in km^2 (% TGA)					
< 4.5 (Deficient)	3.8 (1.4)	96.9 (27.9)	34.4 (19.4)	-	13.5 (3.1)	27.7 (9.8)
4.5 – 10	10.2 (3.9)	93.4 (26.9)	42.8 (24.2)	13.8 (4.7)	26.0 (6.0)	51.0 (18.0)
10 – 25	102.5 (38.9)	110.7 (31.9)	85.2 (48.1)	100.1 (33.8)	188.4 (43.4)	138.9 (49.0)
25 – 50	131.2 (49.7)	13.9 (4.0)	10.6 (6.0)	161.1 (54.3)	184.9 (42.6)	50.4 (17.7)
50 – 100	10.5 (4.0)	-	-	17.6 (5.9)	6.9 (1.6)	6.5 (2.3)
Miscellaneous	5.5 (2.1)	32.3 (9.3)	4.1 (2.3)	3.9 (1.3)	14.3 **(3.3)**	9.1 (3.2)
Total	263.7 (100)	347.1 (100.0)	177.1 (100.0)	296.5 (100.0)	434.0 (100.0)	283.6 (100.0)

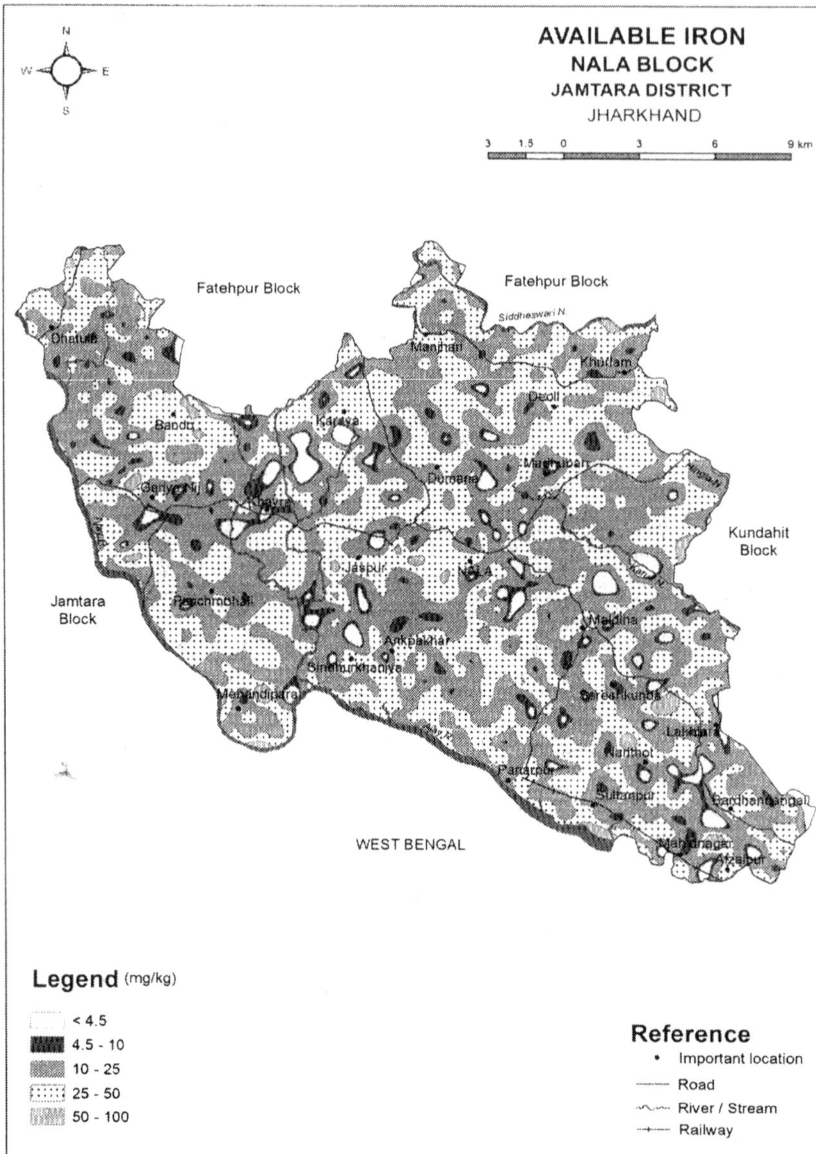

FIGURE 19.7 Available iron status in Nala block of Jamtara district.

The available manganese content in surface soils in these blocks ranges between 0.3 and 119.9 mg kg^{-1}. As per the critical limit of available manganese (<2 mg kg^{-1}), soils of 14.5% area of Jamtara block and 8.2% area of Karmatarn block are deficient in available manganese. The soils of other blocks are sufficient (>2 mg kg^{-1}) in available manganese (Table 19.8). The spatial distribution of available manganese in Karmatarn block is shown in Figure 19.8.

19.4.4.3 Available Zinc

Zinc plays role in protein synthesis, reproductive process of certain plants and in the formation starch and some growth hormones. It promotes seed maturation and production. The available zinc in surface soils ranges between 0.05 and 7.32 mg kg^{-1}. They are grouped and mapped into six classes. As per the critical limit of available zinc (<0.5 mg kg^{-1}), soils of majority area of Jamtara (69.3% of TGA), Karmatarn (78.4% of TGA) and Narayanpur (70.2% of TGA) blocks are deficient in available zinc (Table 19.9). Soils of other blocks (Fatehpur, Kundahit and Nala) are also deficient in zinc covering 9.4 to 34.0% area of TGA. The spatial distribution of available zinc in the soils of Jamtara block is presented in Figure 19.9.

TABLE 19.8 Available Manganese Status

Available Mn (mg kg^{-1})	Fatehpur	Jamtara	Karmatarn	Kundahit	Nala	Narayanpur
	Area in km^2 (% TGA)					
<2 (Deficient)	-	50.3 (14.5)	14.4 (8.2)	-	-	27.0 (9.5)
2–10	15.3 (5.8)	100.7 (29.0)	65.2 (36.8)	8.6 (2.9)	72.9 (16.8)	114.8 (40.5)
10–25	104.4 (39.6)	102.4 (29.5)	80.7 (45.5)	116.4 (39.3)	185.3 (42.7)	110.9 (39.1)
25–50	133.7 (50.7)	41.7 (12.0)	12.7 (7.2)	165.3 (55.7)	140.2 (32.3)	21.8 (7.7)
50–100	4.8 (1.8)	18.0 (5.2)	-	2.3 (0.8)	21.3 (4.9)	-
100–200	-	1.7 (0.5)	-	-	-	-
Miscellaneous	5.5 (2.1)	32.3 (9.3)	4.1 (2.3)	3.9 (1.3)	14.3 **(3.3)**	9.1 (3.2)
Total	263.7 (100)	347.1 (100.0)	177.1 (100.0)	296.5 (100.0)	434.0 (100.0)	283.6 (100.0)

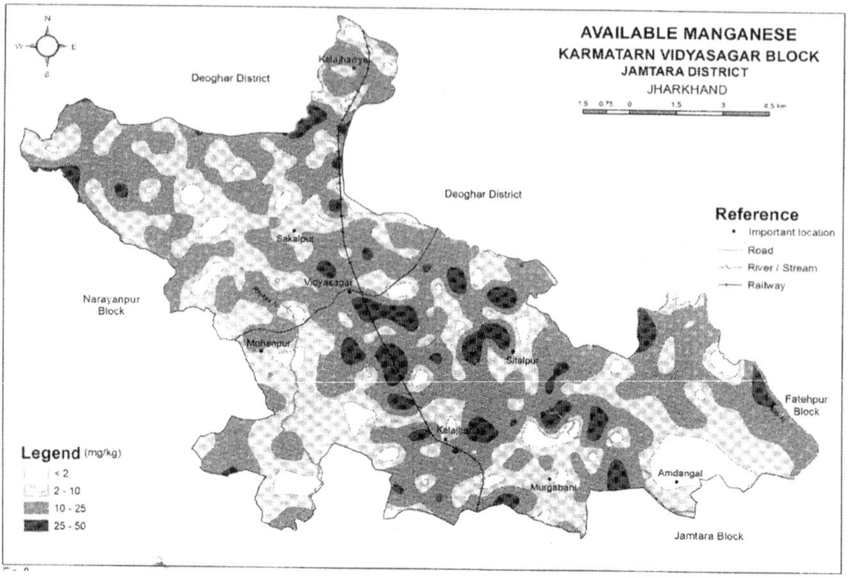

FIGURE 19.8 Available manganese status in Karmatarn Vidyasagar block of Jamtara district.

TABLE 19.9 Available Zinc Status

Available Zn (mg kg^{-1})	Fatehpur	Jamtara	Karmatarn	Kundahit	Nala	Narayanpur
	Area in km^2 (% TGA)					
<0.5 (Deficient)	49.4 (18.7)	240.5 (69.3)	138.8 (78.4)	28.0 (9.4)	147.6 (34.0)	199.1 (70.2)
05–1.0	83.3 (31.6)	60.1 (17.3)	26.1 (14.7)	86.8 (29.3)	134.1 (30.9)	43.0 (15.2)
1.0–2.0	86.2 (32.7)	14.2 (4.1_	8.1 (4.6)	116.5 (39.3)	107.6 (24.8)	23.7 (8.4)
2.0–3.0	26.4 (10.0)	-	-	41.9 (14.2)	20.4 (4.7)	6.7 (2.3)
3.0–5.0	11.3 (4.3)	-	-	16.3 (5.5)	7.8 (1.8)	2.0 (0.7)
>5.0	1.6 (0.6)	-	-	3.1 (1.0)	2.2 (0.5)	-
Miscellaneous	5.5 (2.1)	32.3 (9.3)	4.1 (2.3)	3.9 (1.3)	14.3 **(3.3)**	9.1 (3.2)
Total	263.7 (100)	347.1 (100.0)	177.1 (100.0)	296.5 (100.0)	434.0 (100.0)	283.6 (100.0)

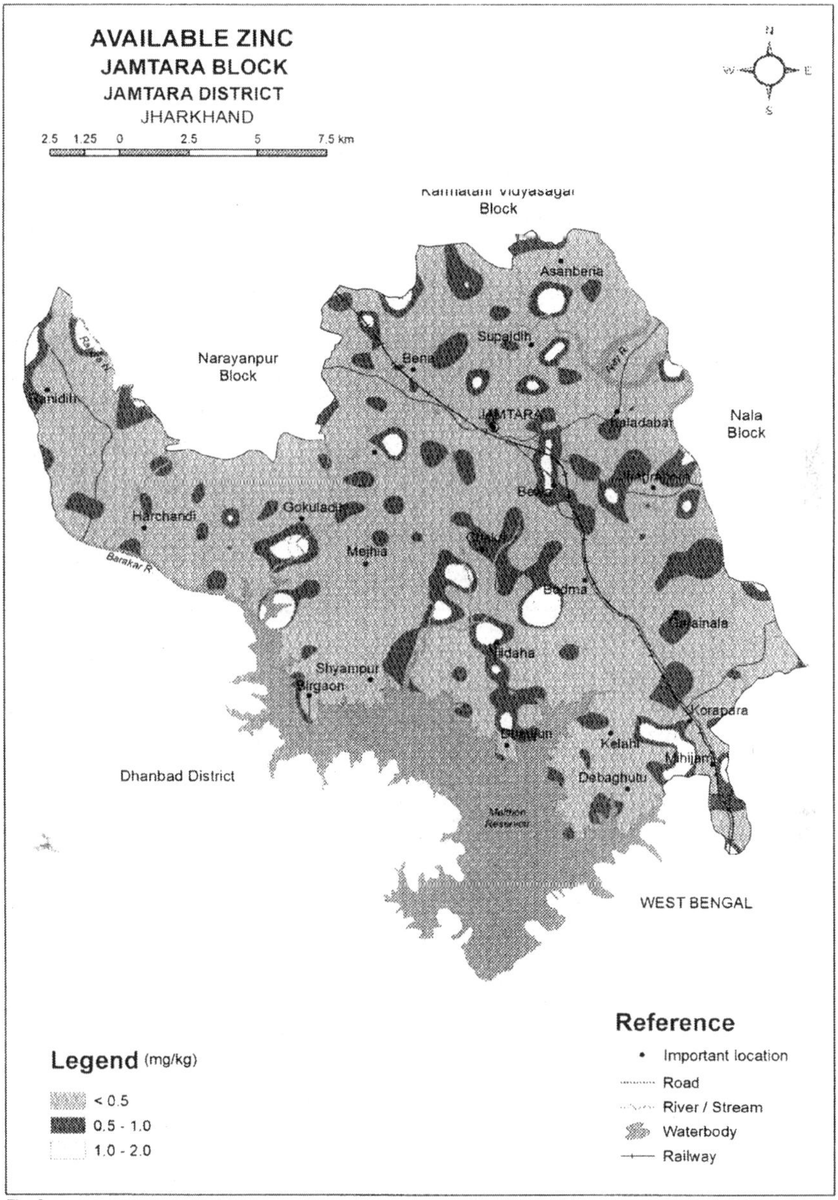

FIGURE 19.9 Available zinc status in Jamtara block of Jamtara district.

19.4.4.4 Available Copper

Copper is involved in photosynthesis, respiration, protein and carbohydrate metabolism and in the use of iron. It stimulates lignifications of all the plant cell walls and is capable of acting as electron carrier in many enzyme systems that bring about oxidation-reduction reactions in plants. The available copper status in surface soils ranges between 0.02 and 8.31 mg kg^{-1}. They are grouped and mapped into six classes. As per the critical limit (<0.2 mg kg^{-1}) of available copper, 31.2, 26.7, 22.2, 9.2, 4.3 and 2.9% area are deficient in available copper in Jamtara, Karmatarn, Narayanpur, Nala, Fatehpur and Kundahit blocks, respectively (Table 19.10). The spatial distribution of available copper in the soils of Jamtara block is shown in Figure 19.10.

19.4.4.5 Available Boron

Boron increases solubility and mobility of calcium in the plant and it acts as regulator of K/Ca ratio in the plant. It is required for development of new meristematic tissue and also necessary for proper pollination, fruit and seed

TABLE 19.10 Available Copper Status

Available Cu (mg kg^{-1})	Fatehpur	Jamtara	Karmatarn	Kundahit	Nala	Narayanpur
	Area in km^2 (% TGA)					
<0.2 (Deficient)	11.3 (4.3)	108.3 (31.2)	47.2 (26.7)	8.4 (2.9)	39.9 (9.2)	63.0 (22.2)
0.2–0.5	20.6 (7.8)	121.8 (35.1)	61.4 (34.7)	20.2 (6.8)	44.3 (10.2)	74.5 (26.3)
0.5–1.0	61.8 (23.4)	72.5 (20.9)	54.1 (30.5)	64.3 (21.7)	99.4 (22.9)	85.7 (30.2)
1.0–2.0	110.2 (41.8)	12.2 (3.5)	10.3 (5.8)	129.4 (43.6)	142.4 (32.8)	45.6 (16.1)
2.0–4.0	53.1 (20.1)	-	-	67.6 (22.8)	85.5 (19.7)	5.7 (2.0)
> 4.0	1.2 (0.5)	-	-	2.7 (0.9)	8.2 (1.9)	-
Miscellaneous	5.5 (2.1)	32.3 (9.3)	4.1 (2.3)	3.9 (1.3)	14.3 **(3.3)**	9.1 (3.2)
Total	263.7 (100)	347.1 (100.0)	177.1 (100.0)	296.5 (100.0)	434.0 (100.0)	283.6 (100.0)

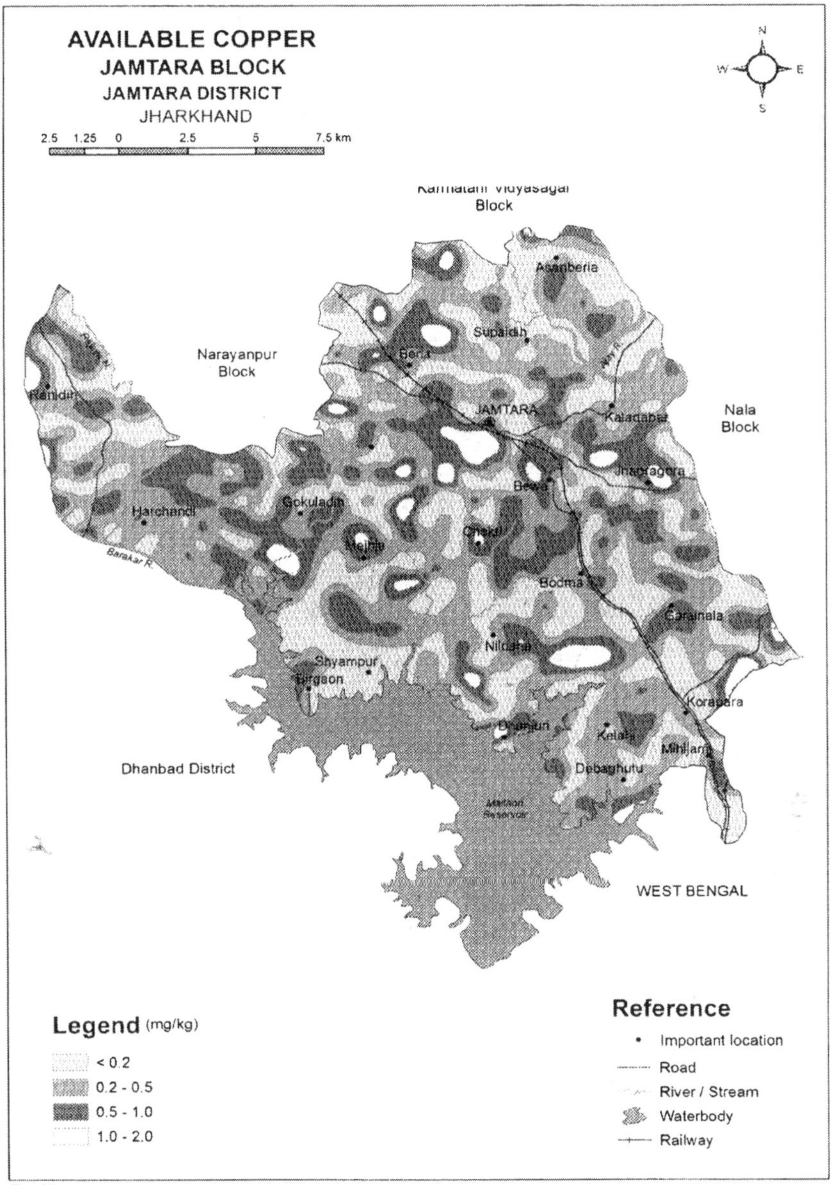

FIGURE 19.10 Available copper status in Jamtara block of Jamtara district.

setting and translocation of sugar, starch and phosphorous, etc. It has role in synthesis of amino acid and protein and regulates carbohydrate metabolism. The available boron content in the soils of different blocks varies from 0.02

to 4.93 mg kg^{-1} and details about area and distribution in different blocks are given in Table 19.11. As per the critical limit of available boron (<0.5 mg kg^{-1}), soils of 40.5, 40.1, 35,5, 24.6, 24.1 and 17.5% area of Jamtara, Karmatarn, Narayanpur, Kundahit, Fatehpur and Nala blocks are deficient in available boron, respectively. The spatial distribution of available boron in the soils of Narayanpur block.

19.5 NUTRIENT STATUS: AT A GLANCE

Soil acidity and nutrient status of Jamtara district is presented in Table 19.12. Data reveals that soils of major area (52.8–85.6%) of all the blocks are strongly to extremely acidic (pH <5.5). The organic carbon and available phosphorus content of the soils are low in major area of all the blocks in the district. Aerial extension of low content of organic carbon (<0.50%) and available phosphorus (<10 kg ha^{-1}) varied from 56.6 to 80.1% and 50.8 to 84.6% area in different blocks, respectively. Available potassium and available sulfur content in the blocks are found low in 28.1 and 24.4% area of the district, respectively. Deficiency of available zinc (DTPA extractable) and available boron (hot water soluble) content are also observed in 44.6 and 29.1% area of the district, respectively. The present study clearly indicates soil acidity to be an "inherent/core problem" of the soils. Low content of soil organic matter and available phosphorus with

TABLE 19.11 Available Boron Status

Available B (mg kg^{-1})	Fatehpur	Jamtara	Karmatarn	Kundahit	Nala	Narayanpur
	Area in km^2 (% TGA)					
<0.50 (Deficient)	63.6 (24.1)	140.6 (40.5)	70.9 (40.1)	72.9 (24.6)	76.0 (17.5)	100.8 (35.5)
0.50–0.75	94.3 (35.7)	111.1 (32.0)	56.7 (32.0)	92.0 (31.1)	131.9 (30.4)	103.1 (36.4)
0.75–1.00	56.1 (21.3)	42.3 (12.2)	28.2 (15.9)	62.7 (21.1)	108.9 (25.1)	43.6 (15.4)
>1.00	44.2 (16.8)	20.8 (6.0)	17.2 (9.7)	65.0 (21.9)	102.9 (23.7)	27.0 (9.5)
Miscellaneous	5.5 (2.1)	32.3 (9.3)	4.1 (2.3)	3.9 (1.3)	14.3 **(3.3)**	9.1 (3.2)
Total	263.7 (100)	347.1 (100.0)	177.1 (100.0)	296.5 (100.0)	434.0 (100.0)	283.6 (100.0)

TABLE 19.12 Nutrient Status – At a Glance

Attributes	Fatehpur (1064)*	Jamtara (1277)	Karmatarn (712)	Kundahit (1231)	Nala (1740)	Narayanpur (1185)	Jamtara district (7209)
				Area in km^2 (% TGA)			
Strongly to extremely acidic soils pH (<5.5)	213.5 (81.3)	185.7 (53.5)	97.0 (54.7)	156.6 (52.8)	371.5 (85.6)	217.6 (76.8)	1241.9 (68.9)
Poor organic carbon status OC (<0.5%)	149.4 (56.6)	240.5 (69.3)	139.8 (79.0)	215.7 (72.8)	296.4 (68.3)	227.1 (80.1)	1268.9 (70.4)
Low availability of plant nutrients							
Available P (<10 kg ha^{-1})	191.7 (72.7)	213.1 (61.4)	105.3 (59.5)	150.7 (50.8)	367.2 (84.6)	165.3 (58.3)	1193.3 (66.2)
Available K (<108 kg ha^{-1})	110.1 (41.7)	80.5 (23.2)	29.9 (16.9)	72.2 (24.4)	149.7 (34.5)	64.7 (22.8)	597.1 (28.1)
Available S (<10 mg kg^{-1})	60.4 (22.9)	83.3 (24.0)	48.9 (27.6)	40.6 (13.7)	118.5 (27.3)	87.7 (30.9)	439.4) (24.4)
Available Zn (<0.5 mg kg^{-1})	49.4 (18.7)	240.5 (69.3)	138.8 (78.4)	28.0 (9.4)	147.6 (34.0)	199.1 (70.2)	803.4 (44.6)
Available B (<0.5 mg kg^{-1})	63.6 (24.1)	140.6 (40.5)	70.9 (40.1)	72.9 (24.6)	76.0 (17.5)	100.8 (35.5)	524.8 (29.1)

*No. of samples analyzed.

**Area in sq.km. and values in parenthesis indicate percent area.

moderate content of available potassium and sulfur as well as deficiency of zinc and boron indicate poor soil fertility status of the district.

19.6 RECOMMENDATIONS

More than 75% soils of Fatehpur, Nala and Narayanpur blocks and 50% soils of Jamtara, Karmatarn and Kundahit blocks pose major challenges in crop production due to high soil acidity which needs to be addressed before resorting to fertilizer use for higher farm profits. Farmers should be encouraged to ameliorate soil acidity problem to sustain soil health and high crop yields.

Soils of all the blocks are of poor fertility as evidenced by low organic matter level and high phosphate deficiency (caused by high soil acidity and P fixation). Potassium deficiency in these soils is widespread (mainly due to its non-application and high crop removal) especially in Fatehpur and Nala blocks. Application of fertilizers and manures as per requirement of crop(s), and cropping sequence based on soil testing in a site specific mode is therefore advocated.

Zinc is deficient in about 70% soils of Jamtara, Karmatarn and Narayanpur blocks (lowlands with continuous rice cropping). Application of 5 kg Zn/ha at transplanting and/or 2 to 3 sprayings of 0.5% $ZnSO_4$ solution in standing crop will help in improving the situation. Boron (35–40%) and Sulfur (24–30%) deficiency are common in soils of Narayanpur, Kamatarn and Jamtara blocks. Farmers need to apply these nutrients based on soil tests.

19.7 CONCLUSIONS

Block level soil nutrient mapping based on geo-referenced soil sampling and GIS based mapping is a unique effort in developing rational fertilizer management practices in Jamtara district. Multi-nutrient deficiencies of phosphorous, potassium, nitrogen, sulfur, boron and soil acidity problems are the reasons for low agricultural productivity/profitability and food security in rainfed conditions of the district. The soil nutrient maps will help the farmers in judicious use of plant nutrients and soil amendments besides reducing the huge nutrient based subsidy burden of the State Government on fertilizers. Promoting proper land use plan based on soil nutrient maps will improve the prospects of agriculture in Jamtara district.

KEYWORDS

- **Agricultural Land Use Planning**
- **Geographic Information System**
- **Global Positioning System**
- **Jamtara District**
- **Macronutrients**
- **Micronutrients**
- **Nutrient Status**
- **Organic Carbon**
- **Soil Acidity**

REFERENCES

Berger, K. C., & Truog, E. (1940). Boron deficiency as revealed by plant and soil tests. *Journal of American Society of Agronomy, 32*(1), 297–301.

FAI. (2011–2012). Fertilizer and Agriculture Statistics, Eastern Region, 36th Ed. 214p.

Follet, R. H., & Lindsay, W. L. (1970). Profile distribution of zinc, iron, manganese and copper in Colorado soils. *Technical Bulletin, Colorado Agricultural Experimental Station, 110*, 79p.

Haldar, A. K., Srivastava, R., Thampi, C. J., Sarkar, D., Singh, D. S., Sehgal, J., & Velayutham, M. (1996). Soils of Bihar for optimizing land use. NBSS Publ. 50b. (Soils of India Series), National Bureau of Soil Survey and Land Use Planning, Nagpur, India, pp. 70+4 sheets soil Map (1:500,000 scale).

Hatcher, J. T., & Wilcox, L. V. (1950). Colorimetric determination of boron using carmine. *Analytical Chemistry, 22*, 567–569.

IARI. (1970). Soil Survey Manual, All India Soil and Land Use Organization, *Indian Agricultural Research Institute*, New Delhi.

Lindsay, W. L., & Norvell, W. A. (1978). Development of a DTPA micronutrients soil test for Zn, Fe, Mn and Cu. *Soil Science Society of America Proceedings, 42*, 421–428.

Mehta, V. S., Singh, V., & Singh, R. B. (1988). Evaluation of some soil test methods for available sulfur in some alluvial soils. *Journal of Indian Society of Soil Science, 36*, 743–746.

NBSS & LUP. (2006). Assessment and mapping of some important soil parameters including soil acidity for the State of Jharkhand (1:50,000 scale) towards rational land use plan, NBSS & LUP Report No. 946, National Bureau of Soil Survey and Land Use Planning, Nagpur, India, 243 pp.

Page, A. L., Miller, R. H., & Keeney, D. R. (1982). Method of Soil Analysis, Part-II, Chemical and Microbiological Properties, *Soil Sci. Soc. Am.* and *Am. Soc. Agron.* Madison, Wisconsin, USA.

Sahoo, A. K., Singh, S. K., Sarkar D., Sarkar, A. K., Agarwal, B. K., Shahi, D. K., Nayak, D. C., Mukhopadhyay, S., & Banerjee, T. (2013). Mapping of Soil Acidity and Nutrient Status at Block Level of Dumka District, Jharkhand, NBSS Report No. 1052B, ICAR-NBSS & LUP, Nagpur, 93 p.

Sen, P., Singh, S. K., Sarkar, D., & Majumdar, A. (2012). Farmer's Advisory Services for Fertilizer and Crop Planning in the State of West Bengal. A Novel Approach to reach out to the farmers. *Fertilizer Marketing News, 43*, 1–7.

Singh, D., Chhonkar, P. K., & Pandey, R. N. (2005). Manual on Soil, Plant and Water Analysis, IARI, New Delhi, 65p.

William, C. H., & Stainbergs, A. (1959). Soil sulfur fraction and chemical indices of available sulfur in some Australian soils. *Australian Journal of Agricultural Research, 10*, 340–352.

CHAPTER 20

SPATIAL DISTRIBUTION OF AVAILABLE NUTRIENTS IN ASHWAGANDHA (*WITHANIA SOMNIFERA*) GROWN SOILS OF FARMERS' FIELDS IN ANDHRA PRADESH

B. VAJANTHA,[1] M. UMADEVI,[2] M. C. PATNAIK,[3] M. RAJKUMAR,[4] M. SUBBARAO,[5] and M. V. S. NAIDU[6]

[1]*Scientist, Agricultural Research Station, Perumallapalle – 517 505, Chittoor, India*

[2]*Director, Water Technology Centre, PJTSAU, Rajendranagar, Hyderabad – 500 030, India*

[3]*Principal Scientist, Agricultural Research Institute, Jayshankar Telangana State Agricultural University, Rajendranagar, Hyderabad – 500 030, India*

[4]*Principal Scientist, Fruit Research Station, Sangareddy, Medak, Telangana – 502 001*

[5]*Principal Scientist and Head, Agricultural Research Station, Perumallapalle – 517 505, Tirupathi, India*

[6]*Professor, S. V. Agricultural College, ANGRAU, Tirupathi – 517 502, India*

CONTENTS

ABSTRACT

To know about the soil fertility status of Ashwagandha growing soils, a survey was conducted during November 2008 in Ananthapur and Kurnool districts of Andhra Pradesh and 34 soil samples were collected at flowering stage of Ashwagandha. The samples were analyzed for their color, texture, pH, electrical conductivity (EC), cation exchange capacity (CEC), organic carbon (OC), available N, P, K, S and soil nutrient index was calculated. The soil samples in both the districts have sandy clay loam to clay loam texture, super active to semi active in CEC activity (0.31 to1.47 cmol(p+) kg^{-1}), neutral to slightly alkaline reaction (7.1 to 7.8), non saline (0.15 to 0.26 dS m^{-1}), low to medium soil OC (0.42 to 0.64%), low available N (185 to 242 kg ha^{-1}), medium available P (12.15 to 18.24 kg ha^{-1}), medium to high available K (235 to 310 kg ha^{-1}), and sufficient available S (10.34 to 18.22 mg kg^{-1}).

20.1 INTRODUCTION

Among the various medicinal plants, *Withania somnifera* Dunal (Wintercherry, Ashwagandha or Asgandh) is an important medicinal plant and its use in Ayurvedic and Unani extends back over 3000 to 4000 years (Atal and Schwarting, 1961). In Ayurveda, the roots of Ashwagandha are known to possess health maintenance and restoration properties, which are similar to ginseng roots, hence it is called as Indian ginseng. In United States market, Ashwagandha acquired considerable significance on account of its large demand due to its reported male sex stimulating properties (Joshi et al., 1981). It is an adoptogenic herb and its roots, seeds and leaves are used in Ayurvedic and Unani medicines. The root drug finds an important place in treatment of rheumatic pain, inflammation of joints, nervous disorders, female disorders, hiccup, cold, cough, ulcers, leprosy, etc. The leaf paste is applied for carbuncles, inflammation and swellings and leaf juice is useful in conjunctivitis. Bark decoction is taken for asthma and applied locally to bed sores. Ashwagandha and its extracts are used in the preparation of herbal tea, powders, tablets and syrups (Nigam and Kandalkar, 1995). Ashwagandha is cultivated over an area of 10,780 ha with a production of 8,429 tons in India. While the annual demand increased from 7028 tons (2001–02) to 9127 tons (2004–05) necessitating the increase in its cultivation and higher production. It stands third with annual growth rate of 9.1% next to Amla and Ashoka. Among the traded medicinal plants in India, Ashwagandha stands second in trading with a worth of Rs. 100–120 million next to Amla (Tripathi et al., 1996). Due to increasing demand for roots in recent times and considering its future demand, there exists much scope for extensive cultivation of the crop in India.

20.1.1 ASHWAGANDHA CULTIVATION IN THE WORLD

Herbal plants are considered as one of the most important sources of medicines since the dawn of human civilization. Out of 80,000 tons of medicinal plants imported by Western countries, India tops the list of exporters to USA and Europe with a share of over 10,000 tons (Praveenkumar et al., 2007). The values of trade in medicinal plants are about Rs. 5,000 crores, while the world trade is about US$ 62 billions. A survey indicated that the use of herbal medicines will reach to the tune of US$ 5 trillion during 2050 (Ashok Kumar, 2003). Currently, the Ayurvedic and herbal products turnover is estimated to be Rs. 25,000 crores (Tikka and Jaimini, 2007).

20.1.2 ASHWAGANDHA CULTIVATION IN INDIA

According to one estimate of botanical survey of India, about 7,500 plants are used for medicinal purposes out of 15,000 plants of our country (Pushpagandan, 1995). There is growing demand for medicines of Ayurveda, Siddha, Unani and Homeopathy both for domestic consumption and export purposes. India exports herbal products and medicines to the tune of Rs. 550 crores annually (Reddy, 2004). The increasing demand for medicinal plant products is much felt by the people and the diversified agro-climate of India is a boon for cultivation and is rightly called the botanical garden of the world. India being one of the 12 mega diversity centers in the world, with this bio-resource wealth, it ranks 10th in the world and 4th in Asia having 15 to 20 thousand plants species with medicinal value of which 30% are considered as endemic to India. Currently, there are about 880 species of medicinal plants in all India trade (Praveenkumar et al., 2007).

Ashwagandha grows in dry and sub-tropical regions. Being hardy and drought tolerant species with its enormous bio compounds, its usage is forever regarded and continuous to enjoy the monopoly in many parts of India. The major Ashwagandha cultivating states are Madhya Pradesh, Rajasthan, Punjab, Uttar Pradesh, Haryana, Gujarat and Maharashtra among which Madhya Pradesh alone has more than 4000 ha area (Misra et al., 1997). *Withania somnifera* Dunal (Winter cherry or Ashwagandha or Asgandh), belonging to *Solanaceae* family is a small woody shrub or herb that grows usually 30 to 50 cm height (maximum of 150 cm). It is an erect growing dicotyledonous plant with fleshy long tap roots. The stem and branches are covered with minute star shaped hairs. Leaves are simple upto 10 cm long, ovate, pedicillate and alternate. Plant bears small (1 cm long), greenish or yellow flowers borne together in short axillary clusters. The fruits or berries are smooth, spherical, red colored with 6 mm diameter enclosed in an inflated and membranous calyx. The fruit has small kidney shaped yellow colored seeds (Nigam and Kandalkar, 1995). Ashwagandha root contains 0.4–1.2% alkaloids, 40–65% starch, 40–65% fibers and minor quantity of oil. The important chemical constituents are alkaloids (withanolides) that are present in roots, leaf and berries. This crop is generally grown in late *kharif* season only on conserved soil moisture and it can be grown on any type of soil with good drainage with 7.5 to 8.0 pH. It requires dry climate for better growth and root development but winter temperatures improve the root quality (Kahar et al., 1991).

20.1.3 SIGNIFICANCE OF THE STUDY

As the quality of root is an important parameter for its marketability, the factors affecting its quality need to be studied and optimized for making Ashwagandha cultivation the most remunerative (Muthumanickam and Balakrishnamurthy, 1999). There is need to study the farmers cultivation practices, method and quantity of fertilizer and manures application adopted by Ashwagandha growing farmers to improve the quantity and quality of root yield. In Andhra Pradesh, the area under Ashwagandha is quite less but increasing year after year due to its drought hardy nature and better performance under moisture stress situations. Considering the medicinal value of the crop, its growing demand and paucity of information on scientific production technology, the survey was carried out during 2008 with an aim to conduct crop nutrition survey in certain Ashwagandha growing tracts of Andhra Pradesh and study the fertility status of surface (0–15 cm) soils and nutrient contents in plant samples.

20.2 STUDY AREA

To know about the soil fertility status of Ashwagandha growing soils, a survey was conducted during November 2008 in Gunthakal mandal of Ananthapur and Alur, Adoni, Aspari and Chippigiri mandals of Kurnool districts in Andhra Pradesh and 34 soil samples were collected at flowering stage of Ashwagandha. The sampling location details (latitude and longitude) were noted using Global Positioning System (GPS) (model – GPSMAP 76 of Garmin company) (Table 20.1) and information regarding to the crops grown, manure and fertilizers schedule used by the farmers were also collected.

20.3 METHODOLOGY

The soil samples collected were analyzed for their color, texture (Piper, 1966), pH (Jackson, 1973), EC (Jackson, 1973), Cation exchange capacity (CEC), OC (Walklcy and Black, 1934), available N (Subbaiah and Asija, 1956), P (Olsen et al., 1954), K (Jackson, 1973), S (Chesnin and Yien, 1951) and soil nutrient index was calculated by using the following formulae to express area wise soil fertility status.

$$\text{Nutrient Index (N1)} = \frac{N_l + 2\,N_m + 3\,N_h}{N_l + N_m + N_h}$$

Rating chart of nutrient index:

NI	**Rate**
<1.5	Low
1.5 to 2.5	Medium
2.5	High

TABLE 20.1 Location Details of Sample Collection Fields

S. No.	Name of the farmer	Village	Mandal	District	Latitude	Longitude
1.	G. V. Kondaiah	G. Kothala	Gunthakal	Ananthapur	E 77°36'35"	N 14°40'47.5"
2.	G. Ramulu	-do-	-do-	-do-	E 77°36'36.8"	N 14°40'49.3"
3.	D. chinnappa	-do-	-do-	-do-	E 77°36'38.3"	N 14°40'56.7"
4.	C. Krishna Reddy	-do-	-do-	-do-	E 77°36'37.4"	N 14°40'50.2"
5.	C. Narasimhulu	-do-	-do-	-do-	E 77°36'17.4"	N 14°40'50.9"
6.	C. Jayaramulu	-do-	-do-	-do-	E 77°36'36.2"	N 14°40'50.8"
7.	S. Srinivasulu	Peddaotur	Alur	Kurnool	E 77°15'44"	N 15°25'39.6"
8.	S. Nagaraju	Peddaotur	-do-	-do-	E 77°13'46.3"	N 15°23'31.9"
9.	K. Lingaji Rao	-do-	-do-	-do-	E 78°16'10.4"	N 15°26'7.2"
10.	K. Ganesh	-do-	-do-	-do-	E 78°15'55.4"	N 15°25'35.6"
11.	K. Adeppa	-do-	-do-	-do-	E 77°15'44.0"	N 15°25'24.6"
12.	G. Ramachandra	-do-	-do-	-do-	E 77°17'15.5"	N 15°26'11.8"
13.	N. Gopalam	Chinna Otur	-do-	-do-	E 77°17'38.3"	N 15°26'28.3"
14.	Sk. Rajak	-do-	-do-	-do-	E 77°17'37.6"	N 15°26'26.3"
15.	G. Govindappa	-do-	-do-	-do-	E 77°18'11.7"	N 15°26'50.8"
16.	V. Ramanjaneyulu	Nagarur	Adoni	-do-	E 77°19'17.5"	N 15°26'49.3"
17.	D. Venkataramana	-do-	-do-	-do-	E 77°19'19.3"	N 14°40'56.7"
18.	V. Ramappa	-do-	-do-	-do-	E 77°19'19.3"	N 14°40'50.2"

19.	V. Krishnappa	Joharapuram	Aspari	-do-	E 77°36'17.4"	N 14°40'50.9"
20.	D. Kajabi	-do-	-do-	-do-	E 77°36'36.2"	N 14°40'50.8"
21.	S. Mahakumaraswamy	-do-	-do-	-do-	E 77°15'44"	N 15°25'39.6"
22.	D. Mohammad Rafi	-do-	-do-	-do-	E 77°13'46.3"	N 15°23'31.9"
23.	D. Sukusab	-do-	-do-	-do-	E 78°16'10.4"	N 15°26'7.2"
24.	D. Basha	-do-	-do-	-do-	E 78°15'55.4"	N 15°25'35.6"
25.	D. Ibraheam	-do-	-do-	-do-	E 77°15'44.0"	N 15°25'24.6"
26.	D. Ismail	-do-	-do-	-do-	E 77°17'15.5"	N 15°26'11.8"
27.	B. Bojjappa	-do-	-do-	-do-	E 77°17'38.3"	N 15°26'28.3"
28.	D. Bhasha	-do-	-do-	-do-	E 77°17'37.6"	N 15°26'26.3"
29.	D. Chalambi	-do-	-do-	-do-	E 77°36'35"	N 14°40'47.5"
30.	M. Bheemanna	Aspari	-do-	-do-	E 77°36'36.8"	N 14°40'49.3"
31.	Srinivasa Chandra	Chippigiri	Chippigiri	-do-	E 77°36'38.3"	N 14°40'56.7"
32.	Ramesh	-do-	-do-	-do-	E 77°36'37.4"	N 14°40'50.2"
33.	A. Sankar	-do-	-do-	-do-	E 77°36'17.4"	N 14°40'50.9"
34.	Govindappa	-do-	-do-	-do-	E 77°36'36.2"	N 14°40'50.8"

where N_l, N_m, N_h are the numbers of soil samples falling in the category of low, medium and high nutrient status and are given weightages of 1, 2 and 3, respectively.

Based on CEC to clay ratio on weight basis, the CEC activity can categorized in to four groups (Soil Survey Staff, 1998).

CEC to clay ratio	activity
>0.6	super active
0.4 to 0.6	active
0.24 to 0.4	semi active
<0.24	subactive

20.4 RESULTS AND DISCUSSION

Data with regard to soil color, texture, pH, EC, soil organic carbon (OC), available N, P, K, S status in surface (0–15 cm) soil samples are presented in Table 20.2. The mandal-wise ranges and means of these parameters are presented in Table 20.3.

20.4.1 SOIL COLOR AND TEXTURE IN ASHWAGANDHA GROWING FIELDS

In general the soils are very dark brown to black color in Ananthapur and Kurnool districts. The soils are dark gray to dark grayish brown in Ananthapur and dark grayish brown to black in Kurnool. In general the soil texture ranged from sandy loam to sandy clay loam in Ananthapur, while sandy loam to clay loam in Kurnool district.

20.4.2 SOIL PH AND EC STATUS IN ASHWAGANDHA GROWING FIELDS

The soils are neutral to slightly alkaline in reaction with the pH ranging from 7.0 to 7.8 with a mean of 7.4. In the soils of Ananthapur district, the pH ranged from 7.3 to 7.8 (neutral to slightly alkaline) with a mean 7.55. In Kurnool district, the pH ranged from 7.0 to 7.8 with a mean of 7.4 (neutral to slightly alkaline). In general, all the soils were found to be non-saline and the EC

TABLE 20.2 Physical and Chemical Properties of Surface Soil Samples Collected From Ashwagandha Growing Farmers' Fields in Ananthapur and Kurnool Districts of Andhra Pradesh During Crop Nutrient Survey

S. No.	Sand (%)	Silt (%)	Clay (%)	Textural class	CEC (c mol (P^+) kg^{-1})	CEC: clay ratio	pH	EC (dS m^{-1})	OC (%)	N (kg ha^{-1})	P (kg ha^{-1})	K (kg ha^{-1})	S (mg kg^{-1})
1.	69.43	4.04	26.53	Sandy clay loam	16.96	0.63 (super active)	7.8	0.15	0.52 (M)	238 (L*)	17.82 (M*)	276 (M)	12.4 (S*)
2.	53.33	18.24	28.43	Sandy clay loam	22.43	0.79 (super active)	7.4	0.18	0.5 (M)	215 (L)	15.35 (M)	265 (M)	11.82 (S)
3.	69.08	6.38	24.54	Sandy clay loam	17.39	0.71 (super active)	7.3	0.22	0.64 (M)	197 (L)	18.24 (M)	280 (M)	12.86 (S)
4.	52.01	6.17	41.82	Sandy clay	24.39	0.58 (active)	7.6	0.18	0.5 (M)	187 (L)	16.16 (M)	292 (H*)	10.34 (S)
5.	75.08	6.04	18.18	Sandy loam	26.69	1.47 (super active)	7.5	0.2	0.5 (M)	195 (L)	17.3 (M)	270 (M)	12.68 (S)
6.	55.01	7.67	37.32	Sandy clay	16.41	0.44 (active)	7.6	0.2	0.45 (L)	200 (L)	18.07 (M)	265 (M)	11.06 (S)
7.	61.63	12.11	26.25	Sandy clay loam	15.87	0.60 (active)	7.4	0.18	0.52 (M)	235 (L)	16.15 (M)	235 (M)	13.04 (S)
8.	50.01	21.42	28.57	Sandy clay loam	18.05	0.63 (super active)	7.4	0.25	0.54 (M)	218 (L)	17.34 (M)	242 (M)	15.78 (S)
9.	74.19	4.69	21.12	Sandy clay loam	21.42	1.01 (super active)	7.5	0.18	0.49 (L)	205 (L)	17.05 (M)	256 (M)	16.02 (S)
10.	42.58	13.65	43.77	Sandy clay	18.78	0.43 (active)	7.7	0.15	0.53 (M)	215 (L)	17.64 (M)	272 (M)	14.15 (S)

TABLE 20.2 (Continued)

S. No.	Sand (%)	Silt (%)	Clay (%)	Textural class	CEC (c mol (P^+) kg^{-1})	CEC: clay ratio	pH	EC (dS m^{-1})	OC (%)	N (kg ha^{-1})	P (kg ha^{-1})	K (kg ha^{-1})	S (mg kg^{-1})
11.	40.81	49.16	10.03	Silty loam	20.3	2.02 (super active)	7.4	0.17	0.52 (M)	197 (L)	17.25 (M)	264 (M)	18.14 (S)
12.	60.97	26.02	13.01	Sandy loam	16.38	1.26 (super active)	7.6	0.17	0.57 (M)	202 (L)	16.15 (M)	270 (M)	15.2 (S)
13.	60.33	8.24	31.43	Sandy clay loam	30.67	0.98 (super active)	7.3	0.2	0.52 (M)	210 (L)	15.18 (M)	288 (H)	15.03 (S)
14.	61.63	15.35	23.02	Sandy clay loam	26.34	1.14. (super active)	7.2	0.18	0.54 (M)	224 (L)	16.34 (M)	265 (M)	12.12 (S)
15	53.12	14.92	31.96	Sandy clay loam	22.86	0.72 (super active)	7.1	0.18	0.47 (L)	232 (L)	17.64 (M)	256 (M)	14.38 (S)
16	37.17	19.21	43.62	Clay loam	24.32	0.56 (active)	7.7	0.24	0.45 (L)	185 (L)	16.34 (M)	270 (M)	10.34 (S)
17	71.78	10.04	18.18	Sandy loam	18.34	1.01 (super active)	7.7	0.2	0.52 (M)	197 (L)	17.7 (M)	265 (M)	14.2 (S)
18	58.61	5.22	43.82	Sandy clay loam	19.26	0.44 (active)	7.6	0.18	0.5 (M)	207 (L)	16.82 (M)	256 (M)	13.18 (S)
19	60.32	18.32	21.36	Sandy loam	17.34	0.81 (super active)	7.3	0.22	0.48 (L)	195 (L)	14.36 (M)	294 (H)	13.46 (S)

20	67.83	8.04	24.13	Sandy clay loam	16.41	0.68 (super active)	7.6	0.17	0.47 (L)	202 (L)	15.82 (M)	272 (M)	14.98 (S)
21	62.32	4.1	33.58	Sandy clay loam	10.34	0.31 (semi active)	7.4	0.2	0.52 (M)	210 (L)	16.34 (M)	310 (H)	14.8 (S)
22	69.53	12.19	18.28	Sandy clay loam	13.04	0.71 (super active)	7.4	0.23	0.5 (M)	225 (L)	17.05 (M)	306 (H)	12.86 (S)
23	57.39	15.04	27.57	Sandy clay loam	16.41	0.60 (active)	7.7	0.18	0.44 (L)	197 (L)	16.72 (M)	287 (H)	18.22 (S)
24	48.37	4.1	47.53	Sandy clay	16.96	0.36 (semi active)	7.3	0.22	0.42 (L)	210 (L)	17.36 (M)	295 (H)	16.7 (S)
25	40.03	13.21	46.53	Sandy clay	17.52	0.38 (semi active)	7.6	0.16	0.46 (L)	218 (L)	18.12 (M)	282 (H)	16.02 (S)
26	57.97	26.02	16.01	Sandy clay	20.85	1.30 (super active)	7.1	0.2	0.52 (M)	236 (L)	16.04 (M)	274 (M)	16.32 (S)
27	63.18	19.71	17.11	Sandy loam	22.16	1.30 (super active)	7.2	0.18	0.57 (M)	224 (L)	17.52 (M)	263 (M)	15.5 (S)
28	62.92	8.56	28.52	Sandy clay	20.62	0.72 (super active)	7.3	0.24	0.54 (M)	218 (L)	16.92 (M)	282 (H)	14.08 (S)
29	42.18	16.65	38.77	Sandy clay	21.85	0.56 (active)	7.5	0.2	0.48 (L)	210 (L)	15.86 (M)	274 (M)	12.87 (S)
30	58.01	13.32	28.57	Sandy clay loam	28.62	1.00 (super active)	7.8	0.23	0.52 (M)	218 (L)	17.05 (M)	289 (H)	14.4 (S)

TABLE 20.2 (Continued)

S. No.	Sand (%)	Silt (%)	Clay (%)	Textural class	CEC (c mol (P^+) kg^{-1})	CEC: clay ratio	pH	EC (dS m^{-1})	OC (%)	N (kg ha^{-1})	P (kg ha^{-1})	K (kg ha^{-1})	S (mg kg^{-1})
31	57.05	16.21	26.74	Sandy clay loam	22.34	0.84 (super active)	7.5	0.18	0.5 (M)	238 (L)	16.82 (M)	294 (H)	14.08 (S)
32	65.85	12.65	21.5	Sandy clay loam	26.85	1.25 (super active)	7.3	0.22	0.47 (L)	224 (L)	17.32 (M)	278 (M)	12.98 (S)
33	60.13	12.12	27.75	Sandy clay loam	18.92	0.68 (super active)	7.6	0.19	0.45 (L)	218 (L)	16.72 (M)	290 (H)	13.11 (S)
34	65.83	10.54	24.13	Sandy clay loam	26.04	1.08 (super active)	7.5	0.19	0.55 (M)	224 (L)	17.84 (M)	285 (H)	14.76 (S)

*M – Medium, *L – Low, *H – High and *S – Sufficient.

TABLE 20.3 Mandal-Wise Characterization of Soil Samples Collected From Ashwagandha Growing Farmers' Field During Crop Nutrition Survey (November 2008)

S. No.	District	Mandal	No of samples	pH	EC (dS m^{-1})	CEC (c mol (P^+) kg^{-1})	CEC to Clay ratio	OC (%)	Available N (kg ha^{-1})	Available P (kg ha^{-1})	Available K (kg ha^{-1})
1	Anantha-pur	Gunthakal	6								
			Range	7.3 – 7.8	0.15–0.22	16.41–24.39	0.44–1.47	0.45–0.64	187–238	15.35–18.24	265–292
			Mean	7.55	0.185	20.40	0.96	0.545	212.5	16.80	278.5
2	Kurnool	Alur	9								
			Range	7.2–7.7	0.15–0.25	15.87–30.67	0.43–1.26	0.49–0.57	197–235	12.15–17.64	235–288
			Mean	7.45	0.2	23.27	0.85	0.53	216	14.89	262
3	Kurnool	Adoni	3								
			Range	7.6–7.7	0.18–0.24	18.34–24.32	0.44–1.01	0.45–0.52	185–207	16.34–17.70	256–270
			Mean	7.65	0.21	21.33	0.73	0.485	196	17.02	263
4	Kurnool	Aspari	12								
			Range	7.1–7.2	0.16–0.26	10.34–28.62	0.31–1.30	0.42–0.57	195–236	14.76–18.12	263–310
			Mean	7.15	0.21	19.48	0.81	0.495	215.5	16.44	287
5	Kurnool	Chippigiri	4								
			Range	7.3–7.6	0.18–0.22	18.92–26.85	0.68–1.25	0.45–0.55	218–238	13.36–17.18	278–294
			Mean	7.45	0.20	22.88	0.96	0.5	228	15.27	286

TABLE 20.3 (Continued)

S. No.	District	Mandal	No of samples	pH	EC (dS m^{-1})	CEC (c mol (P^+) kg^{-1})	CEC to Clay ratio	OC (%)	Available N (kg ha^{-1})	Available P (kg ha^{-1})	Available K (kg ha^{-1})
Kurnool		Overall	28								
		Range	7.1–7.7	0.15–0.26	10.34–30.67	0.31–1.30	0.42–0.57	185–242	12.15–18.12	235–310	
		Mean	7.4	0.21	20.5	0.81	0.495	213.5	15.14	272.5	
Total Range: 7.1–7.8			34								
Mean: 7.45			0.15–0.26	10.34–30.67	0.31–1.47	0.42–0.64	185–242	12.15–18.24	235–310		
			0.21	20.5	0.89	0.53	213.5	15.20	272.5		

ranged from 0.15 to 0.26 dS m^{-1} with a mean of 0.21 dS m^{-1}. In Ananthapur, the EC ranged from 0.15 to 0.22 dS m^{-1} with a mean value of 0.185 dS m^{-1}. In Kurnool it ranged from 0.15 to 0.26 dS m^{-1} with a mean of 0.21 dS m^{-1}.

20.4.3 CEC STATUS IN ASHWAGANDHA GROWING FIELDS

The CEC of soils in the both districts ranged from 10.34 to 30.67 c mol (p^+) kg^{-1} with a mean of 20.50 c mol (p^+) kg^{-1}. The CEC in soils of Anantapur district ranged from 16.41 to 26.69 c mol (p^+) kg^{-1} with mean value of 21.55 c mol (p^+) kg^{-1} and it was 10.34 to 30.67 c mol (p^+) kg^{-1} with mean of 19.48 c mol (p^+) kg^{-1} in Kurnool district. Based on CEC: Clay ratio, CEC activity is ranged from super active to semi active. The soils of Ashwagandha grown fields have the CEC activity ranged from 0.3 to 1.47 with a mean value of 0.89.

20.4.4 ORGANIC CARBON STATUS IN ASHWAGANDHA GROWING FIELDS

The OC content in all the samples were low to medium. It ranged from 0.42 to 0.64% with mean value of 0.53%. In Ananthapur district, the OC content in soil ranged from 0.40 to 0.64% with mean of 0.52%. In Kurnool district it ranged from 0.42 to 0.57% with mean value of 0.49%.

20.4.5 AVAILABLE N STATUS IN ASHWAGANDHA GROWING FIELDS

All the soil samples are low in available nitrogen. In general the range of available N in both the districts was 185 to 238 kg ha^{-1} with mean of 212 kg ha^{-1}. The available N in soils of Ananthapur ranged from 187 to 238 kg ha^{-1} with a mean value of 212.5 kg ha^{-1}. However, in Kurnool district it ranged from 185 to 238 with mean of 212 kg ha^{-1}.

20.4.6 AVAILABLE P STATUS IN ASHWAGANDHA GROWING FIELDS

Among all the soil samples the available P ranged from 12.15 to 18.24 kg ha^{-1} with a mean of 15.20 kg ha^{-1}. The soils are medium in phosphorous content in both the districts. The Ananthapur and Kurnool districts had the available P ranged from 15.35 to 18.24 and 12.15 to 18.12 kg ha^{-1} with mean values of 16.80 and 15.14 kg ha^{-1}, respectively.

20.4.7 AVAILABLE K STATUS IN ASHWAGANDHA GROWING FIELDS

Both the districts are medium to high in available K. The available K in Ananthapur soils ranged from 265 to 292 kg ha^{-1} with mean of 278 kg ha^{-1}. In Kurnool district it ranged from 235 to 310 kg ha^{-1} with mean of 272 kg ha^{-1}.

20.4.8 AVAILABLE S STATUS IN ASHWAGANDHA GROWING FIELDS

The soils in both the districts are found to be sufficient in available sulfur content and it ranged from 10.14 to 18.22 mg kg^{-1} with mean value of 14.18 mg kg^{-1}. The available S in Ananthapur district ranged from 10.34 to 12.86 mg kg^{-1} with mean value of 11.6 mg kg^{-1}. In Kurnool district, it ranged from 10.14 to 18.22 mg kg^{-1} with mean of 14.18 mg kg^{-1}.

The soil samples collected from farmers' fields are low in available N (Menu-Singh et al., 2005). It may be due to the fact that farmers do not apply required dose of N because Ashwagandha is mostly grown as rainfed crop with least cost of cultivation (Kaushal, 2002; Patel, 2001). The medium range of nutrients in soil might be due to application of Farm Yard Manure (FYM), which supplies higher OC that hastens the formation of organic chelates of higher stability with organic ligands, which has lower susceptibility to adsorption, fixation and precipitation in the soil (Maheshwari, 2000; Reddy and Reddy, 1998).

20.5 SOIL NUTRIENT INDEX

The nutrient Index for OC, available N – P – K – S in Ananthapur and Kurnool are presented in Table 20.4. In general, the nutrient Index for OC, available N – P – K – S in Andhra Pradesh was 1.83 – 1 – 2 – 2.33 – 2. The Ananthapur district has 1.56 – 1 – 2 – 2.38 – 2 of OC, N – P – K – S nutrient Index, which indicates that the soils in Ananthapur are low in nitrogen, medium in phosphorous, potassium, sulfur content. The Kurnool district has

TABLE 20.4 Soil Nutrient Index for Ashwagandha Grown Farmers Field in Ananthapur, Kurnool and Andhra Pradesh

S. No.	Nutrient	Nutrient index value			Range
		Ananthapur	Kurnool	Andhra Pradesh	
1	Organic carbon	1.83	1.29	1.56	Medium
2	Available N	1	1	1	Low
3	Available P	2	2	2	Medium
4	Available K	2.33	2.42	2.38	Medium
5	Available S	2	2	2	Medium

the nutrient index of 1.29 – 1 – 2 – 2.4 – 2 for OC, N – P – K – S, which indicated deficiency of nitrogen, and medium status for phosphorous, potassium, sulfur.

20.6 CONCLUSIONS

From crop nutrition survey it was concluded that the soil samples in both the districts have neutral to slightly alkaline in soil reaction. Both the districts have sandy clay, sandy loam, silt loam and clay texture soils. Among the collected soil samples, 16.6% of soils were low in OC content and remaining were medium in range in Ananthapur where as in Kurnool 60% of soils had medium in OC content and 40% were low in content. All the soil samples collected were low in available N and medium in available P. In Ananthapur and Kurnool, 83.3 and 57.1% of soil samples, respectively, have medium available K status and remaining soils are rated as high for available K.

KEYWORDS

- **Ashwagandha**
- **Available Nutrients**
- **Cation Exchange Capacity**
- **Physico Chemical Properties**
- **Soil Nutrient Index**
- **Soil Physical Properties**

REFERENCES

Ashok Kumar, A. (2003). Medicinal plants: Need for financing conservation, collection and cultivation. *Financing Agriculture, 34*(3), 17–21.

Atal, C. K., & Schwarting, A. E. (1961). Ashwagandha, an ancient drug. *Economic Botany, 15*(3), 256–263.

Chesnin, L., & Yein, C. H. (1951). Turbidimetric method of sulfate sulfur in plant materials. *Proceedings of Soil Science Society of America, 15*, 149–151.

Jackson, M. L. (1973). Soil chemical analysis. Prentice Hall of India Private Limited, New Delhi.

Joshi, B. G., Gupta, R. K., Kakade, J. R., & Dahatonde, B. N. (1981). Investigations into the medicinal properties of *Withania dominifera.* Paper presented at *4th All India Seminar on Medicinal and Aromatic Plants* at Madurai, 31st August to 3rd September 1981.

Kahar, L. S., Tomar, S. S., Pathan, M. A., & Nigam, K. B. (1991). Effect of sowing dates and variety on root yield of Ashwagandha. *Indian Journal of Agricultural Sciences, 16*(7), 495.

Kaushal, S. K., Kumar, S., & Upadhyay, R. G. (2002). Response of nitrogen and phosphorus on growth, development and seed yield of Ashwagandha under mid-hill conditions of Himachal Pradesh. *Research on Crops, 3*(2), 376–379.

Soil Survey Staff. (1998). Keys to Soil Taxonomy National Resource Conservation Service USDA.

Maheshwari, S. K., Sharma, R. K., & Gangrade, S. K. (2000). Response of Ashwagandha to organic manures and fertilizers in shallow black soil under rainfed conditions. *Indian Journal of Agronomy, 45*(1), 214–216.

Menu-Singh, Naeem, M., & Nasir khan, M. (2005). Physio morphological response of *Withania somnifera* as influenced by nitrogen fertilization. *Advances in Plant Sciences, 18*(1), 669–673

Misra, H. O., Singh, S., & Kumar, S. (1997). Ashwagandha Cultivation in India. Farm Bulletin No. 5. *Central Institute of Medicinal and Aromatic Plants*, Lucknow.

Muthumanickam, D., & Balakrishnamurthy, G. (1999). Studies on nutritional requirement for Ashwagandha (*Withania somnifera* Dunal) in Shevroy hills of Tamil Nadu. *Journal of Spices and Aromatic Crops, 8*, 179–183.

Nigam, K. B., & Kandalkar, V. S. (1995). Ashwagandha. *Advances in Horticulture, 11*, 335–344.

Olsen, S. R, Cole, C. V., Watanabe, F. S., & Dean, L. A. (1954). Estimation of available phosphorus in soils by extraction with sodium bicarbonate. Circulation from USDA, 939.

Patel, D. H. (2001). Effect of methods of sowing, nitrogen levels and time of harvesting on yield and quality parameters of Ashwagandha var. WS-100. *PhD Thesis*, Gujarat Agricultural University, Anand, 2001.

Piper, C. S. (1966). Soil and Plant Analysis. Hans Publishers, Bombay. pp. 137–153.

Praveenkumar, Channabasappa, K. S., & Hanchinal, S. (2007). Medicinal Plants: Traditional uses properties of *Withania somnifera.* Paper read at 4th All India Seminar on Medicinal and Aromatic Plants at Madurai, 31st August to 3rd September 2007.

Pushpagandan, P. (1995). Ethnobotany in India. A Status Report, Government of India, New Delhi, pp. 83–98.

Reddy, J. (2004). Medicinal Plants Research Scenario in India. Infoconcepts India Inc., January, pp. 25–26.

Reddy, G., & Reddy, M. S. (1998). Effect of organic manures and nitrogen levels on soil available nutrients status in maize – soybean cropping system. *Journal of the Indian Society of Soil Science*, *46*(3), 474–476.

Subbaiah, B. V., & Asija, G. L. (1956). A rapid procedure for the determination of available nitrogen in soils. *Current Science, 25*, 259–260.

Tikka, S. B. S., & Jaimini, S. N. (2007). Medicinal herbs as biofence in arid and semi-arid regions. *Proceedings of the National Symposium on Medicinal and Aromatic Plants for the Economic Benefit of Rural People (MAPER)*, RKVM-IAS, Kolkata, p. 3

Tripathi, A. K., Shukla, Y. N., & Kumar, S. (1996). Ashwagandha (*Withania somnifera* Dunal (*Solanaceae*): a status report. *Journal of Medicinal and Aromatic Plant Sciences, 18*(1), 46–62.

Walkley, A., & Black, C. A. (1934). Estimation of organic carbon by chromic acid titration method. *Soil Science, 37*, 29–38.

CHAPTER 21

MANAGEMENT OF SODIC BLACK CALCAREOUS SOILS: A CASE STUDY FROM CENTRAL INDIA

A. O. SHIRALE,[1] V. K. KHARCHE,[2] R. S. ZADODE,[3] R. N. KATKAR,[4] B. P. MEENA,[1] and A. B. AGE[4]

[1]*Scientist, ICAR-Indian Institute of Soil Science, Babibagh, Bhopal – 462 038, India*

[2]*Associate Dean, College of Agriculture, Nagpur, Dr. Panjabrao Deshmukh Krishi Vidyapeeth, Akola – 444 104, India*

[3]*PhD Scholar, Dr. Panjabrao Deshmukh Krishi Vidyapeeth, Akola – 444 104, India*

[4]*Associate Professor, Dr. Panjabrao Deshmukh Krishi Vidyapeeth, Akola – 444 104, India*

CONTENTS

ABSTRACT

Purna valley is an elongated basin in Vidarbha region of Maharashtra having very deep heavy swell shrink soils. The major problems of these soils are native salinity/sodicity, poor hydraulic conductivity, compact, dense subsoil and incomplete leaching of salts from soil due to severe drainage impairments. Field experiments on cotton (2011) followed by green gram in *kharif* and chickpea in *rabi* season (2012) were conducted on farmers fields in this valley. The pH of experimental sites varied from 8.27–8.34, electrical conductivity (EC) between 0.20–0.25 dS m^{-1}, organic C content between 5.23–5.71 g kg^{-1}, $CaCO_3$ content between 9.67–10.60%, the cation exchange capacity between 52.17–53.47 cmol(p^+) kg^{-1}) and Exchangeable Sodium Percentage (ESP) between 10.39–11.29. The treatments comprised of five different green manures (sunhemp, dhaincha, cowpea, green gram and leucaena loppings), two crop residues (cotton stalk and farm waste as biomulch), gypsum and control. There were nine treatments replicated on three farmers' fields on Vertisols treating each farmer as one replication in randomized block design. The significant improvement in chemical properties of soils was observed under gypsum indicating reduction in pH from the 8.31 to 8.19 and ESP from average initial of 11.15 to 8.11, but simultaneously significant reduction was also observed under organic amendments like dhaincha and sunhemp in terms of reduction in pH upto 8.25 and ESP up to 8.82. Among the crop residues and green manures dhaincha in situ green manuring showed the highest significant decrease in exchangeable Na followed by sunhemp and cowpea. The crop residues and green manures although slow in reclamation were found effective in improving chemical properties of sodic soils gradually in addition to addition of significant amount of biomass resulting into more carbon sequestration.

21.1 INTRODUCTION

The deterioration in soil quality, signifies degradation of certain natural attributes that sustain potential productivity of a soil with a set of management tools and inputs under normal climatic conditions. Soil degradation is antonymous of soil quality. It has been estimated that 23% of all usable land has been affected by degradation to sufficient degree to reduce productivity. Comparable proportion of degraded lands of total geographical area in India is 46%, which is about two times that in the world (Katyal, 2012). Nearly 120 m ha area is under degraded soils in India (Maji et al., 2010). Agricultural production in the arid and semi-arid regions of the world is limited by salinity, sodicity, poor water resources, limited rainfall and loss in soil fertility, constrained to a localized area or sometimes extending over the whole of the basin. Extension of irrigation to the arid and semi-arid regions, however, usually had led to an increase in the area under shallow water tables and to intensify and expanding the hazards of salinity and sodicity. This is because irrigation water brings in additional salts and releases immobilized salts in the soil through evapotranspiration and concentrating dissolved salts in the soil solution. The relative significance of each source in contributing soluble salts depends on the natural drainage condition, soil properties, ground water quality, irrigation water quality, and management practices.

Soil salinity and sodicity are the two major environmental concerns leading to land degradation in irrigated areas of arid and semi-arid regions across the world (Mahmoodabadi et al., 2013). Saline–sodic soils belong to an important category of salt-affected soils characterized by high electrical conductivity (ECe) and Sodium Absorption Ratio/Exchangeable Sodium Percentage (SAR/ESP) (Richards, 1954). Saline–sodic soils are degraded both by the negative effects of salinity on plant growth and of sodicity on soil structure. These soils are high in both soluble sodium and exchangeable sodium, which cause soil swelling and dispersion that leads to poor structure, thus limiting soil water infiltration and permeability (Gharaibeh et al., 2010). In addition to structural losses, chemical and biological properties are also negatively affected by high salinity and sodicity (Ganjegunte et al., 2008). Successful amelioration of a saline–sodic soil involves a two-step procedure with application of a Ca^{2+} source to displace Na^{+} on clay surfaces, which promotes soil flocculation and subsequently followed by leaching to drain out salts from the soil profile. Gypsum is widely accepted as a significant source of Ca^{2+} to soils and its use has been long studied as the most common and a primary chemical remediation method for saline–sodic soils. (Oster et al., 1999). Other chemical reclamation strategies include application of elemental sulfur or sulfuric acid, which solubilize native calcite and provide the required Ca^{2+} to offset Na^{+} on the exchange sites (Amezketa et al., 2005). However, due to increased costs of chemical amendments, phytoremediation with salt tolerant crops has been increasingly adopted, especially in developing countries, as a cheaper alternative to chemical amendments. This technique works on the same principle of native calcite dissolution by building up the partial pressure of CO_2 (PCO_2) in the rhizosphere (Qadir et al., 2007). on the other hand, organic amendments represent an alternative for reclaiming a variety of degraded soils, including salt-affected soils and significant improvements have been reported with respect to soil physical, chemical and biological characteristics after organic matter incorporations (Ghosh et al., 2010). Their contributions towards soil structure development by aggregation and aiding in salt leaching is of prime importance in saline soil amelioration (Lakhdar et al., 2009). While gypsum provides chemical amelioration, the reclamation potential of organic amendments such as composts has been mostly attributed in the literature to the added organic matter, which improves soil structure by promoting soil aggregation through increased soil microbial activity (Roldán et al., 1996). Native $CaCO_3$ mineral, abundant in arid and semiarid region soils, is an important potential Ca^{2+} source, but its low solubility is a main obstacle for the purpose of sodic soil reclamation. The dissolution rate of $CaCO_3$ depends on pH value, CO_2 partial pressure, and its hydrolysis reaction in soil solution (Plummer et al., 1978). Under the alkaline environments, the CO_2 concentration in soil solution is the main factor controlling $CaCO_3$ dissolution (Plummer et al., 1978). Organic matter content, temperature, and moisture content in soil affect CO_2 production. The CO_2 partial pressure in soil air usually is much higher than that in the atmosphere due to plant root respiration and microbial oxidation. The elevated partial pressure of CO_2 within plant root zone and the generation of protons released by roots of certain plant species can assist in the increase of dissolution rate of native $CaCO_3$ (Qadir et al., 2005), and consequently enhance the $Na+$ replacement on exchangeable sites of soils (Robbins, 1986). Several studies have indicated that calcareous sodic soils can be improved by growing crops or trees with or without the application of chemical amendments (Chorom and Rengasamy, 1997). It has also been proved that some organic sources can be applied to reclaim soils affected by salts and sodium (Wichern et al., 2006). The importance of maintaining high levels of soil organic matter has been well established. Soil organic matter improves soil structure and aggregation, increases hydraulic conductivity and promotes higher nutrient levels and greater cation exchange capacity (Jalali and Ranjbar, 2009). Li and Keren (2009) carried out a laboratory experiment in order to study the effects of corn stalks application on reclamation of calcareous sodic soils. The results showed that the reclamation efficiency of calcareous sodic soils increased by adding corn stalks. The addition of organic matter in conjunction with gypsum has been successful in reducing the adverse effects on soil properties associated with the presence of sodium. Vance et al. (1998) found that applying organic matter and gypsum on the soil surface had higher effect on reducing the dispersion of soil particles and on soil electrical conductivity compared to the exclusive gypsum application.

Because there is an increased need to reclaim saline and sodic soils the impact of organic matter on maintaining soil health and crop production should be examined. Many attempts have been made to reclaim the sodic soil with the help of organic matter, but we here examine as many possible locally available organic residues and green manures in comparison to gypsum on sodic black calcareous soils. While the specific effect of organic matter on ESP, has been the subject of several studies its effects on soil exchangeable cations, soluble cations and anions due to the use of different organic matter having varying composition is not well documented. Therefore, this study was conducted to evaluate the effects of two crop residues (composted cotton stalk and farm waste as biomulch) and five green manures (sunhemo, dhaincha, leucaena loppings, cowpea and green gram) as well as the effects of gypsum as a chemical amendment on soil sodicity and cation and anion concentration in soluble and exchangeable phase.

21.1.1 MANAGEMENT OF SODIC BLACK SOILS: GLOBAL SCENARIO

It is known that several organic materials, such as green manures, crop residues, farmyard manures (FYM), agro-industrial by-products and composts can be used as amendments to enhance and sustain the overall soil fertility. The same amendments could likely be considered for soil remediation in the salt-affected areas due to their high organic matter content. In fact, organic matter has several beneficial effects on agricultural fields, such as the slow release of nutrients, soil structure improvement, and the protection of soils against erosion. Selected studies (from literature of the last 10–15 years) are summarized in Table 21.1, focusing on the effects of application of organic matter (i.e., different organic waste materials) in rehabilitation of sodic soils.

21.1.2 MANAGEMENT OF SODIC BLACK SOILS: INDIAN SCENARIO

Many attempts have been made in India for reclamation of sodic soils (Patel and Singh, 1991). Gypsum treatment was more effective in removing leachate Na^+ and Ca^{2+}+Mg^{2+} and cumulative removal of CO_3^{2-} + HCO_3^- was higher

TABLE 21.1 Effect of Different Amendment Application on Soil Sodicity (Global Scenario)

Organic material used	Sodicity levels	Effects	Reference
Sesbania aculeate green manure	pH 10.3, ESP 86.4	Green manuring enhanced the process of reclamation by decreasing the pH value and exchangeable Na of soil.	Swarup (1991)
Sesbania–wheat–sesbania	ESP-49.0	Reduction in ESP upto 28 after one year of experiment	Ilyas et al. (1993)
Cotton gin crushed compost and poultry manure	ESP-15.7 EC 9 mS·cm^{-1} pH-8	Improving soil structure, reducing (by 50%) ESP and increasing different enzyme activities	Tejada et al. (2006)
Corn stalk application	ESP-19 EC-9.7	Recorded 78% decrease after incubation for 60 days and leaching with 6.5 pore volume with distilled water.	Li and Keren (2009)
Poultry and sheep manure	pH-7.5 ESP-10.10	Application of sheep and poultry manure to soils caused a greater adsorption of cations such as calcium, magnesium and potassium than sodium (Na^+). Conversely, the addition of organic amendments led to increased Na^+ leaching and a lower ESP.	Jalali and Ranjbar (2009)
Compost (animal wastes and plant residues)	ESP 34–37 EC-4.03–5.11 mS·cm^{-1} pH-8.62–8.75	Decreased EC and SAR of the saturation extracts of the soils. Organic amendments co-applied with chemical amendments seemed to have a high value for reducing soil pH, soil salinity, and soil sodicity	Mahdy (2011)
Mixture of green waste compost, sedgepeat and furfural residue	ESP -15.8 EC- 3.69 mS·cm^{-1} pH -7.75	Decreased EC, and ESP and increased organic carbon	Wang et al. (2014)

with pressmud treatment than gypsum and pyrite treatment under percolated conditions. Attempts reporting reclamation of sodic soils with the help of organic amendments such attempts are very few but yet signify the superiority of organic amendments over use of gypsum. Singh (1974) reported that removal of exchangeable Na was more pronounced where Sesbania aculeate was added as a source of organic matter as compared with barley straw due to wider C:N ratio of barley straw. The application of FYM @ 6% significantly decreased pHs, exchangeable Na and ESP and enhanced CEC, exchangeable Ca + Mg and organic carbon of soil (Kesvan and Gupta, 1986). Bharambe et al. (2001) studied management of salt affected Vertisols with subsurface drainage and crop residue incorporation under soybean-wheat cropping system on sodic Vertisol and revealed that the physical properties were improved and salinity (ECe) and sodicity (ESP) of soil reduced considerably below critical limit of salinity hazards with incorporation of crop residue such as sugarcane trash @ 5 t ha^{-1} or green manuring with dhaincha over a period of four years. Maurya et al. (2009) evaluated pyrite gypsum and organic amendments compost, digested sludge and sulphitation pressmud for reclamation of sodic soils. The results showed that conjunctive use of gypsum and sulphitation pressmud was more effective in reducing pH, EC, exchangeable Na^+ and ESP over other treatments and significantly improved exchangeable Ca^{2+}, Mg^{2+}, K^+, CEC, available N, P, K status, organic carbon and water holding capacity of soil.

21.2 GEOGRAPHICAL SETTINGS

Purna valley is an east west elongated basin with slight convexity to south, occupying the part of Amaravati, Akola and Buldana districts of Vidarbha region in Maharashtra state of India and extends from 20°45′ to 21°15′ N latitude and 75°15′ to 77°45′ E longitude covering with an area of 2.74 lakh ha in 892 villages. The Purna River originates at an elevation of 900 m in Gawilgarh hills of Satpura in Baitul district of Madhya Pradesh and drains for about 240 km before meeting to Tapi River. The Kutasa village selected for present study is 32 km from Akola and 10 km away from the Purna River. The climate is characterized by three distinct seasons *viz.,* summer being hot and dry from March to May, monsoon characterized as warm and rainy from June to October and winter with dry mild cold from November to February. Most of the rainfall is received from south-west monsoon during

June to October. The mean annual minimum and maximum temperature during cotton, green gram and chickpea growing season are 18.8 and 32.7, 20.4 and 33.7°C, respectively. The Purna valley is a faulted basin bounded on the north by the scarp of the Satpudas and on the south by the scarp of the Ajanta plateau. It is filled with boulders, pebbles, gravels, sands and clays derived entirely from the salts surrounding the valley. The total thickness of the deposits is 140 to 420 meters below the ground level. The river Purna is a perennial stream, which drains in the south facing scarp of Gavilgarh hills in Betul district of Madhya Pradesh. The river banks are dissected, and the right bank is at a higher elevation than the left bank. The aggregated valley has many streams developing a sub-parallel drainage. The river Katepurna is the most important tributary on the left bank, three other rivers on the left bank, are Morna, Mun and Shahanur.

21.3 MATERIALS AND METHODS

21.3.1 PROBLEMS OF THE STUDY AREA

Purna valley is a unique tract of Vertisols and associated soils in Vidarbha, region of Maharashtra in India having combination of three fold problems like the native salinity/sodicity, poor drainability and poor quality ground water. The unique feature of salt affected soils of Purna valley is that though salinity is widely reported in this tract, presence of salts on surface is hardly seen. However, use of well water which is of poor quality makes the situation more problamatic. Farmers in the Valley are therefore compelled to forego irrigation. The precipitation of calcium in the form of calcium carbonate immobilizes calcium and magnesium in the valley soils and dominance of sodium affects physical and chemical properties of soil adversely. Under such situation, it may enhance clay dispersion, destabilization of soil structure and breaking of soil capillary network, which ultimately affect water transmission characteristics of soils (Sagare et al., 2000). The Purna valley soils have severe limitations for their sustainable use owing to the development of adverse physical conditions especially poor internal drainage even at ESP 5 (Balpande et al., 1996). The $CaCO_3$ in clay fractions, the exchangeable Ca/Mg ratio, ESP and saturated hydraulic conductivity (SHC) were reported as the yield influencing factors in soils of Purna valley (Kadu et al., 2003).

21.3.2 EXPERIMENT DESIGN AND TREATMENTS

The field experiments on cotton (2011) followed by leguminous green gram and chickpea (2012) were conducted on selected farmers' fields in Purna valley. The experimental sites were characterized for the chemical and physical properties. The treatments comprised of five different green manures (sunhemp, dhaincha, cowpea, green gram and leucaena loppings), two crop residues (composted cotton stalk and mulching with farm waste as biomulch), gypsum and control. We utilized locally available organic material for this experiment based on their composition (Table 21.2). There were nine treatments replicated on three farmers' fields and the design of experiment was randomized block design, where each farmers' field was treated as one replication. During first year cotton was grown in *kharif* and various green manuring crops (sunhemp, dhaincha, cowpea and green gram) were sown in between two rows of cotton which were buried subsequently at 30, 45, 40 and 40 days after sowing in soil. The addition of biomass and composition of green manuring crops and crop residues were measured before turning into soil. The cotton stalk residues were decomposed using decomposing culture (*Trichoderma viridae*) and mixed into particular soils before sowing. Gypsum application was made to the respective treatment plots uniformly by mixing in the top (10 cm) soil layer. The recommended dose of fertilizer was applied to cotton as per soil test common to all treatments. During second year green gram was grown in *kharif* followed by chickpea in *rabi* season with recommended dose of fertilizer. The crop residues from both the crops grown in second year were incorporated into soil of respective treatments after harvest of the crops so as to ascertain their residual effect.

21.3.3 SOIL SAMPLING AND ANALYTICAL METHODS

Initial composite soil sample was collected during 2011 at each site and soil samples were taken with an auger (4 cm diameter) from 0–20 cm at each site after harvest of each crop under each treatment in all the three replications (2011–12 and 2012–13). Thus three initial composite samples from three sites and from nine treatments and three replications after harvest of the three crops (total 71) were collected. The samples were air-dried and ground to pass through 2 mm sieve before analyzing. Soil samples for biological properties were collected during peak growth stages from rhizosphere region of crop plant. Soil pHs and ESP were analyzed following standard methods (Richards, 1954). Exchangeable sodium was expressed as a percentage of CEC (i.e., ESP. The cation exchange capacity (CEC) was determined using 1N sodium acetate and 1N ammonium acetate as described in USDA Handbook No. 60 (Richards, 1954). The exchangeable Ca^{2+} and Mg^{2+} were determined by leaching the soils in 1N KCl TEA, buffer solution (pH 8.2) and titrating the leachate with standard EDTA solution using murexide and EBT as an indicator (Jackson, 1973). Exchangeable sodium and potassium were determined by leaching the soil with 1N ammonium acetate (pH 7) solution, Na^+ and K^+ from the leachate were estimated by using flame photometer (Page et al., 1982).

21.3.4 STATISTICAL ANALYSIS

The data were analyzed by using SAS statistical software (ver. 9.2; SAS Institute, Cary, NC). One way analysis of variance (ANOVA) was carried out with the ANOVA procedure in SAS enterprise guide 4.2 and the Fisher least significant differences (LSD).

21.4 RESULTS AND DISCUSSION

21.4.1 SOIL pH AND ECE

The highest reduction in pH was recorded due to gypsum application (drop in pH by 0.1 units). However, among the organics sunhemp, dhaincha and cowpea were found equally effective in reducing pHs (drop by 0.04 units) after two years. The significant decrease was noted in pHs at the end of second year under gypsum application due to reclamation. In calcareous saline-sodic and sodic soils, pH usually does not descend to a greater extent (Nelson and Oades, 1998) due to neutralization of H^+ ions by the dissolution of calcite as: $CaCO_3 + H^+ \leftrightarrow Ca^{2+} + HCO_3$.

Plummer et al. (1978) opined that CO_2 concentration in soil solution is the main factor controlling $CaCO_3$ dissolution. The reduction in pH (Table 21.2) owing to application of organic amendments could be ascribed to acidifying effect of organic amendments. Organic acids produced during the course of decomposition of organic amendments increase the availability of Ca and neutralize HCO_3 and CO_3 that otherwise tie up some of Ca to form

TABLE 21.2 Changes in pH, ECe, Organic Carbon and ESP Due Various Treatments (Mean of Three Replicates)

Treatments	pH	ECe (dS m^{-1})	Organic carbon (g kg^{-1})	ESP (%)
Cotton				
Control (no residue, no green manure)	8.36^{a}	0.94^{b}	5.46^{b}	11.30^{a}
Sunhemp in-situ green manuring	8.30^{ab}	0.94^{b}	6.03^{ab}	10.67^{cb}
Dhaincha in-situ green manuring	8.32^{ab}	0.96^{ab}	6.30^{a}	10.43^{c}
Leucaena green leaf green manuring	8.33^{ab}	0.95^{b}	6.00^{ab}	10.92^{ab}
Cowpea in-situ green manuring	8.31^{ab}	0.95^{ab}	6.13^{ab}	10.70^{cb}
Green gram in-situ green manuring	8.29^{ab}	0.95^{ab}	5.73^{ab}	11.01^{ab}
Composted cotton stalk residue	8.31^{ab}	0.95^{b}	5.86^{ab}	11.00^{ab}
Biomulch (mulching with farm waste)	8.30^{ab}	0.94^{a}	5.80^{ab}	11.08^{ab}
Gypsum @ 2.5 t ha^{-1}	8.27	1.00	5.56	10.22^{c}
LSD (P = 0.05)	0.078	0.05	0.81	0.48
Green gram (Residual effect of first year treatment)				
Control (no residue, no green manure)	8.36^{a}	0.95^{ab}	5.43^{c}	11.31^{a}
Sunhemp in-situ green manuring	8.26^{cb}	0.92^{cb}	6.33^{ab}	8.88^{e}
Dhaincha in-situ green manuring	8.27^{cb}	0.91^{c}	6.53^{a}	9.07^{ed}
Leucaena green leaf green manuring	8.30^{ab}	0.93^{cb}	6.13^{cab}	9.64^{dc}
Cowpea in-situ green manuring	8.28^{ab}	0.91^{c}	6.36^{a}b	9.30^{dc}
Green gram in-situ green manuring	8.27^{ab}	0.92^{cb}	5.96^{cab}	9.84^{bc}
Composted cotton stalk residue	8.27^{cb}	0.92^{cb}	6.00^{cab}	9.83^{bc}
Biomulch (mulching with farm waste)	8.29^{ab}	0.93^{cab}	5.90^{cab}	10.46^{b}
Gypsum @ 2.5 t ha^{-1}	8.19^{e}	0.96^{a}	5.56^{cb}	8.76^{e}
LSD (P = 0.05)	0.08	0.03	0.80	0.67
Chickpea (Residual effect of first year treatment)				
Control (no residue, no green manure)	8.35^{a}	0.93^{a}	5.64^{c}	11.22^{a}
Sunhemp in-situ green manuring	8.24^{cb}	0.88^{cdb}	6.47^{cab}	8.98^{dc}
Dhaincha in-situ green manuring	8.25^{cb}	0.85^{d}	6.79^{a}	8.82^{dc}
Leucaena green leaf green manuring	8.30^{ab}	0.87^{cd}	6.37^{cab}	9.42^{bc}
Cowpea in-situ green manuring	8.26^{cab}	0.89^{cb}	6.60^{ab}	9.134^{c}
Green gram in-situ green manuring	8.30^{ab}	0.89^{bc}	6.33^{cab}	9.54^{bc}
Composted cotton stalk residue	8.26^{cb}	0.90^{cab}	6.19^{cab}	9.58^{bc}
Biomulch (mulching with farm waste)	8.27^{cab}	0.91^{ab}	6.10^{cab}	10.16^{b}
Gypsum @ 2.5 t ha^{-1}	8.18^{c}	0.89^{cb}	5.82^{cb}	8.11^{d}
LSD (P=0.05)	0.09	0.03	0.87	1.00

*Data are mean values of three replicates; means with different letters in the same column differ significantly at P = 0.05 according to Fisher LSD.

lime precipitates. Choudhary et al. (2011) observed the significant decrease in pH due to application of dhaincha and gypsum either alone or in combination with each other. Drop in pH under organic amendments treatments was significant ($P > 0.05$) after harvest of green gram and chickpea. This might be due to enhanced proton (H^+) release in the rhizosphere in the case of certain N_2- fixing crops (Qadir et al., 2003) and subsequent amalgamation of residues into soil collected from both the crops in respective treatments.

The electrical conductivity of saturation paste extract was influenced significantly and it was slightly decreased. However, it was well within permissible limits of salinity hazards. During second year after harvest of green gram it was further increased up to 0.96 dS m^{-1} (Table 21.2). Further after harvest of *rabi* chickpea crop it was found to be decreased in all treatments. The application of organic materials also increases the release of salts into soil solution as a result of mineral dissolution due to increase in partial pressure of carbon dioxide and organic acids which leads to very slight increase in electrical conductivity. Similar results were also reported by Choudhary et al. (2011) and Kaur et al. (2008) who showed that application of organic and inorganic amendments increase electrical conductivity initially but decrease the same subsequently due to creation of favorable soil physical condition.

21.4.2 SOIL ORGANIC CARBON

Organic carbon in soil amplified significantly due to addition of organic amendments. The significant improvement in SOC was recorded under dhaincha by 20.39% followed by cowpea and sunhemp by 17.02 and 14.71% over control and gypsum (Table 21.2). Wong et al. (2009) reported significant upsurge in SOC due to addition of organic material over gypsum application in sodic soil. Improvement in SOC due to gypsum was meager during first year of experiment but it was enhanced during second year of experiment by 3.19% over control. Slight increase in values of SOC under gypsum at the end of experiment was attributed to the residues of green gram and chickpea added to the soil. Dalal et al. (2011) reviewed the effect of salinity and sodicity on organic carbon dynamics and reported that salinity and sodicity decreased with the SOC status and stocks in the salt affected soils Thus, gypsum application must be accompanied by organic amendments and phytoremediation to form and stabilize soil aggregates and thus protect SOM. Also potential to accumulate significant amounts of SOC in salt affected landscapes is high. The results thus suggest the superfluous benefit of addition of biomass and in turn carbon to the soil by the practices like crop residue utilization and green manuring over only gypsum application. The different treatments of organic amendments followed the sequence dhaincha > cowpea > sunhemp > leucaena loppings > green gram cotton stalk > biomulch in respect of their carbon sequestration potential. The potential of carbon sequestration by green manures and crop residues over control has also been reported by Surekha et al. (2013) in Vertisols at Hyderabad.

21.4.3 EXCHANGEABLE SODIUM PERCENTAGE

Amendment of sodic soil with gypsum @ 2.5 t ha^{-1} drops down ESP from initial value of 11.15 up to 8.11 (Table 21.2 and Figure 21.1). Drop in ESP under gypsum application was obvious due appreciable quantity of soluble Ca which facilitates replacement of Na ions, converting Na^+ clay into Ca^{2+} clay. Amending soil with green manures also recorded significant reduction in ESP while crop residue treatments recorded modest reduction in ESP. The composition of organic amendments for reclamation clearly shows that all green manures (sunhemp, dhaincha, cowpea and green gram) adds substantial amount of biomass besides this also adds huge amount of calcium in soil which in turn helps in reducing sodicity. Swarup (1991) and Jagadeeswaran et al. (2002) revealed the significant decrease in ESP due to application of dhaincha as green manure. The ESP of experimental soil drop notably under gypsum as well as organic amendment treatments due to inclusion of leguminous green gram (*kharif*) and chickpea (*rabi*) crops during second year of experiment which creates acidic environment in root zone through release of H^+ proton during N_2 fixation and solubilizes native calcium carbonate which

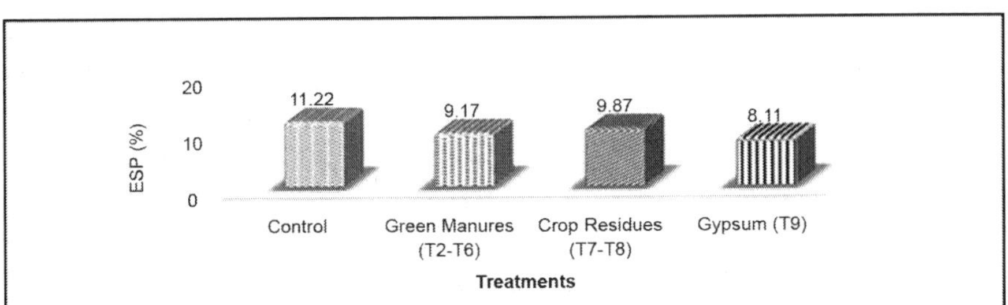

FIGURE 21.1 Average reduction in exchangeable sodium percentage at the end of experiment.

in terns helps in reclamation. Reduction in sodicity due to phytoremidiation with N_2 fixing crop plants has been reported by Mubarak and Nortcliff (2010) and Quadir et al. (2003). The burying of residues from both the crops after their harvest at respective treatments enhanced the reclamation process.

The decomposition of organic matter produces organic acids that solubilizes native calcium carbonate resulting into faster removal of exchangeable sodium and acceleration of the reclamation of calcareous sodic soil. This is also further helpful in reducing the pH of sodic soils which is reflected in the data earlier. The potential of different green manures in regaining sodic soil followed the sequence dhaincha>sunhemp>cowpea>leucaena>green gram.

21.4.4 EXCHANGEABLE CATIONS

The application of gypsum @ 2.5 t ha^{-1} recorded the highest significant improvement in exchangeable Ca^{2+} by 17.01%, Mg^{2+} by 17.45% and K^+ by 18.0% over the control which is due to direct addition of calcium from gypsum and faster reclamation. The application of crop residues and green manures also enhanced the Ca^{2+}, Mg^{2+} and K^+ cations notably at the end of second year (Table 21.3). Among the crop residues and green manures application of dhaincha (by 7.89, 12.0 and 18.0%) and sunhemp (by 6.0, 10.52 and 17%) Ca^{2+}, Mg^{2+}, and K^+, respectively. Jagadeeswaran et al. (2002) reported enhancement in cation like Ca and Mg due to application of sesbania green manure. The significant improvement in Ca^{2+}, Mg^{2+} and K^+ under the organic amendments (T2–T7) might be due to solubilization of calcium carbonate and magnesium carbonate along with addition of potassium which enhance the availability of these cations (Quadir et al., 2007). It has been confirmed in previous studies that Ca^{2+} improves soil structure through cationic bridging with clay particles and soil organic matter (David and Dimitrios, 2002). In addition, Ca^{2+} can inhibit clay dispersion and the associated disruption of aggregates by replacing Na^+ and Mg^{2+} in clay and aggregates, thereby adding to aggregate stability (Zhang and Norton, 2002). Previous studies (Yuan et al., 2007) have shown that Mg^{2+} could be a deleterious ion when its concentration is higher than calcium.

Keren (1996) showed that adsorbed magnesium has adverse effects on infiltration rate for montmorillonitic non-calcareous and calcareous soils. High exchangeable Mg^{2+} is sometimes associated with conditions of low soil conductivity and crusting, similar to those characterizing sodic soils

TABLE 21.3 Changes in Exchangeable Cations Due Various Treatments (Mean of Three Replicates)

Treatments	Ca (cmol(p⁺) kg)	Mg (cmol (p+)kg)	Na (cmol (p+) kg)	K (cmol (p+)kg)
Cotton				
Control (no residue, no green manure)	27.53[b]	18.13[b]	5.99[a]	0.96[c]
Sunhemp in-situ green manuring	30.40[a]	19.30[ab]	5.70[dcb]	1.05[ab]
Dhaincha in-situ green manuring	28.80[ab]	22.63[a]	5.62[cd]	1.09[a]
Leucaena green leaf green manuring	28.20[ab]	18.76[b]	5.78[cb]	1.04[cab]
Cowpea in-situ green manuring	28.00[ab]	19.13[b]	5.70[dcb]	1.04[cab]
Green gram in-situ green manuring	28.53[ab]	18.60[b]	5.88[ab]	1.00[cb]
Composted cotton stalk residue	28.20[ab]	18.60[b]	5.85[ab]	0.99[cb]
Biomulch (mulching with farm waste)	28.26[ab]	18.33[b]	5.88[ab]	0.99[cb]
Gypsum @ 2.5 t ha^{-1}	29.53[ab]	20.06[ab]	5.53[d]	1.02[cab]
LSD (P = 0.05)	2.41	3.46	0.18	0.081
Green gram (Residual effect of first year treatment)				
Control (no residue, no green manure)	27.86[a]	18.20[d]	6.00[a]	0.97
Sunhemp in-situ green manuring	28.93[a]	21.20[a]	5.04[cb]	1.11[ab]
Dhaincha in-situ green manuring	30.06[a]	20.06[cab]	4.97[c]	1.13[ab]
Leucaena green leaf green manuring	28.53[a]	19.26[cd]	5.25[cb]	1.08[ab]
Cowpea in-situ green manuring	29.06[a]	19.73[cb]	5.09[cb]	1.10[ab]
Green gram in-situ green manuring	28.73[a]	19.06[cd]	5.35[cb]	1.05[cb]
Composted cotton stalk residue	28.60[a]	19.13[cd]	5.30[cb]	1.09[ab]
Biomulch (mulching with farm waste)	28.46[a]	18.73[cd]	5.61[ab]	1.04[cb]
Gypsum @ 2.5 t ha^{-1}	30.73[a]	21.06[ab]	4.81[c]	1.14[a]
LSD (P = 0.05)	3.13	1.44	0.58	0.09
Chickpea (Residual effect of first year treatment)				
Control (no residue, no green manure)	27.86[b]	18.33[d]	5.97a	1.00
Sunhemp in-situ green manuring	29.26[b]	20.26[cab]	4.92[dcb]	1.17[b]
Dhaincha in-situ green manuring	30.06[b]	20.53[ab]	4.84[dc]	1.18[a]
Leucaena green leaf green manuring	29.00[b]	20.06[cab]	5.13[cb]	1.11[a]
Cowpea in-situ green manuring	29.60[b]	20.13[cab]	5.01[dcb]	1.13[a]
Green gram in-situ green manuring	29.00[b]	19.66[cdb]	5.20[cb]	1.10[a]
Composted cotton stalk residue	28.53[b]	20.06[cab]	5.18[cb]	1.11[a]
Biomulch (mulching with farm waste)	28.80[b]	18.86[cd]	5.46[ab]	1.09[ab]
Gypsum @ 2.5 t ha^{-1}	32.60[a]	21.53[a]	4.46[d]	1.18[a]
LSD (P=0.05)	2.20	1.61	0.60	0.10

*Data are mean values of three replicates; means with different letters in the same column differ significantly at P = 0.05 according to Fisher LSD.

(Yuan et al., 2007). Initially, the Mg^{2+}/Ca^{2+}ratio were higher than one, which probably led to a deleterious effect on soil aggregate stability by increasing clay dispersion (Zhang and Norton, 2002). Any deleterious impact of Mg^{2+} on soil physical, chemical and biological properties (Clark et al., 2007) might have been offset by the beneficial effect of increased organic matters. The highest significant reduction in Na^+ was observed under the application of gypsum (by 33.85%) followed by dhaincha (25.29%) and sunhemp (17.58%). Sodicity will decline if there is a ready source of favorable cations, particularly Ca^{2+}, to reduce the concentration of Na^+. Because of organic acids formation (Wong et al., 2009) due to organic amendments application, a higher amount of Ca^{2+} was supplied from the soil $CaCO_3$ source to the solution phase. Similar observations were found in sodicity reduction of alkaline subsoil with 20 t ha^{-1} of green shoots of common vetch (Harris and Rengasamy, 2004). The exchangeable cations like Ca^{2+}, Mg^{2+} and K^+ were increased under both organic and chemical amendments and both are effective in removing Na from soil.

21.4.5 SOLUBLE CATIONS

Changes in bivalent cations (Ca^{2+} and Mg^{2+}) were strongly dependent on the applied amendments. Although amended soil appeared to have the highest soluble Ca^{2+} concentrations, application of gypsum itself or incorporated green manures increased soluble Ca^{2+}, strongly. The results also indicated that the exclusive gypsum application resulted in further soluble Ca^{2+} compared to organic amendments (Table 21.4). By adding gypsum to the soil containing $CaCO_3$, Ca^{2+} should act as a common ion, therefore its solubility decrease resulted in increase in soluble as well as exchangeable Ca^{2+}. Comparing the effects of gypsum and organic amendments, it was found that the organic amendments also showed comparable improvement in soluble Ca^{2+} like that of gypsum application. Results from the application of organic amendments, dhaincha addition, showed the highest improvement in soluble Ca^{2+}. In the case of Mg^{2+}, the gypsum and organic amendment treatment showed the increase in Mg^{2+} concentration. In addition, it was observed that gypsum application amplified soluble Mg^{2+} concentration compared to the control and organic amendments. Increase in soluble Mg^{2+} indicates replacement of exchangeable Ca^{2+} with Mg^{2+} in exchange complex. Elrahaman et al. (2012) reported increase in exchangeable Ca^{2+} and Mg^{2+} in solution phase with addition of organic matter.

TABLE 21.4 Changes in Soluble Cations Due Various Treatments (Mean of Three Replicates)

Treatments	Ca (me Lit^{-1})	Mg (me Lit^{-1})	Na (me Lit^{-1})	K (me Lit^{-1})
Cotton				
Control (no residue, no green manure)	2.66^{d}	3.70^{a}	4.80^{a}	0.56^{d}
Sunhemp in-situ green manuring	3.13^{bc}	3.96^{a}	4.23^{cab}	0.66^{cb}
Dhaincha in-situ green manuring	3.40^{b}	4.06^{a}	4.03^{cb}	0.71^{ab}
Leucaena green leaf green manuring	2.86^{dc}	3.86^{a}	4.30^{cab}	0.66^{cb}
Cowpea in-situ green manuring	2.93^{dc}	4.03^{a}	4.46^{cab}	0.65^{cb}
Green gram in-situ green manuring	2.93^{dc}	4.00^{a}	4.63^{ab}	0.60^{cd}
Composted cotton stalk residue	3.13^{bc}	3.90^{a}	4.63^{ab}	0.61^{cd}
Biomulch (mulching with farm waste)	3.06^{c}	3.76^{a}	4.60^{cab}	0.61^{cd}
Gypsum @ 2.5 t ha^{-1}	3.80^{a}	4.20^{a}	3.93^{c}	0.73^{a}
LSD (P = 0.05)	0.31	1.20	0.66	0.06
Green gram (Residual effect of first year treatment)				
Control (no residue, no green manure)	2.40^{e}	3.73^{a}	4.86^{a}	0.57^{c}
Sunhemp in-situ green manuring	3.60^{cb}	4.30^{a}	4.06^{cb}	0.72^{ab}
Dhaincha in-situ green manuring	3.80^{b}	4.06^{a}	3.93^{cb}	0.76^{a}
Leucaena green leaf green manuring	3.20^{d}	3.93^{a}	4.20^{cab}	0.73^{ab}
Cowpea in-situ green manuring	3.33^{dc}	4.10^{a}	4.10^{cab}	0.68^{b}
Green gram in-situ green manuring	3.20^{d}	4.13^{a}	4.20^{cab}	0.68^{b}
Composted cotton stalk residue	3.40^{dc}	3.90^{a}	4.16^{cab}	0.67^{b}
Biomulch (mulching with farm waste)	3.40^{dc}	3.76^{a}	4.33^{ab}	0.66^{b}
Gypsum @ 2.5 t ha^{-1}	4.26^{a}	4.23^{a}	3.43^{e}	0.77^{a}
LSD (P = 0.05)	0.32	1.11	0.79	0.08
Chickpea (Residual effect of first year treatment)				
Control (no residue, no green manure)	2.43^{a}	3.83^{a}	4.83^{a}	0.57d
Sunhemp in-situ green manuring	3.90^{bc}	4.56^{a}	3.76^{bc}	0.76^{cab}
Dhaincha in-situ green manuring	3.93^{b}	4.50^{a}	3.66^{bc}	0.78^{ab}
Leucaena green leaf green manuring	3.60^{dc}	4.13^{a}	4.03^{b}	0.76^{cab}
Cowpea in-situ green manuring	3.56^{dc}	4.33^{a}	3.80^{bc}	0.73^{cb}
Green gram in-situ green manuring	3.50^{d}	4.23^{a}	4.00^{b}	0.71^{c}
Composted cotton stalk residue	3.50^{d}	4.00^{a}	3.96^{b}	0.71^{c}
Biomulch (mulching with farm waste)	3.63^{dbc}	4.06^{a}	4.18^{b}	0.70^{c}
Gypsum @ 2.5 t ha^{-1}	4.76^{a}	4.63^{a}	3.16^{c}	0.82^{a}
LSD (P=0.05)	0.34	1.05	0.66	0.07

*Data are mean values of three replicates; means with different letters in the same column differ significantly at P = 0.05 according to Fisher LSD.

Chaganti and Crohn (2015) also showed that application of green waste compost recorded high Ca^{2+} and Mg^{2+} in leachate over biosolid compost.

The application of gypsum and organic amendments were effective in reducing soluble Na^+ concentration in saturation paste extract in both the years. Among the different treatments application of gypsum recorded significantly lower value of 3.17 me L^{-1} after second year (2012–2013). Among the organic treatments dhaincha recorded lowest value of Na (3.67 me L^{-1}) followed by sunhemp (3.77 me L^{-1}) on an average during second year gypsum treatment reduced soluble Na^+ by 34.36% and dhaincha by 24.01%. The reduction in soluble Na content in saturation paste extract in gypsum and organic treatments might be due to removal of Na due to reclamation process. The potential of different amendments in removing Na^+ follows the order: gypsum>dhaincha>sunhemp>cowpea>leucaena>green gram. Low levels of Na^+ in organic matter treated soil might be due to fact that soil with addition of organic matter exhibit strong preferential adsorption of Na^+ (Pratt and Grover, 1964). Mahmoodabadi et al. (2013) reported the reduction in soluble Na^+ due to addition of pistachio residue and gypsum alone or in combination. Inclusion of N fixing leguminous crops (green gram and chickpea) in cropping system further reduce the Na^+ concentration due to H^+ released by N_2-fixing phytoremediation crops may react with calcite in a way similar to H+ released as a result of H_2CO_3 dissociation into H^+ and HCO_3^-. Qadir et al. (2003) found that H^+ release by N_2-fixing crops could increase the rate of Na^+ removal from calcareous sodic soils.

The increase in soluble K in saturation paste extract was noticed in all treatments except control in both the years of experiment. The highest value of soluble K was observed under gypsum (0.82 me L^{-1}) over control (0.57 me L^{-1}) followed by, sunhemp and leucaena loppings. The increase in K in organic amendments was due to addition of K through biomass. Kharche et al. (2010) studied integrated use of gypsum and organic amendments and bio-inoculant for reclamation of sodic swell shrink soils of Mula command area and reported that integrated as well as combined use of gypsum and organic amendments increased soluble Ca^{2+}, Mg^{2+} and Na^+concentration in saturation paste extract. Similar results were also reported by Murtaza et al. (1999) and Sharma et al. (2001).

21.4.6 SOLUBLE ANIONS

Soluble carbonates were removed significantly due to incorporation of gypsum as well as organic amendments. The significant removal of bicarbonates due to gypsum and organic amendments was due to improvement in infiltration rate. Gypsum application recorded 71.66% reduction while among organic amendments application of dhaincha recorded higher reduction by 43.57% over control (Table 21.5). The significant reduction in bicarbonate in soluble phase due to sesbania leaves (dhaincha) and gypsum application has been reported by Sharma et al. (2001) who showed that gypsum and organic amendments application were most effective in removing CO_3^{2-} and HCO^-_3, respectively. As regards to Cl^- and SO_4^{2-} they are also reduced significantly due to organics matter incorporation and gypsum application. This may be due to increasing leachability of soluble and exchangeable Na^+ throughout the soil profile.

Beheiry et al. (2005) reported that addition of organic manures decreased soil salinity and they attributed to improvement in physical properties of the soil, which facilitate the leaching of salts outside from the root zone.

21.5 RECOMMENDATIONS

It was observed that the highest amount of carbon sequestration occurs in soils with incorporation of crops like dhaincha and sunhemp by adopting a simple practice of sowing one line of dhaincha/sunhemp in between two rows of cotton for eventual in situ decomposition in soil. This practice has been found to increase SOC stock in spite of gradual reduction in pH, ESP and harmful cation like Na^+ and anions. Besides reclamation of sodic soils, utilization of crop residues and green manures add tremendous amount of biomass and nutrient, which can help in overcoming multi-nutrient deficiencies.

21.6 CONCLUSIONS

It can thus be concluded that the use of crop residues and green manuring is beneficial for reclamation of sodic black calcareous soils along with additional benefit of nutrient recycling and increasing soil carbon sequestration. Among the different crop residues and green manures tested in present investigation dhaincha in situ green manuring and sunhemp in situ green manuring for reducing pH and ESP along with removal of HCO_3^-, SO_4^{2-} and Cl^- leads to creation of favorable soil environment for better crop growth. Use of crop residues and green manures showed gradual reclamation as compared to amendments like gypsum.

TABLE 21.5 Changes in Soluble Anions Due Various Treatments (Mean of Three Replicates)

Treatments	HCO_3^- (me Lit^{-1})	Cl^- (me Lit^{-1})	SO_4^{2-} (me Lit^{-1})
Cotton			
Control (no residue, no green manure)	8.33^{a}	1.60^{a}	4.66^{a}
Sunhemp in-situ green manuring	5.83^{cb}	1.20^{ab}	4.30^{ab}
Dhaincha in-situ green manuring	5.00^{c}	0.93^{cb}	4.05^{cb}
Leucaena green leaf green manuring	6.66^{cab}	1.20^{ab}	4.41^{ab}
Cowpea in-situ green manuring	5.83^{cb}	1.06^{cab}	4.41^{ab}
Green gram in-situ green manuring	5.83^{cb}	1.20^{ab}	4.41^{ab}
Composted cotton stalk residue	7.50^{ab}	1.33^{ab}	4.35^{ab}
Biomulch (mulching with farm waste)	8.33^{a}	1.46^{ab}	4.46^{a}
Gypsum @ 2.5 t ha^{-1}	2.50^{d}	0.53^{c}	3.89^{c}
LSD (P = 0.05)	2.02	0.54	0.39
Green gram (Residual effect of first year treatment)			
Control (no residue, no green manure)	8.26^{a}	1.60^{a}	4.67^{a}
Sunhemp in-situ green manuring	5.26^{cb}	1.10^{ab}	4.24^{cb}
Dhaincha in-situ green manuring	4.93^{c}	0.86^{cb}	3.99^{cd}
Leucaena green leaf green manuring	5.86^{cb}	1.10^{ab}	4.37^{ab}
Cowpea in-situ green manuring	5.83^{cb}	1.03^{cb}	4.34^{cab}
Green gram in-situ green manuring	5.93^{cb}	1.16^{ab}	4.33^{cab}
Composted cotton stalk residue	6.73^{cab}	1.16^{ab}	4.31^{cab}
Biomulch (mulching with farm waste)	7.06^{ab}	1.26^{ab}	4.42^{ab}
Gypsum @ 2.5 t ha^{-1}	2.46^{d}	0.50^{c}	3.75^{d}
LSD (P = 0.05)	1.87	0.54	0.37
Chickpea (Residual effect of first year treatment)			
Control (no residue, no green manure)	8.33^{a}	1.73^{a}	4.64^{a}
Sunhemp in-situ green manuring	5.03^{cb}	1.03^{b}	4.05^{cb}
Dhaincha in-situ green manuring	4.70^{c}	0.90^{b}	3.86^{dc}
Leucaena green leaf green manuring	5.60^{cb}	1.00^{b}	4.12^{cb}
Cowpea in-situ green manuring	5.53^{cb}	0.96^{b}	4.22^{b}
Green gram in-situ green manuring	5.73^{cb}	1.06^{b}	4.22^{b}
Composted cotton stalk residue	6.56^{cab}	1.06^{b}	4.22^{b}
Biomulch (mulching with farm waste)	6.90^{ab}	1.10^{b}	4.29^{b}
Gypsum @ 2.5 t ha^{-1}	2.36^{d}	0.60^{b}	3.67^{d}
LSD (P = 0.05)	1.90	0.58	0.32

*Data are mean values of three replicates; means with different letters in the same column differ significantly at P = 0.05 according to Fisher LSD.

KEYWORDS

Carbon Sequestration
Crop Residue Management
Green Manuring
Organic Amendments
Reclamation
Sodic Vertisols

REFERENCES

Amezketa, E., Aragues, R., & Gazol, R. (2005). Efficiency of sulfuric acid, mined gypsum, and two gypsum by-products in soil crusting prevention and sodic soil reclamation. *Agronomy Journal, 97*, 983–989.

Balpande, S. S., Deshpande, S. B., & Pal, D. K. (1996). Factors and processes of soil degradation in vertisols of Purna valley, Maharashtra. India. *Land Degradation and Development, 7*, 1–12.

Beheiry, G., Gh, S., & Soliman, A. A. (2005). Wheat productivity in previously organic treated calcareous soil irrigated with saline water. *Egyptian Journal of Basic and Applied Sciences, 20*, 363–376.

Bharambe, P. R., Shelke, D. K., Jadhav, G. S., Vaishnava, V. G., & Oza, S. R. (2001). Management of salt affected Vertisols with sub-surface drainage and crop residue incorporation under soybean-wheat cropping system. *Journal of the Indian Society* of *Soil Science, 49*, 24–29.

Chaganti, V. N., & Crohn, D. M. (2015). Evaluating the relative contribution of physiochemical and biological factors in ameliorating a saline–sodic soil amended with composts and biochar and leached with reclaimed water. *Geoderma,* 259–260, 45–55.

Choron, M., & Rengasamy, P. (1997). Carbonate chemistry, pH, and physical properties of an alkaline sodic soil as affected by various amendments. *Australian Journal of Soil Research, 35*, 149–161.

Choudhary, O. P., Ghuman, B. S., Singh, B., Thuy N., & Buresh, R. J. (2011). Effect of long term use of sodic irrigation, amendments and crop residues on soil properties and crop yields in rice-wheat cropping system in a calcareous soil. *Field Crop Research, 121*, 363–372.

Clark, G. J., Dodgdhun, N., Sale, P. W. G., & Teng, C. (2007). Changes in chemical and biological properties of Sodic clay subsoil with addition of organic amendments. *Soil Biology and Biochemistry, 39*, 2806–2817.

Dalal, R. C., Wong, V. N. L., & Sahrawat, K. L. (2011). Salinity and sodicity affect organic carbon dynamics in soil. *Bulletin of the Indian Society of Soil Science*, 28, 95–117.

David, R., & Dimitrios, P. (2002). Diffusion and cation exchange during the reclamation of saline structured soils. *Geoderma, 107*, 271–279.

Elrahaman, S. H. A., Mustafa, M. A. M., Taha, T.A, Elsharawy, M. A. O., & Eid, M. A. (2012). Effect of different amendments on soil chemical characteristics, grain yield and elemental content of wheat plants grown on salt-affected soil irrigated with low quality water. *Annals of Agricultural Science*, *57*, 175–182.

Ganjegunte, G., King, L., & Vance, G. (2008). Cumulative soil chemistry changes from land application of saline–sodic waters. *Journal of Environmental Quality, 37*, 128–138.

Gharaibeh, M., Eltaif, N., & Shra'ah, S. (2010). Reclamation of a calcareous saline sodic soil using phosphoric acid and by product gypsum. *Soil Use Management, 26*, 141–148.

Ghosh, S., Lockwood, P., Hulugalle, N., Daniel, H., Kristiansen, P., & Dodd, K. (2010). Changes in properties of sodic Australian Vertisols with application of organic waste products. *Soil Science Society of America Journal, 74*, 153–160.

Harris, M. A., & Rengasamy, P. (2004). Sodium affected subsoils, gypsum, and green manure: inter-actions and implications for amelioration of toxic red mud wastes. *Environmental Geology*, *45*, 1118–1130.

Ilyas, M., Miller, R. W., & Qureshi, R. H. (1993). Hydraulic conductivity of saline-sodic soil after gypsum application and cropping. *Soil Science Society of America Journal, 57*, 1580–5.

Jackson, M. L. (1973). *Soil Chemical Analysis,* 2nd ed.; Prentice Hall of India Pvt. Ltd., New Delhi.

Jagadeeswaran, R., Singaram, P., & Murugappan, V. (2002). Industrial waste ferrogypsum- a substitute for gypsum to alleviate sodicity. *Australian Journal of Experimental Agriculture*, *4*, 309–313.

Jalali, M., & Ranjbar, F. (2009). Effects of sodic water on soil sodicity and nutrient leaching in poultry and sheep manure amended soils. *Geoderma*, *153*, 194–204.

Kadu, P. R., Vaidya, P. H., Balpande, S. S., Satyavathi, P. L. A., & Pal, D. K. (2003). Use of hydraulic conductivity to evaluate the suitability of Vertisols for deep-rooted crops in semi-arid parts of central India. *Soil Use and Management*, *19*, 208–216.

Katyal, J. C. (2012). World soil day: Thou shalt not waste soil but harness quality management practices. *Journal of the Indian Society* of *Soil Science, 60*, 251–260.

Kaur, J., Choudhary, O. P., & Singh, B. (2008). Microbial biomass carbon ans some soil properties as influenced by long term sodic water irrigation, gypsum and organic amendments. *Australian Journal of Soil Research,* 2008, *46*, 141–151.

Keren, R. (1996). Reclamation of sodic-affected soils. *In*: *Soil Erosion, Conservation and Rehabilitation;* Agassi; M. Ed.; Marcel Dekker Inc., New York, 1996.

Kesvan, S. P., & Gupta, R. K. (1986). Amelioration of a sodic soil with organic matter. *Journal of the Indian Society* of *Soil Science, 34*, 442–443.

Kharche, V. K., Patil, S. R., Belur, S. V., & Ghogare, N. S. (2010). Integrated use of gypsum, spent-wash press-mud compost and bio-inoculants for reclamation of sodic swell shrink soils of Maharashtra. *Journal of Soil Salinity and Water Quality*, *2*, 110–115.

Lakhdar, A., Rabhi, M., Ghnaya, T., Montemurro, F., Jedidi, N., & Abdelly, C. (2009). Effectiveness of compost use in salt-affected soil. *Journal of Hazardous Materials, 171*, 29–37.

Li, F. H., & Keren, R. (2009). Calcareous sodic soil reclamation as affected by corn stalk application and incubation: a laboratory study. *Pedosphere*, *19*, 465–475.

Mahdy, A. M. (2011). Soil properties and wheat growth and nutrients as affected by compost amendment under saline water irrigation. *Pedosphere*, *21*, 773–781.

Mahmoodabadi, M., Yazdanpanah, N., Sinobas, L. R., Pazira, E., & Neshat, A. (2013). Reclamation of calcareous saline sodic soil with different amendments (I): redistribution of soluble cations within the soil profile. *Agricultural Water Management, 120*, 30–38.

Maji, A. K., Reddy, G. P. O., & Sarkar, D. (2010). *Degraded and wastelands of India: Status and spatial distribution*. Directorate of Information and Publications of Agriculture, ICAR, Krishi Anusandhan Bhavan I, Pusa, New Delhi, 158 p.

Maurya, B. R., Ram, H., & Prasad, S. S. (2009). Impact of soil amendments on properties of the salt affected rice soil. *Journal of the Indian Society* of *Soil Science, 57*, 385–388.

Mubarak, A. R., & Nortcliff, S. (2010). Calcium carbonate solubilization through H-proton release from some legumes grown in calcareous saline sodic soils. *Land Degradation and Development, 21*, 24–31.

Murtaza, G., Tahir, M. N., Ghafoor, A., & Qadir, M. (1999). Calcium losses from gypsum and ferm yard manure treated saline sodic soil during reclamation. *International Journal of Agriculture and Biology, 90*, 1560–8530.

Nelson, P. N., & Oades, J. M. (1998). *Organic matter, sodicity and soil structure*. In:*Sodic Soils: Distribution, Properties, Management and Environmental Consequences*; Sumner, M. E.; Naidu, R., Eds.; Oxford University Press, New York.

Oster, J. D., Shainberg, I., & Abrol, I. P. (1999). Reclamation of salt-affected soil. *Agricultural Drainage, Agronomy Monograph, 38*, 315–346.

Page A. L., Miller, R. H., & Keeney, D. R. (1982). *Methods of Soil Analysis Part II Chemical and Microbiological Properties*, 2nd Edn.; ASA, Madison, Wisconsin, USA.

Patel, K. P., & Singh, B. (1991). A comparative study on the effect of gypsum, pressmud and pyrite on leachate composition, soil properties and yield of rice and wheat on sodic soil. *Journal of the Indian Society of Soil Science, 39*, 154–159.

Plummer, L. N., Wigley, T. M. L., & Parkhurst, D. L. (1978). The kinetics of calcite dissolution in CO_2-water system at 5 to 60°C and 0.0 to 1.0 atm CO_2. *American Journal of Science, 278*, 179–216.

Pratt, P. F., & Grover, B. L. (1964). Monovalent-divalent cation exchange equilibria in soils in relation to organic matter and type of clay. *Soil Science Society of America Proceedings, 28*, 32–35.

Qadir, M., Noble, A. D., Oster, J. D., Schubert, S., & Ghafoor, A. (2005). Driving forces for sodium removal during phytoremediation of calcareous sodic and saline–sodic soils: a review. *Soil Use Management, 21*, 173–180.

Qadir, M., Oster, J., Schubert, S., Noble, A., & Sahrawat, K. (2007). Phytoremediation of sodic and saline–sodic soils. *Advances in Agronomy, 96*, 197–247.

Qadir, M., Steffens, D., Yan F., & Schubert, S. (2003). Proton release by N-fixing plant roots: a possible contribution to phytore-2 mediation of calcareous sodic soils. *Journal* of *Plant Nutrition* and *Soil Science, 166*, 14 –22.

Richards, L. A. (1954). *Diagnosis and improvement of saline and alkali soils USA Handbook No. 60*, Oxford and IBH Publishing Co. Calcutta.

Robbins, C. W. (1986). Sodic calcareous soils reclamation as affected by different amendments and crops. *Agronomy journal, 78*, 916–920.

Roldán, A., Albaladejo, J., & Thornes, J. (1996). Aggregate stability changes in a semiarid soil after treatment with different organic amendments. *Arid Land Research Management, 10*, 139–148.

Sagare, B. N., Thakare, S. K., & Sonune, B. A. (2000). Saline soil of Purna river basin of Vidarbha region. Department of Agricultural chemistry and soil sciences, Dr. Panjabrao Deshmukh Krishi Vidyapeeth, Akola (M.S.), Extension report, 84.

Sharma, D. R., Minhas, P. S., Sharma, D. K. (2001). Response of rice-wheat to sodic water irrigation and gypsum application. *Journal of the Indian Society* of *Soil Science, 49*, 324–327.

Singh, N. T. (1974). Physico-chemical changes in sodic soils incubated at saturation. *Plant and Soil, 40*, 303–311.

Surekha, K., Rao, K. V. Shobha Rani, N. Latha P. C., & Kumar R. M. (2013). Evaluation of Organic and Conventional Rice Production Systems for their Productivity, Profitability, Grain Quality and Soil Health, *Agrotechnology,* S11, 006. doi: 10.4172/2168-9881.S11-006.

Swarup, A. (1991). Long term effect of green manuring (sesbania aculeate) on soil properties and sustainability of rice and wheat on a sodic soil. *Journal of the Indian Society of Soil Science, 39*, 777–780.

Tejada, M., Garcia, C., Gonzalez, J. L., & Hernandez, M. T. (2006). Use of organic amendment as a strategy for saline soil remediation: Influence on the physical, chemical and biological properties of soil. *Soil Biology and Biochemistry, 38*, 1413–1421.

Vance, W. H., Tisdell, J. M., & McKenzie, B. M. (1998). Residual effects of surface application of organic matter and calcium salts on the sub-soil of a red-brown earth. *Australian Journal of Experimental Agriculture, 38*, 595–600.

Wang, X., Fang, C., Yu, Z., Wang, J., Peng, D., & Jingjing, K. (2014). Seasonal variations of particulate and dissolved organic carbon in Bosten Lake, Xinjiang, *Journal of Lake Sciences, 26*, 552–558.

Wichern, J., Wichern, F., & Joergensen, R. G. (2006). Impact of salinity on soil microbial communities and the decomposition of maize in acidic soils. *Geoderma, 137*, 100–108.

Wong, V., Dalal, R., & Greene, R. (2009). Carbon dynamics of sodic and saline soils following gypsum and organic material additions: a laboratory incubation. *Applied Soil Ecology*, 41, 29–40.

Yuan, B. C., Xu, X. G., Li, Z. Z., Gao, T. P., Gao, M., Fan, X. W., & Deng, J. M. (2007). Microbial biomass and activity in alkalized magnesic soils under arid conditions. *Soil Biology and Biochemistry*, 39, 3004–3013.

Zhang, X. C., & Norton, L. D. (2002). Effect of exchangeable Mg on saturated hydraulic conductivity, disaggregation and clay dispersion of disturbed soils. *Journal of Hydrology*, 260, 194–205.

CHAPTER 22

IMPACT OF LAND CONFIGURATION AND INTEGRATED NUTRIENT MANAGEMENT ON PRODUCTIVITY OF RAINFED COTTON IN VERTISOLS OF CENTRAL INDIA

V. V. GABHANE,[1] RUPALI GHOGARE,[2] M. B. NAGDEVE,[3] M. M. GANVIR,[4] and A. B. TURKHEDE[5]

[1]*Associate Professor, AICRP for Dryland Agriculture, Dr. Panjabrao Deshmukh Krishi Vidyapeeth, Akola – 444 104, India*

[2]*PG Student, Dr. Panjabrao Deshmukh Krishi Vidyapeeth, Akola – 444 104, India*

[3]*Chief Scientist, AICRP for Dryland Agriculture, Dr. Panjabrao Deshmukh Krishi Vidyapeeth, Akola – 444 104, India*

[4]*Assistant Professor, College of Agriculture, Dr. Panjabrao Deshmukh Krishi Vidyapeeth, Akola Akola – 444 104, India*

[5]*Agronomist, AICRP for Dryland Agriculture, Dr. Panjabrao Deshmukh Krishi Vidyapeeth, Akola – 444 104, India*

CONTENTS

ABSTRACT

A field study was conducted to know the effect of land configuration and nutrient module on soil fertility and productivity of rainfed cotton in Vertisols at Research field of AICRP for Dryland Agriculture, Dr. Panjabrao Deshmukh Krishi Vidyapeeth, Akola (MS) during 2008–09 to 2012–13. Land configuration comprised of flat bed, ridges and furrows, opening of furrow after two rows and opening of furrow after each row while nutrient modules comprised of five treatments involving Recommended Dose of Fertilizer (RDF) 50:25:25 NPK kg ha^{-1} through chemical fertilizers, Farmyard Manure (FYM) @ 10t ha^{-1} + PSB + *Azotobacter*, 50% RDF + FYM @ 5t ha^{-1} + PSB + *Azotobacter*, vermicompost @2.5t ha^{-1} + PSB + *Azotobacter* and glyricidia @ 10t ha^{-1} + PSB + *Azotobacter*. Twenty treatment combinations were executed in split plot design with three replications. The significantly higher seed cotton yield (969 kg ha^{-1}) was recorded in land treatment opening of furrow after each row followed by ridges and furrows (942 kg ha^{-1}) which were found to be at par with each other. Among the nutrient modules, significantly higher seed cotton yield (1003 kg ha^{-1}) was recorded with the application of 50% RDF + FYM @ 5 ha^{-1} + PSB + *Azotobacter* followed by 968 kg ha^{-1} with 50: 25: 25 NPK kg ha^{-1} through RDF, which were found to be at par with each other. However, for improvement in soil fertility status, integrated application of 50 % RDF + FYM @5t ha^{-1} + PSB + *Azotobactor* was found superior to rest of the treatments. Hence, it is concluded that integrated application of 50 % RDF (25:12.5:12.5 NPK kg ha^{-1}) + FYM @ 5t ha^{-1} + PSB + *Azotobactor* and opening of furrow after each row at 30–40 days after sowing results in higher cotton productivity with improvement in fertility status of Vertisols under rainfed conditions.

22.1 INTRODUCTION

Cotton (*Gossypium* sp.) is an important cash crop, globally known as 'King of fibers' and plays vital role in the economy of the farmers as well as the country and is popularly known as 'White gold.' It generates employment opportunities to millions not only in production and trade, its contribution in the foreign exchange is also significant, and there exists large potential for export of raw cotton and value added products. Cotton is an important fiber crop of global significance, which is, cultivated in tropical and sub-tropical regions of more than seventy countries the world over. The major producers of cotton are China, India, USA, Pakistan, Uzbekistan, Argentina, Australia, Greece, Brazil, Mexico, and Turkey. These countries contribute about 85% to the global cotton production. World cotton production is estimated at 119.17 m bales of 480 lb in 2014–15 (USDA, 2015). Cotton area decreased to the tune of 1.47 m ha compared to 2013–14. The early estimates of USDA indicates that India has displaced China and become the leading producer of cotton and also maintaining largest area under cotton and second largest exporter of cotton next to United States. Globally 80% of the agricultural land area is rainfed, which generates 65 to 70% staple foods but 70% of the population inhabiting in these areas are poor due to low and variable productivity. India ranks first among the rainfed agricultural countries of the world in terms of both extent and value of produce.

Integrated nutrient management (INM) is the system, which supplies the nutrients from all possible sources like chemical fertilizer, organic manures, green manure and biofertilizers. But in case of organic manures, the nutrient requires more time to become available to plants. Use of chemical fertilizer is also included in the system. It is proved that the continuous use of chemical fertilizers affect the soil health, which can be considered as major limitation of the system. When INM through chemical fertilizers and different organic sources are applied on long term basis, they show beneficial impact on soil quality (Swarup, 2010). Green manure like dhaincha, glyricidia as a source of organic manure increases the total nitrogen, available phosphorus and microbial population in the soil and improves per unit of the water stable aggregate in soil. There is also growing interest in role of micro-organisms, their beneficial effect, their role as soil health indicators and factors that influence their abundance and adversity.

22.1.1 INTEGRATED NUTRIENT MANAGEMENT IN COTTON: GLOBAL CONTEXT

Vertisols mostly under cotton occur principally in hot environments, in the semi arid tropics with marked alternating wet and dry seasons, they are generally found on sedimentary plains as the result of thousands of years eroding the clay content out of the surrounding hills. They can also be found on level land and in depressions. Smaller areas of Vertisols are found on hill slopes and piedmont plains. Their natural climax vegetation is savannah grassland. Vertisols cover an estimated 340 m ha, or about 3% of the world's cultivatable soils, and are found mainly in Africa; in the Sudan's Gezira cotton fields and Southern black soil plains; in South Africa; Ethiopia; and Tanzania. In Asia, they are found extensively in the Indian Deccan Plateau; Vertisols cover large areas of Australia. There are vast areas of Vertisols in Russia.

22.1.2 INTEGRATED NUTRIENT MANAGEMENT IN COTTON: INDIAN CONTEXT

Cotton is an important commercial crop of India, grown by four million farmers in an area of 7.4 m ha. India occupies the foremost position in acreage, which is almost 25% of the global cotton area. However, the productivity is very low in India and its production contribution is only 9% compared to 22% of China and 19.4% of USA (FAO, 2006). Excessive vegetative growth, boll shedding, imbalanced use of organic and inorganic fertilizers, and poor agronomic practices largely attribute to low productivity of cotton in India. It plays a key role in the National economy in terms of generation of direct and indirect employment in the Agricultural and Industrial sectors. Textiles and related exports of which cotton constitutes nearly 65% account for nearly 33% of the total foreign exchange earnings of our country which at present is around 17 billion dollars with a potential for a significant increase in the coming year.

In India it is grown over an area of 116.14 lakh ha with the production of 334 lakh bales and productivity of 489 kg ha^{-1}. In Maharashtra, the cotton grown area is 41.46 lakh ha with production of 74 lakh bales and productivity is 303 kg ha^{-1}. Vidarbha, which is famous for cotton crop, occupies an area of 14.9 lakh ha with 27.4 lakh bales and the productivity of 312 kg ha^{-1} (Anonymous, 2013). Cotton is cultivated in three distinct agro-ecological regions (north, central and south) of the country. The productivity level of cotton in India varies from zone to zone. In India, there are nine major cotton-growing states which fall under three zones *viz.*, the North Zone (Punjab, Haryana and Rajasthan), the Central Zone (Maharashtra, Madhya Pradesh and Gujarat), and the Southern Zone (Andhra Pradesh, Karnataka and Tamil Nadu). Nearly 65% of the cotton crop is cultivated under rainfed conditions in the country.

The northern zone is almost totally irrigated, while the percentage of irrigated area is much lower in the central and southern zones. The lowest being in the central zone which has nearly 60% of cotton area of the country. Under the rainfed growing conditions rainfall ranges from <400 to >900 mm coupled with aberrant precipitation patterns over the years leading to large-scale fluctuations in production. In the irrigated tract canal and well irrigation are resorted including the use of micro-irrigation system. Central zone comprises primarily rainfed tract of Madhya Pradesh, Maharashtra and Gujarat. Predominant area is under black soils, which is subjected to runoff, erosion, soil and nutrient losses. Cotton productivity is low (444 kg lint per hectare) due to uncertainty and vagaries of monsoon. Moisture stress, salinity, soil degradation problems are often encountered. Farmers in this area are resource poor and therefore not in a position to invest more. Cultivation is done traditionally with bullock drawn implements and by manual labor. In Vidarbha, cotton is grown predominantly as a rainfed crop. As such in Vidarbha region about 89% cultivable land is under rainfed farming and rainfed cotton crop production has direct bearing on agrarian economy of region. It is mostly grown on black cotton soil, i.e., Regur/Vertisols. Swelling and shrinkage processes occur in all soils but Vertisols and their intergrades show a greater expression of this phenomenon.

Vertisols cover an area of 26 m ha, and are predominant in the states of Madhya Pradesh (10.7 m ha), Maharashtra (5.6 m ha), Karnataka (2.8 m ha), Andhra Pradesh (2.2 m ha), and Gujarat (1.8 m ha) (Bhattacharyya et al., 2013). Vertisols in India are distributed across varying rainfall regimes ranging from 590 mm in Rajkot to 980 mm in Rewa. The predominant crops grown in these regions are groundnut, soybean, cotton, maize, and pigeon pea during the rainy season and safflower, chickpea, sunflower, and sorghum during the post rainy season. The cotton crop is generally grown in medium to deep black clayey soil, but is also grown in sandy and sandy loam soil through supplemental irrigation by farmers.

Land configuration and nutrient management have significant effect on growth and yield of crop. Land configuration treatments have significant effect on soil moisture conservation and increasing water holding capacity which help in availability of water and nutrients to crop and boost up the overall production. Gokhale et al. (2011) conducted a field experiment on clay soil and showed that opening of furrow in alternate rows recorded significantly higher seed cotton yield as compared to cotton + straw mulching. Sonune et al. (2013) conducted study on effect of tillage and manuring on soil properties and productivity of rainfed cotton on Vertisol and showed that application of 15 t FYM/ha along with 50% recommended dose of fertilizer recorded significantly higher seed cotton yield with improvement in soil fertility. The integrated plant nutrient supply system has emerged as the most logical concept for managing soil fertility. Integrated use of organic manures and chemical fertilizers has been found to be promising in arresting the decline in productivity through the correction of marginal deficiencies of some secondary and macronutrients and its beneficial influence on physical, chemical and biological properties of soil. No single source of plant nutrients either chemical fertilizer or organic manure or green manure or biofertilizer can meet the nutrient requirement of growing crop. This implies that integrated use of plant nutrients is essential. Due to high cost of chemical fertilizers, farmers are facing problems to cultivate their land to the desired level. Therefore, efforts need to be made to follow INM approach for efficient utilization of chemical fertilizers through supplementation with organic manures and biofertilizers. Gabhane et al. (2014) reported that the integrated application of FYM @ 10 t/ha along with 50% RDF was beneficial in improving soil quality and sustaining the rainfed cotton productivity in Vertisols under semi-arid conditions of Maharashtra.

22.1.3 SIGNIFICANCE OF THE STUDY

Recently, Sharma et al. (2005, 2008), while conducting soil quality assessment studies in rainfed semi-arid tropics (SAT) Alfisols, reported significant influence of long-term use of soil restorative practices such as conservation tillage, farm-based residue application, and conjunctive nutrient management on predominant physical, chemical, and biological soil quality indicators. However, similar information is lacking in rainfed cotton-growing Vertisols and related soils. According to Hullugalle et al. (2007), much of the published research on soil quality changes caused by crop rotations in cotton-based farming systems in Vertisols is available for irrigated regions, but the information for rainfed regions is very scanty. In view of the above, it was hypothesized that by following appropriate soil and nutrient-management practices on a long-term basis, soil properties could be improved in cotton-growing rainfed Vertisols of the SAT region of India. With this consideration, an experiment was conducted during 2008–09 to 2012–13 to study the impact of land configuration and INM on productivity of rainfed cotton in Vertisols under semi arid conditions in Maharashtra.

22.2 STUDY AREA

The field experimental site is located in Akola district of eastern Maharashtra of the Deccan Plateau representing a hot semi-arid ecoregion (20° 32′ to 20° 35′ N latitude and 77° 07′ to 77° 10′ E longitude. The region is located in Central Maharashtra Plateau Zone, which is assured rainfall zone with average annual rainfall of 700 to 900 mm in 43–55 rainy days. The zone receives rains in two peaks during June to September and October to December. The major share is from South-West monsoon (June to September) with 688.1 mm accounting to 84.8% followed by North-East monsoon (October to December) with 81.1 mm accounting to 9.99% and remaining rainfall accounting only 2.35 and 2.82% from winter and summer (i.e., January to February and March to May, respectively. In the study area, March to June is the hot period while November to February is the cold period. The soils are derived from trap rock and have varying depth depending upon their physiography. Most of the soils are calcareous, highly base saturated, fairly well drained, well supplied with potash, moderate to low in phosphate, but low in organic matter content and slightly alkaline. Soils deep and clay in texture pose problem of waterlogging.

22.3 METHODOLOGY

22.3.1 EXPERIMENTAL SITE CHARACTERISTICS

The soils of the study area represent Vertisols soil order with medium and deep clay loams to heavy clays, calcareous, and with lime concretions at varying depths. Predominantly, these soils are susceptible to water erosion. The soils are neutral to slightly alkaline in pH (7.8), low in organic carbon (5.2 g kg^{-1}), low in available nitrogen (178 kg ha^{-1}) low in phosphate (17.8 kg ha^{-1}) with high P-fixation capacity, and high in potash (333.6 kg ha^{-1}).

A field experiment was conducted on the research field of AICRP for Dryland Agriculture, Dr. Panjabrao Deshmukh Krishi Vidyapeeth, Akola since 2008–09 to 2012–13 with the cotton crop. The experiment was planned in split plot design with four land configuration treatments *viz*., L_1: Flat bed, L_2: Ridges and furrows, L_3: Opening of furrow after every two rows, L_4: Opening of furrow after each row (*30–40 days after sowing) in main plots and five nutrient management modules in sub plots as M_1: RDF 50:25:25 kg NPK ha^{-1} through chemical fertilizer, M_2: FYM @ 10 t ha^{-1} + PSB + *Azotobacter*, M_3: 50% RDF + FYM @ 5 t ha^{-1} + PSB + *Azotobacter,* M_4: Vermicompost @ 2.5 t ha^{-1} + PSB + *Azotobacter*, M_5: *Glyricidia* @ 10 t ha^{-1} + PSB + *Azotobacter,* with three replications.

22.3.2 SOIL ANALYSIS AND SOIL MOISTURE DETERMINATION

The plot wise soil samples collected after harvest of crop were analyzed for various physical and chemical properties as per standard methods (Black, 1965; Jackson, 1973).The depth wise (0–15 cm and 15–30 m) soil samples were also collected from each plot at various critical growth stages of cotton (emergence, square formation, flowering, boll development and first picking) for soil moisture determination. The test of statistical significance for the experimental data was carried out as per procedure described by Gomez and Gomez (1984). The sustainable yield index (SYI) was computed on the basis of the yield obtained from 2008–09 to 2012–13 and annual rainfall (Vittal et al., 2002). The regression of mean yield with annual rainfall was derived, and then the estimate of error (σ) was computed. Based on the following equation, SYI was calculated. SYI = Y- σ/Y max where, Y is the estimated average yield of a practice across the years, σ is its estimated standard error based on average rainfall of five years and Ymax is the observed maximum yield in the experiment during the years of cultivation.

22.4 RESULTS AND DISCUSSION

22.4.1 SEED COTTON YIELD

The pooled data (2008–09 to 2012–13) in respect of seed cotton yield as influenced by land configuration was found to be significant (Table 22.1). The significantly higher seed cotton yield (969 kg ha^{-1}) was recorded in land treatment opening of furrow after each row followed by ridges and furrows (942 kg ha^{-1}), which were found to be at par with each other. The lowest seed cotton yield (799 kg ha^{-1}) was recorded in flat bed treatment. Similar observations were also reported by Sagare et al. (2001).

TABLE 22.1 Effect of Land Configuration and Nutrient Module on Yield of Cotton

Treatment		Seed Cotton Yield (kg ha^{-1})						SYI
(A)	**Land Configuration**	2008–09	2009–10	2010–11	2011–12	2012–13	Pooled Mean	
L_1	Flat bed	722	789	946	607	905	799	0.45
L_2	Ridges and Furrows	922	948	1017	696	1140	942	0.54
L_3	Opening of furrow after every two rows	844	855	987	668	938	847	0.50
L_4	Opening of furrow after each row	972	970	1144	704	1149	969	0.56
	SE (M) ±	32	27	48	27	22	22	-
	CD at 5%	112	94	NS	NS	75	67	-
(B)	**Nutrient Module**							
M_1	RDF 50:25:25 kg NPK ha^{-1} through chemical fertilizer	972	940	1116	637	1203	968	0.52
M_2	FYM @ 10 t ha^{-1} + PSB + Azotobacter	807	856	1007	676	951	855	0.49
M_3	50% RDF + FYM@5 t ha^{-1} + PSB + Azotobacter	931	975	1263	734	1128	1003	0.56
M_4	Vermicompost @ 2.5 t ha^{-1} +PSB + Azotobacter	817	811	841	633	934	783	0.47
M_5	Glyricidia @10 t ha^{-1} + PSB + Azotobacter	799	868	893	663	950	835	0.50
	SE (m) ±	38	39	39	26	41	27	-
	CD at 5%	116	112	114	NS	118	78	-
(C)	**Interaction**							
	SE (M) ±	76	78	78	53	83	54	-
	CD at 5%	232	NS	226	NS	NS	NS	-

The effect of nutrient module on seed cotton yield was also found to be significant. The significantly higher seed cotton yield (1003 kg ha^{-1}) was recorded with the application of 50% RDF + FYM @ 5 ha^{-1} + PSB + *Azotobacter* followed by 968 kg ha^{-1} in 50: 25: 25 kg NPK ha^{-1} through chemical fertilizer (RDF) which were found to be at par with each other. The lowest seed cotton yield (784 kg ha^{-1}) was recorded with application of vermicompost @ 2.5 t ha^{-1} + PSB + *Azotobacter*. Application of inorganic fertilizer with FYM created maximum nutrient availability to crop, which resulted in higher yield. In central India (Nagpur), at the end of eight cycles of cotton sorghum two-year rotation, seed cotton yield was enhanced by 20.2–22% in the NPK + FYM plots compared to NPK alone plots (Blaise et al., 2003). Manna et al. (2005) noticed that balanced application of plant nutrients has been proved to enhance crop yield and organic matter content of the soil. The increase in seed cotton yield due to integrated use of FYM and chemical fertilizers attributed to efficient utilization of nutrient from soil (Bandopadhyay et al., 2010), improved physicochemical properties of soil, besides direct addition of nutrients from FYM into the soil (Behra et al., 2007). Similar observations were also reported by Padole et al. (1998) and Solunkhe and Fatak (2011). The interaction effect of land configuration and nutrient module on seed cotton yield was found to be non significant.

22.4.2 SUSTAINABLE YIELD INDEX

Five years data indicated that the higher sustainable yield index was observed with the opening of furrow after each row 30–40 days after sowing and integrated application of 50% RDF + FYM 5t ha^{-1} + PSB + *Azotobactor* (0.56), indicating the significant role of opening of furrow and INM in sustaining the yield of cotton on Vertisols under rainfed conditions. Nayak et al. (2012) also observed highest SYI under rice- wheat system with NPK +FYM over NPK alone or NPK + crop residue.

22.4.3 ECONOMICS OF COTTON CULTIVATION

The effect of land configuration on economics of cotton cultivation indicates that the higher gross monetary and net monetary returns were recorded in land treatment opening of furrow after each row followed by ridges and furrows, which were found to be at par with each other (Table 22.2). The higher

B:C ratio (2.16) was also recorded in land treatment opening of furrow after each row.

The effect of various nutrient modules on economics of cotton cultivation reveals that higher gross monetary and net monetary returns were recorded with the application of 50% RDF + FYM @ 5 ha^{-1} + PSB + *Azotobacter* followed by 50:25:25 kg NPK ha^{-1} through chemical fertilizer (RDF), which were found to be at par with each other. Similar observations were also recorded by Rao and Janawade (2006) with application of sunhemp + 100% RDF. Kumar and Yadav (2007) also observed that application of 100% RDF+10 t FYM ha^{-1} to cotton produced significantly higher net return (Rs. 4354 ha^{-1}) and B:C ratio (1.27). Singh et al. (2013) studied the effect of different rainwater conservation practices with INM through different sources on seed cotton yield and economics of Bt cotton (*Gossypium hirsutum* L.)

TABLE 22.2 Effect of Land Configuration and Nutrient Module on Economics of Cotton

Treatment		GMR (Rs ha^{-1})	NMR (Rs ha^{-1})	B:C Ratio
(A)	**Land Configuration**			
L_1	Flat bed	39575	17452	1.80
L_2	Ridges and Furrows	46972	24256	2.09
L_3	Opening of furrow after every two rows	42747	21008	1.99
L_4	Opening of furrow after each row	49182	26334	2.16
	SE (M) ±	912	912	-
	CD at 5%	2810	2810	-
(B)	**Nutrient Module**			
M_1	RDF 50:25:25 kg NPK ha^{-1} through chemical fertilizer	48435	27893	2.35
M_2	FYM @ 10 t ha^{-1} + PSB + *Azotobacter*	42809	17595	1.70
M_3	50% RDF + FYM@5 t ha^{-1} + PSB + *Azotobacter*	50088	26873	2.16
M_4	Vermicompost @ 2.5 t ha^{-1} +PSB + *Azotobacter*	40222	17354	1.76
M_5	Glyricidia @10 t ha^{-1} + PSB + *Azotobacter*	41541	21598	2.08
	SE (m) ±	1155	1155	-
	CD at 5%	3355	3355	-
(C)	**Interaction**			
	SE (M) ±	2310	2310	-
	CD at 5%	NS	NS	-

in Vertisols and results revealed that the intercropping system in Bt cotton hybrid with greengram (*Phaseolus radiatus* L.) enhanced the seed cotton yield and net returns (Rs 30842 ha^{-1}) with B:C ratio of 2.35 when compared to flat bed.

22.4.4 SOIL CHEMICAL PROPERTIES

22.4.4.1 Soil Reaction (pH)

The data presented in Table 22.3 regarding effect of land configuration on soil pH was found to be non significant. It ranged from 7.82 to 7.85 indicating that soil was slightly alkaline in reaction. Similar findings were also reported by Sagare et al. (2001). The effect of nutrient module on soil pH

TABLE 22.3 Effect of Land Configuration and Nutrient Module on Soil Properties

Treatment		pH (1:2.5)	EC (dS m^{-1})	OC (g kg^{-1})
(A)	**Land Configuration**			
L_1	Flat bed	7.84	0.15	5.83
L_2	Ridges and Furrows	7.85	0.13	5.85
L_3	Opening of furrow after every two rows	7.82	0.13	5.85
L_4	Opening of furrow after each row	7.85	0.14	5.96
	SE (M) ±	0.032	0.031	0.16
	CD at 5%	NS	NS	NS
(B)	**Nutrient Module**			
M_1	RDF 50:25:25 kg NPK ha^{-1} through chemical fertilizer	7.89	0.15	5.28
M_2	FYM @ 10 t ha^{-1} + PSB + *Azotobacter*	7.79	0.14	6.41
M_3	50% RDF + FYM@5 t ha^{-1} + PSB + *Azotobacter*	7.83	0.13	6.38
M_4	Vermicompost @ 2.5 t ha^{-1} +PSB + *Azotobacter*	7.84	0.14	5.68
M_5	Glyricidia @10 t ha^{-1} + PSB + *Azotobacter*	7.85	0.14	5.62
	SE (m) ±	0.049	0.017	0.24
	CD at 5%	NS	NS	0.69
(C)	**Interaction**			
	SE (M) ±	0.097	0.033	0.47
	CD at 5%	NS	NS	NS
	Initial value(2008–09)	7.8	0.20	5.2

was found to be non significant. The higher pH value (7.89) was recorded with the application of 50:25:25 kg NPK ha^{-1} through chemical fertilizer (RDF). The lowest pH value (7.79) was recorded with the application of FYM @ 10 t ha^{-1} + PSB + *Azotobacter* due to application of organic fertilizers, which results in production of certain organic acids which decrease the pH of soil. Similar results were observed by Chaudhary et al. (1981) who showed a slight but non-significant decrease in soil pH due to addition of FYM which may be attributed to the decomposition of organic matter. Interaction effect of land configuration and nutrient module on soil pH was found to be non significant.

22.4.4.2 Electrical Conductivity

The electrical conductivity is a measure of soluble salt concentration in soil. Higher amount of salts in the soil restrict the nutrient uptake and thus affect the plant growth. The data (Table 22.3) in respect of electrical conductivity as influenced by land configuration was found to be non significant. The lowest value (0.13 dS m^{-1}) of electrical conductivity was recorded in land treatment ridges and furrows and land treatment opening of furrow after every two rows. The higher value (0.15 dS m^{-1}) of electrical conductivity was recorded in land treatment flat bed. Similar observations were also reported by Sagare et al. (2001). The data in respect of electrical conductivity as influenced by nutrient module was found to be non significant. The lowest value (0.13 dS m^{-1}) of electrical conductivity was recorded with the application of 50% RDF + FYM @ 5 t ha^{-1} + PSB + *Azotobacter*. The higher value (0.15 dS m^{-1}) of electrical conductivity was recorded with the application of 50: 25: 25 kg NPK ha^{-1} through chemical fertilizer (RDF). The interaction effect of land configuration and nutrient module was found to be non significant.

22.4.4.3 Organic Carbon

Organic matter is an indication of organic carbon (OC) fraction of soil formed due to microbial decomposition of organic residue. The data (Table 22.3) pertaining to OC content of soil as influenced by land configuration was found to be non significant. The highest OC (5.96 g kg^{-1}) was recorded in land treatment opening of furrow after each row while the lowest OC (5.83 g kg^{-1}) was observed in land treatment flat bed. Lower OC with intensive tillage is frequently reported in the literature, and appears to be due to rapid microbial decomposition by incorporation of crop residues during tillage (Doran et al., 1994).

The OC in soil was significantly influenced due to nutrient module. The higher (6.41 g kg^{-1}) OC was recorded with the application of FYM @ 10 t ha^{-1} + PSB + *Azotobacter* followed by the application of 50% RDF + FYM @ 5 t ha^{-1} + PSB + *Azotobacter* (6.38 g kg^{-1}) which were found to be at par with each other. The lowest organic carbon (5.28 g kg^{-1}) was recorded with the application of 50:25:25 kg NPK ha^{-1} through chemical fertilizer (RDF). Significant improvements in OC was seen in the NPK + FYM compared to NPK alone plots after a 10-year of cotton cropping on Vertisols (Venugopalan and Pundarikakshudu, 1999). Tolanur and Badanur (2003) also reported that organic carbon content in soils increased significantly with incorporation of FYM or green manuring in combination with 50% RDN through fertilizer as compared with recommended dose of fertilizers. Similar observations were also reported by Singh et al. (2004) and Goud and Konde (2007). The higher gain of organic carbon (1.21 g kg^{-1}) was recorded with the application of FYM @ 10 t ha^{-1}+ PSB + *Azotobacter* followed by 50 % RDF + FYM @ 5 t ha^{-1} + PSB + *Azotobacter* (1.18 g kg^{-1}) over the initial value (5.2 g kg^{-1}). These results suggest that soil carbon levels can be improved substantially even under semi-arid regions by regular addition of organic manures along with mineral fertilizers. The interaction effect of land configuration and nutrient module was found to be non significant.

22.4.5 AVAILABLE NUTRIENTS

22.4.5.1 Available Nitrogen

The data in respect of available nitrogen as influenced by land configuration was found to be non significant (Table 22.4). However, numerically higher available nitrogen (178.9 kg ha^{-1}) was recorded in land treatment opening of furrow after each row while the lowest available nitrogen (176.5 kg ha^{-1}) was observed in land treatment opening of furrow after every two rows. The effect of nutrient module on available nitrogen was found to be significant. The highest available nitrogen (186.0 kg ha^{-1}) was recorded with the application of 50% RDF + FYM @ 5 t ha^{-1} + PSB + *Azotobacter* followed by RDF 50:25:25 kg NPK ha^{-1} through chemical fertilizer (183.9 kg ha^{-1}) which were

TABLE 22.4 Effect of Land Configuration and Nutrient Module on Available Nutrient Status of Soil

Treatment		Available nutrients (kg ha^{-1})			
(A)	**Land Configuration**	**N**	**P**	**K**	**S**
L_1	Flat bed	177.2	15.2	338.9	16.3
L_2	Rides and Furrows	178.1	15.2	342.6	16.5
L_3	Opening of Furrow after every two rows	176.5	15.0	341.9	16.4
L_4	Opening of furrow after each row	178.9	15.7	342.7	16.6
	SE(m)+	2.4	0.20	3.1	0.2
	CD at 5%	NS	NS	NS	NS
(B)	**Nutrient Module**				
M_1	RDF 50:25 kg NPK ha^{-1} through chemical fertilizer	183.9	17.0	361.2	16.6
M_2	FYM @ 10 t ha^{-1}+ PSB + *Azotobacter*	172.4	14.5	329.4	16.8
M_3	50% RDF + FYM@ 5 t ha^{-1} + PSB + *Azotobacter*	186.0	16.7	366.7	17.8
M_4	Vermicompost @ 2.5 t ha^{-1} + PSB + *Azotobacter*	170.3	14.1	323.7	15.8
M_5	Glyricidia @ 10 t ha^{-1} + PSB + *Azotobacter*	175.7	14.1	326.6	15.4
	SE (m)+	2.3	0.4	5.7	0.3
	CD at 5%	6.9	1.2	16.6	0.8
(C)	**Interaction**				
	SE (m)+	4.7	0.8	11.5	0.5
	CD at 5%	NS	NS	NS	NS
	Initial value (2008–09)	178	17.8	333.6	-

found to be on par with each other. The lowest available nitrogen (170.3 kg ha^{-1}) was recorded with the application of vermicompost @ 2.5 t ha^{-1} + PSB + *Azotobacter*. The increase in available nitrogen due to organic materials application could also be attributed the greater multiplication of soil microbes which could convert organically bound N to nitrogen inorganic form. Similar findings were also reported by Malewar and Hasnabade (1995) and Katkar et al. (2002). The increase in available N might be due to direct addition of N through FYM and green manure to the available pool of soil (Tolanur and Badanur, 2003). The gain of available nitrogen (8.0 kg ha^{-1}) was recorded with the application of 50 % RDF + FYM @ 5 t ha^{-1}+ PSB + *Azotobacter* followed by 50:25:25 kg NPK ha^{-1} through chemical fertilizer (RDF) over the initial value (178 kg ha^{-1}). Interaction effect of land configuration and nutrient module on available nitrogen was found to be non significant.

22.4.5.2 Available Phosphorus

The data in respect of available phosphorus as influenced by land configuration was found to be non significant (Table 22.4). However, numerically highest amount of available phosphorus (15.7 kg ha^{-1}) was recorded in land treatment opening of furrow after each row while the lowest available phosphorus (15.0 kg ha^{-1}) was observed in land treatment opening of furrow after every two rows. The effect of nutrient module on available phosphorus was found to be significant. The higher available phosphorus (17.0 kg ha^{-1}) was recorded with the application of 50: 25:25 kg NPK ha^{-1} through chemical fertilizer (RDF) followed by 50% RDF + FYM @ 5 t ha^{-1} + PSB + *Azotobacter* (16.7 kg ha^{-1}) which were found to be at par with each other. The lowest available phosphorus (14.1 kg ha^{-1}) was recorded with the application of vermicompost @ 2.5 t ha^{-1} + PSB + *Azotobacter* and glyricidia @ 10 t ha^{-1} + PSB + *Azotobacter*. This could be attributed to the effect of applied fertilizer and mineralization of organic sources or through solubilization of the nutrients from the native sources during the process of decomposition. Similar observations were also reported by Malewar and Hasnabade (1995), Bellaki et al. (1998) and Babhulkar et al. (2000). The lower loss of available phosphorus (–0.73 kg ha^{-1}) was recorded with the application of 50:25:25 kg NPK ha^{-1} through chemical fertilizer (RDF) over the initial value (17.8 kg ha^{-1}). Interaction effect of land configuration and nutrient module on available phosphorus was found to be non significant.

22.4.5.3 Available Potassium

The effect of land configuration on available potassium was found to be non significant (Table 22.4). However, numerically higher available potassium (342.7 kg ha^{-1}) was recorded in land treatment opening of furrow after each row while the lowest available potassium (338.9 kg ha^{-1}) was recorded in flat bed treatment. The effect of nutrient module on available potassium was found to be significant. The higher amount of available potassium (366.7 kg ha^{-1}) was recorded with the application of 50% RDF + FYM @ 5 t ha^{-1} + PSB + *Azotobacter* followed by RDF 50:25:25 kg NPK ha^{-1} through chemical fertilizer (361.2 kg ha^{-1}) which were found to be at par with each other. The lowest available potassium (323.7 kg ha^{-1}) was recorded with the application of glyricidia @ 10 t ha^{-1} + PSB + *Azotobactor*. Singh et al. (2001) explained that significant increase in available K content was due to either FYM or green manure along with fertilizers N, suggesting that FYM and green manure helped to maintain the supply of K by releasing the K from reserve source. Similar observations were also recorded by Padole et al. (1998) and Katkar et al. (2002). The higher gain of available potassium (33.1 kg ha^{-1}) was recorded with the application of 50 % RDF + FYM @ 5 t ha^{-1} + PSB + *Azotobacter* followed by 50:25:25 kg NPK ha^{-1} through chemical fertilizer (RDF) (27.6 kg ha^{-1}) over the initial value (333.6 kg ha^{-1}). Interaction effect of land configuration and nutrient module on available potassium was found to be non significant. The increase in availability of NPK was attributed to N mineralization besides addition (Tolanur and Badanur, 2003), solubilization of native P and reduction of P fixation and direct addition of K to the available pool of soil with reduction in K fixation and release of K (Sharma et al., 2001).

22.4.5.4 Available Sulfur

The effect of land configuration on available sulfur was found to be non significant (Table 22.4). The numerically higher amount of available sulfur (16.6 kg ha^{-1}) was recorded in land treatment opening of furrow after each row while the lowest available sulfur (16.3 kg ha^{-1}) was recorded in flat bed treatment. The available sulfur was found to be significantly influenced by nutrient module. The significantly higher amount of available sulfur (17.8 kg ha^{-1}) was recorded with the application of 50% RDF + FYM @ 5 t ha^{-1} + PSB + *Azotobacter*. The lowest available sulfur (15.4 kg ha^{-1}) was recorded with the application of glyricidia @ 10 t ha^{-1} + PSB + *Azotobacter*. Similar findings were reported by Hegde (1996). Interaction effect of land configuration and nutrient module on available sulfur was found to be non significant.

22.4.6 DTPA EXTRACTABLE MICRONUTRIENTS IN SOIL

22.4.6.1 Available Iron

The effect of land configuration on available iron was found to be non significant (Table 22.5). The numerically higher available iron (7.91 mg kg^{-1})

was recorded in land treatment opening of furrow after each row while the lowest available iron (7.77 mg kg^{-1}) was recorded in land treatment opening of furrow treatment after every two rows. The available Fe was found to be significantly influenced by nutrient module. The higher available iron (9.01 mg kg^{-1}) was recorded with the application of 50% RDF + FYM @ 5 t ha^{-1} + PSB + *Azotobacter* followed by FYM @ 10 + ha^{-1} + PSB + *Azotobacter* (8.64 mg kg^{-1}) which were found to be at par with each other. The lowest available iron (6.53 mg kg^{-1}) was recorded with the application of vermicompost @ 2.5 t ha^{-1} + PSB + *Azotobacter*. Similar findings were reported by Babaria and Patel (1980) and Lohakare (1980). Interaction effect of land configuration and nutrient module on available iron was found to be non significant.

22.4.6.2 Available Manganese

The effect of land configuration on available manganese was found to be non significant (Table 22.5). The numerically higher amount of available manganese (12.69 mg kg^{-1}) was recorded in opening of furrow after every two rows treatment. The lowest available manganese (12.59 mg kg^{-1}) was observed in flat bed treatment. The effect of nutrient module on available manganese was found to be significant. The higher amount of available manganese (13.06 mg kg^{-1}) was recorded with the application of 50% RDF + FYM @ 5 t ha^{-1} + PSB + *Azotobacter* followed by FYM @ 10 t ha^{-1} + PSB + *Azotobacter* (12.94 mg kg^{-1}) which were found to be at par with each other. The lowest amount of available manganese (12.14 mg kg^{-1}) was recorded with the application of glyricidia @ 10 t ha^{-1} + PSB + *Azotobacter*. Similar findings were also reported by Prasad and Singh (1980) that continuous use of FYM and chemical fertilizer increased the in available manganese in soil. Interaction effect of land configuration and nutrient module on available Mn was found to be non significant.

22.4.6.3 Available Zinc

The effect of land configuration on available zinc was found to be non significant (Table 22.5). The numerically higher amount of available zinc (0.63 mg kg^{-1}) was recorded in opening of furrow after each row. The lowest amount (0.57 mg kg^{-1}) of available zinc was recorded in land treatment

TABLE 22.5 Effect of Land Configuration and Nutrient Module on DTPA Extractable Micronutrients in Soil

Treatment		DTPA extractible micronutrients (mg kg^{-1})			
		Fe	**Mn**	**Zn**	**Cu**
(A)	**Land Configuration**				
L_1	Flat bed	7.86	12.59	0.59	2.18
L_2	Ridges and Furrows	7.84	12.64	0.60	2.19
L_3	Opening of furrow after every two rows	7.77	12.69	0.57	2.19
L_4	Opening of furrow after each row	7.91	12.63	0.63	2.18
	SE (M) ±	0.35	0.04	0.02	0.007
	CD at 5%	NS	Ns	NS	NS
(B)	**Nutrient Module**				
M_1	RDF 50:25:25 kg NPK ha^{-1} through chemical fertilizer	7.99	12.77	0.58	2.18
M_2	FYM @ 10 t ha^{-1} + PSB + *Azotobacter*	8.64	12.94	0.65	2.21
M_3	50% RDF + FYM@5 t ha^{-1} + PSB + *Azotobacter*	9.01	13.06	0.66	2.23
M_4	Vermicompost @ 2.5 t ha^{-1} +PSB + *Azotobacter*	6.53	12.14	0.54	2.14
M_5	Glyricidia @10 t ha^{-1} + PSB + *Azotobacter*	7.07	12.30	0.56	2.17
	SE (m) ±	0.34	0.07	0.02	0.007
	CD at 5%	0.98	0.21	0.05	0.019
(C)	**Interaction**				
	SE (M) ±	0.68	0.14	0.038	0.013
	CD at 5%	NS	NS	NS	NS

opening of furrow after every two rows. The effect of nutrient module on available zinc was found to be significant. The higher amount of available zinc (0.66 mg kg^{-1}) was recorded with the application of 50% RDF + FYM @ 5 t ha^{-1} + PSB + *Azotobacter* followed by application of FYM @ 10 t ha^{-1} + PSB + *Azotobacter* (0.65 mg kg^{-1}) which were found to be at par with each other. This may be due to addition of Zn through FYM in the soil. The lowest amount of available zinc (0.54 mg kg^{-1}) was recorded with the application of vermicompost @ 2.5 t ha^{-1} + PSB + *Azotobacter*. Similar findings were also noticed by Indulkar and Malewar (1996) that application of FYM resulted in significantly higher availability of zinc. Interaction effect of land configuration and nutrient module on available zinc was found to be non significant.

22.4.6.4 Available Copper

Effect of land configuration on available copper was found to be non significant (Table 22.5). The numerically higher available copper (2.19 mg kg^{-1}) was recorded in land treatments ridges and furrows and land opening of furrows after every two rows. The lower amount of available copper (2.18 mg kg^{-1}) was recorded in flat bed and opening of furrow after each row. The effect of nutrient module on available copper was found to be significant. The significantly higher amount of available copper (2.23 mg kg^{-1}) was recorded with the application of 50% RDF + FYM @ 5 t ha^{-1} + PSB + *Azotobacter* followed by FYM @ 10 t ha^{-1} + PSB + *Azotobacter* (2.21 mg kg^{-1}), which were found to be at par with each other. The lowest amount of available copper (2.14 mg kg^{-1}) was recorded with the application of vermicompost @ 2.5 t ha^{-1} + PSB + *Azotobacter*. Similar observations were also recorded by Bellaki and Badanur (1997) and Bellaki et al. (1998). Interaction effect of land configuration and nutrient module on available copper was found to be non significant.

22.4.7 SOIL MOISTURE

The data on effect of land configuration and nutrient module on soil moisture status (Table 22.6) indicate that there was slight difference in the soil moisture status of various treatments. However, among the land configuration treatments, slightly higher values of soil moisture were recorded in ridges and furrow and opening of furrow after each row treatments, whereas, among the nutrient modules application of FYM @10t ha^{1}+PSB= *Azotobactor* (M_2) and incorporation of *glyricidia* @10t ha^{-1}+PSB=*Azotobactor* (M_5) recorded slightly higher amount of soil moisture during the crop growth period. In general, the soil moisture status decreased after square formation.

TABLE 22.6 Soil Moisture at Different Growth Stages of Cotton (cm^3 cm^{-3})*

Treatment	Soil depth (cm)	Emergence	Square formation	Flowering	Boll development	First picking
L_1M_1	A	34.9	39.0	34.2	31.4	27.2
	B	35.5	38.6	32.2	33.4	29.0
L_1M_2	A	36.2	37.5	35.5	30.6	28.0
	B	36.8	37.3	35.3	33.9	28.0
L_1M_3	A	34.9	38.7	37.9	34.1	29.8
	B	35.6	38.4	31.5	25.8	24.9
L_1M_4	A	34.2	34.2	35.5	30.8	21.7
	B	35.2	35.1	35.3	27.5	28.3
L_1M_5	A	36.9	40.3	35.6	35.2	29.1
	B	37.0	38.6	38.3	31.3	26.7
L_2M_1	A	37.5	37.2	36.8	31.1	23.8
	B	37.7	38.7	35.5	27.5	33.5
L_2M_2	A	37.7	42.3	40.4	31.7	24.8
	B	37.9	41.6	36.2	30.6	27.9
L_2M_3	A	37.2	38.0	39.3	32.0	26.9
	B	38.6	39.3	36.3	34.9	24.6
L_2M_4	A	37.0	38.3	38.7	33.1	27.3
	B	37.9	36.6	33.8	26.9	26.7
L_2M_5	A	39.6	38.0	36.2	28.7	27.3
	B	40.7	39.2	31.5	26.3	25.5
L_3M_1	A	34.9	37.9	33.1	32.7	31.3
	B	35.5	38.2	35.8	31.7	30.4
L_3M_2	A	37.5	40.3	37.2	32.9	26.5
	B	38.3	38.0	37.9	26.5	31.3
L_3M_3	A	35.5	38.2	38.9	30.4	24.2
	B	36.5	34.8	49.2	26.3	26.7
L_3M_4	A	38.3	42.0	37.7	35.9	33.9
	B	39.6	42.8	36.9	27.6	31.1
L_3M_5	A	35.2	37.9	35.9	29.7	25.6
	B	35.6	39.6	38.2	35.1	31.1
L_4M_1	A	36.9	41.1	36.6	28.3	26.5
	B	38.6	38.3	35.2	29.8	28.3
L_4M_2	A	35.2	39.3	35.8	31.3	32.1
	B	37.7	37.7	34.6	27.3	28.0

TABLE 22.6 (Continued).

Treatment	Soil depth (cm)	Emergence	Square formation	Flowering	Boll development	First picking
L_4M_3	A	35.6	39.3	36.8	35.3	27.6
	B	36.8	38.6	36.2	33.7	27.2
L_4M_4	A	35.3	40.0	36.2	30.6	29.3
	B	36.9	40.1	36.5	29.8	29.3
L_4M_5	A	37.7	37.5	37.6	30.4	22.9
	B	38.0	40.3	40.0	34.1	26.2

*A = 0–15 cm; B = 15–30 cm.

22.5 CONCLUSIONS

It is concluded that integrated application of 50% RDF (25:12.5:12.5 NPK kg ha^{1}) + FYM @ 5 t ha^{-1} + PSB + Azotobactor and opening of furrow after each row at 30–40 days after sowing resulted in sustainable cotton productivity and improvement in fertility status of Vertisols under rainfed conditions in Central India.

KEYWORDS

- **Cotton**
- **Integrated Nutrient Management**
- **Land Configuration and Nutrient module**
- **Soil Fertility**
- **Soil Moisture**
- **Sustainable Yield Index**
- **Vertisols**

REFERENCES

Anonymous, (2013). Area and Production of Cotton. www.cci.co.in.

Babaria, C. J., & Patel, C. L. (1980). Effect of application of FYM and sulfur on the availability of Fe in medium black calcareous soil at different level of moisture regimes. *Journal of the Indian Society of Soil Science, 28*(2), 302–306.

Babhulkar, P. S., Wandile, R. M., Badole, W. P., & Balpande, S. S. (2000). Residual effect of long term application of FYM and fertilizers on soil properties (Vertisols) and yield of soybean. *Journal of the Indian Society of Soil Science, 48*(1), 89–92.

Bandyopadhyay, K. K., Misra, A. K., Ghosh, P. K., & Hati, K. M. (2010). Effect of integrated use of farmyard manure and chemical fertilizers on soil physical properties and productivity of soybean. *Soil & Tillage Research, 110*, 115–125.

Behra, U. K., Sharma, A. R., & Pandey, H. N. (2007). Sustaining productivity of wheat–soybean cropping system through integrated nutrient management practices on the Vertisols of central India. *Plant & Soil, 297*, 185–199.

Bellaki, M. A., & Badanur, V. P. (1997). Long term effect of integrated nutrient management on properties of Vertisol under dryland agriculture. *Journal of the Indian Society of Soil Science, 45*(3), 438–442.

Bellaki, M. A., Badanur, V. P., & Setty, R. A. (1998). Effect of long term nutrient management on some important properties of Vertisols. *Journal of the Indian Society of Soil Science, 46*(2), 176–180.

Bhattacharyya, T., Pal, D. K., Mandal, C., Chandran, P., Ray, S. K., Sarkar, Dipak., Velmourougane, K., Srivastava, A., Sidhu, G. S., Singh, R. S., Sahoo, A. K., Dutta, D., Nair, K. M., Srivastava, R., Tiwary, P., Nagar, A. P., & Nimkhedkar, S. S. (2013). Soils of India: historical perspective, classification and recent advances. *Current Science, 104*(10), 1308–1323.

Black, C. A. (Ed.). (1965). Methods of Soil Analysis, Parts 1 & 2, American Society of Agronomy, Madison, Wisconsin, USA.

Blaise, D., Singh. J. V., Venugopalan, M. V., & Mayee, C. D. (2003). Effect of continuous application of manures and fertilizers on productivity of cotton sorghum rotation. *Acta Agronomica Hungarica, 51*, 61–67.

Chaudhary, M. C., Singh, J. P., & Narwal, R. P. (1981). Effect of long term application of P, K and FYM on some soil chemical properties. *Journal of the Indian Society of Soil Science, 29*, 81–85.

Doran, J. W., Coleman, D. C., Bezdicek, D. F., & Stewart, B. A. (Eds). (1994). Defining Soil Quality for a Sustainable Environment. Science Society of America Special Publication, 35. Soil Science Society of America, Madison, WI. (608–273–8080).

FAO. (2006). FAO Statistical Database [Internet]. Food and Agriculture Organization of the United Nations [cited 2007, Dec 21], Available from: http://faostat.fao.org.

Gabhane, V. V., Sonune, B. A., & Katkar, R. N. (2014). Long-term effect of tillage and integrated nutrient management on soil quality and productivity of rainfed cotton in Vertisols under semi-arid conditions of Maharashtra. *Indian Journal of Dryland Agricultural Research and Development, 29*(2), 71–77.

Gokhale, D. N., Shinde, V. S., Gadade, G. D., Sawaraonkar, G. L., & Zade, K. K. (2011). Sustaining rainfed Bt. Cotton (*Gossypium hirsutum* L.) productivity through moisture conservation and integrated nutrient management techniques. *Journal of Cotton Research and Development, 25*(2), 197–201.

Gomez, K. A., & Gomez, A. A. (1984). *Statistical Procedures for Agricultural Research*, John Wiley & Sons, New York, pp. 241–266.

Goud, V. V., & Konde, N. M. (2007). Effect of rotations and continuous application of manures and fertilizers on soil properties under dry farming conditions. *Journal of the Indian Society of Soil Science, 18*, 45–49.

Hegde, D. M. (1996). Long term sustainability of productivity in an irrigated sorghum-wheat system through integrated nutrient supply. *Field Crop Research, 48*(2–3), 167–175.

Hullugalle, N. R., Weaver, T. B., Finlay, L. A., Hare, J., & Entwistle, P. C. (2007). Soil properties and crop yields in a dryland Vertisol sown with cotton-based crop rotations. *Soil and Tillage Research, 93*, 356–369.

Indulkar, B. S., & Malewar, G. U. (1996). Influence of various zinc sources on availability of N, P and Zn in sorghum-wheat cropping system. *Journal of Soils and Crops*. 1996, *6*(2), 139–141.

Jackson, M. L. (1973). *Soil Chemical Analysis*. Prentice Hall Publication Pvt. Ltd., New Delhi, India.

Katkar, R. N., Turkkhede, A. B., Solanke, V. M., Wankhade, S. T., & Patil, M. R. (2002). Effect of integrated management of organic manures and fertilizer on soil properties and yield of cotton. *Journal of Cotton Research and Development, 16*(1), 89–92.

Kumar, J., & Yadav, M. P. (2007). Effect of integrated nutrient management on yield attributes, yield and economics of cotton (*Gossypium hirsutum*) in Central Plain Zone of Uttar Pradesh. *Research on Crops, 8*(2), 347–349.

Lohakare, V. P. (1980). Effect of long term manuring and fertilization on micronutrient status of Vertisol. M.Sc. Thesis (Unpub.), Dr. PDKV, Akola (M.S.).

Malewar, G. U., & Hasnabade, A. R. (1995). Effect of long-term application of fertilizer and organic sources on some properties of Vertisol. *Journal of Maharashtra Agricultural Universities, 20*(2), 285–286.

Manna, M. C., Swarup, A., Wanjari, R. H., Ravankar, N. H., Mishra, B., Saha, M. N., Singh, Y. V., Shahi, D. K., & Swarup, P. A. (2005). Long-term effect of fertilizer and manure application on soil organic carbon storage, soil quality, and yield sustainability under sub-humid and semi-arid tropical India. *Field Crops Research, 93*, 264–280.

Nayak, A. K., Gangwar, B., Shukla, A. K., Mazumdar, S. P. Kumar, Anjani., Raja, R., Kumar, Anil Kumar K., Rai, P. K., & Mohan, U. (2012). Long term effect of different integrated nutrient management on soil organic carbon and its fraction and sustainability of rice-wheat system in Indo gangetic plains of India. *Field Crop Research, 127*, 129–139.

Padole, V. R., Deshmukh, P. W., Nikesar, R. J., & Bansode, N. V. (1998). Effect of organics and inorganics on yield and quality of cotton grown on Vertisol. *PKV Research Journal, 22*(1), 6–8.

Prasad, B., & Singh, A. P. (1980). Changes in soil properties with long term use of fertilizer, lime and farm yard manure. *Journal of the Indian Society of Soil Science, 28*(4), 465–468.

Rao, S., & Janawade, A. D. (2006). Studies on integrated nutrient management in irrigated hybrid cotton. *Journal of Cotton Research and Development, 20*(2), 212–215.

Sagare, B. N., Rewatkar, S. S., & Babhulkar, V. P. (2001). Effect of land configuration and gypsum levels on dynamics of soil properties and productivity of cotton grown in sodic Vertisols. *Journal of the Indian Society of Soil Science, 49*(2), 377–379.

Sharma, K. L., Grace, J. K., Mandal, U. K., Gajbhiye, P. N., Srinivas, K., Korwar, G. R., Ramesh, V., Ramachandran, Kausalya, & Yadav, S. K. (2008). Evaluation of long-term soil management practices using key indicators and soil quality indices in a semi-arid tropical Alfisol. *Australian Journal of Soil Research, 46*, 368–377.

Sharma, K. L., Mandal, U. K., Srinivas, K., Vittal, K. P. R., Mandal, B., Grace, J. K., & Ramesh, V. (2005). Long-term soil management effects on crop yields and soil quality in a dryland Alfisol. *Soil and Tillage Research, 83*, 246–259.

Sharma, M. P., Bali, S. V., & Gupta, D. K. (2001). Soil fertility and productivity of rice (*Oryza sativa*)-wheat (*Triticum aestivum*) cropping system in an Inceptisol as influenced by integrated nutrient management. *Indian Journal of Agricultural Sciences, 71*(2), 82–86.

Singh, J., Shilpa, B., & Venugopalan, V. (2013). Interactive effect of moisture conservation and integrated nutrient management on yield and nutrient utilization efficiency of rainfed Bt cotton (*Gossypium hirsutum*). *Indian Journal of Agricultural Sciences, 83*(10), 1069–1074.

Singh, J., Venugopalan, M. V., Mayee, C. D., Deshmukh, M. S., & Tandulkar, N. R. (2004). Effect of manure and cotton based cropping systems on productivity and micro nutrient availability in rainfed Vertisols. *Journal of Indian Society of Cotton Improvement, 29*(2), 100–105.

Singh, L. R., Verma, N. S., & Lohia, S. L. (2001). Effect of continuous application of FYM and chemical fertilizers on some soil properties. *Journal of the Indian Society of Soil Science, 28*, 171–172.

Solunkhe, P. S., & Fatak, S. U. (2011). Influence of organic and inorganic sources of nutrients on production of desi cotton under different plant protection measures. *Journal of Cotton Research and Development, 25*(1), 42–45.

Sonune, B. A., Gabhane, V. V., Nandanwar, V. S., & Rewatkar, S. S. (2013). Effect of tillage and manuring on soil properties and productivity of rainfed cotton on Vertisol. *Journal of Cotton Research and Development, 27*(2), 234–237.

Swarup, A. (2010). Integrated plant nutrient supply and management strategies for enhancing soil quality, input use efficiency and crop productivity. *Journal of the Indian Society of Soil Science, 58*, 25–31.

Tolanur, S. I., & Badanur, V. P. (2003). Effect of integrated use of organic manure, green manure and fertilizer nitrogen on sustaining productivity of *rabi* sorghum-chickpea system and fertility of a Vertisol. *Journal of the Indian Society of Soil Science, 51*(1), 25–31.

USDA. (2015). World Agricultural Supply and Demand Estimates, January 12, pp 26.

Venugopalan, M. V., & Pundarikakshudu, R. (1999). Long-term effect of nutrient management and cropping system on cotton yield and soil fertility in rainfed Vertisols. *Nutrient Cycling in Agroecosystems, 55*, 159–164.

Vittal, K. P. R., Maruthi Sankar, G. R., Singh, H. P., & Samra, J. S. (2002). Sustainability of Practices of Dryland Agriculture: Methodology and Assessment. All India Coordinated Research Project for Dryland Agriculture, Central Research Institute for Dryland Agriculture, Indian Council of Agricultural Research, Hyderabad, p. 100.

CHAPTER 23

IMPACT OF TANK SILT ON SOIL QUALITY AND CROP PRODUCTIVITY IN RAINFED AREAS: A CASE STUDY FROM CENTRAL INDIA

V. K. KHARCHE,[1] S. M. PATIL,[2] D. V. MALI,[3] S. M. JADHAO,[3] A. O. SHIRALE,[4] and R. N. KATKAR[3]

[1]*Professor, Dr. Panjabrao Deshmukh Krishi Vidyapeeth, Akola – 444 104, India*

[2]*MSc Scholar, Dr. Panjabrao Deshmukh Krishi Vidyapeeth, Akola – 444 104, India*

[3]*Assistant Professor, Dr. Panjabrao Deshmukh Krishi Vidyapeeth, Akola – 444 104, India*

[4]*PhD Scholar, Dr. Panjabrao Deshmukh Krishi Vidyapeeth, Akola – 444 104, India*

CONTENTS

ABSTRACT

Good agricultural practices that are simple, cost effective and farmer friendly need to be identified for judicious management of land resources. Specifically the shallow and marginal soils have numerous constraints for crop production in rainfed areas and need restoration by using locally available resources. In this view the present study was carried out to ascertain the physical and chemical characteristics of tank silt and its effects on various soil properties and crop yields. The tank silt from various dams in Akola district is being lifted and applied to the soil by farmers. Ten farmers' fields were selected in Akola district where the tank silt was applied. The tank silt as well as the soil sample collected before and after application of tank silt were analyzed for various properties. The tank silt was found to contain considerable amount of silt (30.16 to 50.35%) and clay (21.33 to 41.47%) and had higher water retention and available water capacity (10.60 to 17.07%). It was further observed that there was improvement in the silt and clay content, saturated hydraulic conductivity (sHC) and water retention of the soil and slight decrease in bulk density with considerable increase in the available water capacity of soil. The pH and electrical conductivity of tank silt was not found to cause any harm to the soil and values of pH and electrical conductivity were within the permissible limit for tank silt applied soil. The tank silt was found to contain considerable organic carbon (OC) (4.8 to 9.0 g kg^{-1}) and increased the OC of soil by (4 to 51%). There was substantial content of available N, P, K, S, micronutrients and exchangeable cations in the tank silt along with high cation exchange capacity (20.00 to 32.53 cmol (p^+) kg^{-1}) thereby improving nutrient status of soil considerably. The tank silt application recorded sequestration of considerable carbon in soil as evidenced by the soil OC stock thus, indicating its usefulness in crop productivity, soil health as well as mitigating adverse effects of climate change. Tank silt has potential to improve soil health and can also serve as potential source of nutrients for partially substituting chemical fertilizers. It can be concluded that tank silt is good source of organic matter, silt and clay, major, secondary and micro nutrients with high cation exchange and water holding capacity and improves the physical and chemical properties of soils.

23.1 INTRODUCTION

Tank silt is the deposited suspended matter or eroded soil in tank, which comes along with surface runoff caused due to intensive rainfall. Tank sediment is invariably available free of cost. Addition of tank sediments to cultivated fields improves physical and chemical properties of soil, which results in good crop growth and yield. The practice also minimizes cost on the other external inputs such as manures, fertilizers, etc. Tanks thus serve as good trap for eroded soil generating large quantities of accumulated sediment. Poor management practices in catchment result in silting of water bodies and significant reduction of storage capacity. Silt deposit not only reduces the storage capacity but also groundwater recharge, eutrophication of tanks and most importantly higher release of carbon to atmosphere through silt mediated anaerobic decomposition of OC. Good practices such as desilting and application of silt to agricultural fields are therefore imperative. Keshavamurthy and Kotur (1996) reported that the nutritive value of tank sediment and farm yard manure (FYM) was compared at Indian Institute of Horticulture Research (IIHR), Bangalore. The total P in tank sediment was 0.26% while that of FYM was 0.21%. Continued mining by crops and reduced application of organic manures in rainfed areas has resulted in deficiency of several nutrients particularly that of micronutrients. Recycling of tank silt provides a win-win situation to both, improvement in soil health and renovation of the tanks.

Soil is the non-renewable resource. The per capita availability of land in the country is declining. The challenge of increasing in food production to satisfy the demands of growing population is to be met by increasing the per hectare crop productivity. However, due to increasing land degradation the food security and sustainability of agriculture is threatened by severe land degradation. Due to the effects of climate change very intense rains received within short span of time are causing severe soil loss and complete removal of soil at places. The area of shallow soils is relatively more in rainfed areas of central India. The concept of soil hybridization is therefore very pertinent in view of above-mentioned issues. The potential sources of nutrients as alternative to the conventional sources are also necessary. The tank silt lifted from the dams can serve as the best available resources to restore the coarse textured shallow soils. It can also improve the soil properties and serve as the cheapest source of nutrients.

Over exploitation of ground waters through bore wells in semi-arid regions of central India have neglected the tanks and this also resulted into deepening of water levels causing severe water scarcity (Osman et al., 2009). Poor management of catchments and inadequate soil and water conservations measures have caused accelerated soil erosion resulting in reduction of storage capacity of dams, which also caused significant soil loss that is posing the greatest threat to the sustained agricultural crop production in these impoverished soils.

23.1.1 IMPACT OF TANK SILT ON SOIL QUALITY AND CROP PRODUCTIVITY: GLOBAL CONTEXT

The importance of silt and its manure properties have been studied. Krishnappa et al. (1998) reported the impacts of tank desiltation on agricultural production and economics. There are studies that report the economics of desilting and subsequent application on croplands. Interest in evaluating the quality and health of soil resources has been stimulated by increasing awareness that soil is a critically important component of the earth's biosphere, functioning not only in the production of food and fiber but also in ecosystems function and the maintenance of local, regional, and global environmental quality (Glanz, 1995).

Green revolution has virtually transformed low external input to high external input agriculture. Soil is considered as pool of nutrients present in both available and reserve forms. Depletion occurs when nutrients do not get replenished from the reserve pool. Soil is not an eternal supplier of all the nutrients when exploited indiscriminately. In most soils, the deficiency of boron and zinc is widely noticed (Rego et al., 2005, Sahrawat et al., 2008).

23.1.2 IMPACT OF TANK SILT ON SOIL QUALITY AND CROP PRODUCTIVITY: INDIAN CONTEXT

The practice of addition of silty loam tank sediment to clay soils resulted in increased sand and silt content (Ramesh, 2001). Improvement in clay content will not only retain higher moisture but will also reduce the losses of nutrients through leaching because of improved cation exchange capacity (Wassan, 2002). Application of tank silt @ 12.5 t/ha + cow dung @ 12.5 t/ha + 50% recommended dose of fertilizers (RDF) recorded higher uptake of nitrogen (230 kg/ha), phosphorous (26.55 kg/ha) and potassium (60.7 kg/ha) compared to other treatments (Shankaranarayana, 2001). Application of tank silt increased the water holding capacity, clay content and nutrient availability in the soil. It also enhanced the soil moisture content. Addition of 100 t/ha of tank silt over 2 or 3 years was found to improve light soils in Andhra Pradesh (Wassan, 2002). Annandurai et al. (2005) reported that, the mixture of tank silt and press mud when amended to soils improved the physical condition of soil by increasing the hydraulic conductivity, organic matter content and water holding capacity and reducing pH and bulk density of soils. There is a growing need and feedback from the large number of farmers in the rain fed areas and state agricultural departments that the study should be undertaken to generate detailed information on the contents of tank silt and its advantages and drawbacks if any.

Invariably tank sediments have higher nutrient value over their respective cultivated catchment soil (Anonymous, 2003). Tank sediments can be used preferably in the fields of respective catchment to build up their productivity. Several research reports *viz.,* Keshavamurthy and Kotur (1996), Shankaranarayana (2001) and Ramesh (2001) observed higher crop yield and improvement in soil health with the addition of tank sediments to the soil. Based on the characterization of tank silt, its addition to marginal soils can be regarded as good agricultural practice for enhancing soil quality. In this context, the present investigation has been carried out to characterize the tank silt and ascertain its effect on soil properties.

23.1.3 IMPORTANCE OF THE STUDY

It is very essential to ascertain the effect of tank silt on soil properties and in turn on soil quality. It is also equally important to assess the composition of the tank silt because of the varied nature of the silt collected from diverse combinations of vegetation, landform, slope, etc. The catchment area and the local conditions govern the type and nature of tank silt. Many times the components like more soluble salts, concentration of cations like sodium are predominant in the eroded sediments. The resultant tank silt may not be useful for addition to soil. Therefore, the systematic studies are necessary to characterize tank silt and its effect on soil quality.

23.2 STUDY AREA

The tank silt from the bed of various dams is being lifted out and applied to the field by number of farmers in rainfed areas of Maharashtra in central India as a part of the Mahatma Phule Land and Water Management Mission implemented by State Agriculture Department. Tank silt was collected at selected sites in Akola district of Maharashtra in semi-arid region of central India. Akola is situated in between 20°42′ N and 77°02′ E longitude at an altitude of 307.4 m above mean sea level. The study area represents Agro-Ecological Region (AER) 6, which is hot semi-arid on the AER map of the country developed by National Bureau of Soil Survey and Land Use Planning (NBSS & LUP), Nagpur (Challa et al., 1995). The climate is subtropical semi-arid with four main seasons during the year, from June to September (southwest monsoon), October and November (post monsoon) period. The period from December to February is mild winter and March to May hot summer, the distribution of rainfall is erratic. The average annual rainfall is 780 mm. The soil moisture regime is mainly ustic and soil temperature regime is hyperthermic.

23.3 MATERIALS AND METHODS

23.3.1 SOIL SAMPLING

The soil samples were collected from selected ten farmers' fields where the tank silt was applied. The soil samples were also collected from each site before application of tank silt. The soil sample before and after application of tank silt at each site were collected from the 0–20 cm depth. The details regarding the quantity of tank silt applied and crops grown were also recorded at each site.

23.3.2 LABORATORY ANALYSIS

The tank silt as well as soil samples were analyzed for physical and chemical properties. The physical properties like bulk density was determined by method suggested by Blake and Hartge (1986), sHC and water retention was determined by method given by Klute (1986). The chemical properties *viz.*, soil pH and EC (Jackson, 1973), OC (Nelson and Sommers, 1982), calcium carbonate (Piper, 1966), exchangeable Ca and Mg (Lanyon and Headl, 1982), exchangeable Na, K and CEC (Page, 1982), available N (Subbiah and Asija, 1956), available P (Watanabe and Olsen, 1965), available K (Hanway and Heidel, 1952), available S (Chesnin and Yien, 1950) and micronutrients *viz.,* Zn, Cu, Fe and Mn was determined by method described by Lindsay and Norvell (1978). The data were statistically analyzed as per Gomez and Gomez (1990) using method of analysis of variance and means were tested at 5% level of significance and were used for comparison.

23.4 RESULTS AND DISCUSSION

23.4.1 EFFECT OF TANK SILT APPLICATION ON PHYSICAL PROPERTIES OF SOILS

The bulk density and sHC of soil varied from 1.34 to1.86 Mg m^{-3} and 0.43 to 0.64 cm hr^{-1} while that of tank silt was found to vary from 1.42 to 1.81 Mg m^{-3} and 0.69 to 0.87 cm hr^{-1} (Table 23.1). The variation in the bulk density and sHC of tank silt can be attributed to the variation in the mechanical composition of tank silt. However, it was further observed that the bulk density was slightly reduced, which results in improvement in sHC due to tank silt application. Similar results were also reported by Jeyamangalam et al. (2012). The variation in bulk density and lack of definite trend is due to varied mechanical components. The water retention was observed to improve due to application of tank silt which may be attributed to the addition of clay and silt fractions to the soils through tank silt. Higher clay content not only retains higher moisture but also reduces the losses of nutrients through leaching because of improved cation exchange capacity. The available water capacity of the soil was improved by 9.23–39.82% due to tank silt. Anonymous (2011) also reported improvement in water retention of soil due to tank silt application in Osmanabad district. The increase in clay and silt content of soil due to tank silt application has also been reported by Krishanappa et al. (1998) in Kolar district of Karnataka.

The soil hybridization caused due to adoption of this practice can enhance soil quality with minimum cost. Besides its transport cost, the tank silt abundantly is available in the water reservoirs, dams, rivers due to considerable siltation, which can be used for augmenting the soil health especially shallow and skeletal soils. The increase in silt and clay has special significance because of better to nutrients availability.

TABLE 23.1 Effect of Tank Silt Application on Physical Properties of Soil

Site	Bulk density (Mg m^{-3})			%	Saturated Hydraulic conductivity (cm hr^{-1})			%	Available water capacity (%)			%
	Tank silt	Soil	Tank silt applied soil	Increase/ decrease	Tank silt	Soil	Tank silt applied soil	Increase/ decrease	Tank silt	Soil	Tank silt applied soil	Increase/ decrease
1	1.42	1.63	1.73	6.25	0.84	0.54	0.77	42.59	17.07	9.01	11.36	26.08
2	1.43	1.72	1.76	2.32	0.87	0.59	0.67	13.55	10.60	10.17	11.57	13.76
3	1.81	1.66	1.81	9.03	0.76	0.43	0.54	25.58	15.62	8.28	10.81	30.55
4	1.77	1.71	1.52	–11.76	0.82	0.52	0.67	28.84	10.82	7.94	10.18	28.21
5	1.62	1.85	1.83	0.00	0.77	0.59	0.67	13.55	18.20	10.17	11.14	9.23
6	1.61	1.34	1.82	35.5	0.69	0.52	0.54	3.84	13.15	10.83	12.30	13.57
7	1.76	1.62	1.61	0.00	0.81	0.64	0.78	21.87	14.54	11.69	13.54	15.82
8	1.63	1.71	1.50	12.28	0.79	0.60	0.68	13.33	15.26	11.18	11.07	-0.98
9	1.75	1.86	1.60	3.97	0.82	0.64	0.71	10.93	11.78	9.90	11.30	14.14
10	1.74	1.61	1.62	0.00	0.76	0.57	0.65	14.03	11.04	7.96	11.13	39.82
Average	1.62	1.64	1.66		0.79	0.56	0.66		13.80	9.71	11.44	
S.E. (m) ±	0.04	0.04	0.03		0.01	0.02	0.02		0.86	0.42	0.28	
C.V. (%)	8.12	8.71	7.07		6.40	11.31	12.01		19.80	13.95	7.97	

23.4.2 EFFECT OF TANK SILT APPLICATION ON CHEMICAL PROPERTIES OF SOIL

The pH of tank silt was alkaline and varied from 8.0 to 8.6 (Table 23.2). The soil pH was also alkaline and application of tank silt did not cause much change in soil pH. The pH of tank silt applied soil was also alkaline (8.0 to 8.4). The electrical conductivity of tank silt was variable and ranged from 0.29 to 0.44 dS m^{-1}. The EC of soils was also low. The EC of tank silt applied soils was slightly increased, however, it was below the limit of causing salinity hazard to soil.

The OC content was found to vary from 4.2 to 13.0 g kg^{-1}. The OC content in soil was low to moderate 4.9 to 9.70 g kg^{-1}. It was also further observed that the OC content was increased due to application of tank silt to soil (5.1–11.0 g kg^{-1}). The increase in OC from 4 to 51% has been observed in the tank silt applied soil. It thus becomes apparent that application of tank silt is a soil restorative practice for enhancing soil quality due to addition of enough OC. Considerable increase in OC due to addition of tank silt has also been observed by Osman et al. (2009). The tank silt application recorded considerable sequestration of carbon in soil as evidenced by the soil OC stock thus, indicating its usefulness in crop productivity, soil health as well as mitigating adverse effects of climate change.

The tank silt contains considerable quantity of $CaCO_3$ (12.01–15.01%) and recorded slight increase in $CaCO_3$. Thus, it is important to make use of tank silt with little caution so as to ensure the calcium carbonate accumulation in soil. Anonymous (2011) also reported considerable contents of calcium carbonate in the soils applied with tank silt. The exchangeable calcium was increased by 10.26–90.2% due to tank silt in different soils (Table 23.3). However, magnesium was also increased by 3.8–51% suggesting that the fine soil particles in the tank silt adsorbed sufficient divalent basic cations, which can be recycled in the soil again by adopting the practice of lifting of tank silt from dams and its application in the soils. Deshmukh and Deve (1993) reported considerable increase in calcium and magnesium of tank silt applied soil. The exchangeable sodium in the tank silt was also low varying from 0.27 to 1.31 cmol (p^+) kg^{-1}. The tank silt applied soil also recorded exchangeable sodium of 0.91 to 1.91 cmol (p^+) kg^{-1} suggesting that these values are sufficiently below the levels to cause soil sodicity. However, these levels need to be monitored subsequently if the frequency of tank silt application is increased. Exchangeable potassium

TABLE 23.2 Effect of Tank Silt Application on Chemical Properties of Soil

Site	pH				EC (dS m^{-1})				$CaCO_3$ (%)				Organic carbon (g kg^{-1})			
	Tank silt	Soil	Tank silt applied soil	% Increase/ decrease	Tank silt	Soil	Tank silt applied soil	% Increase/ decrease	Tank silt	Soil	Tank silt applied soil	% Increase/ decrease	Tank silt	Soil	Tank silt applied soil	% Increase/ decrease
1	8.4	8.1	8.3	2.2	0.4	0.4	0.4	–2.4	14.0	9.3	13.0	39.5	9.0	6.4	6.7	4.6
2	8.2	8.1	8.3	2.5	0.3	0.3	0.4	24.2	12.2	9.3	11.9	27.2	5.8	6.1	6.8	11.4
3	8.5	7.9	8.1	1.7	0.3	0.3	0.4	32.2	14.1	8.0	11.0	36.3	5.2	6.4	7.0	9.3
4	8.6	8.4	8.4	–0.3	0.3	0.3	0.4	7.6	13.7	7.8	12.8	63.9	8.1	9.7	11.0	13.4
5	8.5	8.1	8.3	2.0	0.4	0.4	0.4	00	14.1	10.6	15.8	48.8	6.6	7.9	8.9	12.6
6	8.6	8.1	8.3	2.5	0.4	0.4	0.4	5.0	14.5	11.8	17.0	43.4	8.3	6.4	7.9	23.4
7	8.0	8.3	8.1	–2.2	0.3	0.3	0.4	5.1	15.0	10.1	12.9	27.5	5.7	6.1	6.8	11.4
8	8.2	7.9	8.0	0.5	0.2	0.2	0.3	50.0	13.0	13.3	16.0	19.7	4.2	5.2	7.9	51.9
9	8.4	8.0	8.1	1.3	0.4	0.4	0.4	4.7	14.0	11.8	14.8	25.6	7.8	6.6	7.2	9.0
10	8.3	8.1	8.2	1.8	0.3	0.4	0.4	–2.3	12.0	9.9	14.1	42.4	4.8	4.9	5.1	4.0
Average	8.3	8.1	8.2		0.3	0.3	0.4		13.7	10.2	13.9		7.3	6.5	7.5	
S.E. (m) ±	0.06	0.04	0.04		0.01	0.02	0.00		0.30	0.54	0.61		0.94	0.43	0.49	
C.V. (%)	2.3	1.9	1.5		15.0	17.6	7.1		7.0	16.9	13.8		40.9	20.7	20.8	

TABLE 23.3 Effect of Tank Silt Application on Exchangeable Cations of Soil

Site	Ca (cmol (p$^+$) kg^{-1})				Mg (cmol (p$^+$) kg^{-1})				Na (cmol (p$^+$) kg^{-1})				K (cmol (p$^+$) kg^{-1})			
	Tank silt	Soil	Tank silt applied soil	Increase/ decrease (%)	Tank silt	Soil	Tank silt applied soil	Increase/ decrease (%)	Tank silt	Soil	Tank silt applied soil	Increase/ decrease (%)	Tank silt	Soil	Tank silt applied soil	Increase/ decrease (%)
1	10.7	16.7	19.2	15.0	4.5	6.7	7.7	16.0	1.3	1.0	1.9	87.2	0.4	0.1	0.9	378.9
2	13.6	19.6	22.7	15.8	3.2	5.4	6.8	26.4	0.9	0.9	0.9	7.6	0.5	1.0	1.0	–2.7
3	12.9	24.0	26.5	10.2	3.7	5.3	8.1	51.0	0.2	0.7	0.9	15.1	0.2	0.9	0.7	–27.5
4	12.9	18.6	20.7	11.2	4.9	6.9	7.2	5.0	0.6	1.4	1.9	36.4	0.4	0.2	0.9	220.6
5	14.5	20.7	24.6	19.1	4.7	8.9	9.2	3.8	0.7	0.9	1.0	9.1	0.3	0.5	1.3	124.1
6	9.2	22.5	26.1	16.1	3.9	5.7	7.7	5.7	0.6	0.7	0.9	37.0	0.8	1.2	1.0	–14.0
7	11.6	20.8	24.9	19.7	3.8	5.9	7.4	24.3	0.7	1.0	1.1	4.6	0.7	0.9	1.3	36.0
8	10.9	18.6	20.6	10.7	3.6	7.2	8.6	19.7	0.8	1.5	1.5	2.6	0.5	1.2	1.3	10.7
9	17.5	24.3	27.5	10.2	5.4	7.8	8.8	13.0	0.6	0.8	1.0	20.2	1.0	1.2	1.0	–19.3
10	15.7	26.2	29.7	13.0	5.8	7.7	8.5	11.1	0.9	0.9	1.0	12.3	1.0	0.9	1.1	15.0
Average	12.9	21.2	24.2		4.3	6.7	8.0		0.7	1.0	1.2		0.6	0.8	1.	
S.E. (m) ±	0.79	0.95	1.06		0.26	0.37	0.24		0.08	0.08	0.12		0.08	0.12	0.06	
C.V. (%)	19.2	14.1	13.8		19.3	17.2	9.4		34.5	24.7	31.0		41.1	44.4	18.74	

content of the tank silt varied from 0.29 to 1.02 cmol(p^+) kg^{-1} which caused slight improvement in the exchangeable potassium in the tank silt applied soil. The cation exchange capacity of tank silt ranged from 20.0 to 32.53 cmol (p^+) kg^{-1} (Table 23.4). The variation is also due to varied mechanical composition of tank silt and tank silt application recorded slight improvement (8.4 to 23.01%).

23.5 EFFECT OF TANK SILT APPLICATION ON AVAILABLE NUTRIENT STATUS

23.5.1 MAJOR AND SECONDARY NUTRIENTS

The available nitrogen and phosphorus in tank silt varied from 100.97 to 204.80 and 10 to 25.8 kg ha^{-1} (Table 23.4 and Figure 23.1). They were considerably improved by tank silt application and varied from 129 to 239 kg ha^{-1} and 17.92 to 27.9 kg ha^{-1}. The increase in available nitrogen in the tank silt applied soil was to the extent of 2 to 32% while considerable increase in available phosphorus was also observed and it was to the tune of 2.7 to 44% in different soils amended with silt from various tanks. Similar results were also reported by Shankaranarayana (2001). The available potassium content of tank silt was considerably high ranging from 138.0 to 267.2 kg ha^{-1} indicating that tank silt is a good source of potassium (Figure 23.2). It was also further observed that the soils, which were initially low in available potassium improved their potassium status almost by 8.3 to 115%. This has got a special significance in the context of huge imports of potassium fertilizers in our country. Osman et al. (2009) also reported increase in potassium status of soils due to tank silk application in Patancheru in Andhra Pradesh. The available sulfur content of tank silt varied from 6.2 to 30.1 mg kg^{-1}. The soils had sulfur content ranging from 4.6 to 22.1 mg kg^{-1} indicating that most of the soils are deficient to marginal in sulfur (Figure 23.3). However, the available sulfur content was considerably improved due to tank silt application and it was increased by 14.17 to 58.44%. The practice of tank silt application has thus got the significance in view of recently observed wide spread sulfur deficiency emerging in soils of Maharashtra (Katkar et al., 2013). Similarly the increase in available S was reported by Sen and Asija (1954).

TABLE 23.4 CEC and Available Nitrogen of Tank Silt, Soil and Tank Silt Applied Soil

		CEC cmol (p^+) kg^{-1}				Available nitrogen (kg ha^{-1})			
S. No.	**Name of the farmer/ sites**	**Tank silt**	**Soil**	**Tank silt applied soil**	**Increase/ decrease (%)**	**Tank Silt**	**Soil**	**Tank silt applied soil**	**Increase/ decrease (%)**
1	Shri. Shivkumar Agrawal	20.9	27.7	30.2	9.0	100.9	115.4	158.7	37
2	Shri. Dipak Dhore (I)	23.6	29.2	34.2	16.9	129.8	158.7	173.1	9
3	Shri. Dipak Dore (II)	26.7	34.8	39.1	12.3	115.4	144.3	190.4	32
4	Shri. Mangesh Umbarkar	32.5	28.1	32.6	16.2	103.8	118.2	129.8	9
5	Smt. Sulochana Raut	27.8	32.1	38.1	18.6	132.7	148.6	158.7	6
6	Shri. Mahadevrao Wankhede	21.6	32.2	39.6	23.0	145.6	152.9	157.2	2
7	Shri. Ambadas Fatkar	29.8	33.3	38.0	14.2	160.1	154.3	167.3	8
8	Mr. Anvar Khan	20.4	30.1	35.6	17.9	171.6	187.5	200.5	6
9	Shri. Ashvin Dalvi	24.1	36.1	42.0	16.5	158.6	201.9	219.3	8
10	Shri. Niranjan Bombatkar	20.0	39.2	42.5	8.4	204.8	196.2	239.4	22
	Average	24.7	32.3	37.0		142.3	157.8	179.4	
	S.E. (m) ±	1.36	1.16	1.26		10.2	9.4	10.4	
	C.V. (%)	17.4	11.4	10.7		22	18	18	

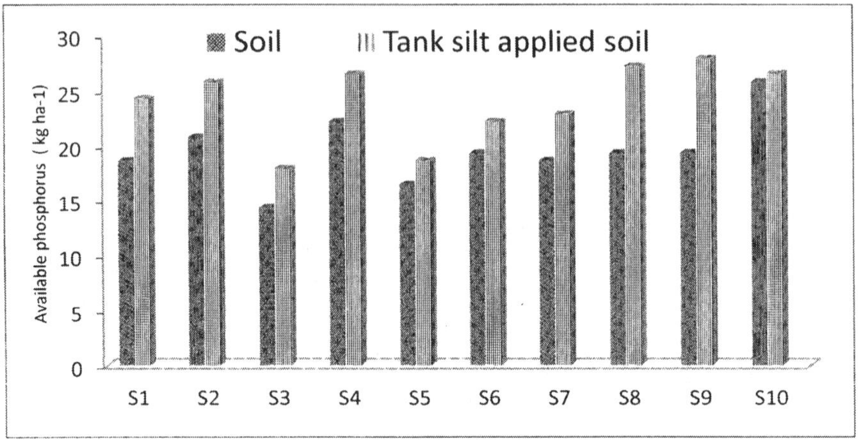

FIGURE 23.1 Effect of Tank silt application on soil available phosphorus.

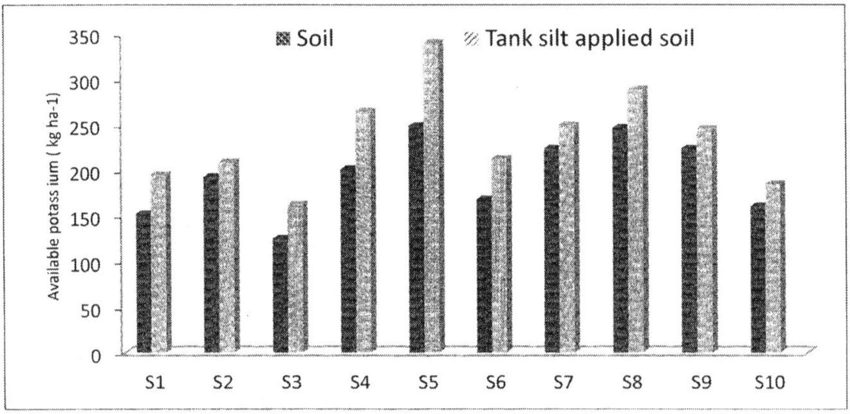

FIGURE 23.2 Effect of tank silt application on available potassium.

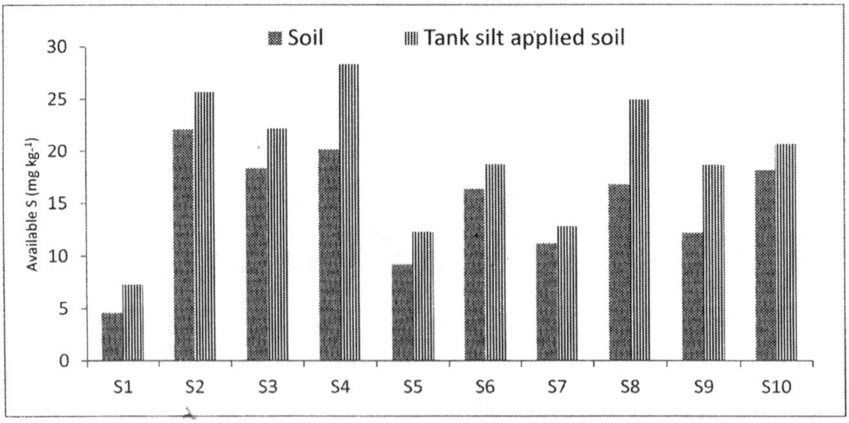

FIGURE 23.3 Effect of tank silt application on soil available sulfur.

23.5.2 MICRONUTRIENTS

The tank silt contains sufficiently higher quantity of micronutrients (Table 23.5). The present soils under study showed considerable improvement after tank silt application. The available zinc in soil and tank silt ranged from 0.48 to 0.81 mg kg^{-1} and 0.80 to 1.48 mg kg^{-1}. Consequently, the zinc content of soil after application of tank silt was increased (0.67 to 1.10 mg kg^{-1}) thus making the soil sufficient in zinc. In view of the widespread zinc deficiency in the predominantly black swell shrink alkaline soils of Maharashtra the practice of tank silt application appears to be the most promising option available to correct the deficiency of zinc. Similar findings on micronutrient contents of tank sediments were reported by Ramesh, (2001) in Dharwad district of Karnataka. The crop yields were also improved (Table 23.6) on the tank silt applied fields as compared to previous year yield before application of tank silt. Application of silt back to the agricultural fields forms an improved agricultural management practice that enhances and protects the soil quality resulting in improved production capacity of soils.

Water harvesting for life saving irrigation to crops at their critical growth stages in semi-arid regions of central India is necessary and this practice can be easily adopted by the farmers with dual advantage of fulfilling the needs of protective irrigation and restoring soil quality through recycling of tank silt in their fields. De-silting activity needs greater support from the Government for rejuvenating the tanks as well as for improving crop productivity in rainfed areas.

It further becomes clear that there is no detrimental effect on soil provided the sites of tank silt application must be monitored for the parameters like exchangeable sodium and calcium carbonate. The composition of tank silt is variable and will depend upon the characteristics of the catchment areas from where the sedimentation is transported. Accordingly its impact on soil properties may vary. Soil testing at such soils needs to be carried out periodically in order to monitor the changes in soil quality. The tank silt can be preferably used for addition in shallow coarse texture soils in the black soil regions of semiarid ecosystems of central India however; it may not be used in deep heavy soils to avoid poor drainage.

Tank silt is a good source of plant essential nutrients *viz*., N, P, K, S, Ca, Mg, Fe, Mn, Zn, Cu along with considerable amounts of silt and clay, OC with good water and nutrient holding capacity thereby improving soil physical and chemical properties beneficial for enhancement in soil quality.

TABLE 23.5 Effect of Tank Silt Application on Micronutrient Status of Soil

	Zn (mgkg^{-1})				Cu (mgkg^{-1})				Fe (mgkg^{-1})				Mn (mgkg^{-1})			
Site	**Tank silt**	**Soil**	**Tank silt applied soil**	**Increase/ Decrease (%)**	**Tank silt**	**Soil**	**Tank silt applied soil**	**Increase/ Decrease (%)**	**Tank silt**	**Soil**	**Tank silt applied soil**	**Increase/ Decrease (%)**	**Tank silt**	**Soil**	**Tank silt applied soil**	**Increase/ Decrease (%)**
1	0.8	0.4	0.6	39	4.2	3.1	6.2	99	10.0	8.3	10.1	21.0	15.0	9.9	11.0	11.5
2	1.0	0.6	0.9	54	5.6	2.9	4.9	70	7.7	9.1	12.1	33.2	12.1	8.7	10.2	17.9
3	0.8	0.7	0.8	9	3.9	2.2	5.1	135	9.4	6.4	10.0	57.1	10.7	7.8	8.0	1.9
4	0.9	0.5	0.7	51	4.2	2.7	4.6	69	10.0	9.0	10.9	20.8	8.6	9.1	9.9	8.5
5	1.2	0.6	0.9	46	4.2	3.2	5.2	60	11.6	7.2	9.5	31.4	7.6	9.4	9.9	4.9
6	1.4	0.7	1.1	56	6.3	1.8	3.8	105	10.2	6.7	9.3	44.3	10.0	5.3	7.6	44.1
7	0.8	0.5	0.7	30	5.2	3.7	5.6	52	8.9	9.1	12.3	35.7	11.2	8.1	10.6	30.8
8	0.9	0.7	0.7	2	5.2	2.8	4.9	74	9.9	8.2	11.2	36.6	9.9	7.2	10.1	12.7
9	1.0	0.8	1.0	23	6.1	4.1	6.2	52	11.3	7.9	10.0	27.0	8.4	9.1	9.9	8.7
10	0.9	0.4	0.7	59	5.9	3.6	5.7	57	11.0	9.0	11.2	23.6	10.3	8.8	9.9	12.5
Average	**0.09**	**0.06**	**0.08**		5.1	3.0	5.2		10.0	8.1	10.7		10.4	8.3	9.7	
S.E. (m) ±	**0.06**	**0.03**	**0.04**		0.28	0.21	0.23		0.36	0.32	0.32		0.67	0.41	0.34	
C.V. (%)	**228**	**135**	**148**		17	22	13		11	12	9		20	15	11	

TABLE 23.6 Effect of Tank Silt Application on Crop Yields

S. No.	Name of the farmer	Before tank silt application (2011–12) Crop	Yield ($q\ ha^{-1}$)	After tank silt application (2012–13) Crop	Yield ($q\ ha^{-1}$)
1	Shri. Shivkumar Agrawal	–	–	Cotton (Drip irrigation; 5)	25.0
2	Shri. Dipak Dhore (I)	Soybean	17.5	Soybean	25.0
3	Shri. Dipak Dhore (II)	Chickpea	10.0	Chickpea	17.5
4	Shri. Mangesh Umbarkar	Wheat	13.8	Cotton (Well irrigation; 4)	17.5
5	Smt. Sulochana Raut	Cotton (Well water irrigation; 5)	13.8	Cotton (Well irrigation; 6)	15.0
6	Shri. Mahadevrao Wankhede	Soybean	14.0	Cotton (Protective irrigation, 2)	10.0
7	Shri. Ambadas Fatkar	Bhendi	50.0	Cotton (Well irrigation; 5)	15.0
8	Mr. Anvar Khan	Cotton (Sprinkler irrigation 4)	12.5	Cotton (Sprinkler irrigation; 5)	25.0
9	Shri. Ashvin Dalvi	Soybean	12.5	Pigeon pea	25.0
10	Shri. Niranjan Bombatkar	Chickpea	10.0	Soybean	20.0

23.6 IMPLICATIONS

In view of the considerable contents of plant essential nutrients and potential to enhance soil quality as well as carbon sequestration the addition of tank silt can be regarded as good agricultural practice specially for coarse textured shallow, skeletal and marginal soils. The option of tank silt application is useful for restoring marginal soils by way of soil hybridization resulting into considerable improvement in their production potential.

23.7 CONCLUSIONS

The tank silt was found to contain considerable amount of silt (30.16 to 50.35%) and clay (21.33 to 41.47%) and had higher water retention and available water capacity (10.60 to 17.07%). There was improvement in the silt and clay content, sHC and water retention of the soil and slight decrease in bulk density with considerable increase in the available water capacity of soil. The pH and electrical conductivity of tank silt were within the permissible limits for tank silt applied soil. The tank silt was found to contain considerable OC (4.8 to 9.0 g kg^{-1}) and increased the OC of soil by (4 to 51%). There were substantial contents of available N, P, K, S, micro nutrients and exchangeable cations in the tank silt alongwith high CEC (20.00 to 32.53 cmol (p^+) kg^{-1}) improving thereby nutrient status of soil. It thus becomes apparent that tank silt has considerable potential to improve soil health and can also serve as potential source of nutrients for partially substituting chemical fertilizers.

KEYWORDS

- **Carbon Sequestration**
- **Marginal Soil**
- **Shallow soil**
- **Soil Health**
- **Soil Restoration**
- **Tank silt**

REFERENCES

Annadurai, B., Arunachalam N., & Mahalingam, K. (2005). Effect of tank silt and pressmud mixture, as amendment on the physical properties. *Journal of Soils and Crops, 15*(1), 26–29.

Anonymous. (2003). Annual Report of Integrated Tank Development Project, University of Agricultural Sciences, Dharwad, p. 14.

Anonymous. (2011). Effect of tank silt application on yield of soybean under very shallow soil of Agriculture College farm, Osmanabad. Project AGRESCO-SSAC. MAU, Parbhani.

Blake, G. R., & Hartge, K. H. (1986). Bulk density, *In*: *Methods of Soil Analysis.* Part-1.Klute A. (Ed.). American Society of Agronomy, Inc. Madison, Wisconsin USA, pp. 371–373.

Challa, O., Vadivelu, S., & Sehgal, J. (1995). Soils of Maharashtra for optimizing Land Use. NBSS Publ. No. 54, Soils of India Series 5, NBSS & LUP, Nagpur.

Chesnin, L., & Yien, C. H. (1950). Turbidimetric determination of available sulfur. *Proceedings of Soil Science Society of America, 14*, 149–151.

Deshmukh, D. D., & Deve, D. V. (1993). Nodulation in groundnut under different package of practice during winter. *Legume Research, 16*(4), 127–130.

Glanz, J. T. (1995). Saving Our Soil: Solutions for Sustaining Earth's Vital Resource. Johnson Books, Boulder, Co.

Gomez, K. A., & Gomez, A. A. (1990). *Statistical Procedures for Agricultural Research*, 2nd Edn. John Willy and Sons, New York.

Hanway, J. J., & Heidel, H. (1952). Soil Analysis Methods, as used in Iowa State. College Soil Testing Laboratory, Iowa, Agriculture. *57*, 1–37.

Jackson, M. L. (1973). *Soil Chemical Analysis* 2nd Edn. Prentice Hall of India Pvt. Ltd., New Delhi. pp. 69–182.

Jeyamangalam, F. M., Annadurai, B., & Arunachalam, N. (2012). Effect of tank silt as organic amendment on physical properties of their soil using Groundnut. *Journal of Soils and Crops, 22*(1), 10–14.

Katkar, R.N, Kharche, V. K., Dey, P., Lakhe, S. R., Jadhao, S. D., Chaudhari, R. D., Damre, P. R., Ikhe, U. D., & Sonune, B. A. (2013). GIS Based Soil Fertility Assessment for Soil Health Sustenance. Research publication Tech. Bulletin; Dr. PDKV/325/2013 Dr. PDKV, Akola. p.155.

Keshavamurthy, S. V., & Kotur, S. C. (1996). A comparison of tank silt, FYM and SSP as a source of P to Ney Poovan banana. Extended Summary: International Conference on Managing Natural Resources, p. 238.

Klute, A. (1986). Water retention. Laboratory Methods. In: *Methods of, Soil Analysis* A. Klute (Ed) Part 1. 2nd ed. Agron. Monograph 9. Madison, Wisconsin. pp. 635–662.

Krishnappa, A.M, Ranganna, B, Ramanna Gowda, P and Arun Kumar, Y. S. (1998). Impact and economics of tank desiltation in southern Districts with special reference to Kolar. Booklet: University of Agricultural Sciences, Bengaluru.

Lindsay, W. L., & Norvell, W. A. (1978). Development of a DTPA soil test for Zn, Fe, Mn and Cu. *Soil Science Society of America Journal, 42*, 421–428.

Lanyon, L. E., & Headl, W. R. (1982). Magnesium, calcium, strontium and barium. *In*: Methods of soil analysis. Part 22nd ed. Page, A. L., Miller, R. H., & Keeney, D. R. (Eds). Agronomy No. 9 American Society of Agronomy, Madison WI, pp. 247–262.

Nelson, D. W., & Sommers, L. E. (1982). Total carbon, organic carbon and organic matter. *In: Methods of Soil Analysis.* Part 2 Chemical and microbiological properties. (A. L. Page, R. H. Miller, & Dr. Keeney, Eds.), American Society of Agronomy, Madison, WI, USA. pp. 539–579.

Osman, M., Wani, S. P. Vineela, C., & Murali, R. (2009). Quantification of Nutrients Recycled by Tank silt and its Impact on soils and Crop: A Pilot Study in Warangal District of Andhra Pradesh. International Crops Research Institute for the Semi-Arid Tropics. Patancheru – 502 324, Andhra Pradesh, India.

Page, A. L., Miller, R. H. & Keeney, D. R. (1982). *Methods of Soil Analysis*, IInd edition part II. Chemical and microbiological properties ASA, Madison, Wisconsin, USA.

Piper, C. S. (1966). *Soil and Plant Analysis*, University of Adelaide, Australia.

Ramesh, N. R. (2001). Characterization of tank sediments of Dharwad district. M. Sc. (Agri.) Thesis, University of Agricultural Sciences, Dharwad.

Rego, T. J., Wani, S. P., Sahrawat, K. L., & Pardhasaradhi, G. (2005). Macro-benefits from boron, zinc and sulfur application in India SAT: A Step for Gray to Green Revolution in agricultural. Global Theme on Agroecosystem Report No. 16, Patanchuru – 502 324, Andhra Pradesh, India: International Crops Research Institute for the Semi-Arid Topics, 24 pp.

Sahrawat K. L., Rego T. J., Wani S. P., & Pardhasaradhi, G. (2008). Sulfur, boron and zinc fertilization effect on grain and straw Quality of Maize and Sorghum Grown in Semi-Arid Tropics Region of India. *Journal of Plant Nutrition, 31*, 1578–1584.

Sen, A., & Asija, G. L. (1954). Action of tank silts on soils. *The Indian Journal of Agricultural Sciences, 24*, 51–62.

Shankaranarayana, V. (2001). Agronomic investigations to achieve increased and sustainable productivity in *kharif* groundnut. PhD Thesis, University of Agricultural Sciences, Dharwad.

Subbiah, B. V., & Asija, G. L. (1956). A rapid procedure for the estimation of available nitrogen in soils. *Current Science, 25*, 259–260.

Wassan (2002). Voices for the ground: Issues in Tank desiltation under 'Neeru-Meeru,' Secunderabad: Watershed Support Services and Activities Network.

Watanabe, F. S., & Olsen, S. R. (1965). Test of ascorbic acid method for determining phosphorus in water and sodium bicarbonate extracts of soils. *Proceedings of Soil Science Society of America, 29*, 677–678.

CHAPTER 24

IMPACT OF TREATED DOMESTIC SEWAGE WATER IRRIGATION ON SOIL PROPERTIES, MAIZE YIELD AND PLANT UPTAKE: A CASE STUDY FROM NAGPUR CITY, CENTRAL INDIA

A. R. MHASKE,[1] S. M. TALEY,[2] and R. N. KATKAR[3]

[1]*Associate Professor, College of Agriculture, Dr. Panjabrao Deshmukh Krishi Vidyapeeth, Nagpur – 444 104, India*

[2]*Head, Department of Soil Water Conservation Engineering, Dr. Panjabrao Deshmukh Krishi Vidyapeeth, Akola – 444 104, India*

[3]*Associate Professor, Dr. Panjabrao Deshmukh Krishi Vidyapeeth, Akola – 444 104, India*

CONTENT

ABSTRACT

In the present study, maize crop was irrigated with well water and domestic waste water, treated through phytorid sewage treatment plant. Recommended NPK dose of fertilizers was applied for maize crop in both the treatments. The use of the treated domestic waste water has showed improvement in the uptake of pollutants by the plants, yields were improved and the nutrient status was also better as compared to the application of well water. The use of the treated domestic waste water showed improvement in the physico-chemical properties of the soil, along with the nutrient status as compared to the application of well water. The pH of the soil in treated sewage water was 7.78, whereas, it was 7.93 in well water. Electrical conductivity (EC) and organic carbon (OC) was 0.34 dS m^{-1} and 5.32 g kg^{-1}, respectively in treated domestic sewage water whereas they were 0.28 dS m^{-1} and 5.24 g kg^{-1}, respectively in well water. NPK were 268.4, 16.99 and 440.1 kg ha^{-1}, respectively in treated domestic waste water, while 266.5, 16.36 to 434.0 kg ha^{-1}, respectively in well water. The physical properties of the soil like bulk density (BD) and saturated hydraulic conductivity (sHC) observed 1.34 Mg m^{-3} and 1.50 cm hr^{-1}, respectively in treated domestic waste water and 1.35 Mg m^{-3} and 1.52 cm hr^{-1}, respectively in well water. The content of micro-nutrients and heavy metals in soil due to irrigation of treated domestic waste water were well below the safe limits in soil. The findings give applicable advice to commercial farmers and agricultural researchers for proper management and use of domestic waste water for agricultural purpose.

24.1 INTRODUCTION

Population in India is growing by geometric progression, whereas, food production is growing by arithmetic progression. The population increase has not only increased the fresh water demand but also increased the volume of waste water generated. Treated or recycled waste water (RWW) appears to be the only water resource that is increasing as other sources are dwindling. Increasing need for water has resulted in the emergence of domestic waste water application for agriculture and its relative use. Therefore, there is an urgent need to conserve and protect fresh water and to use the water of lower quality for irrigation as reported by Al-Rashid and Sherif (2000).

Water is vital but severely limited resource in most countries. Increasing industrial development activities, urbanization and population growth in consuming more and more water day to day throughout the world. Consequently the reuse of waste water for agriculture is highly encouraged (Al-Salem, 1996; Mohammad and Mazahreh, 2003). The reuse of waste water for agricultural irrigation purposes reduces the amount of water that needs to be extracted from water resource thereby reducing the burden on the fresh water (USEPA. 1992; Gregory, 2000). It is the potential solution to reduce the fresh water demand for zero water discharge avoiding the pollution load in the receiving sources. It is the necessity of the present era to think about the existing urban waste water disposal infrastructure, waste water agriculture practices, quality of water used, the health implications and the level of institutional awareness of waste water related issues (Rutkowski et al., 2006). Due to ever increasing population huge volume of domestic waste water is being produced in cities. Indiscriminate disposal of such water is a cause for pollution of air, soil and groundwater supplies (Yadav et al., 2002).

24.1.1 USE OF SEWAGE WATER IRRIGATION: GLOBAL CONTEXT

Xu et al. (2010) studied the Impact of long-term reclaimed waste water irrigation on agricultural soils in California for various lengths of time (3, 8, and 20 years, respectively). The no effluent- irrigated plot served as the control and provided reference "background" values. The soil pH values were significantly ($p < 0.05$) lowered in plots with 20-year irrigation to a depth of 140 cm, while EC was elevated for all three plots compared with control. Organic matter (OM), total carbon and total nitrogen contents increased in the top 10-cm soil layers in plots with 8 and 20 year effluent irrigation. Irrigation with effluents also increased both the total and EDTA-extractable metals in the fields.

Cost of treatment of domestic waste water for recycling is too high to be generally feasible in developing countries like India. However, such waste water with nutrient load could be used as irrigation water to certain crops, tree and plants that may lead to increase in agricultural produce and plantation. It has a potential to supply (organic) carbon nutrients, Nitrogen, Phosphorus, Potassium (NPK) and (inorganic) micro nutrients to support crop/plant growth (Weber et al., 1996).

Nutrients present in treated waste water increased dry matter and grain yield appreciably as compared to yields obtained using fresh water irrigation.

Treated municipal waste water in crop irrigation saved 45% and 94% in the cost of the fertilization programs for wheat and alfalfa, respectively. Overall profit for wheat and alfalfa were 14% and 28%, respectively higher than the control. Irrigation with treatment municipal waste water (TMWW) did not influence the chemical composition of soil and plants to hazardous limits. Hence, irrigation with TMWW is safe and is a potential source of supplemental irrigation not only to meet growing crop water needs but also for increased agricultural production Ali (2010).

Irrigation by TMW had a significant positive effect on cotton yield, plant height, internodes distance, number of bolls per m^2, boll weight, leaf area index (LAI), dry matter as well as dry leaf weight. There were no significant differences observed in cotton yield. TMW treatment exerted no significant effect on fiber quality parameters, but cultivation of cotton in non agricultural soil increased lint percentage and while reducing other characteristics related to fiber quality (Alikhasi et al., 2012).

Seven crops including celery, wheat, maize millet, apple, rapeseed and yellow beans irrigated by treated sewage water did not differ distinctly from crops that did irrigated with good quality water. Leachates at different depths did not show alarming levels of the constituents. With treated sewage irrigation a slight increase in OC of the soil was observed.

24.1.2 USE OF SEWAGE WATER IRRIGATION: INDIAN CONTEXT

Many research reports related to use of sewage/waste water in different conditions and different agro-climatic conditions in India could be found. Sewage effluent-irrigation resulted in reduction of pH and increase in OC as compared to irrigation through tube well water. P in the sewage-effluent-irrigation increased to 180% and a 20% decrease was discernible in available K status. DTPA-extractable Zn, Cu, Fe, Mn, Ni and Pb contents were increased due to long-term use of sewage effluents over tube well water irrigation. The concentration of metals in all crops were below the generalized critical levels of phytotoxicity except Fe (Datta et al., 2000).

Use of treated domestic sewage water irrigation for wheat and black gram under different fertilizer levels improved the characters like number of tillers/plant, panicle length (cm), number of grains/panicle and grain weight over the tube well irrigated crop. This resulted in increase in yield to the tune of 3.63% with sewage water irrigation (Pradhan et al., 2001). Long term studies also indicate that sewage water could be safely used for irrigation.

Municipal sewage being used for about three decades showed the enrichment of soils with both OM and nutrients without excessive accumulation of any toxic elements in soils and plants. Thus, the efficient use of such domestic sewage can effectively increase water resource for irrigation and may prove to be a boon for agricultural production. However, traces of some of the toxic ions like Ni, Cd and Pb were noticed in plants and that NO_3 in some well waters should be a matter of concern and indicates the need for treatment of sewage when used for irrigation (Yadav et al., 2002).

According to Singh et al. (2012) sewage water irrigation increased the yield of *rabi* crops with total N, P, K and OC content of soil compared to irrigation with well water. In the field irrigated with sewage water, the pH of soil slightly decreased. Similarly OC in sewage water was higher than well water. The extractable metals after harvest were found to be slightly higher than that of well water. The sewage water contains large amount of nutrients and therefore could be used as a source of irrigation.

Ladwani et al. (2012) observed that application of domestic water increased the yield of crops compared to irrigation with groundwater; it also increases total N, P, K and OC content of soil.

24.1.3 IMPORTANCE OF THE STUDY

The use of waste water in agriculture is gaining importance, because of its value as a potential irritants and a nutrient donor. Use of waste water for irrigation makes it possible to conserve the limited water resources for crop production and also prevents pollution of water bodies, as soil is a very good sink. Application of waste water to agricultural lands may promote the growth of crops and conserve water and nutrients. But the indiscriminate use of the sewage effluents for irrigation to agricultural crops may cause soil and groundwater pollution problems in the long run when they are not properly handled before and after their application to land. Present investigation deals with effect of application of treated sewage water on: (i) growth parameters and increase in the yield of the crop; (ii) pollutant uptake of the crop; (iii) increase in the nutrient status of the soil; and (iv) physical, chemical and biological properties of the soil including heavy metal accumulation in comparison to well water.

This investigation is carried out at College of Agriculture, Nagpur. Increased population has increased the waste water generation. As per Central Pollution Control Board (CPCB) recent report 38,258 MLD waste water is generated in different cities and towns of India. If the similar efforts are carried out in different cities and towns then waste water can be converted into water resource with nutrient reducing the load on fresh water and doses of costly fertilizers for Agriculture in sustainable environment. This simple solution will reduce the problem of disposal of generated waste water and also risk of the health hazards in the cities and towns.

24.2 MATERIAL AND METHODS

24.2.1 EXPERIMENTAL SET-UP

An experimental set up was prepared for conducting the study to investigate the effects of irrigation of treated domestic sewage water (TDSW) and waste water (WW) on soil and maize crop. Field experiment was carried out at Agriculture College Farm, Nagpur. Phytorid based sewage treatment plant was installed and commissioned during June–2012 on the domestic waste water open drain passing through the Agriculture College Farm Maharajbag under Dr. Panjabrao Deshmukh Krishi Vidyapeeth (Agricultural University) at Nagpur. The treatment plant has capacity of 100 m^3 per day. Experiment on maize was conducted adopting factorial, completely randomized design with treatments of TDSW and WW. Each treatment was replicated ten times. Maize was sown by dibbling method with the application of TDSW and WW. In Nagpur the temperature ranges from 10–28.6°C to 30.7–44.5°C in winter and summer, respectively with annual rainfall 1145 mm and humidity from 10 to 88%. Crop experiment details are as given below.

Crop: Maize

Variety: PKV Shatak	**Plot Size: 2 m x 2 m**
Statistical Design: CRD	No. of irrigations: 8
Replications = 10	Quantity of water: 5 ha cm/irrigation
Spacing: 60 cm × 30 cm	Fertilizer dose: 120:60:30
Date of sowing: 13/12/2012	Date of harvesting: 10/4/2013

24.2.2 SAMPLING OF TREATED DOMESTIC SEWAGE AND WELL WATER

Samples of treated domestic waste water from treatment plant were collected two times during the study period in pre-sowing and after harvesting field crops and analyzed in the laboratory for their physico-chemical parameters.

24.2.3 DATA COLLECTION AND ANALYSIS

24.2.3.1 Analysis of Treated Domestic Sewage and Well Water

The pH of the samples was determined using the pH meter, by calibrating the pH meter using the buffer solutions of known pH values (Potentiometric method-1985). Electrical conductivity (EC) was determined using the conductivity Meter (Conductrometric Method) calibrated with conductivity standard (0.01 m KCl with conductivity 1413 μ Scm^{-1} APHA 1985). Biological oxygen demand was determined by Winkler titration method APHA (1985). Chemical oxygen demand was determined by Reflux method (APHA, 1985). Total dissolved solids and Total suspended solids were determined by Gravimetric method (APHA, 1985). Phosphate was determined by Vanadomolybdate phosphoric acid (colorimetric) method (APHA, 1985). Available nitrogen was estimated by Kjeldhal method. Total Nitrogen was determined using the Kjeldahl procedure (Bremner and Mulvaney, 1982). Turbidity was determined by Nephelometric method (APHA, 1985). Micronutrient and Heavy Metal analysis was done by Atomic Absorption Spectrophotometric method Page et al. (1982).

24.2.3.2 Soil Sample Collection and Analysis

The composite surface soil sample (0–15 cm) was collected from experimental site prior to the start of the field experiment. After harvest of the crops, soil samples were collected for different treatment, air dried and ground to pass through 2 mm sieve and stored in plastic bottles before analysis. The samples were analyzed for different physical and chemical properties as per the standard procedure. International pipette method was used to determine the individual soil fraction (i.e., sand, silt and clay (Piper, 1966). The pH of the samples was determined using the pH meter; by calibrating the pH meter using the buffer solutions of known pH values (Potentiometric method-1985). Electrical conductivity (EC) was determined using the Conductivity Meter (ELICO Conductivity Bridge). Soil OC was estimated by Walkley-Black method (Jackson, 1967), phosphate was determined by extracting the soil with Olsen's reagent 0.5 m $NaHCO_3$ of pH 8.5 and in the extract, available P was estimated calorimetrically (Jackson, 1967). Available potassium estimated by leaching the soil with ammonium acetate and the determination of potassium by using flame photometer as per the standard method, available nitrogen was estimated by Kjeldhal method. Total Nitrogen was determined using the Kjeldahl procedure (Bremner and Mulvaney, 1982). Bulk Density was determined by oven dry weight and volume of the core (Black, 1965). sHC was determined by constant water head method (Kult and Dricksen 1986). Available micronutrients and heavy metals were estimated as per procedure described by Atomic Absorption Spectrophotometric method (Page et al., 1982).

24.2.3.3 Plant Sample Collection and Analysis

The plant samples of maize were collected from each plot in both the treatments of TDSW irrigation and WW irrigation. Different standard methods of analysis were used for the analysis of the plant samples to assess the pollutant taken up by the plants. Each root sample was placed in a plastic sieve and was flushed with low pressure tap water, till the complete removal of the soil particles, then root, leaf, head and sprout samples, were washed with deionized water followed by cleaning with the dilute solution of 0.005% HCL, and they were thoroughly washed, by means of a special detergent (alconox 0.1%), and rewashed repeatedly (four times) with distiled water, left to drain on the filter paper, and dried in a ventilated oven at 70°C. They were then ground by means of special hammer mill, and were ready for chemical analysis. The plant samples were dried, mineralized by washing 1.0 g in a muffle furnace at 500°C for 10–12 hrs and the ash was dissolved in 50:50 (v/v) H_2O: HCL solution (Page et al., 1982). The micronutrient (Zn, Fe, Mn and Cu) and heavy metals (Pb, Cd, Co and Cr) were determined by Atomic Absorption Spectrophotometer.

24.2.3.4 Statistical Analysis

Experimental data were statistically analyzed and the mean was calculated against the critical difference at 5% level of significance. CD values were not reported where there was no significant effect of the treatment (Panse and Sukhatme, 1978).

24.3 RESULTS AND DISCUSSION

24.3.1 QUALITY TREATED DOMESTIC SEWAGE WATER AND WELL WATER

The quality of irrigation waters was assessed for its suitability for irrigation with respect to their pH, EC, sodium absorption ratio (SAR), carbonate, bicarbonate, calcium, magnesium, sodium, biological oxygen demand (BOD), chemical oxygen demand (COD), nitrogen, phosphate and potassium. The irrigation water from the two resources was slightly alkaline in reaction. The pH of the treated domestic sewage water was (7.1) slightly lower than the well water, whereas salt content (EC 0.602 dS m^{-1}) was higher than that of well water. Calcium was the dominant cation followed by magnesium and sodium, although the sodium content was slightly higher in treated domestic sewage water. The SAR of both the water resources is much less than the critical limit. Carbonate and bicarbonate of the treated domestic sewage water was slightly higher than the well water, whereas chloride content was 2 to 3 fold higher in treated domestic sewage water than well water. on the basis of SAR, both the water was suitable for irrigation. on the basis of BOD and COD, the treated domestic sewage water was rated as suitable for irrigation purpose when compared with the prescribed limit of 100 and 250 mg L^{-1} for BOD and COD, respectively. In case of all the major nutrients (NPK) they were slightly higher in treated domestic waste water but not significantly more than well water. Nitrogen was three times higher, phosphorus was five times more and potassium was slightly higher in sewage water. Characteristics of two water sources used (i.e., TDSW and WW are depicted in Table 24.1.

TABLE 24.1 Characteristics of Treated Domestic Sewage Water and Well Water Used for Irrigation (Adapted from Mhaske, A.R., Balpande S.S., Khodpage R.M., & Raut, M. (2016). Effects of treated domestic waste water irrigation on soil properties, nutrient availability and crop yield in Indian context Ecology Environment & Conservation, 22(4): 1967-1975. With permission.)

Parameter	TDSW	Well water
pH	7.1±0.92	7.5±0.84
EC (dSm^{-1})	0.602	0.412
SAR	0.656	0.615
Carbonates (CO_3) (mg L^{-1})	0.57	0.30
Bicarbonates (HCO_3–) (mg L^{-1})	3.81	3.18
Chlorides (Cl–) (mg L^{-1})	3.68	1.48
Calcium (mg L^{-1})	4.12	2.68
Magnesium (mg L^{-1})	1.42	0.72
Sodium (Na) (mg L^{-1})	1.09	0.80
BOD (mg L^{-1})	4.14	1.62
COD (mg L^{-1})	15	5.78
TDS (mg L^{-1})	399	278
Nitrogen (mg L^{-1})	3.7	1.1
Phosphate (mg L^{-1})	1.3	0.26
Potassium (mg L^{-1})	0.32	0.22

24.3.2 MAIZE YIELD

24.3.2.1 Effect of Treated Domestic Sewage Water Irrigation on Growth and Green Fodder Yield of Maize

The data reported in Table 24.2 revealed that the plant height and the green fodder yield parameters were significantly higher in irrigation with treated domestic sewage water over well water.

The maize height 30 DAS was recorded 60.79 cm and increased up to 227.5 cm at harvesting (90 DAS) stage in case of irrigation with treated domestic sewage water. However, the height was observed comparatively less (55.06 cm) at 30 DAS and at harvest (189.3 cm) 90 DAS in case of irrigation with well water as a result, the increase in green fodder yield was

TABLE 24.2 Effect of Treated Domestic Sewage Irrigation on Green Fodder Yield of Maize

	Plant height (cm)					
Treatment	**30 DAS**	**60 DAS**	**At harvest 90 DAS**	**Green fodder yield per plant (g)**	**Green fodder yield (kg plot^{-1}) at harvest stage 90 DAS**	**Green fodder yield (t ha^{-1})**
T1-TDSW	60.79	99.57	227.5	135.5	13.80	34.5
T2-WW	55.06	88.86	189.3	124.70	11.80	29.5
SE (m) ±	1.41	1.20	5.41	0.61	0.33	0.82
CD at 5%	4.20	3.58	16.08	1.81	0.97	2.43
F Test	Sig	Sig	Sig	Sig	Sig	Sig

observed to thc tunc of 16.94% ovcr irrigation with well water. Emonger and Ramolemana (2004) in their study reported 8.62% increase in maize yield by irrigation with treated domestic sewage water as compared to the well water. The favorable effect on enhanced plant height, weight and finally yield might be due to higher nutrient contents in treated domestic sewage water as compared to well water. Monteiro de Paula et al. (2011) also reported similar result in case of sugarcane.

24.3.2.2 Effect of Irrigation with Treated Domestic Sewage Water on Grain and Dry Fodder Yield of Maize

The data pertaining to growth parameters, grain and fodder yield reported in Table 24.3 revealed significant increase in the plant height at 30, 60 and at 90 DAS (at harvest) due to irrigation with treated domestic sewage water.

The increase in growth was found to the tune of 4.18 to 15.24% over irrigation with well water. Maize cob length was observed higher in treated domestic sewage water as compared to well water. The grain yield (53.44q ha^{-1}) and fodder yield (77.49q ha^{-1}) recorded in irrigation with treated domestic sewage water was significantly superior over irrigation with well water (48.50 and 70.15q ha^{-1}), respectively (Antolin et al., 2005). The increase in the grain yield was observed to the tune of 10.19% and the dry fodder 10.46%. Higher dry fodder and grain yield might be recorded due to higher nutrient content in treated domestic sewage water. Similar trend of results was reported by Pradhan et al. (2001). In Bostwana irrigation with treated

TABLE 24.3 Effect of Treated Domestic Sewage Irrigation on the Grain and Dry Fodder Yield of Maize (Adapted from Mhaske et al., 2016)

Treatment	Plant height(cm) 30 DAS	60 DAS	At harvest 90 DAS	Length of cob in (cm) at harvest	Diameter of stem at harvest (cm)	Diameter of cob at harvest (cm)	Grain yield per plant (g)	Grain yield (q ha^{-1})	Dry fodder yield (q ha^{-1})
T1-TDSW	57.31	96.70	222.30	27.16	2.68	5.60	89.30	53.44	77.49
T2-WW	52.01	85.87	192.90	25.16	2.33	5.16	84.40	48.50	70.15
SE(m)	1.43	1.21	5.24	0.52	0.05	0.06	1.27	0.81	1.37
CD at 5% level	4.26	3.60	15.57	1.54	—	—	3.76	2.39	4.06
F Test	*	*	*	*	NS	NS	*	*	*

*Significant.

sewage effluent increased maize yield by 8.62% over irrigation with well water (Emonger and Ramolemana, 2004).

24.3.3 CONTENT OF MICRONUTRIENTS AND HEAVY METALS IN MAIZE PLANT

24.3.3.1 Micronutrients

The data in Table 24.4 revealed that the concentration of the micronutrients in the maize plant was not influenced significantly due to irrigation with treated domestic sewage water. The Figure 24.1 revealed that the observed concentration of micronutrients like Zn, Fe, Cu, and Mn in the plant irrigated with the treated domestic sewage water was 69, 407, 14.40 and 22.60

TABLE 24.4 Concentration of Micronutrients and Heavy Metals in Maize

	Micronutrients content (mg kg^{-1})				Heavy metals content (mg kg^{-1})			
Treatments	**Zn**	**Fe**	**Cu**	**Mn**	**Co**	**Cd**	**Cr**	**Pb**
T1-TDSW	69	407	14.40	22.60	0.63	0.66	0.93	1.24
T2-WW	65	369	13.90	19.80	0.61	0.61	0.86	1.19
Percentage increase over, T2	6.15	10.29	3.59	14.14	3.28	8.19	8.13	4.20
Safe limit	1–400	500	20–1000	5–20	0.02–1	0.1–2.4	0.03–1.4	0.2–20
'F' test	NS	NS	NS	NS	NS	NS	NS	NS
SE m(±)	5.19	8.58	2.20	2.31	0.06	0.06	0.08	0.10

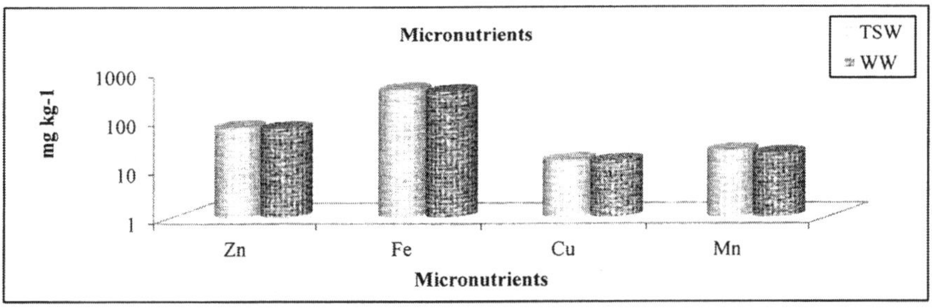

FIGURE 24.1 Micronutrients content in maize. (Adapted from Mhaske, A.R., Balpande S.S., Khodpage R.M., & Raut, M. (2016). Effects of treated domestic waste water irrigation on soil properties, nutrient availability and crop yield in Indian context Ecology Environment & Conservation, 22(4): 1967-1975. With permission.)

mg kg^{-1}, respectively, which were slightly higher over the irrigation by well water. By and large accumulation of micronutrients due to irrigation with treated domestic sewage water was well below the critical limits prescribed for the phytotoxicity of these micronutrients. Similar findings are in line with the results reported earlier by Madyiwa et al. (2002) for irrigation with domestic sewage water.

24.3.3.2 Heavy Metals Content in Maize

The data on heavy metals concentration in the maize plant reported in Table 24.4 revealed that the content of heavy metals *viz.,* Co, Cd, Cr and Pb in the maize plant were not accumulated much due to irrigation with treated domestic sewage water. The content of Co, Cd, Cr and Pb were observed to the tune of 0.63, 0.66, 0.93 and 1.24 mg kg^{-1}, respectively, in plants irrigated with domestic sewage water were slightly higher as compared to plants irrigated with well water (Figure 24.2).

The heavy metals content in maize due to irrigation with treated domestic sewage water was lower than the critical limits prescribed for the phytotoxicity of these metals. The finding corroborates with the results reported earlier by Azad et al. (1986) for the soils irrigated with domestic sewage water in Punjab.

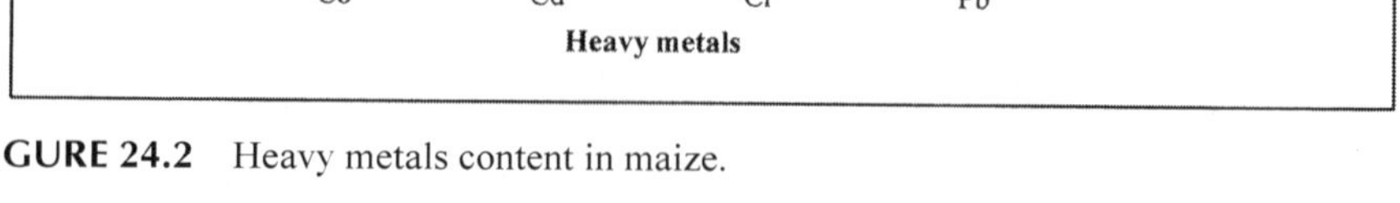

FIGURE 24.2 Heavy metals content in maize.

24.3.4 PHYSICAL AND CHEMICAL CHARACTERISTICS OF SOILS

The experimental soil was classified under the clay textural class (clay-52.4%, silt-29.2% and sand-19.4%), which has considerable amount of smectite clay minerals. The calcium carbonate content of the soil was slightly calcareous (4.34%). The treated domestic sewage water irrigation applied for a season had no significant effect on a clay soil in terms of the physical properties. The pH of the pre-sowing soil was slightly alkaline (7.9), electrical conductivity (0.32 dS m^{-1}). The nutrient status was moderate. Available micronutrients and heavy metal content in soil are depicted in Figure 24.3. The initial status of micronutrients and heavy metals were within the safe limits for crop growth.

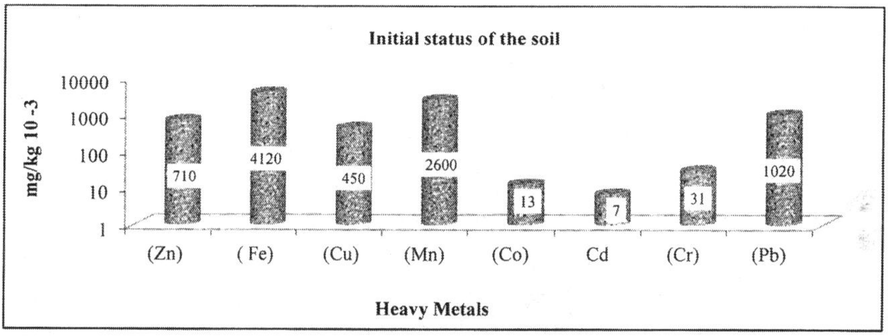

FIGURE 24.3 Available Micronutrients and heavy metal in Soil. (Adapted from Mhaske et al., 2016).

24.3.4.1 Changes in the Soil Status After Harvest of the Maize Crop Chemical Properties

The results in relation to the chemical properties of soil after harvest of the maize crop (Table 24.5) showed that the pH of the soil was 7.85 after

TABLE 24.5 Changes in the Physico-Chemical Properties of the Soil Irrigated with Treated Domestic Sewage and Well Water After Harvest of the Maize Crop (Adapted from Mhaske et al., 2016)

Treatments	pH	EC	OC	Available micronutrients ($kg\ ha^{-1}$)			Bulk density	Saturated Hydraulic conductivity
		(dS m^{-1})	(g kg^{-1})	N	P	K	($Mg\ m^{-3}$)	($cm\ hr^{-1}$)
T1-TDSW	7.85	0.36	5.40	285.20	17.21	432	1.33	1.49
T2-WW	7.94	0.30	5.26	269.10	16.98	429	1.36	1.54
Initial status	7.91	0.32	5.21	270.20	15.23	417.28	1.35	1.52
'F' test	NS	NS	NS	NS	NS	NS	NS	NS
SE m(±)	0.07	0.03	0.06	5.69	0..93	5.40	0.04	0.04

irrigation with treated domestic sewage water, whereas, it was 7.94 in well water irrigation. Al Omran et al. (2011) reported similar findings of lower pH due to use of treated sewage effluent as compared to well irrigation.

EC of soil was increased slightly with treated domestic sewage irrigation (0.36 dS m^{-1}) as compared to well water irrigation (0.30 dS m^{-1}) to maize. This slight increase in EC might be attributed to content of the salts in sewage. Kalavrouziotis et al. (2008) also reported similar findings due to use of treated sewage water irrigation in soil as compared to normal irrigation water and initial status value. Yadav et al. (2002) also reported similar results.

The OC content was not influenced statistically but irrigation with treated domestic sewage water to maize increased OC content (5.40 g kg^{-1}) as compared to well water irrigation (5.26 g kg^{-1}). The increase in OC content in soil due to treated domestic sewage water irrigation was noticed to the extent of 3.65% over initial organic content of the soil while 2.70% higher as compared to the soil irrigated with well water. Wang et al. (2007) reported similar finding of increases in OC in soil due to use of sewage water for irrigation.

The available nitrogen in the soil was increased slightly by irrigation with treated domestic sewage water (285.2 kg ha^{-1}) and slightly reduced in treatment by well water irrigation (269.1 kg ha^{-1}). The increase in available nitrogen was noticed to the extent of 5.63% by treated domestic sewage irrigation over the initial value of available nitrogen in soil after harvest of the maize crop.

The increase in available phosphorous in soil after harvest of the maize crop was observed in treatment of irrigation with treated domestic sewage water (17.21 kg ha^{-1}) which was to the tune of 11.55% over initial.

The available potassium in soil after harvest of the maize crop in the treatment of irrigation with the treated domestic sewage water was 432 kg ha^{-1}, however well water irrigation increased its availability in the soil (429 kg ha^{-1}).

The increase in available potassium was noticed to the extent of 3.53% by treated domestic sewage irrigation over the initial status. These finding are consistent with reports by Ladwani et al. (2012), Badelesha et al. (1997), and Mandal et al. (2008).

24.3.4.2 Impact on Soil Physical Properties

The data pertaining to the physical properties of soil *viz.*, bulk density and sHC are presented in Table 24.5.

The bulk density of soil was not influenced significantly due to irrigation with treated domestic sewage irrigation. It was 1.33 Mg m^{-3} in plot irrigated with treated domestic sewage water which reduced to 1.5% over initial after harvest of the maize crop. Lado and Meni (2002) also observed the decrease in bulk density of soil in treated sewage irrigation.

The sHC was found to decrease slightly in soils irrigated with treated domestic sewage water (1.49 cm hr^{-1}) while with well water irrigation it was 1.54 cm hr^{-1}. The decrease in sHC due to treated domestic sewage irrigation was registered to the tune of 1.97% over initial. The findings are consistent with the Vinten et al. (1983), Viviani and Lovino, (2004), and Lado et al. (2002) in the review article on treated domestic sewage irrigation effect on soil hydraulic properties in the arid and semiarid zones.

24.3.4.3 Impact on Soil Micronutrients

The data pertaining to micronutrient content in soil after harvest of maize (Table 24.6) indicated that there was no significant difference due to irrigation with treated domestic sewage water.

The available Zn, Fe, Cu, and Mn were found 0.74, 4.31, 0.56, 2.71 mg kg^{-1}, respectively, which were slightly higher as compared to irrigation with well water. The increase of Zn, Fe, Cu, and Mn was to the extent of 5.71, 3.60, 16.67 and 3.44%, respectively as depicted in Figure 24.4. Al Omron

TABLE 24.6 Extractable Micronutrients and Heavy Metal Status in Soil Irrigated Treated Sewage and Well Water After Harvest of the Maize

	Micronutrients content (mg kg^{-1})				Heavy metals content (mg kg^{-1})			
Treatments	**Zn**	**Fe**	**Cu**	**Mn**	**Co**	**Cd**	**Cr**	**Pb**
Treated water	0.74	4.31	0.56	2.71	0.018	0.012	0.036	1.15
Well water	0.70	4.16	0.48	2.62	0.016	0.010	0.033	1.04
Initial status	0.71	4.12	0.45	2.60	0.013	0.007	0.031	1.02
Safe limit	2.00	10.00	5.00	5.00	2.000	0.500	1.000	5.00
'F' test	NS	NS	NS	NS	NS	NS	NS	NS
SE m (±)	0.05	0.10	0.04	0.07	0.030	0.000	0.000	0.03

et al. (2011) reported similar results of increase of micronutrients in the soil irrigated with sewage water.

24.3.4.4 Heavy Metals

The heavy metal accumulation content in the soil after harvest of maize crop (Table 24.6) revealed that the extractable heavy metals *viz.,* Co, Cd, Cr and Pb did not accumulate much in plots irrigated with treated domestic sewage water after harvest of the maize crop. The content of Co, Cd, Cr and Pb was observed to be 0.018, 0.012, 0.036 and 1.15 mg kg^{-1}, respectively in domestic sewage water irrigated plots. The accumulation of Co, Cd, Cr and Pb was noticed to the tune of 12.5, 20.0, 9.09, and 10.58% as depicted in Figure 24.4 but accumulation of the heavy metals in soil due to irrigation of treated domestic sewage was well below the safe limits in soil. Emongor and Ramolemana (2004) reported similar findings in use of treated sewage water for irrigation of the crop.

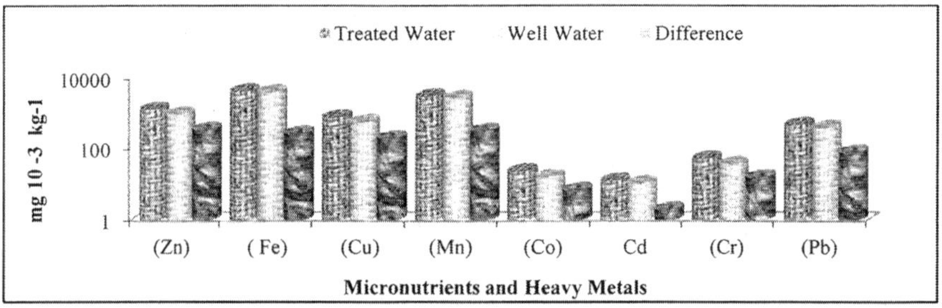

FIGURE 24.4 Available micronutrients and heavy metals in soil after harvesting maize. (Adapted from Mhaske, A.R., Balpande S.S., Khodpage R.M., & Raut, M. (2016). Effects of treated domestic waste water irrigation on soil properties, nutrient availability and crop yield in Indian context Ecology Environment & Conservation, 22(4): 1967-1975. With permission.)

24.3.4.5 Microbial Count in Soil After Harvest of Maize

The data pertaining to the microbial count *viz.,* bacterial population, fungal population and antinomycetes population in soil after harvest of cotton are presented in Table 24.7. The results revealed that the bacterial, fungal and antinomycetes population in soil after harvest of crop was slightly higher in soil irrigated with treated domestic sewage water as compared to irrigation with well water.

TABLE 24.7 Changes in the Microbial Properties of the Soil Irrigated with Treated Domestic Sewage Water and Well Water After Harvest of the Maize

Treatments	Bacterial population (cfu x 10^6)	Fungal population (cfu x 10^4)	Antinomycetes population (cfu x 10^4)
T1-TDSW	41.00	8.89	29.67
T2-WW	24.67	3.04	10.33
SE (m)±	0.54	0.22	1.55
CD at 5%	1.44	0.579	4.09

24.4 CONCLUSIONS

Sewage waste water treated with Phytorid technology proved to be effective in physical, chemical and biological treatments to reduce pollutant load well below the phytotoxicity limits. The concentrations of micronutrients (Zn, Fe, Cu, and Mn) and heavy metals like Co, Cd, Cr and Pb in maize irrigated with treated domestic sewage water was observed well below the critical limits prescribed for the phytotoxicity of these metals. Its application increased the yield of crops compared to irrigation with well water. It also increased total N, P, K and OC content of soil.

KEYWORDS

- **Crops**
- **Plant Uptake**
- **Soil**
- **Soil Contamination**
- **Treated Sewage Water Irrigation**

REFERENCES

Al Omron, A. M., El-Maghraby, S. E., Nadeem, M. E. A., El-Eter, A. M., Al-Mohani, H. (2011). Long term effect of irrigation with the treated sewage effluent on some soil properties of Al-Hassa Governorate. *Saudi Arabia Journal of Saudi Society of Agricultural Sciences, 11(1),* 15–18.

Ali A. Al-Jaloud (2010). Reuse of wastewater for irrigation in Saudi Arabia and its effect on soil and plant. Natural Resources and Environment Research Institute (NRERI), King

Abdul-Aziz City for Science and Technology (KACST), Riyadh Kingdom of Saudi Arabia, Email aljaloud@kacst.edu.sa 19th World Congress of Soil Science, Soil Solutions for a Changing World 1–6 August 2010, Brisbane, Australia. Published on DVD. 163–166.

Alikhasi, M., Kouchakzadeh, M., & Baniani, E. (2012). The effect of treated municipal wastewater Irrigation in non-agricultural soil on cotton plant. *Journal of Agricultural Science and Technology*, *14*, 1357–1364.

Al-Rashid, M. F., & Sherif, M. M. (2000). Water resources in the GCC countries: an overview, *Water Resources Management, 14*(1), 59–73.

Al-Salem, S. S. (1996). Environmental considerations for wastewater reuse in agriculture, *Water Science and Technology, 33*(10–11), 345–353.

Anonymous. (1985). APHA (American Public Health Association), Standard methods for the examination of water and wastewater. 19th ed., Washington, DC.

Antolın, M., Carmen, I., Pascual, C., Garci, A., Polo, M., & Sanchez-Diaz (2005). Growth, yield and solute content of barley in soils treated with sewage sludge under semiarid Mediterranean conditions. *Field Crops Research, 94*, 224–237.

Azad, A. S., Sekhon, G. S., & Arora, B. R. (1986). Distribution of cadmium, nickel and cobalt in sewage water treated soils. *Journal of the Indian Society* of *Soil Science*, *34*, 619–621.

Badelesha, H. S., Chhabra, R., & Ghuman, B. S. (1997). Changes in soil chemical properties and plant nutrient content eucalyptus irrigated with sewage water. *Journal of the Indian Society* of *Soil Science*, *45*(2), 358–364.

Black, C. A. (Ed). (1965). Methods of soil analysis, part 1&2, American Society of Agronomy, Madison, Wisconsin, USA.

Bremner, J. M., & Mulvaney, C. S. (1982). Methods of Soil Analysis, part 2 Chemical and Microbiological Properties, 595–624.

Datta, S. P., Biswas, D. R., Sarang, N., Ghosh, S. K., & Rattan, R. R. (2000). Effect of long-term application of sewage effluents on organic carbon, bioavailable phosphorus, potassium and heavy metal status of soils and content of heavy metals in crops grown thereon. *Journal of the Indian Society* of *Soil Science*, *48*(4), 836–839

Emongor, V. A., & Ramolemana. (2004). Treared sewage effluent (water) potential to be used for horticultural production in Botswana, *Physics and Chemistry of the Earth*, 29, 1101–1108.

Gregory, A. (2000). Strategic direction of water recycling in Sydney, Proceeding of the First Symposium Water Recycling, Australia Adelaide, 19–20. pp. 35–41.

Jackson, M. L. (1967). Soil Chemical Analysis. Prentice Hall of India Pvt. Ltd., New Delhi pp. 205.

Kalavrouziotis, I. K., Robolas, P., Koukoulakis, P. H. Papadopoulos, A. H. (2008). Effects of municipal reclaimed wastewater on the macro and micro elements status of soil Brassica oleracea var. Italica and B. oleracea var. gemmifera. *Agricultural Water Management, 95*(4), 419–426.

Klute, A., & C. Dirksen, (1986). Hydraulic Conductivity and Diffusivity: Laboratory Methods. In: A.Klute (ed.). Methods of Soil Analysis. Part 1. Physical and Mineralogical Methods. SSA Book Series, 5.

Lado, M., & Meni, B. H. (2002). Treated domestic sewage irrigation effects on soil hydraulic properties in arid and semiarid zones: A review metal contents of soils under vegetables in Harare, Zimbabwe. Agriculture, Ecosystems and Environment, *107*, 151–165

Ladwani, K. D., Ladwani, K. D., Manik, V. S., & Ramteke, D. S. (2012). Impact of domestic waste water irrigation on soil properties and crop yield. *International Journal of Scientific and Research Publications, 2*(10), 1–7.

Madyiwa, S., Chimbari, M., Nyamangara, J., & Bangira, C. (2002). Cumulative effects of sewage sludge and effluent mixture application on soil properties of a sandy soil under a mixture of star and kikuyu grasses in Zimbabwe. *Physics and Chemistry of the Earth, 27*, 747–753.

Mhaske, A.R., Balpande S.S., Khodpage R.M., & Raut, M. (2016). Effects of treated domestic waste water irrigation on soil properties, nutrient availability and crop yield in Indian context Ecology Environment & Conservation, 22(4): 1967-1975.

Mandal, U. K., Warrington, D. N., Bhardwaj. A. K., Bar-Tal. A., Kautsky. L., Minz, D., & Levy, G. J. (2008). Evaluating impact of irrigation water quality on a calcareous clay soil using principal component analysis, *Geoderma, 144*, 189–197.

Mohammed, M. J., & Mazahreh, N. (2003). Changes in soil fertility parameters in response to irrigation of forage crops with secondary treated wastewater, *Communications* in *Soil Science and Plant Analysis, 34*(9 &10) pp. 1281–1294.

Monteiro de Paula, Nogueira, S. F., Montes, C. R., & Melfi, A. J. (2011). Unburned sugarcane irrigated with treated sewage: effects on carbon and nitrogen exportation and crop yield. São Paulo State Research Aid Foundation (FAPESP), National Council for Scientific and Technological Development (CNPq), Municipal Sewage Plant of the municipality of Piracicaba, São Paulo, Brazil (SEMAE – Piracicaba).

Page, A. L., Miller, R. H., & Keeney, D. R. (Eds.) (1982). Methods of Soil Analysis, Part-2: Chemical and microbiological properties. *American Society of Agronomy and Soil Science Society of America.* Madison, Wisconsin, 1159 p.

Panse, V. G., & Sukhatme, P. V. (1978). Statistical Methods for Agricultural Workers. I. C. A. R., New Delhi, India.

Piper, C. S. (1966). "*Soil and Plant Analysis, IVth edition.*" University of Acelicide, Adcitada, Australia, 135–200.

Pradhan, S. K., Sarkar, S. K., & Prakash, S. (2001). Effects of sewage water on the growth and yield parameters of wheat and black gram with different fertilizer levels. *Journal of Environmental Biology*, *22*(2), 133–6.

Rutkowski, T., Raschid-Sally, L., & Buechler, S. (2006). Wastewater irrigation in the developing world—Two case studies from the Kathmandu Valley in Nepal," *Agricultural Water Management,* 88, pp. 83–91.

Singh, P. K., Deshbhratar, P. B., & Ramteke, D. S. (2012). Effects of sewage wastewater irrigation on soil properties, crop yield and environment. *Agricultural Water Management. 103*, 100–104.

U.S. Environmental Protection Agency. (1992). Offices of water and wastewater and compliance (Ed.) Guidelines for water reuse. U.S. EPA, Washington. WA State Water Strategy.

Vinten, A. J. A., Mingelgrin, U., & Yaron, B. (1983). The effect of suspended solids in wastewater on soil hydraulic conductivity: II. Vertical distribution of suspended solids. *Journal of Alliance of Crop, Soil and Environmental Science Societies, 47*(3), 408–412.

Viviani, G., & Iovino, M. (2004). Wastewater reuses effects on soil hydraulic conductivity. *Journal of Irrigation and Drainage Engineering, 130*, 476–484.

Wang, J., Wang, G., & Wanyan, H. (2007). Treated waste water irrigation effect on soil, crop and environment: Waste water recycling in the loess area of China. *Journal of Environmental Sciences. 19*, 1093–1099.

Weber, B., Avnimelech, Y., & Juanico, M. (1996). Salt enrichment of municipal sewage new prevention approaches in Israel, *Environmental Management*, *20*(4), 487–495.

Xu, J., Wu, L., & Chang, A. C., & Zhang, Y. (2010). Impact of long-term reclaimed wastewater irrigation on agricultural soils: A preliminary assessment. *Journal of Hazardous Materials*, *183*, 780–786.

Yadav, R. K., Goyal, B., Sharma, R. K., Dubey, S. K., & Minhas, P. S. (2002). Post-irrigation impact of domestic sewage effluent on composition of soils, crops and groundwater—A case study: *Environment International. 28*(6), 481–486.

PART IV

LAND USE PLANNING AND LIVELIHOOD SECURITY

CHAPTER 25

STATUS OF LAND USE PLANNING IN INDIA

N. G. PATIL[1] and S. K. SINGH[2]

[1]*Principal Scientist, ICAR-National Bureau of Soil Survey and Land Use Planning, Amravati Road, Nagpur – 440 033, India*

[2]*Director, ICAR-National Bureau of Soil Survey and Land Use Planning, Amravati Road, Nagpur – 440 033, India*

CONTENTS

ABSTRACT

Land use planning (LUP) research in India is reviewed in this article. It is noted that most of the work in the recent past is based on edaphic approach. The LUP research in India began with zonation of the country into different

Agro-ecological regions (AER) using meteorological data in conjunction with soil survey data. Evaluation of soils to grow different crops is the most extensively reported topic after AER zonation. Third area of interest was observed to be farm or village based studies wherein prevailing crops were evaluated and/or alternative crops were suggested based on land evaluation. The land use planning however remained an academic exercise with little attention to practical constraints or adoption. Drawing from the practical experience, it is argued that paradigm shift is required from LUP based on edaphic factors to more inclusive LUP based on water, livestock, forest and common property resources. Now, geospatial tools have completely transformed the land use planning capabilities. It is anticipated that many problems like scale limitations in acquisition of varied data, bringing out uniformity and consistency, co-ordination amongst different related agencies, etc., could be overcome through remedial measures suggested in this review.

25.1 INTRODUCTION

During the last decade 'land use' has become the most debated topic in India. In fact it could be termed a decade of land conflicts as numerous protests have been reported from different parts of the country. Farmers have opposed acquisition of their lands for infrastructural development, industry, special export zones, etc. The decade has also witnessed subtle shift from fertility based demand for land to location based demand that has triggered price rise in the country and successive Governments have been forced to revise the land acquisition policies effected from time to time. The significant spurt in demand for land for different purposes is now causing rapid changes in the way lands are managed in the country where land is a scarce resource. It is therefore essential to know our land resources, their potential, constraints and come out with plans aimed at optimal utilization while ensuring that the demands for higher food production are met. It is notable that the conversion of land use is mostly irreversible when agriculture land is acquired for other purpose. It is therefore feared that indiscriminate conversion of land use could be at the peril of food production in India. Prime lands of the country therefore need to be protected. There are many such issues like producing enough fodder for the livestock, managing our water resources efficiently, ensuring green cover, etc., It calls for a scientific land use planning. Are we ready to take up the challenge?

25.1.1 CURRENT NEEDS AND FUTURE DEMANDS

Under the assumption of 3.5% growth in per capita Gross Domestic Product (GDP) (low income growth scenario), demand for food grains (including feed, seed, wastage and export) is projected in the year 2020 at the level of 256 mt comprising 112 mt of rice, 82 mt of wheat, 39 mt of coarse grains and 22 mt of pulses (Agriculture Policy Vision 2020, Planning Commission). The demand for sugar, fruits, vegetables, and milk is estimated to grow to a level 33,77,136 and 116 mt, respectively. The demand for meat is projected at 9 mt, fish 11 mt and eggs 77.5 billion. Future increases in the production of cereals and non-cereal agricultural commodities will have to be essentially achieved through increases in productivity, as the possibilities of expansion of area and livestock population are minimal. To meet the projected demand in the year 2020, country must attain a per hectare yield of 2.7 tons for rice, 3.1 tons for wheat, 2.1 tons for maize, 1.3 tons for coarse cereals, 2.4 tons for cereal, 1.3 tons for pulses, 22.3 tons for potato, 25.7 for vegetables, and 24.1 tons for fruits. The production of livestock and poultry products must be improved 61% for milk, 76% for meat, 91% for fish, and 169% for eggs by the year 2020 over the base year 1999. Average yields of most crops in India are still rather low (Paroda and Kumar 2000). Keeping this in view, the land use plans have to be made for achieving higher productivity and sustainable food security.

25.1.2 STATUS OF LUP RESEARCH

Land-use planning means the systematic assessment of physical, social and economic factors in such a way as to encourage and assist land users in selecting options that increase their productivity, are sustainable and meet the needs of society (FAO, 1993).

The FAO definition further elaborates that land-use planning is the systematic assessment of land and water potential, alternatives for land use and economic and social conditions in order to select and adopt the best land-use options. Its purpose is to select and put into practice those land uses that will best meet the needs of the people while safeguarding resources for the future.

Land use planning research in India emerged as an offshoot of extensive soil surveys that were taken up after 1980. Till then soil information of the country was limited to traditional classification such as red soils, black soils, etc. The National Bureau of Soil Survey and Land Use Planning (NBSS &

LUP) is entrusted with research, training, correlation, classification, mapping and interpretation of soil information and work done in India is broadly confined to the research work by NBSS & LUP that could be broadly divided into four categories in a chronological order.

1. During the early years, zonation of the country based on length of growing period criteria was focused.
2. Development of soil suitability criteria for major crops of the country followed by crop experiments to evaluate the developed criteria including soil attributes
3. LUP at different planning levels/units like village, watershed, district based on soil distribution, topography and climate.
4. Customized LUP like identifying suitable area for commercial crops like rubber, tea, etc.

During the early years, soil survey information was used for zonation of the country into agro-ecological regions (AER) based on length of growing period (LGP) criteria (Sehgal et al., 1992, 1993, 1995). Realizing the limitations of crop diversity and narrower LGP, the AERs were further subdivided into 60 sub-regions (Velayutham et al., 1999). The zonation work mostly assisted planners and administrators in taking informed decisions on specific target programrs like drought mitigation, explicit works in drought prone area. Besides, the agro ecological sub region map of the country also assisted in prioritization of research issues that were central to the zone or sub zone. The administrators could gauge the magnitude and scale of schemes to be implemented because of the information available.

The next step was to asses if current land use was based on scientific rationale or not. This required land evaluation with a specified objective. The crops grown by farmers usually reflect the wisdom passed on from generation to generation and socio-economic factors operating at the time of reference. However, there have been rapid changes especially during the last three decades and influence of socio-economic factors appears to have become overriding *vis-a-vis* edaphic factors. Irrespective of these changes, the primary step in land use planning is to evaluate lands for their ability to support prevailing use. Therefore, it is logical to note that the LUP research in the country subsequently focused on developing soil suitability criteria for major crops of the country. It was essential to develop such criteria without which it was not possible to judge the prevailing crop choices. Publication named 'Soil Suitability Criteria" by Naidu et al. (2006) compiles all the research work in this regard.

Once the criteria had been fixed, their testing in the field was taken up in a series of studies. Such studies evaluated different soils for their ability to grow a chosen crop. The findings enabled to ascertain the feasibility of continuing with the same crop and experimenting with an alternate crop option. Expectedly this phase lasted very long as the criteria were evaluated in different growing conditions for different crops, different agro-ecological regions and management (Annual Report, NBSS & LUP, 2014).

The research efforts in general identified three broad categories of state intervention: (i) introduction of new crops and livestock components based on soil suitability and potential for enhanced agricultural productivity; (ii) introduction of new varieties commiserate with the soil information/properties; and (iii) adoption or changes in land management techniques. These findings facilitated the planners to assess the potential of different crops and possible changes in land use in different agro-eco systems.

In concurrent as well as subsequent efforts, land use plans were developed for different management units like village, watershed, district, etc. The bureau has implemented soil based land use plans for number of villages in the country and demonstrated that agricultural productivity is intrinsically linked to soil status and enhanced productivity can be achieved with scientific utilization of soils. Efficacy of LUP has been demonstrated (Charurvedi et al., 2015; Hajare et al., 2002; Hirekerur et al., 1986) on 2180 ha area of seven different villages. The last decade has witnessed a major change in the way land use plans are prepared. The Indian experience showed that farmers do not adhere to the land use plans or grow crops according to soil suitability irrespective of the location and findings of soil suitability studies. The reasons for non-compliance were varied, for instance socio-economic compulsions (market availability, family needs, cash crop), management issues (spatial distance from village, labor availability, land size), resource availability (credit, tools). Further, top-down approach adopted by the researchers was not readily acceptable. Departing from the usual approach, inclusive land use planning was emphasized during the recently concluded National Agriculture Innovation Project (NAIP) implemented by the bureau (NAIP 2009 to 2014). Moreover, in addition to the soil resources, common property resources such as community water tanks, pasture lands, community lands, Non Timber Forest Produce (NTFP) were also included for arriving at a land use plan for a village. Crop diversification was also promoted with active participation of the farmers. Landless villagers participated in the process as they had stakes in common property resources. For example, in a study undertaken in Gondia district (Maharashtra), land use planning in tribal village primarily aimed at optimal use of soil and water resources and use of common property resources especially for landless villagers. Community nursery utilized available water in organized way and facilitated early transplanting of rainfed rice that led to early harvest with 50% increase in yields. The farmers realized mean additional income of Rs. 14,525/hectare (Table 25.1). It also opened new possibility of *rabi* crop raised on residual soil moisture (Figure 25.1). The system has implications for 11.6 m ha. rainfed paddy area of the country.

TABLE 25.1 Rice Yield in Gondia Clusters Before and After Soil Based Land Use During 2009–14

Year	(q/ha)	Area (ha)	Total Yield (q)	Income (Rs/ha)
2009–10	29.8	66.8	1990.64	35760
2010–11	27.8	181.2	5037.36	33360
2011–12	20.1	212.6	4273.26	24120
2012–13	28.4	240	6818.00	34080
2013–14	23.3	300	7005.00	30355
Mean	**25.88**	**200.12**	**5024.85**	**31535**

Baseline 16.2 q/ha—income Rs.17,000/ha (Before soil based interventions).

Source: NAIP (2014).

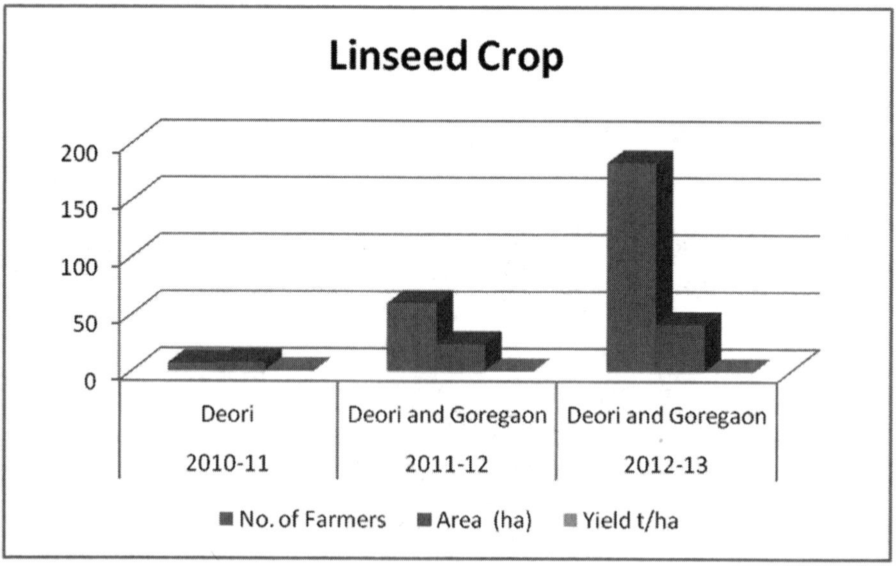

FIGURE 25.1 Linseed production in project clusters.

A farming system of rice-utera linseed – NTFP was developed to provide livelihood requirements.

In another participatory land use planning experiment conducted in Dhule district of Maharashtra, soil suitability for the onion crop in the district showed that it could be grown in approximately 1900 km^2 land. Field trials also suggested that the onion crop could achieve higher economic returns (Rs. 100,000 per ha).

25.2 TECHNOLOGIES USED IN LAND USE PLANNING

Traditional land use planning relied on many different sources of physical/printed information like topographic maps, aerial photographs, soil survey manuals, botanical surveys, flood zone maps, drainline maps, hydrology maps, etc. Obviously it was difficult to comprehend, interrelate and correlate all these information for land use decisions. Further the information was static and lost relevance after some time. Periodicity of aerial photographs was not frequent that hampered utility. Shift from printed information to digital information and integration using Geographic Information System (GIS) enables land-use planners to handle, analyze and correlate multiple data layers to one location and visualize trends and patterns. The GIS also facilitates georeferencing of old tabular soil information to produce interpretive maps, helping the user in visual representation of the tabular data.

Technological advances are also aiding hitherto difficult embedding of multi scale data and simulating different scenarios of land use and agricultural productivity. Under National Agriculture Innovation Project (Component 4) based on saturated hydraulic conductivity (sHC) database and length of growing period estimation, AESRs of IGP were modified from the existing 17 to a total 29 units. These modifications were accomplished using tools in GIS platform, climate data and expert opinion. Using district-level rice yield data, rice-growing areas in the IGP were divided into four regions such as low, medium, medium high and high representing areas with rice yield of <1000, 1000–2500, 2500–4000 and >4000 kg ha^{-1}, respectively. These assessments show that nearly 33% area produces medium high to high category yields, while medium level yield is observed in 63% of the total rice-growing areas (Patil et al., 2014). Since the data on sHC were not readily available, pedotransfer functions were developed (Tiwari et al., 2014) for estimating the same for the soils of two food-growing zones of India (the Indo-Gangetic Plains) and the black soil region (BSR). The soil information systems are thus being developed with the help of modern data mining techniques. New tools like artificial neural networks, principal component analysis have been shown to enhance value of the acquired data and prediction of unknown soil properties (Patil et al., 2011, 2013; Tiwary et al., 2014).

Now research in land use planning could benefit from a large variety of tools such as: Information Management, System Analysis, Decision Support Systems, Multicriteria Analysis, Geographic Information Systems, Remote Sensing, Computer Image Analysis, Sensors, Modeling Technique, Neural Network Technology, Land Evaluation, etc. All these tools have to be considered under a broad and integrated approach related to food and other agricultural commodity production, rational land use planning, water saving, resource conservation, environmental impacts and socio-economic effects (Riveira and Maseda, 2006). Given these complexities, it is obvious that land use planning at local level under the broader framework at state or country level is an appropriate approach as it is indeed difficult to conceive a local plan for a remotely connected centralized or apex agency.

It is however important that local focus is not lost while using the technology. The soil suitability to support a crop of choice was recently examined by Venugopalan et al. (2014) using greater data on climate and soils. Computer model Infocrop could satisfactorily capture subtle differences in soil variables and weather patterns prevalent in the Bench Mark locations spread over 16 agro-ecological sub-regions (AESRs) resulting in a wide range of mean simulated rainfed cotton yields (482–4393 kg ha^{-1}). It is apparent that though the model put greater data demand, it could also perform simulations with greater accuracy. However, the testing was limited to only two crops leaving a major information gap indicating similar work to be taken up in different local units or AESRs for a variety of crops.

25.3 PARADIGM SHIFT IN LAND USE PLANNING

Apart from technological changes, there have been major changes in the way land use plans are prepared now. It is now well accepted fact that farmers do not adhere to the land use plans prescribed by land use planners, instead they have their preferences and justifications for growing crops that they deem suitable for the quality of land holdings and many other socio-economic parameters. Emphasis on edaphic factors is thus now shifted to more integrated approach based on farmers' considerations and his/her interaction with the realm of environment. The reasons for non-compliance to external planner's LUP were varied ranging from socio-economic issues like market access, market availability, family needs, education, management issues (spatial distance from village, labor availability, land size), resource availability (credit, tools). The prescriptive LUP is negated by many factors such as lack of institutional support to implement LUP, lack of coordination among disparate agencies working for the same goal without synergy. Now a land use planner must consider these constraints and formulate LUP with knowledge that individual farmer or land holder is fully at liberty to make land use decisions and thus make provisions for possible alternatives that may not be necessarily scientific. Mutually agreed land use plans or in other words participatory land use planning is the only option to ensure compliance (even if it partial) of LUP.

State ownership of forest lands and rigid laws governing use of forest lands leave little scope for any change in use. Thus land use plan is merely a suggestive/consultative document with no legislative support. *For example*, a water reservoir suggested in the LUP may be located in forest department owned land or it could be revenue department's property. The property rights in such case could become a major hindrance in effective LUP. If the location happens to be a land parcel owned by marginal farmer, his/her livelihood will be lost.

While socio-economic factors have great relevance in India, the ecological/environmental factors have little or no attention. The issues like management of salt affected soils, ground water pollution due to agro-chemicals, loss of bio-diversity have not so far been considered in land use planning. Most of the developed countries have developed criteria for identification of the prime lands (lands that are of national importance for food security) and delineated the areas to be protected for sustained agricultural productivity. In Indian context such studies are limited. Recently Naidu et al. (2014) reported delineation of prime lands for the state of Andhra Pradesh. Such exercise however needs to be taken up at national level. Naidu et al. (2014) used 1:250K soil survey data for this purpose. Such delineation is not expected to assist the decision makers much in land acquisition that occurs at local level. Because of the scale limitations, a polygon indicating prime land may contain large swathes of problem soils or soils with major limitations. As of now we have to contend with the available data. As soil information on larger scale becomes available, delineation of prime lands will require revisiting. Preparedness of any management unit (village/watershed/ district/region) to overcome challenges of drought in short term and climate change in long term hinges on how smartly we use the available information and append it with more data using modern tools.

25.4 LACK OF INSTITUTIONAL MECHANISM

It is notable that implementation of land use plans in general has been an activity of only one institute in India i.e., NBSS and LUP. Apart from the work in hundred odd villages over the period of three decades, there is no notable or reported LUP implementation activity in India. The study villages represent geographically minuscule area with likely inadequate representation of the landscape. Implementation of land use plan for relatively bigger administrative/management units like block, tehsil or district has not been reported so far in India primarily because it is not feasible for a researcher to pursue execution of LUP in the absence of institutional support. There is a perceptible lack of interest in state agricultural universities and research centers of ICAR. The district level land use planning has thus been restricted to preparation of plans and then leaving implementation to the administrators and line departments. In the near future also the same conditions are expected to prevail leaving little scope for any changes in approach to implementation of LUP unless the plethora of rules and regulations are amended to achieve synergy benefits of scientific LUP. However, it is not the lack of institutions but the lack of linkages between different development departments and institutions that appears to be the main hurdle in formulation of effective land use plans. For instance, clearance of proposal to build a water reservoir by irrigation or water supply department in area owned by forest department could take years because of the processes and multi-point checks that the proposal has to undergo. It may not be advisable to create specialized institutions as strengthening of existing structure through sensitizing, imparting skills, raising awareness level and better coordination that would enhance the utility of land use plans prepared with scientific rationale.

25.5 HUMAN RESOURCES DEVELOPMENT

Land use planning is a specialized work and one needs to acquire a set of skills. Unfortunately such highly skilled manpower is not available in the country. Given the high variability in socio-economic conditions and challenges that India faces, expertise in the field of LUP can only be developed in situ. National Bureau of Soil Survey and Land Use Planning can help the nation to develop required manpower. Practical training of land use planners, administrators, researchers, and decision makers is very limited. In fact there is a complete lack of awareness at all the tiers of governance. A revenue officer like District Magistrate or Sub Divisional Magistrate is empowered to allow conversion of land use without any scientific criteria in judging the proposed conversion. He/she is well ill informed about the soil quality or the potential agricultural production that is likely to be lost. Neither there is socio-economic consideration nor an ecological consideration. Thus the immediate concern would be to force a rethink on such policies. The recent opposition to land acquisition bill amendments and draft land use policy (2013 – available at website of Department of Land Records) of the central Government are the welcome changes. The necessity of economic feasibility and social acceptability of land use decisions has been now highlighted. However, these are only small steps.

25.6 NEED FOR A NEW APPROACH

The potential of land for various uses depends on both biophysical and socio-economic conditions. However, as discussed earlier, studies in India and abroad have been restricted to evaluation of soils, terrain and climate for intended and existing land uses. Other factors like socio-economic conditions have been rarely analyzed. Implementation of LUP is not possible without coordination amongst different stake holders, institutional support and legal framework. Because of the land ownership issues and different agencies involved, agricultural scientists in India have confined research to agricultural land use planning and thus, landscape as a unit of planning is under emphasized. Moreover, evaluation of physical environment is also limited. For instance, land use plans are routinely prepared for rainfed agriculture. However, it is essential to incorporate water resources as a critical input and develop land use plans that consider optimum utilization of water resources at different planning scales. The edaphic approach is too sectoral and lacks multi-disciplinarity. In many parts of India, rural livelihood depends upon livestock, fish farming, non timber forest produce and landless population depends upon common property resources available in the village. These facets of LUP have not received much research attention. The Table 25.2 encapsulates research gaps, resultant effects and implications for the future research work and policy decisions.

TABLE 25.2 Research Gaps, Resultant Effects and Implications for the Future Research Work and Policy Decisions

Existing research gap/constraint/problem	Effect/Consequences /Implications	Remedial Suggestions
Data required for land use planning are not available or available at multi scales	Inconsistent land use plans. Decision makers are not confident resulting in sub optimal utilization of resources.	A protocol for data collection to be developed and followed for national, regional and micro level land use planning. Georeferencing of all the data to be made mandatory.
Land use conflicts and concern about losing fertile lands to non-agriculture uses.	Significant loss of agricultural production.	Most suitable lands for high agriculture production to be delineated to facilitate decisions after appropriate evaluation of potential loss.
Landless population is neglected in all the land use plans/studies because LUP is viewed as a process of using land especially agricultural land. The latest *National Sample Survey Organization study (1999) on the role of land, water and forest commons in the life and economy of rural Indians has revealed that CPRs provide as much as 58% of fuel wood requirements and up to 25% of fodder requirements. It also provides evidence of large-scale depletion of CPRs, with CPR lands in rural India declining by almost 2% every five years (Goswami, 2011).*	Major part of the population is left out of the development process. A large majority of over 75 billion rural population of India are dependent on CPRs for their livelihood (Pradhan and Patra, 2011) and yet the issue of land use planning in CPRs has remained neglected mainly due to the protected nature of these resources, where no change of land use is possible (as in case of forest), or the possibility of no modifications in its characteristic (as in case of village ponds, common grazing land).	Encourage policy that aims at protection and optimal utilization of common property resources such as pasture lands, water bodies, community forests, etc. Develop LUPs encompassing growth of non-timber forest produce species on common lands.
Socio-economic factors are ignored with disproportionately higher emphasis on soil resources, their quality and ability to support an agricultural crop.	Low acceptability of land use plans by administrators, decision makers, planners, etc.	Shift to integrated land use policy formulations based on land information, socio-economics and environment concerns.

Available water resources in the planning unit are not taken into account. LUP is confined to rainfed agriculture. Implicit to such planning is an assumption that irrigated agriculturists need not be offered or do not need any land use planning advice and crop choice is entirely decided by amount of water available.	The perspective of raising less water consuming crops for food security is not addressed. Similarly "What if" simulations to facilitate policy decisions such as offering incentives to farmers cultivating less water-intensive crops are lacking.	All future studies, policy analysis to include water resources data as an integral part.
Wide difference between expert opinion and field reality/farmer's opinion about crop choice. For instance most of the land in Yavatmal district of Maharashtra is termed unsuitable to cultivate cotton and yet the existing cotton cropped area exceeds 60%.	The low input, drought resistant cereals like sorghum and millets were fast replaced by cotton and soybean during the last decade. The district is also infamous for cotton farmers' suicides. There was a spread of intra-hirsutum hybrids from 40–45% of the cotton area in 2002 to 92% in 2012, a concomitant decline in the area under *desi* (*G. arboretum* and *G. herbaceum*) and Egyptian (*G. barbadense*) cotton from 31% to 3% during the corresponding period (Venugopalan et al., 2015). Its immediate fall out have been a skewed market surplus of long staple cotton and a huge shortage of both short staple cotton needed for non spinning application and extra long staple cotton needed for premium textiles.	Weaning away the farmers from cotton cultivation especially in land parcels unsuitable for growing cotton crop is a major challenge. It calls for defining area to be covered under each type of cotton annually with targeted production and allocating suitable lands for this purpose. Major cotton producing federal states need to prepare a joint policy to keep the area under cotton crop optimal. A paradigm shift in approach would be required as input costs differ across regions. Development of Decision Support System that simulates different scenarios at national scale would be the first step in formulating land use policy for cotton.
Similar wide difference between expert opinion and field reality/farmer's opinion about crop choice.	A mismatch between suggested land use/crop choice and farmers' choice.	By extension above (cotton related) exercise could also be useful for major cereal crops, pulse crops of the country.

TABLE 25.2 (Continued).

Existing research gap/constraint/problem	Effect/Consequences /Implications	Remedial Suggestions
Acceptability of Land Use Plans.	Because of several factors the land use plans prepared by experts do not find acceptability amongst the planners, policy makers and stakeholders. Existing institutional mechanism does not support or provide legislative backing to the land use plans and hence LUP is by and large academic exercise with limited and sporadic execution.	National priorities for land use and crop choice need to be defined and mechanism to be set up to implement decisions taken in national interest. The local land use plans must be in sync with national land use plans. Guidelines may be issued for preparing for formulation of land use plans that consider national objectives such as food security, livelihood to landless population, etc.
The plans do not indicate economic value of existing and proposed land use or comparison with neighboring unit. For example, existing productivity of cotton in nearby village/district to gauge potential of introducing the crop, or national production so that decision maker can decide if the crop is already in excess or information on market/infrastructure.	Non-acceptability of LUP by stakeholders.	Economic evaluation to be made an integral part of Land Use Planning.
Provisions for supporting livestock is not made in LUP	Decision maker, administrators are not able to take informed decisions.	Information on livestock to be analyzed for better land use plans
Land use plans do not address Bio-diversity or environment protection and ecological services issues.	No information on potential loss or implications resulting in ill informed decisions	Integration of eco-services as an integral part of LUP and economic evaluation of such services to be inducted in all simulations/analyzes. Eco-sensitive zones to be delineated and evaluated for the ecological services they provide.

Negligible research interest in LUP outside NBSS and LUP due to lack of skills and access to required data	Lack of well defined land use policy.	Raising awareness levels, Human resources development for LUP research. Geo-portal will overcome the inadequacies partially by providing soil data. Other data also need to be made accessible.
Lack of linkages with other natural resources organizations	To link the land use priorities of prime agricultural lands with non- agricultural land use.	It will help to develop integrated land use plan for diversified land use types.

25.7 CONCLUSIONS

The review of LUP research in India suggested that there are many limitations. To overcome these, we recommend: (i) Georeferencing of all the relevant data; (ii) Prime lands of the country need to be delineated immediately; (iii) Common property resources, water resources, livestock resources, and socio-economic factors to be made integral part of LUP; (iv) National priorities for land use and crop choice need to be defined and mechanism to be set up to implement decisions taken in national interest. The local land use plans must be in sync with national land use plans. Guidelines may be issued for formulation of land use plans that consider national objectives such as food security, livelihood to landless population, etc.; (v) Integration of eco-services as an integral part of LUP and economic evaluation of such services to be inducted in all simulations/analyzes. Eco-sensitive zones to be delineated and evaluated for the ecological services they provide; and (vi) Raising awareness levels, Human resources development for LUP research.

KEYWORDS

- **Agro-Ecological Regions**
- **Common Property Resources**
- **Decision Support Systems**
- **Geographic Information System**
- **Length of Growing Period**
- **Pedotransfer Functions**
- **Saturated Hydraulic Conductivity**

REFERENCES

Chaturvedi, A., Hajare, T. N., Patil, N. G., Chaturvedi, A., Mungole, A., & Kamble, R. (2015). Land use planning issues in management of common property resources in a backward tribal area. Land Use Policy *42*, 806–812.

FAO. (1993). Guidelines for land use planning, FAO Development series 1. Food and Agriculture Organization of the United Nations, Rome.

Goswami, B. (2011). Dependence of the poor on commons (Common Property Resources CPR). Available at http://infochangeindia.org/agenda/enclosure-of-the-commons/dependence-of-the-poor-on-commons.html

Hajare, T. N, Nair, K. M., & Velayutham, M. (2002). Soil Based Agro Technology Transfer in Sukli (Distt. Nagpur), NBSS Publ. 87, ISBN, 81-85460-64-7.

Hirekerur, L. R., & Sehgal, J. L. (1986) The Soils of Mondha Village (Nagpur) for Agro- technology Transfer, NBSS Publ. 11, 65 p.

http://dolr.nic.in Last accessed March 2015.

Naidu, L. G. K., Dharumarajan, S., Lalitha, M., Srinivas, S., Ramamurthy, V., & Singh, S. K. (2014). Categorization and delineation of prime and marginal lands of Andhra Pradesh for different uses. *Agropedology, 24*(2), 253, 261

Naidu, L. G. K., Ramamurthy, V., Challa O., Hegde, R., & Krishnan, P. (2006) Manual Soil-Site Suitability Criteria for Major Crops NBSS Publication no. 129, 118 pp.

NAIP (2014). www.naip.icar.org.in/…/NBSSLUP.

National Sample Survey Organization (1999). Common Property Resources in India Report No. 452(54/31/4). National Sample Survey Organization. Department of Statistics and Program Implementation, Government of India.

NBSS & LUP (2014). Annual Report, NBSS & LUP, Nagpur.

Paroda, S., & Kumar, P. (2000). Food Production and Demand in South Asia. Agril. Econ. Res. Rev. *13*(1), 1–24.

Patil, N. G., Pal, D K., Mandal, C., & Mandal D. K. (2011). on Describing Soil Water Retention Characteristics of Vertisols and Pedotransfer Functions Based on Nearest Neighbor and Neural Networks Approach to Estimate AWC *Journal of Irrigation and Drainage Engineering, 138*(2), 177–184.

Patil, N. G., Tiwary, P., Bhattacharyya, T., Chandran, P. et al. (2014). Natural resources of the Indo-Gangetic Plains: a land-use planning perspective. *Current Science 107*, 1537–1549.

Patil, N. G., Tiwary, P., Pal, D., Bhattacharya, T., Sarkar, D., Mandal, C., Mandal, D., Chandran, P., Ray, S., Prasad, J., Lokhande, M., & Dongre, V. (2013). Soil Water Retention Characteristics of Black Soils of India and Pedotransfer Functions Using Different Approaches. *Journal of Irrigation and Drainage Engineering, 139*(4), 313–324.

Planning Commission, Govt. of India, http://www.planningcommission.nic.in/reports/genrep/bkpap2020/24_bg2020.pdf.

Pradhan, A. K., & Patra, R. (2011). Common property resources in rural India: dependence, depletion and current status. The IUP Journal of Managerial Economics. *9*(1), 6–20.

Riveira, I. S., & Maseda, R. C. (2006). A review of rural land-use planning models Environment and Planning B: Planning and Design *33*(2), 165–183.

Sehgal, J. (1995). Soil resource mapping and agro-ecological zoning for land use planning. India perspective Indian farming, p. 15–21.

Sehgal, J., Mandal, D.K and Mandal, C. (1995). Agro-Ecological Subregions of India (Map). NBSS and LUP. (ICAR) Publication 24, Nagpur.

Sehgal, J. L., Mandal, D. K., Mandal, C., & Vadivelu, S. (1992). Agro-Ecological Regions of India. NBSS Publ. 24, 1992, 130 p. ISBN: 81-85460-15-9.

Sehgal, J. L., Mandal, D. K., Mandal, C., & Yadav, S. C. (1993). India – Growing Period for Crop Planning. NBSS & LUP Publ. 39, 1993, ISBN: 81-85460-19-1.

Tiwary, P., Patil, N. G., Bhattacharyya, T., et al. (2014). Pedotransfer functions: a tool for estimating hydraulic properties of two major soil types of India. *Current Science 107*, 1431–1439.

Velayutham, M., Mandal, D. K., Mandal, C., & Sehgal, J. (1999). Agro-Ecological Subregions of India for Planning and Development. NBSS & LUP, Publ. No. 35, 372 p.

Venugopalan, M. V., Tiwary, P., Ray, S. K., Patil, N. G., et al. (2014). Info Crop-cotton simulation model – its application in land quality assessment for cotton cultivation. *Current Science, 107*, 1512–1518.

www.planningcommission.nic.in/reports/genrep/index.php (Last accessed March 2015).

CHAPTER 26

LAND RESOURCE INVENTORY TOWARDS VILLAGE LEVEL AGRICULTURAL LAND USE PLANNING

V. RAMAMURTHY,[1] S. K. SINGH,[2] S. CHATTARAJ,[3] G. P. OBI REDDY,[4] and S. C. RAMESH KUMAR[1]

[1]*Principal Scientist, ICAR-National Bureau of Soil Survey and Land Use Planning, Regional Centre, Bangalore-560 024, India*

[2]*Director, ICAR-National Bureau of Soil Survey and Land Use Planning, Amravati Road, Nagpur – 440 033, India*

[3]*Scientist, ICAR-National Bureau of Soil Survey and Land Use Planning, Amravati Road, Nagpur – 440 033, India*

[4]*Principal Scientist, ICAR-National Bureau of Soil Survey and Land Use Planning, Amravati Road, Nagpur – 440 033, India*

CONTENTS

ABSTRACT

Land resource is finite and competing demands for land are infinite. Arable land is shrinking because of diversion of agricultural lands to other non-agricultural uses and agriculture related activities are being taken up on marginal lands. This has resulted in land degradation and land resource management is considered as one of the priority areas for achieving sustainable food security by raising land productivity. The concept of using the land for suitable utilization lies within the land use planning (LUP) process, which aims at optimizing the use of land while sustaining its potential by avoiding resource degradation. It has been recognized that the land assessment and its reliability for land use decisions depend largely on the quality of soil information. Efforts were made to develop regional level land use plans by using land resource information generated at small scale (1:250,000 and 1:50,000). However, the efforts could not yield desired results at village level due to unavailability of large scale land resources database. Stakeholders seldom adopted these land use plans due to lack of site-specific information. Land Resource Inventory (LRI) at large scale (1:10,000) provides required information to prepare sustainable land use plan at village level, which sets the path for using right land use and right agro-techniques on each parcel of land. In India, LUP at local level are governed by farmers own requirement and market prices rather than land suitability criteria alone. LUP aims to encourage and assist land users in selecting options that increase their productivity, are sustainable and meet the needs of society. The systematic evaluation and planning of land resources requires basic data and information about the land, the people and the organization of administration and service. Participatory land use planning (PLUP) approach helps greatly in developing site-specific land resource management options to improve the land productivity and to minimize land degradation.

26.1 INTRODUCTION

In recent years, village level LUP has become widely accepted felt need as a means to resolve land use conflicts and enhance sustainable utilization and management of natural resources. Important natural resources at the village include soils, plants, water, minerals and sunshine. In a predominantly rural society of India, people who own productive lands are considered relatively rich and wealthy; landless people are poor, underfed, and often remain dependent on others for their livelihood. Land is scarce resource in India, even though the country has a land area of about 328 m ha, which is the seventh largest land area in the world. India is burdened with a population of 1210 million as per the 2011 census, which grew from 345 million in 1947 with a growth rate of 1.76 in the last decade. Population density has increased from 117 per sq.km in 1951 to 368 in 2011.

The rapid population increase currently experienced in the country has put land resources under enormous pressure. Competition for land among the different uses is becoming acute and conflicts arising out of this competition are more frequent and more complex. Land is crucial for all developmental activities, for natural resources, ecosystem services and for food security. As a consequence of various developmental endeavors like intensive farming, tourism, development of infrastructural facilities, etc., present arable land is shrinking because of diversion of prime agricultural lands to other non-agricultural uses and agriculture related activities are being taken up on marginal lands. Due to these activities ecological balances is being impaired by soil erosion, siltation of dams, shortage of ground water, land and water pollution, water logging, etc.

During the last two decades Indian agriculture has been facing major challenges like deceleration in growth rate, inter-sectoral and inter-regional inequality, declining input use efficiency, degradation of natural resources, etc., with consequent adverse effects on food and nutritional security and food inflation. Over the past few decades, agriculture has changed from a traditional low-tech and ecologically-benign sector into a modern high-tech industrial sector. In the past, agricultural land use served mainly a monosectoral purpose, *viz.,* the production of foodstuffs in order to meet the multifaceted demand for nutrients in various forms. Agriculture contributes around 14% to India's Gross Domestic Product, but absorbs nearly 60% of the country's working population. About three-fourth of the total population draw their livelihood from agriculture. In recent decades, the industrial development and urbanization are major drivers of economic growth in India. The XII Five Year Plan provides that the country needs to reach an economic growth rate of at least 8% in the next five years in order to significantly increase the quality of life for its citizens, reducing poverty and fostering environmentally sustainable development. The sustainability movement has emphasized the need for environmentally-benign modes of agricultural production, so that the agricultural sector – the biggest land use consumer – would also serve the broadly accepted policy objective of sustainable land use in an ecologically vulnerable world. Consequently, agriculture is becoming a center piece in the worldwide sustainability debate, as food, energy, ecology and land use are concentrated here in one sector.

Therefore, better-directed efforts are needed to preserve our resource *viz.*, land, water and soil biota. In view of these, the LUP process forms an important exercise in this direction (Bauer, 1973). Proper planning of land and its resources allows for rational and sustainable use of land catering to various needs, including social, economic, developmental and environmental needs. Proper LUP based on sound scientific, and technical procedures, and land utilization strategies, supported by participatory approaches empowers people to make decisions on how to appropriately allocate and utilize land and its resources comprehensively and consistently catering to the present and future demands. These decisions depend largely on the quality of land resource information (Bogaert and D'Or, 2002; FAO, 1976; Salehi et al., 2003). Land resource information at large scale provides site-specific information, which is need of the hour to improve the productivity of crops and efficiency of inputs besides conserving precious natural resources.

26.2 STATUS OF LAND RESOURCES IN INDIA

The area available for agriculture, forestry, pasture and other bio mass production in India is 262 m ha and the net sown area is 140 m ha (Table 26.1). The remaining area is not suitable for agriculture due to inaccessibility of the terrain or harsh nature of the climate.

The degradation of land resources is taking place at an alarming rate and not all the cultivated lands at present are highly productive. A number of surveys and assessments have been carried out by different agencies over the past several decades on the extent and type of land degradation occurring in the country. The major surveys/estimates are summarized in Table 26.2. Estimates prepared by different agencies vary considerably from 53 m ha to 239 m ha. (Planning Commission, Tenth Five-Year Plan documents). Sehgal and Abrol (1994) reported that about 57% (187.8 m ha) of the land resources in the country are subjected to different types of degradation and threatening the sustainability of the resource base (Table 26.2).

TABLE 26.1 Land Utilization Pattern in India During 2012–13

S.No.	Land use	Area (m ha)	Area (% to TGA)
1	Total geographical area (TGA)	328.73	-
2	Forest	70.00	22.89
3	Area under non agricultural uses	26.45	8.6
4	Barren and un-cultural land	17.28	5.6
5	Permanent pastures and grazing lands	10.24	3.35
6	Miscellaneous tree crops and groves	3.16	1.03
7	Culturable wasteland	12.58	4.11
8	Old Fallow lands	11.00	3.6
9	Current fallows	15.28	5.0
10	Net sown area	139.93	45.76

Source: Directorate of Economics and Statistics, Ministry of Agriculture (2012–13).

However, recently ICAR and NAAS (2010) reported harmonized estimates on land degradation to be about 120.72 m ha (Table 26.3), which was worked out through reconciliation of datasets from various sources *viz.*, NBSS & LUP, CSWCR&TI, CAZRI, CSSRI, NRSC, FSI, NAAS).

A close look at the present health of the soil and water resources of India reveals a failure of the land use policy. About 120 m ha are threatened by

TABLE 26.2 Estimates of Soil Degradation in India by Various Agencies (area in m ha)

Agency	Estimated area (m ha)	Criteria for delineation
National Commission on Agriculture (1976)	148.09	Based on secondary data
Ministry of Agriculture (1978)	175.00	Based on NCA estimates
Society for Promotion of Wastelands Development (1984)	129.58	Based on secondary collected data
National Remote Sensing Agency (1985)	153.28	Mapping on 1:1 million scale based RS techniques
Ministry of Agriculture (1985)	173.64	Land degradation statistics of the states
Ministry of Agriculture (1994)	107.43	Elimination of duplication of data
NBSS & LUP (ICAR) (1994)	187.70	Mapping of 1:4.4 million scale based on GLASOD guidelines

Source: Gajbhiye and Sohanlal (2006).

TABLE 26.3 Extent of Degraded and Wastelands in India

Degradation type	Area (in m ha)	Open forest (<40% canopy) (mha)
Water erosion	73.27	9.30
Wind erosion	12.40	-
Chemical degradation soils		
Exclusively Salt affected soils	5.44	
Salt affected and water eroded soils	1.20	0.10
Exclusively acidic soils (pH<5.5)	5.09	-
Acidic and water eroded soils	5.72	7.13
Physical degradation		
Mining and industrial waste	0.19	
Water logging	0.88	
Total	104.19	16.53
Grand Total area (Arable land and open forest)	120.72	

Source: ICAR and NAAS (2010).

various types of degradations like salinity, alkalinity, ravine and gully erosion areas, areas under ravages of shifting cultivation, and desertification. The highest proportion of degradation is caused by soil erosion (9.86%). There are also specific problems of land degradation due to open-cast mining operations.

Currently, India produces about 257 m tons of food grains (2014–15), whereas, by the end of the decade, it is estimated that the demand for food shall rise to 307 m tons. Further, LUP in the country has so far not been comprehensive and adequate, particularly to deal the competitive demands by various sectors. Thus, there is a need to protect agricultural areas that are essential for food security including the prime agricultural lands, command areas, double cropped land and other lands that are essential for livelihood of rural population. Thus, conserving prime agricultural lands by proper LUP at different levels assumes mammoth importance.

26.3 NEED OF LRI AT LARGE SCALE

LRI plays a vital role in resource management. It assists in the planning for future land use, particularly agriculture, because it assesses the land resource and its potential for sustainable agricultural production.

The LRI provides two sets of data:

i) LRI – inventory of five physical factors (rock, soil, slope, erosion type and severity, and vegetation) which is the basis of assessing land resources.
ii) Land Use Capability classification (LUC) – evaluation of the potential for sustained agriculture production (land use) in the long term.

In the early years the use of the soil information was mainly focused at national and regional level but in recent years this has expanded to be increasingly at district, block and village. Information about the land resources and its contribution to food production for an ever-growing population, biodiversity, urban and rural infrastructures, is now required at village level. To achieve a quantum jump in agricultural productivity at reduced cost on a sustainable basis and to usher second green revolution in India, natural resources like soil and water need to be efficiently utilized by developing site specific soil management strategies. Soil specific management provides the required input on each soil type and prevents over and under application of inputs resulting from blanket field applications, which are currently followed.

Spatial variation of soil properties causes uneven patterns in soil fertility and crop growth, and decreases the use efficiency of inputs applied uniformly at the field scale (Bhatti et al., 1991; Larson and Robert, 1991; Miller et al., 1988). Application of variable rather than uniform rates of N has been proposed to avoid application of excessive N where it will not be utilized by crops (Carr et al., 1991; Mulla et al., 1992). In order to apply fertilizer at variable rate or management of application, a methodology needs to be developed to divide farmlands into management zones/units that have similar soil characteristics, management responses and climate.

Delineation of spatial variability and mapping at village level is possible at larger scale, i.e., 1:5000 or 1:10,000. Soil series, the lowest category in soil taxonomy, which is considered the most homogeneous unit for management interpretations (Soil Survey Staff 2000). It is the basic units of soil classification as well as mapping at large scale. The pedon represents the sampling unit for soil series. The soils within a series may have similar properties but these may not be identical. Differences in characteristics like slope, texture, stoniness, degree of erosion and other features known as soil phases separate them each other. These have developed on similar parent material and comparable climatic and geo-morphic environments. Since soil series represents comparatively uniform edaphic characteristics of an area occupying larger extent helps in formulating land use plans and for natural resource conservation and management. In this direction, NBSS & LUP and ICRISAT in collaboration tried to identify suitability of Vertisols and associated soils for improved cropping systems in Central India based on 10 benchmark soil series (NBSS & LUP-ICRISAT, 1991). The study suggested possible cropping systems for different benchmark soil series of Vertisols based on soil resource inventory, farmers interviews and with limited field testing. Similarly, Gaikwad et al. (1986) and Yadav et al. (1985) suggested soil-site characters in relation to crop productivity as case studies based at soil order level, i.e., Vertisols. Management strategies suggested at soil order level may not applicable to soil series and within series. Naidu et al. (1986) assessed productivity and potentiality of eight extensively occurring soil series of Delhi. The soil series are grouped into three productivity classes namely good, average and poor on the basis of morphological, physico-chemical and soil environmental factors. Among the parameters considered for productivity and potentiality assessment, texture and soil moisture were found to be pre-dominant factors governing the rating indices. Hence, management strategies need to be developed at phase level by considering site characteristics.

The soil series are useful in developing inter relationship among properties to predict soil qualities and crop management strategies including leaching, run-off potential, drainage, terraces and diversions, grassed water ways, capability classes and sub-classes, crop yield, wind break and rangeland suitability.

Khakural et al. (1992) reported that when soil series are most distinct, then individual tillage system could be developed for specific soil series. Several articles in the literature indicate an increase in profits from applying plant nutrients to soil as compared to general application in the larger area (Bechman, 1992; Richter, 1991). Munson and Runge (1990) suggested that the site specific testing would help in identification of soil sensitivity to nutrient leaching or erosion and run-off.

At North Dakota, a study was carried out to know the response of barley and wheat to fertilizer management and nutrient grid method based on soil series (at phase level). Although grid sampling produced significantly higher yields, the extra cost of soil sampling and management caused it to have the lowest net returns (Wibawa, 1991).

Although India has made much advances in agricultural research, but still the blanket recommendations are very much in practice for adoption over large area. These blanket recommendations are no more useful to enhance productivity gains, which were witnessed between 1965 and 1985. Now, to enhance growth rate in productivity, precision agriculture technologies have to be developed. To realize the agronomic potential of existing soil type without soil degradation or to improve the potential of soils on sustainable basis. For assessing the productivity of soils in an area, one needs to have knowledge on the different soil types and their properties. The research results and predictions can be enhanced using the site-specific information. Land resource information at phase/series level is more important for prioritizing integrated soil management strategies to develop farm/village level LUP for optimizing natural resource, conservation and management.

The concept of using the land for suitable utilization lies within the LUP process, which aims at optimizing the use of land while sustaining its potential by avoiding resource degradation. It has been recognized that the land assessment and its reliability for land use decisions depend largely on the quality of soil information. Efforts were made to develop regional level land use plans by using land resource information generated at small scale (1:250,000 and 1:50,000). However, the efforts could not yield desired results at village level due to unavailability of large-scale (site-specific) land resources database. Stakeholders seldom adopted these land use plans due to lack of site-specific information. LRI at large scale (1:10,000) provides required information to prepare sustainable land use plan at village level, which sets the path for using right land use and right agro-techniques on each parcel of land. Also LRI information could be utilized by government departments to implement various developmental programs in the country (*viz.*, IWMP, RKVY, NHM, MGNREG, etc.).

26.3.1 APPROACH

Soils are normally mapped based on physiography/landform-soil relation. The accuracy of soil map largely depends on how precisely and accurately the physiography/landform units are delineated. Soil maps are prepared at different scale *viz.*, small scale (1:250,000, 1:1M, or smaller), medium scale (1:100,000, 1:63,360, 1:50,000) and large scale (1:25,000, 1:10,000, or larger), depending upon the purpose and requirement of user agencies.

The application of satellite remote sensing data products for small and medium scale of soil mapping are widely accepted (Soil Survey Staff, 1995) but the same have not been used in large-scale soil mapping frequently due to their coarse resolution. Large-scale soil mapping done so far following conventional methods that are time consuming, expensive and have low repetitive value especially in difficult and inaccessible terrain. However, with the availability of high resolution LISS-IV data (better than 6 m) from IRS-P6 satellite and panchromatic mode (2.5 m resolution) from Cartosat-1 IRS satellite, it is now possible to utilize these data for quick and precise LRI on 1: 10,000 scale.

26.3.2 METHODOLOGY

The methodology proposed for LRI on 1:10,000 scale is essentially a six tier approach comprising: (i) generation of orthorectified Cartosat merged LISS-IV using digital terrain database, (ii) landform/physiography analysis based on interpretation of high resolution satellite data, (iii) field characterization of soils for landform-soil relationship and mapping, (iv) laboratory characterization of soil, (v) development of land resource information system, and (vi) Decision Support System (DSS) for LUP (Figure 26.1).

26.3.3 DATASETS REQUIRED

Stereo pair of Cartosat-1 data (2.5 m resolution) provides the opportunity for precise and accurate delineation of landform units as it can be used to generate digital elevation model (DEM), generation of contours (10 m) and deriving information on slope and other terrain features of land. The high resolution IRS LISS-IV data (5.8 m) provides multispectral information about the object which helps in better delineation of land use/land cover, vegetation condition, etc., that may provide indirect inferences about the soils. Keeping this in view the datasets like Stereo pair Cartosat-1 digital data, IRS-P6 LISS IV digital data, Survey of India (SOI) toposheets (1:50,000 scale or larger), cadastral maps, other collateral data *viz.* global spatial datasets, SRM reports, district reports, geology maps, available LULC maps, etc., are proposed.

26.3.4 SATELLITE DATA PROCESSING

***Geo-referencing*:** As the SOI topomaps on 1: 10,000 scales are not available, it is proposed that all the spatial datasets *viz.*, Cartosat-1 data, LISS-IV

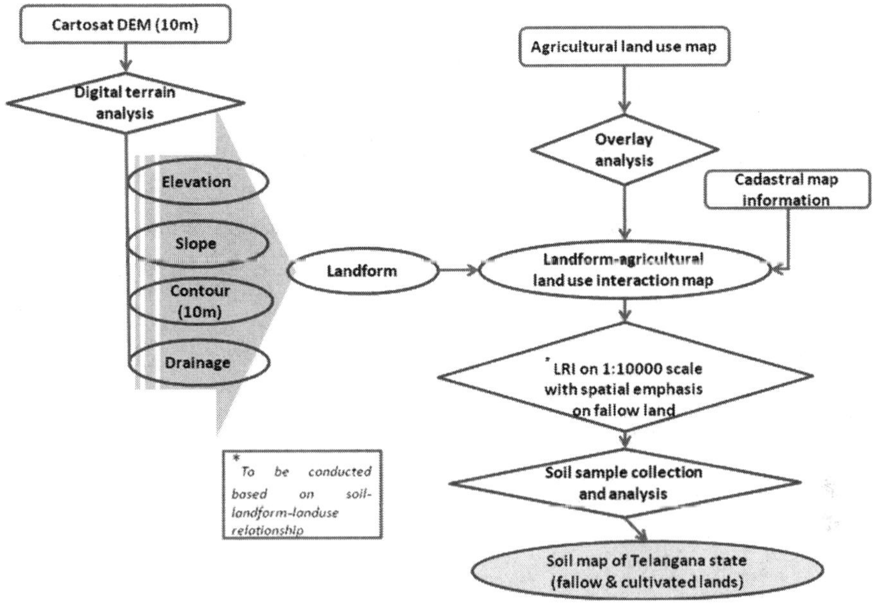

FIGURE 26.1 Land Resource Inventory methodology at 1:10000 scale.

data, Cadastral maps, etc., were georeferenced with the Global datasets using Global coordinate system (GCS) and WGS-84 datum.

Generation of DEM (10 m): It is proposed to use digital elevation model (DEM) of 10 m resolution as it provides better visualization features for precise delineation of landform units on 1:10,000 scale. As the DEM (10m) is not readily available, the same need to be generated from the stereo pair Cartosat-1 data.

***Generation of ortho-rectified Cartosat merged LISS-IV data*:** IRS LISS-IV data (5.8 m resolution) was fused with high resolution Cartosat-1 data (2.5) to generate ortho-rectified cartosat merged LISS-IV data which was used for on-screen digitization of different landform units.

26.3.5 SATELLITE DATA INTERPRETATION

26.3.5.1 Land Use/Land Cover (LULC) Mapping

LULC map of the area was generated based on digital classification of ortho-rectified IRS-P6 LISS-IV data. Normalized difference vegetation index

(NDVI) map was be generated to account for the variation, if any, within same land use/land cover classes.

26.3.5.2 Generation of Digital Terrain Attributes

Cartosat DEM (10 m resolution) was used to generate spatial maps of different terrain attributes *viz.*, slope, aspect, hill shade, terrain wetness index, topographic position index, etc.

26.3.5.3 Landform Mapping

Hybrid approach (involving both visual and digital techniques) was adopted to generate landform maps using orthorectified Cartosat merged LISS-IV data and terrain features derived through Cartosat DEM.

26.3.5.4 Pre-Field Landform Map

The pre-field landform map was prepared by integrating spatial maps of landform, slope, land use/land cover, NDVI, and other relevant maps (e.g., physiographic regions, geology, AESR, etc.). Proposed tentative legend for landform map is shown in Figure 26.2.

26.3.6 FIELD WORK FOR GROUND-TRUTHING

Intensive traversing has been done in the area to verify the pre-field physiographic units and make any correction, if needed. Transects (approx. 8–10) were drawn in such a way that all major physiographic units occurring in the block/tehsil are covered. Intensive observations (profiles/mini-pits/augur bores) were taken in each physiographic unit to establish the soil series and

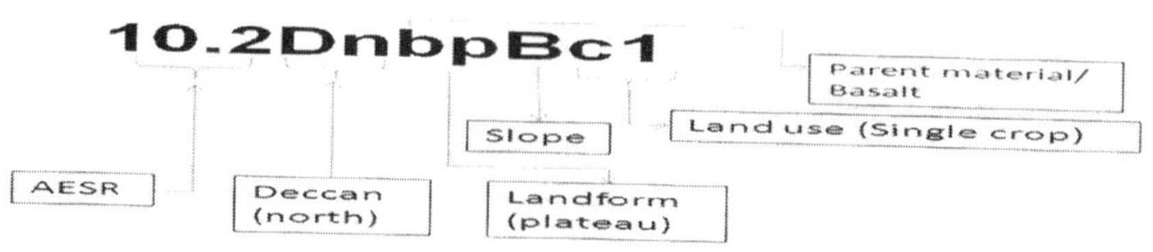

FIGURE 26.2 Interpretation legend for landform unit.

develop landform-soil relationship in the area. Horizon-wise soil samples were collected from the representative soil series for laboratory analysis.

After establishing the landform-soil relationship, detailed field-work was undertaken in each village of the block/tehsil. Soil profiles (approx. 8–10) were studied on all dominant physiographic units of the village and horizon-wise soil samples were collected for detailed laboratory characterization.

26.3.6.1 Soil Mapping

The process of soil mapping begins with the checking of physiographic map units in the field. Simultaneously, soils were studied in different physiographic units for developing correlation between physiographic units and soil composition to ensure correct stipulation of soil of the map units. Soil mapping unit essentially phases of soil series or soil complexes. Soil map generated at village level for Kokarda village, Kalmeshwar thesil of Nagpur, Maharashtra is shown in Figure 26.3. Dominant phases of soil series (e.g., surface texture, slope, erosion, salinity/sodicity, stoniness, flooding, etc.), influencing land use (agricultural productivity) and soil management were shown in each soil mapping unit. Necessary guidelines need to be framed to maintain the quality and accuracy of datasets developed.

26.4 VILLAGE LEVEL LAND USE PLANNING (VLUP)

Village land-use planning is the process of evaluating and proposing alternative uses of natural resources in order to improve the socio-economic conditions of villagers. It is believed that this process only becomes effective when it is carried out in a participatory way, which means that the principal

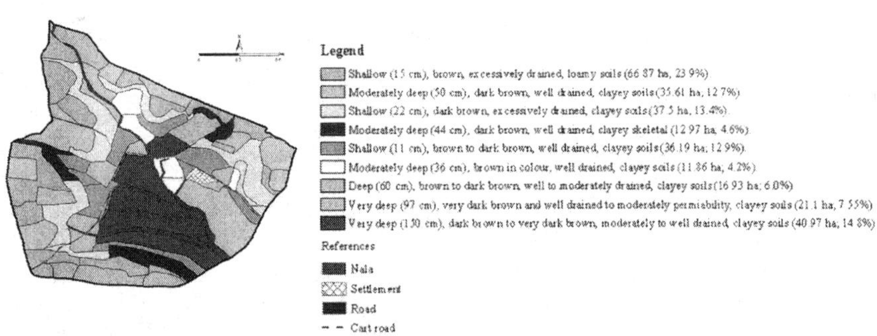

FIGURE 26.3 Soil map of Kokarda Village.

users of land, the villagers, are fully involved. To ensure full participation it is important to consider the different socio-economic groups in a village (including gender) which have different interests and expectations. The optimal use of these natural resources depends mainly on: the potential of people to utilize and manage them; their priorities; the socio-economic conditions and; the carrying capacity of the natural resources.

Objective of LUP is "Land use planning creates the preconditions required to achieve a type of land use that is environmentally sustainable, socially desirable and economically sound. It thereby activates social processes of decision making and consensus building concerning the utilization and protection of private, communal or public areas" (GTZ, 1995).

LUP is an iterative process based on the dialog amongst all stakeholders aiming at the negotiation and decision for a sustainable form of land use in rural areas as well as initiating and monitoring its implementation (GTZ, 1999).

In India, LUP at local level are governed by farmers own requirement and market prices (Velayutham et al., 2001) rather than land suitability criteria (Ramamurthy et al., 2000) which is followed in developed countries. The land use plans suggested by national and state land use boards and research institutes are seldom adopted by local communities/stakeholders. The reasons for non-adaption are the initiatives usually come from government officials or others outside the local community and the techniques, resources and skills suggested rely heavily upon innovations developed at research stations. Moreover, suggested plans developed from soil survey and land capability assessments (Dhanorkar et al., 2013; Patil et al., 2011), focuses upon the relationship between land use and its environmental compliance alone. The socio-economic and political factors at the household, community and national levels, which influence land use, are often neglected. Also, there is a tendency to focus on land use per se and to neglect the details of land management and husbandry. Such land use plans have limited replicability because it involves considerable manpower and technical resources like maps and field staff and farmers do not easily comprehend the technicalities in this top down approach. As a result the implementation of LUP is often difficult.

It has now been endorsed that successful LUP depends on the participation of farmers. Thus, PLUP is considered as an important tool for sustainable resource management by local communities (Amler et al., 1999; Fagerstrom et al., 2003; Oltheten, 1999; Sawathvong, 2003). The aim of PLUP is to strike a balance between technical approach and farmer's requirements to maintain natural resources in sustainable manner. The plan should blend with bio-physical, socio-economic, gender, policy, equity, community participation and institutionalized management of common property resources on a village basis.

26.4.1 PRINCIPLES OF LUP

There are eleven principles of LUP, they are:

1. LUP is orientated to local conditions in terms of both method and content
2. LUP considers cultural viewpoints and builds up on local environmental knowledge
3. Land use planning takes into account traditional strategies for solving problems and conflicts
4. LUP assumes a concept which understands rural development to be a "bottom-up" process based on self-help and self-responsibility.
5. LUP is a dialog, creating the prerequisites for the successful negotiation and cooperation among stakeholders.
6. LUP is a process leading to an improvement in the capacity of the participants to plan and take actions.
7. LUP requires transparency. Therefore, free access to information for all participants is a prerequisite.
8. The differentiation of stakeholders and the gender approach are core principles in land use planning.
9. LUP is based on interdisciplinary cooperation.
10. LUP is an iterative process; it is the flexible and open reaction based on new findings and changing conditions.
11. LUP is implementation-orientated.

26.4.2 PARTICIPATORY LAND USE PLANNING (VILLAGE/WATERSHED/FARM LEVEL)

After several failures of top-down approach, 'bottom-up approach' has become the jargon of implementation of developmental plans in recent days, where farmer is the main player in the development or identification of problem, planning, implementation of interventions and evaluation. The other agencies act as facilitators in the program implementation (Figure 26.4). For the success of such a planning process, the need for changes in land use or action to prevent some unwanted change must be desired and accepted by all the stakeholders and there must be a political will.

PLUP is "an iterative process based on dialog among all participants involved, and aiming to reach decisions on a sustainable form of land use in rural areas. It also includes the initiation of and support to appropriate implementation measures" (GTZ, 1995). In the Indian context, it is essentially a collective consensus on the use of resources.

PLUP should be methodologically and contextually oriented to local conditions and should be built on local environment using indigenous knowledge for problem and conflict management. It is based on the assumption that development is a process brought about 'from bottom' and based on self-help and collective responsibility. It is an interdisciplinary task and is intended to improve the participant's capacity to prioritize, plan, implement and evaluate. This requires transparency of information. The focus should be planning for and by the people.

Three types of land use options are possible through PLUP:

(1) Scientifically optimal, socio-economically tenable and accepted by partners/stake holders;
(2) Socio-economically desirable, scientifically permissible and accepted by partners;
(3) Socio-economically acceptable, scientifically feasible and desired by partners.

Using the above options, PLUP aims to achieve the following objectives:

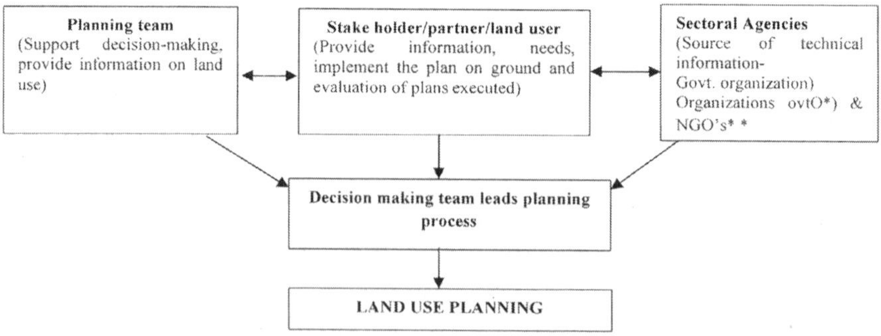

FIGURE 26.4 Players in participatory land use planning and their roles (in parentheses).

- Optimization of production from different farming systems to improve farm income;
- Controlling of soil and water erosion/degradation;
- Restoration of degraded lands to their primary production potential;
- Utilization of wastelands;
- Optimization of use of surface/ground water resources;
- Provision for food, fodder, fiber and fuel security;
- Environment security.

An important aspect of the PLUP process is the participation of villagers in managing and protecting natural resources, participation is key to sustainable development.

26.4.3 INFORMATION REQUIRED FOR VLUP

The local planning unit may be the village/Panchayat/Watershed. At this level, it is easiest to fit the plan to the people, making use of local people's knowledge and contributions. Where planning is initiated at the district level, the program of work to implement changes in land use or management has to be carried out locally. Alternatively, this may be the first level of planning, with its priorities drawn up by the local people. Local-level planning is about getting things done on particular areas of land – what shall be done where and when, and who will be responsible. At local level, the plan is very detailed, and it is possible for all participants to take part directly in the decision-making process. The data sets required for local planning is presented in Table 26.4.

The information to be collected for LUP can be broadly categorized as:

1) Bio-physical information
    - Land use/cover
    - Soil variability
    - Soil fertility status
    - Soil constraints and potential
    - Soil degradation
    - Climatic information
2) Socio-economic information
    - Socio-economic constraints and potential
    - Adoption and yield gap of crops
    - Economics of land use types

TABLE 26.4 Data Set for Village/Village Panchayat/Watershed Level Land Use Plan

Management units	Base data needed				Quality of Plan	Implementing Agency	Goals
	Soil	Climate	Land Use	Social and Economic			
• Iso grow area maps	• Soil series-phases (1:10,000)	Weekly data • Rainfall • PET • Temp	Major LUTS • Rainfed • Irrigated • Forestry • Barren • Flood area • Habitation • Specific crop information like varieties, management, etc.	Farm family wise • Capital availability • Socio-Economic status • Needs	Quantitative (management unit wise crop, variety & management details)	• Dept of Agri./ horticulture • Dept. of Forestry • Village Panchayats • Village level workers • Farming Community	Village/ Farm/ Watershed development

- Farming constraints and potentials
- Stakeholders priorities

3) Environmental information

The required information and data can be sub categorized as:

- Population, demographic trends, migration;
- Actors and institutions;
- Land and other resources, including information on biodiversity, conservation values, ecosystem services, etc.;
- Environment, e.g., availability of water, climate trends, natural hazards;
- Past, present and future land use;
- Production and trends; Infrastructure;
- Social services;
- Topographic references; administrative boundaries.

26.4.4 HOW TO GET THE INFORMATION?

It is assumed that PLUP begins with LRI.

- Information on current land use, water resources, forest resource and soil resources and land degradation could be documented during LRI.
- Information documented during LRI could be cross checked from secondary sources and information on population of the village and other means of livelihood could be either collected from secondary sources such as Panchayat records or line departments.
- Socio-economic information may be collected by visiting the house-holds and conducting structured interview. Data sources could be formal or informal, physical or digital. Questions related to socio-economic information should not be fixed, use Participatory Rural Appraisal (PRA) tools for collecting maximum information.

26.4.5 PROCESS OF VLUP

Understanding the sequence of steps (or the process) in PLUP before attempting the fieldwork is most important. In the PLUP process there are procedures to follow, methods to apply and practices or tasks to undertake. PLUP involves different stages in planning and implementation. The process of PLUP is outlined in Figure 26.5 and stages are described below.

***Stage 1: Preparation for Implementing LUP*:** Prepare implementation teams at local level and conduct training for team members in Participatory Land Use and Management approaches. This will include preparation of survey and mapping equipment and materials. Villagers also need to be well informed, and must have the implementation activities and methods of the LUP process and policies, regulations and objectives explained to them properly.

***Stage 2: Survey and mapping of village boundary and forest and agricultural land use zones/units*:** This necessitates the determination of village boundaries and the preparation of boundary agreements, followed by the drawing of a village base map. Study the land resources of the village and document village landmarks and topographic features to establish village reference points and to identify and map village land-use zones/units.

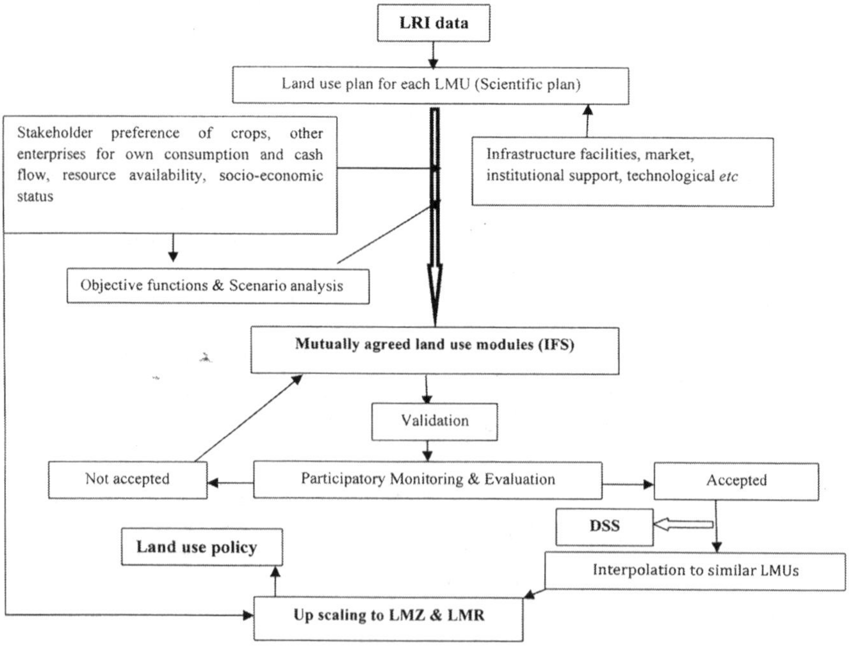

FIGURE 26.5 Conceptual model for village level land use planning.

Stage 3: Data collection and analysis: Information on village land tenure, land use and land claims including traditional village agreements and farming systems need to be collected. In addition, information on socio-economic conditions and villagers' perceived problems and needs are to be documented by employing PRA tools and techniques. This information should then be summarized and analyzed to determine agricultural land allocation criteria.

Stage 4: Village land-use plans: Conduct staff and villager awareness training on the definition, objectives and activities in agricultural land use planning. Evaluate land resources of each farmer, match with farmers needs and prepare mutually agreed land use plans considering resources available locally. Select and demonstrate with participating families suitable land use options based on the above (Figure 26.6). Year-end monitoring should be undertaken in order to facilitate planning and expansion of demonstration activities prior to adopting ongoing land use plans.

Stage 5: Land use and management plan: Use the land-use zoning/unit map prepared in stage 2 to discuss land use management with villagers. It is important to reach agreement on appropriate land uses for each of the land-use zones/unit. It is then necessary to conduct a village meeting to discuss about land management options.

Stage 6: Agricultural land allocation records: When land allocation is completed in each village, a temporary land-use certificate (TLUC) record book needs to be established at the Village panchyat. The book should include all TLUCs cross-referenced, coded and checked against the village land use map.

Stage 7: Monitoring and evaluation: Conduct field monitoring of land-use practices with the village LUP/LA committee and villagers. Make a

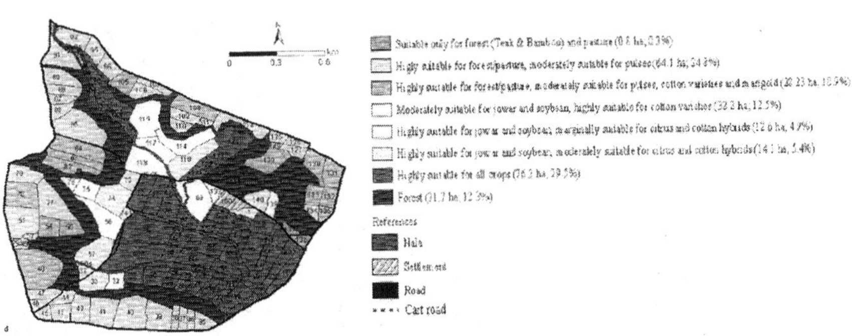

FIGURE 26.6 Land use plan map of Kokarda village.

report and feedback monitoring results to villagers and the LUP/LA committee. Use the results to prepare a follow up activity plan to address issues and problems identified during monitoring work and to improve the whole LUP procedure.

26.4.6 ADVANTAGES OF VLUP

- Local targets, local management and local benefits. People will be more enthusiastic about a plan seen as their own, and they will be more willing to participate in its implementation and monitoring;
- More popular awareness of land-use problems and opportunities;
- Plans can pay close attention to local constraints, whether these are related to natural resources or socio-economic problems;
- Better information is fed upwards for higher levels of planning

28.4.7 DISADVANTAGES OF VLUP

- Local interests are not always the same as regional or national interests;
- Difficulties occur in integrating local plans within a wider framework;
- Limited technical knowledge at the local level means technical agencies need to make a big investment in time and labor in widely scattered places;
- Local efforts may collapse because of a lack of higher-level support or even obstruction.

26.5 IMPACT OF VLUP ON LAND USE AND LIVELIHOOD SECURITY

LRI based LUP was implemented in Kokarda and Kaniyadol villages of Kalmeshwar thesil of Nagpur from 2000–2005 under Institute-Village linkage program (Figure 26.6). The mean annual rainfall of the area is 976 mm received mostly from south-west monsoon, from second fortnight of June to October. About 96% of the total rainfall occurred during June to September in 67 rainy days with length of growing period workout to be about 150–170 days. Delayed onset, early withdrawal and prolonged dry spells, are the common characteristics of the monsoon. These aberrations have a severe impact on the crop production and productivity.

The soils are shallow to moderately deep and deep, well drained/moderately well drained, with gravely clay loam/sandy clay loam soils (Figure 26.3). Shallow soils occur on hill slopes and deep soils in valley bottom. Shallow soils, are under *kharif* monocrops. Deep soils are cultivated to *kharif* and *rabi* crops.

Before implementation of soil based land use plans, about 26% of the area was under sorghum, 22% under cotton, 17% under soybean and 13% of the area was under orange orchards. The small and marginal farmers were mostly the wage earners. The cattle and goat rearing formed an integral part of their livelihood. Dairying was supplementary enterprise to most of the farmers.

After LRI, methodology outlined in Figure 26.5 was followed to prepare and implement LUP of the village. At the end of the project impact assessment was carried out and it was found that socially acceptable, technically feasible and economically viable technologies were by and large, considered sustainable.

- Due to VLUP, area under cultivation increased in the village. The uncultivable land (12%) and current fallows (34%) were brought under cultivation by way of fodder production, afforestation and agro forestry. There was 5% increase in net sown area.
- Changes in land use and cropping pattern were significant due to VLUP. Before the implementation of the VLUP, maximum area was under cotton hybrids and sorghum. After five years, cotton hybrids have shifted to deep soils and varieties to shallow soils. Similarly, sorghum and citrus has been shifted to soybean. This indicates that productivity level and market forces influence the land use rather than soil suitability alone.
- The productivity of dryland and irrigated crops has increased in the range of 14 to 48% in grain crops and 400% in fodder crops.
- The Crop Productivity Index (CPI) of rainfed crops is higher than irrigated crops.
- The income of farmers was increased by about 32% after implementation of the VLUP.
- In a short period, VLUP has been able to motivate large number of farmers to adopt modern farm technology and help them in raising farm production and consequently income.

- The VLUP helped in generating knowledge and providing skills about new production technology among adopted and also other farmers of the area.
- The VLUP provided an excellent mechanism for feedback information for generation of refined, low cost and effective technology and develop interpretive base for soil units mapped.
- The VLUP has generated demand for better services and supplies of inputs and creation of physical facilities essential for raising production and livelihood.

26.6 CONCLUSIONS

Land resource is finite and competing demands for land are infinite. Present arable land is shrinking because of diversion of prime agricultural lands to other non-agricultural uses and agriculture related activities. The concept of using the land for suitable utilization lies within the LUP process, which aims at optimizing the use of land while sustaining its potential by avoiding resource degradation. It has been recognized that the land assessment and its reliability for land use decisions depend largely on the quality of soil information. Efforts were made to develop regional level land use plans by using land resource information generated at small scale (1:250,000 and 1:50,000). However, the efforts could not yield desired results at village level due to unavailability of large-scale land resources database. Stakeholders seldom adopted these land use plans due to lack of site-specific information. LUP at local level are governed by farmers own requirement and market prices rather than land suitability criteria alone. LUP aims to encourage and assist land users in selecting options that increase their productivity, are sustainable and meet the needs of society. The systematic evaluation and planning of land resources requires basic data and information about the land, *viz.*, land use/cover, soil variability, soil fertility status, soil constraints and potential, soil degradation, climatic information, socio-economic constraints and potential, adoption and yield gap of crops, economics of land use types and stakeholders priorities. Proper implementation of VLUPs based on LRI, showed significant change in land use, productivity improvement of crops besides improving economic capability of farmers. LRI at large scale (1:10,000) provides required information to prepare sustainable land use plan at village level, which sets the path for using right land use and right agro-techniques on each parcel of land.

KEYWORDS

- **Integrated Land Use Plan**
- **Land Evaluation**
- **Land Use**
- **Livelihood**
- **Mutually Agreed Plan**
- **Participatory Land Use Planning**
- **Stakeholders**

REFERENCES

Amler, B., Betke, D., Eger, H., Ehrich, C., Hoesle, U., Kohler, A., Kösel, C., Lossau, A. V., Lutz, W., Müller, U., Schwedersky, T., Seidemann, S., Siebert, M., Trux, A., & Zimmermann, W. (1999). Land Use Planning Methods, Strategies and Tools GTZ.

Bauer, K. W. (1973). The use of soils data in regional planning, *Geoderma, 10*, 1–26.

Benchman, T. J. (1992). Farming by soil type cuts costs, boosts profits. *Prairie Farmer* (Jan 7), 12–13.

Bhatti, A. U., Mulla, D. J., & Frazier, B. E. (1991). Estimation of soil properties and wheat yields on complex eroded hills using geostatistics and Thematic Mapper images. *Remote Sensing of Environment*, *37*, 181–191.

Bogaert. P., & D'Or. (2002). Estimating soil properties from thematic soils maps. The Bayesian Maximum Entropy Approach. *Soil Science Society of America Journal, 66*, 1492–1500.

Carr, P. M., Carlson, G. R., Jacobsen, J. S., Nielsen, G. A., & Skogley, E. O. (1991). Farming soil, not fields: A strategy for increasing fertilizer profitability. *Journal of Production Agriculture*, *4*, 57 61.

Dhanorkar, B. A., Koyal, Arti., Mohekar, D. S., Naidu, L. G. K., Reddy, R. S., & Sarkar, D. (2013). Soil resource assessment for crop planning in Medak district, Andhra Pradesh. *Agropedology*, *23*(1), 23–29.

Directorate of Economics & Statistics (2012–13), Ministry of Agriculture, Government of India. http://eands.dacnet.nic.in/ (Accessed on 15th December, 2015).

Fagerstrom, M. H. H., Messing, I., & Wen, Z. M.. (2003). A participatory approach for integrated conservation planning in a small catchment in Loess Plateau, China-Part I. Approach and methods. *Catena*, 54, 255–269.

FAO. (1976). *A Frame work for Land Evaluation.* Food and Agricultural Organization. Soils Bulletin No. 32, Rome. Italy.

FAO. (1993). Guidelines for land use planning. FAO Development Series No.1. Rome.

Gaikwad, S. T., Tamgadge, D. B., & Meshram, R. S. (1986). Landscape and soils relationship and its impact on land use planning. *Journal of Maharashtra Agriculture Universities*, *11*, 141–143.

Gajbhiye and Sohanlal (2006). Land utilization. In Handbook of Agriculture, 5th Edition, Directorate of Information and Publications in Agriculture, ICAR, New Delhi-12 Larson, W. E., & P. C. Robert. (1991). Farming by soil. *In*: *Soil Management for Sustainability.* R. Lal and F. J. Pierce (ed.) Soil Water Conservation Society, Ankey, IA. p. 103–112.

GTZ. (1995). *Landnutzungsplanung. Strategien, Instrumente, Methoden*. Eschborn.

GTZ. (1999). *Land Use Planning. Methods, Strategies and Tools*. Eschborn.

ICAR and NAAS. (2010). Degraded and wastelands of India: Status and spatial distribution. Project report, Indian Council of Agricultural Research (ICAR), New Delhi. p. 158.

Khakural, B. R., Lemme, G. D., Schumacher, T. E., & Lindstrom, M. J. (1992). Tillage systems and landscape position: effects on soil properties. *Soil Tillage Research, 25*, 43–52.

Larson, W. E., & Robert, P. C. (1991). Farming by Soil. *In*: *Soil Management for Sustainability,* Lal, R., F. J. Pierce and I. A. Ankeny (Eds.). Soil Water Conservation Society, Ankeny, IA, pp. 103–112.

Miller, M. P., Singer, M. J., & Nielsen, D. R. (1988). Spatial variability of wheat yield and soil properties on complex hills. *Soil Science Society of American Journal, 52*, 1133–1141.

Mulla, D. J., Bhatti, A. U., Hammond, M. W., & Benson, J. A. (1992). A comparison of winter wheat yield and quality under uniform versus spatially variable fertilizer management. *Agriculture Ecosystem Environment, 38*, 301–311.

Munson, R. D., & Runge, C. F. (1990). *Improving Fertilizer and Chemical Efficiency Through "High Precision Farming*." Centre for International Food and Agric. Policy, Univ. of Minnesota, St. Paul.

Naidu, L. G. K., Verma, K. S., Jain, S. P., Rana, K. P. C., & Sidhu, G. S. (1986). An appraisal of the productivity potential of dominant soils of Delhi Territory. *Journal of Indian Society of Soil Science, 34*, 821–829.

NBSS & LUP-ICRISAT. (1991). The suitabilities of Vertisols and associated soils for improved cropping systems in Central India, NBSS &LUP, Nagpur and ICRISAT, Patancheru, India, 61 pp.

Oltherten, T. M. P. (1999). *Participatory approaches to planning for community forestry. Results and lesson from case study in Asia, Africa and Latin America*. Forest, tree and people program- Forestry department.

Patil, P. L., Vinay, L., & Dasog, G. S. (2011). Land evaluation of Bhanapur micro-watershed in Northern Dry Zone of Karnataka. *Agropedology, 21*(2), 10–16.

Ramamurthy, V., Patil, N. G., Bankar, W. V., Gajbhiye, K. S., & Venugopalan, M. V. (2000). Land use planning: Farmers' perception and priorities–A case study. Paper presented in National Symposium on Perspectives and policies for land use planning, held at NBSS & LUP, 27-30, Dec., 2000

Richter, S. (1991). *Applying plant food by bits and bytes*. Cooperative Partners (March/April), 1–4.

Salehi, M. H., Eghbal, M. K., & Khademi, H. (2003). Comparison of soil variability in a detailed and a reconnaissance soil map in Central Iran. *Geoderma, 111*, 45–46.

Sehgal, J., & Abrol, I. P. (1994). Soil degradation in India: assessment and impact. *In*: *8th International Soil Conservation Conference: Soil and Water Conservation, Challenges and Opportunities,* pp. 15–18.

Soil Survey Staff. (1995). Soil Survey Mannual. Washington DC, USDA.

Soil Survey Staff. (2000). *Soil Survey Manual*, USDA Hand Book No.18 (Indian print).

Swathvong, S. (2003). Participatory land use management planning in biodiversity conservation areas of Lao PDR. Acta-University-Agriculturae- Sueciae-Silvestria, No. 267, 44 pp.

Velayutham, M., Ramamurthy, V., & Venugopalan, M. V. (2001). Agricultural land use planning from theoretical perspectives to participatory action plan in the Indian context. *The Land, 6*(4), 45–60.

Wibawa, W. D. (1991). *Variable Rate of Fertilization Based on Yield Goal, Soil Fertility and Soil Series*. Unpublished MS thesis, North Dakota State University Library, Fargo.

Yadav, S. C., Mishra, M. K., Choube, S. D., & Saxena, R. K. (1985). Crop response under different fertility levels on various soil units. *JNKVV Research Journal, 19*, 93.

CHAPTER 27

A DECADE OF Bt COTTON IN INDIA: LAND USE CHANGES AND OTHER SOCIO-ECONOMIC CONSEQUENCES

M. V. VENUGOPALAN,[1] A. R. REDDY,[2] K. R. KRANTHI,[3] M. S. YADAV,[4] VANDANA SATISH,[5] and DHANASHREE PABLE[6]

[1]*Principal Scientist, ICAR-Central Institute for Cotton Research, Nagpur – 400 010, India*

[2]*Principal Scientist, ICAR-Agricultural Technology Application Research Institute, Zone V, Hyderabad – 500 059, India*

[3]*Director, ICAR-Central Institute for Cotton Research, Nagpur – 400 010, India*

[4]*Chief Technical Officer, ICAR-Central Institute for Cotton Research, Nagpur – 400 010, India*

[5]*Senior Technical Assistant, ICAR-Central Institute for Cotton Research, Nagpur – 400 010, India*

[6]*Senior Research Fellow, ICAR-Central Institute for Cotton Research, Nagpur – 400 010, India*

CONTENTS

ABSTRACT

Bt hybrids were commercially released in India in 2002 and within ten years time their area increased to 10.4 m ha, occupying over 90% of the cotton growing area. Bt hybrids along with associated agro-techniques propelled India to become the second largest producer (390.9 lakh bales), consumer (306 lakh bales) and the largest exporter (118 lakh bales) of raw cotton in 2013. The land use changes accompanying Bt-cotton revolution is the unprecedented increase in cotton acreage from 7.8 m ha in 2002 to 12.6 m ha in 2014. Other structural changes include the spread of intra-*hirsutum* hybrids to 92% in 2012, a concomitant decline in the area of Diploid and Egyptian cotton from 31% to 3%. This caused a glut of long staple cotton and a shortage of both short staple cotton and extra long staple cotton. The gain in area under cotton in the post-Bt era is at the expense of pulses, oilseeds and coarse cereals that are vital for our food security. The attractive price of cotton and the relatively lower risks with Bt cotton encourage farmers to cultivate hybrid cotton even in soils and climates that are not ideal for cotton. Other consequences include the emergence of a vibrant private sector seed industry, switch over to mono/sole cotton system; re-emergence of cotton leaf curl virus disease in north India, resurgence of secondary pests, etc. Consequently, the average productivity has plateaued between 500 and 560 kg lint/ha during the last five years but the production costs have escalated. Decrease in raw cotton exports that lead to a huge surplus in the national cotton stocks and the continued supply glut is unlikely to push cotton prices upwards in the near future. Therefore, future land use policies including cotton policy need to be thoughtfully formulated keeping in view the past and emerging trends.

27.1 INTRODUCTION

Cotton is the most important cash crop of India and has immense potential to sustain the livelihood of farmers and several other stakeholders along the cotton value chain. About 37% of the world's cotton area is in India and it contributes about 25% to the global cotton production. India is the only county to grow all the four cultivated species of cotton *viz., Gossypium hirsutum* (American upland cotton), *G. arboreum* (diploid, Asiatic cotton), *G. herbaceum* (Levant cotton) and *G. barbadense* (Egyptian cotton). Currently, Intra – *hirsutum* hybrids occupy more than 90% of the cotton area and all the current hybrids are Bollgard II Bt hybrids.

27.1.1 Bt COTTON IN GLOBAL SCENARIO

Genetically modified (GM) crops are engineered to confer one or both of the two traits *viz.* herbicide tolerance and insect resistance (Bt). Four crops *viz.,* soybean (50%), corn (30%), cotton (14%) and canola (5%) account for 99% of the global GM crop area. In 2014, GM cotton was planted in 25.1 m ha, which accounts for 68% of the 37 m ha of the global cotton planted. Bt cotton is essentially an operational yield enhancing or more appropriately, a yield protecting technology, intended to protect the cotton crop against the damage by bollworms. Bt cotton was first introduced in USA in 1996. Subsequently, it was also introduced into other countries like Australia (1996), Argentina (1997), China (1997), Mexico (1998), South Africa (1998), Colombia (2002) and then into India in 2002. Today it is globally cultivated in around 25 mha, about 40% of which is in India. China, Pakistan and USA are other major adopters of Bt transgenic cotton (James, 2012).

27.1.2 Bt COTTON IN INDIAN SCENARIO

Genetically modified Bt cotton technology is unarguably the fastest of all technologies adopted by Indian cotton farmers. The area under Bt cotton reached 10.8 m ha in 2012, a phenomenal increase from the area planted in 2002, when the technology was first commercialized in India. This period of exponential adoption of Bt cotton can be divided into 2 distinct phases. During the first phase from 2002–2007 corresponding to the 10th Five-year plan, the area under Bt cotton increased from 0.3 to 62 lakh ha, and the productivity of cotton in India increased steadily from 302 to 554 kg lint/ha. During the second phase 2008–2012, corresponding to the 11th plan the coverage under Bt cotton steadily rose to 108 lakh ha, 93% of total cotton area of 116 lakh ha, but there was no further gain in productivity.

Unlike other countries, Bt gene is embedded in hybrids in India. Several Bt hybrids, predominantly intra-*hirsutum* and a few *Gossypium hirsutum* x *Gossypium barbadense* have been released by the Genetic Engineering Approval Committee (GEAC), Ministry of Environment and Forests, Government of India (Choudhary and Gaur, 2015). These hybrids belong to six Bt cotton events- Mon 531 (*cry*1*Ac*), Mon 15985 (*cry*1*Ac*+*cry*2*Ab2*), Event 1 (*Cry*1Ac), GFM event (fusion of *cry*1*Ab*+*cry*1*Ac*), BNLA 601 (*cry*1*Ac*) and MLS 9124 (*cry*1*Ac*). During 2012–13, 90% of the Bt cotton area was planted with Bollgard II (*cry*1Ac+*cry*2Ab2, Mon 15985).

Bt hybrids along with associated agro-techniques allowed India to expand cotton acreage thereby emerging as the second largest producer (390.9 lakh bales), consumer (306 lakh bales) and the largest exporter (118 lakh bales) of raw cotton in 2013–14. The production of cotton seed oil, an important by-product also increased substantially. The most obvious land use change accompanying this technological revolution is the unprecedented increase in cotton acreage from 7.8 m ha in 2002 to 12.2 m ha in 2011 and further to 12.6 m ha in 2014, a quantum jump of 62%. Other structural changes in the cotton scenario include the spread of intra-*hirsutum* hybrids from 40–45% of the cotton area in 2002 to 92% in 2012, a concomitant decline in the area under diploid (*G. arboreum* and *G. herbaceum*) and Egyptian (*G. barbadense*) cotton from 31% to 3% during the corresponding period. Its immediate fall out has been a skewed market surplus of long staple cotton and a huge shortage of both short staple cotton needed for non-spinning applications and extra long staple cotton needed for premium textiles. The gain in area under cotton in the post Bt era is at the expense of groundnut, soybean, sorghum and pearl millet in Maharashtra; sunflower, groundnut and sorghum in Andhra Pradesh; pearl millet, sorghum, pigeon pea and groundnut in Gujarat, coarse cereals in South Rajasthan and finger millet in South Karnataka. The high Minimum Support Price (MSP) of cotton compared to its competing crops and the low risks associated with Bt cotton are luring farmers to cultivate hybrid cotton beyond its preferred niches. Consequently, the average productivity hovered between 500 and 568 kg/ha during the last 5 years. The implications of these land use changes on profitability of cotton production and the long-term impact on food security need introspection. The subsequent sections discuss these aspects in detail.

27.2 CHANGES IN AREA UNDER COTTON AFTER THE ADOPTION OF Bt HYBRIDS

Transgenic cotton was first commercialized globally in 1996 and India adopted it for commercial cultivation six years later in 2002. Only 3 hybrids, Bt MECH 12, Bt MECH 162 and Bt MECH 184 developed by Maharashtra Hybrid Seed Company (MAHYCO) containing *cry1Ac* gene were planted in 29,307 ha (0.5% of the cotton area) in 2002. The GEAC approved the cultivation of Bt cotton in the cotton growing states of Maharashtra, Madhya Pradesh and Gujarat in Central zone and Tamilnadu, Karnataka and Andhra Pradesh in South zone. In the year 2005 the approval was further extended to the states of Punjab, Haryana and Rajasthan in North zone but commercial sowing of cotton in this zone began in 2006. In the same year, the first two-gene event MON15985, commonly known as Bollgard II® sourced from Monsanto containing two genes *cry1Ac* and *cry2Ab*, was developed by MAHYCO and approved for commercial cultivation. By the 6th year, in 2007, Bt cotton was planted in 3.2 m ha, registering a 210 fold increase (Figure 27.1). At the end of the first decade, in the year 2012, 10.8 m ha or 93% of the cotton area was planted under Bt cotton. The area further increased to 11.6 m ha in 2013 (James, 2014). In the states of Punjab (96%), Haryana (95%), Maharashtra (91%), Madhya Pradesh (92%) and Andhra Pradesh (98%), the adoption of Bt cotton hybrids exceeded 90%. Thus, the technology is claimed to be widely accepted among farmers of India.

Another parallel revolution accompanying the adoption of Bt cotton was a steady increase in area sown under cotton from 7.8 m ha in 2002 to 12.2 m ha in 2012 (Figure 27.1). The area further increased to 12.7 m ha in 2014. Further introspection reveals that the average annual growth rate in area under cotton, which was 0.57% during the 10th plan period (2002–03 to 2006–07) corresponding to the first phase of a decade of Bt cotton rose to 5.97% during the 11th plan period (2007–08 to 2011–12) corresponding

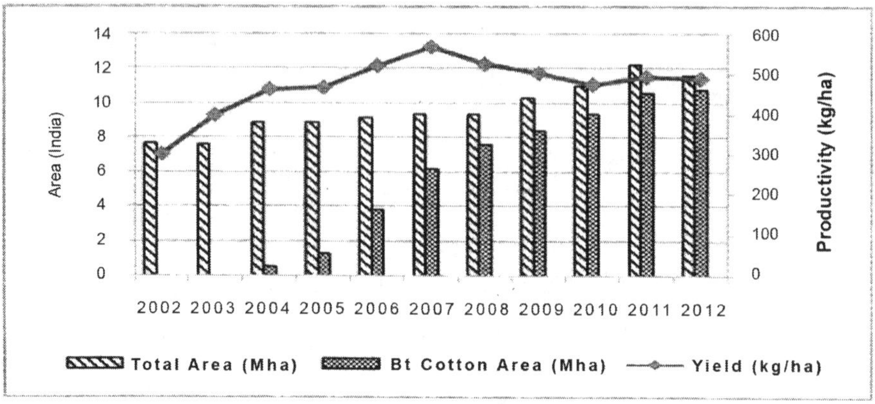

FIGURE 27.1 Area under cotton and Bt cotton hybrids along with productivity of cotton in India during the period 2002 to 2012 (Source: ISAAA, 2014).

to the second phase of Bt cotton in India (Figure 27.2). Only four crops-soybean (4%), pigeon pea (3.1%), chickpea (2.3%) and maize (2.2%), registered a growth rate of over 2% during the 11th plan period (2007–08 to 2011–12). The share of cotton to gross cropped area increased from 4.1% in 2002 to 6.0% in 2012 making it the third most important crop in terms of area after rice and wheat.

The gain in area under cotton has not been uniform across the 10 cotton growing states. Major gains occurred in the states of Maharashtra, Gujarat and Andhra Pradesh (Figure 27.3). The share of Maharashtra, Gujarat and

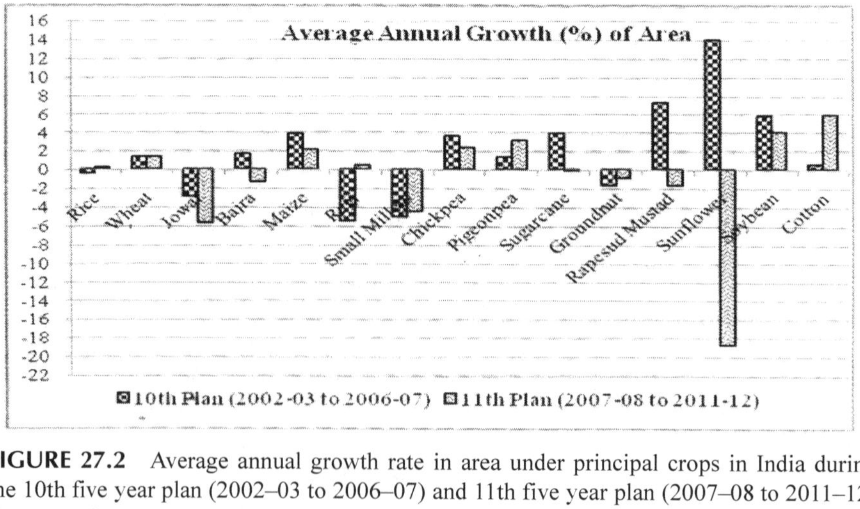

FIGURE 27.2 Average annual growth rate in area under principal crops in India during the 10th five year plan (2002–03 to 2006–07) and 11th five year plan (2007–08 to 2011–12) (Source: Directorate of Economics & Statistics, MoA).

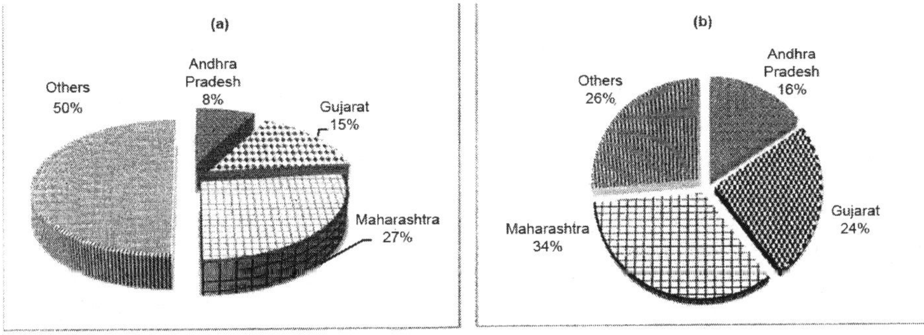

FIGURE 27.3 Distribution of area under cotton in India among important states (a) 2002 (b) 2011.

Andhra Pradesh to the national cotton area increased from 27%, to 34%, 15% to 24% and 8% to 16%, respectively. In 2011, ten years after the introduction of Bt cotton, these 3 states collectively accounted for 75%[th] of the cotton area as against 50% in 2002. The combined area under cotton in the remaining states also increased from 24.6 lakh ha to 31.66 lakh ha.

27.3 LAND USE CHANGES AFTER THE INTRODUCTION OF Bt COTTON

According to Food and Agriculture Organization (FAO), land use is characterized by arrangements, activities and inputs people undertake in a certain land cover type to produce, change or maintain it. A change in the agricultural land use indicates a change in the proportion of area under different crops at different time intervals. All the changes in cropping pattern are essentially changes in agricultural land use. The cropping pattern in several states has undergone significant changes after the introduction of Bt cotton hybrids and a general trend indicates a shift in the area allocated to cereals, pulses and oilseeds in favor of cotton to varying degrees in different states. The total area under sorghum registered a 51% decline during the last two decades, primarily at the expense of soybean in the first decade and cotton in the next.

Significant gain in the area under cotton was observed in the states of Andhra Pradesh, Gujarat and Maharashtra. Hence a detailed analysis of these states along with that of Madhya Pradesh was done to understand the land use (cropping pattern) changes in these states following the introduction of

Bt hybrids. To understand the relationship between the area under cotton and the area of other crops, a correlation analysis was carried out. Crops grown in the same season with highest negative correlation coefficients were considered as competing crops to cotton. Area changes and compounds growth rates were also worked out to understand the changes occurring in the area under cotton as well as in the area of competing crops. Significant gain in the area under cotton was observed in the states of Andhra Pradesh, Gujarat and Maharashtra. Hence, a detailed analysis of these states along with that of Madhya Pradesh was done to understand the land use (cropping pattern) changes in these states following the introduction of Bt hybrids. Data from the Directorate of Economics and Statistics, Department of Agriculture and Cooperation, Ministry of Agriculture, Government of India (eands.dacnet.nic.in) for the period 2002 to 2011 was used for the analysis.

27.3.1 ANDHRA PRADESH

In Andhra Pradesh, there was an inverse relationship between the area under cotton and the area under sunflower, jute & mesta, finger millet, niger, linseed, sorghum, pearl millet. Among these crops, jute and mesta need moist climate and are not grown in areas where cotton is grown and hence do not compete with cotton. The area of sugar cane, sesamum and groundnut was also negatively correlated with cotton area. In Andhra Pradesh, cotton area increased by 10.75 lakh ha with a compound growth rate of 9.83% per annum during the first decade of adoption of Bt cotton. The increase in area is mainly due to the extension of cotton area to non-traditional region of Telangana (Maharana et al., 2011). During the same period all the above mentioned crops registered a negative growth in area (Table 27.1). Highest negative growth was observed in the area of linseed followed by sunflower, jute & mesta and pearl millet. The highest decline in area was observed in groundnut followed by sorghum and sunflower. The correlation coefficient of groundnut with cotton area is lowest. But the area of sorghum and sunflower has high negative correlation with cotton area indicating the possibility of cotton replacing these two crops (Figure 27.4(a)). Finger millet, niger, linseed and pearl millet are the other crops which lost most of their area in favor of cotton. Reddy (2011) also observed a diversion in the area allotted to coarse cereals and pulses (green gram and black gram) towards cotton and other high value crops in Andhra Pradesh.

TABLE 27.1 Change in Area, Crop Growth Rate and Correlation of Cotton Area with Area of Other Crops in Andhra Pradesh During the Post-Bt Era

Crop	Area (000 ha)			Compound growth rate (CGR)	Correlation coefficient
	2002	2011	Change		
Cotton	803.3	1879.0	1075.70	9.83	
Sunflower	127.0	17.0	–110.00	–16.72	–0.93
Jute & Mesta	81.0	26.0	–55.00	–12.16	–0.90
Finger millet	71.0	42.0	–29.00	–7.07	–0.89
Niger	15.0	7.0	–8.00	–8.63	–0.85
Linseed	4.0	0.0	–4.00	–44.82	–0.83
Sorghum	264.0	103.0	–161.00	–9.51	–0.80
Pearl millet	87.0	43.0	–44.00	–10.48	–0.66
Sugar cane	232.5	204.0	–28.50	–2.24	–0.58
Sesamum	115.0	72.0	–43.00	–6.65	–0.40
Groundnut	1271.0	1057.0	–214.00	–2.05	–0.24

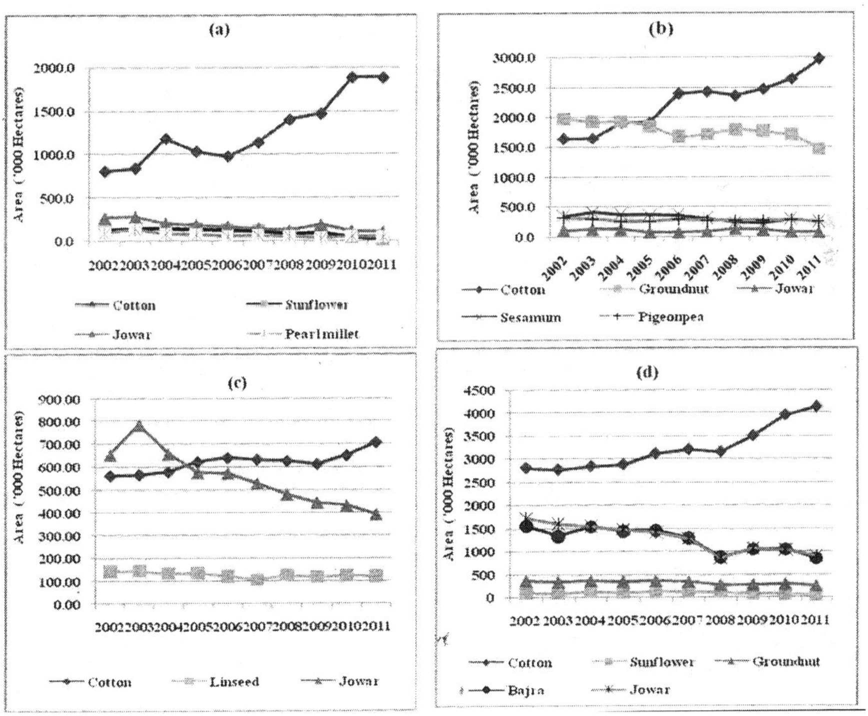

FIGURE 27.4 Trend in area under cotton and competing crops during the decade after the introduction of Bt hybrids (2002 to 2011) in the states of (a) Andhra Pradesh (b) Gujarat (c) Madhya Pradesh and (d) Maharashtra.

27.3.2 GUJARAT

In Gujarat there was a high negative correlation between the area under cotton with the area of groundnut, sesamum, finger millet and maize (Table 27.2). The area of pigeon pea and sorghum was also negatively correlated with cotton area. The area under cotton increased by 13.27 lakh ha with a compound growth rate of 6.63% per annum during post Bt era. Earlier Mehta (2012) reported that during the post reform period, about 12.2 lakh ha of area has been diverted from other crops mainly coarse cereals (pear millet, sorghum) and pulses towards cotton. Ramandeep Kaur and Kiran (2015) also reported that between 2000 and 2011, the area under cotton in Gujarat increased by 85.9%. Groundnut was the main crop which showed the highest decrease in area during the same period and has high negative correlation with cotton area. Hence, most of the groundnut area in Gujarat was replaced by cotton (Figure 27.4(b)). Sesamum, finger millet and maize were other important crops which have negative growth in area and have high negative correlation with cotton. Probably these crops were also replaced by cotton in Gujarat. The gain in area under cotton was primarily in the traditional groundnut belt of the Saurashtra region.

27.3.3 MADHYA PRADESH

In Madhya Pradesh the area under cotton increased from 5.59 lakh ha in 2002 to 7.06 lakh ha in 2011 with a growth rate of 2.08% per annum (Table 27.3).

TABLE 27.2 Change in Area, Crop Growth Rate and Correlation of Cotton Area with Area of Other Crops in Gujarat During the Post-Bt Era

	Area ('000 ha)				
Crop	**2002**	**2011**	**Change**	**CGR**	**Correlation coefficient**
Cotton	1634.8	2962.0	1327.2	6.63	
Groundnut	1972.6	1454.0	–518.6	–2.48	–0.94
Sesamum	344.1	247.0	–97.1	–5.45	–0.77
Finger millet	24.6	16.0	–8.6	–4.39	–0.72
Maize	464.5	387.0	–77.5	–2.32	–0.66
Pigeon	313.0	244.0	–69	–1.47	–0.55
Sorghum	104.4	68.0	–36.4	–3.37	–0.50

TABLE 27.3 Change in Area, Crop Growth Rate and Correlation of Cotton Area With Area of Other Crops in Madhya Pradesh During the Post-Bt Era

Crop	Area ('000 ha)				Correlation coefficient
	2002	2011	Change	CGR	
Cotton	559.3	706.0	146.7	2.08	
Finger millet	0.6	0.3	–0.3	–5.09	–0.88
Sunflower	0.7	0.1	–0.6	–17.15	–0.86
Sorghum	651.8	390.9	–260.9	–6.66	–0.83
Linseed	142.5	120.3	–22.2	–2.09	–0.64

In this state the area of finger millet, sunflower, sorghum and linseed showed high negative correlation with the area under cotton during the post Bt era but the area under finger millet and sunflower was negligible. Among these crops the area under sorghum declined drastically by 2.61 lakh ha while the area of linseed decreased by 0.22 lakh ha. It is evident that the gain in area under cotton was at the expense of a decline in area under sorghum and linseed (Figure 27.4(c)).

27.3.4 MAHARASHTRA

In Maharashtra an additional 13.25 lakh ha area was diverted to cotton area during the period from 2002 to 2011 (Table 27.4). During this period cotton area showed a positive growth rate of 4.53% per annum. Sorghum showed highest decrease in area followed by pearl millet, groundnut, sesamum and sunflower. The area under these crops showed high negative correlation with the area under cotton during the post Bt era. Hence most of the area of these crops was diverted to cotton (Figure 27.4(d)). The other crops whose area showed negative growth and high negative correlation with cotton area include finger millet, linseed and castor. Additionally, during the period, 2007–2010, cotton has replaced soybean crop in Vidarbha region of Maharashtra (Deokate and Bandgar, 2013). Sabesh et al., 2014a established a trade-off of area under oilseeds and pulses with cotton. They also observed that there was a change in the cropping pattern in Marathwada and Khandesh regions of Maharashtra in favor of cotton.

Changes in cropping pattern after the introduction of Bt cotton is also evident in other cotton growing states. In Karnataka, cotton is spreading

TABLE 27.4 Change in Area, Crop Growth Rate and Correlation of Cotton Area With Area of Other Crops in Maharashtra During the Post-Bt Era

Crop	Area ('000 ha)				Correlation coefficient
	2002	2011	Change	CGR	
Cotton	2800	4125	1325.0	4.53	
Sesamum	115	46	–69.0	–10.63	–0.81
Sunflower	97	42	–55.0	–6.71	–0.80
Groundnut	351	238	–113.0	–4.14	–0.80
Sorghum	1718	896	–822.0	–7.27	–0.79
Pearl millet	1547	838	–709.0	–6.36	–0.78
Finger millet	146	130	–16.0	–2.22	–0.74
Linseed	54	31	–23.0	–6.79	–0.70
Castor	30	8	–22.0	–12.70	–0.66
Niger	54	37	–17.0	–6.28	–0.46

to non-traditional areas, into the southern districts of Mysore, Shivamogga, Chamarajanagar and Davenagere (Pavithra and Kunnal, 2013) at the expense of sorghum and finger millet. In Rajasthan cotton area is increasing in the southern districts of Ajmer, Nagaur, Pali and Jodhpur whereas the area in the traditional cotton belt of Sriganganagar and Hanumangarh is declining. The latter two districts used to contribute about 85% of state's cotton production. But during 2012 season, about 50% of the total production of cotton in Rajasthan was from the southern districts.

Subtle changes in the area under cotton cultivation are also evident in Pakistan, where cotton is exclusively grown under irrigated conditions. In the districts of Vehari, Multan, Bahawalpur (cotton zone) and Jhang, TTS, Faisalabad (central zone), the area under Bt cotton increased at the expense of wheat and sugarcane. Majority of the small farmers (70%) left their land as fallow instead of cultivating wheat in order to grow Bt cotton in the months of February-March (Sabir et al., 2011).

India today has an exportable surplus of cotton and export of raw cotton is not a lasting solution. It is also currently self sufficient in cereals but is dependent on huge imports for meeting the domestic requirement of pulses and oilseeds. India imports 3.0 to 3.5 m tones of pulses annually. During 2011–12, the import was 3.3 m tons valued at Rs 8767 crore. During the same year, India imported 8.4 m tons of edible oil, almost half of its domestic requirement, at a cost of Rs. 46242 cores. By the end of 2016–17, the demand for edible oil is projected at 16.64, m tons and this would require a production of 59 m tons of oilseed crops (at the current crop composition). The demand for coarse cereals (sorghum and millets) for nutritional security and for providing dry folder for cattle is immense. During the decade of introduction of Bt cotton, the reduced risk (of bollworms) and attractive prices have increased the area under cotton, disproportionately to the demand, at the expense of coarse cereals, pulses and oilseeds.

27.4 CHANGE IN SPECIES COMPOSITION

India is the only country which cultivates all the four cultivable species of cotton *viz.*, *G. hirsutum*, *G. arboreum*, *G. herbaceum* and *G. barbadense* on a commercial scale. Asiatic cotton belonging to the diploid *G. arboreum* and *G. herbaceum* are inherently resistant to several biotic and abiotic stresses. At the dawn of India's independence in 1947, 97% of the cotton area was under *desi* varieties and only 3% was under *G. hirsutum* varieties. To meet the requirement of the domestic textile mills during the 1950's and 1960's there was a conscious effort to partly replace *desi* cotton with *G. hirsutum* cotton. Introduction of hybrids cotton in 1970 hastened this process (Figure 27.5) and introduction of Bt hybrids further accelerated the pace of replacement of *desi* cotton specially from Punjab, Haryana and Rajasthan of North zone and Madhya Pradesh in the Central zone. The area under *G. hirsutum* cotton increased post independence to 41% by 1965 and to 75% by 2000 (of which 35% was under varieties and 40% was under hybrids). By 2010, more than

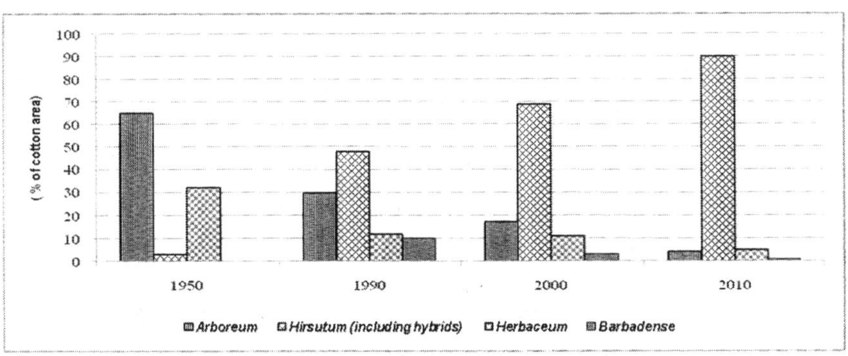

FIGURE 27.5 Relative composition of various species of cotton grown in India.

90% of the area was covered by intra- *hirsutum* Bt hybrids (Ramasundaram et al., 2011). It is also note worthy to note that during 1996, about 45% of the cotton area was under hybrid cotton but majority of this was occupied by hybrids from the public sector. In 2004, hybrids were grown in 6 m ha, (two third of the cotton area) of which 5 m ha (83%) were sown to propriety hybrids of the private sector (Murugkar et al., 2007). Popular varieties from the public sector including F 1378, LH 1556, F 846, F1054, H 1098, RST 9, RS 875, LRA 5166, LRK 516, PKV 081, PKV Rajat, MCU 5VT, MCU 5, LK 861, Surabhi, Sahana, G. cot 11, G cot 13 (*G. hirsutum*) and LD 327, HD 107, RG 8, G 27, PA 255, AKA 7, AKA 8, AKA 840, Maljari (*G. arboreum*) to cite a few have become almost extinct following the introduction of Bt cotton hybrids (Ramasundaram 2013). Around 1167 Bt hybrids were released by the private sector and over 98% of these are based on Monsantos Cry gene events (James, 2014).

27.5 CHANGES IN FIBER QUALITY PROFILE: POST-Bt ERA

During 1997, the share of medium long staple cotton (26 mm to 32 mm) to the total fibers produced was 64.3% (Kranthi et al., 2012). Adoption of Bt cotton hybrids shifted the balance in favor of long staple cotton (27.5 to 32 mm) and by 2011–12, 77% of the cotton produced was under this category as against 38% of the total cotton in 2002. This shift has been at the expense of short staple (*desi* cotton) and Extra long staple cotton. The production of long staple cotton has increased more than five times during the period 2002–03 to 2011–12 (Singh et al., 2013). The production of short staple cotton has been witnessing a steady decline during the last decade from around 9.5 lakh bales in 2001–03 to 6.0 lakh bales in 2006–07 and further to 4.0 lakh bales in 2011–12 (Figure 27.6). The annual requirement of short staple cotton, a trait unique to Asiatic cotton varieties, is around 10–15%. It is worth mentioning that surgical cotton is an important end use product of short staple, coarse cotton.

27.6 CHANGES IN IMPORT-EXPORT SCENARIO

During the period from mid 1990's to the first half of 2000's, domestic consumption of cotton out-stripped the production warranting high imports. During the period 2002 to 2011 the consumption of cotton continued to

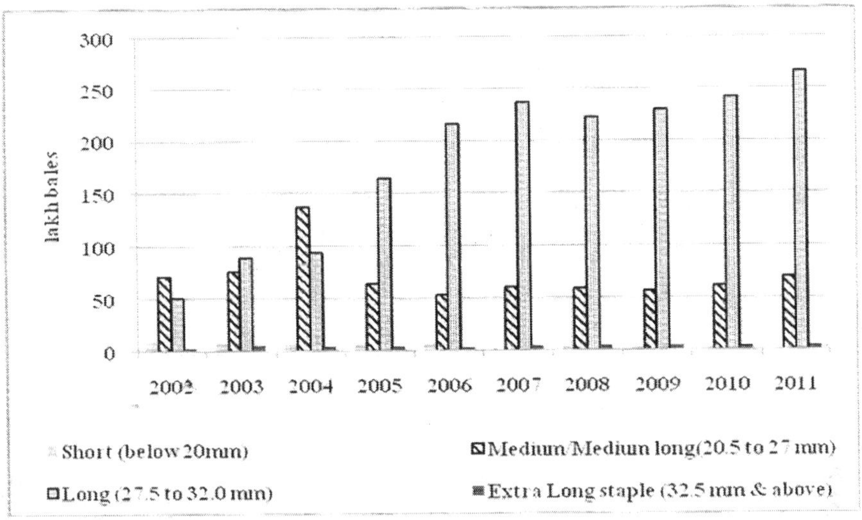

FIGURE 27.6 Staple wise cotton production in India during the decade (2000 to 2011).

increase from 154 lakh bales to 230 lakh bales. India was the third largest importer of cotton in 2002–03, importing 17.7 lakh bales of raw cotton. After the introduction of Bt cotton from 2005–06. India became a net cotton exporter (Figure 27.7) and today it is the second largest exporter of raw cotton. During the period from 2001 to 2011 the domestic consumption of cotton increased from 159 lakh bales to 230 lakh bales (of 170 kgs). The value

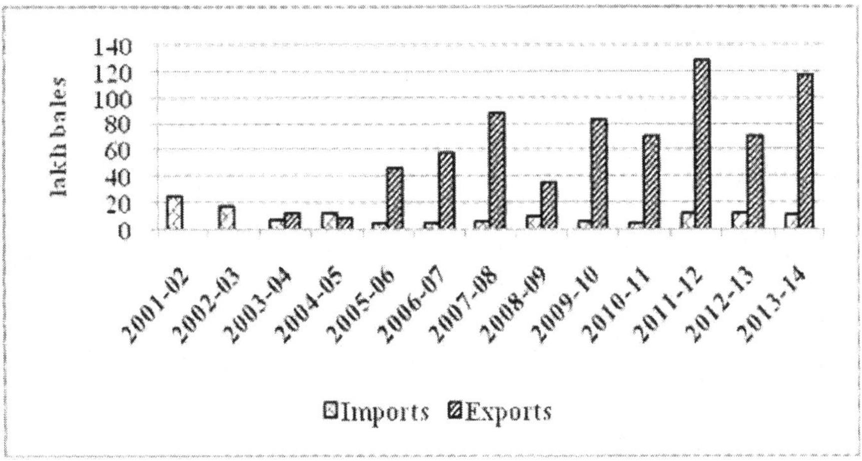

FIGURE 27.7 Import and export of cotton by India during the period 2001–02 to 2013–14.

of raw cotton exported in 2011–12 was Rs. 21623 crores. This transformation from being a chronic importer to a major and regular exporter is largely attributable to the increase in production following the large scale adoption of Bt cotton. During the period 2005 to 2013, India continued to import 5 to 15 lakh bales of cotton annually, mostly in the extra long staple category. India emerged as the second largest exporter of raw cotton after USA, and it caters to the markets of Bangladesh, Pakistan and China.

27.7 PRODUCTIVITY OF COTTON DURING THE Bt COTTON ERA

Bt cotton is essentially not an yield enhancing technology, rather it is a loss minimizing technology aimed to protect the host cultivar against losses due to key lepidopteron pests- the American bollworm *Helicoverpa armigera* (Hubner), the spotted bollworm, *Erias vitella* (Fab) and *Erias insulana* (Boisd) and the pink bollworm; *Pectinophora gossypiella* (Saunders). By the mid 1990s cotton farmers in India were spending >43% of the variable costs of production on insecticides, around 80% of that being used for the control of bollworm in general and *Helicoverpa armigera* (American bollworm) in particular (ICAC, 1998). Bt cotton hybrids were introduced as a strategy to effectively manage bollworms. Within a period of 10 years the productivity of cotton increased from 302 kg in 2002 to 489 kg/ha in 2012.

The compound annual growth rate in productivity was only 0.5% during the entire adoption period from 2004 to 2012 (Table 27.5). A closer scrutiny of the data furnished in Figure 27.1 also indicates that there was a 153% increase (from 302 to 463 kg/ha) in the first 3 years, when only 6% area was under Bt cotton. The highest yield during the decade (2002–2012) was 568 kg lint/ha in 2007, when the area under Bt hybrids was only 6.2 m ha, 66% of the total area under the crop. During the subsequent years, yields have declined despite progressive increase in area under Bt hybrids, release of new Bt hybrids, the introduction of the superior Bollgard II, approval of new Bt events, increased use of fertilizes and insecticides. Yield gains have been negative in the states of Gujarat and Madhya Pradesh during the post Bt era (Table 27.5). Is the spread of hybrids into domains not inherently suitable for their cultivation a cause of this decline remains unanswered?.

The compound growth rate in area and productivity of cotton in India after the large scale adoption of Bt cotton, during the period 2004 and 2012 was 3.7 and 0.5%, respectively and together they contributed to rise in cotton production at the rate of 4.3% (Table 27.6).

TABLE 27.5 Compound Annual Growth Rate of Area, Production and Productivity in India During the Post Bt Era (2004 to 2012)

States	Area	Production	Productivity
Punjab	–0.1	2.4	2.5
Haryana	–0.1	5.6	5.8
Rajasthan	0.3	4.8	4.4
Gujarat	2.9	2.2	–0.7
Madhya Pradesh	4.8	4.5	–0.3
Maharashtra	0.7	1.5	0.8
Andhra Pradesh	8.5	11.2	2.5
Karnataka	–0.9	6.3	7.2
Tamilnadu	–0.3	1.1	1.4
All India	3.7	4.3	0.5

Source: Sabesh et al. (2014b).

TABLE 27.6 Relationship Between Area Under Bt Cotton (m ha) and Yield (kg/ha) During 2002–2013

State	Years	Linear Regression Equation	R^2
Punjab	9	y = 640–67.002x	0.013
Tamil Nadu	10	y = 704+597.72x	0.038
Madhya Pradesh	11	y = 539–78.554x	0.102
Andhra Pradesh	12	y = 524+41.252x	0.168
Gujarat	12	y = 600+61.71x	0.178
Maharashtra	12	y = 217+34.739x	0.679
Haryana	9	y = 401+508.07x	0.732
Rajasthan	8	y = 381+617.72x	0.759
Karnataka	10	y = 254+501.54x	0.859

A regression analysis of the area under Bt cotton in different states and the lint yields (Table 27.6) indicated that the R^2 (coefficient of determination), which is an estimate of goodness of fit was relatively high for Maharashtra, Haryana, Rajasthan and Karnataka where the base yields during the period when the Bt cotton hybrids was introduced were low. In states like Punjab, Tamil Nadu, Madhya Pradesh, Andhra Pradesh and Gujarat the goodness of fit was 0.013 to 0.0178 indicating a poor relationship between expansion in area under Bt cotton and yields.

Results of a survey by Mayee and Choudhary (2013) highlighted that only 24% of the respondents in Maharashtra realized higher yields after the adoption of Bt cotton. In contrast, 79% of the respondents in Andhra Pradesh and 100% of the respondents in Punjab realized enhancement in yield with the adoption of this technology.

27.8 SOCIO-ECONOMIC IMPACTS OF BT HYBRIDS

Around 220 studies were made on the socio-economic impact of Bt cotton in India until 2010 (Kaphengst et al., 2011). Most of the studies on impact analysis considered data collected from experimental fields, from farmers fields, opinions or experiences expressed by farmers from different locations and time. Impact based on wide macro level analysis are few (Ramasundaram et al., 2014). Many of these studies have not been scientifically robust either due to non-availability of proper checks/controls (like iso-genic non-Bt line), difficulty to distinguish the effect of the trait alone from other confounded effects, unequal sample size of Bt and non-Bt farmers, and limitations of recall method. Nevertheless several studies have quantified positive economic benefits of Bt cotton technology and a spectrum is presented in Table 27.7.

TABLE 27.7 Published Reports Highlighting the Benefits of Bt Cotton in India

Study period	% yield increase	% spray Reduction	% Profit	Reference
2002–03	45–63	66	50	Bennett (2006)
2004	31	39	88	Gandhi and Namboodiri (2006)
2006	46	55	110	Ramgopal (2006)
2004–05	32	25	83	Dev and Rao (2007)
2006–07	43	21	134	Subramanian and Qaim (2010)
2006–07	43	21	70	Sadashivappa and Qaim (2009)
2002–08	24	–	50	Kathage and Qaim (2012)
2012	98	83	–	Mayee and Choudhary (2013)
–	34	30	–	Prathyusha et al. (2015)
2002–04	25		30	Ramasundaram et al. (2007)
2004–05	43	36	120	Gandhi and Namboodiri (2009)
2007–08	30	–	151	Kiresur and Ichangi (2011)
2011	–	50	57	Benagi et al. (2015)
2011–12	100	43	–	Reddy and Narala (2015)

The economic benefits resulted from more effective control of bollworms resulting in increased yields, saving in the cost of insecticides and higher profits (higher yield or better prices). In Burkina Faso, where Bt cotton was released in 2009, based on the results of field trials and producer surveys, Vitale et al. (2011) reported an increase in cotton yields to an extent of 21.3 % and income by $106/ha. In India, the extent of benefit widely varied across states and across growing conditions. Wherever, the prescribed norms of cultivation, for instance, maintenance of refugia are not followed, the variation was more pronounced (Sadashivappa and Qaim, 2009). Brookes and Barfoot (2008) concluded that among the different Bt cotton growing countries, India experienced the highest advantage of the Bt trait amounting to 54% increase in yield, whereas in the other countries it ranged between 0% and 27%. Gruère and Sengupta (2011) reported that due to the adoption of Bt cotton, the increase in yield was 34–42% and the increase in net returns was 50–94%. The reduction in spray of pesticide was 30–36% with an associated cost reduction in plant protection to the extent of 35–52%.

Jost et al. (2008) opined that the profitability of transgenic cotton was more closely associated with yield and not with the technology. Ramasundaram and Vennila (2013) also opined that the recent gains in cotton production cannot be solely attributed to Bt gene technology. Other factors including coverage under hybrids seed, increased fertilizer use, favorable monsoon, higher prices cotton leading to more investment in inputs and expansion in irrigation have also contributed to yield gains (Grure and Sun, 2012). Treatment of the Bt hybrid seeds with imidacloprid was also instrumental in protecting the crop against sap-sucking pests during the vulnerable vegetative stage of the crop leading to higher yield realization.

The impact of Bt technology has plateaued during the recent years. Recent reports have expressed concern over the beneficial impact of Bt cotton particularly under rainfed conditions. Kuruganti (2009) argued that high growth rate of yield in Gujarat was due to low incidence of target pest, consistent good monsoon, increase in area under irrigation, higher fertilizer use and more importantly shift in favor of hybrids. Gutierrez et al. (2015) concluded that Bt cotton was economical under irrigated conditions. On the contrary, the cost of Bt seed and insecticides increase the risk of farmers under low yielding rainfed environments. The results of study by Romeu–Dalmau et al. (2015) also highlighted that the potentials of Bt cotton is constrained under rainfed conditions (in Maharashtra) and under such situations Asiatic cotton (*G. arboreum*) would be a better alternative.

27.9 CHANGE IN INPUT USE SCENARIO DURING THE POST-Bt PERIOD

The impact of Bt cotton technology on the input *viz.,* seeds, fertilizers and pesticides use was analyzed by the using the data derived from the http://eands.dacnet.nic.in maintained by the Directorate of Economic and Statistics (DES) and the Commission for Agricultural Costs and Prices (CACP) *viz.,* http://cacp.dacnet.nic.in/, Ministry of Agriculture, Government of India. The trend in the usage of seeds and fertilizers is presented in Table 27.8.

Across different states, during 2001 the quantity of seed used for sowing ranged from 4.3 kg/ha in Maharashtra to 16.2 kg/ha in Rajasthan. Farmers in the states like Punjab, Haryana, Rajasthan where hybrids were not grown commercially prior to the introduction of Bt cotton adopted 15–20 kg seeds/ha as the recommended plant population was higher. Adoption of Bt cotton resulted in a progressive reduction in the quantity of seed used across all the cotton growing states. Suresh et al. (2013) reported that the reduction at the All India level was from 7 kg/ha in 2001 to 3 kg/ha in 2009. A similar trend is reflected across different states (Table 27.8) and the sharpest decline was observed in Punjab. The current seed rate ranges from 1.13 kg/ha in Madhya Pradesh to 6.68 kg/ha in Rajasthan. Incidentally, farmers of certain areas in Rajasthan continue to grow non-Bt cultivars sown with higher seed rate. Since the seeds of Bt cotton are costly, farmers dibble single seed at fixed row to row and plant to spacing to provide a plant population of 12,000–18,000 plants/ha and hence the seed rate is low.

TABLE 27.8 Changes in Input Use and Production Cost During the Bt Cotton Period

	Seed Quantity (kg/ha)			Fertilizer Quantity (kg/ha)		
State	**2001**	**2006**	**2011**	**2001**	**2006**	**2011**
Andhra Pradesh	4.5	1.6	1.7	139.7	226.4	237.2
Gujarat	4.4	3.5	1.7	81.0	124.9	215.1
Haryana	13.3	9.3	2.3	59.5	106.9	134.4
Karnataka	7.9	4.2	2.0	59.9	92.1	164.2
Madhya Pradesh	5.1	1.6	1.1	93.7	173.1	92.2
Maharashtra	4.3	3.0	1.8	103.3	106.0	273.2
Punjab	12.7	4.0	2.8	76.3	145.7	231.0
Rajasthan	16.2	14.5	6.9	65.8	78.7	119.2
Tamil Nadu	10.0	7.1	6.0	129.7	209.8	276.4

The primary benefit from Bt cotton is in the protection of cotton plants from bollworm damage and thereby reducing the insecticide usage for bollworm control. After the introduction of Bt cotton, there was a significant reduction in insecticide used from 13176 metric tons (1.5 kg/ha) in 2001 to 4623 metric tons (0.5 kg/ha in 2006). After 2006, the use of insecticides progressively increased to reach 11,598 metric tons (0.97 kg/ha) in 2013 (Kranthi, 2014). Further, while the insecticide usage for the control of bollworms declined from 9410 metric tons in 2001 to 121 metric tons in 2013 but the usage for the control of sucking pests increased from 3312 metric tons in 2001 to 11366 metric tons in 2013 (Kranthi, 2014) because of increasing infestation of whiteflies, leaf hoppers and thrips and resistance built up in these insect pests against insecticides. The release of a large number of hybrids from 62 in 2006 to 1167 hybrids in 2014 without proper screening for tolerance against pests and diseases is another reason.

The use of chemical fertilizers in cotton cultivation increased steadily at the rate of 8.1% per annum during the first decade of Bt cotton. The increase between 2001 and 2011 was 3 fold in Punjab, 2.5 to 3.0 fold in Maharashtra, Karnataka and Gujarat, 2.0 to 2.5 fold in Tamilnadu and Haryana and 1.5 to 2.0 fold in the remaining states except Madhya Pradesh (Table 27.8).

Farmers applied more fertilizers with the hope of reaping additional benefits particularly in areas with assured rainfall/ supplementary irrigation. There has been a reduction in partial factor productivity of fertilizers and increase in the cost of production (Suresh et al., 2013). Productivity trend in Figure 27.8 indicates that there was a significant reduction in seed cotton yield per kg of fertilizer applied in the states of Andhra Pradesh, Haryana, Punjab, Tamilnadu and Maharashtra between the period 2001 and 2011. This decline in partial factor productivity for fertilizers could be an indicator of a decline in soil health or a nutrient imbalance which might be a reason for yield stagnation.

27.10 COST OF COTTON CULTIVATION DURING THE POST-Bt PERIOD

While the gains in yield and profits following the adoption of Bt cotton have been widely reported, the increase in expenditure has often been over looked. Taking the country as a whole, the nominal cost of cultivation (at current prices) increased @ 9.49% per annum during triennium ending

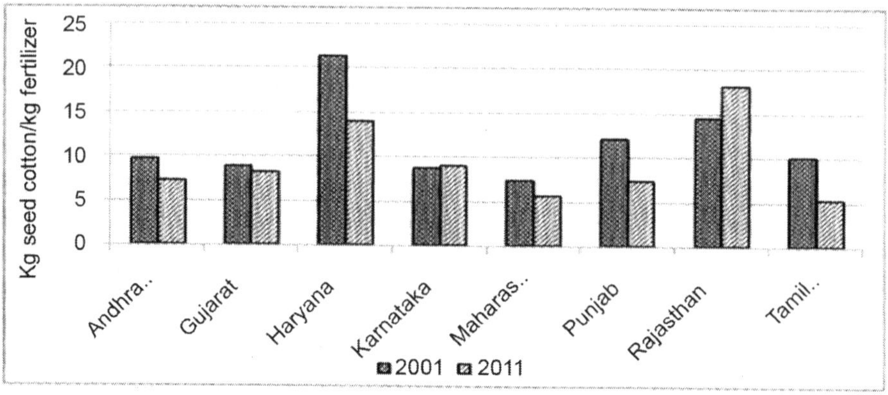

FIGURE 27.8 Partial factor productivity of fertilizers in cotton production during 2001 and 2011.

2001–02 to triennium ending 2009–10. In absolute terms, the increase was from Rs. 16,037/ha to Rs. 35,275/ha during the same period (Suresh et al., 2013). Although a similar increasing trend was observed in different states, the magnitude of change differed (Table 27.9). The rate of increase in cost was slow during the period 2002 to 2006 (with the exception of Gujarat), but rose steadily during the period 2006 to 2011. During the entire period (2001 to 2011), the cost of cultivation increased ten fold in Gujarat, four fold in Karnataka, Madhya Pradesh, Haryana and Maharashtra and two to three fold in other states. Sabesh et al. (2014a) reported that the cost of cultivation in Maharashtra increased from Rs. 20,860/ha in 2000–03 to Rs. 52,583/ha in 2010–11 at a compounded average growth rate of 12.25%.

Seeds, fertilizer (including manures), insecticides and labor account for roughly half of the total production cost with the share of labor component being the highest (Table 27.10). Contrary to the popular belief, seeds account for only 4–9% of the total cost. However, the proportional expenditure on seeds increased rapidly in the states of Punjab, Rajasthan, Haryana where hybrids were hitherto not cultivated. Roughly 10% of the expenditure was incurred on fertilizers in most of the states except in the states of the north zone where the proportional expenditure was slightly lower. Between 2001 and 2011, a significant reduction in the proportionate share on the cost in insecticides was noticed in the states of Andhra Pradesh, Gujarat, Haryana, Karnataka, Punjab and Rajasthan but not so in the states of Maharashtra, Madhya Pradesh and Tamilnadu. The expenditure on labor gradually increased over the 10-year period in most of the states except

TABLE 27.9 Cost of Cotton Cultivation and Proportional Share of Different Components to Total Cost

State	Year	Cost of Cultivation[a] (Rs/ha)	Percentage share to total cost			
			Seed	Fertilizer[b]	Insecticide	Human Labor
Andhra Pradesh	2001	23,401	6.3	8.8	9.2	25.6
	2006	27,184	8.0	14.7	4.8	29.6
	2011	61,634	6.0	10.4	4.4	31.4
Gujarat	2001	5,475	12.6	31.6	21.9	26.6
	2006	34,999	5.9	7.4	8.5	22.8
	2011	58,038	5.1	10.5	4.2	34.5
Haryana	2001	14,760	2.6	4.9	11.7	42.7
	2006	29,779	4.4	4.6	6.8	30.8
	2011	62,330	7.3	3.4	3.2	35.1
Karnataka	2001	11,006	5.6	9.3	5.7	30.8
	2006	15,802	16.1	12.5	4.5	26.8
	2011	44,735	5.2	10.2	2.0	31.3
Madhya Pradesh	2001	11,009	6.1	12.2	3.4	27.4
	2006	28,745	5.8	11.4	6.2	23.5
	2011	41,861	9.0	6.0	4.6	27.4
Maharashtra	2001	17,234	5.3	11.8	3.6	26.3
	2006	21,669	8.8	9.2	1.4	24.7
	2011	61,907	6.1	11.0	3.6	32.5
Punjab	2001	25,772	2.9	3.5	27.8	22.7
	2006	36,866	8.9	4.9	4.6	26.5
	2011	66,698	8.7	4.8	8.5	28.5
Rajasthan	2001	13,104	3.4	6.9	9.8	38.4
	2006	21,381	3.1	7.5	8.7	34.8
	2011	54,296	7.3	5.6	4.5	40.9
Tamil Nadu	2001	31,845	1.6	5.8	2.9	41.6
	2006	29,422	4.7	9.3	3.5	43.9
	2011	61,319	3.9	11.4	2.6	49.6

[a]All expenses in cash and kind incurred in production by owner + interest on value of owned capital assets (excluding land) + rental value of owned land and rent paid for leased in land + input value of family labor. [b]Including manure.

TABLE 27.10 R^2 (Coefficient of deTermination) Values of Linear and Quadratic Equations Between Seed Cotton Yield (kg/ha) and Costs (Rs/ha) Between 2000 and 2011

State	R^2 values	
	Linear	**Quadratic**
Punjab	0.28	0.45
Haryana	0.83	0.89
Rajasthan	0.74	0.75
Madhya Pradesh	0.54	0.58
Maharashtra	0.44	0.74
Gujarat	0.59	0.86
Karnataka	0.61	0.74
Tamilnadu	0.61	0.61
Andhra Pradesh	0.21	0.44

Haryana. Suresh et al. (2013) reported that in 2009–10, human labor was the single largest component, accounting for 44% of the total cost of cultivation.

Sabesh et al. (2014b) also concluded that the cost of cultivation was high in Andhra Pradesh, Maharashtra, Gujarat and Madhya Pradesh. Though the damage due to bollworms was reduced, the damage due to sucking pests rose after 2006–07 and this in turn escalated the cost of production in some states.

Ideally, if the returns are in proportion to the investment in inputs as reflected by the cost of cultivation, a linear relationship between the cost of cultivation and yield is expected. But this has not been the case with Bt cotton hybrids in most of the states (Table 27.10). R^2 values, indicating the proportion of variability explained through linear or quadric models between input costs and yields, are in favor of the latter model. A typical relationship for Gujarat state is also depicted in Figure 27.9.

It is evident from the data in Table 27.10 and Figure 27.9 that increase in the expenditure incurred on cotton cultivation is not translated into seed cotton yields. The reasons could either be technology fatigue or improper stewardship or placement of technology. For instance on marginal soils, Bt hybrids needing higher amount of inputs may be a poor choice of technology in comparison to varieties. Often farmers are continuously over investing in inputs in Bt cotton fields in certain rainfed regions of Maharashtra where productivity enhancement is not in proportion with the investment

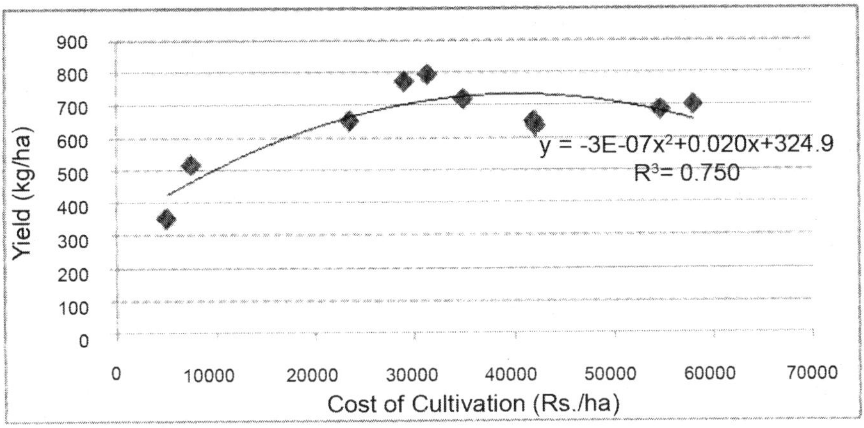

FIGURE 27.9 Relationship between lint yield and cost of cotton cultivation in Gujarat between 2000 and 2011.

(Ramasundaram et al., 2011). Under such circumstances, short duration early maturing, sucking pest and drought tolerant varieties preferably with Bt gene would be a better choice.

27.11 OTHER SIGNIFICANT CHANGES IN THE POST-Bt ERA

- Seeds of Bt cotton hybrids are entirely produced by the private sector. The onus of its marketing and agro-technology transfer is also carried out mostly by private sector companies. Nearly 40 million packets @ Rs. 950/packet – sold every year and the total sales are to the tune of Rs. 3800 crore. About 44 companies are involved in the production and marketing of the seeds of Bt cotton hybrids. The turnover for the Indian biotechnology seed industry is estimated at US$ 551 million. As cotton is the only permitted biotechnology crop in India, biotech cotton contributes the lion's share to this turnover and growth (James, 2012).
- The increase in cotton production leads to an increase in the production and consumption of cotton seed by-products-oil and seed meal. The production of cotton seed oil increased from 0.46 m tons in 2002 to 1.5 m tons in 2013 (Choudhary and Gaur, 2015).
- In India, the Plant Varieties Protection and Farmers Rights Act empowers the farmers to retain his/her own produce for seed purpose

but the Bt technology is available only in Hybrids and the seeds can therefore not be re-used. This makes farmers rights infructuous.

- Short staple varieties of *Desi* cotton, *Gossypium arboreum,* with high micrornaire and low ash content are ideal for surgical use. The rapid adoption and diffusion of Bt cotton hybrids has virtually wiped out the *G. arboreum* cotton from their traditional tracts in North and Central India, causing huge shortage of cotton for the manufacture of surgical cotton.
- The expansion of rainfed cotton into red soil areas of Telangana and Karnataka and into shallow black soils of Maharashtra, sub-humid areas of eastern Vidarbha (Maharashtra) both of which are not suitable for the cultivation of hybrid cotton is a cause of concern.
- Breakdown in the seed production chain of the public sector bred non-GM varieties and the fear of contamination with GM cotton has adversely affected the organic cotton production system in recent years.

27.12 CONCLUSIONS

An estimated 93% of the area under cotton is occupied by Bt hybrids and perhaps this has reached a saturation point. The Bt technology has thus far been able to protect cotton against bollworms, the primary purpose for which it was evolved and released. It is also evident that lint yields have not increased after 2007, despite increase in the dose of farm inputs. New events carrying more potent *cry* genes coding for Bt toxins that can either reduce pesticides usage or delay resistance build up are not in the anvil, at least in the near future. The rise in area under cotton in India at the expense of coarse cereals, pulses and oilseeds, all being vital components of food/ nutritional security, needs re -thinking and consideration, It is imperative to sustain yields, reduce cost of cultivation and improve farm profits. Proper stewardship in the management of cotton based production system holds the key. At this juncture, it is important to identify regions where other factors of production like soil-site sustainability, moisture availability, biotic stresses like leaf curl virus, etc.; are preventing the expression of the potential of Bt hybrids and wean these area away from hybrids. A Bt version of popular early maturing, compact sucking pest tolerant varieties available in the public sector would be a benefiting alternative to farmers in such resource poor regions that are not conducive to hybrids. Emphasis can also be given to the cultivation of suitable pulses and leguminous oilseed crops as intercrops in cotton to economize N use, improve farm income and augment pulses/ oilseeds production. Farmers need to adopt, yield enhancing technologies (like high density planting), cost saving measures and switch over to more sustainable production systems.

Production technologies for the next generation must be sustainable, in harmony with nature and in consonance with local production systems. At macro level, a huge surplus in the national cotton stocks and the continued supply glut is unlikely to push cotton prices further in the near future. An immediate increase in domestic mill consumption is also unlikely. Safeguarding the cotton farmers against low, uncertain and volatile prices, will need sustained Government intervention in the market. Future land use policies including cotton policy needs to be thoughtfully formulated keeping in view the past and emerging trends of Bt- Cotton.

KEYWORDS

- **Bt Hybrids**
- **Cost of Production**
- **Cotton Import-Export**
- **Cropping Pattern**
- **Fiber Quality**
- **Input Use**
- **Productivity**
- **Socio-Economic Impact**
- **Species Composition**

REFERENCES

Benagi, V. I., Naik, L. K., Rafee, C. M., Udikeri, S. S., Gopala, M., Basavaraja, H., & Patil, B. C. (2015). Impact evaluation of Bt cotton in Karnataka. University of Agricultural Sciences, Dharwad and Department of Agriculture, Government of Karnataka.

Bennett, R. M., Kambhampati, U., Morse, S., & Ismael, Y. (2006). Farm-level economic performance of genetically modified cotton in Maharashtra, India. *Review of Agricultural Economics, 28*(1), 59–71.

Brookes, G., & Barfoot, P. (2008). Global impact of biotech crops: socioeconomic and environmental effects, 1996–2006. *AgBioForum, 11*, 21–38.

Choudhary, B., & Gaur, K. (2015). *Biotech Cotton in India, 2002 to 2014*; ISAAA Series of Biotech Crop Profiles; ISAAA: Ithaca, NY.

Deokate, T. B., & Bandgar, D. P. (2013). Changes in land-use and cropping pattern in Maharashtra. *Agricultural Economics Research Review,* 26, 257.

Dev, S. M., & Rao, N. C. (2007). *Socio-economic Impact of Bt Cotton.* Centre for Economic and Social Studies (CESS) Monograph No. 3. CESS, Hyderabad.

Gandhi, V. P., & Namboodiri, N. V. (2009). *Returns and Economics of Bt Cotton vis-a vis Traditional Cotton Varieties in the State of Maharashtra in India.* Centre for Management in Agriculture, Indian Institute of Management, Ahmedabad.

Gandhi, V. P., & Namboodiri, N. V. (2006). *The adoption and economics of Bt cotton in India: Preliminary results from a study.* Indian Institute of Management, Ahmedabad.

Gruère, G., & Sengupta, D. (2011). Bt Cotton and farmer suicides in India: An evidence-based assessment. *Journal of Development Studies, 47*(2), 316–337.

Grure, G. P., & Sun, Y. (2012). *Measuring the Contribution of Bt Cotton Adoption to India's Cotton Yields Leap.* International Food Policy Research Institute (IFPRI), 2012.

Gutierrez, A. P., Ponti, L., Herren, H. R., Baumgartner and Kenmore, P. E. (2015). Deconstructing Indian cotton; weather, yields and suicides. *Environmental Sciences Europe, 27*(12).

http://cacp.dacnet.nic.in/, Commission for Agricultural Costs and Prices (CACP), Ministry of Agriculture, Government of India, Accessed on Accessed on 25 July, 2015.

http://eands.dacnet.nic.in, Directorate of Economic and Statistics (DES), Accessed on Accessed on 25 July, 2015.

ICAC. (1998). *Survey of the Costs of Production of Raw Cotton.* ICAC, Washington.

James, C. (2012). *Global Status of Commercialized Biotech/GM Crops, 2012.* ISAAA Brief No. 44, ISAAA: Ithaca, NY.

James, C. (2014). *Global Status of Commercialized Biotech/GM Crops, 2014.* ISAAA Brief No. 49. ISAAA: Ithaca, NY.

Jost, P., Shurley, D., Culpepper, S., Roberts, P., Nichols, R., Reeves, J., & Anthony, S. (2008). Economic comparison of transgenic and nontransgenic cotton production systems in Georgia. *Agronomy Journal, 100*(1), 42–51.

Kaphengst, T., Nadja, E. B., Clive, E., Robert, F., Sophie, H., Stephen, M., & Nataliya, S. (2011). Assessment of the economic performance of GM crops worldwide; Report to European Commission, March, p. 149.

Kathage, J., & Qaim, M. (2012). Economic impacts and impact dynamics of Bt (*Bacillus thuringiensis*) cotton in India. *PNAS, 109*, 11652–11656.

Kiresur, V. R., & Ichangi, M. (2011). Socio-economic impact of *Bt* cotton- A case study of Karnataka. *Agricultural Economics Research Review, 24*(1), 67–81.

Kranthi, K. R., Venugopalan, M. V., & Yadav, M. S. (2012). Hybrid Cotton Revolution in India. *Seed Times, 5*(4), 29–38.

Kranthi, K. R. (2014). Cotton Production Systems – Need for a Change in India. *Cotton Statistics & News* Weekly, *38*, 4–7.

Kuruganti, K. (2009). Bt Cotton and the myth of enhanced yields. *Economic and Political Weekly, 44*, 29–33.

Maharana, L., Dash, P. P., & Krishnakumar, K. N. (2011). A comparative assessment of BT and non-BT cotton cultivation on farmers livelihood in Andhra Pradesh. *Journal of Biosciences Research, 2*(2), 99–111.

Mayee, C. D., & Choudhary, B. (2013). Adoption and uptake pathways of Bt cotton in India. *Indian Society for Cotton Improvement*. pp. 120.

Mehta, N. (2012). Performance of crop sector in Gujarat during high growth period: some explorations. *Agricultural Economics Research Review, 25*(2), 195–204.

Murugkar, M., Ramaswami, B., & Shelar, M. (2007). Competition and monopoly in the Indian cotton seed market, *Economic and Political Weekly, 42*(37), 3781–3789.

Pavithra, B. S., & Kunnal, L. B. (2013). Performance of cotton crop in non-traditional areas of Karnataka: An economic analysis. *Karnataka Journal of Agricultural Sciences, 26*(2), 243–246.

Prathyusha, T., Vasantha, R., & Supriya, K. (2015). Consequences of Cultivation of Bt Cotton as Perceived by farmers of Andhra Pradesh, India. *International Research Journal of Social Sciences, 4*(6), 7–14.

Ramgopal, N. (2006). *Economics of Bt cotton vis-à-vis traditional cotton varieties – Study in Andhra Pradesh*, Agro-Economic Research Centre, Andhra University, Andhra Pradesh.

Ramandeep Kaur M. M., & Kiran, G. S. (2015). Analysis of trends in area, production and yield of important. *International Journal of Agronomy and Agricultural Research, 7*(1), 86–92.

Ramasundaram, P., & Vennila, S. (2013). A decade of Bt cotton experience in India: Pointers for transgenics in pipeline. *Current Science, 104*(6), 697–698.

Ramasundaram, P., Suresh, A., & Chand, R. (2011). Manipulating technology for surplus extraction: The case of Bt cotton in India. *Economic and Political Weekly, 43*(46), 23–26.

Ramasundaram, P., Suresh, A., Samuel, J., & Wankhade, S. (2014). Welfare Gains from Application of First Generation Biotechnology in Indian Agriculture: The Case of Bt Cotton. *Agricultural Economics Research Review, 27*(1), 73–82.

Ramasundaram, P., Vennila, S., & Ingle, R. K. (2007). Bt Cotton Performance and Constraints in Central India. *Outlook on Agriculture, 36*(3), 175–80.

Reddy, A. A. (2011). Sources of agricultural growth in Andhra Pradesh, India: Scope for small farmer participation. *The Indian Economic Journal, 59*(3), 87–108.

Reddy, A. R., & Narala, A. (2015). Impact evaluation of Bt cotton in Maharashtra. ICAR-Central Institute for Cotton Research, Nagpur, Maharashtra, India.

Romeu–Dalmau, C., Bonsall, M. B., Willis, K. J., & Dolan, L. (2015). Asiatic cotton can generate similar economic benefits to Bt cotton under rainfed conditions. *Nature Plant, 1*(6), 1–5.

Sabesh, M., Prakash, A. H., & Bhaskaran, G. (2014b). Shift in Indian Cotton Scenario due to shift in Cotton Technology, *Cotton Research Journal, 6*(1), 75–82.

Sabesh, M., Ramesh, M., Prakash, A. H., & Bhaskaran, G. (2014a). Is there any shift in cropping pattern in Maharashtra after the introduction of Bt cotton? *Cotton Research Journal, 6*(1), 63–70.

Sabir, H. M., & Khan, M. B. (2011). Bt Cotton and its impact on cropping pattern in Punjab. *Pakistan Journal of Social Sciences, 31*(1), 127–134.

Sadashivappa, P., & Qaim, M. (2009). Bt cotton in India: Development of benefits and the role of government seed price interventions. *AgBio Forum, 12*(2), 172–83.

Singh, S. P., Jena, P. C., & Singh, N. K. (2013). Cotton Production and Environmental Sustainability in India. CUTS International, Jaipur.

Subramanian, A., & Qaim, M. (2010). The impact of Bt cotton on poor households in rural India. *Journal of Development Studies, 46*(2), 295–311.

Suresh, A., Ramasundaram, P., Samuel, S., & Wankhade S. (2013). "Impact of Technology and Policy on Growth and Instability of Agricultural Production: The Case of Cotton in India, " *Indian Journal of Agricultural Sciences*, *83*(8), 939–48.

Vitale, J., Ouattarra, M., & Vognan, G. (2011). Enhancing sustainability of cotton production systems in West Africa: A summary of empirical evidence from Burkina Faso. *Sustainability, 3*, 1136–1169.

CHAPTER 28

LAND EVALUATION FOR RAINFED COTTON: A CASE STUDY FROM CENTRAL INDIA

DHANASHREE PABLE,[1] S. CHATTERJI,[2] T. K. SEN,[2] M. V. VENUGOPALAN,[3] and J. D. GIRI[4]

[1]*PhD Scholar, ICAR-National Bureau of Soil Survey and Land Use Planning, Amravati Road, Nagpur – 440 033, India*

[2]*Principal Scientist, ICAR-National Bureau of Soil Survey and Land Use Planning, Amravati Road, Nagpur – 440 033, India*

[3]*Principal Scientist, ICAR-Central Institute for Cotton Research, Nagpur – 400 010, India*

[4]*Senior Scientist, ICAR-National Bureau of Soil Survey and Land Use Planning Amravati Road, Nagpur – 440 033, India*

CONTENTS

ABSTRACT

Land evaluation (LE) is the assessment of the suitability of land for man's use in agriculture, forestry, engineering, hydrology, regional planning, recreation, etc. Land suitability evaluation is an approach in LE that concerns the appraisal and grouping of specific areas of land in terms of their suitability for defined uses. The present investigation was undertaken to evaluate selected soils of Akola district for rainfed cotton. Nine soil profiles, representative of these areas were identified for the present study. Identification of minimum datasets (MDS) of land parameters governing cotton yield was done using expert system and available literature keeping in view the importance that they exert on crop production. Various parameters identified were, rainfall, depth, clay and exchangeable sodium percent (ESP). A number of methods are used for LE. A fuzzy modeling-based method has been applied in the present study. The method based on fuzzy modeling gives additional information on the "position" of a land unit within the suitability class relative to the neighboring suitability class and hence overcomes the limitation of abrupt class boundary of conventional method. The soil suitability classes were computed for each pedon based on the four soil attributes. The suitability classes for the crop were identified by placing the composite land index (CLI) values (of the land units) in a set of equally-spaced classes on a 0–100 scale, with a 10 unit gradation. Pedons 1, 3, 5, 6, 7, belong to class I, whereas, the CLI was observed to be the highest in P1 (92.96) and the lowest observed in P6 (90.60). Pedons 2, 4 and 9 belong to class II, the highest CLI was found in P9 (86.08) and the lowest CLI was found in P4 (85.50). Pedon 8 belongs under class III, which has the lowest CLI of 79.73.

28.1 INTRODUCTION

Land, the mother earth supports and nourishes all life supporting systems. It has been meeting the basic needs of food, fodder, fuel, fiber and other raw materials since its origin. Land is precious and every inch of it is to be used judiciously. The demands on the finite land resources are increasing exponentially due to the ever-growing population. It has been estimated that India's population may reach 1.6 billion by about 2050 AD. This mammoth growth is making the man to land ratio more unfavorable. The per capita cultivable land holding is expected to decline from 0.5 ha (1951–52) to 0.08 ha (by 2050 AD). The food production although increased from 52 m tones (in 1950) to almost 190 m tones (in 1990s), but this has been largely due to expansion in cultivated area and use of high levels of inputs. The significant growth of agriculture has been at the cost of soil quality and risk of soil degradation (Abrol and Sehgal, 1994). According to a study conducted by ICAR and NAAS (2010), about 121 m ha of soils in the country are under different kinds and degrees of degradation and these are getting further deteriorated thereby increasing the risk of jeopardizing our food security system. on the top of it, many more issues concerning environment, sustainability, carrying capacity of our land resources, etc., are emerging. All these demand our focused attention on developing alternative land use options for sustaining the ever-increasing population. It is, therefore, imperative that the land resources should be evaluated in terms of their suitability for different agricultural uses.

LE is concerned with the assessment of land performance for specified land utilization purposes (Rossiter, 1996). Such evaluation is essential in the process of land use planning because it may guide decisions on land utilization in such a way that the resources of the environment are optimally used and a sustained land management is achieved. In other words, LE is the assessment of the suitability of land for man's use in agriculture, forestry, engineering, hydrology, regional planning, recreation, etc. Land suitability evaluation is an approach in LE that concerns the appraisal and grouping of specific areas of land in terms of their suitability for defined uses (FAO, 1976).

The term 'Land Evaluation' was first used in 1950 at the Amsterdam Congress of the International Society of Soil Science, where Visser (1950) presented a paper entitled, 'The trend of the development of LE in the future. Stewart (1968) defined LE as assessment of suitability of land for man's use in agriculture, forestry, engineering, hydrology, regional planning, recreation, etc. Beck (1981) suggested that LE is to bring out an understanding of relationship between conditions of land and manner in which it is utilized. It is concerned with performance of land as it affects the change in land quality. LE is matching of the ecological and management requirements of relevant kinds of land use with land qualities of land use planning. LE should be regarded as an integrated process. The matrix of intractions between governance, different sectoral policies, science and technology, and investment and finance provides a background for understanding people's strategies in managing the assets to which they have access (FAO, 1976).

The process of assessment of land performance when used for specified purpose involving the execution and interpretation of surveys and studies of all aspects of land in order to identify and make a comparison of promising kinds of land use in terms applicable to the objectives of the evaluation (FAO, 1976). The main objective of LE is to put at the disposal of user (farmers, planners, or governmental departments), the relevant information about land resources that is necessary for planning, development and management decision. The end result of LE survey is thus to have a number of clear recommendations with alternatives on appropriate type of land use together with their consequences (FAO, 1993).

In the parametric method, the suitability classes are defined as discrete groupings, separated by strict class definitions or fixed class limits. Land units that have a degree of suitability somewhat intermediate between classes can, however, only be classified in one single suitability class. The method based on fuzzy modeling, on the other hand, gives additional information on the "position" of a land unit within the suitability class relative to the neighboring suitability class. In general terms, the traditional LE systems follow a Boolean or rule-based approach adapted to the principle of maximum limiting factors. There is a growing concern regarding failure of this method to incorporate the inexact or fuzzy nature of much of the land resource data. In recent years, there has been marked interest in the use of fuzzy modeling-based methodology (a mathematical approach) in LE, and it can be considered as a new phase in the quantification trend. The use of strict Boolean algebra with a simple true/false logic in combination with a rigid, exact model is often inappropriate for LE because of the continuous nature of soil variation, the uncertainties associated with describing the phenomenon itself or in the measurements made on it, or because of inexactness of much of the land resource data (Burrough, 1989; Chatterji, 2000).

The uncertainty that is inherent in any evaluation process involves both data and model ambiguity; this ambiguity includes measurement error, inherent soil variability, soil instability, conceptual ambiguity, over-abstraction or even simple ignorance of key factors that can impact LE. Due to the wide range of factors that make up LE and its inherent uncertainty, we believe that a unique approach must be taken to address LE. It was proposed that randomness and uncertainty of soil quality (SQ) be dealt with by using fuzzy modeling (Jager, 1995; Pedrycz and Gomide, 1998; Ross, 1995). This approach provides the basis for analysis of systems characterized by a high degree of uncertainty, nonlinearly and complexity. The concept of fuzzy modeling is a generalization of conventional set theory; the concept of belongingness to a set has been modified to include partial degrees of membership (i.e., values along the continuum between 0 and 1, encoded as a fuzzy membership function (MF). An MF is the central concept of the fuzzy set theory, where MF represents the relationship of an element to a set. An MF of a fuzzy set is expressed on a continuous scale from 1 (full membership) to 0 (full non-membership).

LE is even more challenging issue in rainfed cotton growing environs of Maharashtra in view of various constraints *viz.*, inadequate rainfall, shallow depth, sodicity, poor fertility and shrink swell nature of soils at many of the places. An yield gap of 840 kg/ha of cotton was reported in Maharashtra by Aggarwal et al. (2008). The physical constraints that limit cotton yield and result in its yield gap also include, among others, unsuitable soils. The yield of cotton has either plateaued or started declining in Maharashtra, one of the reasons for which could be soil constraints (Sadanshiv et al., 2012). After the green revolution in wheat and rice during the late 1960s and early 1970s, if there is any other crop that has registered a phenomenal growth during the last 13–14 years, it is cotton. Cotton production in India has increased manifold, from 15.8 m bales in 2001–02, through 30.7 m bales in 2007–08 to 33.4 m bales in 2012–13 (Cotton Advisory Board, 2013). This has been primarily the result of introducing Bt (*Bacillus thuringiensis*) technology in cotton.

There is also growing threat from land degradation on cotton productivity and ecosystem sustainability. Despite this, only few studies have been carried out to assess the land for cotton sustainability in this region. The study reports fuzzy modeling based LE of soils of Akola district for rainfed cotton.

28.1.2 LAND EVALUATION FOR RAINFED COTTON: GLOBAL SCENARIO

Cotton is an important fiber crop of global significance, which is, cultivated in tropical and sub tropical regions of more than seventy countries the world over. The major producers of cotton are China, India, USA, Pakistan, Uzbekistan, Argentina, Australia, Greece, Brazil, Mexico, and Turkey. These countries contribute about 85% to the global cotton production. India, during 2012–13, was the second largest producer of cotton in the world after China, accounting for about 18% of the world cotton production. With 11–12 m ha,

India has the distinction of having the largest area under cotton cultivation in the world constituting about 25% of the world area (36.01 m ha) under cotton cultivation. The yield per hectare during 2012–13 in the country is projected to be 489 kg/ha (CCI, 2013).

28.1.3 LAND EVALUATION FOR RAINFED COTTON: INDIAN SCENARIO

Over the past few years, India has achieved significant growth in cotton production. About a decade ago, India was barely self sufficient to meet its cotton requirement but is now poised to overtake China to become the world's biggest producer of cotton this year (2013–14). Bt cotton since its release in India in 2002 by Genetic Engineering Approval Committee (GEAC) replaced more and more conventional cotton area. There was an exponential increase in Bt cotton area from 29,000 ha in 2002 to 11.16 m ha in 2010 accounting for a staggering 92% of the total cotton area in India. It is estimated that Bt cotton would cover an area of 15.0 m ha by 2015 (Choudhary and Gaur, 2010). The production increased from 13.6 m ton in 2002–03 to 33.40 m ton during 2012–13 (CCI, 2013) following introduction of Bt cotton.

Cotton production and productivity in India have increased considerably over the years. The production increase in subsequent years was due to increase in productivity, which was spectacular after 1966–67, when the All India Coordinated Cotton Improvement Project was launched by the Indian Council of Agricultural Research (ICAR). The factors responsible for the phenomenal increase in production are the development of improved technology by the scientists, its dissemination by the extension workers and proper adoption by the cotton producing farmers. The advent of hybrids during 1970's resulted in quantum jump in production and brought about change in cotton scenario. More importantly, the introduction of Bt cotton since 2001 has revolutionized the cotton production scenario in India.

India, during 2012–13, was the second largest producer of cotton in the world after China. In India, it is grown in about 11–12 m ha area of shrink-swell soils under rainfed conditions, and alluvium derived soils under irrigated conditions which constitute about 25% of the world area (36.01 m ha) under cotton cultivation. The yield per hectare during 2012–13 in the country is projected to be 489 kg ha^{-1} (CCI, 2013). Cotton is cultivated in three distinct agro-ecological regions (North, Central and South) of the country. India in the recent years has emerged as the largest producer (73702 Mt) of organic cotton in the world, which has good emerging prospects (Barik, 2010). India is the only country to grow all the four species of cultivated cotton, *Gossypium arboreum* and *Gossypium herbaceum* (Asian cotton), *Gossypium barbadence* (Egyptian cotton) and *Gossypium hirsutum* (American upland cotton), besides hybrid cotton. *Gossypium hirsutum* represents 90% of the hybrid cotton in India and all the current Bt Cotton hybrids are *Gossypium hirsutum*. In India, cotton is mostly grown in Maharashtra, Gujarat, Andhra Pradesh, Haryana, Madhya Pradesh, Karnataka and Tamil Nadu states which accounts for nearly 98% of total cotton cultivated area (Basavanneppa and Biradar, 2003).

The major states growing cotton in 2013 in order of hectarage were Maharashtra (4.13 m ha) representing almost half of the total area growing cotton, or 40%, of all cotton area in India in 2012, followed by Gujarat (2.36 m ha or 20%), Andhra Pradesh (2.14 m ha or 16%), Northern Zone (1.56 m ha or 15%), Madhya Pradesh (608,000 ha or 8%), and the rest in Karnataka, Tamil Nadu and other states. During 2012–13, the production and productivity of cotton in Maharashtra were 80.0 lakh bales and 329 kg ha^{-1}, respectively (CCI, 2013).

The majority (90%) of cotton in the state of Maharashtra is rainfed, and this area is expected to increase in the coming years (Singh et al., 2012) In India, Vidarbha region of Maharashtra (16.33 lakh ha) is the main cotton-producing region (Mehetre et al., 2000). In Vidarbha region of Maharashtra it is grown predominantly in Vertisols and associated soils. The reasons for low productivity of cotton in Vidarbha include erratic distribution of rainfall, imbalanced fertilizer use, poor quality seed, low organic carbon and low adoption of improved agro-techniques and decline in soil health (Katkar, 2008).

The district of interest in Vidarbha in the present investigation is Akola district. It lies in the western part of Nagpur Division of Maharashtra state. The total geographical area of Akola district is 540 ha (in '000) of which during 2012–13, area under cotton cultivation was 168.3 ha (in '000), production was 2689 M ton lint (in '00) and productivity was 4.18 q ha^{-1} (Lint) (Anonymous, 2013). The district faces a number of biophysical constraints (including those of soils) to cotton production often resulting in large yield gap. However, not much work has been undertaken on ascertaining the suitability of cultivating cotton in soils of Akola district.

The baseline (reference) information obtained on various soil functions of the study area could be translated into a workable scale for farmers and land managers to identify whether a soil is being aggraded or degraded. The concept of Minimum Data Sets (MDS) refers to a set of specific soil measurements which is considered as a basic requirement for systematically assessing SQ and has to be developed region specific. MDS constitutes mutually exclusive and locally relevant indicators. The use of MDS followed by its validation through correlation with sustainable yield index reduced redundancy of datasets and helped in enhancing the pragmatic value of CSIs. Further, most of the researchers in the country have used only the MDS of physical and chemical LE indicators. The MDS of microbiological indicators for assessing soil quality has, however, not been often recommended. The present study overcomes this limitation by also recommending MDS of biological soil indicators for the study area.

There is urgent need to adopt appropriate soil and crop management practices in these cotton growing areas which would reduce soil degradation at desirable level so as to enhance and sustain cotton productivity. This study, therefore, aimed at LE for rainfed cotton in selected soils of Akola district using fuzzy modeling-based method undertaken with the objectives to characterize the selected soils in terms of their physical and chemical properties.

28.2 STUDY AREA

The study area constitutes cotton growing soils of Akola district of Maharashtra, India. It is situated at Deccan plateau in between 19° 51′ to 21° 16′ N latitude and 76° 38′ to 77° 44′ E longitude at an altitude of 285 m above sea level. It lies in the western part of the Nagpur Division of Maharashtra state. Nine cotton-growing sites were selected for profile studies based on variation in soil depth (shallow, moderately deep, deep and very deep), after traversing the area and their location details are given in Table 28.1.

The geology of the region (Akola district) is marked by Deccan trap, which covers major part of the district and is characterized by basalt lava flows, which are generally dark gray, hard and compact. The entire district, forming part of the Purna valley, is a basin shape depression in the Deccan traps filled by river deposits. The northern edge of the basin in contact with Satpura hills is supposed to be a fault line and covered by boulders and debris derived from the disintegration of the traps.

TABLE 28.1 Site location of Selected Pedons of Akola District

Pedon No.	Village	Site location	Elevation (above MSL)	Soil depth
Pedon 1	Dongargaon	20°42"13" N 76°56"55" E	296 m	Very deep (150+ cm)
Pedon 2	Sonori	20°43"23" N 76°50"25" E	315 m	Shallow (43 cm)
Pedon 3	Mana	20°46"20" N 76°28"30" E	307 m	Very deep (150+ cm)
Pedon 4	Vallabhnagar	20°50"17" N 76°59"18" E	273 m	Deep (143 cm)
Pedon 5	Rohana	20°52"26" N 76°55"21" E	276 m	Deep (120 cm)
Pedon 6	Popatkhed	20°57"05" N 77°22"89" E	417 m	Moderately deep (91cm)
Pedon 7	Kanheri (Sarap)	20°38"37" N 77°2"20" E	300 m	Very deep (150+ cm)
Pedon 8	Mandva Budrukh	20°38"47" N 76°47"47" E	300 m	Very deep (150+ cm)
Pedon 9	Thar	21°0"29" N 76°51"52" E	282 m	Very deep (150+ cm)

Akola district has been classified under hot, moist semi-arid ecosubregion, medium to high AWC and LGP of 120–150 days. In general, more than 90% of rainfall is received during July to September. The average rainfall is 707.3 mm. The weather from November to February is cool and high temperature is experienced from March to June. The average annual temperature is 27.3°C. May is the hottest month with average maximum temperature being 42.4°C and January is the coldest month with average minimum temperature being 12.3°C. The average relative humidity was recorded as high as 71.3% during the rainy season.

28.3 METHODOLOGY

28.3.1 SOIL ANALYSIS

Horizon-wise soil samples were collected for determining physical and chemical properties. The soil samples were air dried at room temperature. The samples were ground using wooden mortar and pestle and sieved through <2 mm sieve. The processed samples of <2 mm size were labeled and stored in polythene bags for subsequent physical and chemical analyzes. The methods of analysis followed in laboratory are described below.

A minimum dataset (MDS) comprising climatic parameter (rainfall), physical parameters (depth, and clay), chemical (ESP) was developed for LE (Table 28.2). Soil samples were analyzed for these physical and chemical properties following standard procedures. Particle-size distribution (sand, silt and clay) was determined as per international pipette method (Jackson, 1979) and ESP was determined by formula as the ratio of exchangeable Na to cation exchange capacity (CEC) by Jackson (1967). After analyzing them, profile weighted mean (PWM) (0–150 cm) was determined using formula

$$PWM - X = \frac{\sum DX}{\sum Dp}$$

where, PWM value was determined as the ratio of summation of depth of an individual horizon (D) multiplied by any parameter of soil (X) and total depth of the profile (Dp).

TABLE 28.2 Minimum Datasets Identified for Land Evaluation

MDSs Component	Justification
Depth	Cotton is a long duration deep-rooted crop, the top roots of cotton grow upto 1.2 to 1.5 m depth. Under rainfed conditions and being a long duration crop (6–7 months) the crop has to depend upon the moisture available in the soil profile during post monsoon period (September-January). For providing adequate rooting depth and for supplying moisture depth becomes an important parameter for MDS.
Rainfall	Being a long duration crop rainfed (6–7 months) it has to depend upon the soil moisture (rainfall) available in the soil profile.
Clay	Clay by virtue of its high moisture holding capacity and direct correlation between yield of cotton crop.
ESP	High sodium percentage leads to structural decline e.g., dispersion of soil aggregates into individual soil particles leading to reduce water availability, low saturated hydraulic conductivity (sHC), low permeability and reduced crop yield as a result of reduced availability of other nutrients.

28.3.2 LAND EVALUATION

The identification of MDS: Selection of the attributes for developing MDS comprising physical and chemical properties of soils (that best represent soil functions) was carried out using expert knowledge (Chatterji, 2000; Kadu et al., 2003; Kundu et al., 2012; Pable et al., 2014; Venugopalan et al., 2009). The MDS for cotton comprised the characteristics namely, soil depth, clay, rainfall and ESP. The methodological approaches of the aforementioned methods are discussed below in brief.

This fuzzy modeling based method consists of three steps like generation of membership values for the soil characteristics, determination of weights for the membership values, and combination of weighted membership values to produce a composite land index (CLI) (Burrough, 1989; Chatterji, 2000).

a. $$\sum \mu(z) = A \frac{1}{1 + a_i\ (z_i - c_i)^2 \text{ for } 0 \leq z \leq £} \quad (1)$$

where A is the soil characteristic set; a is the dispersion index that determines the shape of the function, c (called the ideal point or standard index) is the value of the property z at the center of the set and £ is the maximum value that z can take.

b. The Joint Membership function (JMF) for each pedon and for each parameter was computed using the convex combination rule, which is a linear weighted combination of membership values of each land characteristic A_i

$$I = \sum_{i=1}^{N} w_i\ \mu\ A_i \quad (2)$$

where I is the joint membership function w_i are the weights of the memberships value μA_i.

c. To ensure that weights sum up to unity, the rank r_i of a land characteristic, A was converted to weight W_i using the equation:

$$W_i = \frac{r_i}{\sum_{i=1}^{n} r_i} \quad (3)$$

Equation (2) shows that the choice of weights W_i is crucial in the determination of the overall land suitability index. Davidson et al. (1994) suggest that this choice should be based on data and knowledge of the relative importance of differentiating land characteristics to crop growth.

d. Simple ranking procedure was used to rate land characteristics from 1 (least important) to 4 (most important) in deriving weights. This ranking was based on literature (Kadu et al., 2003; Naidu et al., 2006; NBSS & LUP, 1994; Sys 1985;), which identified the relative importance of a particular parameter to the cultivation of cotton crop.
e. The composite land index (CLI) of a SQ parameter was determined as the average of the aggregated JMF values of the parameters for a particular pedon which in concept and for all practical purposes holds the same implication as that by SQI.
f. The suitability classes for the crops were identified by placing the (CLI) values in a set of equally-spaced classes on a 0–100 scale, with a 20 unit gradation. CLI lying between 100–80 comes under class I, CLI lying between 80–60 comes under class II, CLI lying between 60–40 comes under class III, CLI between 40–20 comes under class IV and class V has CLI values ranging from 20–0.

28.4 RESULTS AND DISCUSSION

28.4.1 MDSS PROPERTIES

Rainfall is of vital importance since it affects the climatic condition and also largely determines the success or failure of rainfed agriculture. Proper distribution of precipitation is more important than the total rainfall. In the study area, the mean annual rainfall is 707.30 mm. The rainy season is from June to September and during this period about 90% of the total rainfall is received. Highest rainfall is received during the month of July and lowest during February. This indicates that there is a water deficiency during pre-summer and summer periods (December to May) in the area. The climatic and soil parameters *viz.*, rainfall, depth, clay and ESP were identified as indicators for cotton. Rainfall is an important parameter for cultivation of cotton. Depth is an important parameter for cotton which has deep rooting system. Clay soils (Vertic properties) are preferable as it can hold more water. High sodium percentage leads to structural decline (i.e., dispersion of soil aggregates into individual soil particles leading to reduced water availability, low permeability and reduced crop yields as a result of reduced availability of other nutrients.

The study pedons showed variation in soil depth and were grouped under various categories *viz.*, shallow (<50 cm), moderately shallow (50–75 cm), moderately deep (75–100 cm), deep (100–150 cm) and very deep (>150) as per Soil Survey Manual (Soil Survey Staff, 1998). In cotton growing soils of Akola district, pedon 2 (43 cm) was shallow; pedon 6 (91 cm) was moderately deep; pedons 4 (143 cm) and 5 (120 cm) were deep; pedons 1, 3, 7, 8 and 9 (150+ cm) were very deep.

The soils, in general have high amounts of clay in most of the profiles. The clay content increased with depth in all the pedons and ranged from 25.02 to 69.98%. It is advocated that as the immediate weathering product of basalt is smectite, hence more clay was observed in soils developed on basalt. The ESP of soils varied from 1.11 (P5) to 13.83 (P9). The high amount of sodium in soils cause poor physical properties of cracking clay soils (Balpande et al., 1996). Kadu et al. (1993) observed that an ESP of > 5% may adversely affect the hydraulic properties of Vertisols. The profile weighted means of all the MDS's parameters are given in Table 28.3.

28.4.2 LAND EVALUATION BY USING FUZZY MODELING-BASED METHOD

28.4.2.1 Computation of Membership Functions

The ranks and statistics of standard indices, the dispersion indices and weightages (computed through ranking approach) for cotton are presented in Table 28.4. The same were required for developing membership functions.

TABLE 28.3 Weighted Means of MDS Parameters

Pedon No.	Growing period rainfall (mm)	Depth	Texture (% clay)	ESP
Pedon 1	791	150	57.6	4.65
Pedon 2	791	43	31.6	2.53
Pedon 3	954	150	58.8	2.61
Pedon 4	791	143	66.3	7.70
Pedon 5	791	120	62.3	4.17
Pedon 6	891	91	65.7	2.25
Pedon 7	809	150	50.7	6.97
Pedon 8	680	150	64.6	5.69
Pedon 9	848	150	68.0	7.25

Standard indices (Xi) for the MDS components for cotton were finalized on the basis of their point/range value in the highly suitable class (Naidu et al., 2006) and knowledge of experts on the soils in relation to the crop (NBSS & LUP, 1994). The standard indices considered for depth were >90 cm, rainfall >750 mm and >1300 mm for clay >27.5 and ESP <5.

A value of dispersion index (ai) 0.0002 developed for depth (Table 28.4), implies that the various soil units have their depth belongingness to the ideal value scattered within a band 0.0002 measure. Similar interpretation of dispersion indices holds good for other properties as well. Ranking was assigned to each parameter based on the relative importance of that parameter to the cultivation of cotton crop and was based on local experts' knowledge (Chatterji et al., 2002; Kadu et al., 2003). Depth, being the most important parameter for cotton, was assigned the highest rank of 4, rainfall 3, clay 2 and ESP was assigned the lowest rank of 1. The weightages derived from the ranking were 0.4 for depth, 0.3 for rainfall, 0.2 for clay and 0.1 for ESP.

The membership functions of the relevant parameters for the given pedons for cotton are presented in Table 28.5. The membership functions (or

TABLE 28.4 Ranks, Values of Different Indices for Selected Parameters (MDS) and Weightages for Computing Membership Functions

Selected Parameters	Rank	Standard Index (X_i)	Dispersion Index (a_i)	Weightage (W_i)
Depth (cm)	4	> 90	0.0002	0.4
Rainfall (mm)	3	>750 and <1300	0.0001	0.3
Clay (%)	2	< 27.5	0.0006	0.2
ESP	1	<5	0.131	0.1

TABLE 28.5 Membership Functions (Ei) of MDS Constituents

Pedon No.	Depth	Rainfall	Clay	ESP
P1	1.00	1	0.65	1.00
P2	0.69	1	0.99	1.00
P3	1.00	1	0.63	1.00
P4	1.00	1	0.53	0.51
P5	1.00	1	0.57	1.00
P6	1.00	1	0.53	1.00
P7	1.00	1	0.76	0.66
P8	1.00	0.66	0.55	0.94
P9	1.00	1	0.50	0.60

values) indicate the degree of suitability at a given location with respect to a given land characteristic. on a 0–1 scale, any parameter having a membership of 1 (highly suitable class) in any land unit suggests that the parameter has the complete belongingness to a particular class. The membership value of depth for pedon 1 is 1.0, which implies that the parameter has a belongingness of 100% to that class. Similarly, membership value of rainfall for pedon 8 is 0.66, clay for pedon 2 is 0.99, ESP for pedon 8 is 0.94 which implies that the parameter has a belongingness of 66, 99 and 94% to that class, respectively.

The joint membership function (JMF) is a sum of the product of membership functions and weightage (Table 28.6). The composite land index (CLI) of land parameter is the average of the aggregated JMF value of the parameters for a particular pedon.

The suitability classes (Table 28.6) for the crops were identified by placing the CLI values (of the land units) in a set of equally-spaced classes on a 0–100 scale, with a 10 unit gradation. CLI lying between 100–90 comes under class I, CLI lying between 90–80 comes under class II, CLI lying between 80–70 comes under class III, CLI between 70–60 comes under class IV, CLI lying between 60–50 comes under class V, CLI lying between 50–40 comes under class VI, CLI lying between 40–30 comes under class VII, CLI lying between 30–20 comes under class VIII, CLI lying between 20–10 comes under class XI and CLI lying between 10–0 comes under X.

TABLE 28.6 Joint Membership Functions (JMFs), Composite Land Indices (CLI) and Land Classes for Cotton in Fuzzy Modeling

Pedon No.	Depth	Rainfall	Clay	ESP	CLI	Land Classes
P1	0.40	0.30	0.13	0.10	92.96	I
P2	0.28	0.30	0.20	0.10	87.55	II
P3	0.40	0.30	0.13	0.10	92.60	I
P4	0.40	0.30	0.11	0.05	85.50	II
P5	0.40	0.30	0.11	0.10	91.49	I
P6	0.40	0.30	0.11	0.10	90.67	I
P7	0.40	0.30	0.15	0.07	92.11	I
P8	0.40	0.20	0.11	0.09	79.73	III
P9	0.40	0.30	0.10	0.06	86.08	II

It was observed (Table 28.5) that pedons 1, 3, 5, 6, and 7, belong to class I; pedons 2, 4, and 9 placed in class II and pedon 8 comes under class III. The composite land index (CLI) was observed to be highest in P1 (92.96), whereas the lowest CLI was observed in P8 (79.73).

28.5 CONCLUSIONS

The use of the fuzzy technique is helpful for land suitability evaluation, especially in applications in which subtle differences in land are of a major interest. This fuzzy model approach helps in overcoming limitations of abrupt boundary of land classes thus enables classification of land units on a continuous scale. Fuzzy modeling based method providing quantification of the belongingness of a characteristic to a particular class and taking into account the difference in factor weighting was the basis for LE as well as water limited potential yield. It was found to be a sound technique for evaluating suitability of selected soils for rainfed cotton. The successful application of this method provides us to suggest that whenever we have such datasets (as used in the present investigation), we could use this method for reliably assessing and monitoring land suitability. Having mentioned this, more studies will, however, need to be conducted before this could be recommended as a standard method of LE. The fuzzy modeling-based method is intuitive based on sound logic and is thus of much pragmatic value.

KEYWORDS

- **Composite Land Index**
- **Expert Knowledge**
- **Fuzzy Modeling**
- **Joint Membership Functions**
- **Land Evaluation**
- **Minimum datasets**
- **Rainfed Cotton**

REFERENCES

Abrol, I. P., & Sehgal, J. L. (1994). Degraded lands and their rehabilitation in India. *In: Soil Resilience and Sustainable Land Use.* D. J. Greenland and I. Szabolcs (Eds). CAB international Wallingford, Oxon, UK, pp. 129–144.

Aggarwal, P. K., Hebbar, K. B., Venugopalan, M. V., Rani, S., Bala, A., Biswal, A., & Wani, S. P. (2008). Quantification of yield gaps in rainfed rice, wheat, cotton and mustard in India. Global Theme on Agro-ecosystems-Report no. 43. Patancheru 502324, Andhra Pradesh, India. International Crops Research Institute for the Semi-Arid Tropics, pp. 36.

Anonymous, (2013). Yield data of Akola district. District agriculture officer report for period, 2012–2013.

Balpande, S. S., Deshpande, S. B., & Pal, D. K. (1996). Factors and processes of soil degradation in Vertisols of Purna valley, Maharashtra, India. *Land Degradation and Development, 7,* 313–324.

Barik, A. (2010). Emerging prospects of organic carbon cotton cultivation in India. *Cotton Reserch Journal, 1,* 18–36.

Basavanneppa, M. A., & Biradar, D. P. (2003). Productivity and profitability of split application of NPK nutrient on hybrid cotton. *Madras Agricultural Journal, 90*(1–3), 86–90.

Beck, K. J. (1981). From Soil survey interpretation to land evaluation. Part I from the past to present soil survey and land evaluation. *1*(2), pp. 5.

Burrough, P. A. (1989). Fuzzy mathematical methods for soil survey and land evaluation. *Journal of Soil Science, 40*(3), 477– 492.

CCI. (2013). Cotton Corporation of India. The Cotton Corporation of India Ltd. Online available at www. Cotcrop. Gov. in/share. Aspx.

Chatterji, S. (2000). Evaluation of suitability of some benchmark soil of India for rice cultivation using Boolean and Fuzzy logic: A comparative study. *Journal of the Indian Society of Soil Science, 38*(4), 809–813.

Chatterji, S., Bhattacharya, S., & Shekinath, D. E. (2002). Fuzzy estimation of minimal set of land characteristics for cultivating cotton & soybean in some soil series of Nagpur Dist. Maharashtra. *Journal of the Indian Society of Soil Science, 50*(1) 94–97.

Choudhary, B., & Gaur, K. (2010). Bt Cotton in India: A Country Profile. ISAAA Series of Biotech Crop Profiles. Pub: The International Service for the Acquisition of Agribiotech Applications (ISAAA).

Cotton Advisory Board. (2013). Cotton Association of India. Available online at www.caionline.in

Davidson D. A., Theocharpolulos, S. P., & Bloksma, R. J. (1994). A land evaluation project in Greece using GIS and based on Boolean and Fuzzy set methodologies. *International Journal of Geographic Information System, 8*(4), 369–384.

FAO. (1976). A Framework for Land Evaluation Soils bull. *32,* FAO, Rome.

FAO. (1993). Guideline: Land evaluation for extensive grazing, Soil Bulletin, *58,* FAO, Rome, 158.

Jackson, M. L. (1967). Soil chemical analysis, Prentice Hall India Pvt. Ltd., New Delhi.

Jackson, M. L. (1979). Soil chemical analysis – Advanced Course, 2nd Edition., University of Wisconsin, MD., WI.

Jager, R. (1995). Fuzzy Logic in Control. Delft TU press, Delft, Belgium. pp. 123.

Kadu, P. R., Pal, D. K., & Deshpande, S. B. (1993). Effect of low exchangeable sodium on hydraulic conductivity and drainage in shrink-swell soils of Purna Valley, Maharashtra. *Clay Research, 12,* 65–70.

Kadu, P. R., Vaidya, P. H., Balpande, S. S., Satyavati P. L. A., & Pal, D. K. (2003). Use of hydraulic conductivity to evaluate the suitability of Vertisols for deep-rooted crops in semi-arid parts of central India. *Soil Use and Management*, *19*, 208–216.

Katkar, R. N. (2008). Effect of crop residue management on soil and crop productivity and water use efficiency of cotton under differential moisture regimes. PhD Thesis, submitted to Dr. PDKV, Akola. Maharashtra (Unpub).

Kundu, S., Vassanda, M., Saha, J. K., Rajendrian, S., Hati, K. M., Bisawas, A. K., Reddy, K. S., & Subbarao, A. (2012). Assessing soil health of Vertisol of AESR 10.1 using selected physical, chemical and biological attributes of soils. *Journal of the Indian Society of Soil Science, 60*(4), 281–287.

Mehetre, S. S., Patil, V. R., Ghadge, S. B., Pawar, J. R., & Mahajan, S. V. (2000). Trends and Growth Rates of Area, Production and Productivity of 'White Gold' in different district of Maharashtra over different time period. *Journal of Maharashtra Agricultural University*, 279–284.

ICAR and NAAS. (2010). Degraded and Wastelands of India: Status and Spatial Distribution, Indian Council of Agricultural Research-National Academy of Agricultural Sciences, New Delhi. p.158.

Naidu, L. G. K., Ramamurthy, V., Challa, O., Hegde, R., & Krishnan, P. (2006). Manual on soil site suitability criteria for major crops. NBSS Publ. No. 129 NBSS & LUP. Nagpur. p.118.

NBSS & LUP. (1994). Soil-sit suitability criteria for different crops. *In*: *Proceeding of National Meet on Soil-Site Suitability Criteria for Different Crops*. Feb. 7–8, 31p.

Pable D., Chatterji, S., Venugopalan, M. V., Sen, T. K., Giri, J. D., & Sarkar D. (2014). Soil quality assessment using Fuzzy modeling- A case study in rainfed cotton growing agroecological subregions of Vidarbha, Maharashtra, *Cotton Research Journal 5*(2), 126–131.

Pedrycz, W., & Gomide, F. (1998). An introduction to fuzzy sets: Analysis and Design. MIT Press, Cambridge, USA.

Ross, T. J. (1995). Fuzzy Logic with engineering applications. McGraw–Hill, New York, USA.

Rossiter, D. G. (1996). A theoretical framework for land evaluation, *Geoderma, 72*, 165–190.

Sadanshiv, N. S., Chatterji, S., Sen, T. K., Venugopalan, M. V., Tiwary, P., Wagh N. S., & Chaturvedi, A. (2012). Application of a crop simulation model for quantification of yield gap of Cotton in Wardha district, Maharashtra. *Agropedology, 22*(2), 74–79.

Singh, J., Babar, S., Abraham., S., Venugopalan M. V., & Mujumdar, G. (2012). Fertilization of High Density, Rainfed Cotton Grown on Vertisols of India. *Better Crops, 96*, 26–27.

Soil Survey Staff. (1998). Soil Taxonomy: a basic system of soil classification for making and interpreting soil surveys. USDA Handbook No. 436, US Government printing office, Washington, DC.

Stewart, A. R. (1968). Land evaluation, *In*: *Land Evaluation, Paper of a CSIRO Symposium,* G. A. Stewart (Ed.). Macmillan of Australia, Molebourne, pp. 1–10.

Sys, C. (1985). Land Evaluation Part I, II and III. State Univ. Ghent Publ., Belgium. 343 p.

Venugopalan, M. V., Sankarnarayan, K., Blaise, D., Nalayini, P., Prahraj C. S., & Gangaiah. B. (2009). Bt cotton (*Gossipium* sp.) in India and its agronomic requirements – A review. *Indian Journal of Agronomy, 54*(4), 343–360.

Visser, W. C. (1950). The Trend of Development of Land Evaluation in the Future. Trans. 4th International Congress on Soil Science. (Amsterdam), *1*, 334–336.

CHAPTER 29

NUTRIENT OPTIMA-BASED PRODUCTIVITY ZONALITY AND RATIONALE OF FERTILIZER USE IN CITRUS

A. K. SRIVASTAVA[1] and S. N. DAS[2]

[1]*Principal Scientist, (Soil Science), ICAR – Central Citrus Research Institute, Nagpur, Maharashtra – 440 010, India*

[2]*Director, Maharashtra Remote Sensing Applications Centre, Nagpur, Maharashtra – 440 010, India*

CONTENTS

ABSTRACT

In an long term experiment, the orange fruit yield difference of 30.2 and 48.9 kg/tree initially observed on shallow soil (Typic Ustorthent) and deep soil (Typic Haplustert) in an orchard size of 11 ha, reduced to respective fruit yield of 62.7 and 68.5 kg/tree with corresponding fertilizer does (g/tree) of 1200 N – 600 P – 600 K – 75 Fe – 75 Mn -75 Zn – 30 B, and 600 N – 400 P – 300 K – 75 Fe – 75 Mn –75 Zn – 30 B, suggesting the necessity of fertilizer application on variable rate application for rationality in fertilizer using SSNM as rationale of fertilizer use. Optimum grid size soil mapping using specific nutrient-based spatial variograms as a interpretation tool popularly known as DRIS developed for Khasi mandarin of northeast India and Nagpur mandarin of central India aided in measuring the magnitude of changes in pool of soil available nutrients before and after fertilization. Grid sampling is integrated into Global Positioning System (GPS) based soil sampling and nutrient-mapping that in turn uses a Geographic Information Systems (GIS) to employ variable rate technology (VRT) for fertilizer. Hence, it is necessary to realize the full impact of SSNM, unless the usefulness of grid soil/leaf sampling, production zone vis-à-vis management zone strategies, aerial imagery vis-a-vis canopy reflectance is exploited in tandem with ultimate aim of rationalizing fertilizer use with productivity maximization.

29.1 INTRODUCTION

Globally, soil nutrient deficits are estimated at an average of 18.7 N, 5.1 P, and 38.8 (kg/ha/year) with an annual total nutrient deficit of 5.5 Tg (1 Tg = 10^{12} g) N, 2.3 Tg P and 12.2 Tg K coupled with a total potential global production loss of 1136 Tg/year considering four major (rice, wheat, maize, and barley) cereal crops (Tan et al., 2005). These figures indirectly sound an alarm in respect of highly nutrient responsive perennial crops as well (although, such statistics pertaining to perennial crops are missing), warranting the importance of balanced fertilization.

Considering 60% of the world's arable lands having mineral deficiencies or elemental toxicity problems, citrus, by the virtue of its perennially have emerged as world's leading fruit crop. Nutrient management-based production system of citrus like any other fruit crop is inherently complex to understand due to large variation in nutrient-use-efficiency. Citrus is considered evergreen in nature, blessed with nutrient conservation mechanism, to facilitate an increased carbon return per unit of invested nutrient due to comparatively longer duration of photosynthesis eventually leading to higher nutrient-use-efficiency. Perennial fruit trees play an important role in carbon cycle of terrestrial ecosystems and sequestering atmospheric CO_2 (Carbon sequestration in biomass of citrus trees ranges from 23.9 tons CO_2 ha^{-1} for young trees to 109 tons CO_2 ha^{-1} for mature trees). 4R nutrient management concept, exploiting criteria of right choice of fertilizers at right dose at right stage of right crop has been the most pivotal driving force towards improved nutrient-use-efficiency. However, citrus by the virtue of their perennial nature of woody framework (Nutrients locked therein), extended physiological stages of growth, differential root distribution pattern (root volume distribution), and growth stages from the point of view of nutrient requirement and preferential requirement of some nutrients by specific fruit crop, collectively make them nutritionally more efficient than the annual crops.

Our initial efforts were to: standardize optimum fertilizer requirement as recommended doses of fertilizers (RDF) through multi-location experiments, scheduling nutrient application across crop phenophases, scheduling fertigation by partitioning both nutrient and water requirement across critical growth stages, backed up by development of cultivar specific nutrient diagnostics, to be in a position to practice advanced methods of nutrient management. Attempts were later made to address these issues with combined use of concepts like Site Specific Nutrient Management (SSNM) and soil fertility spatial variogram-based fertilizer recommendations as decision support tool. The success of SSNM during the last 10 years has been prominently realized on a number of crops *viz.,* cereals, black gram, avocado, citrus, etc., to cite few success stories (Srivastava and Singh et al., 2008c). Recent review on the subject introduced a model to answer three questions pertaining to SSNM in production of perennial crops, namely, which input factors of crop production are limiting yield; what action should be taken to remove the limiting factors; and what is the potential gain in revenue from taking the action. The suggested model captured the essence of the law of the minimum in yield and revenue increase only if the limiting nutrients are appropriately adjusted (Srivastava, 2013c). Success of SSNM depends on correctness of measurement and understanding on variability in available supply of nutrients, which can be summarized in three steps *viz.,* (i) assessing variation, (ii) managing variation, and (iii) evaluation, but there is hardly any comprehensive coverage addressing the perennial crops. Spatial maps are fundamental to SSNM addressing variation either soil fertility or leaf nutrient composition because they represent either the spatial state of a growing condition. With new advances in technology, grid sampling for developing precision variogram is increasing. The first step in the process is to divide large fields into small zones using a grid. Next, a representative location within the grid is identified for precision soil sampling. (Srivastava et al., 2010) Grid sampling is integrated into Global Positioning System (GPS) based soil sampling and nutrient-mapping that in turn uses a Geographic Information Systems (GIS) to employ variable rate technology (VRT) for fertilizer applications (Srivastava and Singh, 2010). Optimum grid size soil mapping using specific nutrient-based spatial variograms as a interpretation tool popularly known as DRIS developed for Khasi mandarin of northeast India and Nagpur mandarin of central India aided in measuring the magnitude of changes in pool of soil available nutrients before and after fertilization (Srivastava and Singh, 2008b; Srivastava, 2013b).

In another long term experiment, the large fruit yield difference of 30.2 and 48.9 kg/tree initially observed on shallow soil (Typic Ustorthent) and deep soil (Typic Haplustert) in an orchard size of 11 ha, reduced to respective fruit yield of 62.7 and 68.5 kg/tree with corresponding fertilizer does (g/tree) of 1200 N – 600 P – 600 K – 75 Fe – 75 Mn – 75 Zn – 30 B, and 600 N – 400 P – 300 K – 75 Fe – 75 Mn –75 Zn – 30 B, suggesting the necessity of fertilizer application on variable rate application for rationality in fertilizer using SSNM as rationale of fertilizer use (Srivastava et al., 2009). The success of VRT depends to a large extent on the quality of fertility management maps. The ratio between nugget semivariance and total semivariance or still was used to define different classes of spatial dependence for leaf nutrient. If ratio was ≤25%, the leaf nutrient was considered to be strongly spatially dependent, or strongly distributed in patches; if ratio was between 26 and 75%, the leaf nutrient was considered weakly spatially dependent; if the ratio was 100%, or the slope of the semivariogram was close to zero, the leaf nutrient was considered as not being spatially correlated (pure nugget). Studies on the variations between ground truth yield versus two yield estimation methods [inverse distance weighted (IDW) and density grid method] using Pearson coefficient method further aided in better interpretation. Our studies generating databank on leaf analysis and fruit yield through exploration of 7 states across northeast India were analyzed through combined application of diagnosis and recommendation integrated system (DRIS) to determine leaf nutrient optima and GIS, which delineated major production zones facing minimum nutrient constraints through spatial variogram of nutrient constraints (Srivastava et al., 2006). In this background, the present review attempts to highlight certain pertinent issues on the necessity of developing SSNM concept as rationale of fertilizer use to harness better fertilizer use efficiency with emphasis on Citrus.

29.2 DEVELOPMENT OF RATIONALE FOR FERTILIZER USE

Of late, there has some distinct headways towards development of rationale of fertilizer use. In this regard, exploiting spatial variability has huge practical significance. Characterizing spatial variability of soil physico-chemical properties is a fundamental element of: (i) soil quality assessment, (ii) modeling non-point source pollutants in soil, and (iii) site-specific crop management. The heterogeneity of soil physico-chemical properties has been known since the classic study of Nielsen et al. (1973), which characterized the spatial variability of soil-water properties for a 150 ha field at the University of California's West Side Field Station in the San Joaquin Valley. The protocols developed by Corwin and Lesch (2005) for site specific evaluation of soil properties (EC appraisal) comprised of eight general steps: (i) site description, and EC_a, survey design; (ii) EC_a data collection with mobile GPS-based equipment; (iii) soil sampling design; (iv) soil core sampling; (v) laboratory analysis; (vi) calibration of EC_a to EC_e; (vii) spatial statistical analysis; and (viii) GIS database development and graphic display. Siqueira et al. (2010) suggested that relief may be considered an integrating factor that expresses the interaction of various soil and plant attributes. The study further analyzed the potential use of landforms to predict the variability of soil and orange attributes, with large spatial variability in soil and temporal variability in orange quality.

29.3 SOIL FERTILITY BASED SPATIAL VARIOGRAM AS A DECISION SUPPORT TOOL

A total of 56 grid based plants identified within the Nagpur mandarin orchard using two layered variograms (spatial variogram for fruit yield and other soil test values for various nutrients and delineating different management zones using DRIS-based analysis and formulated variable rates of fertilizer doses depending upon the soil test value of various grids (Srivastava, 2012b) These variable rates of fertilizer doses comprised of:

- 200 g N – 100 g P_2O_5 – 200 g K_2O – 100 g $FeSO_4$ – 100 g $MnSO_4$ – 100 g $ZnSO_4$
- 400 g N – 150 g P_2O_5 – 300 g K_2O – 200 g $FeSO_4$ – 200 g $MnSO_4$ – 300 g $ZnSO_4$
- 600 g N – 200 g P_2O_5 – 400 g K_2O – 300 g $FeSO_4$ – 300 g $MnSO_4$ – 300 g $ZnSO_4$

After the application of variable rates of fertilizers, the difference in yield responses over initial values, and likewise differences in soil test values with respect to $KMnO_4$-N, Olsen-P, NH_4OAc-K, DTPA-Fe, DTPA-Mn, DTPA-Cu and DTPA-Zn were computed and summarized as below highlighting the magnitude of different types of responses.

Response variability	Yield (qha^{-1})	N (mg kg^{-1})	P (mg kg^{-1})	K (mg kg^{-1})	Zn (mg kg^{-1})	Fe (mg kg^{-1})	Mn (mg kg^{-1})
Fertilizer doses: 200 g N – 100 g P_2O_5 – 200 g K_2O – 100 g $FeSO_4$ – 100 g $MnSO_4$ – 100 g $ZnSO_4$							
Initial	85.79	114.0	12.00	147.0	0.82	11.64	8.72
Final	163.74	116.8	12.78	152.3	0.88	12.42	9.68
Difference	37.45	2.8	0.78	5.3	0.06	0.77	0.96
Fertilizer doses: 400 g N – 150 g P_2O_5 – 300 g K_2O – 200 g $FeSO_4$ – 200 g $MnSO_4$ – 200 g $ZnSO_4$							
Initial	90.63	106.8	9.95	120.5	0.66	8.65	5.97
Final	102.30	113.3	11.36	128.9	0.78	10.35	8.12
Difference	11.68	6.63	1.40	8.41	0.12	1.70	2.15
Fertilizer doses: 600 g N – 200 g P_2O_5 – 400 g K_2O – 300 g $FeSO_4$ – 300 g $MnSO_4$ – 300 g $ZnSO_4$							
Initial	47.82	98.5	8.83	110.3	0.51	8.58	6.11
Final	85.79	117.3	11.07	138.9	0.87	11.29	9.24
Difference	37.95	19.01	2.23	28.9	0.35	2.70	3.30

Source: Srivastava (201 d).

The changes in area distribution either with regard to fruit yield or with regard to different nutrients ($KMnO_4$-N, Olsen-P, NH_4OAc-K, DTPA-Fe, DTPA-Mn and DTPA-Zn)as a result of variable doses of fertilizer application (Table 29.1) further revealed some distinct reorientation. Out of different grid sizes, the grid size upto 40 × 40 m^2, 42.42–43.02% area of the identified orchard (Total area of 11,184 sq.m) was observed to possess optimum fruit yield. While, sampling with both grid size of 60 × 60 m^2, as high as 68.33% area was observed having optimum yield, suggesting further the

TABLE 29.1 Distribution of Area Statistics (Sq. m) of Nutrients Under Different Grid Size Sampling at Village Ladgaon, Katol, Nagpur (Total Area: 11,184 sq.m)

	Grid size (m)					
	20 x 20		40 x 40		60 x 60	
Nutrient indices	Area (m^2)	% age	Area (m^2)	% age	Area (m^2)	% age
Iron						
Deficient	0.0	0.0	0.0	0.0	0.0	0.0
Low	7881.5	70.47	8780.0	78.50	8313.0	74.33
Optimum	3302.5	29.53	2404.0	21.50	2871.0	25.67
High	0.0	0.0	0.0	0.0	0.0	0.0
Excess	0.0	0.0	0.0	0.0	0.0	0.0
Manganese						
Deficient	711.8	6.34	400.3	3.58	108.5	0.97
Low	4926.7	44.05	4424.5	39.56	6561.0	58.67
Optimum	5545.0	49.60	6359.3	56.86	4514.5	40.36
High	0.0	0.0	0.0	0.0	0.0	0.0
Excess	0.0	0.0	0.0	0.0	0.0	0.0
Zinc						
Deficient	0.0	0.0	0.0	0.0	0.0	0.0
Low	1456.5	13.02	494.0	4.42	564.3	5.04
Optimum	9727.5	86.98	10690.0	95.58	10619.8	94.96
High	0.0	0.0	0.0	0.0	0.0	0.0
Excess	0.0	0.0	0.0	0.0	0.0	0.0
$KMNO_4$-N						
Deficient	0.0	0.0	0.0	0.0	0.0	0.0
Low	25.5	0.23	0.0	0.0	15.8	0.14
Optimum	11158.5	99.77	11184.0	100	11168.3	99.86
High	0.0	0.0	0.0	0.0	0.0	0.0
Excess	0.0	0.0	0.0	0.0	0.0	0.0
Olsen-P						
Deficient	0.0	0.0	0.0	0.0	0.0	0.0
Low	0.0	0.0	0.0	0.0	0.0	0.0
Optimum	11177.0	99.94	11164.5	99.83	11184.0	100
High	7.0	0.06	19.5	0.17	0.0	0.0
Excess	0.0	0.0	0.0	0.0	0.0	0.0
NH_4OAc-K						
Deficient	0.0	0.0	0.0	0.0	0.0	0.0

TABLE 29.1 (Continued)

	Grid size (m)					
	20 x 20		**40 x 40**		**60 x 60**	
Nutrient indices	**Area (m^2)**	**% age**	**Area (m^2)**	**% age**	**Area (m^2)**	**% age**
Low	8268.8	73.93	8097.0	72.40	8180.3	73.14
Optimum	2915.3	26.07	3087.0	27.60	3003.8	26.86
High	0.0	0.0	0.0	0.0	0.0	0.0
Excess	0.0	0.0	0.0	0.0	0.0	0.0
Fruit yield						
Deficient	62.3	0.56	3.3	0.03	3.0	0.02
Low	6377.3	57.02	6369.3	56.95	7663.5	68.53
Optimum	4744.5	42.42	4811.5	43.02	3517.5	31.45
High	0.0	0.0	0.0	0.0	0.0	0.0
Excess	0.0	0.0	0.0	0.0	0.0	0.0

Source: Srivastava (2013d).

optimum grid size of 40 × 40 m^2, using all the grid sizes (20 × 20 m^2 to 60 × 60 m^2) no significant area under any category was observed to be significantly influence since most of the orchard area 99.77 to as much as 10% area was observed within the optimum range. Across all the three grid size sampling from 20 × 20 m^2 to 60 × 60 m^2, an area ranging from 99.83 to as high as 100% area was observed to fall within optimum range. On the other hand, zoning of NH_4OAc-K based on variograms developed through soil sampling via three grid sizes of 20 × 20 m^2, 40 × 40 m^2 and 60 × 60 m^2, 73.93%, 72.40% and 73.14% area were observed within low range. While rest of the 26.07%, 27.60% and 26.86% area, respectively showed NH_4OAc-K within optimum range at grid size of 20 × 20 m^2, 40 × 40 m^2 and 60 × 60 m^2. Under grid size of 20 × 20 m^2, 40 × 40 m^2 and 60 × 60 m^2, respectively, 29.54%, 21.50% and 25.67% area of the orchard was observed displaying optimum range of DTPA-Fe. While other 70.47%, 78.50% and 74.33% area of the orchard showed low level of DTPA-Fe under grid size sampling of 20 × 20 m^2, 40 × 40 m^2 and 60 × 60 m^2. DTPA-Mn showed a distant variation under differing zone when compared across three-grid size sampling. Under the grid size sampling 20 × 20 m^2, 6.34%, 44.05 and 49.60% area of the orchard was registered to fall within deficient, low and optimum range, respectively. Likewise under grid size of 40 × 40 m^2, 3.5%, 39.56% and 56.86% area of the orchard, respectively, showed low, optimum and optimum range of DTPA-Mn. These changes were well corroborated with changes in fruit yield zones (Figure 29.1).

29.4 DEVELOPMENT OF FERTILIZER PREDICTION MODELS

Basic data developed for targeted yield using prediction model:

Nutrients	Parameters		
	NR	CS	CF
Nitrogen	4.49	81.5	34.3
Phosphorous	0.27	65.0	5.55
Potassium	1.10	21.1	64.9
Zinc	0.01	30.4	1.23
Iron	0.04	8.69	2.06
Manganese	0.01	1.98	1.30

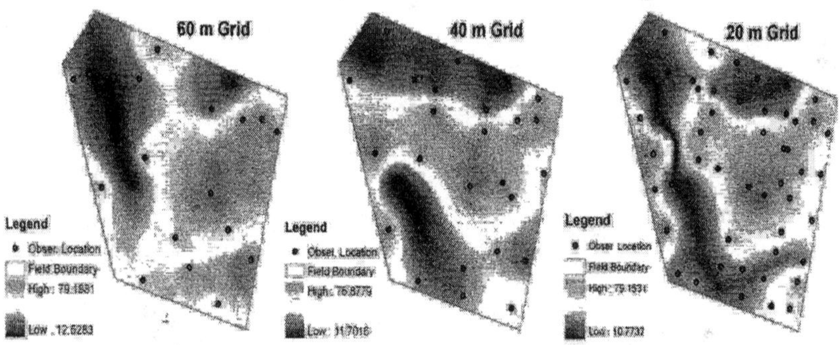

a. Spatial variation in FRUIT YIELD based on actual FRUIT YIELD data at different grid sampling.

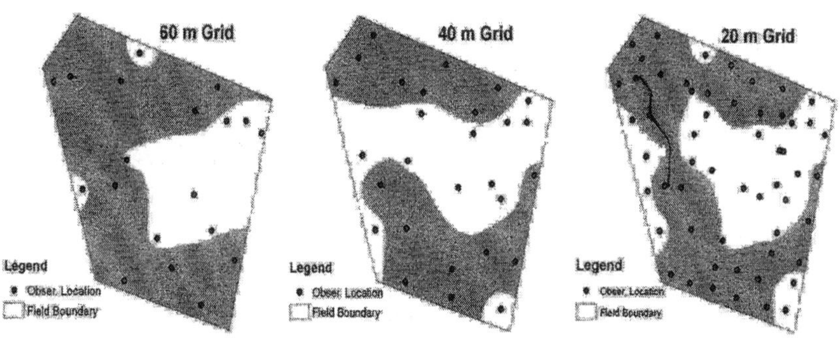

b. Spatial variation in FRUIT YIELD based on DRIS-based different FRUIT YIELD classes at different grid sampling

FIGURE 29.1 Spatial variation in fruit yield. (Adapted from Srivastava, A. K. 2012b and Srivastava, A. K. 2013d).

In the above table, NR (Nutrient requirement) computed as the amount of nutrient required (kg) to produce one quintal of fruits, using formula: NR = Uptake of nutrient (kg ha^{-1})/Fruit yield (q ha^{-1}) CS (Percent contribution from soil to total nutrient uptake) computed as the amount of per cent contribution of soil nutrient, using formula: CS = Uptake of nutrient (kg ha^{-1}) by fruits × 100/Available soil test value for nutrient (kg ha^{-1}) from control treatment CF (Percent contribution from fertilizer to total uptake) computed as the amount of per cent contribution of fertilizer nutrient, using the formula: CF = Uptake of nutrient (kg ha^{-1}) from fruit crop)-(Avail. STV from nutrient (kg ha^{-1}) × CS/100 × 100/Fertilizer nutrient applied (kg ha^{-1}). The Following predication equations for computing fertilizer doses were developed:

i. Fertilizer Nitrogen = 13.09 (Targeted Fruit Yield) – 2.37 (Soil test value for Nitrogen);
ii. Fertilizer Phosphorous = 4.08 (Targeted Fruit Yield) – 26.83 (Soil test value for Phosphorous);
iii. Fertilizer Potassium = 1.69 (Targeted Fruit Yield) – 0.39 (Soil test value for Potassium);
iv. Fertilizer Zinc = 0.98 (Targeted Fruit Yield) – 24.73 (Soil test value for Zinc);
v. Fertilizer Iron = 1.94 (Targeted Fruit Yield) – 4.21 (Soil test value for Iron);
vi. Fertilizer Manganese = 0.77 (Targeted Fruit Yield) – 1.52 (Soil test value for Manganese).

Using these fertilizer prediction equation, ready reckoner (Table 29.2) developed for different soil test values-based fertilizer recommendation in relation to different target fruit levels, have shown good promise to rationalized the soil test-crop response based fertilizer recommendations.

29.5 SITE SPECIFIC NUTRIENT MANAGEMENT

Future gains in productivity and input use efficiency will require soil and crop management technologies that are knowledge-intensive, and are tailored to specific characteristics of individual farms or fields to manage the variability that exists between and within them (Tiwari, 2007a). The SSNM approach is one such option that has been tried successfully in India using

TABLE 29.2 Soil Test-Based Fertilizer Recommendation in Relation to Different Targeted Fruit Yield of Nagpur Mandarin as a Ready Reckoner

Soil test value (kg/ha)	Targeted fruit yield (q/ha)				
	100	150	200	250	270
Soil nitrogen test-based fertilizer recommendation					
100	107.2	172.65	238.10	303.55	369.0
200	101.27	166.72	232.17	297.62	363.07
300	95.35	160.80	226.25	291.70	357.15
400	89.42	154.87	220.32	285.77	345.30
500	83.50	148.95	214.4	279.85	315.22
Soil phosphorous test-based fertilizer recommendation					
5	345.85	585.85	825.85	106.5	130.5
10	211.7	451.7	691.7	931.7	117.1
15	77.55	317.55	557.55	797.55	103.7
20	–56.6	183.4	423.4	663.4	903.4
25	–190.8	49.25	289.25	529.25	769.25
Soil potassium test-based fertilizer recommendation					
100	130.10	214.70	299.30	383.90	468.50
200	91.00	175.60	260.20	344.80	429.40
300	51.90	136.50	221.10	305.70	390.30
400	12.80	97.40	182.00	266.60	351.20
500	–26.30	58.30	142.90	227.50	312.10
Soil zinc test-based fertilizer recommendation					
0.89	75.84	124.84	173.84	222.84	242.44
1.79	53.68	102.68	151.68	200.68	220.28
2.68	31.53	80.53	129.53	178.53	198.13
3.58	9.37	58.37	107.37	156.37	175.97
4.48	–12.79	36.21	85.21	134.21	153.81
Soil iron test-based fertilizer recommendation					
8.96	156.28	253.28	350.28	427.88	486.08
13.4	137.42	234.42	331.42	409.02	467.22
17.9	118.56	215.56	312.56	390.16	448.36
22.4	99.70	196.70	293.70	371.30	429.50
26.8	80.84	177.84	274.84	352.44	410.64
Soil manganese test-based fertilizer recommendation					
11.2	59.98	98.48	136.98	175.48	190.88
20.1	46.36	84.86	123.36	161.86	177.26

TABLE 29.2 (Continued)

Soil test value (kg/ha)	Targeted fruit yield (q/ha)				
	100	150	200	250	270
29.1	32.74	71.24	109.74	148.24	163.64
38.0	19.12	57.62	96.12	134.62	150.02
44.8	8.90	47.40	85.90	124.40	139.80

Source: Srivastava (2013 d).

different approaches. A plant-based SSNM approach developed in the early 1990s by the International Rice Research Institute in collaboration with partners has been adopted by IPNI, India. The scientific principles of SSNM were compiled into a practical into a practical guidebook to nutrient management for rice (Dobermann et al., 2003a, 2003b). Numerous experimental results have proved that application of SSNM technologies to improve productivity in the country is a necessity. However, large scale implementation of SSNM technologies has proved elusive due to lack of adequate soil testing infrastructures. Large variation in tree canopy and subsequently, the tree-to-tree yield difference are common in many of the large sized citrus orchards. Knowing the required nutrients for all stages of growth, and understanding the soil's ability to supply those needed nutrients are critical to profitable crop production. The recommendations on fertilizer application may not, however, produce the same magnitude of yield response when practiced in an orchard of large area, because of its inability to accommodate variation in soil fertility status.

Spatial variability studies (Figure 29.2) carried out in an orchard identified two different soil types, *viz.*, S_1 (Typic Ustorthent) and S_2 (Typic Haplustert). Comparison of SSNM versus conventional fertilizer response as RDF (T_{16}) and farmers practices (T_7) showed a significant increase in fruit yield alongwith fruit quality due to SSNM treatment which doubled after 3 years of experimentation on both the soil types (Table 29.2) compared to either recommended practices RDF (T_{16}) or farmers' practices (T_7) suggesting better fertilizer-use-efficiency through SSNM. On the other hand, such studies also showed that the current farmers' practices have a vast potential of improvement. SSNM induced significant changes in leaf nutrient concentrations in response to differential SSNM based fertilization. Best SSNM treatments (T_9 and T_6 on S_1 and S_2 soil type, respectively) elevated leaf N, P, K and Zn concentrations significantly compared to either farmers' practices (T_7) or RDF (T_{16}), the recommended practices in vogue in the region. All the nutrients

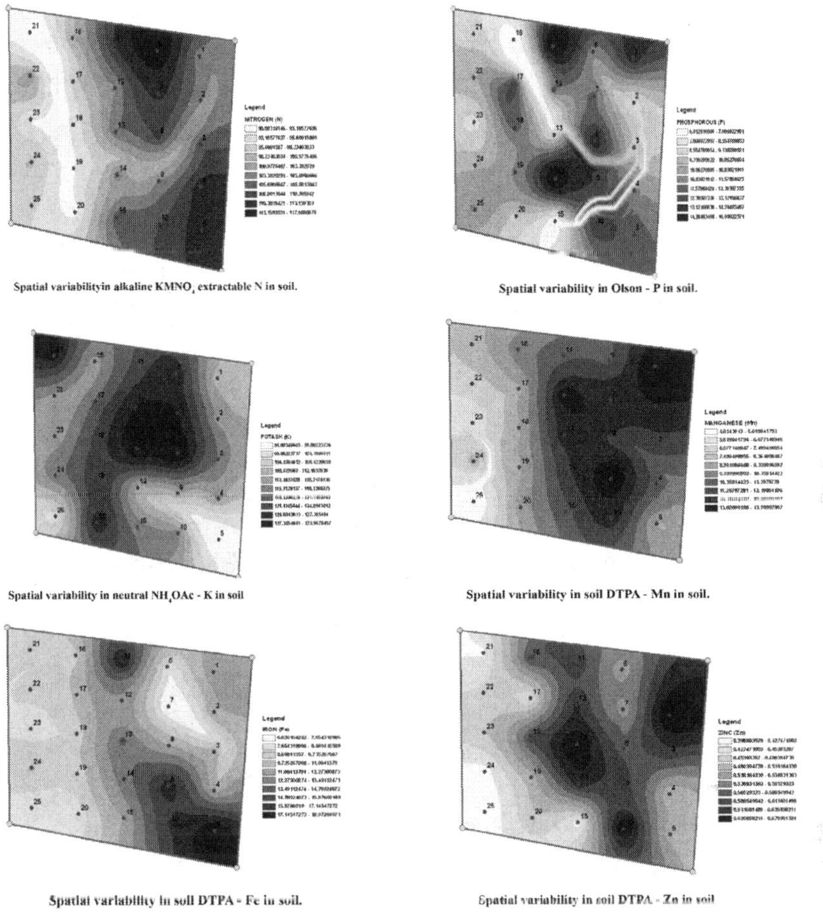

FIGURE 29.2 Types of spatial variability studies. (Srivastava, A. K. 2013c)

were within the deficient range in trees fertilized with farmer's practices, which attained an optimum level after 2 years through site-specific fertilizer treatments. Application of K increased leaf Zn concentration irrespective of soil type or whether any micronutrient was included in the treatment. However, the effect of K was greatest when co-applied with the micronutrients, and the effect increased as K supply increased. Hence, K application improved the efficacy of applied Zn, because of the similar metabolic pathways are involved with regard to their biochemical transformation. According to Tucker et al. (1995), as much as 30% saving in granular fertilizer cost was estimated for Valencia grove if variable N rates were implemented on per tree basis, ranging from 129 to 240 kg N/ha/year. For comparison purposes, the eastern half of the grove received the full uniform rate of 240 kg N/ha/year, ensuring that the granules are accurately placed adjacent to the tree. Application of variable fertilizer rate technology in this grove saved in nitrogen equivalent to the 32 to 43% reduction of N rates achieved through use of fertigation and foliar sprays of urea (Lamb et al., 1999). These observations suggested that SSNM-based fertilization taking advantage of spatial variability in deficient nutrients and accordingly tailoring anticipated fertilizer requirement within an orchard, could play a vital role in not only rationalizing the fertilizer use but reducing the production cost as well.

Fruit yield potential was observed to be much higher on S_2 soil type (39.2–68.5 kg/tree) than on S_1 soil type (27.9–52.7 kg/tree), irrespective of treatments. However, maximum fruit yield of 52.7 kg/tree and 68.5 kg/tree was observed with treatments T_9 and T_6, respectively, displaying the difference in response behavior of two contrasting soil types. Treatments involving micronutrients, T_1 (58.9 and 58.9 kg/tree) and T_{11} (39.3 and 56.1 kg/tree) showed higher fruit yield over treatments T_7 (25.1 and 39.2 kg/tree) and T_{15} (30.0 and 45.5 kg/tree), respectively. The response of P was not evident with treatment T_2 (30.6 kg/tree), but became significant with treatment T_4 having higher P (39.2 kg/tree) over no P though treatment T_3 (27.9 kg/tree) on S_1 soil type with no response of P at all on S_2 soil type. Increasing doses of K from K_0 (T_5) to K_{900} (T_8), fruit yield maximized with treatment T_8 (49.9 kg/tree) and T_4 (58.0 kg/tree), respectively, on S_1 and S_2 soil type. While at higher N+P (N_{1200} + P_{600}) level, increasing K from K_0 (T_{13}) to K_{1500} (T_{12}), fruit yield maximized with treatment T_9 (52.7 and 60.9 kg/tree) on both the soil types. Analysis of tree size of 3040 trees space of 40-acre grove showed a skewed distribution with 51.1% trees having 25–100 m^3/tree size classes and a median size of 82 m^3/tree. At a uniform fertilization rate of 240 kg N/ha/year, the leaf N concentration of 12 trees with different canopy sizes that were randomly sampled in the grove showed optimal levels (2.4–2.6%) in the large trees and excess levels (> 3%) in the medium to small trees (Tucker et al., 1995). From the regression line, trees with excess N had canopies < 100 m^3/tree, and constituted 62% of the grove. Under such conditions, variable rate fertilization can, therefore, save production costs, reduce N leaching and increase yields per variable acre (Schumann et al., 2003). All the three fruit quality parameters were significantly affected by different treatments on both the soil types (Table 29.1). The maximum juice content of 48.6% and 49.8% was observed with treatments T_8 and T_6, respectively, in S_1 and S_2 soil types. On S_1 soil type, the juice content increased by 6.7–6.8% considering T_5–T_8 involving application of K_2O from 0 to 900 g/tree along with N_{600} + P_{400} + M_1S_1 or K_2O from 0 to 1500g/tree along with N_{1200} + P_{600} + M_1S_1. While on S_2 soil type, increase in juice content was comparatively higher (5.8%) at lower nutrient level of N_{600} + P_{400} + M_1S_1 (T_5–T_8) compared to higher nutrient level of N_{1200} + P_{600} + M_1S_1 (T_9–T_{13}). However, juice content was not significantly influenced by micronutrient application, since treatments, T_7 versus T_1 and T_{15} versus T_{11} were statistically at par with each other irrespective of soil type. These observations suggested that on deep S_2 soil type, higher K-reserve coupled with available water content, fruit quality development could be achieved at a much lower nutrient dose. Total soluble soils (TSS) on the other hand observed the reverse pattern of response and remained unaffected by micronutrient or secondary nutrient fertilization. There was a decrease in total soluble solids from 9.6% with T_5 (N_{600} + P_{400} + K_0 + M_1S_1) to as low as 8.2% with treatment T_8 (N_{600} + P_{400} + K_{900} + M_1S_1) on S_1 soil type likewise, increasing K application from 600 g/tree to 1500 g/tree with N_{1200} + P_{600} (T_9–T_{13}) produced no significant change in TSS (8.2%).

29.6 VARIABLE RATE FERTILIZER APPLICATION

Variable rate fertilization is one of the most effective techniques for rationale use of fertilizers executed by matching the fertilizer rate with tree requirement on a per tree size basis (Miller et al., 2005). Site specific management of 17-year-old 'Valencia' grove (2980 trees) in Florida using automated sensor system equipped with differential global positioning system and variable rate delivery of fertilizers (135–170 kg N/ha/year) on a tree size basis (0–240 m^3/tree), achieved a 38–40% saving in granular fertilizers cost. While, conventional uniform application rate of 270 kg N /ha/year showed that trees with excess nitrogen (>3%) had canopies less than 100 m^3 with lower fruit yield and inferior quality (Zaman et al., 2005). In another long term experiment, the large fruit yield difference of 30.2 and 48.9 kg/tree initially observed on shallow soil (Typic Ustorthent) and deep soil (Typic Haplustert) in an orchard size of 11 ha, reduced to respective fruit yield of 62.7 and 68.5 kg/tree with corresponding fertilizer does (g/tree) of 1200 N – 600 P – 600 K – 75 Fe – 75 Mn -75 Zn – 30 B, and 600 N – 400 P – 300 K – 75 Fe – 75 Mn –75 Zn – 30 B, suggesting the necessity of fertilizer application on variable rate application for rationality in fertilizer use (Srivastava et al., 2006; Srivastava, 2013c).

Analysis of tree size of 3040 trees space of 40-acre grove showed a skewed distribution with 51.1% trees having 25–100 m^3/tree size classes and a median size of 82 m^3/tree. At a uniform fertilization rate of 240 kg N/ha/year, the leaf N concentration of 12 trees with different canopy sizes that were randomly sampled in the grove showed optimal levels (2.4–2.6%) in the large trees and excess levels (> 3%) in the medium to small trees (Tucker et al., 1995). From the regression line, trees with excess N had canopies < 100 m^3/tree, and constituted 62% of the grove. Under such conditions, variable rate fertilization can, therefore, save production costs, reduce N leaching, and increase yields per variable acre (Schumann et al., 2003). A 30% saving in granular fertilizer cost was estimated for this 'Valencia' grove if variable N rates were implemented on a per tree basis ranging from 129 to 240 kg N/ha/year. For comparison purposes, the eastern half of the grove received the full uniform rate of 240 kg N/ha/year. No fertilizer was allocated by spreader to skips or resets of one-to-three year age. Due to a very restricted root system, new resets should be fertilized individually, usually by hand (Tucker et al., 1995), ensuring that the granules are accurately placed adjacent to the tree. Application of variable fertilizer rate technology in this grove saved in nitrogen equivalent to the 32 to 43% reduction of N rates achieved through use of fertigation and foliar sprays of urea (Lamb et al., 1999). Elprince (2009) suggested multivariable fertilizer recommendation models allowing necessary adjustments in variable rate fertilizer applications. Later, these logistic models were cross validated and combined in a GIS to derive N and K fertilization class maps using kriged-interpolated data sets of the significant site variables. Chan et al. (2002) studied the effect of boundary determination, interpolation method and GPS location error for determining variable rate nitrogen application map based upon yield maps. A general linear model for mean absolute error approximated the error effects.

29.7 STRATEGIES AND OPPORTUNITIES

The fall out of a generalized fertilizer s recommendation over large areas of small-scale farming leads to the possibility of over or under-application of nutrients, by and large, with its economic and environmental consequences (Srivastava 2013a, 2013e). The more apparent consequences of falling productivity and nutrient efficiency, multi-nutrient deficiencies, increasing pace of nutrient mining and falling farm income are highlighted by earlier researchers (Ghosh et al., 2004; Srivastava and Singh, 2004a, 2005,

2008b; Tiwari, 2007a). The environmental impacts are not very apparent yet, probably because of the generally low nutrient application rates, except few crops of very high commercial importance. The SSNM, on the other hand, is an approach for feeding crops with nutrients, as and when they are needed, considering inherent spatial variability associated with fields under crop production. The SSNM also avoids indiscriminate use of fertilizers by preventing excessive/ inadequate rates of fertilization, ensuring that all the required nutrients are applied at proper rates and in proper ratios commensurating with the crop's nutrient demands.

The work on precision citriculture in Florida (Whitney et al., 1999) is carried out with the following objectives: (i) develop a system to map citrus yields using conventional manual harvest labor and electronically record harvester identity associated with each citrus container loaded in the grove; (ii) develop a system to measure and map tree location, canopy volume, and height in citrus grove; (iii) determine the feasibility of using GPS/GIS for variable rate application of fertilizers and pesticides, and for monitoring, tracking, and controlling grove equipment; and (iv) determine what GPS/GIS information is most valuable and how it may be used effectively to improve management of production and harvesting operations. The increased availability of hyperspectral imaging sensors and advanced analysis tools like partial least squares regression and spectral mixing techniques will further facilitate extend this concept at canopy level. The newer approaches using laser induced fluorescence have considerable potential for pre-visual identification of nutrient and water stress in addition to detection of optimum levels of growth and yield under different fertilization rates under orchard conditions (Srivastava and Singh, 2001, 2002) (Table 29.3).

The most important step towards the calibration of site specific fertilizer requirement is the estimation of the indigenous nutrient supplies, which we define as the cumulative amount of nutrient originating from all indigenous sources (Srivastava et al., 2007). There are several approaches to determine indigenous nutrient supply. Thus far, the most popular method in India has been soil testing as it proved to a rapid and reliable indicator for many nutrients. However, the staggering number of land holdings in India and the meager soil testing infrastructure pose a major challenge to wide-scale adoption of SSNM technologies in India. The major challenges for SSNM research and extension in future will be two-fold: i. to retain the demonstrated potential of the approach and ii. to build upon what has already been achieved while reducing the complexity of the technologies as it is disseminated to farmers (Johnston et al., 2009). However, Zhang et al. (2010) suggested that the regionalized maps

TABLE 29.3 Response of Different Site Specific Nutrient Management-Based Treatments on Canopy Growth, Fruit Yield and Quality of Nagpur Mandarin on Typic Ustorthent (S_1) and Typic Haplustert Soil (S_2) Soil Types (Pooled data)

Treatments	Canopy volume (m^3)*		Fruit yield (kg/tree)		Fruit quality parameters (%)					
					Juice Content		TSS		Acidity	
	S_1	S_2	S_1	S_2	S_1	S_2	S_1	S_2	S_1	S_2
T_1 ($N_{600} + P_{400} + K_{600} + M_1S_1$)	3.9 (25.7)	3.5 (19.3)	37.4	58.9	45.7	45.5	8.2	8.1	0.56	0.77
T_2 ($N_{600} + P_{200} + K_{600} + M_1S_1$)	3.7 (27.3)	2.7 (21.3)	30.6	57.2	44.5	41.6	8.5	7.6	0.64	0.68
T_3 ($N_{600} + P_0 + K_{600} + M_1S_1$)	3.4 (29.5)	3.1 (23.0)	27.9	57.1	44.1	42.4	9.1	8.4	0.60	0.75
T_4 ($N_{600} + P_{600} + K_{600} + M_1S_1$)	4.6 (28.7)	2.9 (20.6)	39.2	58.0	44.7	43.7	8.8	7.9	0.63	0.68
T_5 ($N_{600} + P_{400} + K_0 + M_1S_1$)	4.2 (31.7)	2.4 (21.4)	33.4	55.3	41.9	42.4	9.6	8.7	0.56	0.64
T_6 ($N_{600} + P_{400} + K_{300} + M_1S_1$)	4.7 (32.6)	5.4 (21.7)	33.9	68.5	44.9	49.8	9.3	8.6	0.58	0.67
T_7 ($N_{600} + P_{400} + K_{600} + M_0S_0$)	3.8 (27.0)	2.6 (19.6)	25.1	39.2	45.2	46.5	8.6	7.8	0.62	0.81
T_8 ($N_{600} + P_{400} + K_{900} + M_1S_1$)	5.7 (28.5)	4.3 (20.5)	49.9	48.7	48.6	48.2	8.2	7.9	0.75	0.82
T_9 ($N_{12\,00} + P_{600} + K_{600} + M_1S_1$)	6.6 (31.2)	3.7 (23.4)	52.7	60.9	45.4	42.7	8.9	8.8	0.55	0.62

T_{10} (N_{1200} + P_{600} + K_{900} + M_1S_1)	6.6	3.3	41.8	50.4	42.6	43.6	8.6	8.2	0.59	0.71
	(29.1)	(23.4)								
T_{11} (N_{1200} + P_{600} + K_{1200} + M_1S_1)	5.8	3.9	39.3	56.1	44.9	44.8	8.5	8.1	0.62	0.76
	(30.0)	(20.5)								
T_{12} (N_{1200} + P_{600} + K_{1500} + M_1S_1)	4.6	4.3	36.3	56.3	48.4	46.3	8.2	7.6	0.80	0.86
	(28.4)	(19.7)								
T_{13} (N_{1200} + P_{600} + K_0 + M_1S_1)	3.8	3.3	33.3	46.5	41.6	42.5	8.2	9.1	0.51	0.66
	(26.6)	(19.5)								
T_{14} (N_{1200} + P_{600} + K_{1200} + M_0 S_1)	4.5	2.9	33.9	46.3	43.2	43.7	9.5	8.5	0.64	0.77
	(28.1)	(21.2)								
T_{15} (N_{1200} + P_{600} + $K_{12\,00}$ + M_0S_0)	3.9	2.9	30.0	45.5	44.6	43.8	9.6	8.2	0.63	0.74
	(27.2)	(23.2)								
T_{16} (N_{600} + P_{200} + K_{100}) – RDF	3.5	3.0	31.5	43.7	44.0	43.1	8.6	8.5	0.57	0.68
	(24.5)	(23.7)								
LSD (*p=0.05*)	1.2	0.8	8.0	8.1	3.1	2.2	0.5	0.6	0.09	0.08

M_1 stands for application of 300 g each of $ZnSO_4$, $FeSO_4$, $MnSO_4$ and 100 g borax/tree;

Mo stands for no application of micronutrient fertilizers;

S_1 stand for application of 400 g $MgSO_4$/tree and 100 g elemental S/tree;

So stands for no application of Mg and S;

RDF stands for recommended doses of fertilizers;

*Expressed as increase over initial value as given in parenthesis.

are practical alternative to site-specific soil nutrient management approaches in areas where it is not practical to implement SSNM due to small field size or other constraints to use intensive soil sampling and chemical analyzes.

ACKNOWLEDGMENT

First author is thankful to International Plant Nutrition Institute, Gurgaon, Haryana for sanctioning a series of research projects namely, (i) Site specific nutrition management in Nagpur mandarin grown in central India, (ii) Site specific nutrient management in mosambi sweet orange (*Citrus sinensis* Osbeck), and (iii) Development of soil fertility map as a decision support tool for fertilizer recommendation in citrus, from which majority of work has originated. We place on record the guidance of Dr. K.N. Tiwari, Ex-Director, IPNI, Gurgaon, Haryana in various capacities. The cooperation and guidance received from Dr. A.K. Joshi, General Manager, Regional Remote Sensing Centre – Central, Nagpur is also thankfully acknowledged.

KEYWORDS

- **Diagnosis and Recommendation Integrated System**
- **Geographic Information Systems**
- **Global Positioning System**
- **Recommended Doses of Fertilizers**
- **Site Specific Nutrient Management**
- **Total Soluble Soils**
- **Variable Rate Technology**

REFERENCES

Chan, C. W., Schueller, J. K., Miller, W. M., Whitney, J. D., Wheaton, T. A., & Cornell, J. A. (2002). Error sources on yield-based fertilizer variable rate application maps. *Precision Agriculture 3*, 81–94.

Corwin, D. L., & Lesch, S. M. (2005). Characterizing soil spatial variability with apparent soil electrical conductivity I. Survey protocols. *Computers and Electronics in Agriculture, 46*, 103–33.

Dobermann A., Witt C., Abdulrachman S., Gines H. C., Nagarajan R., Son, T. T., Tan, P. S., Wang, G. H., Chien, N. V., Thoa, V. T. K., Phung, C. V., Stalin, P., Muthukrishnan, P.,

Ravi, V., Babu, M., Simbahan, G. C., & Adviento, M. A. A. (2003a). Soil fertility and indigenous nutrient supply in irrigated rice domains of Asia. *Agronomy Journal, 95*, 913–923.

Dobermann A., Witt C., Abdulrachman S., Gines, H. C., Nagarajan, R., Son, T. T., Tan, P. S., Wang, G. H., Chien, N. V., Thoa, V. T. K., Phung C. V., Stalin P., Muthukrishnan P., Ravi V., Babu M., Simbahan, G. C., & Adviento, M. A. A. (2003b). Estimating indigenous nutrient supplies for site-specific nutrient management in irrigated rice. *Agronomy Journal, 95*, 924–935.

Elprince, Adel, M. (2009). Prediction of soil fertilization maps using logistic modeling and a geographical information system. *Soil Science Society of America Journal, 73*, 2032–2042.

Ghosh, P. K., Bandopadhyay, K. K., Mishra, A. K., & Subbarao, A. (2004). Balanced fertilization for maintaining soil health and sustainable agriculture. *Fertilizer News, 49*, 13–24.

Johnston, A. M., Khurana, H. S., Majumdar, K., & Satyanarayana, T. (2009). Site Specific Nutrient Management – Concept, Current Research and Future Challenges in Indian Agriculture. *Journal of the Indian Society of Soil Science, 57*, 1–10.

Lamb, S. T., Graham, W. D., Harrison, C. B., & Alva, A. K. (1999). Impact of alternative citrus management practices and groundwater nitrate in the Central Florida ridge. Field investigation, *Transactions of American Society of Agriculture Engineers, 42*, 1653–1668.

Miller, W. M., Schumann, A. W., Whitney, J. D., & Buchanon, S. (2005). Variable rate applications of granular fertilizer for citrus test plots. *Applied Engineering in Agriculture, 21*(5), 795–702.

Nielsen, D. R., Biggar, J. W., & Erh K. T. (1973). Spatial variability of field-measured soil-water properties. *Hilgardia, 42*(7), 215–259.

Schumann, A. W., Fares, A., Alva, A. K., & Paramasivam, S. (2003). Response of 'Hamlin' orange to fertilizer source, annual rate and irrigated area. *Proceedings of Florida State Horticulture Society, 116*, 256–260.

Siqueira, D. S., Marques, Jr. J., & Pereira, G. T. (2010). The use of landforms to predict the variability of soil and orange attributes. *Geoderma, 155*(1–2), 55–66.

Srivastava, A. K. (2006). Shyam Singh, & Tiwari, K. N. Site specific nutrient management for Nagpur mandarin (*Citrus reticulata* Blanco) *Better Crops, 88*, 22–25.

Srivastava, A. K. (2007). Shyam Singh, & Tiwari, K. N. Diagnostic tools for citrus: Their use and implications in India. *Better Crops – India, 1*, 26–29.

Srivastava, A. K. (2012b). Development of soil fertility map as a decision support tool for fertilizer recommendation in citrus. Annual Report 2011–12. National Research Centre for Citrus, Nagpur, Maharashtra, pp. 48–52.

Srivastava, A. K. (2013a). Nutrient deficiency symptomology in citrus: An effective diagnostic tool or just an aid for post–mortem analysis. *Agricultural Advances, 2*, 177–94.

Srivastava, A. K. (2013b). Recent developments in diagnosis and management of nutrient constraints in acid lime. *Scientific Journal of Agricultural* (*Agricultural Advances*), *2*, 86–96.

Srivastava, A. K. (2013c). Site specific nutrient management in citrus. *Scientific Journal of Agricultural (Agricultural Advances), 2,* 1–15.

Srivastava, A. K. (2013d). Development of soil fertility map as a decision support tool for fertilizer recommendation in citrus. Annual Report 2012–13. National Research Centre for Citrus, Nagpur, Maharashtra, pp. 40–42.

Srivastava, A. K. (2013e). Nutrient diagnostics in citrus: Are they applicable to current season crop. *Agrotechnology, 2*, 104–105. (doi: 10.4172/2168P-9881.1000e104).

Srivastava, A. K., & Shyam Singh. (2001). Development of optimum soil property limits in relation to fruit yield and quality of *Citrus reticulata* Blanco cv. Nagpur mandarin. *Tropical Agriculture*, *78*, 174–181.

Srivastava, A. K., & Shyam Singh. (2002). Soil analysis based diagnostic norms for Indian citrus cultivar. *Communications in Soil Science and Plant Analysis, 33*, 1689–1706.

Srivastava, A. K., & Shyam Singh. (2004a). Soil and plant nutritional constraints contributing to citrus decline in Marathawada region, India. *Communications in Soil Science and Plant Analysis, 35*, 2537–2550.

Srivastava, A. K., & Shyam Singh. (2005). Diagnosis of nutrient constraints in citrus orchards of humid tropical India. *Journal of Plant Nutrition, 29*, 1061–1076.

Srivastava, A. K., & Shyam Singh. (2008a). Analysis of citrus orchard efficiency in relation to soil properties. *Journal of Plant Nutrition, 30*, 2077–2090.

Srivastava, A. K., & Shyam Singh. (2008c). Citrus nutrition research in India: Problems and prospects. *Indian Journal of Agricultural Sciences, 78*, 3–16.

Srivastava, A. K., & Shyam Singh. (2010). Citrus production zones for citrus industry. *ICAR News.* 16, 7.

Srivastava, A. K., Huchche, A. D., Lallan Ram, & Shyam Singh. (2001a). Pre-and post-harvest response of Nagpur mandarin (*Citrus reticulata* Blanco) to K fertilization in Typic *Haplustert. Journal of Potash Research. 17*, 93–97.

Srivastava, A. K., Kohli, R. R., Dass, H. C., Huchche, A. D., & Lallan Ram. (1999). Evaluation of the nutritional status of Nagpur mandarin (*Citrus reticulata* Blanco) by foliar sampling. *Tropical Agriculture,72*, 93–98.

Srivastava, A. K., Shyam Singh., Diware, V. S., & Haramandeep Singh. (2009). Site-specific nutrient management in 'Mosambi' sweet orange. *Better Crops – India, 3*, 10–11.

Srivastava, A. K., Shyam Singh; Das, S. N., & Harmandeep Singh. (2010). Delineation of productivity zones in mandarin orchards using DRIS and GIS. *Better Crops – South Asia, 4*, 13–15.

Tan, Z. X., Lal, R., & Wiebe, K. D. (2005). Global soil as nutrient depletion and yield reduction. *Journal of Sustainable Agriculture. 26*, 123–146.

Tiwari, K. N. (2007a). Breaking yield barriers and stagnation through site-specific nutrient management. *Journal of the Indian Society of Soil Science, 55*, 444–454.

Tiwari, K. N. (2007b). Reassessing the role of fertilizers in maintaining food, nutrition and environmental security. *Indian Journal of Fertilizer, 3*(1), 33–48 & 51–52.

Whitney, J. D., Miller, W. M., Wheaton, T. A., Salyani, M., & Schueller, J. K. (1999). Precision farming applications in Florida citrus. *Applied Engineering in Agriculture, 15*(5), 399–303.

Zaman, Q. U., Schumann, A. W., & Miller, W. M. (2005). Variable rate nitrogen application in Florida citrus based on ultrasonically-sensed tree size. *Applied Engineering in Agriculture, 21*(3), 331–336.

Zhang, Q., Yang, Z., Li, Y., Chen, D., Zhang, J., & Chen, M. (2010). Spatial variability of soil nutrients and GIS-based nutrient management in Yongji County, China. *International Journal of Geographical Information Science, 24*(7), 965–981.

CHAPTER 30

IMPACT OF MAHATMA GANDHI *NREGS* ON LAND USE PATTERN AND NATURAL RESOURCE MANAGEMENT IN THE DROUGHT-PRONE RAYALASEEMA REGION OF ANDHRA PRADESH, INDIA

V. SURESH BABU[1] and G. RAJANIKANTH[1]

[1]*Associate Professor, National Institute of Rural Development and Panchayati Raj (NIRD&PR), Hyderabad – 500 030, India*

CONTENTS

ABSTRACT

The Ministry of Rural Development (MoRD), Government of India, has launched a Rights Based and People-centric wage employment program known as Mahatma Gandhi National Rural Employment Guarantee Scheme (Mahatma Gandhi NREGS), which essentially combines both social security and rejuvenation of agriculture by creating durable community and individual assets. The Scheme incorporated a component, known as category IV works, to improve the productivity of marginal and small farms by providing irrigation, land development and development of plantations. Implementation of category IV works is the adoption of block cum group approach in Chittoor and individual approach in Anantapur has resulted in higher investment by the beneficiaries. While 85% of the beneficiaries made own investment (Rs. 16,000/acre) in Chittoor, only 39% of the Anantapur beneficiaries (Rs. 1630/acre) did so. Works in Chittoor focused more on development of waste land as a block and converging with other schemes to cover activities not permissible under Mahatma Gandhi NREGS and beneficiaries not eligible under category IV works. The impact was an increase in cropping intensity and continuous employment in agriculture, increase in labor absorption in agriculture and improvements in the livelihoods. The results are divergent because of the variations in agro-climatic conditions in the study area. The category IV works under Mahatma Gandhi NREGS need to be restricted to agriculturally backward areas through participatory planning approach at Gram Panchayat level with focus on macro plan for land use and resource development and a micro plan to cover individual farmers.

30.1 INTRODUCTION

Even after six decades of planned development, India has not been able to achieve its long-cherished goal of providing sustainable livelihoods even for the marginal and small farmers, let alone the rural landless. This failure could be largely ascribable to the stagnancy in crop yields. Since 60% of the cropped area has been cultivated under rain-fed conditions, weather shocks have escalated the risks in the recent period. This has led to decline in the private investment in agriculture. The National Sample Survey (NSS) survey on the Situation Analysis of Farmers (2003) showed that nearly 60% of the Indian farmers want to leave agriculture as the returns are low and risk is high. To ameliorate the situation, the central and state governments have taken up a number of initiatives like comprehensive area development, technology missions, watershed development programs.

With a view to improve the deteriorating rural conditions of the rural labor, the Ministry of Rural Development, Government of India, has launched a Rights Based and People-centric wage employment program known as Mahatma Gandhi National Rural Employment Guarantee Scheme (Mahatma Gandhi NREGS). The program is able to reach about 50 million rural households in the country. It aims at improving the assets of the poor, especially scheduled castes and scheduled tribes and reducing the adverse impact of drought. It is essentially a program which combines both social security and rejuvenation of agriculture by creating durable community and individual assets.

The individual assets are classified as category IV works in the list of permissible categories of works. The guidelines specify both types of works that can be covered and the eligibility conditions. Three types of works namely, (1) provision of irrigation facility, (2) development of plantations, and (3) land development to arrest soil erosion and to conserve moisture, can be taken up on private lands. These interventions are intended to bring the fallow and cultivable waste lands under cultivation and improve the productivity of the land presently under cultivation, besides reducing production risks. The order of priority for selecting the beneficiaries is fixed as: (1) SC/ST farmers, (2) beneficiaries of land reforms, (3) beneficiaries of Indira Awas Yojana, and (4) marginal and small farmers below the poverty line (Anonymous, 2008 and 2013a).

Indian Institute of Science, Bangalore (Anonymous, 2013b), reported that implementation of Mahatma Gandhi NREGS works such as water conservation and harvesting works, drought proofing, irrigation provisioning and improvement works and renovation of traditional water bodies have contributed to improved ground water levels, increased water availability for irrigation, increased area irrigated by ground and surface water sources and finally improved drinking water availability for humans and livestock. Further, land development works such as land leveling, conservation bench terracing, contour and graded bunding, field bunding, pasture development, silt application and drought proofing have contributed to improved soil organic carbon (SOC) content, reduced surface runoff and reduction in soil erosion. It is also reported that assessed land development and water related Mahatma Gandhi NREGS works contributed directly or indirectly to increasing crop production as well as reducing the risk of crop failure. The major limitation of this rapid assessment study conducted in five states during 2012, was no baseline or benchmark or pre-Mahatma Gandhi NREGS scenario data or information for comparison with the post-Mahatma Gandhi NREGS implementation scenario.

Dandekar et al. (2010) indicated that they have found overall positive impacts of the assets created in Sikkim State under Mahatma Gandhi NREGS. They have foreseen a second round of positive multiplier effect of NREGS program; however, the mechanism to ensure durability and sustainability of the assets created under this program was not achieved.

Works on individual lands accounted for about 12.0% of the total works carried out in 2011– 12. Since the inception of the scheme, 20.0 lakh works, accounting for 14% of the total works, were taken up under category IV. However, their share in the total expenditure will be small as the unit costs are quite low as compared to the costs of general works.

Category IV works have been highly concentrated in some states. Madhya Pradesh is at the top with 5.78 lakh works, accounting for 31% of the total works (Table 30.1). Uttar Pradesh occupies second place with 3.71 lakh works, which account for 15% of the total works. Andhra Pradesh stands at the third place with 3.26 lakh works. These works formed only 9% of the total works. Rajasthan took up 2.00 lakh works which accounted for 30% of the total works. These four states accounted for 74% of the total Category IV works at the national level (Figure 30.1).

Category IV works are important for several reasons. Firstly, lands left fallow will be brought under cultivation or productivity of lands presently under cultivation will be improved. Secondly, these works will reduce the incidence as well as the duration of migration as labor requirement on own land will rise. Thirdly, the dependence of the small and marginal farmers on the scheme will come down as the labor requirement for their own land will increase. It is reported that 50% of the beneficiaries have not sought employment under the scheme after category IV work.

TABLE 30.1 Distribution of Category IV Works and Their Share in Total Works

State	No. of Works (2006–07 to 2011–12)	Percentage of All Works
Madhya Pradesh	5,78,050	31
Uttar Pradesh	3,71,200	15
Andhra Pradesh	3,25,758	9
Rajasthan	2,00,260	30
All Other States	5,24,809	NA
All India	20,00,077	14

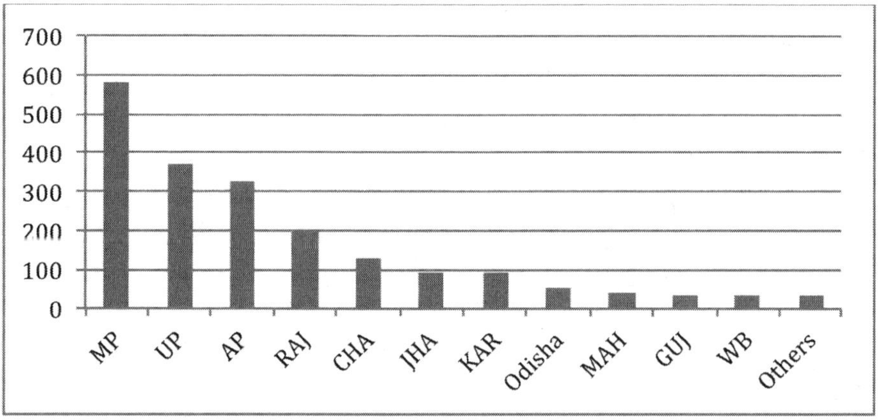

FIGURE 30.1 Distribution of Category IV Works across States ('000).

30.2 OBJECTIVES

In view of the importance of the Category IV works in terms of their absolute number and because of their importance for improving agriculture and reducing the dependence on MGNREGS, this study is taken up to examine the impact of these works on agricultural development and improvement in the incomes of the beneficiaries. The following are the objectives of the study:

- To examine the supplementary investments made by the farmers after these works.
- To study the changes in cropping pattern resulting from these works.
- To study the impact of the Category IV works on the yields of crops.
- To identify the difficulties in the implementation of the works.
- To examine the farmers' views on the implementation of the works.
- To study the impact of these works on migration.
- To provide suggestions for making these works more effective.

30.3 METHODOLOGY

In Andhra Pradesh, two districts were selected where Drought Prone Area Program (DPAP) and Desert Development Program (DDP) are under implementation namely Anantapur and Chittoor. Anantapur district is a DDP district. In Andhra Pradesh, there are 11 DPAP districts comprising of 94 blocks, of which, Chittoor district was selected for the study purpose (http://www.cwc.nic.in). These districts were selected for having large number of Category IV works. From each selected district, two blocks and from each block, two Gram Panchayats are selected for a detailed study. The selected blocks and Gram Panchayats are based on the condition that they had sufficient beneficiaries of category IV works of MGNREGS. The particulars of states, districts, blocks/Mandals and Gram Panchayats/villages selected for the study are shown in Table 30.2.

From each sample village, 25 beneficiaries were selected and a detailed questionnaire was canvassed. In each village, focus group discussions were conducted to elicit the views of the beneficiaries regarding the importance of the works taken up vis-à-vis the problems of agriculture in the village and impact of these works. The total samples from the two districts were 200. Information was also collected from the Secretary of the Gram Panchayat regarding conditions of agriculture, works taken up in the village under category IV.

30.4 RESULTS AND DISCUSSION

Incidence of poverty, estimated from the below poverty line (BPL) survey data, is very high in Anantapur, and Chittoor districts. The incidence is more than 90% in these districts. Anantapur has negligible area under irrigation, while Chittoor has less than 50% area under irrigation. The two districts in the study area can be classified into two categories. In Category I district with 30 to 45% of the gross cropped area in the *rabi* season. They require moderate intervention through MGNREGS in *rabi* season, but the requirement is moderate as agricultural activities are also carried out at a moderate level [Category II district with low share of *rabi* (less than 30%)]. This district requires high intervention of MGNREGS for six months. Chittoor falls under Category I, while Anantapur is under Category II. Anantapur district

TABLE 30.2 States and Districts Selected for the Study

State	Districts	Blocks/Mandals	Villages
Andhra Pradesh	1. Anantapur	1. Kalyanadurg	1. Garudapalem; 2. Golla
		2. Nallamada	1. Gopepalli; 2. Kurmala
	2. Chittoor	1. Punganuru	1. Aredigunta; 2. Chandramakulapalli
		2. Sadum	1. Kammmvaripalli; 2. Gongivaripalli

needs employment in winter and summer seasons. The agricultural situation in sample districts is described below in detail.

30.4.1. AGRICULTURE DEVELOPMENT

All the indicators of agricultural development during the latest period 2005–08 are presented in Table 30.3. The ranking of the districts on the value of agricultural output per hectare indicates that Chittoor in Andhra Pradesh occupies top position. The value of output per hectare in Chittoor and Anantapur ranges between Rs. 11,600 to Rs. 6,500 during 2005–08.

While land productivity indicates level of development of land resources, the value of output per worker indicates the capacity of crop sector to provide adequate livelihood. Anantapur occupies top position with Rs. 5,000 per worker and Chittoor being a backward district records per worker income around Rs. 3,000. Thus, agricultural development with high land productivity does not necessarily ensure adequate livelihood for the dependent workers. Glaring contradiction in these two indicators is observed in Chittoor district. Chittoor has high land productivity, but low labor productivity due to a large size of the labor force.

Irrigation is one of the factors responsible for agricultural development. Chittoor has 47.0% irrigation ratio and Anantapur has extremely low irrigation ratio (13.3%).

Irrigation intensity is another indicator of agricultural development. This indicator does not always reflect correct position as low irrigation ratio may have high irrigation intensity. Anantapur has low irrigation ratio, but its irrigation intensity is not very low. With three times irrigation ratio, Chittoor has lower irrigation intensity than Anantapur. Thus, irrigation intensity becomes a good indicator only when considered along with irrigation ratio.

Cropping intensity reveals the intensity with which land is used. Except in the case of plantation crops and annual crops like sugarcane, cropping intensity is positively associated with agricultural development. Cropping intensity is very low in both the districts namely, Anantapur and Chittoor with less than 110%. Thus, higher cropping intensity may not guarantee higher value of agricultural output if crop yields are low.

Mechanization and fertilizer use are also important indicators of agricultural development (Table 30.3). Chittoor has high use of fertilizer about 125 kg. Anantapur exhibits low use of fertilizer 70 kg/ha.

30.4.2 LAND USE PATTERN

The analysis focuses mainly on four categories of land use namely, net area sown, forest area, fallow land and culturable waste. These four categories have a significant influence on livelihoods of the people. Anantapur district has a high proportion of area under crops and low proportion of area under forest. Net area sown accounts for 57.6% as against 40.7% at the state level. Only 10.3% the area is under forest as against 22.7% at the State level. The district is similar to the state in other indicators. Chittoor has very low proportion of area under crops and high proportion of area under forest. Net sown area accounts for only 25.0% and forest area 29.8%. The share of fallow land is also high at 17.5% as against 12.5% at the state level (Table 30.4).

30.4.3 RAINFALL AND IRRIGATION

Anantapur receives precipitation of about 700 mm per annum (Average of rainfall received during 2005–2010). Chittoor is in the high rainfall zone with annual precipitation exceeding 1000 mm (Average of rainfall received during 2005–2010). Anantapur has very low irrigation cover of 12.7% and major source of irrigation in this district is tube well, which accounts for 72.2% of the net irrigated area. In the absence of sufficient rainfall, wells will become dry and actual irrigation may be much less than what is reported (Table 30.5).

30.4.4 RELEVANCE OF CATEGORY IV WORKS AND SUGGESTIONS

Category IV works under MGNREGS are useful for agricultural development as most of the holdings are small and marginal and the owners have limited capacity to invest on land development. However, difference arises in the nature of works to be emphasized under varying agro-climatic conditions. The areas may be classified into low rainfall and low irrigation areas, low rainfall with moderate irrigation and high rainfall/irrigation areas. Land development has to be a common approach for all the areas. In the low rainfall and low irrigation areas, moisture conservation works need priority and dry land horticulture is suitable for these areas. When rainfall is moderate or high, irrigation sources can be developed under MGNREGS. In the

TABLE 30.3 Indicators of Agricultural Development in Sample Districts: 2005–08

District	Output at 1990–93 prices		Fertilizer/ha. (kg)	Gross Irrigation (%)	Crop Intensity (%)	Irrigation Intensity (%)	Tractors/ 10,000 ha NAS
	Per ha.	Per worker					
Anantapur	6585	5111	70	13.3	106	126	28
Chittoor	11,624	3085	125	47.0	110	122	137

NAS: Net Area Sown.

TABLE 30.4 Land Use Pattern in Sample Districts

District	Forest	Not Available for Cultivation	Other Uncultivable waste*	Fallow	NAS	Total	Reported Area (lakh ha)
Anantapur	10.3	16.3	3.4	12.5	57.6	100.0	19.1
Chittoor	29.8	20.5	7.2	17.5	25.0	100.0	15.2
AP	22.7	17.8	5.4	13.5	40.7	100.0	275.0

*Includes pastures and other grazing land, land under miscellaneous crops and groves and culturable waste.

TABLE 30.5 Rainfall, Irrigation and Sources of Irrigation

S. No.	District	Average Rainfall for 6 years (2005–10)	(%) of NIA	Source-wise Percentage Share in Net Irrigated Area					
				Canal	Tank	Tube Well	Other Well	Others	Total
1	Anantapur	718 mm	12.7	14.3	5.5	74.5	4.4	1.2	100.0
2	Chittoor	1045 mm	40.4	5.7	11.1	72.2	12.0	0.1	100.0

Source: www.imd.gov.in.

irrigated areas, there is not much need for these works and the focus should be on providing drainage and solving water logging problems. However, decisions about the classification have to be based on ground realities rather than aggregate indicators. The approach adopted for the component of category IV in MGNREGS should be different from the approach adopted for general and community works. In this case, agricultural development plans for a small geographical area should be prepared with focus on the development of land and water resources on the lands of small and marginal farmers and weaker sections. The unit of planning should be a Block or Mandal and cluster approach should be adopted.

The plans have to be prepared under technical guidance and people's participation, especially small and marginal farmers. The focus should be more on employment generation after the development of lands rather than immediate employment creation.

30.4.5 AGRICULTURE IN THE STUDY AREA: RAINFALL AND IRRIGATION

The two Mandals in Anantapur district receive a very low rainfall of about 550 mm per annum (Table 30.6). Nallamada has slightly higher rainfall than Kalyandurg, but it exhibits greater fluctuations. In the three years from 2006–07, rainfall varied between 450 mm and 1059 mm in Nallamada, but the variation in Kalyanadurg is only between 670 mm and 955 mm. The two Mandals in Chittoor receive higher precipitation than their counterparts in Anantapur. Punganur has an annual rainfall of 858 mm and Sodum 884 mm. Just as in the case of Nallamada, Sodum has higher rainfall, but greater fluctuation. During the three years from 2006–07, annual rainfall varied between 560 mm and 934 mm in Punganur, whereas, it varied between 663 mm and 1036 mm in Sodum.

Besides low rainfall, the two Mandals in Anantapur have very low irrigation ratio – 5.4% in Nallamada and 3.1% in Kalyanadurg (Table 30.7). Thus, the study area in Anatapur is highly drought-prone. Even the low irrigation is provided by tube wells in Kalyanadurg and dug wells and tube wells in Nallamada. Dug wells are not feasible in Kalayandurg as the ground water level is very deep. Bore wells are dug even up to 800 feet. The two Mandals of Chittoor have higher irrigation ratio than their counterparts in Anantapur. Both the Mandals have one-third of the net area sown under irrigation. But they differ in the sources of irrigation. While Sodum has tank irrigation

TABLE 30.6 Trends in Rainfall in Sample Areas of Anantapur District

	Anantapur		Chittoor	
Rainfall (mm)	**Kalyandurg Mandal**	**Nallamada Mandal**	**Punganur Mandal**	**Sodum Mandal**
Normal (50 yrs.)	544	577	858	884
2006–07	671	450	560	663
2007–08	954	1059	934	1036
2008–09	696	880	795	725

TABLE 30.7 Irrigation Ratio and Sources in Sample Areas of Anantapur: 2008–09

	Anantapur		Chittoor	
Item	**Kalyandurg**	**Nallamada**	**Punganur**	**Sodum**
Irrigation Ratio (Net)	3.1	5.4	31.9	33.9
Canal	–	–	–	–
Tank	–	4.2	–	17.0
Dug Well	–	36.6	7.7	4.6
Tube Well	100.0	59.2	92.3	78.4
All Sources	100.0	100.0	100.0	100.0

along with tube wells, Punganur depends completely on wells. However, ground water level has steeply declined in both the Mandals. It is reported that all the dug wells are dried up.

Land use pattern in the study area indicates that a high proportion of area is brought under cultivation in Anantapur district 71.2% in Kalyanadurg and 67.3% Nallamada block (Table 30.8). The two Mandals in Chittoor district have a lower proportion area under crops – 34% in Punganoor and 28% in Sodum. Is there any scope for expansion of area further? Four categories of land will be useful for productive use. They are: (1) permanent pastures, (2) miscellaneous tree crops and groves, (3) culturable waste, and (4) fallows (both current and other fallows). The total area under these four categories is found to be about 10% of the reporting area in Anantapur district and 25% in Chittoor district. However, a close examination of the available land is needed for taking up any developmental activities on these lands.

Cropping pattern in the study area indicates that groundnut is the major crop accounting for about 90% of the cropped area in both the Mandals of Anantapur. Groundnut is the major crop in the two Mandals of Chittoor as well, but rice and sugarcane, fruits and vegetables also account for significant extent of area (Table 30.9). Thus, the sample area in Chittoor is more developed than the sample area in Anantapur.

30.4.6 INCIDENCE OF POVERTY

Incidence of poverty is calculated from the data on the BPL cards issued in the village. It is significant to note that almost all the beneficiary households are BPL in both the districts (Table 30.10).

TABLE 30.8 Land Use Pattern in Study Area: 2008–09 (Percentages)

Item	Kalyandurg	Nallamada	Punganur	Sodum
Forest	9.5	2.5	10.0	10.0
Barren and Unculturable	3.5	19.0	19.0	23.0
Non-agricultural Uses	4.2	1.1	15.0	13.0
Permanent Pastures	–	–	4.0	2.0
Miscellaneous Tree Crops	–	0.2	3.0	1.0
Culturable Waste	0.1	1.5	1.0	1.0
Other Fallow	1.1	3.9	7.0	15.0
Current Fallow	11.4	8.3	15.0	23.0
Net Area Sown	71.2	67.3	34.0	28.0
Net Area Sown ('000 ha)	34.9	15.0	10.1	6.3
Scope for area expansion (%)	11.5	10.0	23.0	27.0

TABLE 30.9 Cropping Pattern in Sample Mandals: 2008–09 (%)

	Anantapur		Chittoor	
Crop	Kalyandurg	Nallamada	Punganoor	Sodum
Groundnut	92.2	88.8	43.7	28.8
Sunflower	1.1	1.4	–	–
Rice	–	–	4.4	16.7
Sugarcane	–	–	14.8	10.6
Ragi	–	–	1.6	–
Fruits & Vegetables	–	–	19.4	39.3

TABLE 30.10 Estimated Incidence of Head Count of Poverty (%)

Caste	Anantapur	Chittoor	Combined
SC	100.0	100.0	100.0
Others	100.0	97.6	98.6
Overall	**100.0**	**98.8**	**99.4**

30.4.7 SIZE OF LANDHOLDING

The size of operated area is only 3.29 acres per household and hardly one–fourth of the land has got irrigation facilities (Table 30.11). Most of this facility is due to intervention through MGNREGS category IV works. The size of landholding is distinctly lower for SC beneficiaries as compared to 'Others.' While the holding size is 3.92 acres for 'Others,' it is only 2.8 acres for SC. The two districts differ in the irrigation ratio. While irrigation ratio is negligible in Anantapur, it is high at 49% in Chittoor. In Chittoor, irrigation ratio is higher for SC as compared to 'Others' and this is mainly due to category IV works in Chittoor. Culturable wasteland was developed adopting Block Approach. Though the development process was initiated earlier, category IV works have strengthened the development process by converging with the initiatives of other schemes/ departments.

30.4.8 CATEGORY IV WORKS: APPROACH AND TYPES

The two districts selected for study are highly backward as they receive very low precipitation. Irrigation facility is completely absent in Anantapur district. Chittoor with more rainfall and irrigation facility is better placed.

The study area in Anantapur depends completely on groundnut and there was no crop during the last four years due to failure of monsoon. MGNREGS provided good opportunity for the livelihood and saved them from large scale distressed migration. The area in Chittoor is slightly more developed than Anantapur because of relatively higher rainfall and better ground water availability. But water level is going down due to the failure of monsoon.

TABLE 30.11 Average Size of Operated Area and Irrigated Area by Caste

	Anantapur		Chittoor		Combined	
Caste Group	Operated Area (Ac)	Irrigated Area (%)	Operated Area (Ac)	Irrigated Area (%)	Operated Area (Ac)	Irrigated Area (%)
SC	2.91	6.1	2.64	57.8	2.79	27.3
Others	4.18	1.3	3.72	43.3	3.92	24.0
Overall	3.39	3.8	3.19	49.2	3.29	25.5

The following works were implemented in Anantapur district:

1. Farm bunds
2. Bush clearance
3. Boulder removal
4. Mahatma Gandhi Van Nursery
5. Dry Land Horticulture

The works are taken up by the labor groups. Each group consists of a *mate* and workers are employed on the lands selected for improvement. Land owner has little role in carrying out the work and he will not participate if he/she is not a member of the group, which asked for work. As per the guidelines, the land owner is expected to join the other workers in executing. The results of Andhra Pradesh indicate departure from the guidelines.

Horticulture is the major activity taken up in Andhra Pradesh. The state implemented the works as a package and horticulture is combined with land development and border plantations. Horticulture beneficiaries account for 53.8% of the beneficiaries and remaining 46.2% got development works (Table 30.12).

It is found that in both the districts, the proportion of horticulture beneficiaries is higher among SC communities than others. Only a small proportion of beneficiaries (30% in Anantapur and 3% in Chittoor) got land development work. The higher concentration of horticulture works for scheduled castes.

The extent of land benefitted and the average cost per acre are shown in Table 30.13. The average extent of land benefited by the category IV work is 3.0 acres. It is slightly higher in Anantapur (3.33 acres) than in Chittoor (2.90 acres). In the case of silt application, the same land is treated three times in three consecutive years, but it is considered only once in calculating the average extent of land benefited per household.

TABLE 30.12 Distribution of Beneficiaries by Types of Category IV Work Undertaken (%)

	Anantapur			Chittoor			Combined		
Type of Work	**SC**	**Other**	**All**	**SC**	**Other**	**All**	**SC**	**Other**	**All**
Horticulture	46.2	16.0	31.4	43.8	8.8	25.8	44.8	11.9	28.2
Land Development	30.8	80.0	54.9	3.1	73.5	39.4	15.5	76.3	46.2
Horticulture + Others	23.0	4.0	13.7	53.1	17.7	34.8	39.7	11.8	25.6

TABLE 30.13 Average Extent of Land Benefitted and Cost of Category IV Work

	Anantapur			Chittoor			Combined		
Type of Work	**SC**	**Others**	**All**	**SC**	**Others**	**All**	**SC**	**Others**	**All**
Land (ac/hld)	2.91	3.76	3.33	2.84	2.95	2.90	2.87	3.30	3.09
Cost ('000/hld)	208	74	142	278	111	192	247	95	170
Cost ('000/ac)	71	20	43	98	38	66	86	29	55

The extent of land benefitted is less for SC beneficiaries. While the average extent of land benefitted is 3.3 acres for other communities, it is 2.87 acres for SC beneficiaries. The average cost of work per beneficiary is found to be Rs. 1.70 lakh. For SC community, the amount spent is Rs. 2.47 lakh, while it is 0.95 lakh for others. This is because of the difference between the works given to these two communities. The unit cost of horticulture is more than that of all other works and inclusion of other components will also add to the cost. The cost of horticulture development is Rs. 1.18 lakh/acre for three years. The support will end with three years. If a farmer has five acres, the cost of development comes to Rs. 5.90 lakh.

Average cost of development is more in Chittoor because of the dominance of horticulture works. The cost per beneficiary is Rs. 1.92 lakh in Chittoor and Rs. 1.42 lakh in Anantapur. This higher cost is partly due to the larger extent of land per beneficiary. Average cost of category IV works per acre of land also shows the same variation between the two districts and between the two groups. Cost per acre is Rs. 86,000 for SC beneficiaries and Rs. 29,000 for other beneficiaries. Similarly, cost of work per acre is Rs. 43,000 in Anantapur and Rs. 66,000 in Chittoor.

Once public investment is made for land development, farmers have to make complementary investments. The benefit of public investment will be more if these investments are coming forward. It is found that 62% of the beneficiaries made their own investment after the completion of Category IV works. For instance, the activity taken up by the government will not cover removal of heavy boulders if it requires the use of JCB machine. This activity is undertaken by the beneficiaries. It is found that about 60% of the beneficiaries made their own investment and there is no difference between the two communities. However, the amount of investment is higher for other communities than SC. While SC communities invested Rs. 7,000 per beneficiary, investment of other communities is about Rs. 12,000 per beneficiary (Table 30.14).

TABLE 30.14 Details of Making Own Investment and Amount of Investment Per Household

	Anantapur			Chittoor			Combined		
Item	**SC**	**Others**	**All**	**SC**	**Others**	**All**	**SC**	**Others**	**All**
Beneficiaries Invested (%)	41.2	35.5	39.0	87.2	82.9	85.0	61.1	62.5	61.7
Amount (Rs.)	1751	1432	1630	13,762	19,854	16,884	6956	11922	9163

There is significant difference in the proportion of beneficiaries making own investment as well as investment per beneficiary between the two districts. While 85% of the beneficiaries made own investment in Chittoor, only 39% of the Anantapur beneficiaries did so. Similarly, the amount of investment per beneficiary is very low at Rs. 1630 in Anantapur, while it is an impressive amount of Rs. 16,000 in Chittoor. The variation in own investment is mainly due to the nature of works undertaken. Works in Chittoor focused on development of waste land as a cooperative venture, but works were not found in the study area of Anantapur. All works in Anantapur are on individual lands and the average cost of the work is also low. Accordingly, the requirement of supplementary investment is low.

Investment made by the beneficiaries as a sequel to the category IV work is classified into four broad categories – pump set and pipes, land development, purchase of plants and fencing. The results on this parameter are presented in Table 30.15. Land development and purchase of plants are the two

TABLE 30.15 Percentage of Beneficiaries by Type of Own Investment After Category IV Work

Type of Work	Anantapur			Chittoor			Combined		
	SC	**Others**	**All**	**SC**	**Others**	**All**	**SC**	**Others**	**All**
Pump set and Pipes	–	–	–	–	40.0	11.1	18.2	–	6.7
Land Development	16.7	50.0	33.0	53.8	60.0	55.6	54.5	42.1	46.7
Purchase of Plants	50.0	50.0	50.0	46.2	–	33.3	27.3	47.4	40.0
Fencing	33.3	–	16.7	–	–	–	–	10.5	6.7
Total	100.0	100.0	100.0	100.0	100.0	100.0	100.0	100.0	100.0

items on which 87% of the beneficiaries have made investment. Actually, there is no need for the beneficiaries to pay for the plants because the project allows for replacement also. Beneficiaries have to invest when mortality exceeds the permissible limits.

Category IV works is expected to improve crop yields or cropping intensity or both. But in drought prone areas, these changes depend heavily on monsoon conditions. The results show that there is no increase in the cropping intensity of the beneficiaries in Anantapur district. In fact, there is decline in the gross cropped area by 9.1% and the extent of decline is more for SC beneficiaries (Table 30.16). It is reported that there is no good monsoon after the land development activities. The decline would have been still more but for the interventions through Category IV works. The decline is not very high because horticultural plantations took place and intercrop is also supported under the program. The agricultural situation in Anantapur is very bad and marginal and small farmers are incapable of taking risk. In the absence of proper interventions, there will be heavy distress migration in terms of number of persons as well as duration from the area. It is necessary to support the livelihoods of these people who are actively participating in MGNREGS.

Convergence Model of Category IV Works in Chittoor District of Andhra Pradesh

A new model is adopted in Chittoor district under Category IV works. In this model, a block of about 167 acres of waste land is identified near Chandramakulapalli village in Punganoor Mandal. The land was barren and uneven and it was about to be taken by a mineral water producing company. A group of 53 landless laborers from scheduled castes formed an association and started working on the land in 2006. They started land leveling activity by mobilizing community labour at the rate of one person from each family per day. If any household fails to contribute labor, they had to pay Rs. 150 per day. An amount of Rs. 30,000 was provided for land leveling from Comprehensive Land Leveling Programme (CLDP) of Government of Andhra Pradesh. Seven bore wells were sanctioned under RIDF of NABARD and five were successful. Road is sanctioned under MGNREGS. Initially, they were carrying water from nearby area. Drip irrigation equipment is provided after one year. Plantation is developed around the boundary with teak, sapota and neredu plants. Vegetables are grown. Mango plantation and intercrop are supported under MGNREGS. Additional bore wells are dug under Andhra Pradesh Micro-Irrigation Project. The programme is successful because of people's initiative, technical support from departments and financial support in a convergence mode under various programmes. The same experiment is carried out in other parts of the district also. While land required development, water resources contributed for the success of the programme. The beneficiaries have taken up dairying also. They need wire fencing and electricity. One important aspect is mutual cooperation and support among the members. They need some assistance for electricity and wire fencing to protect from forest pigs. Each beneficiary has three acres of land. They need not depend on MGNREGS in future. Full crop of horticulture will be realized after two more years.

TABLE 30.16 Change in Gross Cropped Area Per Household

	Gross Cropped Area in Acres								
	Anantapur			Chittoor			Combined		
Period	SC	Other	All	SC	Other	All	SC	Other	All
Before Cat. IV Work	2.96	3.96	3.45	1.16	2.42	1.81	1.97	3.07	2.53
After Cat. IV Work	2.29	4.02	3.14	2.70	2.91	2.81	2.52	3.38	2.95
Change (%)	–22.7	1.5	–9.1	132.2	20.5	55.4	27.8	10.1	17.0

Application of fertilizer is one of the ways of increasing productivity. Only 20% of the beneficiaries reported increase in fertilizer use. Chittoor is ahead of Anantapur in this aspect. Households reporting increase are more in Chittoor than in Anantapur (Table 30.17). All the three varieties namely, urea, DAP and manure have been increased.

Productivity of land before and after the category IV work is computed using the same prices for both the periods. It is found that productivity increased after the category IV works are implemented. Though area declined marginally, productivity increased in both the districts. As expected, increase is more in Chittoor district than in Anantapur district. While productivity

TABLE 30.17 Farmers Reporting Increase in Fertilizer Use and Extent of Increase

	Anantapur			Chittoor			Combined		
Item	**Bef.**	**Aft.**	**Inc. (%)**	**Bef.**	**Aft.**	**Inc. (%)**	**Bef.**	**Aft.**	**Inc. (%)**
Reporting Increase (%)	–	–	13.7	–	–	24.2			19.7
Use of Urea (kg/hld)	147	245	66.7	–	265	NC	64	256	300.0
Use of DAP (kg/hld)	343	490	42.9	833	3693	343.2	620	2297	270.7
Manure Value (Rs.)	235	596	153.3	205	402	96.3	280	486	123.1

Bef.: Before; Aft.: After; Inc.: Increase.

increased from Rs. 5000 to Rs. 10,500 in Chittoor district, the corresponding increase in Anantapur is in the range of Rs. 8,000 to Rs. 9,600. But increase is more in Chittoor. Productivity increase is more for other communities as compared to SC (Table 30.18). This is because horticulture projects have not yet reached the full productivity stage.

Since the district is under drought since 2010, information on yield is taken from the beneficiaries covered before this year. In one year, monsoon conditions were slightly better and yield information belongs to that year. The situation in Chittoor is slightly different. Groundnut is an intercrop for most of the beneficiaries and hence yield levels will not be improved as the plantation grows. Groundnut yield has gone up from 2.5 quintals to 3.1 quintals/acre in Anantapur (Table 30.19).

TABLE 30.18 Change in Land Productivity

	Value of Output Per Acre (Rs.)								
	Anantapur			Chittoor			Combined		
Period	**SC**	**Others**	**All**	**SC**	**Others**	**All**	**SC**	**Others**	**All**
Before	6880	9255	8214	3990	5679	4939	5409	7434	6547
After	7680	11076	9588	6747	13191	10368	7205	12153	9985
Change (%)	11.6	19.7	16.7	69.1	132.3	109.9	33.2	63.5	52.5

TABLE 30.19 Changes in Yield Per Acre

	Anantapur		Chittoor		Combined	
Crop	**Before**	**After**	**Before**	**After**	**Before**	**After**
Groundnut	250	310	229	232	242	282
Millets	1267	1240	–	–	1267	1240

Income from crop production is estimated by calculating the value of output at constant (2011–12) prices and taking 50% of output as input. Since labor input consists of family labor, It is assumed that 50% of the labor input comes from family labor. The value is estimated from estimated labor requirement for the crops grown. Income from labor and other sources are estimated from the days of outside employment and days of participation in MGNREGS in 2009–10 and 2011–12. Income from other sources is taken directly from the reported figures. As it is not high it will not create much problem in estimation. Per capita income is found to be Rs. 11,900 before the category IV works and Rs. 13,600 after the category IV works (Table 30.20). There is an increase of 14.5% in per capita income. This low increase in income is partly due to crop failure and partly due to new horticulture plantations. Mango plantations are expected to give an income of Rs. 500

TABLE 30.20 Change in Per Capita Annual Income

	Income (Rs.)								
	Anantapur			Chittoor			Combined		
Period	SC	Others	All	SC	Others	All	SC	Others	All
Before	8616	12092	10183	12836	13831	13323	10829	13071	11887
After	8978	13095	10834	14407	17571	15956	11825	15616	13614
Increase (%)	4.2	8.3	6.4	12.2	27.0	19.8	9.2	19.5	14.5

PLATE 1 Dryland horticulture crops grown in Kalyandurg mandal, Anantapur district.

to Rs. 1000/tree after six years and total income on this count comes to Rs. 35,000 to Rs. 70,000/acre. In the present calculations only the increase realized is taken into account. Dryland horticulture mango crop grown in Kalyandurg mandal, Anantapur district is shown in plate 1. The gain to SC farmers is found to be less than that of the other group because most of the SC farmers were given horticulture, which has not yet attained full growth. While the income of other communities increased by 19.5%, the gain for SC is only 9.2%. In the context of agro-climatic conditions of these two districts, mango plantations give more stable income than seasonal crops.

Income from different sources is shown in Table 30.21. Share of income from crop production increased from 20% to 26% income share of income from non-MGNREGS income declined from 22% to 19%. Thus, the share of crop income increased and if both family labor income and crop income are added, the share of income from agriculture comes to 34%.

30.4.9 IMPACT OF CATEGORY IV WORKS ON EMPLOYMENT

Changes in employment days in agriculture are computed by taking the labor use per acre in each crop and employment content of each crop. As there is not much improvement in crop production, the gain in employment is also not much (Table 30.22). But the potential will be higher. It is to be noted that employment generation will be a challenge in these areas because horticulture development is not going to generate much employment. Increase in employment in marketing will not benefit these farmers. There is a need to create employment opportunities outside crop production for these beneficiaries. All the MGNREGA workers participating seriously in the Scheme have to be consulted about changes in their employment needs to find out solutions for sustainable employment.

TABLE 30.21 Income from Different Sources (%)

	Anantapur		Chittoor		Combined	
Share of Income	**Before**	**After**	**Before**	**After**	**Before**	**After**
Share of Agriculture	31.0	34.0	12.4	21.8	19.7	26.3
Family Labor	11.5	10.9	5.1	6.5	7.6	8.1
NREGS	16.6	16.6	17.0	16.9	16.8	16.8
Wage	14.9	14.0	26.7	22.3	22.1	19.3
Other Income	26.1	24.5	38.8	32.4	33.8	29.5

TABLE 30.22 Change in Employment in Agriculture per Household

	Person Days of Employment								
	Anantapur			Chittoor			Combined		
Period	**SC**	**Others**	**All**	**SC**	**Others**	**All**	**SC**	**Others**	**All**
Before	101	136	118	44	82	64	70	105	87
After	83	138	110	83	93	88	83	112	98
Change (%)	–18.4	1.5	–7.1	88.7	13.1	38.5	19.0	6.7	11.6

30.4.10 IMPACT OF CATEGORY IV WORKS ON MIGRATION

One of the major objectives of MGNREGS is to arrest distress migration from rural areas. Implementation of category IV works and other MGNREGS works helped in the generation of employment to the beneficiaries. But agricultural employment has not yet gone up partly due to monsoon failure and partly due to gestation period of horticulture crops. The state is providing more than 100 days of employment and in fact, the major demand of the beneficiaries is to enhance the employment days to 150 days. The demand is justified in view of the recurring drought in these areas. It is found that 100 days ceiling is reached by middle of June for a family with two workers. Agricultural work starts from the month of June if monsoon is normal. In the case of monsoon failure, there will be distress migration due to the absence of agricultural activities in the area.

The proportion of households reporting migration has not changed after the implementation of category IV works and remained around 10%. There is no significant variation between the two communities and between the two districts (Table 30.23). But the proportion of migrating households is higher in Anantapur as compared to Chittoor. It is reported that 20% of the households are migrating even after providing employment for 100 days/family. This is because MGNREGS can retain them for 50 days if the family has two workers. The problem of migration is not severe in Chittoor and

TABLE 30.23 Households Migrating for Work Before and After Category IV Work (%)

	Anantapur		Chittoor		Combined	
Caste Group	**Before**	**After**	**Before**	**After**	**Before**	**After**
SC	15.4	11.5	6.3	9.4	10.3	10.3
Overall	19.6	19.6	3.0	4.5	10.3	11.1

the general MGNREGS has resulted in decline in employment. In fact, the state has been proactive in providing more than 100 days of employment in drought-hit areas.

30.4.11 USEFULNESS OF WORK

Information about perceptions about the usefulness of the work is collected from the beneficiaries. Since the works are taken up on the basis of the discussions in Gram Sabha, answers for this question should be interpreted carefully. If they report that they are not of much use, it should be treated that they could not realize the benefit due to external factors like drought. A small proportion of the workers reported that works taken up are not of much use – 32% in Anantapur and 10% in Chittoor (Table 30.24). The answers are compatible with the climatic conditions in the area. Further, the convergence program with the initiative of the beneficiaries made a lot of difference about the perceptions of the beneficiaries about the works. The block approach in Chittoor is also responsible for uniform perceptions among the beneficiaries about the usefulness of the works. However, there are variations in the benefits derived by the beneficiaries due to differences in the access to water resources.

The beneficiaries of category IV works have worked in MGNREGS for about 100 days in each year during the last three years. Employment generated is more than 100 days in Chittoor and 86 days in Anantapur. Participation of SC is slightly higher than that of other communities, but the difference is not much. This high participation of others clearly indicates the need for MGNREGS in the drought prone areas.

The need for MGNREGS is clearly understood by examining the distribution of households by days of employment in MGNREGS. A very small proportion of households worked for less than 100 days. It is observed that only 10% of the households worked for less than 50 days and another 11%

TABLE 30.24 Percentage of Beneficiaries Expressing Usefulness of Work (%)

Level of Utility of Works Taken Up	Anantapur			Chittoor			Combined		
	SC	Others	All	SC	Others	All	SC	Others	All
Highly Useful	64.7	74.2	68.3	89.7	90.2	90.0	75.6	83.3	79.0
Not of Much Use	35.3	25.8	31.7	10.3	9.8	10.0	24.4	16.7	21.0

worked for more than 50 days but less than 100 days. The proportion of households working for more than 100 days is about 80%.

The participation of SC is higher than that of other communities. The proportion of households with less than 100 days of participation is only 15% among SC communities as compared to 27% among other communities. Thus, 85% of the SC households participated for more than 100 days and 12% of the beneficiaries participated for more than 125 days. This is mainly due to the severe drought in these areas and lack of alternative employment opportunities (Table 30.25).

Though drought situation is same in both the districts, participation is more in Chittoor as compared to Anantapur. While 22% of SC beneficiaries worked for more than 125 days in Chittoor, there are no such beneficiaries in Anantapur. The same pattern is observed during the last three years.

30.5 SUMMARY AND CONCLUSIONS

Most of the beneficiaries are below the poverty line and belong to the category of marginal and small farmers. They have good housing conditions, but are educationally backward. Their demographic structure is also progressive. Chittoor is more developed than Anantapur as shown by socio-economic indicators *viz.*, education, housing, demographic indicators and irrigation ratio. The variation between the two districts is partly due higher irrigation facilities and partly due to the model adopted. The pattern of development of the two caste groups indicates that development percolates first to general population before it percolates to the weaker sections.

Monsoon failure during the last four years affected Anantapur more than the Chittoor. Even in the same district, the effect of drought differs between the two Mandals.

Development of horticulture is the major focus of category IV works. The work is taken up in a comprehensive manner covering land development, soil and moisture conservation and horticulture plantation. The risk of monsoon failure will have less impact on the livelihood of the people after the gestation period of the plantation. Now the benefit is realized mostly in terms of wages. The land development works will result in higher income during normal monsoon period. The staggered trenches are helpful for improving the soil moisture.

Two models are followed in the implementation of category IV works in the state. One is the individual approach in which the lands of beneficiaries

TABLE 30.25 Percentage Distribution of Beneficiaries by Days of Participation in MGNREGS

Days of Participation	Anantapur			Chittoor			Combined		
	SC	Others	All	SC	Others	All	SC	Others	All
2009–10									
< 50	7.7	16.0	11.7	6.2	11.8	9.0	6.9	13.6	10.3
50 to 100	11.5	12.0	11.7	6.2	14.7	10.6	8.6	13.6	11.1
100 to 125	80.8	72.0	76.6	65.6	58.8	62.1	72.4	64.3	68.3
Above 125	–	–	–	22.0	14.7	18.2	12.1	8.5	10.3
Total	100.0	100.0	100.0	100.0	100.0	100.0	100.0	100.0	100.0
Average Days	89	84	86	112	98	105	102	92	97
2010–11									
< 50	7.7	12.0	9.8	6.3	5.8	6.0	6.9	8.5	7.7
50 to 100	11.5	20.0	15.6	18.8	23.5	21.2	15.5	22.1	18.8
100 to 125	80.8	68.0	74.5	35.3	58.8	45.5	53.4	62.7	58.1
Above 125	–	–	–	48.8	11.8	27.3	24.1	6.8	15.4
Total	100.0	100.0	100.0	100.0	100.0	100.0	100.0	100.0	100.0
Average Days	90	81	86	115	102	109	104	93	99
2011–12									
< 50	7.7	20.0	13.8	6.3	5.8	6.0	6.9	11.9	9.4
50 to 100	7.6	8.0	7.9	12.5	14.7	13.7	10.3	11.9	11.1
100 to 125	84.7	72.0	78.3	28.1	38.2	33.3	53.4	52.5	53.0
Above 125	–	–	–	53.1	41.3	47.0	29.3	23.7	26.5
Total	100.0	100.0	100.0	100.0	100.0	100.0	100.0	100.0	100.0
Average Days	92	76	84	119	115	117	107	99	103

are improved individually. This is the approach observed in the study area of other states also. The other is Group Approach for a Block of Land. This approach is found in Chittoor district. The initiative of the beneficiaries is very high and departmental support is available in a convergence mode. This is found to be more effective and beneficial with several externalities. However, land and motivated people must be available for adopting this approach.

The beneficiaries made complementary investment to maximize the benefits by way of buying pump sets, investing on land development activities and putting fencing for the land. The major deficiency observed in the implementation is the supply of plants for border plantation. Supply of too tender plants resulted in high mortality after undergoing a lot of trouble in watering them.

30.6 SUGGESTIONS

Analysis of the field data and information collected through Focused Group Discussions suggest some aspects for consideration by the implementing agency. In some cases, a thorough examination of these aspects mentioned may be necessary before acting on them.

30.6.1 CATEGORY IV WORKS IN ANDHRA PRADESH

1. The study area in Anantapur receives a very low rainfall and there is no irrigation. Chittoor receives slightly higher rainfall and irrigation ratio is moderate at 33%.
2. A variety of works have been taken up in the state. The main among them are: farm bunds, bush clearance, boulder removal, Mahatma Gandhi Van Nursery and dry land horticulture. These works are taken up by the labor groups. Land owner has no role in carrying out the work as he can participate only when he is a member of the group which has taken up the work. This aspect is a departure from the guidelines.
3. Horticulture beneficiaries account for 53.8% of the beneficiaries and remaining 46.2% got development works. The average cost of work per beneficiary is found to be Rs. 1.70 lakh. The cost of horticulture development is Rs. 1.18 lakh/acre for three years. The support will end after three years, two years before the yield of horticulture commences.

4. Farmers made necessary investment after the public investment is made. It is found that 62% of the beneficiaries made such investment. For instance, the norms will not permit the use of JCB for the removal of boulders. Beneficiaries spent wherever big boulders are left in the field. It is found that about 60% of the beneficiaries made investment after the work is completed.
5 The difference between the two districts in the implementation of category IV works is the adoption of block cum group approach in Chittoor and individual approach in Anantapur. This has resulted in higher investment by the beneficiaries. While 85% of the beneficiaries made own investment in Chittoor, only 39% of the Anantapur beneficiaries did so. Similarly, the amount of investment per beneficiary is very low at Rs. 1630 in Anantapur, while it is an impressive amount of Rs. 16,000 in Chittoor.
6. Works in Chittoor focused more on development of waste land as a block and converging with other schemes to cover activities not permissible under MGNREGS and beneficiaries not eligible under category IV works.

30.6.2 IMPACT ON CROP PRODUCTION

1. Mango plantations have not yet reached production stage, but they have attained healthy growth. The benefit from moisture conservation works is not yet realized because of monsoon failure during last five years. The rainfall was good in 2005–06 and moderate in 2009–10. In all the other years, there was severe shortfall of rainfall. But farmers appreciate the works taken up and feel that the full benefit will be realized when monsoon is normal.
2. Though gross cropped area declined marginally, productivity increased in both the districts. As expected, increase is more in Chittoor district than in Anantapur district. The average improvement per acre is found to be Rs. 6500 to Rs. 10,000.
3. There is an increase of 12% in per capita income. The low increase in income is partly due to crop failure and partly due to new horticulture plantations. Mango plantations are expected to give an income of Rs. 500 to Rs. 1000/tree after six years and total income on this count comes to Rs. 35,000 to Rs. 70,000/acre.

30.6.3 IMPACT ON EMPLOYMENT

As there is not much improvement in crop production, the gain in employment is also not much. Employment increased by 11.6% after the works and the increase occurred only in Chittoor. Anantapur witnessed slight decline in employment after the works due to monsoon failure. It is to be noted that employment generation will be a challenge in these areas because horticulture development is not going to generate much employment. Increase in employment in marketing will not benefit these farmers. There is a need to create employment opportunities outside crop production for these beneficiaries. Proper planning is needed to shift them to self-employment activities.

30.6.4 IMPACT ON MIGRATION

The state is providing more than 100 days of employment and beneficiaries want enhancement to 150 days. The demand is justified in view of the recurring drought in these areas. It is found that 100 days ceiling is reached by middle of June for a family with two workers. Agricultural work starts from the month of June if monsoon is normal. If monsoon fails, there will be distress migration. There is not much migration because of effective implementation of MGNREGS in the state. The proportion of beneficiaries reporting migration lower in Chittoor than in Anantapur because of the group and block approach in this district.

KEYWORDS

- **Category IV Works**
- **Distress Migration**
- **Durable Assets**
- **Employment**
- **Land Development**
- **Mahatma Gandhi NERGS**
- **Productivity**

REFERENCES

Anonymous, (2008). Operational Guidelines, 2008. The National Rural Employment Guarantee Act, 2005 (3rd Edition). Ministry of Rural Development, Government of India. pp. 27.

Anonymous, (2013a). Operational Guidelines, 2013. Mahatma Gandhi National Rural Employment Guarantee Act, 2005 (4th Edition). Ministry of Rural Development, Government of India. pp. 55.

Anonymous, (2013b). Environmental Benefits and Vulnerability Reduction through Mahatma Gandhi National Rural Employment Guarantee Scheme. Report synthesized by Indian Institute of Science in collaboration of Ministry of Rural Development, Government of India and Deutsche Gesellschaft für Internationale Zusammenarbeit (GIZ). pp. i–iv.

Dandekar, A., Bhandari, V., & Modi, P. (2010). An Impact Assessment Study of the Usefulness and Sustainability of the Assets Created under Mahatma Gandhi NREGA in Sikkim. Institute of Rural Management, Anand. Report submitted to The Secretary, Rural Management and Development Department, Government of Sikkim.

www.cwc.nic.in, Central Water Commission, Government of India (Accessed on 15th December, 2015).

INDEX

B

D

E

F

G

M

N

Q

U

V